T0190327

Lecture Notes in Computer Science

Lecture Notes in Artificial Intelligence 14180

Founding Editor

Jörg Siekmann

The series Lecture Notes in Artificial Intelligence (LNAI) was established in 1988 as a topical subseries of LNCS devoted to artificial intelligence.

The series publishes state-of-the-art research results at a high level. As with the LNCS mother series, the mission of the series is to serve the international R & D community by providing an invaluable service, mainly focused on the publication of conference and workshop proceedings and postproceedings.

Xiaochun Yang · Heru Suhartanto ·
Guoren Wang · Bin Wang · Jing Jiang · Bing Li ·
Huaijie Zhu · Ningning Cui
Editors

Advanced Data Mining and Applications

19th International Conference, ADMA 2023
Shenyang, China, August 21–23, 2023
Proceedings, Part V

Editors
Xiaochun Yang
Northeastern University
Shenyang, China

Heru Suhartanto
The University of Indonesia
Depok, Indonesia

Guoren Wang
Beijing Institute of Technology
Beijing, China

Bin Wang
Northeastern University
Shenyang, China

Jing Jiang
University of Technology Sydney
Sydney, NSW, Australia

Bing Li
Agency for Science, Technology
and Research (A*STAR)
Singapore, Singapore

Huaijie Zhu
Sun Yat-sen University
Guangzhou, China

Ningning Cui
Anhui University
Hefei, China

ISSN 0302-9743 ISSN 1611-3349 (electronic)
Lecture Notes in Artificial Intelligence
ISBN 978-3-031-46676-2 ISBN 978-3-031-46677-9 (eBook)
https://doi.org/10.1007/978-3-031-46677-9

LNCS Sublibrary: SL7 – Artificial Intelligence

This Springer imprint is published by the registered company Springer Nature Switzerland AG
The registered company address is: Gewerbestrasse 11, 6330 Cham, Switzerland

Paper in this product is recyclable.

Preface

The 19th International Conference on Advanced Data Mining and Applications (ADMA 2023) was held in Shenyang, China, during August 21–23, 2023. Researchers and practitioners from around the world came together at this leading international forum to share innovative ideas, original research findings, case study results, and experienced insights into advanced data mining and its applications. With the ever-growing importance of appropriate methods in these data-rich times, ADMA has become a flagship conference in this field. ADMA 2023 received a total of 503 submissions from 22 countries across five continents. After a rigorous double-blind review process involving 318 reviewers, 216 regular papers were accepted to be published in the proceedings, 123 were selected to be delivered as oral presentations at the conference, 85 were selected as poster presentations, and 8 were selected as industry papers. This corresponds to a full oral paper acceptance rate of 24.4%. The Program Committee (PC), composed of international experts in relevant fields, did a thorough and professional job of reviewing the papers submitted to ADMA 2023, and each paper was reviewed by an average of 2.97 PC members. With the growing importance of data in this digital age, papers accepted at ADMA 2023 covered a wide range of research topics in the field of data mining, including pattern mining, graph mining, classification, clustering and recommendation, multi-objective, optimization, augmentation, and database, data mining theory, image, multimedia and time series data mining, text mining, web and IoT applications, finance and healthcare. It is worth mentioning that ADMA 2023 was organized as a physical-only event, allowing for in-person gatherings and networking. We thank the PC members for completing the review process and providing valuable comments within tight schedules. The high-quality program would not have been possible without the expertise and dedication of our PC members. Moreover, we would like to take this valuable opportunity to thank all authors who submitted technical papers and contributed to the tradition of excellence at ADMA. We firmly believe that many colleagues will find the papers in these proceedings exciting and beneficial for advancing their research. We would like to thank Microsoft for providing the CMT system, which is free to use for conference organization, Springer for their long-term support, the host institution, Northeastern University, for their hospitality and support, Niu Translation and Shuangzhi Bo for their sponsorship. We are grateful for the guidance of the steering committee members, Osmar R. Zaiane, Chengqi Zhang, Michael Sheng, Guodong Long, Xue Li, Jianxin Li, and Weitong Chen. With their leadership and support, the conference ran smoothly. We also would like to acknowledge the support of the other members of the organizing committee. All of them helped to make ADMA 2023 a success. We appreciate the local arrangements, registration and finance management from the local arrangement chairs, registration management chairs and finance chairs Kui Di, Baoyan Song, Junchang Xin, Donghong Han, Guoqiang Ma, Yuanguo Bi, and Baiyou Qiao, the time and effort of the proceedings chairs, Bing Li, Huaijie Zhu, and Ningning Cui, the effort in advertising the conference by the publicity chairs and social network and social media coordination chairs, Xin Wang, Yongxin

Tong, Lina Wang, and Sen Wang, and the effort of managing the Tutorial sessions by the tutorial chairs, Zheng Zhang and Shuihua Wang, We would like to give very special thanks to the web chair, industry chairs, and PhD school chairs Faming Li, Chi Man Pun, Sen Wang, Linlin Ding, M. Emre Celebi, and Zheng Zhang, for creating a successful and memorable event. We also thank sponsorship chair Hua Shao for his sponsorship. Finally, we would like to thank all the other co-chairs who have contributed to the conference.

August 2023

Xiaochun Yang
Bin Wang
Jing Jiang

Organization

Chair of the Steering Committee

Xue Li	University of Queensland, Australia

Steering Committee

Osmar R. Zaiane	University of Alberta, Canada
Chengqi Zhang	Sydney University of Technology, Australia
Michael Sheng	Macquarie University, Australia
Guodong Long	Sydney University of Technology, Australia
Xue Li	University of Queensland, Australia
Jianxin Li	Deakin University, Australia
Weitong Chen	Adelaide University, Australia

Honor Chairs

Xingwei Wang	Northeastern University, China
Xuemin Lin	Shanghai Jiao Tong University, China
Ge Yu	Northeastern University, China

General Chairs

Xiaochun Yang	Northeastern University, China
Heru Suhartanto	The University of Indonesia, Indonesia
Guoren Wang	Beijing Institute of Technology, China

Program Chairs

Bin Wang	Northeastern University, China
Jing Jiang	University of Technology Sydney, Australia

Local Arrangement Chairs

Kui Di	Northeastern University, China
Baoyan Song	Liaoning University, China
Junchang Xin	Northeastern University, China

Registration Management Chairs

Donghong Han	Northeastern University, China
Guoqiang Ma	Northeastern University, China
Yuanguo Bi	Northeastern University, China

Finance Chair

Baiyou Qiao	Northeastern University, China

Sponsorship Chair

Hua Shao	Shenyang Huaruibo Information Technology Co., Ltd., China

Publicity Chairs

Xin Wang	Tianjin University, China
Yongxin Tong	Beihang University, China
Lina Wang	Wuhan University, China

Social Network and Social Media Coordination Chair

Sen Wang	University of Queensland, Australia

Proceeding Chairs

Bing Li	Agency for Science, Technology and Research (A*STAR), Singapore
Huaijie Zhu	Sun Yat-sen University, China
Ningning Cui	Anhui University, China

Tutorial Chairs

Zheng Zhang	Harbin Institute of Technology, Shenzhen, China
Shuihua Wang	University of Leicester, UK

Web Chair

Faming Li	Northeastern University, China

Industry Chairs

Chi Man Pun	University of Macau, China
Sen Wang	University of Queensland, Australia
Linlin Ding	Liaoning University, China

PhD School Chairs

M. Emre Celebi	University of Central Arkansas, USA
Zheng Zhang	Harbin Institute of Technology, Shenzhen, China

Program Committee

Meta Reviewers

Bohan Li	Nanjing University of Aeronautics and Astronautics, China
Can Wang	Griffith University, Australia
Chaokun Wang	Tsinghua University, China
Cheqing Jin	East China Normal University, China
Guodong Long	University of Technology Sydney, Australia
Hongzhi Wang	Harbin Institute of Technology, China
Huaijie Zhu	Sun Yat-sen University, China
Jianxin Li	Deakin University, Australia
Jun Gao	Peking University, China
Lianhua Chi	La Trobe University, Australia
Lin Yue	University of Newcastle, Australia
Tao Shen	University of Technology Sydney, Australia

Wei Emma Zhang	University of Adelaide, Australia
Weitong Chen	Adelaide University, Australia
Xiang Lian	Kent State University, USA
Xiaoling Wang	East China Normal University, China
Xueping Peng	University of Technology Sydney, Australia
Xuyun Zhang	Macquarie University, Australia
Yanjun Zhang	Deakin University, Australia
Zheng Zhang	Harbin Institute of Technology, Shenzhen, China

Reviewers

Abdulwahab Aljubairy	Macquarie University, Australia
Adita Kulkarni	SUNY Brockport, USA
Ahoud Alhazmi	Macquarie University, Australia
Akshay Peshave	GE Research, USA
Alex Delis	Univ. of Athens, Greece
Alexander Zhou	Hong Kong University of Science and Technology, China
Baoling Ning	Heilongjiang University, China
Bin Zhao	Nanjing Normal University, China
Bing Li	Institute of High Performance Computing, A*STAR, Singapore
Bo Tang	Southern University of Science and Technology, China
Carson Leung	University of Manitoba, Canada
Changdong Wang	Sun Yat-sen University, China
Chao Zhang	Tsinghua University, China
Chaokun Wang	Tsinghua University, China
Chaoran Huang	University of New South Wales, Australia
Chen Wang	Chongqing University, China
Chengcheng Yang	East China Normal University, China
Chenhao Zhang	University of Queensland, Australia
Cheqing Jin	East China Normal University, China
Chuan Ma	Zhejiang Lab, China
Chuan Xiao	Osaka University and Nagoya University, Japan
Chuanyu Zong	Shenyang Aerospace University, China
Congbo Ma	University of Adelaide, Australia
Dan He	University of Queensland, Australia
David Broneske	German Centre for Higher Education Research and Science Studies, Germany

Dechang Pi	Nanjing University of Aeronautics and Astronautics, China
Derong Shen	Northeastern University, China
Dima Alhadidi	University of New Brunswick, Canada
Dimitris Kotzinos	ETIS, France
Dong Huang	South China Agricultural University, China
Dong Li	Liaoning University, China
Dong Wen	University of New South Wales, Australia
Dongxiang Zhang	Zhejiang University, China
Dongyuan Tian	Jilin University, China
Dunlu Peng	University of Shanghai for Science and Technology, China
Eiji Uchino	Yamaguchi University, Japan
Ellouze Mourad	University of Sfax, Tunisia
Elsa Negre	LAMSADE, Paris-Dauphine University, France
Faming Li	Northeastern University, China
Farid Nouioua	Université Mohamed El Bachir El Ibrahimi de Bordj Bou Arréridj, Algeria
Genoveva Vargas-Solar	CNRS, France
Gong Cheng	Nanjing University, China
Guanfeng Liu	Macquarie University, Australia
Guangquan Lu	Guangxi Normal University, China
Guangyan Huang	Deakin University, Australia
Guannan Dong	University of Macau, China
Guillaume Guerard	ESILV, France
Guodong Long	University of Technology Sydney, Australia
Haïfa Nakouri	ISG Tunis, Tunisia
Hailong Liu	Northwestern Polytechnical University, China
Haojie Zhuang	University of Adelaide, Australia
Haoran Yang	University of Technology Sydney, Australia
Haoyang Luo	Harbin Institute of Technology (Shenzhen), China
Hongzhi Wang	Harbin Institute of Technology, China
Huaijie Zhu	Sun Yat-sen University, China
Hui Yin	Deakin University, Australia
Indika Priyantha Kumara Dewage	Tilburg University, The Netherlands
Ioannis Konstantinou	University of Thessaly, Greece
Jagat Challa	BITS Pilani, India
Jerry Chun-Wei Lin	Western Norway University of Applied Sciences, Norway
Jiabao Han	NUDT, China
Jiajie Xu	Soochow University, China
Jiali Mao	East China Normal University, China

Jianbin Qin	Shenzhen University, China
Jianhua Lu	Southeast University, China
Jianqiu Xu	Nanjing University of Aeronautics and Astronautics, China
Jianxin Li	Deakin University, Australia
Jianxing Yu	Sun Yat-sen University, China
Jiaxin Jiang	National University of Singapore, Singapore
Jiazun Chen	Peking University, China
Jie Shao	University of Electronic Science and Technology of China, China
Jie Wang	Indiana University, USA
Jilian Zhang	Jinan University, China, China
Jingang Yu	Shenyang Institute of Computing Technology, Chinese Academy of Sciences
Jing Du	University of New South Wales, Australia
Jules-Raymond Tapamo	University of KwaZulu-Natal, South Africa
Jun Gao	Peking University, China
Junchang Xin	Northeastern University, China
Junhu Wang	Griffith University, Australia
Junshuai Song	Peking University, China
Kai Wang	Shanghai Jiao Tong University, China
Ke Deng	RMIT University, Australia
Kun Han	University of Queensland, Australia
Kun Yue	Yunnan University, China
Ladjel Bellatreche	ISAE-ENSMA, France
Lei Duan	Sichuan University, China
Lei Guo	Shandong Normal University, China
Lei Li	Hong Kong University of Science and Technology (Guangzhou), China
Li Li	Southwest University, China
Lin Guo	Changchun University of Science and Technology, China
Lin Mu	Anhui University, China
Linlin Ding	Liaoning University, China
Lizhen Cui	Shandong University, China
Long Yuan	Nanjing University of Science and Technology, China
Lu Chen	Swinburne University of Technology, Australia
Lu Jiang	Northeast Normal University, China
Lukui Shi	Hebei University of Technology, China
Maneet Singh	IIT Ropar, India
Manqing Dong	Macquarie University, Australia

Mariusz Bajger	Flinders University, Australia
Markus Endres	University of Applied Sciences Munich, Germany
Mehmet Ali Kaygusuz	Middle East Technical University, Turkey
Meng-Fen Chiang	University of Auckland, New Zealand
Ming Zhong	Wuhan University, China
Minghe Yu	Northeastern University, China
Mingzhe Zhang	University of Queensland, Australia
Mirco Nanni	CNR-ISTI Pisa, Italy
Misuk Kim	Sejong University, South Korea
Mo Li	Liaoning University, China
Mohammad Alipour Vaezi	Virginia Tech, USA
Mourad Nouioua	Mohamed El Bachir El Ibrahimi University, Bordj Bou Arreridj, Algeria
Munazza Zaib	Macquarie University, Australia
Nabil Neggaz	Université des Sciences et de la Technologie d'Oran Mohamed Boudiaf, Algeria
Nicolas Travers	Léonard de Vinci Pôle Universitaire, Research Center, France
Ningning Cui	Anhui University, China
Paul Grant	Charles Sturt University, Australia
Peiquan Jin	University of Science and Technology of China, China
Peng Cheng	East China Normal University, China
Peng Peng	Hunan University, China
Peng Wang	Fudan University, China
Pengpeng Zhao	Soochow University, China
Philippe Fournier-Viger	Shenzhen University, China
Ping Lu	Beihang University, China
Pinghui Wang	Xi'an Jiaotong University, China
Qiang Yin	Shanghai Jiao Tong University, China
Qing Liao	Harbin Institute of Technology (Shenzhen), China
Qing Liu	Data61, CSIRO, Australia
Qing Xie	Wuhan University of Technology, China
Quan Chen	Guangdong University of Technology, China
Quan Z. Sheng	Macquarie University, Australia
Quoc Viet Hung Nguyen	Griffith University, Australia
Rania Boukhriss	University of Sfax, Tunisia
Riccardo Cantini	University of Calabria, Italy
Rogério Luís Costa	Polytechnic of Leiria, Portugal
Rong Zhu	Alibaba Group, China
Ronghua Li	Beijing Institute of Technology, China
Rui Zhou	Swinburne University of Technology, Australia

Weijia Zhang	University of Newcastle, Australia
Weiwei Ni	Southeast University, China
Weixiong Rao	Tongji University, China
Wen Zhang	Wuhan University, China
Wentao Li	Hong Kong University of Science and Technology (Guangzhou), China
Wenyun Li	Harbin Institute of Technology (Shenzhen), China
Xi Guo	University of Science and Technology Beijing, China
Xiang Lian	Kent State University, USA
Xiangguo Sun	Chinese University of Hong Kong, China
Xiangmin Zhou	RMIT University, Australia
Xiangyu Song	Swinburne University of Technology, Australia
Xianmin Liu	Harbin Institute of Technology, China
Xianzhi Wang	University of Technology Sydney, Australia
Xiao Pan	Shijiazhuang Tiedao University, China
Xiaocong Chen	University of New South Wales, Australia
Xiaofeng Gao	Shanghai Jiaotong University, China
Xiaoguo Li	Singapore Management University, Singapore
Xiaohui (Daniel) Tao	University of Southern Queensland, Australia
Xiaoling Wang	East China Normal University, China
Xiaowang Zhang	Tianjin University, China
Xiaoyang Wang	University of New South Wales, Australia
Xiaojun Xie	Nanjing Agricultural University, China
Xin Cao	University of New South Wales, Australia
Xin Wang	Southwest Petroleum University, China
Xinqiang Xie	Neusoft, China
Xiuhua Li	Chongqing University, China
Xiujuan Xu	Dalian University of Technology, China
Xu Yuan	Harbin Institute of Technology, Shenzhen, China
Xu Zhou	Hunan University, China
Xupeng Miao	Carnegie Mellon University, USA
Xuyun Zhang	Macquarie University, Australia
Yajun Yang	Tianjin University, China
Yanda Wang	Nanjing University of Aeronautics and Astronautics, China
Yanfeng Zhang	Northeastern University, China
Yang Cao	Hokkaido University, China
Yang-Sae Moon	Kangwon National University, South Korea
Yanhui Gu	Nanjing Normal University, China
Yanjun Shu	Harbin Institute of Technology, China
Yanlong Wen	Nankai University, China

Yanmei Hu	Chengdu University of Technology, China
Yao Liu	University of New South Wales, Australia
Yawen Zhao	University of Queensland, Australia
Ye Zhu	Deakin University, Australia
Yexuan Shi	Beihang University, China
Yicong Li	University of Technology Sydney, Australia
Yijia Zhang	Jilin University, China
Ying Zhang	Nankai University, China
Yingjian Li	Harbin Institute of Technology, Shenzhen, China
Yingxia Shao	BUPT, China
Yishu Liu	Harbin Institute of Technology, Shenzhen, China
Yishu Wang	Northeastern University, China
Yixiang Fang	Chinese University of Hong Kong, Shenzhen, China
Yixuan Qiu	The University of Queensland, Australia
Yong Zhang	Tsinghua University, China
Yongchao Liu	Ant Group, China
Yongpan Sheng	Southwest University, China
Yongqing Zhang	Chengdu University of Information Technology, China
Youwen Zhu	Nanjing University of Aeronautics and Astronautics, China
Yu Gu	Northeastern University, China
Yu Liu	Huazhong University of Science and Technology, China
Yu Yang	Hong Kong Polytechnic University, China
Yuanbo Xu	Jilin University, China
Yucheng Zhou	University of Technology Sydney, Australia
Yue Tan	University of Technology Sydney, Australia
Yunjun Gao	Zhejiang University, China
Yunzhang Huo	Hong Kong Polytechnic University, China
Yurong Cheng	Beijing Institute of Technology, China
Yutong Han	Dalian Minzu University, China
Yutong Qu	University of Adelaide, Australia
Yuwei Peng	Wuhan University, China
Yuxiang Zeng	Hong Kong University of Science and Technology, China
Zesheng Ye	University of New South Wales, Sydney, Australia
Zhang Anzhen	Shenyang Aerospace University, China
Zhaojing Luo	National University of Singapore, Singapore
Zhaonian Zou	Harbin Institute of Technology, China
Zheng Liu	Nanjing University of Posts and Telecommunications, China

Zhengyi Yang	University of New South Wales, Australia
Zhenying He	Fudan University, China
Zhihui Wang	Fudan University, China
Zhiwei Zhang	Beijing Institute of Technology, China
Zhixin Li	Guangxi Normal University, China
Zhongnan Zhang	Xiamen University, China
Ziyang Liu	Tsinghua University, China

Contents – Part V

Data Mining

Applications (Including Industry Track Papers)

Data Mining

Heterogeneous Line Graph Neural Network for Link Prediction

Yiyang Sun[1], Yuncong Zhao[1], Longjie Li[1,2(✉)], and Hu Dong[1]

[1] School of Information Science and Engineering, Lanzhou University, Lanzhou 730000, China
{sunyy20,zhaoyc20,ljli,dongh20}@lzu.edu.cn
[2] Key Laboratory of Media Convergence Technology and Communication, Lanzhou 730000, Gansu, China

Abstract. Heterogeneous network link prediction is an important network information mining problem. Existing link prediction methods for heterogeneous networks typically require predefined meta-paths with prior knowledge. To address the problem, we propose a new model, named Heterogeneous Line Graph Neural Network (HLGNN), in this paper. Firstly, we design a line graph transformation module to encapsulate node features and transform the heterogeneous network into a heterogeneous line graph. Then, we propose an intra-type aggregation component to collect the same type of edges. As we have aggregated node information in each type, we design an inter-layer aggregation to combine messages from multiple node types. Finally, we put the aggregation results into a multilayer perceptron to achieve link prediction. The experimental results show that, compared to the state-of-the-art baselines, the proposed method achieves superior performance.

Keywords: Link prediction · Heterogeneous networks · Graph neural networks · Line graph

1 Introduction

Link prediction is one of the key problems in network information mining, which can predict whether there is an edge between two nodes based on the attributes of the nodes and the structural information of the network [1]. As a consequence, link prediction can be utilized for recommendation system [2], implicit relation mining [3], knowledge graph completion [4], protein interaction prediction [5], and so on.

A homogeneous network is composed of a single type of nodes and edges. However, in the real world, there are many kinds of complex systems that cannot be adequately described by homogeneous networks. In recent years, the study of heterogeneous networks has become a consistently interesting research topic owing to its practical implications in a wide range of applications.

In earlier times, some existing researches suggested heuristic methods for link prediction, which extract the structural features of a network to measure

X. Yang et al. (Eds.): ADMA 2023, LNAI 14180, pp. 3–14, 2023.
https://doi.org/10.1007/978-3-031-46677-9_1

the connectivity between nodes. For example, the Common neighbors index [6] computes the number of shared neighbors between two nodes as their similarity. The Resource Allocation index [7] further discriminates the contributions of different common neighbors by penalizing the influence of common neighbors with large degrees. The Katz index [8] counts all paths connecting two nodes, simultaneously assigning smaller weights to longer paths. However, these methods rely on the heuristics of the existence of links, and thence are not applied to networks with different characteristics.

As a deep learning technique on graph data, graph neural networks (GNNs), such as GCNs (Graph Convolutional Networks) [9], GATs (Graph Attention Networks) [10], and GraphSAGE [11], have been considerably successful in many graph-based tasks. These GNNs can effectively generate the embeddings of nodes by aggregating the information of their neighbors. Using the embeddings of nodes, we can solve the task of link prediction in homogeneous networks. However, due to the diverse types of edges and nodes in heterogeneous networks, link prediction in heterogeneous networks cannot be commendably handled by these methods.

Inspired by the success of GNNs in homogeneous networks, the study of heterogeneous graph neural networks (HGNNs) has attracted extensive attention from researchers. For the complexity of heterogeneous networks, one major line of HGNNs is to define and leverage metapaths [12] to preserve semantics and benefit heterogeneous structure modeling. HGNNs, such as HAN (Heterogeneous Graph Attention Network) [13] and MAGNN (Metapath Aggregated Graph Neural Network) [14], generally choose a number of specific metapaths and aggregate information on these metapaths. However, it is obvious that the information of heterogeneous networks cannot be completely represented by the fixed metapaths. Other methods, such as HetGNN [15] and THGNN (Topic-aware Heterogeneous Graph Neural Network) [16], further improve heterogeneous graph neural networks by using special attributes or contents associated with nodes, such as text, image, and video. However, not all heterogeneous networks in practical application have such rich attribute content information.

In addition, similar to GNNs, HGNNs usually propagate and aggregate features within various neighborhoods via different mechanisms [10,11]. This way may lead to homophily [17]. However, in heterogeneous networks, there also may be some 'opposites attract' situations leading to heterophily. As in protein networks, there are cases where proteins of different types are more likely to react. Currently, link prediction methods for heterogeneous network rarely consider heterophily.

To address the above challenges, we propose a heterogeneous line graph neural network, named HLGNN, to learn comprehensive link representations for link prediction. HLGNN first applies line graph transformation to generate heterogeneous edge features and then transforms the heterogeneous network into heterogeneous line graph. Next, HLGNN conducts intra-type aggregation on the transformed line graph. By aggregating the information of neighbors according to their types, intra-types aggregation can carry latent but more fine-grained type-level semantics than aggregation by metapaths. Meanwhile, HLGNN imports the principle of heterophily from homogeneous networks. Then, HLGNN applies inter-layers aggregation to combine the messages that obtained from different types. In this way, HLGNN can get more comprehensive information in heterogeneous networks.

In summary, this work makes the following major contributions.

(1) We propose an end-to-end heterogeneous network link prediction method without utilizing metapaths.

(2) We propose a line graph transformation that can transform the link prediction task to the binary classification of nodes in the transformed line graph.

(3) We introduce the idea of dealing with heterophily in heterogeneous networks and propose a new aggregation method based on node type in the transformed heterogeneous line graph. Therefore, we can distinguish the contribution of different node type.

(4) We conduct experiments to evaluate the performance of the proposed model. Experimental results demonstrate that our proposed model outperforms the state-of-the-art baselines.

2 Related Work

Heterogeneous network link prediction task is mostly completed by heterogeneous graph embedding methods that originally aim to deal with the scenarios with nodes and/or edges of various types through projecting nodes into a low-dimensional vector space. Heterogeneous graph embedding-based link prediction methods settle the task through calculating the similarity of the embeddings of target nodes. Metapath2vec [18] first generates a sequence of nodes by performing random walk on a certain metapath [19]. Then, it puts the sequence into the skip-gram model to produce node embeddings. However, metapath2vec can only take advantage of one metapath and may ignore productive features of the networks. HERec [20] generates random walk sequences that are guided by selected metapaths. Then HERec fuses the result of different metapaths to perform embeddings. Although HERec can gather more metapath information, it discards node attributes. GATNE [21] considers both the structure and attribute information of heterogeneous neighbors, but ignores the multiple relations among different types of nodes. HAN [13] is a HGNN, which extends the attention mechanism into heterogeneous networks to automatically learn the importance of neighbors on different metapaths. MAGNN [14] proposes both intra-metapath aggregation and inter-metapath aggregation to capture the information from the metapath connecting to the node and fuse the latent features obtained from multiple metapaths. THGNN [16] is a HGNN designed for link prediction, which expands the text content of specific nodes to import topic-aware semantics in heterogeneous networks.

In heterogeneous networks link prediction, the above methods usually lean node embeddings and then calculate the possibility of link existence. Therefore, they might loss the potential association of nodes. On the other hand, these methods depend on the predefined metapaths, which cannot completely retain all the semantic information in heterogeneous networks. In conclusion, an end-to-end link prediction method for heterogeneous networks that does not rely on metapath is a necessary.

3 Preliminaries

Definition 1. ***(Line Graph)*** *Given a graph $\mathcal{G} = (\mathcal{V}, \mathcal{E})$, the line graph transformed from $\mathcal{G}$ is denoted as $L(\mathcal{G})$. Let $e \in \mathcal{E}$ be an edge in $\mathcal{G}$, then there is a node, marked as v_e, in $L(\mathcal{G})$. Given two edges $e_1, e_2 \in \mathcal{E}$, if e_1 and e_2 share the same end-node, then there is an edge between nodes v_{e_1} and v_{e_2} in $L(\mathcal{G})$.*

Definition 2. ***(Heterogeneous Network)*** *A heterogeneous network, is defined as $\mathcal{G} = (\mathcal{V}, \mathcal{E}, T_v, T_e)$, consists of a set of nodes $\mathcal{V}$ and a set of edges $\mathcal{E}$. The node set $\mathcal{V}$ and the edge set $\mathcal{E}$ are associated with a node-type mapping function $\Phi : \mathcal{V} \rightarrow T_v$ and an edge-type mapping function $\Psi : \mathcal{E} \rightarrow T_e$, respectively. T_v denotes the set of node types and T_e represents the set of edge types, where $|T_v| + |T_e| > 2$.*

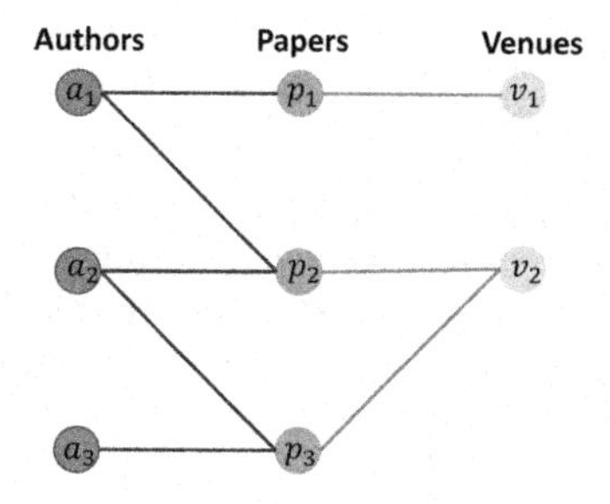

Fig. 1. An example heterogeneous network with three types of nodes (i.e., authors, papers, and venues).

Figure 1 illustrates an example heterogeneous network, which includes three types of nodes (i.e., authors, papers, and venues) and two kinds of edges (i.e., authors-papers and papers-venues).

According to the definition of line graph [22,23], we define the heterogeneous line graph as follows.

Definition 3. ***(Heterogeneous Line Graph)*** *Given a heterogeneous network $\mathcal{G} = (\mathcal{V}, \mathcal{E}, T_v, T_e)$, $HL(\mathcal{G}) = (\mathcal{V}^l, \mathcal{E}^l, T_v^l, T_e^l)$ is the heterogeneous line graph transformed from $\mathcal{G}$. According to Definition 1, a node in $HL(\mathcal{G})$ is generated from an edge in $\mathcal{G}$ and an edge in $HL(\mathcal{G})$ is derived from two edges from $\mathcal{G}$ those share a common end-node. T_v^l denotes the set of node types and T_e^l represents the set of edge types in $HL(\mathcal{G})$, such that $T_v^l = T_e$ and $T_e^l = T_e \times T_e$.*

Figure 2 shows an example of heterogeneous line graph.

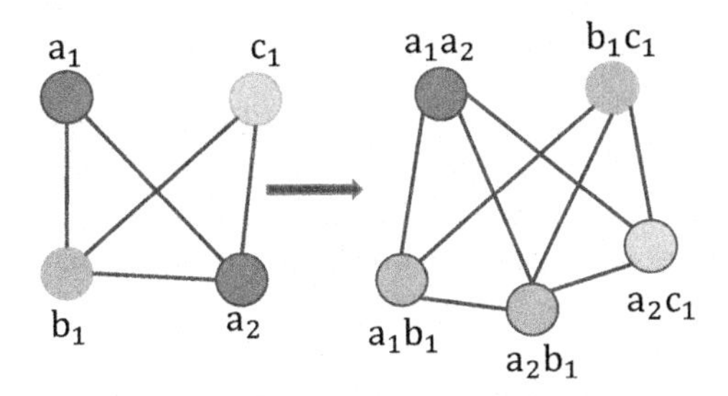

Fig. 2. An example of heterogeneous line graph.

4 Methodology

In this section, we describe the proposed heterogeneous line graph neural network (HLGNN), which is designed for link prediction in heterogeneous networks. HLGNN consists of three major components: line graph transformation, intra-type aggregation, and inter-layer aggregation. Figure 3 illustrates the framework of HLGNN.

Line Graph Transformation. Given a heterogeneous network, HLGNN generates features for edges and transforms the heterogeneous network into a heterogeneous line graph.

Intra-Type Aggregation. In this part, to discriminate the contributions of nodes with different types to link prediction task, HLGNN aggregates the information from different types nodes separately.

Inter-layer Aggregation and Model Training. After aggregating the features of the neighbors of different types, HLGNN combines the embeddings in different hops with concatenation. Then, we train the prediction model using the embeddings of all sampled nodes.

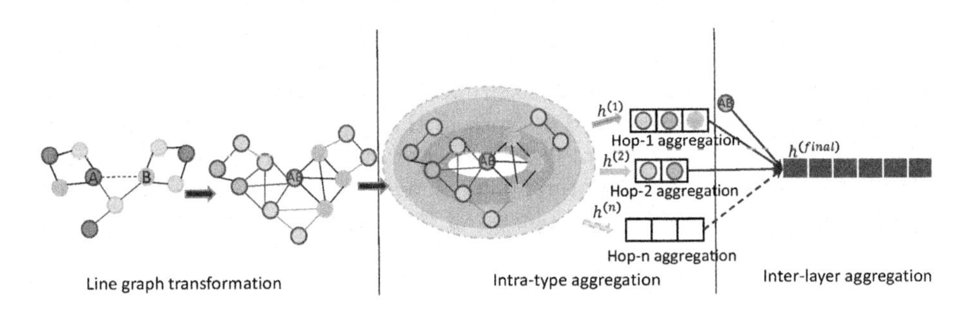

Fig. 3. The overall framework of the proposed model HLGNN.

4.1 Line Graph Transformation

Heterogeneous networks are usually composed of many types of nodes and edges, and different types of nodes generally have different features, which means that different types of nodes may have various levels of vector dimensions. For example, in an academic network, the features of authors are usually determined by their research fields, while the features of papers are decided by their keywords. In general, the dimensions of different types of nodes are not the same. Even if their dimensions are equal, the meanings of the corresponding features are diverse.

To overcome these challenges, we adopt the line graph transformation component to transform the sets of node pairs, which are located in different dimensions, into the sets of edges with features in the same potential dimensions. To this end, we compute the features of edges and project them into the same potential space before aggregating the information. The operation is given as follows [22]:

$$\mathbf{h}_e = \mathbf{W}_{v1,v2}(\mathbf{X}_{v1}||\mathbf{X}_{v2}), \tag{1}$$

where $\mathbf{X}_{v1} \in \mathbb{R}^{dv1}, \mathbf{X}_{v2} \in \mathbb{R}^{dv2}$ are the feature vectors of nodes $v1$ and $v2$, $\mathbf{W}_{v1,v2} \in \mathbb{R}^{de \times (dv1+dv2)}$ is the parametric matrix for the edge that consists of $v1$ and $v2$, $\mathbf{h}_e \in \mathbb{R}^{de}$ is the projected feature vector of the edge, and $||$ denotes the vector concatenation operator. In addition, $dv1$, $dv2$, and de are the corresponding dimensions.

According to Eq. (1), we can concatenate the features of original nodes as the features of edges. But the features might be different when we change the sequence of the nodes. To overcome these limitation, we design a particular ordering for node types in a heterogeneous network. For instance, in Fig. 1, we have the node types of *authors*, *papers*, and *venues*. We define the edge types as *author-paper* and *paper-venue*. Afterward, we can only use the edges like (a_1, p_1), (a_2, p_3), and (p_2, v_2).

Then we convert the heterogeneous network into a heterogeneous line graph. The transformation rules are as in Definition 3. In addition, we attach a type to each node in the heterogeneous line graph, which is the type of the edge that generates the node. Line graph transformation has addressed the diversity of vector dimensions of different edge types in the origin graph and successfully converted the heterogeneous graph to line graph. Through this operation, the task of link prediction is transformed into a node classification task on the line graph, which can reduce the complexity of the calculation and preserve more information.

4.2 Intra-type Aggregation

From the previous module, we get a transformed heterogeneous line graph. In this step, we shall zoom into the intra-type aggregation block of HLGNN.

Given a transformed heterogeneous line graph $HL(\mathcal{G})$, the intra-type aggregation can learn more fine-grained semantic and structural information of the target node, than the metapath-based aggregation, by separating the effects of the individual types of neighbors. Let v be a target node in $HL(\mathcal{G})$, if the distance from node u to v is k, we say u is a k-hop neighbor of v. Suppose there are n node types in the line graph, i.e., $t_1, t_2, \cdots, t_n$, intra-type aggregation employs a type-specific encoder to transform all the k-hop neighbors' feature vectors into a single vector, which is

$$\mathbf{h}_v^{(k)} = \mathbb{C}_{i=1}^{n}(agg\{\mathbf{h}_u | t(u) = t_i\}), \tag{2}$$

where $\mathbf{h}_u$ denotes the feature vector of node u that is a k-hop neighbor of v, $t(u) = t_i$ means that the type of u is t_i. To distinguish different effects of node type, inspired by H2GCN [24], we set $\mathbb{C}(\cdot)$ as concatenation and $agg(\cdot)$ as degree-normalized average of neighbors' feature vectors. Then the equation can be rewritten as:

$$\mathbf{h}_v^{(k)} = \|_{i=1}^{n} \left(\sum_{u,t(u)=t_i} \frac{\mathbf{h}_u}{\sqrt{d_{v,k}}\sqrt{d_{u,k}}} \right), \tag{3}$$

where $d_{v,k}$ is the number of v's k-hop neighbors. Using Eq. (3), we gather the information from each type of nodes through the degree-normalized average in the specific type, and then separate the various contributions of different types by the concatenation operation.

4.3 Inter-layer Aggregation and Model Training

From the previous aggregation, we get the information of k-hop neighbors. Each hop of neighbors' embeddings contain type-wise information. In this step, we shall fuse the information from different hops. Suppose we consider h-hop neighbors in the proposed model, we obtained $\mathbf{h}_v^{(1)}, \mathbf{h}_v^{(2)}, \cdots, \mathbf{h}_v^{(h)}$ for node v from the last step. To combine all of the h embeddings of v, a naïve approach is to take an element-wise mean of these vectors. Cause of the weak performance of the mean pooling operation, as in H2GCN [24] and SGC [25], we do not use this approach. Since the effects of neighbors in different hops may be diverse, we combine the h embeddings of v with simple concatenation to compute the final embedding of v learned from $HL(\mathcal{G})$, which is

$$\mathbf{h}_v^{(final)} = \|_{k=1}^{h} \mathbf{h}_v^{(k)}. \tag{4}$$

After getting the final embedding of node v in the transformed line graph, we perform the downstream task through a multilayer perceptron (MLP) classifier. In the heterogeneous line graph $HL(\mathcal{G})$, the downstream task is node classification; mapped to the original heterogeneous network, the task is link prediction. For a linked node pair in the original graph, we give the ground-truth label of 1. Correspondingly, the class label of the node in the line graph generated from the node pair is 1. For a non-linked node pair in the original graph, we assign the ground truth label of 0. Accordingly, the class label of the corresponding node in the line graph is 0. By entering the embedding $\mathbf{h}_v^{(final)}$ into MLP, we obtain the output of v as

$$z_v = \sigma(\mathbf{W} \cdot \mathbf{h}_v^{(final)}), \tag{5}$$

where $\sigma(\cdot)$ is the sigmoid function and $\mathbf{W}$ is a weight matrix. The output z_v is the predicted label of node v.

Referring to negative sampling [26], we choose all the linked node pairs as positive samples and sample the same number of non-linked node pairs as negative samples. The overall training process is optimized by minimizing the cross-entropy loss.

5 Experiments

In this section, we experimentally evaluate the performance of the proposed HLGNN model. The experiments aim to address the following two questions:

RQ1. How does HLGNN perform the link prediction task in heterogeneous network?

RQ2. How do the three major components, i.e., line graph transformation, intra-type aggregation, and inter-layer aggregation, affect the performance of HLGNN?

5.1 Networks

Table 1. Statistics of datasets.

Dataset	Node	Edge	Metapath
DBLP	# author (A): 4,057 # paper (P): 14,328 # term (T): 7,723 # venue (V): 20	# A-P: 19,645 # P-T: 85,810 # P-V: 14,328	A-P-A A-P-T-P-A APVPA
Amazon	# movie (M): 1,754 # brand(B): 293 # user(U): 2,476 # category(C): 95	# M-U: 34,042 # M-B: 734 # M-C: 4,913	U-M-U U-M-B-M-U U-M-C-M-U
Last.fm	#user(U): 1,892 #artist(A): 17632 #tag(T): 1,088	#U-U: 12717 #U-A: 92,834 #A-T: 23,253	U-A-U U-A-T-A-U

In our experiments, we use three commonly used heterogeneous networks, i.e., DBLP, Amazon, and Last.fm, to evaluate the performance of HLGNN. Simple statistics of the three networks are summarized in Table 1.

DBLP: This network is a subset of DBLP, composed of authors (A), papers (P), terms (T), and venues (V). On this network, we predict both author-paper links and co-author hyper-links.

Amazon: It is a movie review dataset extracted from Amazon. The node types in this network are movies (M), brands (B), users (U), and categories(C). On this network, we choose to predict U-M edges to simulate movie recommendation.

Last.fm: This network is extracted from a British online music website that consists of the relationship between listeners and singers. It is also a widely used recommendation dataset, consisting of users, artists, and artist tags. We predict user-artist links on Last.fm.

In experiments, the node pairs with different existence of links are divided into training, validation, and testing sets by 70%, 10%, 20%, respectively.

5.2 Baselines

To evaluate the performance of the proposed HLGNN, we choose a collection of baselines, which are Deepwalk [27], node2vec [28], metapath2vec [18], HERec [20], GAT [10], H2GCN [24], HAN [13], MAGNN [14], and THGNN [16]. Among these baselines, Deepwalk and node2vec are node embedding methods designed for homogeneous networks, while metapath2vec and HERec are node embedding methods applied to heterogeneous networks. GAT and H2GCN are two homogeneous GNNs; HAN, MAGNN, and THGNN are three heterogeneous GNNs.

For the four node embedding methods, we set the window size to 5, walk length to 100, walks per node to 40, and the number of negative samples to 5. For GAT, HAN, and MAGNN, we set the number of attention head to 8. For THGNN, we set the same metapath as MAGNN and the topic-aware factors is 8. For all baselines, the parameters we set are based on the common settings of most articles. For our HLGNN, we optimize it by setting the learning rate to 0.005, the hop parameter h to 2, and the number of epochs to 210.

5.3 Link Prediction

We perform the experiment of link prediction to answer the question of RQ1 in this subsection. The results, measured by *area under the ROC curve* (AUC) and *average precision* (AP), are reported in Table 2. The results are the average of 10 independent runs. We can easily observe that HLGNN achieves the best performance in most cases. In general, the performance of HAN, MAGNN, and THGNN is superior to that of other baselines because they are heterogeneous GNNs. Specially, among all the baselines, THGNN obtains the best performance since it considers a more fine-grained semantics in heterogeneous networks. Although GAT and H2GCN are homogeneous GNNs, their results are better than those of metapath2vec and HERec that are heterogeneous node embedding methods. The reason is metapath2vec and HERec are not end-to-end methods, i.e., the embeddings learned in these methods are independent of link prediction. However, both metapath2vec and HERec outperform DeepWalk and node2vec because the latter are homogeneous node embedding methods.

Table 2. Experiment results (%) of link prediction on three networks. The optimal value of each network id in bold.

Datasets	DBLP(A-A)		DBLP(A-P)		Amazon		Last.fm	
Metrics	AUC	AP	AUC	AP	AUC	AP	AUC	AP
DeepWalk	75.54	74.22	78.83	78.09	76.01	78.35	81.88	79.37
node2Vec	63.35	62.87	63.92	63.57	61.05	59.73	65.04	66.32
metapath2vec	76.60	75.92	73.42	72.81	78.12	78.34	84.32	83.69
HERec	76.92	75.83	78.72	78.69	80.65	78.58	85.51	83.46
GAT	85.35	85.08	83.25	84.31	83.29	83.43	87.13	86.73
H2GCN	86.98	86.35	84.59	83.93	81.47	82.09	88.40	88.17
HAN	87.44	89.81	85.60	86.38	84.18	84.14	80.65	88.03
MAGNN	88.56	90.78	85.74	86.80	83.92	83.25	92.89	92.14
THGNN	**90.40**	90.22	86.18	87.62	85.60	84.86	93.28	93.57
HLGNN	89.92	**91.69**	**88.00**	**87.80**	**86.24**	**85.74**	**93.89**	**94.06**

Table 3. Experimental results (%) of ablation study. The optimal value of each network id in bold.

Variant	DBLP(A-A)		Amazon		Last.fm	
	AUC	AP	AUC	AP	AUC	AP
HLGNN	**89.92**	**91.69**	**86.24**	**85.75**	**93.89**	**94.06**
HLGNN-1	87.97	89.36	83.79	83.38	92.14	92.70
$\text{HLGNN}_{\text{intra-mean}}$	84.61	85.73	81.52	81.03	89.95	89.49
$\text{HLGNN}_{\text{intra-linear}}$	86.50	87.24	84.68	84.47	90.91	91.28
$\text{HLGNN}_{\text{inter-linear}}$	87.15	88.62	83.25	81.86	91.87	92.18

5.4 Ablation Study

In this subsection, to answer RQ2, we verify the effectiveness of the components of HLGNN by comparing the results of HLGNN with those of four variants of HLGNN. The results are outlined in Table 3. Here, HLGNN-1 means only aggregating the information of 1-hop neighbors. In HLGNN, we aggregate the information of both 1-hop and 2-hop neighbors. $\text{HLGNN}_{\text{intra-mean}}$ is a variant of HLGNN. When combining the feature vectors of different types of neighbors, mean pooling is used in $\text{HLGNN}_{\text{intra-mean}}$. $\text{HLGNN}_{\text{intra-linear}}$ is an extension of $\text{HLGNN}_{\text{intra-mean}}$ by appending it with a linear transformation. In $\text{HLGNN}_{\text{inter-linear}}$, we use a linear transformation after mean pooling of the embeddings of different hops.

From results in Table 3, we can find that HLGNN outperforms all variants, which proves the effectiveness of the components in HLGNN. By comparing the results of HLGNN and HLGNN-1, we conclude that, besides 1-hop neighbors, 2-hop neighbors can also provide useful information. In addition, the comparison results of HLGNN and $\text{HLGNN}_{\text{intra-mean}}$ indicate that, when combining the feature vectors of different types of neighbors, our approach is superior to the mean pooling operation. Similarly, the comparison results of HLGNN and $\text{HLGNN}_{\text{inter-linear}}$ prove that, when computing the final embedding of a node, the simple concatenation is better than the mean pooling.

6 Conclusion

In this paper, we propose a hetereogeneous line graph neural network, namely HLGNN, to address the problem of hetereogeneous network link prediction. The proposed HLGNN contains three major components, i.e., line graph transformation, intra-type aggregation, and inter-layer aggregation, to deal with the limitations in heterogeneous network link prediction. Additionally, we define the conception of heterogeneous line graph, which can preserve more information in hetereogeneous networks and reveal the relationship between edges. We import the idea of heterophily into heterogeneous networks and propose intra-type aggregation and inter-layer aggregation to perform fine-grained aggregation. Experimental results manifest that HLGNN outperforms the state-of-the-art methods in link prediction task.

Acknowledgments. This study was supported in part by the Science and Technology Program of Gansu Province (Nos. 21JR7RA458 and 21ZD8RA008), and the Supercomputing Center of Lanzhou University.

References

1. Liben-Nowell, D., Kleinberg, J.: The link-prediction problem for social networks. J. Am. Soc. Inform. Sci. Technol. **58**(7), 1019–1031 (2007)
2. Adamic, L.A., Adar, E.: Friends and neighbors on the web. Soc. Networks **25**(3), 211–230 (2003)
3. Lü, L., Zhou, T.: Link prediction in complex networks: a survey. Phys. A **390**(6), 1150–1170 (2011)
4. Nickel, M., Murphy, K., Tresp, V., Gabrilovich, E.: A review of relational machine learning for knowledge graphs. Proc. IEEE **104**(1), 11–33 (2016)
5. Airoldi, E. M., Blei, D. M., Fienberg, S. E., et al.: Mixed membership stochastic blockmodels. J. Mach. Learn. Res., 1981–2014 (2008)
6. Newman, M.E.J.: Clustering and preferential attachment in growing networks. Phys. Rev. E **64**(2), 25102 (2001)
7. Zhou, T., Lü, L., Zhang, Y.: Predicting missing links via local information. Europ. Phys. J. B. **71**(4), 623–630 (2009)
8. Katz, L.: A new status index derived from sociometric analysis. Psychometrika **18**(1), 39–43 (1953)
9. Kipf, T.N., Welling, M.: Semi-supervised classification with graph convolutional networks. In: ICLR (2017)
10. Velickovic, P., Cucurull, G., Casanova, A., et al.: Graph attention networks. In: ICLR (2018)

11. Hamilton, W., Ying, Z., Leskovec, J.: Inductive representation learning on large graphs. In: NeurIPS, pp. 1024–1034 (2017)
12. Sun, Y., Han, J., Yan, X., et al.: Pathsim: Meta path-based top-k similarity search in heterogeneous information networks. Proc. VLDB Endowment **4**(11), 992–1003 (2011)
13. Wang, X., Ji, H., Shi, C., et al.: Heterogeneous graph attention network. In: WWW, pp. 2022–2032 (2019)
14. Fu, X., Zhang, J., Meng, Z., et al.: MAGNN: metapath aggregated graph neural network for heterogeneous graph embedding. In: WWW, pp. 2331–2341 (2020)
15. Zhang, C., Song, D., Huang, C., et al.: Heterogeneous graph neural network. In: KDD, pp. 793–803 (2019)
16. Xu, S., Yang, C., Shi, C., et al.: Topic-aware heterogeneous graph neural network for link prediction. In: CIKM, pp. 2261–2270 (2021)
17. McPherson, M., Smith-Lovin, L., Cook, J.M.: Birds of a feather: homophily in social networks. Ann. Rev. Sociol. **27**(1), 415–444 (2001)
18. Dong, Y., Chawla, N.V., Swami, A.: Metapath2vec: scalable representation learning for heterogeneous networks. In: KDD, pp. 135–144 (2017)
19. Sun, Y., Han, J.: Mining heterogeneous information networks. Principles Methodol. **14**(2), 20–28 (2012). Morgan & Claypool Publishers
20. Shi, C., Hu, B., Zhao, W.X., et al.: Heterogeneous information network embedding for recommendation. IEEE Trans. Knowl. Data Eng., 357–370 (2019)
21. Cen, Y., Zou, X., Zhang, J., et al.: Representation learning for attributed multiplex heterogeneous network. In: KDD, pp. 1358–1368 (2019)
22. Cai, L., Li, J., Wang, J., et al.: Line graph neural networks for link prediction. IEEE Trans. Pattern Anal. Mach. Intell. **44**(9), 5103–5113 (2021)
23. Chen, Z., Li, L., Bruna, J.: Supervised community detection with line graph neural networks. In: ICLR (2019)
24. Zhu J., Yan Y., Zhao L., et al.: Beyond homophily in graph neural networks: current limitations and effective designs. In: NeurIPS, pp. 7793–7804 (2020)
25. Wu, F., Souza, A., Zhang, T., et al.: Simplifying graph convolutional networks. In: ICML, pp. 6861–6871 (2019)
26. Mikolov, T., Chen, K., Corrado, G., et al.: Efficient estimation of word representations in vector space. In: ICLR (2013)
27. Perozzi, B., Al-Rfou, R., Skiena, S.: Deepwalk: online learning of social representations. In: KDD, pp. 701–710 (2014)
28. Grover, A., Leskovec, J.: Node2vec: scalable feature learning for networks. In: KDD, pp. 855–864 (2016)

Improving Open-Domain Answer Sentence Selection by Distributed Clients with Privacy Preservation

Weikuan Wang, Tao Shen, Michael Blumenstein, and Guodong Long(✉)

University of Technology Sydney, Ultimo, Australia
weikuan.wang@student.uts.edu.au,
{tao.shen,Michael.Blumenstein,guodong.long}@uts.edu.au

Abstract. Open-domain answer sentence selection (OD-AS2), as a practical branch of open-domain question answering (OD-QA), aims to respond to a query by a potential answer sentence from a large-scale collection. A dense retrieval model plays a significant role across different solution paradigms, while its success depends heavily on sufficient labeled positive QA pairs and diverse hard negative sampling in contrastive learning. However, it is hard to satisfy such dependencies in a privacy-preserving distributed scenario, where in each client, fewer in-domain pairs and a relatively small collection cannot support effective dense retriever training. To alleviate this, we propose a brand-new learning framework for **P**rivacy-preserving **D**istributed O**D-AS2**, dubbed PDD-AS2. Built upon federated learning, it consists of a client-customized query encoding for better personalization and a cross-client negative sampling for learning effectiveness. To evaluate our learning framework, we first construct a new OD-AS2 dataset, called Fed-NewsQA, based on NewsQA to simulate distributed clients with different genre/domain data. Experiment results show that our learning framework can outperform its baselines and exhibit its personalization ability.

Keywords: Information Retrieval · Federated Learning · Personalization

1 Introduction

Open-domain answer sentence selection (OD-AS2) aims to fetch relevant sentences from a large-scale collection given a query, which is also known as long answer in open-domain question answering (OD-QA). Its interest is growing from both academia and industry [18] as it reaches a balanced granularity between coarse-grained passages [25] and fine-grained phrases [18]. Such balanced-granular answers can relieve crowdsourcing burdens and satisfy most real-world scenarios.

Advanced by surging pre-trained language models [5], representation learning entered a new era and rendered dense retrieval as a significant prerequisite across

X. Yang et al. (Eds.): ADMA 2023, LNAI 14180, pp. 15–29, 2023.
https://doi.org/10.1007/978-3-031-46677-9_2

different solution paradigms (e.g., '*retrieval & read*') to OD-AS2. Built upon a dual-encoder (a.k.a. bi-encoder, two-stream encoder), dense retrieval represents both questions from users and sentences in the collections as dense vectors in the same semantic space, and measures question-sentence relevance via a lightweight metric, e.g., doc-product [10,17].

As training an effective dense retrieval model requires sufficient data – both human-created positive question-answering pairs and a large-scale collection to support negative mining, it remains a formidable challenges to directly apply the dense retrieval to the real-world industrial scenarios, e.g., in-house data inquiry, individual email searches, and personal intelligent assistants. The corpus (i.e., the labeled QA pairs and collections) in each client is usually too scarce and biased to train an effective model, while the corpus from each client cannot be uploaded to a central server for standard distributed learning for a privacy-preserving purpose.

To this end, we propose a new learning framework for **P**rivacy-preserving **D**istributed **OD-AS2**, called PDD-AS2. In particular, built upon a prevailing federated learning (FL) framework, FedAvg [24], PDD-AS2 alleviates the data-scarcity problem along with two significant directions. On the one hand, our framework learns generic representation across clients via FL. On the other hand, we present a client-customized query encoding for personalization and client-specific query distribution. In line with dynamic hard negatives and query-side fine-tuning, it will significantly improve the model's effectiveness. To evaluate our learning framework, PDD-AS2, we propose to construct a new distributed OD-AS2 dataset based on NewsQA [33] w.r.t. news story's genre.

In the experiments, we show that our PDD-AS2 framework can improve the performance of our baseline by 5%-15%. Clients with insufficient training data benefit from the model aggregation greatly.

The main contributions of this work can be summarized as

- We highlight a promising setting of open-domain answer sentence selection (OD-AS2) for real-world industrial applications and propose a privacy-preserving distributed OD-AS2 (PDD-AS2) learning framework towards both personalization and effectiveness.
- We propose a key technique, i.e., client-customized query encoding method to effectively learn PDD-AS2 framework.
- We construct a new distributed OD-AS2 dataset upon, dubbed Fed-NewsQA to evaluate the effectiveness of our framework and its baselines.

2 Related Work

2.1 Open-Domain Question Answering

Open-domain question answering (OD-QA) answers a given question using a collection of documents. It does not require a specified context. Compared with Machine Reading Comprehension (MRC), which is another popular task in NLP, OD-QA is more in line with human behaviors. MRC can only retrieve answers from a given context. Therefore, OD-QA has a very promising future in industry applications.

Traditional OD-QA systems often consists of a multi-stage method, i.e., query analysis, context retrieval, and answer retrieval [1,12,26]. DrQA [4] is the first work to incorporate neural MRC models into OD-QA. A new diagram of OD-QA is proposed. This diagram is a two-stage retriever-reader diagram, which combines IR methods like TF-IDF with a neural MRC model. Nowadays, the retriever-reader diagram is studied in many works [11,20–22,30] and proved to outperform significantly traditional methods in performance and efficiency.

However, the two-stage structure of this retriever-reader has a big problem in practical use. Whenever the model receives a query, it requires a complex and heavy reader model to encode several or even dozens of long contexts in real time, which is unacceptable in practice.

2.2 Dense Retrieval

Dense retrieval has recently become a popular topic in industry and academia due to its advantages of both latency and performance. The key to the success of dense retrieval is its leverage of negative samples to train the model. The early stage of research only uses random negatives to train dense retrieval models [14]. Recently, researchers applied hard-negatives to train the model. Hard negatives refer to samples that are semantically similar to positive samples but are in fact negatives. Some studies [37] demonstrate that most of the boost in the training phase come from these hard negatives. Some researchers use BM25 to retrieve hard negatives [7,17]. Some others use static hard-negatives fixed during the entire training or an epoch [10,35]. [37] propose a dynamic hard-negative method which called query-side fine-tuning.

However, insufficient training data would result in severe performance degradation. [17] shows around a 10% performance difference in top-5 passage retrieval due to an insufficient number of negative samples. [27] found that it is beneficial to increase the number of random negatives in the mini-batch. When using only 10% of training data, the normal dense retrieval model's performance can drop by 20% [23]. In this work, we propose an open-domain question answering method empowered by Federated learning to alleviate the problem. Also, we further explore the potential of query-side fine-tuning for personalization.

2.3 Answer Sentence Selection

The Answer Sentence Selection task was defined by [34]. This task aims to select a sentence that correctly answers the question from a set of sentence candidates. This task has been studied in many works [8,31,32,36]. However, in a typical AS2 task, the model is required to select sentences from several candidates. In our Open-domain Sentence Selection setting, the number of candidates can scale up to one million, which significantly increases the task's difficulty.

2.4 Federated Learning

Federated learning was proposed by [24] as a privacy-preserving solution to leverage personal data on different clients. All the training data is stored locally on each client. Each client uses local data to train its own model locally. After each round of training or a certain training time, these clients allow other clients to learn from the training data of this client with privacy protection by sharing the model weights or gradients.

Recently, some researchers have applied Federated learning to different NLP tasks [9,13,15]. In these scenarios, user data are scattered in different devices (e.g., cell phones) or different facilities (e.g., banks, hospitals). Moreover, these data cannot be uploaded to the central server due to privacy concerns related to items such as users' input method records, medical records, etc. However, the combination of Federated learning of open-domain question answering has not been studied yet.

3 Methodology

In this section, we first introduce the preliminaries of our work. Then we present our proposed client-customized query encoding and cross-client negative sampling in our PDD-AS2 framework. Later, we detail the training process of our PDD-AS2 framework and our proposed Fed-NewsQA benchmark for evaluating our framework.

3.1 Preliminary

Task Formulation. In line with existing works [8,17,31,37], we first formulate open-domain answer sentence selection (OD-AS2) under distributed setting as follows: For each client $c^i \in \mathbb{C}$ with its large-scale sentence collection $\mathbb{S}^i = \{s_1^i ... s_n^i\}$, it aims to fetch potential answer sentence(s) s_k^i from $\mathbb{S}^i$ that answers a given query $q \in \mathbb{Q}$. In the OD-AS2 setting, the sentence set $\mathbb{S}^i$ contains sentences from all passages in c^i. If no confusion is caused, we omit the superscript 'i' for a specific client in the remainder.

Usually, a query q and its answer sentence s_q^+ are often provided as positive training data in each client. Hence, it is necessary to sample a set of negative for q to construct negative samples, i.e.,

$$\mathbb{N}_q = \{d | d \sim P(\mathbb{S})\}, \tag{1}$$

where $P(\cdot)$ denotes a distribution over $\mathbb{S}$. For simplicity, we omit the query-specific subscript indicator, q.

Then, a contrastive learning framework is usually employed to learn an efficient retrieval model. Formally, a representation learning module is first used to embed q and each $s \in \{s^+\} \cup \mathbb{N}$ and then derive a probability distribution over $\{s^+\} \cup \mathbb{N}$. Specifically,

$$P(\{s^+\} \cup \mathbb{N} | q; \Theta) = 1/Z \tag{2}$$

$$exp(< \text{Enc}(q; \Theta^{(q)}), \text{Enc}(s; \Theta^{(s)}) >$$

where $\Theta = \{\Theta^{(q)}, \Theta^{(s)}\}$, Z denotes softmax normalization term, Θ parameterizes a text encoder for a single vector representation, $<,>$ denotes a lightweight relevance metric (say, a dot product) for their similarity score. Here, $\Theta^{(q)}$ and $\Theta^{(s)}$, whether tied or not, compose a dual-encoder structure for efficient dense retrieval. Lastly, the training loss of contrastive learning can be defined to optimize Θ, i.e.,

$$L^{(\mathrm{ct})}(\mathbb{Q};\Theta) = -\sum_{q\in\mathbb{Q}} \log P(\mathrm{s} = s^{+}|q, \{s^{+}\} \cup \mathbb{N};\Theta), \tag{3}$$

where $P(\cdot|q;\Theta)$ denotes the probability distribution over $\{s^{+}\}\cup\mathbb{N}$ for q by Eq. (2).

Subsequently, considering the distributed setting of OD-AS2, the overall training loss can be defined as

$$L(\{\mathbb{Q}^{i}\}_{i};\{\Theta^{i}\}_{i}) = \sum_{i} L^{(\mathrm{ct})}(\mathbb{Q}^{i};\Theta^{i}). \tag{4}$$

However, directly optimizing Eq. (4) cannot deliver a satisfactory performance for each client i since both labeled question-answering pairs and the collection are too scarce to effectively learn. Therefore, we adopt a popular federated learning method, FedAvg [24], as the backbone of our framework. It will leverage the training data distributed in each client in a privacy-preserving way. We denote the weight of global model as Θ^{global}. For each $c \in \mathbb{C}$ with model weight Θ^{i}, we update Θ^{i} with a learning rate of α locally by

$$\Theta^{i} = \Theta^{i} - \alpha\nabla L(\mathbb{Q}^{i};\Theta^{i}), \tag{5}$$

where L is the loss function of local training objective defined in Eq. 4. After local updates, each client sends their weights Θ^{i} to the central server. Central server aggregate the weights by

$$\Theta^{global} = \sum_{i=1}^{k} \frac{|\mathbb{D}_i|}{\sum_{i=1}^{k}|\mathbb{D}_i|}\Theta^{i}, \tag{6}$$

where k is the number of clients, $\mathbb{D}_i$ denotes the volume of the dataset on each client. Note that our PDD-AS2 framework is also compatible with other federated learning methods.

3.2 Fed-Negative: Cross-Client Negatives

However, federated learning cannot fulfill the needs of negative samples in terms of quality and quantity for some clients with few document collections. Building on this problem, we propose fed-negative: a cross-client negative sampling method inspired by dynamic negative sampling for introducing more diverse negative samples. Given a client c, we first encode q into representations by $\mathrm{Enc}(q;\Theta)$. Then we select a subset of clients from the whole client set as

$$C_s = Select(\{C\}), c \notin C_s, \tag{7}$$

where the select function can be based on network condition or geographical distance estimated by the client's region. Then we send the query representation $\mathrm{Enc}(q;\Theta)$ to each client in C_s.

Once each client receives the query, they perform a similarity search on their own sentence embedding matrix to retrieve the top n sentence embeddings and send them back to c. c chooses the top n negatives from all negatives based on the similarity score as

$$N^{fed} = TopK(\{(N_{c_k})\}), c_k \in C_s \tag{8}$$

where N_{c_k} is the negative set of q sampled in client c_k.

3.3 Client-Customized Query Encoding

On top of fed-negative, we propose client-customized query encoding inspired by query-side fine-tuning. We aim to provide each client with a personalized query encoder to resolve miscellaneous queries. For this purpose, we personalize $\text{Enc}(q; \Theta)$ with local training while fixing the $\text{Enc}(s; \Theta)$. $\text{Enc}(s; \Theta)$ shares a global weight among all clients. In this stage, we utilize our proposed fed-negative method to generate diverse negative samples.

Training Objective. To learn a personalized query encoder, we apply the constrative loss defined in Eq. 4. Formally, given a query q and its gold answer s^+, we first sample the negative set N^{fed} defined by Eq. 8. Therefore, we only update the weight Θ of the query encoder with the loss function defined in Eq. 4

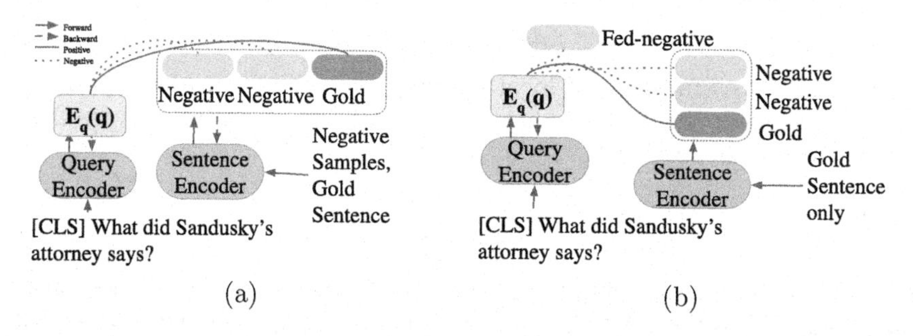

Fig. 1. (a) Train query encoder $\text{Enc}(q; \Theta)$ and sentence encoder $\text{Enc}(s; \Theta)$ with Static hard-negative sampling (b) Personalize the query encoder $\text{Enc}(q; \Theta)$ with fed-negative

3.4 Training Pipeline of PDD-AS2

Finally, we introduce the overall training pipeline of our PDD-AS2 framework. As shown in Fig. 1, we organize our training procedure into two stages, adapted from some prevailing works [17,37]: (Stage 1) ***Federated Static negative training***: we train the encoders with static hard negative sampling N^{static} under FedAvg. Due to the instability of the model in the early training stage, we initially sample BM25 negatives N^{BM25} to warm up the model, following the approach of some works [6,37]. We update both $\text{Enc}(q; \Theta)$ and $\text{Enc}(s; \Theta)$ by $\mathcal{L}$ defined in Eq. 4.

The overview of the federated framework is illustrated in Algorithm 1. (Stage 2) ***Query encoder personalization***: Continuing from the first stage, we sample N^{fed}, as defined in Sect. 3.2, to train a client-customized query encoder following the method described in Sect. 3.3.

3.5 Fed-NewsQA: A Multi-client OD-AS2 Benchmark

To better evaluate our method in a distributed setting, we propose a multi-client OD-AS2 benchmark based on NewsQA. Recent open-domain question answering works often use datasets such as SQuAD [28], TREC [34], WebQuestions [2], Natural Questions [3] in their experiments. However, we propose to use NewsQA [33] as our original dataset for two main reasons.

Algorithm 1. PDD-AS2: Federated Static negative training

1: **Input:** Clients set $\mathbb{C}$, Training set D_i on client c_i, global model weight Θ^{global}, learning rate α
2: **Initialize:** global model Θ^{global};
.
3: **for** $r = 0, 1, \ldots, R$ **do**
4:   **for** Client $c_i \in \mathbb{C}$ **in parallel do**
5:     Initialize local model $\Theta^i \leftarrow \Theta$.
6:     **for** batch b in D_i **do**
7:       Send queries $q_b \in b$ to other clients $c_j \in \mathbb{C}$
8:       Receive negative samples $\mathbb{N}_{q_b}$
9:       $\Theta^i \leftarrow \Theta^i - \eta \nabla \mathcal{L}(s^+ \cup \mathbb{N}; q; \Theta^i)$
10:     **end for**
11:   **end for**
12:   Server updates Θ by global aggregation
13: **end for**

First, to better mimic the difference between each client's personal documents and the data scarcity problem in the real-world cases, we propose to split the dataset into different genres for simulating different clients. Among all these datasets, we find that NewsQA meets our requirements perfectly. We split the dataset into different genres directly from the web-link of each passage. We choose ten genres from NewsQA since the remaining genres do not have enough number of samples in the dev/test set. Each of these genres represents a different client in our Federated setting. The statistics of each genre are shown in the Fig. 2.

Second, NewsQA significantly outnumbers some other datasets on the distribution of the more difficult reasoning questions, such as SQuAD [33]. We believe inferencing and reasoning queries are essential for open-domain question answering in real-world cases.

3.6 Retrieval Schemes

Our model is compatible with two retrieval schemes: sentence-level retrieval and passage-level retrieval. For sentence-level retrieval, we retrieve the top sentences follow the probability distribution defined in Eq. 2. For passage-level retrieval, based on the fact that sentences are extracted from their source passages, we retrieve the passage with highest relevance score as

$$f(p,q) := \max_{s \in p}\{< \text{Enc}(q;\Theta), \text{Enc}(s;\Theta) >\}, \forall s \in \mathbb{S}, \quad (9)$$

where $s \in p$ represents the set of sentences in a given passage p. The additional cost of sorting sentence scores can be ignored [19]. Therefore, the inference speed of our sentence-based passage retrieval is the same as for sentence retrieval.

4 Experiments

4.1 Setup

Baselines. We conduct experiments[1] to compare the performance of our method with several dense retrieval methods, including: (1) dense retrieval trained with random negative [14] (2) dense retrieval trained with BM25 negative [7]; (3) dense retrieval trained with STAR [37]. In personalization stage, we compare our proposed fed-negative to dynamic hard-negatives in [37].(4) a simple sparse retriever constructed by BM25.

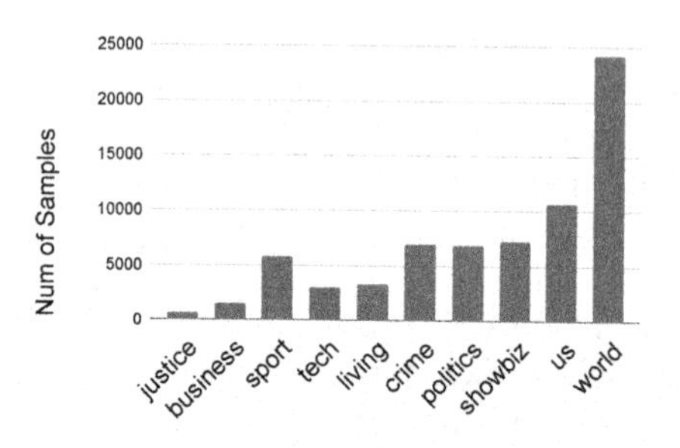

Fig. 2. Statistics of each genre in our Benchmark

Implementation. We use pre-trained DistilBERT [29] by Hugging Face as our model. We use AdamW with a learning rate of 3e-5. We use Faiss [16] to perform the similarity search. We use open-sourced BM25 model in training. Queries and sentences are truncated to a maximum of 32 tokens and 512 tokens, respectively. We represent query embeddings simply using the $[CLS]$ token, and we represent sentence embeddings using the average pooling of word embeddings in the sentence.

The details of our training procedure is described as follows: In the federated static negative training, we pair each query with BM25 negatives and gold-negatives with a batch size of 8 in the warm-up stage. Then we replace them with static hard-negatives. To demonstrate the influence of numbers of negatives, we also experiment with settings with different numbers of negatives. We enable in-batch negative in this stage. We implemented vanilla FedAvg as our Federated learning framework. We aggregate local weights after each epoch.

In the personalized query encoder training, we pair each query with dynamic hard negatives or fed-negatives with a batch size of 32. To demonstrate the influence of numbers of negatives, we also experiment on settings with different numbers of negatives. We enable in-batch negatives in this stage.

[1] We will make our data and codes public.

We report two levels of metrics in our experiments: sentence-level and passage-level. The retrieval procedure of both levels is defined in Sect. 3.6. In both levels, we report the MRR@10, Recall@1,20,100 scores.

4.2 Experiment Results

Table 1. Results on our Fed-NewsQA Benchmark.

Models	Sentence-level Retrieval				Passage-level Retrieval			
	MRR@10	**R@1**	**R@20**	**R@100**	**MRR@10**	**R@1**	**R@20**	**R@100**
Upper Bound								
Central-training	0.338	0.284	0.629	0.781	0.502	0.447	0.553	0.821
Sparse Retriever								
BM25	0.172	0.152	0.343	0.533	0.343	0.288	0.345	0.598
Dense Retriever								
dense retrieval-Random Neg	0.194	0.171	0.466	0.62	0.376	0.323	0.401	0.702
dense retrieval-Bm25 Neg	0.188	0.151	0.475	0.639	0.353	0.303	0.388	0.679
dense retrieval-STAR	0.232	0.190	0.535	0.679	0.403	0.350	0.421	0.709
Dense Retriever: Ours								
PDD-AS2	0.261	0.217	0.546	0.695	0.429	0.395	0.479	0.745
+client-customized query encoding	0.289	0.232	0.556	0.711	0.445	0.414	0.489	0.75
+client-customized query encoding with fed-negative	**0.309**	**0.252**	**0.577**	**0.72**	**0.458**	**0.431**	**0.504**	**0.762**

The main result of our experiments is shown in Table 1. We conclude with two main findings from the results. First, compared with the dense retrieval baselines trained on a single client, our PDD-AS2 outperformed all other methods. This is because the number of documents in some clients is very restricted. Our method can leverage training data on each client in a privacy-preserving way. Therefore, our federated method can achieve better performance than non-Federated methods.

Second, our personalization method with fed-negative can outperform the method with local dynamic hard negatives. This is because the scarcity of training data in some clients can lead to a much worse hard-negative sampling result. Compared with static hard negative sampling, the training of the client-customized query encoder introduces far more negative samples, strengthening the need for hard negatives in terms of quality and quantity. Our method alleviates the problem by leveraging diverse hard negatives on other clients in a privacy-preserving way.

4.3 Influence of Numbers of Negatives

We explore the influence of *num_negatives* in our setting. We experiment with the combinations of different numbers of negatives used in each method. The result of different *num_negatives* is shown in Table 2. We show the impact of *num_negatives* on both stages of training separately. The maximum number of hard-negatives we can test in stage 1 training is limited due to GPU RAM cost. For BM25 negative sampling and static hard-negative sampling, we train the model with our PDD-AS2 framework from the beginning of our training procedure. In experiments of stage 2 training with fed-negative, we continue our training from the model weights trained in previous steps, which follows our training procedure.

We have two findings from the results. First, we found that insufficient numbers of negative samples can lead to much worse performance. This is intuitive since the model saw fewer numbers of samples during training. Second, client-customized query encoder training can benefit more from the larger amount of negatives. Our experiment shows that the optimal number for BM25 negative sampling is not very large. BM25 negative sampling cannot leverage the larger amount of negatives effectively. However, due to the limitation of hardware resources, we cannot test on larger numbers of negatives in stage 1 training.

Meanwhile, client-customized query encoder can be steadily improved while feeding much more negatives compared with stage 1 training. This result indicates the need for introducing more hard-negatives with higher quality in stage 2 training, further proving the effectiveness and necessity of our fed-negative. What's more, the computational cost does not scale with the *num_negatives*. As a consequence, client-customized query encoder can benefit from fed-negative with little cost.

Table 2. Different num_negative in Training

Models	Sentence-level Retrieval				Passage-level Retrieval			
	MRR@10	R@1	R@20	R@100	MRR@10	R@1	R@20	R@100
Dense Retriever with BM25 negatives								
num_negative=2	0.143	0.123	0.302	0.489	0.310	0.247	0.311	0.582
num_negative=8	0.172	0.151	0.343	0.533	0.343	0.288	0.345	0.598
Dense Retriever with STAR								
num_negative=2	0.201	0.160	0.506	0.655	0.352	0.305	0.379	0.705
num_negative=8	0.232	0.191	0.535	0.679	0.403	0.350	0.421	0.709
PDD-AS2								
num_negative=2	0.242	0.193	0.516	0.645	0.392	0.354	0.432	0.719
num_negative=8	0.261	0.217	0.546	0.695	0.429	0.395	0.479	0.745
+client-customized query encoding								
num_negative=10	0.272	0.233	0.557	0.705	0.431	0.415	0.487	0.746
num_negative=200	**0.289**	**0.251**	**0.576**	**0.711**	**0.445**	**0.434**	**0.489**	**0.75**

4.4 Influence of Training Data Size

In this section, we first explore whether our PDD-AS2 can effectively handle the data scarcity problem in each client by leveraging data from different clients. In training, we select different ratios of data randomly. We present the sentence-level R@1 score on our Fed-NewsQA in Fig. 4. Compared with single-client training, the PDD-AS2 can achieve higher accuracy in all data ratio settings. Moreover, as the ratio of training data in each client decreases, the data scarcity problem in single-client training becomes more serious. As a consequence, PDD-AS2 can bring about a more significant performance improvement over single-client training.

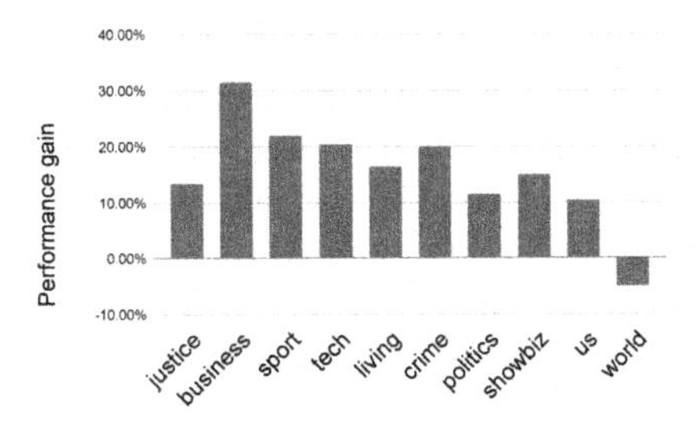

Fig. 3. Difference of performance gain of each client on Sentence R@1

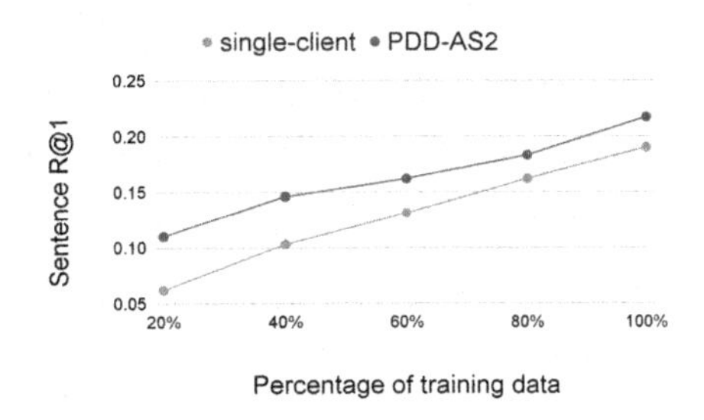

Fig. 4. Difference of performance in training dataset size on Sentence R@1

Table 3. Different k while sampling 10 negatives

Method	Sentence R@1	Passage R@1
k=10	0.121	0.235
k=50	0.202	0.379
k=100	0.211	0.352
k=300	**0.217**	**0.395**

Table 4. Perplexity of gpt-2 on our dataset.

Method	Perplexity
Without training	36.3
CLM without embedding	25.9
CLM with sentence embedding	**25.6**

We also explore to what extent each client benefits from the PDD-AS2. We show the performance improvement in sentence-level R@1 on Fed-NewsQA of each client in Fig. 3. We found that clients with less training data can benefit more from the PDD-AS2 framework. These results indicate that our framework can effectively leverage the training data on different clients. However, performance on some clients with a larger amount of training data decreased when applying our framework, implying the need for personalization in this scenario (Table 3).

4.5 Privacy

When transferring sentence embeddings between clients, one key concern is whether the user's privacy would be leaked. However, no work has been dedicated to restoring private information from mere sentence embeddings. In order to measure the risk involved, we conducted an experiment to detect whether our transmitted sentence embeddings contained information related to the original text.

In this experiment, we used GPT-2, a model that performs well on text generation tasks. In the first part of the experiment, we trained GPT-2 on the language modeling task using our dataset and measured its perplexity on the test set. In the second part of the experiment, we added the sentence embeddings generated by the previously trained sentence encoder in PDD-AS2 to the training and testing procedure. In detail, we feed the sentence embeddings into the GPT-2 as key-value pairs together with the text input. After receiving the input, the model tries to establish the connection between the embedding and the actual sentence it represents through the self-attention structure. Table 4 shows no significant difference in the perplexity between the two groups of experiments. The group with sentence embeddings has slightly lower perplexity on the test set. However, these differences are not statistically significant. To further demonstrate that we cannot obtain private information from the sentence embeddings, we let GPT-2 generate actual sentences directly from their corresponding embeddings without any input and prompts. We show the result in the Table 5.

We found that GPT-2 could not restore the actual sentence using only the sentence embeddings. Sentence embeddings did have an impact on the generated results. However, these effects are seemingly random and irrelevant to the actual sentence.

Table 5. Case study of sentence-embeddings decoding

Original Sentences	Generated Sentences
Four Australian troops have now died in the conflict in Afghanistan.	"It's not the first time that we've had
It made my stomach turn," Bertha Lewis, chief executive officer of ACORN, told reporters at the National Press Club in Washington.	"I think it's important? very important? Very difficult to the one. I think. is, part of me. I the to blame, I don't blame my
Read the story at the WRTV web site	CNN's a great-school program that's not

5 Conclusion

In this paper, we propose a Privacy-preserving Distributed OD-AS2 method, dubbed PDD-AS2. Our method utilizes training data on different clients while eliminating the need to transfer the raw data between clients. The training process of our approach is two-stage. In the first stage, we train both query encoder and sentence encoder with static hard-negatives under a federated framework. In the second stage, we personalize a client-customized query encoder for each client. We also propose a new negative sampling method called fed-negative. In fed-negative, we introduce diverse negatives from other clients to enhance the training. We further test our method on a new Federated Open-domain Sentence Selection benchmark based on NewsQA. This benchmark better mimics real-world cases than other benchmarks in terms of data distribution and query types.

The experiment results show that our method can effectively improve the performance of open-domain answer sentence selection under distributed settings by leveraging training data on different clients in a privacy-preserving way.

We prove that not every client can benefit from Federated learning, which indicates the need for personalization in such a scenario. As a solution, we provide each client with a client-customized query encoder which handles miscellaneous queries.

References

1. Allam, A.M.N., Haggag, M.H.: The question answering systems: a survey (2016)
2. Berant, J., Chou, A., Frostig, R., Liang, P.: Semantic parsing on Freebase from question-answer pairs. In: Proceedings of the 2013 Conference on Empirical Methods in Natural Language Processing, pp. 1533–1544. Association for Computational Linguistics, Seattle, Washington, USA, October 2013. https://aclanthology.org/D13-1160
3. Bird, S., Klein, E., Loper, E.: Natural language processing with Python: analyzing text with the natural language toolkit. O'Reilly Media, Inc. (2009)
4. Chen, D., Fisch, A., Weston, J., Bordes, A.: Reading wikipedia to answer open-domain questions. In: ACL (2017)
5. Devlin, J., Chang, M.W., Lee, K., Toutanova, K.: BERT: Pre-training of deep bidirectional transformers for language understanding. In: Proceedings of the 2019 Conference of the North American Chapter of the Association for Computational Linguistics: Human Language Technologies, Volume 1 (Long and Short Papers), pp. 4171–4186. Association for Computational Linguistics, Minneapolis, Minnesota, June 2019. https://doi.org/10.18653/v1/N19-1423. https://aclanthology.org/N19-1423
6. Gao, L., Callan, J.: Unsupervised corpus aware language model pre-training for dense passage retrieval. In: Proceedings of the 60th Annual Meeting of the Association for Computational Linguistics (Volume 1: Long Papers), pp. 2843–2853. Association for Computational Linguistics, Dublin, Ireland, May 2022. https://doi.org/10.18653/v1/2022.acl-long.203. https://aclanthology.org/2022.acl-long.203
7. Gao, L., Dai, Z., Chen, T., Fan, Z., Van Durme, B., Callan, J.: Complement lexical retrieval model with semantic residual embeddings. In: Hiemstra, D., Moens, M.-F., Mothe, J., Perego, R., Potthast, M., Sebastiani, F. (eds.) ECIR 2021. LNCS, vol. 12656, pp. 146–160. Springer, Cham (2021). https://doi.org/10.1007/978-3-030-72113-8_10
8. Garg, S., Vu, T., Moschitti, A.: Tanda: transfer and adapt pre-trained transformer models for answer sentence selection. In: Proceedings of the AAAI Conference on Artificial Intelligence, vol. 34, pp. 7780–7788 (2020)
9. Ge, S., Wu, F., Wu, C., Qi, T., Huang, Y., Xie, X.: Fedner: privacy-preserving medical named entity recognition with federated learning. ArXiv abs/2003.09288 (2020)
10. Guu, K., Lee, K., Tung, Z., Pasupat, P., Chang, M.W.: Realm: retrieval-augmented language model pre-training. In: Proceedings of the 37th International Conference on Machine Learning. ICML 2020, JMLR.org (2020)
11. Guu, K., Lee, K., Tung, Z., Pasupat, P., Chang, M.W.: Realm: Retrieval-augmented language model pre-training. ArXiv abs/2002.08909 (2020)
12. Harabagiu, S.M., Maiorano, S.J., Pasca, M.: Open-domain textual question answering techniques. Nat. Lang. Eng. **9**, 231–267 (2003)
13. Hardy, S., et al.: Private federated learning on vertically partitioned data via entity resolution and additively homomorphic encryption. ArXiv abs/1711.10677 (2017)

14. Huang, J.T., et al.: Embedding-based retrieval in facebook search. In: Proceedings of the 26th ACM SIGKDD International Conference on Knowledge Discovery & Data Mining, pp. 2553–2561 (2020)
15. Jiang, D., et al.: Federated topic modeling. In: Proceedings of the 28th ACM International Conference on Information and Knowledge Management, CIKM 2019, pp. 1071–1080. Association for Computing Machinery, New York (2019). https://doi.org/10.1145/3357384.3357909. https://doi.org/10.1145/3357384.3357909
16. Johnson, J., Douze, M., Jegou, H.: Billion-scale similarity search with gpus. IEEE Trans. Big Data **7**(03), 535–547 (2021). https://doi.org/10.1109/TBDATA.2019.2921572
17. Karpukhin, V., et al.: Dense passage retrieval for open-domain question answering. In: Proceedings of the 2020 Conference on Empirical Methods in Natural Language Processing (EMNLP), pp. 6769–6781. Association for Computational Linguistics, Online, November 2020. https://doi.org/10.18653/v1/2020.emnlp-main.550. https://aclanthology.org/2020.emnlp-main.550
18. Kwiatkowski, T., et al.: Natural questions: a benchmark for question answering research. Transactions of the Association of Computational Linguistics (2019)
19. Lee, J., Sung, M., Kang, J., Chen, D.: Learning dense representations of phrases at scale. In: Proceedings of the 59th Annual Meeting of the Association for Computational Linguistics and the 11th International Joint Conference on Natural Language Processing (Volume 1: Long Papers), pp. 6634–6647. Association for Computational Linguistics, Online, August 2021. https://doi.org/10.18653/v1/2021.acl-long.518. https://aclanthology.org/2021.acl-long.518
20. Lee, J., Yun, S., Kim, H., Ko, M., Kang, J.: Ranking paragraphs for improving answer recall in open-domain question answering. In: EMNLP (2018)
21. Lee, K., Chang, M.W., Toutanova, K.: Latent retrieval for weakly supervised open domain question answering. ArXiv abs/1906.00300 (2019)
22. Lin, Y., Ji, H., Liu, Z., Sun, M.: Denoising distantly supervised open-domain question answering. In: ACL (2018)
23. Lu, S., et al.: Less is more: Pretrain a strong Siamese encoder for dense text retrieval using a weak decoder. In: Proceedings of the 2021 Conference on Empirical Methods in Natural Language Processing, pp. 2780–2791. Association for Computational Linguistics, Online and Punta Cana, Dominican Republic, November 2021. https://doi.org/10.18653/v1/2021.emnlp-main.220. https://aclanthology.org/2021.emnlp-main.220
24. McMahan, B., Moore, E., Ramage, D., Hampson, S., Arcas, B.A.y.: Communication-efficient learning of deep networks from decentralized data. In: Singh, A., Zhu, J. (eds.) Proceedings of the 20th International Conference on Artificial Intelligence and Statistics. Proceedings of Machine Learning Research, vol. 54, pp. 1273–1282. PMLR (20–22 Apr 2017). https://proceedings.mlr.press/v54/mcmahan17a.html
25. Nguyen, T., et al.: Ms marco: a human generated machine reading comprehension dataset, November 2016
26. Paca, M.: Open-domain question answering from large text collections. Comput. Linguist. **29**, 665–667 (2003)
27. Qu, Y., et al.: RocketQA: an optimized training approach to dense passage retrieval for open-domain question answering. In: Proceedings of the 2021 Conference of the North American Chapter of the Association for Computational Linguistics: Human Language Technologies, pp. 5835–5847. Association for Computational Linguistics, Online, June 2021. https://doi.org/10.18653/v1/2021.naacl-main.466. https://aclanthology.org/2021.naacl-main.466

28. Rajpurkar, P., Zhang, J., Lopyrev, K., Liang, P.: SQuAD: 100,000+ questions for machine comprehension of text. In: Proceedings of the 2016 Conference on Empirical Methods in Natural Language Processing, pp. 2383–2392. Association for Computational Linguistics, Austin, Texas, November 2016. https://doi.org/10.18653/v1/D16-1264.https://aclanthology.org/D16-1264
29. Sanh, V., Debut, L., Chaumond, J., Wolf, T.: Distilbert, a distilled version of bert: smaller, faster, cheaper and lighter. arXiv preprint arXiv:1910.01108 (2019)
30. Seo, M., Lee, J., Kwiatkowski, T., Parikh, A.P., Farhadi, A., Hajishirzi, H.: Real-time open-domain question answering with dense-sparse phrase index. ArXiv abs/1906.05807 (2019)
31. Shen, G., Yang, Y., Deng, Z.H.: Inter-weighted alignment network for sentence pair modeling. In: Proceedings of the 2017 Conference on Empirical Methods in Natural Language Processing, pp. 1179–1189. Association for Computational Linguistics, Copenhagen, Denmark, September 2017. https://doi.org/10.18653/v1/D17-1122. https://aclanthology.org/D17-1122
32. Tran, Q.H., Lai, T., Haffari, G., Zukerman, I., Bui, T., Bui, H.: The context dependent additive recurrent neural net. In: Proceedings of the 2018 Conference of the North American Chapter of the Association for Computational Linguistics: Human Language Technologies, Volume 1 (Long Papers), pp. 1274–1283. Association for Computational Linguistics, New Orleans, Louisiana, June 2018. https://doi.org/10.18653/v1/N18-1115. https://aclanthology.org/N18-1115
33. Trischler, A., et al.: NewsQA: a machine comprehension dataset. In: Proceedings of the 2nd Workshop on Representation Learning for NLP, pp. 191–200. Association for Computational Linguistics, Vancouver, Canada, August 2017. https://doi.org/10.18653/v1/W17-2623. https://aclanthology.org/W17-2623
34. Wang, M., Smith, N.A., Mitamura, T.: What is the Jeopardy model? a quasi-synchronous grammar for QA. In: Proceedings of the 2007 Joint Conference on Empirical Methods in Natural Language Processing and Computational Natural Language Learning (EMNLP-CoNLL), pp. 22–32. Association for Computational Linguistics, Prague, Czech Republic, June 2007. https://aclanthology.org/D07-1003
35. Xiong, L., et al.: Approximate nearest neighbor negative contrastive learning for dense text retrieval. arXiv preprint arXiv:2007.00808 (2020)
36. Yoon, S., Dernoncourt, F., Kim, D.S., Bui, T., Jung, K.: A compare-aggregate model with latent clustering for answer selection. In: Proceedings of the 28th ACM International Conference on Information and Knowledge Management (2019)
37. Zhan, J., Mao, J., Liu, Y., Guo, J., Zhang, M., Ma, S.: Optimizing dense retrieval model training with hard negatives. In: Proceedings of the 44th International ACM SIGIR Conference on Research and Development in Information Retrieval, pp. 1503–1512. SIGIR '21. Association for Computing Machinery, New York (2021). https://doi.org/10.1145/3404835.3462880. https://doi.org/10.1145/3404835.3462880

An Early Stage Identification of Cryptomining Behavior with DNS Requests

Hui Li[1], Yihang Hao[1,2], Mengda Lyu[1], Xiaojie Yu[1], Bo Yang[1,2], and Lizhi Peng[1,2](✉)

[1] Shandong Provincial Key Laboratory of Network Based Intelligent Computing, University of Jinan, Jinan, China
{qinglang,lvengda}@stu.ujn.edu.cn, {plz,yangbo}@ujn.edu.cn
[2] Quancheng Laboratory, Jinan, China

Abstract. The booming of cryptocurrencies in the last decade brought about the burst of cryptomining for obtaining cryptocurrencies in recent years. Only those users with plenty of computing resources are able to gain profits according to the design of block chain. As a result, this brings out more and more criminal attacks to maliciously plunder private and public computing resources through networks. Consequently, the detection of malicious cryptomining behavior is particularly important for network security and management. In this paper, we designed Mining Vanguard, realizing the recognition of mining behavior through the detection of DNS behavior. By constructing a comprehensive feature set that includes both traditional DNS resolution features and morpheme features, we combine network characteristics with semantic characteristics, aiming to achieve early recognition. Through a large number of targeted experiments, it is verified that Mining Vanguard is promising for detecting mining behaviors on the Internet.

Keywords: Cryptomining detection · Network security · Data mining · Machine learning

1 Introduction

Cryptocurrency is a media of exchange that uses cryptographic principles to ensure security and trustworthiness. Unlike traditional banking systems that rely on a centralized regulatory system, cryptocurrencies are based on decentralized consensus mechanism. This mechanism is powered by block chain technology that uses distributed ledgers. Therefore, cryptocurrency transactions need to be verified and added to the distributed public ledger. This process is called crypto-mining.

H. Li and Y. Hao—Contributed equally to this research.

X. Yang et al. (Eds.): ADMA 2023, LNAI 14180, pp. 30–44, 2023.
https://doi.org/10.1007/978-3-031-46677-9_3

Cryptomining needs to connect a large number of computers through the network to jointly calculate a complex mathematical problem as the proof of work (PoW) to connect blocks and maintain the integrity of transactions [13]. In return, the miners will receive a certain amount of cryptocurrency as a reward. Since the first person who solves these problems is the only one to receive rewards, cryptomining has evolved into a highly competitive field where success relies heavily on the miner's hash rate. As a result, a large number of malicious actors have been attracted to illegally use hijacked computing resources to act as "miners" [1], posing a big threat to the security of computing resources. This threat has grown significantly in recent years with the skyrocketing value of cryptocurrencies, consequently, the detection of crypto-mining behaviors become an increasingly important research topic for network security [12]. According to the current research [19], billions of devices are maliciously excavating cryptocurrencies for the benefit of cybercriminals with no awareness of the damage they may cause.

Recent studies of detecting cryptomining behaviors have been carried out using unique characteristics. The majority of the previous studies focus on the host level, i.e., detecting the cryptomining behavior through system calls or web scripts on the browsers [3,6]. Apparently, it is not efficient for large-scale detection. While the other approach works on the network level, detecting mining behavior by analyzing the network communicating patterns [10,18]. For this type of method, existing studies rely on extracting features from network traffic generated by cryptomining behaviors and building machine-learning models. However, such methods generally need a great number of observation samples for detection and they are not able to make the final decisions even at the end of a session. Such efficiency is undoubtedly not enough for online systems to meet the requirements.

According to common sense, almost every network behavior first needs to make DNS requests before establishing network connections as a fundamental step. Consequently, as long as the DNS requests from a specific host are captured and traced, we can infer the network behavior at the very beginning of a network connection. Therefore, this paper proposes a new methodology with several meaningful innovations, namely, Mining Vanguard, for it is designed to accurately detect mining behaviors at a very early stage. On the basis of our research, no previous studies have attempted to detect mining behavior through DNS request analysis. Our contributions are as follows:

- We propose a novel approach to detect the crypto-mining behavior at the early stage by analyzing the DNS requests.
- We adopt the concept of the morpheme to enhance the matching speed of regular expression, and extracted more meaningful semantic information features in a domain name based on morpheme.
- Targeted experiments were carried out to verify the precision of our method on real-world DNS traffic, extracting satisfying results to show that our method is effective.

The remainder of the paper is structured as follows. First, we outline related work in Sect. 2. Then, we present the framework in Mining Vanguard and describe in detail the work in Sect. 3. Implementation details and experimental results are described in Sect. 4. Discussion and future work are provided in Sect. 5.

2 Related Works

2.1 Crypto-Mining Software Detection

Cryptomining software mining usually has two main forms. One is browsers based encrypted mining program. The mining process is run in a script embedded in web content. In order to check such content, many researchers have investigated a large number of attacked websites that were affected with drive-by mining and proposed kinds of mining detecting techniques with so-called "web fingerprints". Most of those approaches such as CoinSpy [7] and MineSweeper [9] are able to work with a browser that to warn the user of the potential risk.

The other one uses malicious mining software, which is run on infected computers connected to the Internet. The pattern of detecting such mining behavior is host based detection. Host based detection technology is to install a program in the target host, i.e., the detector, which is used to detect the mining behavior of the host. The detector mainly uses system call [6] or operation code [3] as features, and inputs the features to train a recognition model such as SVM [5].

Although the accuracy of these two methods is very high, they can only be detected on a single host. But in most cases, we need to detect the mining behavior of multiple hosts on a large scale at the same time, so we need to deploy the detection at the network level.

2.2 Crypto-Mining Traffic Identification

Network traffic identification is a main approach to detect whether there is malicious behavior in the network environment. Many works [4,15] have been proposed in order to ensure network security. DPI (Deep Packet Inspection) is a traditional traffic identification technique, but as most modern traffic is encrypted, DPI methods gradually became ineffective. Thus, people started to use the spatial and temporal characteristics of the traffic, which are not related to the content, as the basis for identification. With the help of machine learning, this kind of method is able to achieve high accuracy [11,16].

In recent years, with the continuous rampant cryptomining, the evolution of identifying cryptomining traffic is very similar to that of traditional traffic identification. At first, content detection is still available since crypto-mining traffic is unencrypted. For instance, Li et al. [10] proposed a passive method to detect Ethereum traffic by examining the Ethereum protocol-related fields in the packets. While Jordi et al. [14] regarded that DPI methods are ineffective in detect analyze encrypted traffic and not sustainable to be deployed in real-world networks with large traffic volumes. Therefore, they designed a method to detect crypto-mining using NetFlow without inspecting the packets' payload.

However, these methods usually need to input more packets into the classifier to obtain high-accuracy classification results. While in real circumstances, it makes no sense to recognize the behavior when the connection has disconnected. Thus, we need to make the decision accurately as soon as possible.

2.3 DNS Behavioral Identification

As an important infrastructure of the Internet, DNS carries the important task of mapping between domain names and IP addresses. Domain name analysis has a significant effect on identifying different network behaviors at the early stage. At present, the main approaches to detect DNS requests generated by botnet or phishing websites are still based on a blacklist. With its defects of high overhead and poor timeliness in maintenance and update, thereby dynamic and real-time methods of analyzing DNS requests gradually become a research hotspot.

Many works have been proposed that mainly focus on analyzing malicious network behavior. For example, Bilge et al. [2] detected malicious domain names by analyzing the time distribution of domain name query requests, the length of TTL, and the language characteristics of domain names. Yadav et al. [21] noticed that Normal domain names are usually readable while malicious domain names tend to be more chaotic so as to avoid supervision and reduce the cost of registration, thus it is easy to discover the statistical difference. In this way, Yadav et al. [20] designed a method to discover malicious domain names by comparing the Bi-gram distribution of the target domain names and the normal ones. Compared with simple cutting, Khaitan and Srinivasan [8,17] enhanced word segmentation with natural language models.

3 Methodology

The main purpose of Mining Vanguard is to achieve early identification of crypto-mining behavior. Since DNS requests are always the first to appear on the link, if only it is determined that a user has the request or intention to connect to the mining pools, we can detect the mining behavior at the early stage.

When it comes to data mining of mining domain names, separating the components of the mining domain names string is an essential process due to the semantic significance that these strings often carry. One of the most significant contributions of our method is the use of linguistic concept "morpheme" as the fundamental basis for segmenting domain names. Additionally, we have proposed several statistical features for domain names that are based on morphemes. Unlike words, morphemes are the smallest meaningful lexical item in a language. Compared with word-based segmentation or mechanically using N-gram as statistics, morpheme-based method is more dynamic and more corresponded with linguistic laws. Take the domain name "internationalreview.com" as an example. After removing the top domain name ".com", the second level

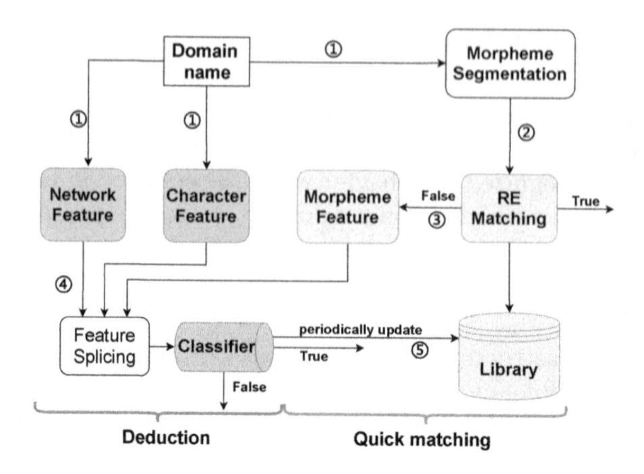

Fig. 1. System architecture and identification procedure.

can be divided into "inter", "nation", "al", "re" and "view" with morpheme segmentation. While for word segmentation, it can only be divided into "inter", "national" and "review", as a result, the partition granularity is quite rough.

The main architecture of Mining Vanguard is shown in Fig. 1. Mining Vanguard consists of two main function modules: matching (the yellow parts) and deduction (the blue parts). When a domain name is captured, (1) perform morpheme segmentation for regular expressions (RE) matching, meanwhile, extract both network resolution features and character features. (2) Execute RE matching, if it is determined that this is a mining domain name, the next operation will not be carried out. Otherwise, (3) morpheme features will be extracted, and then, (4) splice all the features together for classifier making further identification. If it is predicted as a mining domain name, (5) the matching library will be updated with a certain strategy.

3.1 Quick Matching

A large library of the crypto-mining domain name was built at first for quick matching. For the initialized data we used a large number of mining domain names and extracted a RE library. If the REs are matched sequentially, the time required will increase linearly with the number of updated new REs. Suppose the number of the REs is n, then we can easily get the computational complexity of the sequent matching algorithm, $O(n)$. To reduce the computational complexity and speed up the matching procedure, we design a hash storage method based on morpheme information.

Through long-term observation, we found that a group of REs may contain the same morpheme mor_x. When the number of mor_x satisfies:

$$count(mor_x) \geq th \tag{1}$$

where th is the threshold we predefined, mor_x judged to be a frequent morpheme mor_f. Then, this group of REs will be divided into different sequential lists based

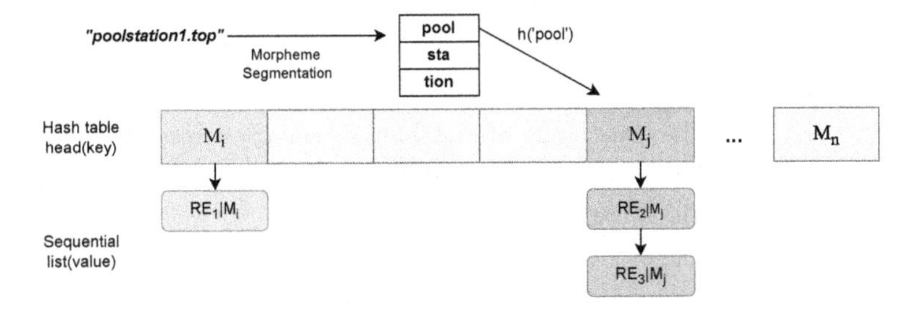

Fig. 2. Architecture of *RE-Hash Table.*

on the different mor_f they contain. Repeat this process until there is no mor_f left, then the other REs are led by mor_n. Afterward, use these groups of REs to build the *RE-Hash table* as shown in Fig. 2.

Hash table is a data structure that provides fast insertion and lookup operations. No matter how many pieces of data there are in the hash table, the time complexity of insertion and lookup is $O(1)$. Based on this, we constructed *RE-Hash table* to store REs. Each node has the structure of $\langle key, value \rangle$, where $key = h(mor_f)$ i.e., the hash result of mor_x, and the $value$ is the RE list led by mor_f. After inserting all the frequent morphemes with their RE list, the rest REs are inserted to the last node in *RE-Hash table.*

let c_i represents the ASCII value of the ith letter in the morpheme information (if the ith letter is empty, c_i is set to zero), we design the hash function of a morpheme as Eq. 2

$$key = (c_1 - 97) \times 27^3 + (c_2 - 97) \times 27^2 \quad +(c_3 - 97) \times 27 + c_4 \tag{2}$$

Figure 2 shows an example of the matching process. The morpheme segmentation of the target domain name is first launched, and then the first morpheme "pool" is extracted and executed hash search. The result is Mi, thus, this domain name needs to sequentially match the RE list $RE_x|M_i$, i.e., the RE list led by "pool". If it is not successfully matched, repeat this operation with the second morpheme till the end. If the domain name being tested is not matched successfully or does not contain mor_f, it will be matched with the regular expression in the last node of the hash table. If a match is still not found, the decision-making process will be handed over to the deduction module.

We prove that our morpheme-based matching method has a sublinear time complexity concerning the number of REs present in the library. For better presentation, we have listed and explained the symbols used in this section in Table 1.

As Fig. 3 shows, the linear function $f(x)$ satisfy $f(x_1 + x_2) = f(x_1) + f(x_2)$ while function $g(x)$ has the quality of sub-linearity, i.e., $g(x_1 + x_2) \leq g(x_1) + g(x_2)$. Therefore, given that an algorithm has a time complexity of $O(n^m)$, if $m = 1$, this algorithm has a linear time complexity with the increase of n. If

Table 1. Symbols and principles used in this paper

Variable	Meaning
$f(l)$	The time usage of matching l regular expressions
L	The original number of REs in the library
ΔL	The increased number of REs
N	The original number of hash table heads
p	$Pr[count(mor_x) \geq th]$, see Sect. 3.1

only we could prove that an algorithm has sub-linearity, we can draw a conclusion that this algorithm has the time complexity of $O(n^m)$ where $0 \leq m \leq 1$.

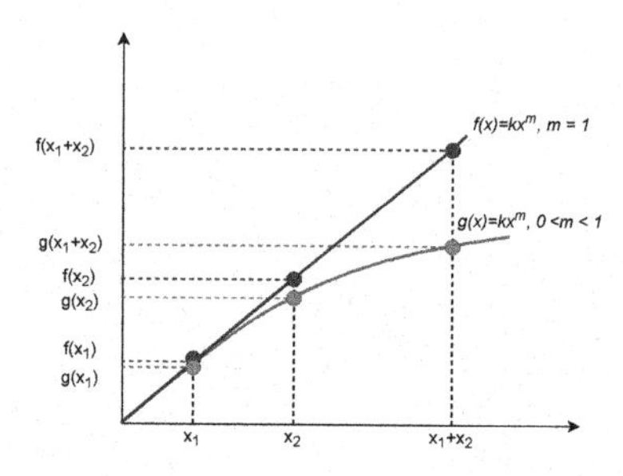

Fig. 3. Illustration of sublinearity.

Proof. Let p stands for the probability of finding frequent morphemes in the update set, i.e.,

$$p = Pr[count(mor_x) \geq th] \tag{3}$$

Since the length of each is different, but the sum of all nodes is equal to L (changed to $(L + \Delta L)$ after updating). Therefore, $f(L)$ can be estimated as:

$$\hat{f}(L) = E[f(L)] = \frac{c}{N} \cdot L \tag{4}$$

where c stands for the time required for single RE matching. After the update,

$$\hat{f}(\Delta L) = p \cdot \frac{c\Delta L}{N+1} + (1-p) \cdot \frac{c\Delta L}{N} = (\frac{p}{N+1} + \frac{1-p}{N}) \cdot c(\Delta L) \tag{5}$$

The same procedure may be easily adapted to obtain to calculate:

$$\hat{f}(L + \Delta L) = (\frac{p}{N+1} + \frac{1-p}{N}) \cdot c(L + \Delta L) \tag{6}$$

Let *Eq.* 6 - *Eq.* 4 - *Eq.* 5, i.e.,

$$\begin{aligned} &\hat{f}(L + \Delta L) - \hat{f}(\Delta L) - \hat{f}(L) \\ &= (\frac{p}{N+1} + \frac{1-p}{N}) \cdot c(L + \Delta L) \quad -(\frac{p}{N+1} + \frac{1-p}{N}) \cdot c(\Delta L) - \frac{c}{N} \cdot L \\ &= (\frac{p}{N+1} - \frac{p}{N}) \cdot cL \leq 0 \end{aligned} \tag{7}$$

Therefore, $\hat{f}(L + \Delta L) \leq \hat{f}(\Delta L) + \hat{f}(L)$ is proved, which means that our method has less time complexity with the RE list growing.

3.2 Feature Extraction and Machine Learning Model

The matching method suffers from detecting zero-day domain names because it can not build the hash table without any prior knowledge or data. Therefore, we designed an unknown mining domain name prediction method based on machine learning. Domain name features are first extracted. As it is shown in Fig. 1, the extracted features of domain name D include three main parts: morpheme features, network features, and character features.

Morpheme Features. In the feature extraction stage, we will continue to use morpheme-based features in depth. And different from the fast matching stage, we have divided morphemes in more detail. When the number of mor_x satisfies:

$$th^{*} > count(mor_x) \geq th \tag{8}$$

we will consider it as normal mining domain name morphemes, and they are defined as weak morphemes.

When the number of mor_x satisfies:

$$count(mor_x) \geq th^{*} \tag{9}$$

It shows that this morpheme is with high frequency in mining domain names, and such a morpheme is defined as a strong morpheme, we focus on analyzing strong morpheme.

- **Number of morpheme**: The number of morphemes containing in a domain name D. Generally, non-mining domain names contain more meaningful morphemes than mining domain names do.
- **Number of strong/weak morphemes**: Strong morphemes are considered more relevant to the implication of mining, while weak morphemes are not.
- **Proportion of strong/weak morphemes**: The proportion of strong/weak morphemes in the whole domain name. A mining domain name may contain more strong morphemes. For non-mining domain names, although they may contain some mining morphemes, they usually account for a low proportion.

Network Features. Network features can better reflect the relationship between domain and IP address, and can contribute effective server information for identification.

- **Number of Resolved IPs:** The number of IP addresses obtained by resolving D on the public DNS server.
- **Time usage of Resolving:** The time taken to resolve D on the public DNS server. If failed, record it as - 1.
- **TLD length:** The length of top-level domain name.
- **TLD encoding:** The sum of ASCII codes for top-level domain names.

Table 2. Individual feature

Feature type	Feature name
Morpheme Features	Number of morpheme
	Number of strong/weak morpheme
	Proportion of strong/weak morpheme
Network Features	Number of resolved IPs
	Time usage of resolving
	TLD length
	TLD encoding
Character Features	Dot number
	Char-Num switch number
	Total length

Character Features. To prevent the mine pool domain name from using the program to randomly generate the domain name in order to escape supervision, the following string features are added to enable the classifier to have a certain recognition ability for the randomly generated domain name.

- **Dot number**: The number of "." in string D.
- **Char-Num switch number**: The switching times of letters and numbers in string D. Non-mining domain names usually have better readability, so they switch letters and numbers less often.
- **Total length**: The total length of the domain name. Due to the high cost of a short domain name, malicious applications prefer a longer one.

Table 2 shows the complete feature set. After extracting the features, a random forest can be adopted to identify mining domain names. Compared with other traditional machine learning methods, the random forest has obvious advantages in model generalization and the capability of dealing with unbalanced problems. This correspondent with the target problem. On one hand, mining domain name requests and the normal ones are highly unbalanced in the real world, on the other hand, the model we trained is expected to be able to accurately predict the probability that a never-met domain name could be a mining one.

When the random forest model gives a prediction, it will also give confidence at the same time. Only samples with confidence greater than 95% will be added to the update cache. After the cache records more than ΔL, we will regularly count the frequent morphemes and create new heads in *RE-Hash table*. Here ΔL should be limited within the range of $(th, 2th)$ to ensure that there is at most one update in ΔL new REs.

4 Experiments

4.1 Data Collection

In this paper, we constructed three datasets to evaluate the effectiveness of Mining Vanguard. Table 3 shows the size of the selected datasets, and we will provide a detailed overview of each dataset below.

Table 3. Categories and examples of datasets.

Dataset	Types	Total	Samples
RE-Qianxin	Regular Expression	337	^.*\\.?f2pool\\.com ^.*\\.?alwayshashing\\.com ^.*\\.?webcoin\\.com
Dataset-I	Mining	5238	ca01.supportxmr.com mun.suprnova.cc pool.0cash.org
	Normal	100000	google.com youtube.com baidu.com
Dataset-II	Mining	2012329	etc2.poolgpu.com btc.f2pool.com bsty.hashlink.eu
	Normal	2086230	bilibili.com drive.wps.com edge.microsoft.com

In RE-Qianxin, we collected regular expressions provided by the Qianxin Threat Intelligence Platform for mining pool matching. The dataset consists of 337 regular expressions, which we used to store all regular expressions in Mining Vanguard's RE-hash as initial data for evaluating the performance of the fast matching module.

Dataset-I aims to evaluate the effectiveness of the deduction module. In this dataset, we collected mining pool lists provided by the Qianxin Threat Intelligence Platform, including large mining pools such as Fish Pool and Spark Pool. Additionally, we selected the top 100,000 accessed domain names by traffic ranking from Alexa as benign domain names.

The data in Dataset-II was all captured in a real environment, and the mining pool domain names were provided by CNCERT. The mining pool domain names were captured twice, with a three-month interval between captures. There were 963,754 mining pool domain names captured in the first capture, and the set of these mining pool domain names was named cert_202112. In the second capture, there were 1,048,575 mining pool domain names captured, and the set of these mining pool domain names was named cert_202202. Additionally, during the monitoring period, we considered all other domains as benign domains.

4.2 RE Matching Speed Test

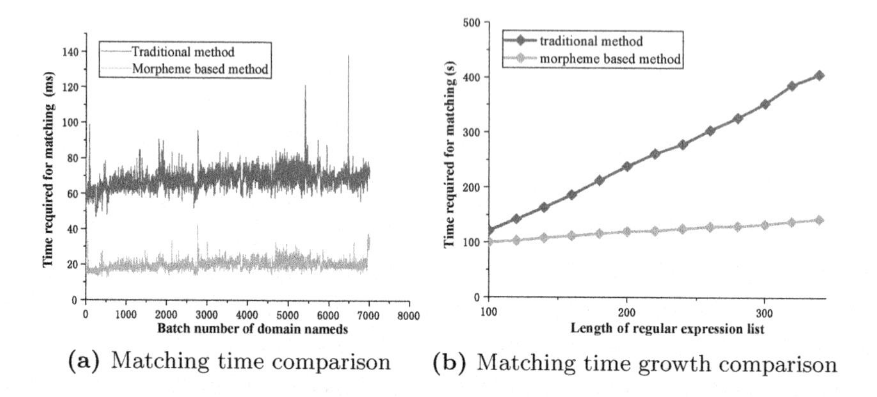

(a) Matching time comparison (b) Matching time growth comparison

Fig. 4. Comparative result of RE matching speed.

In order to verify the efficiency performance of the morpheme-based RE matching method and the traditional sequential method, we compared our proposal with the traditional sequential method by testing the matching time required by both methods. In this experiment, we utilized regular expressions from RE-Qianxin and stored them in an RE-hash data structure. We then used Dataset-I as the domain dataset for matching purposes. Since it is not easy to observe the time for every single RE matching, every 100 domain names are bundled into a batch to reckon the time.

The results in Fig. 4a show that our method is averagely 40 ms faster than the traditional one on every batch. In addition, we compare the time consumption of our proposed method with traditional methods by gradually increasing the number of domain names to be matched. It can be seen from Fig. 4b that our morpheme-based matching method has the advantage of sub-linear time complexity when the regular matching list becomes larger.

4.3 Classification Algorithm Selection and Ablation Evaluations

To evaluate the effectiveness of the selected features and select the most effective classification algorithm, we used Data-I and set the training-to-testing ratio to 7:3. We then evaluated three lightweight classifiers: Support Vector Machine (SVM), Random Forest (RF), and K-Nearest Neighbor (KNN). We conducted 30 independent repetitions of the evaluation and measured performance using three common criteria: precision, recall, and F1-score. Table 4 displays the average results across all 30 runs.

As shown in Table 4, RF outperformed other classifiers by achieving a precision rate of over 95% and a recall rate of over 82% for mining detection. These findings demonstrate that RF has superior performance compared to other classifiers. Based on these results, we selected RF as the classifier for Mining Vanguard.

Table 4. Performance comparison of classifiers.

Type	Algorithm	percision	recall	f1-score
Normal	SVM	0.985	0.997	0.991
	KNN	0.985	0.998	0.991
	RF	**0.991**	**0.998**	**0.994**
Mining	SVM	0.946	0.708	0.810
	KNN	0.954	0.704	0.81
	RF	**0.959**	**0.827**	**0.888**

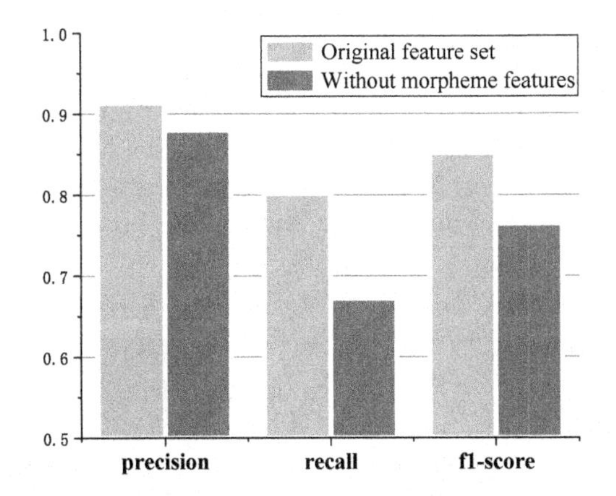

Fig. 5. Ablation test results of morpheme feature.

In order to verify the effectiveness of the morpheme-based features we proposed, we conducted a set of ablation experiments. Two different classifiers are trained independently using the feature set with/without morpheme features. It can be clearly seen from Fig. 5 that the morpheme-based features are able to improve the precision by 3%, recall by 13%, and F1-score by 10%. The experiment result shows that the morpheme-based features can significantly improve performance.

4.4 Real-Network Environment Validation

We deploy Mining Vanguard in a real networking environment to detect mining DNS requests. Figure 6 illustrates the architecture of the platform. We conducted an evaluation of Mining Vanguard's effectiveness in real-world settings using Dataset-II. The experiment was divided into two steps. In the first step, we utilized cert_202112 from Dataset-II as positive samples and normal domain names as negative samples. In the second experiment, we used cert_202202 as positive samples while keeping the negative samples identical to those used in the first step. We designed the experiment in this way to evaluate Mining Vanguard's robustness since the positive samples used in the first and second steps were captured three months apart.

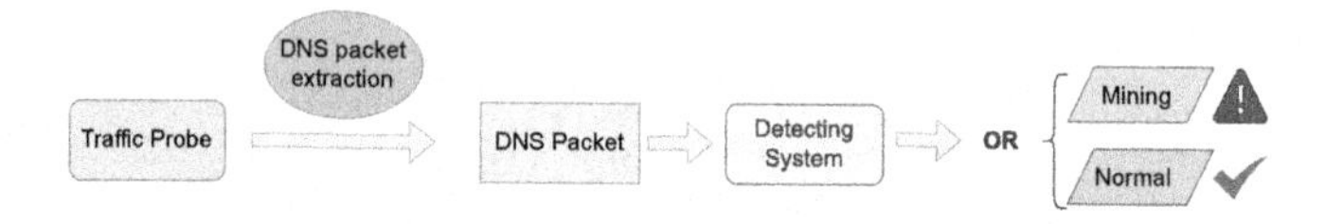

Fig. 6. Flow chart of the testing bed.

As shown in Fig. 7, the performance of Mining Vanguard is excellent. In both cert_202112 and cert_202202, the true positive rate is more than 99.5% and the false negative rate is less than 0.02%. It shows that Mining Vanguard can detect most mining domain names in the real environment and has a low false positive rate. From cert_202112 to cert_202202, the long time span of data does not affect the performance of Mining Vanguard. It shows the Mining Vanguard is robust.

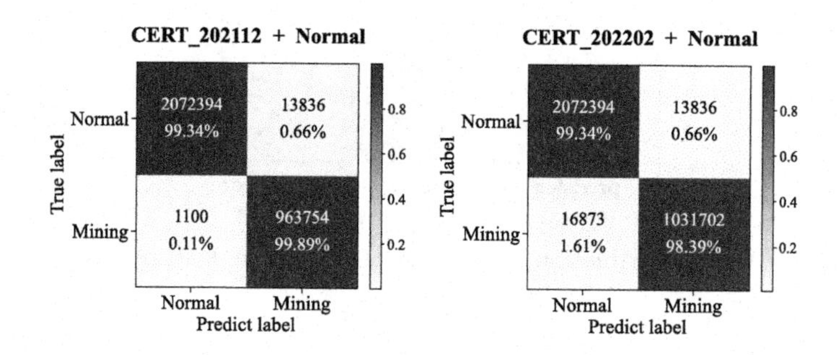

Fig. 7. Result of real-network environment validation.

5 Conclusions

In this paper, we propose a novel method, namely Mining Vanguard, aiming at realizing the earliest recognition of mining behavior by analyzing the DNS requests. While the primary focus of this framework in this paper is early detection of mining behavior, our proposed method of combining static matching and dynamic recognition can serve as a useful approach for detecting other forms of malicious activities. In order to fully utilize the semantic information implied in domain names, we design a fast regular matching algorithm, and several morpheme-based features are adopted to enhance the precision of the machine learning model. Experimental results show that Mining Vanguard is effective enough at present, while extensive work is also required to be done in the future to deal with the problem of encrypted DNS requests such as DoH (DNS-Over-Https) since the content of DNS request is gradually encrypted.

Acknowledgment. This research was partially supported by the National Natural Science Foundation of China under Grant No. 61972176, Shandong Provincial Natural Science Foundation, China under Grant No. ZR2021LZH002, Jinan Scientific Research Leader Studio, China under Grant No. 202228114, Shandong Provincial key projects of basic research, China under Grant No. ZR2022ZD01, Shandong Provincial Key R&D Program, China under Grant No. 2021SFGC0401, and Science and Technology Program of University of Jinan (XKY1802).

References

1. Europe's supercomputers hijacked by attackers for crypto mining, 18 May 2020. https://www.bbc.com/news/technology-58678907
2. Bilge, L., Sen, S., Balzarotti, D., Kirda, E., Kruegel, C.: Exposure: a passive dns analysis service to detect and report malicious domains. ACM Trans. Inf. Syst. Secur. **16**(4), April 2014
3. Darabian, H., et al.: Detecting cryptomining malware: a deep learning approach for static and dynamic analysis. J. Grid Comput. **18**(2), 293–303 (2020)
4. Du, C., Liu, S., Si, L., Guo, Y., Jin, T.: Using object detection network for malware detection and identification in network traffic packets. CMC-Comput. Mater. Continua **64**(3), 1785–1796 (2020)
5. Gangwal, A., Piazzetta, S.G., Lain, G., Conti, M.: Detecting covert cryptomining using HPC. In: Krenn, S., Shulman, H., Vaudenay, S. (eds.) CANS 2020. LNCS, vol. 12579, pp. 344–364. Springer, Cham (2020). https://doi.org/10.1007/978-3-030-65411-5_17
6. Karn, R.R., Kudva, P., Huang, H., Suneja, S., Elfadel, I.M.: Cryptomining detection in container clouds using system calls and explainable machine learning. IEEE Trans. Parallel Distributed Syst. **32**(3), 674–691 (2021). conference Name: IEEE Transactions on Parallel and Distributed Systems
7. Kelton, C., Balasubramanian, A., Raghavendra, R., Srivatsa, M.: Browser-based deep behavioral detection of web cryptomining with CoinSpy. In: Proceedings 2020 Workshop on Measurements, Attacks, and Defenses for the Web. Internet Society (2020)
8. Khaitan, S., Das, A., Gain, S., Sampath, A.: Data-driven compound splitting method for English compounds in domain names. In: Proceedings of the 18th ACM Conference on Information and Knowledge Management, CIKM 2009, pp. 207–214. Association for Computing Machinery, New York (2009)
9. Konoth, R.K., et al.: MineSweeper: an in-depth look into drive-by cryptocurrency mining and its defense. In: Proceedings of the 2018 ACM SIGSAC Conference on Computer and Communications Security, CCS 2018, pp. 1714–1730. Association for Computing Machinery (2018)

10. Li, Z., Hou, J., Wang, H., Wang, C., Kang, C., Fu, P.: Ethereum behavior analysis with netflow data. In: 2019 20th Asia-Pacific Network Operations and Management Symposium (APNOMS), pp. 1–6 (2019)
11. Lotfollahi, M., Siavoshani, M.J., Zade, R.S.H., Saberian, M.: Deep packet: a novel approach for encrypted traffic classification using deep learning. Soft. Comput. **24**(3), 1999–2012 (2020)
12. McAfee: Cloud adoption and risk report: Work from home edition, 1 May 2020. https://www.mcafee.com
13. Mukhopadhyay, U., Skjellum, A., Hambolu, O., Oakley, J., Yu, L., Brooks, R.: A brief survey of cryptocurrency systems. In: 2016 14th Annual Conference on Privacy, Security and Trust (PST), pp. 745–752 (2016)
14. Muñoz, J.Z.i., Suárez-Varela, J., Barlet-Ros, P.: Detecting cryptocurrency miners with netflow/ipfix network measurements. In: 2019 IEEE International Symposium on Measurements Networking (M N), pp. 1–6 (2019)
15. Nari, S., Ghorbani, A.A.: Automated malware classification based on network behavior. In: 2013 International Conference on Computing, Networking and Communications (ICNC), pp. 642–647 (2013)
16. Ren, X., Gu, H., Wei, W.: Tree-rnn: tree structural recurrent neural network for network traffic classification. Expert Syst. Appl. **167**, 114363 (2021)
17. Srinivasan, S., Bhattacharya, S., Chakraborty, R.: Segmenting web-domains and hashtags using length specific models. In: Proceedings of the 21st ACM International Conference on Information and Knowledge Management, CIKM 2012, pp. 1113–1122. Association for Computing Machinery, New York (2012)
18. Sun, P., Lyu, M., Li, H., Yang, B., Peng, L.: An early stage convolutional feature extracting method using for mining traffic detection. Comput. Commun. **193**, 346–354 (2022)
19. Swedan, A., Khuffash, A.N., Othman, O., Awad, A.: Detection and prevention of malicious cryptocurrency mining on internet-connected devices. In: Proceedings of the 2nd International Conference on Future Networks and Distributed Systems, ICFNDS 2018. Association for Computing Machinery, New York (2018)
20. Yadav, S., Reddy, A.K.K., Reddy, A.L.N., Ranjan, S.: Detecting algorithmically generated domain-flux attacks with dns traffic analysis. IEEE/ACM Trans. Networking **20**(5), 1663–1677 (2012)
21. Yadav, S., Reddy, A.K.K., Reddy, A.N., Ranjan, S.: Detecting algorithmically generated malicious domain names. In: Proceedings of the 10th ACM SIGCOMM Conference on Internet Measurement, pp. 48–61. IMC '10. Association for Computing Machinery, New York (2010)

NFAQP: Normalizing Flow Based Approximate Query Processing

Libin Cen, Jingdong Li, Wenjing Yue, and Xiaoling Wang(✉)

School of Computer Science and Technology, East China Normal University, Shanghai, China
{libincen,jdl,wjyue}@stu.ecnu.edu.cn, xlwang@cs.ecnu.edu.cn

Abstract. With the unprecedented rate at which data is being generated, Approximate Query Processing (AQP) techniques are widely demanded in various areas. Recently, machine learning techniques have made remarkable progress in this field. However, data with large domain sizes still cannot be handled efficiently by existing approach. Besides, the accuracy of the estimate is easily affected by the number of predicates, which may lead to erroneous decisions for users in complex scenarios. In this paper, we propose NFAQP, a novel AQP approach that leverages normalizing flow to efficiently model the data distribution and estimate the aggregation function by multidimensional Monte Carlo integration. Our model is highly lightweight - often just a few dozen of KB - and is unaffected by large domains. More importantly, even under queries with a large number of predicates, NFAQP still achieves relatively low approximation errors. Extensive experiments conducted on three real-world datasets demonstrate that NFAQP outperforms baseline approaches in terms of accuracy and model size, while maintaining relatively low latency.

Keywords: Approximate Query Processing · Normalizing Flow · Aggregate function

1 Introduction

Approximate Query Processing (AQP) has been developed for decades, aiming to provide users with the approximation of aggregation functions while keeping low latency and high accuracy. With the explosive growth of data, AQP technology has become widely demanded in data exploration, analysis, visualization, and even business decision-making.

Traditionally, approximate query processing was dominated by sampling-based methods [2,15,20]. The basic idea of those methods is to preserve a small sample of the original table to answer the query to avoid the expensive row scan on the origin table. However, the query accuracy is inherently constrained by sample size. When it comes to queries with low selectivity, insufficient samples often lead to unacceptable results, rendering such approaches dependent on a large sample, further leading to significant space overhead.

X. Yang et al. (Eds.): ADMA 2023, LNAI 14180, pp. 45–60, 2023.
https://doi.org/10.1007/978-3-031-46677-9_4

More recently, there has been a flurry of research that leverages machine learning techniques to process approximate queries. Despite the learning-based AQP approaches having shown promising results, several issues still need to be addressed. First, previous density estimation-based methods [16–18] are mainly based on simple ML models, such as Kernel Density Estimation (KDE) and Mixture Density Network (MDN). However, these ML models cannot efficiently model the high-dimensional distribution of tables to support multi-predicate queries. Second, the RSPN-based method [9] introduces an independence assumption between weakly correlated attributes in the table and recursively decomposes the table into conditionally independent attributes. However, this assumption makes it challenging for RSPN to accurately estimate the probability. Finally, generative model-based methods [22,23] suffer from the constraints imposed by the inefficient encoding (one-hot or binary, typically) on categorical attributes with a large domain, leading to a large model size inevitably.

In this paper, we proposed a Normalizing Flow-based Approximate Query Processing approach, NFAQP, which leverages the normalizing flow(NF) to efficiently address the AQP problem. Normalizing Flow is a highly expressive deep generative model, excelling at handling tasks with high-dimensional data, such as image generation [1,10], speech synthesis [21], graph generation [13], etc. The NF model can be further utilized as a density estimator [19] and capable of efficiently modeling the complex and high-dimensional joint distribution of the table and precisely estimating the probability density of the tuple. Besides, the size of NF model is relatively smaller and unrelated to the domain size. Learning on a table with 6 categorical attributes and each attribute has a domain size of 1000 on average, the input dimension of VAE would be 6000 with one-hot encoding (even 600 with binary encoding), but 6 for the NF model. NFAQP estimates the query result by multidimensional Monte Carlo integration to achieve high accuracy even with low selectivity. For example, with a selectivity of 0.1%, there are only 100 valid tuples for estimating the query for 100k samples, which is usually insufficient for AQP systems. For our approach, the number of sampling points to evaluate the integration is always constant and unaffected by selectivity because the integration only performs in the predicate domain.

The main contributions of this paper are as follows:

- We elaborate on the ways that utilize normalizing flow to address the AQP problem. We illustrate how to train the model to learn the joint distribution and how to answer approximate queries combined with multidimensional Monte Carlo integration.
- We propose NFAQP, a Normalizing Flow-based Approximate Query Processing framework, which leverages the NF model to support the estimation of five aggregation functions. In addition, we adopt the adaptive importance sampling based-integration to AQP to reduce the need for sampling points and accelerate the processing of Group By queries by batch evaluation.
- Extensive experimentation on three real-world datasets has demonstrated that NFAQP significantly reduces the AQP error, especially with a large number of predicates, while achieving low latency and a small model size.

2 Related Work

Approximate Query Processing. The early research of AQP mainly relies on sampling. Among them, online-sampling-based methods [11,15] collect the sample on the fly when the query arrives without requiring pre-existing samples. On the contrary, methods based on offline sampling [2,20] create the sample in advance and preserve it for query processing. Nevertheless, sampling-based approaches can only provide relatively small speedup because the accuracy is highly related to the sample size.

Learning-based AQP has made remarkable progress recently. DBEst [18] combines kernel density estimator and regression model(including XGBoost, LightGBM, etc.) to approximate the query result. Its extended version, DBEst++ [17], utilizes word embedding and mixture density networks to achieve lower latency. However, those models are inefficient to learn a high-dimensional distribution and hard to handle the queries with multiple range predicates. DeepDB [9] proposed the Relational-Sum-Product Network (RSPN) to learn the joint distribution. RSPN recursively decomposes the joint probability into multiple independent probabilities according to the weakly correlated columns. However, the independence assumption between weakly correlated columns limits the accuracy of RSPN, despite its ability to model high-dimensional distributions. The utilization of deep generative models in AQP is first proposed in [23], which directly generates samples with VAE to answer the query. Further, ELECTRA [22] extended it with Conditional VAE (CVAE). By taking predicates as the condition input of CVAE, the model directly generates samples that match the predicates to reduce the error in high-predicate queries. Indeed, generative models are effective ways to fit the high-dimensional distribution. However, as mentioned above, both [22,23] fail to address the low encoding efficiency on attributes with large domain.

Cardinality Estimation. In the research of learning-based query optimization, data-driven cardinality estimation, or selectivity estimation, is similar to approximate query processing to some extent. Both require the use of ML models to learn the complex distributions in the table. Naru [27] factorizes the joint probability into a series of conditional probabilities and fits them by Masked AutoEncoder (MADE) [6]. Besides, it proposes the progressive sampling to infer the selectivity in an autoregressive way. Neurocard [26], further extended it to process the join query and decompose the large-domain columns to reduce the model dimension. DQM-D [8] takes the autoregressive model as a density estimator and estimates the selectivity by integration instead of progressive sampling. However, the autoregressive models also suffer from the large dimension problem as we mentioned before. FACE [25] is the first cardinality estimator that utilizes the normalizing flow. It transforms all columns into continuous float numeric and the NF model can directly learn the joint probability from this format while keeping a small size. Normalizing flow has achieved excellent performance on cardinality estimation. Our work further extended the NF-based query processing framework in FACE [25] to address the AQP problem.

3 Methodology

We proposed NFAQP, a normalizing flow-based approximate query processing Framework. In this section, we first describe the overall architecture of NFAQP and supported queries in Sect. 3.1. In Sect. 3.2, we introduce the background knowledge about utilizing normalizing flow to estimate probability density. After that, we introduce how to train the NF model to learn the joint distribution of the table in Sect. 3.3. Finally, we elaborate on the detail of query processing, the adaptive importance sampling-based integration and the support of queries with Group By In Sect. 3.4. Table 1 shows the notations used in this section.

Table 1. Notations

T	Original table
$\|T\|$	Number of the rows of table
AGG	Aggregate functions
θ	The selectivity of query
N	The number of sample points used to evaluate integration
$\mathbb{R}$	Integration domain
$\boldsymbol{x}$	Tupe
$\Pi_c(\boldsymbol{x})$	The projection of tuple $\boldsymbol{x}$ over column c
p	Probability density
NF	The normalizing flow model
f	The bijective transformation of NF

3.1 Overall Architecture

Figure 1 presents the high-level architecture of NFAQP. In the training stage, the training data sampled from the database will be preprocessed into continuous numeric data, which the NF model can easily learn. The metadata recorded in the preprocessing stage and the trained model will be preserved. In the query processing stage, the SQL will be parsed into the predicates range and Group By values(if contained) to construct the integration domain. After that, sampling points are generated within the integration domain and the NF model evaluates their probability density. Finally, the integrator estimates the approximation of the aggregation function. There are three advantages to our design:

- The normalizing flow model can learn a joint distribution of continuous data while keeping a small number of parameters. For discrete data, it can also be transformed into continuous data after dequantization for easily learning. Thus, we can obtain the distribution of all columns using only a lightweight model.

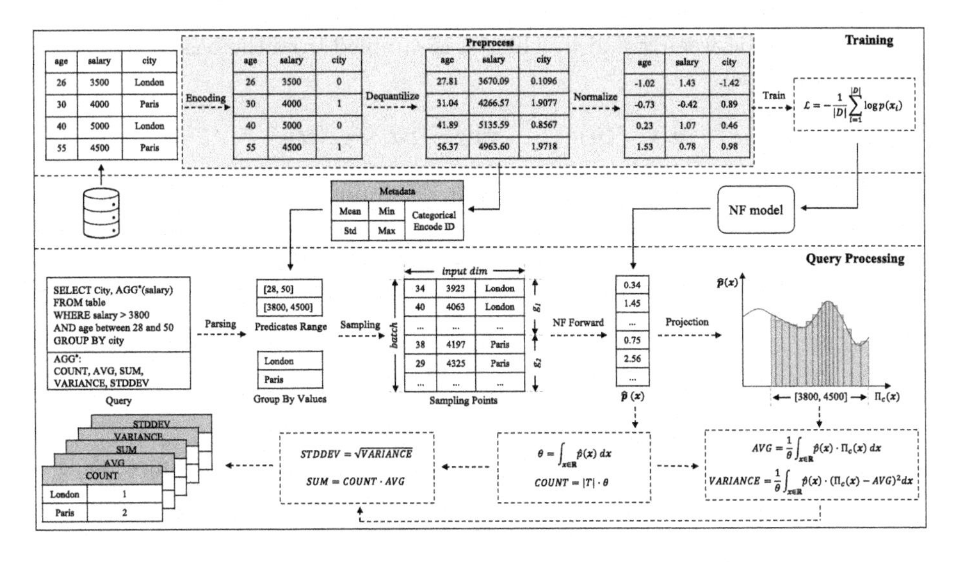

Fig. 1. NFAQP Architecture

- The space overhead is relatively small. We only need to preserve several KBs for the necessary components of the query engine while handling the queries on a table with millions of rows.
- The processing of queries can be accelerated easily. We design a sampling method that treats the sampling points for GROUP BY queries as a batch to reduce the inference time.

NFAQP supports queries whose SQL like:

```
SELECT [g,] AGG(c)
FROM T
[WHERE predicates]
[GROUP BY g]
```

where AGG is the aggregate function. The supported functions include COUNT, AVG, SUM, VARIANCE, and STDDEV. Besides, NFAQP supports point predicate [=], range predicate[>, >=, <, <=, between], and also the query with multiplepredicates. It also supports GROUP BY queries and both the WHERE and GROUP BY clauses are optional.

3.2 Density Estimation via Normalizing Flow

Normalizing flow provides an effective way to model complex target distribution and exact density estimation [12,19]. Given a d-dimensional real vector $\boldsymbol{x}$, which is a random variable sampled from the target distribution, such as the training table T, the basic idea of normalizing flow is to learn the bijective transformation

T that mapping between $\boldsymbol{x}$ and $\boldsymbol{u}$, where $\boldsymbol{u}$ is sampled from the base distribution of NF model $p_u(\boldsymbol{u})$:

$$\boldsymbol{x} = f(\boldsymbol{u}) \qquad \text{where} \quad \boldsymbol{u} \sim p_u(\boldsymbol{u}) \tag{1}$$

Among them, the base distribution $p_u(\boldsymbol{u})$ is a simple known distribution, such as normal distribution, and the transformation f must be invertible and differentiable. Both f and $p_u(\boldsymbol{u})$ contain trainable parameters.

Once the NF model is trained, we can utilize it to estimate the probability density of tuple $\boldsymbol{x}$ in T by:

$$p_x(\boldsymbol{x}) = p(\boldsymbol{u})|\text{det}J_f(\boldsymbol{u})|^{-1} \tag{2}$$

where $\boldsymbol{u}$ can be acquire with the invert transform as $\boldsymbol{u} = f^{-1}(\boldsymbol{x})$ and $J_f(\boldsymbol{u})$ is the Jacobian of transform f. Noticed that when scaling the probability density from $p(\boldsymbol{u})$ to $p(\boldsymbol{x})$, the determinant of Jacobian $|\text{det}J_f(\boldsymbol{u})|^{-1}$ is needed. Thus, for efficiency, f should be carefully designed to have the determinant of its Jacobian easy to acquire.

3.3 Training

We followed the training method for the NF model as described in [25] while saving the metadata for the query processing. Herein, we provide a brief overview of the overall training process.

Preprocessing. Before training, a series of preprocessing steps will be applied to transform the raw data into continuous numeric data first, which can be easily learned by the NF model.

For categorical columns like "city", we encode them to integers and save the encoded ID for each distinct value. After encoding, the training data have been converted into numeric. Next, we perform spline dequantization [25] to all columns to make data more continuous. Specifically, we first collect and sort all the distinct values of columns. After that, we counted the cumulative probability of each distinct value according to their occurrence frequency, corresponding to the points on the cumulative distribution function (CDF). After that, we construct the CDF using Monotone Piecewise Cubic Spline Interpolation on these points. Next, we uniformly sample the CDF points on the y-axis and map them back to the corresponding value on the x-axis, which is the dequantized value. Finally, we store the Min and Max of each column for parsing the predicates into ranges when querying. The effect of dequantilization is to scatter each distinct value v onto the interval $[a, a + \Delta_a)$, where Δ_a is the distance between a and the smallest value bigger than a. Finally, the data will be normalized before being fed into the NF model and the Mean and Std of each column will be stored for mapping the predicate ranges into normalized ranges.

Model Training. The NF model is trained by maximizing negative log-likelihood. Given the training set D and each tuple $\boldsymbol{x}_i \in D$, the loss function is defined as:

$$\mathcal{L} = -\frac{1}{|D|}\sum_{i=1}^{|D|} \log p(\boldsymbol{x}_i) \tag{3}$$

3.4 Query Processing

In this section, we illustrate how to process approximate query by using the trained NF model. We first show a simple case that without groupby:

SELECT AGG(salary) FRRM table
WHERE salary > 3800
AND age BETWEEN 28 AND 45
AND city = "Paris"

Parsing. As we can see, the given query example contains multiple predicates, including range and point predicates. First, as shown in the parsing stage of Fig. 1, the where clause of the SQL will be parsed into predicate ranges to create a bounded integration domain $\mathbb{R}$, where $\mathbb{R}$ is d-dimensional for a table with d columns. For numerical data, we complete its boundary according to the Min and Max stored in the Metadata. Thus, "salary > 3800" would be converted into [3800, 4500]. "age between 28 and 45" are not need to convert since it is already bounded. For the point predicate on categorical data, we find the encoded ID of the predicate value, and the corresponding range is $[\mathrm{ID}, \mathrm{ID}+1]$ since all the categorical columns are encoded to and dequantized as numeric type with bin size of 1. Finally, we obtain a 3-dimensional integration domain $\mathbb{R} = [[3800, 4500], [28, 45], [1, 2]]$ for the above example.

Integration. We already trained a normalizing flow model NF. For any tuple $\boldsymbol{x}$, we estimate its probability density by $\hat{p}(\boldsymbol{x}) = NF(\boldsymbol{x})$. Additionally, we have obtained the integration domain $\mathbb{R}$ after the parsing stage. Now, we illustrate how to approximate the aggregation function.

For COUNT on table T, is formally defined as:

$$COUNT = |T| \cdot \theta \tag{4}$$

where $|T|$ is the size of the table and θ is the selectivity of predicates. We use the multidimensional Monte Carlo integration to approximate θ:

$$\begin{aligned}\theta &= \int_{\boldsymbol{x}\in\mathbb{R}} p(\boldsymbol{x})\, d\boldsymbol{x} \\ &\approx \sum_{i=1}^{N} \hat{p}(\boldsymbol{x}_i) \cdot v_i\end{aligned} \tag{5}$$

In the approximate formulation, the integration domain $\mathbb{R}$ is divided to N hypercubes, where N is the number of sampling points used to evaluate the integration, and v_i is the volume of the i-th hypercube.

The AVG on the aggregate column c is defined as:

$$\begin{aligned} AVG &= \frac{\int_{\boldsymbol{x} \in \mathbb{R}} p(\boldsymbol{x}) \cdot \Pi_c(\boldsymbol{x})\, d\boldsymbol{x}}{\int_{\boldsymbol{x} \in \mathbb{R}} p(\boldsymbol{x})\, d\boldsymbol{x}} \\ &\approx \frac{1}{\theta} \sum_{i=1}^{N} \hat{p}(\boldsymbol{x}_i) \cdot \Pi_c(\boldsymbol{x}) \cdot v_i \end{aligned} \tag{6}$$

where $\Pi_c(\boldsymbol{x})$ is the projection of $\boldsymbol{x}$ on aggregate column c. For instance, if the generated sample point is $\boldsymbol{x}_i = [34, 3923, \text{London}]$ and the aggregate column is "salary", then the projection $\Pi_c(\boldsymbol{x}) = 3923$. We utilize the previously calculated selectivity to scale up the integration form predicate domain to $\frac{1}{\theta}$.

After AVG is obtained, we can further acquire VARIANCE, the variance on column c of table T:

$$\begin{aligned} VARIANCE &= \frac{\int_{\boldsymbol{x} \in \mathbb{R}} p(\boldsymbol{x}) \cdot (\Pi_c(\boldsymbol{x}) - AVG)^2\, d\boldsymbol{x}}{\int_{\boldsymbol{x} \in \mathbb{R}} p(\boldsymbol{x})\, d\boldsymbol{x}} \\ &\approx \frac{1}{\theta} \sum_{i=1}^{N} \hat{p}(\boldsymbol{x}_i) \cdot (\Pi_c(\boldsymbol{x}) - AVG)^2 \cdot v_i \end{aligned} \tag{7}$$

Noticed that the SUM can be defined as the product of AVG and COUNT. Furthermore, the STDDEV is the square of VARIANCE. Thus, we can fully reuse the previous result to approximate those:

$$\begin{aligned} SUM &= |T| \cdot \int_{\boldsymbol{x} \in \mathbb{R}} p(\boldsymbol{x}) \cdot \Pi_c(\boldsymbol{x})\, d\boldsymbol{x} \\ &\approx COUNT \cdot AVG \end{aligned} \tag{8}$$

$$\begin{aligned} STDDEV &= \sqrt{\frac{\int_{\boldsymbol{x} \in \mathbb{R}} p(\boldsymbol{x}) \cdot (\Pi_c(\boldsymbol{x}) - AVG)^2\, d\boldsymbol{x}}{\int_{\boldsymbol{x} \in \mathbb{R}} p(\boldsymbol{x})\, d\boldsymbol{x}}} \\ &\approx \sqrt{VARIANCE} \end{aligned} \tag{9}$$

Adaptive Importance Sampling. As the dimension of integration increases, we will encounter the curse of dimensionality. Even with a 6-dimensional integration and just 20 sampling points assigned to each dimension, the total number of sampling points comes to $20^6 = 64,000,000$, leading to an unacceptable computational cost.

Previous research [8,18,25] has shown the significance of Adaptive Importance Sampling (AIS) [14] for reducing the demand of sampling points. AIS is an extension of Monte Carlo integration which adaptively adjusts the distribution of sampling points, concentrating them in areas with higher probability density. This enables AIS to achieve the same accuracy with much fewer sampling points compared with the naive Monte Carlo. However, AIS has its limitation that typically leads to an underestimation of VARIANCE and STDDEV because the sampling points are concentrated on areas with high probability density, where also close to the position of AVG. Despite this, Considering the trade-off between accuracy and computational cost, AIS remains the best choice for most scenarios, and the resolution of this issue still requires future research.

Our implementation is based on TORCHQUAD [7], a Python module for multidimensional numerical integration on GPU. We extended its implementation for Adaptive Importance Sampling integration to support approximate query processing.

Group by Processing. For the processing of GROUP BY queries, we start with the example showing in Fig. 1:

SELECT AGG(salary)
FRRM table
WHERE salary > 3800
AND age BETWEEN 28 AND 45
GROUP BY City

Once receiving this query, the SQL parser decomposes it into multiple subqueries according to the distinct values of the Group By column, and each subquery represents a group g_i of the original query. Each subquery has the GROUP BY clause removed and replaced with a point predicates to the GROUP BY value. For the above example, it will be decomposed into two subqueries in which "GROUP BY City" is replaced with "City = London" and "City = Paris".

After that, we treat those subqueries as a batch and process them in parallel. Specifically, we concatenate the sample points of those subqueries along the batch dimension, as shown in Fig. 1, and evaluate their probability density in a single forward pass. Typically, the inference of the NF model has not fully utilized the parallel performance of GPU. Therefore, concatenating multiple queries to increase the batch size can improve its efficiency compared to serial evaluation. For the subsequent integration, we extend its parallel version to integrate multiple subqueries simultaneously.

4 Performance Evaluation

4.1 Experiment Setting

Environment. All the experiments were perform on server with 8 core Intel(R) Core(TM) i7-9700K CPU, a Navidia 2080Ti GPU, 64 GB RAM, 1 TB SSD and Ubuntu 18.04.

Datasets. We conduct the experiment on three real-world datasets, which have been widely adopted by existing works [9,18,23]:

- Flights [4], which contains statistical data on domestic flights in the USA for the past few years.
- Beijing PM2.5 [3], an hourly record of air pollutant data from air quality monitoring sites in Beijing.
- TPC-H [24], a widely adopted synthetic benchmark dataset, with our experiments primarily conducted on the LINEITEM table.

To make AQP challenging, we follow the setting of previous work [9,23], scaling Flights and Beijing PM2.5 to 10 million tuples by IDEBench [5] while keeping the relationships between attributes. For TPC-H, we set the scale factor as 1 to generate the data, and the corresponding table contains 6 million tuples.

Baselines. We evaluate NFAQP against a variety of publicly available AQP systems:

- VerdictDB [20], A sampling-based AQP method, deployed as an intermediate component between the database and the user. The sampling rate is set to 1% in our experiment.
- VAE-AQP [23], A deep generative models-based method that generates samples by VAE on the client side directly for answering queries. The setting of VAE is kept default and the number of generated samples during query processing is set to 1% of the origin table size in our experiment.
- DeepDB [9], A Relational-Sum-Product-Network-based method that Leverages RSPN to model the joint probability by decomposing it into independent probability recursively. Noticed that VARIANCE and STDDEV are not supported by DeepDB, we only compare it on COUNT, AVG and SUM. The number of tuples assigned to SPN is set to 1% of the origin table.

The evaluation mainly focuses on the performance of modern AQP technique on multi-predicate queries. DBEst [18] and DBEst++ [17] are not considered in the comparison because their publicly available versions do not fully support the complex workload in our evaluation.

Workload. Inspired by [27], we generate synthetic workload for evaluation. For each query, we randomly choose k attributes that will be applied predicates, where k increases from 1 to the number of attributes of the dataset. For each predicate (op, v), we randomly select the operation type op from $[=, >, >=, <, <=, \text{between}]$ (if the corresponding attribute is categorical, only $=$ will be selected) and the predicates value v sampling from the respective column. After that, we check the validity, ensuring that there exist tuples in the original table satisfying the predicates combination. Additionally, the attribute c to estimate the aggregation function is randomly selected from the numeric attributes, and the Group By attribute(if contained) is randomly selected from the categorical attributes. Finally, we apply all aggregation functions [COUNT, AVG, SUM, VARIANCE, STDDEV] to generate 5 queries. For each dataset, we generate 2000 queries for the evaluation.

Hyper-Parameter Setting. For joint distribution learning, we use a lightweight NF model with 2 coupling layers and each layer with 28 hidden units. For query processing, We set $N = 16000$ as the number of sampling points for the evaluation of AIS-Integration.

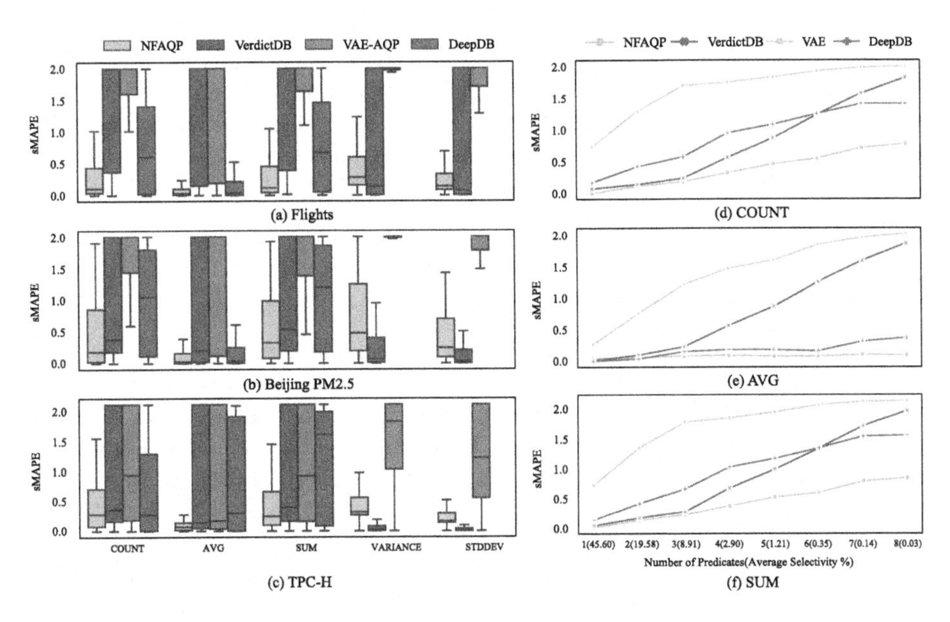

Fig. 2. (a), (b) and (c) denote the overall AQP error on all datasets. (d), (e) and (f) show the average sMAPE under workload with different k on Beijing PM2.5. The percentage % on top shows the average selectivity of queries with corresponding k.

Error Metric. The approximate error of query can be expressed by relative error. However, the relative error is unbounded. When encountering rare difficult queries, some AQP systems may produce inaccurate results that relative error approaching infinity. Therefore, we adopt Symmetric Mean Absolute Percentage Error (sMAPE), which is a bounded metric and lies between 0 and 2 as the error metric for better visualization. For any query, we denote the ground truth of the query answer as g and the approximate result as $\hat{g}$, then sMAPE is defined as:

$$sMAPE = 2 * \frac{|g - \hat{g}|}{|g| + |\hat{g}|} \tag{10}$$

4.2 Comparison of Accuracy

We visualize the distribution of AQP error by box plots, as shown in Fig. 2(a), (b) and (c). The top, middle line, and bottom of each box correspond to the 75th, 50th and 25th percentiles of sMAPE, while the top and bottom whiskers show the 95th and 5th percentiles.

For all datasets, NFAQP consistently achieves the lowest of approximate error on COUNT, AVG and SUM, showing the superiority of the NF model in dealing with the most common queries. For VARIANCE and STDDEV, NFAQP is second to VerdictDB and the reasons contain two folds. First, VerdictDB collects samples by uniform sampling, which does not significantly affect the distribution of variance. Second, as mentioned earlier, AIS integration tends to collect sampling points close to the AVG, resulting in the underestimation of VARIANCE and STDDEV. VAE-AQP also suffers from this issue as the generated samples are common and lack diversity. However, apart from DeepDB, which dose not support querying VARIANCE and STDDEV, NFAQP still achieves the lowest among the learning-based approaches in this part.

4.3 Vary Number of Predicates

To further investigate the performance under the challenging workload, we compared the AQP error of COUNT, AVG and SUM under different predicates on the Beijing PM 25 dataset, as shown in Fig. 2 (d), (e) and (f). The number of predicates k on the x-axis increases from 1 to 8, where 8 is the number of attributes in the dataset, and the percentage next to it shows the average selectivity of the queries with corresponding k. The average sMAPE is recorded on the y-axis. Note that as the number of predicates increases, NFAQP exhibits the least impact. Although in some simple cases with high selectivity, NFAQP is second to DeepDB. However, as k increases, NFAQP achieves a distinct advantage low selectivity while the error of VerdictDB and VAE-AQP has grown to unacceptable. The reasons for that can be attributed in three folds. First, VerdictDB and VAE-AQP are based on samples that are unrelative to the query, no matter they are sampled from data or generated by VAE. In low selectivity, only little tuples satisfy the predicates in the samples that could be used to approximate the query. Second, the independence assumption of DeepDB limits its performance. Finally, NFAQP relies on the sampling points generated within $\mathbb{R}$ to evaluate the query, where $\mathbb{R}$ is the integration domain that satisfies the predicates. Therefore, no matter how low the selectivity is, it always has enough valid sampling points to evaluate the query, ensuring the accuracy.

4.4 Group by Evaluation

The performence on Group By scenario was evaluated on TPC-H. As show in Fig. 3(a), NFAQP achieves the lowest approximate error on COUNT, AVG, SUM, while its advantages are even more significant compared to Non-Group

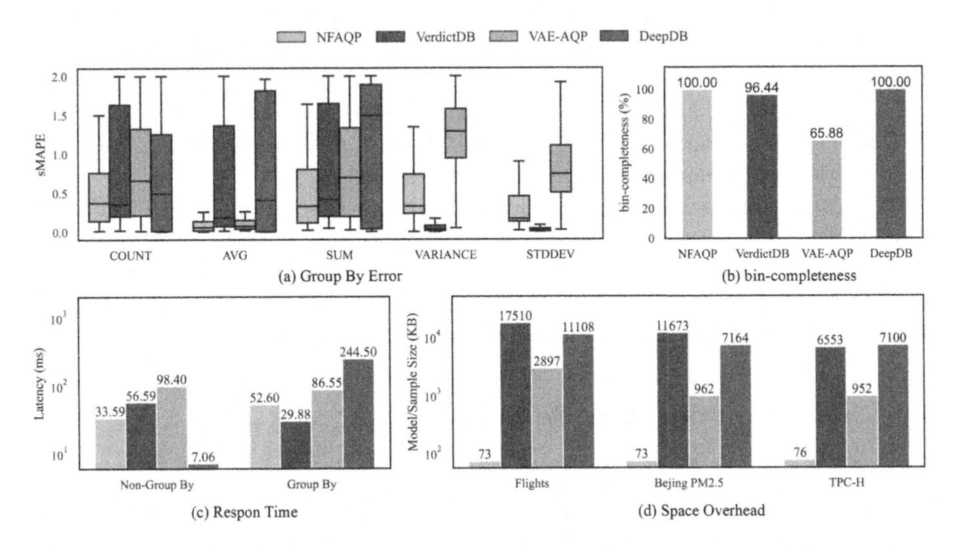

Fig. 3. (a) shows the comparison of AQP error and (b) shows the bin-completeness on Group By queries. (c) denotes the query processing latency and (d) shows the space overhead comparison.

By scenarios. This is because the selectivity of Group By queries is relatively lower (for each group), which other AQP methods are less competitive.

Noticed that under Group By scenario, some AQP methods have too few valid tuples after predicate filtering, making them unable to produce answers for all groups. The Group By error shown Fig. 3(a) is only compute on groups that return by the AQP system. We also compare the bin-completeness as do in [22], which is defined as:

$$bin\text{-}completeness = \frac{|G_{group} \cap \hat{G}_{group}|}{|G_{group}|} \tag{11}$$

where G_{group} is the set of the groups in the ground true while $\hat{G}_{group}$ is in the approximate result. As shown in Fig. 3(b), only NFAQP and DeepDB can answer the queries with complete groups. This is because both VerdictDB and VAE-AQP answer the queries with limited samples, which some groups with low selectivity may not exist in those samples. Different from this, NFAQP directly evaluates the integration on the queried domain. Therefore, no matter how low the selectivity is, it can always answer with complete groups.

4.5 Latency and Space Overhead

We also compare the time and space overhead for all AQP systems. The average query response time is shown in Fig. 3 (c), NFAQP both achieves the second low latency in tow scenarios. In Non-Group By scenarios, DeepDB is very fast, but it drastically increases in latency in Group By scenarios because it does not take measures to accelerate grouping. For VerdictDB and VAE-AQP, the latency is lower in Group By because there are fewer valid tuples in samples under low selectivity, which significant affects their accuracy. Benefiting from the batch acceleration, the latency of NFAQP does not increase much in Group By scenarios while ensuring the approximate error.

We further compare the model size of learning-based approaches (NFAQP, VAE-AQP, DeepDB) and the sample size of the sampling-based approach (VerdictDB). As is shown in Fig. 3 (d), NFAQP achieves the lowest space overhead. Besides, the model size of NF is almost unaffected by the dataset, while VAE is relative to the domain size and RSPN is relative to correlation of attributes.

5 Conclusion and Future Work

This study proposes a noval approximate query processing approach that leverage the normalizing flow to efficiently learn the data's distribution and estimate the aggregation function by multi-dimensional MonteCarlo integration. The experiments on three real-world dataset has demonstrate that our approach has significant advantages in accuracy, especially under low selectivity, while keeping low latency and small model size.

In the future, we will further explore the impact of the model size to achieve a better tradeoff between model size, accuracy, and latency. Besides, this work will be further extended to support the join queries with multi-table.

Acknowledgment. We thank editors and reviewers for their suggestions and comments. This work was supported by National Key R&D Program of China (No. 2021YFC3340700), NSFC grants (No. 61972155 and 62136002), and Shanghai Trusted Industry Internet Software Collaborative Innovation Center.

References

1. Abdal, R., Zhu, P., Mitra, N.J., Wonka, P.: Styleflow: attribute-conditioned exploration of styleGAN-generated images using conditional continuous normalizing flows. ACM Trans. Graph. (ToG) **40**(3), 1–21 (2021)
2. Agarwal, S., Mozafari, B., Panda, A., Milner, H., Madden, S., Stoica, I.: BlinkDB: queries with bounded errors and bounded response times on very large data. In: Hanzálek, Z., Härtig, H., Castro, M., Kaashoek, M.F. (eds.) Eighth Eurosys Conference 2013, EuroSys 2013, Prague, Czech Republic, 14–17 April 2013, pp. 29–42. ACM (2013)
3. Beijing Multi-Site Air-Quality Data Data Set: Beijing multi-site air-quality data data set (2021). https://archive.ics.uci.edu/ml/datasets/Beijing+Multi-Site+Air-Quality+Data. Accessed 9 Apr 2022
4. Bureau of Transportation Statistics: Carrier on-time performance (2019). https://www.transtats.bts.gov/. Accessed 9 Apr 2023
5. Eichmann, P., Zgraggen, E., Binnig, C., Kraska, T.: Idebench: a benchmark for interactive data exploration. In: Proceedings of the 2020 ACM SIGMOD International Conference on Management of Data, pp. 1555–1569 (2020)

6. Germain, M., Gregor, K., Murray, I., Larochelle, H.: Made: masked autoencoder for distribution estimation. In: International Conference on Machine Learning, pp. 881–889. PMLR (2015)
7. Gómez, P., Toftevaag, H.H., Meoni, G.: torchquad: Numerical integration in arbitrary dimensions with PyTorch. J. Open Source Softw. **6**(64), 3439 (2021)
8. Hasan, S., Thirumuruganathan, S., Augustine, J., Koudas, N., Das, G.: Deep learning models for selectivity estimation of multi-attribute queries. In: Proceedings of the 2020 ACM SIGMOD International Conference on Management of Data, pp. 1035–1050 (2020)
9. Hilprecht, B., Schmidt, A., Kulessa, M., Molina, A., Kersting, K., Binnig, C.: DeepDB: learn from data, not from queries! Proc. VLDB Endow. **13**(7), 992–1005 (2020)
10. Ho, J., Chen, X., Srinivas, A., Duan, Y., Abbeel, P.: Flow++: improving flow-based generative models with variational dequantization and architecture design. In: International Conference on Machine Learning, pp. 2722–2730. PMLR (2019)
11. Kandula, S., et al.: Quickr: lazily approximating complex adhoc queries in bigdata clusters. In: Proceedings of the 2016 International Conference on Management of Data, pp. 631–646 (2016)
12. Kobyzev, I., Prince, S.J., Brubaker, M.A.: Normalizing flows: an introduction and review of current methods. IEEE Trans. Pattern Anal. Mach. Intell. **43**(11), 3964–3979 (2020)
13. Kuznetsov, M., Polykovskiy, D.: Molgrow: a graph normalizing flow for hierarchical molecular generation. In: Proceedings of the AAAI Conference on Artificial Intelligence, vol. 35, pp. 8226–8234 (2021)
14. Lepage, G.P.: Adaptive multidimensional integration: vegas enhanced. J. Comput. Phys. **439**, 110386 (2021)
15. Li, F., Wu, B., Yi, K., Zhao, Z.: Wander join: online aggregation via random walks, pp. 615–629 (2016)
16. Lin, C., Li, J., Wang, X., Lu, X., Zhang, J.: WFApprox: approximate window functions processing. In: Nah, Y., Cui, B., Lee, S.-W., Yu, J.X., Moon, Y.-S., Whang, S.E. (eds.) DASFAA 2020. LNCS, vol. 12112, pp. 72–87. Springer, Cham (2020). https://doi.org/10.1007/978-3-030-59410-7_5
17. Ma, Q., Shanghooshabad, A.M., Almasi, M., Kurmanji, M., Triantafillou, P.: Learned approximate query processing: Make it light, accurate and fast. In: CIDR (2021)
18. Ma, Q., Triantafillou, P.: Dbest: revisiting approximate query processing engines with machine learning models. In: Proceedings of the 2019 International Conference on Management of Data, pp. 1553–1570 (2019)
19. Papamakarios, G., Nalisnick, E., Rezende, D.J., Mohamed, S., Lakshminarayanan, B.: Normalizing flows for probabilistic modeling and inference. J. Mach. Learn. Res. **22**(1), 2617–2680 (2021)
20. Park, Y., Mozafari, B., Sorenson, J., Wang, J.: VerdictDB: universalizing approximate query processing. In: Proceedings of the 2018 International Conference on Management of Data, pp. 1461–1476 (2018)
21. Prenger, R., Valle, R., Catanzaro, B.: Waveglow: a flow-based generative network for speech synthesis. In: ICASSP 2019–2019 IEEE International Conference on Acoustics, Speech and Signal Processing (ICASSP), pp. 3617–3621. IEEE (2019)
22. Sheoran, N., et al.: Conditional generative model based predicate-aware query approximation. In: Proceedings of the AAAI Conference on Artificial Intelligence, vol. 36, pp. 8259–8266 (2022)

23. Thirumuruganathan, S., Hasan, S., Koudas, N., Das, G.: Approximate query processing for data exploration using deep generative models. In: 2020 IEEE 36th international conference on data engineering (ICDE), pp. 1309–1320. IEEE (2020)
24. TPC: TPC Benchmark^{TM} H (Transaction Processing Performance Council Benchmark^{TM} H) (2022). https://www.tpc.org/tpch/. Accessed 13 Apr 2022
25. Wang, J., Chai, C., Liu, J., Li, G.: Face: a normalizing flow based cardinality estimator. Proc. VLDB Endow. **15**(1), 72–84 (2021)
26. Yang, Z., et al.: Neurocard: one cardinality estimator for all tables. Proc. VLDB Endow. **14**(1), 61–73 (2020)
27. Yang, Z., et al.: Deep unsupervised cardinality estimation. Proc. VLDB Endow. **13**(3), 279–292 (2019)

Efficient Blockchain Data Trusty Provenance Based on the W3C PROV Model

Zhongming Yao[1], Zhiqiong Wang[2,3(✉)], Liang Wen[1], Kun Hao[2], and Junming Xu[1]

[1] School of Computer Science and Engineering, Northeastern University, Shenyang 110819, China
yaozming@stumail.neu.edu.cn

[2] College of Medicine and Biological Information Engineering, Northeastern University, Shenyang 110819, China
{wangzq,haokun}@bmie.neu.edu.cn

[3] Key Laboratory of Big Data Management and Analytics (Liaoning Province), Northeastern University, Shenyang 110819, China

Abstract. Data provenance can effectively ensure the correctness of data mining results. To realize efficient and trusty data provenance, we propose an efficient blockchain data trusty provenance architecture based on the W3C PROV model. First, we integrate the W3C PROV model into the blockchain system to offer a unified provision information standard so as to form convincing assessments about its quality, reliability, or trustworthiness. Second, we design multi-bucket indexes within blocks and a skip list index among blocks to improve the query efficiency of the provenance information on blockchain. Finally, we design hash-linked list indexes to improve the verification efficiency of the completeness and correctness of the provenance information and propose a verifiable data query algorithm based on the above index. Experiential results show that our proposed architecture performance is effective and efficient.

Keywords: Blockchain · Data provenance · W3C PROV Model · Index

1 Introduction

Data mining is finding helpful information in a large amount of data using computer technology. Data mining can help us see patterns, trends and laws hidden in data, and predict future trends [7]. Data provenance refers to tracking and recording all changes and operations of data in the whole life cycle to determine the source, usage, and transmission path of data. In data mining, we need to know the source, data quality, and data processing methods, which must be obtained through data provenance [13]. Therefore, effective and trusty data

X. Yang et al. (Eds.): ADMA 2023, LNAI 14180, pp. 61–76, 2023.
https://doi.org/10.1007/978-3-031-46677-9_5

provenance is crucial for data mining. Data provenance traces the flow path of the raw data by recording the related metadata, also called provenance information, to determine the historical information of the raw data. These metadata record the evolution of the whole life cycle of the raw data, covering the processes of generation, distribution, and modification. By tracing the raw data through these metadata, we can reproduce the historical events, flow paths, related sources, and information related to data leakage. This information facilitates accountability.

Traditional data provenance has the problem of being vulnerable to tampering, forgery, and malicious attacks. In addition, traditional data provenance also has the problems of information islands and difficulty in sharing, which leads to worse information flow and more complicated cooperation. Therefore, traditional data provenance lacks sufficient credibility and transparency. Motivated by the academic attention to blockchain technology, there is an idea to use it for data provenance. Because blockchain system provides complete data history and immutable data storage, as well as adopts the decentralized network structure to realize information sharing and cooperation. However, this idea is non-trivial and presents three main challenges.

- *How to introduce the existing excellent data provenance model into the blockchain system?* Many excellent provenance models exist in traditional data provenance, such as the W3C PROV model[1]. The provenance results of using mature provenance models for data provenance are more convincing to the public. Still, there has been little work to integrate mature provenance models into blockchain systems [1,3–5,8]. In addition, these efforts only carry the mature provenance model to the blockchain but do not combine the two organically.
- *How to ensure the query efficiency of provenance information on blockchain according to provenance requirements?* The lack of effective indexes in traditional blockchain leads to low query efficiency. There are efforts to improve the efficiency of data queries by building indexes on blockchain, which are aimed at common data query requirements [12,14,16–18]. Therefore, the current blockchain lacks an effective index designed according to the characteristics of provenance information, which cannot meet the actual query requirements in the data provenance scenario.
- *How to efficiently ensure the completeness and correctness of the query results of provenance information?* The block-chain structure and hash information on blockchain can ensure data's integrity. Each piece of data in the blockchain can be verified by using the previous block hash and Merkel tree root field. However, provenance information generally involves many pieces of data. It takes a lot of work to verify them in the general blockchain, which cannot effectively guarantee the completeness and correctness of the query results of provenance information.

To tackle the above challenges, we propose an efficient blockchain data trusty provenance architecture based on the W3C PROV model. In this architecture,

[1] https://www.w3.org/TR/prov-overview/.

we integrate the W3C PROV model into the blockchain system to realize data provenance based on blockchain. And we design the indexes within and among the blocks to support the efficient query of the provenance information. In addition, we propose an efficient verifiable query algorithm for the architecture and introduce a hash structure to support trusty verification of provenance information query on the premise of ensuring efficiency. The contributions of our paper are summarized below.

- We propose an efficient blockchain data trusty provenance architecture based on the W3C PROV model by integrating the W3C PROV model into the blockchain system to make the data provenance more convincing for the public.
- We design the multi-bucket indexes within blocks and the skip list index among blocks to improve the query efficiency of the provenance information on blockchain.
- We propose an efficient verifiable query algorithm based on our architecture. Specifically, we design a hash-linked list index to realize the overall verification of provenance information so as to improve the verification efficiency of the completeness and correctness of provenance information.
- We design detailed experiments. The experiments show that the on-chain data query speed is indeed improved after optimizing the structure. In addition, the experiments also demonstrate that this architecture applies to the data provenance verification process.

In this paper, Sect. 2 introduces the related work of blockchain and data provenance based on blockchain. Section 3 presents the preliminary knowledge and defines the problems. Section 4 proposes a data provenance architecture based on blockchain and introduces the indexes in the architecture. Section 5 proposes an efficient verifiable query algorithm. Section 6 evaluates the architecture and analyzes the experimental results. Section 7 summarizes the paper.

2 Related Work

2.1 Blockchain

In 2008, when Satoshi Nakamoto introduced the concept of Bitcoin in the Bitcoin white paper, blockchain technology started to come into people's view. With the rise of Bitcoin, more and more people have begun to pay attention to and study blockchain technology. Many scholars [6] summarized the architecture, data structure, consensus mechanism, and smart contracts in blockchain technology in a comprehensive and detailed way. They compared Bitcoin, Ethereum, and Hyperledger Fabric platforms from different aspects and summarized the advantages and disadvantages of blockchain technology and the future development trend of blockchain.

Specifically, for blockchain data queries, Zhang et al. [16,17] propose a data authentication structure called GEM^2-tree for range queries to reduce the overhead of using smart contracts. Then they optimized the GEM^2-tree not to sacrifice much query performance. Subsequently, the team optimized the keyword query to reduce the contract overhead. Xu et al. [14] propose a processing framework for verifiable queries, chain, which allows lightweight users to verify query results obtained from untrusted parties. They also developed two new indexes to aggregate data within and between blocks to ensure efficient verification of query results and speed up subscription queries by inverted prefix trees. Zhu et al. [18] propose and implement a blockchain database called SEBDB to add relational data semantics to the blockchain platform and implement a SQL-like language to support easy application development. To develop block-level, table-level, and hierarchical multi-level indexes are proposed to ensure data query efficiency. For blockchain data storage, Xu et al. [15] introduce the concept of consensus unit to solve the storage problem, which organizes different nodes into a unit and allows them to store at least one copy of blockchain data in the system. Based on this, they further define the block allocation optimization problem to utilize the storage space and minimize the query cost fully. Qi et al. [10] propose a new storage engine to enhance storage scalability by integrating erasure encoding and Byzantine Fault Tolerance consensus protocol, which can increase the overall storage capacity when more nodes join the blockchain.

2.2 Provenance Based on Blockchain

There is also an amount of current work in blockchain provenance. Demichev et al. [5] propose a new approach to design a fully decentralized data management system for distributed environments with distributed groups of users. The system is suitable for distributed environments with groups of users that are administratively unrelated or loosely related in the case of a partial or complete lack of trust between the groups of users. Tang et al. [11] propose Prv^{2}chain, a framework that allows users to privately store records of their linked sources, protect data on the chain of blocks, and the association of data. They also design the record ID generation algorithm to allow users to perform on-chain tree traversal efficiently. Liu et al. [9] propose a blockchain-based architecture for secure and efficient distribution network sources in the Internet of Things. They introduced a unified source query model to avoid storing and querying the entire source data directly on the blockchain. Chen et al. [2] propose a decentralized data-sharing platform that provides a secure, correct source record using a networked blockchain named ProvNet. All good shared records are collected and stored in a tamper-proof web-based blockchain called a block. In addition to the above work, some scholars have also studied data provenance based on blockchain [1,3,4,8]. The above work either does not adopt the existing advanced model, or the data provenance efficiency needs to be higher.

3 Preliminaries

3.1 Blockchain

Blockchain, also known as a distributed ledger, is a database in which multiple mutually untrusted sections jointly maintain the same global state, which has the advantages of decentralization, data tamper-proof, and data provenance.

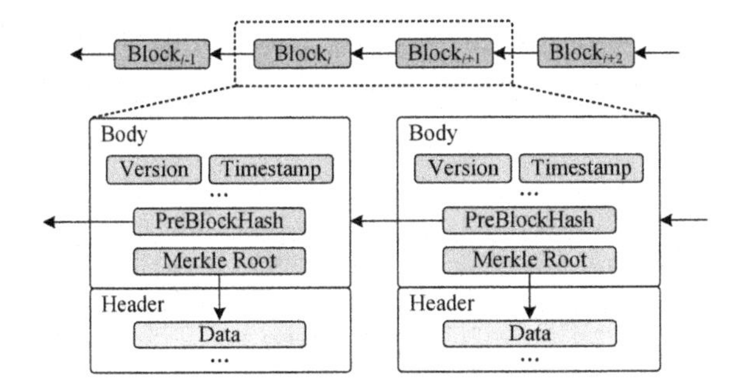

Fig. 1. Blockchain structure.

In the blockchain, all nodes are equal, and each node redundantly stores a complete copy of the blockchain, and there is no centralized hardware or governing body. It is only possible to tamper with the data stored in the blockchain for the malicious nodes if their computing power reaches more than half of the total computing power. In addition, blockchain uses a pre-block hash to connect all blocks first and last and uses a Merkle tree structure to store data to achieve the tamper-proof data feature. As shown in Fig. 1, a block in the blockchain consists of a block header and a block body. The block header includes fields such as version number, timestamp, difficulty target, the hash value of the previous block, and Merkle tree root. And the block body stores the detailed data.

3.2 W3C PROV Model

Provenance is information about entities, activities, and people involved in producing a piece of data, which can be used to form assessments about its quality, reliability, or trustworthiness. The PROV Family of Documents defines a model, corresponding serializations, and other supporting definitions to enable the interchange of provenance information in heterogeneous environments. And the W3C PROV data model is its core. As shown in Fig. 2, the W3C PROV data model is a generic data model for provenance that allows domain and application-specific representations of provenance to be translated into such a data model and interchanged between systems. Thus, heterogeneous systems can export their native provenance into such a core data model, and applications that need to make sense of provenance can import, process, and reason. The W3C PROV data model has a modular design and is structured according to six components covering various facets of provenance:

- Component 1: *entities* and *activities*, and the *time* at which they were created, used, or ended;
- Component 2: *derivations* of entities from others;
- Component 3: *agents* bearing responsibility for entities that were generated and activities that happened;
- Component 4: *bundles* mechanism to support provenance of provenance;

- Component 5: properties to link entities that refer to the same thing;
- Component 6: *collections* forming a logical structure for its members.

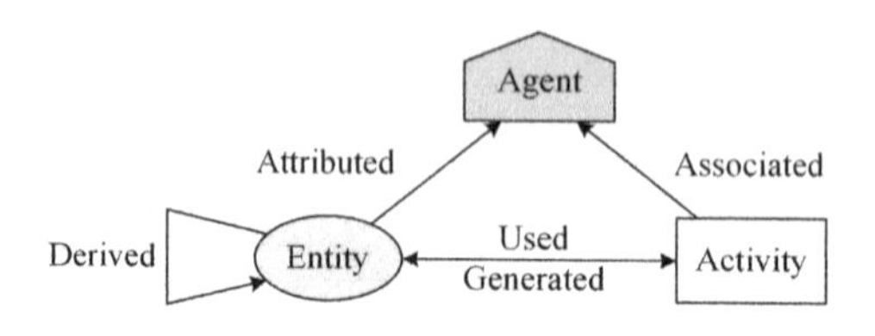

Fig. 2. The W3C PROV data model.

According to official documents[2], an entity is a physical, digital, conceptual, or other kinds of thing with some fixed aspects. An activity occurs over a while and acts upon or with entities. A derivation is a transformation of an entity into another, an update of an entity resulting in a new one, or the construction of a new entity based on a pre-existing entity. An agent bears some form of responsibility for an activity taking place, an entity's existence, or another agent's activity. A bundle is a named set of provenance descriptions and is an entity allowing provenance of provenance to be expressed. A collection is an entity that provides a structure to some constituents who must be entities.

3.3 Problem Definition

In the scenario of the data provenance based on blockchain, on-chain and off-chain data can be defined as follows.

Definition 1. *The on-chain data is denoted as* $BC = \{B_1, B_2, \cdots\}$*, where* $B_i = \{H_i, O_i\}$ *denotes the block. The* H *denotes the block header, and the* $O = \{o_1, o_2, \cdots\}$ *denotes the block body, where* o *denotes a data object in the block.*

Definition 2. *The off-chain data is denoted as* $D = \{d_1, d_2, \cdots\}$*, where* $d_i = \{v_i, w_i\}$ *denotes a piece of data in a certain version. The* v *denotes the version of the data, and the* w *denotes the value of the corresponding version* v*.*

According to the above definitions, the problem in this paper can be defined as follows.

Definition 3. *Suppose the user initiates data provenance for a data according to the keyword* k*. In that case, this architecture returns the data* $S_{d_k} = \{d_k^1, d_k^2, \cdots\}$ *and provenance information* $prov = \{p_1, p_2, \cdots\}$ *where* p *corresponds to a data object* o*.*

[2] https://www.w3.org/TR/2013/REC-prov-dm-20130430/.

4 Architecture

4.1 Architecture Overview

Our architecture consists of four modules, including the off-chain database, blockchain, index structure, and W3C PROV data model modules, as illustrated in Fig. 3.

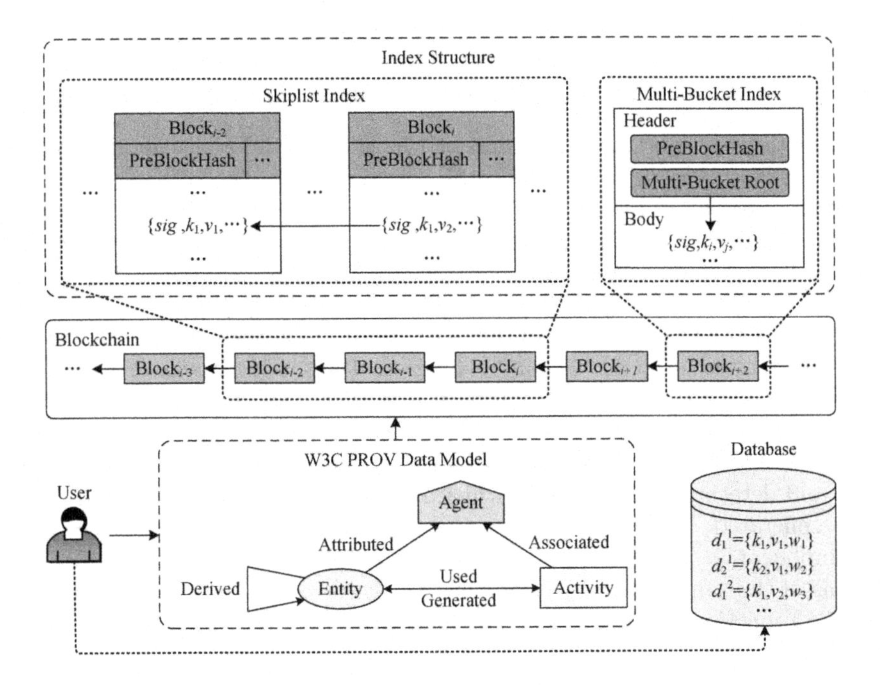

Fig. 3. The architecture.

The off-chain database module stores the raw data and the historical versions of the raw data. The blockchain module stores the provenance information. The index structure module includes the multi-bucket indexes within the blocks and the skip list index among blocks. The W3C PROV data model module processes the provenance information into a unified standard form.

In our architecture, the users perform operations on the raw data. Then the blockchain network processes and stores the provenance information about the data operations through the W3C module while modifying and recording the version of the raw data in the off-chain database. When the users want to perform data provenance, they request data access from the blockchain network, which queries the data on blockchain by the index structure and the data in the off-chain database.

The details of integrating the W3C PROV model into the blockchain system are described in Sect. 4.2. And the details of the index structure are described in Sect. 4.3.

4.2 W3C PROV Model Integration

To provide a unified provenance information standard so as to form convincing assessments about its quality, reliability, or trustworthiness, we integrate the W3C PROV data model into the blockchain system by defining data objects o on blockchain in detail. The data object o is defined as follows.

Definition 4. *The data object* o *is denoted as* $o = \{k, v, a, t, sig_u\}$*, where* k *denotes the keyword of the data object,* v *denotes the version of the data object,* v *denotes the activity on the data object,* t *denotes the time of the activity, and* sig_u *denotes the signature of the blockchain node performing the activity.*

We realize the partial integration of the W3C PROV data model by Definitions 1, 2, 3, and 4. Specifically, the raw data d on the off-chain database corresponds to the *entities* in Component 1. The t field in the data object on blockchain corresponds to the *time* in Component 1. The a field in the data object on blockchain corresponds to the *activities* in Component 1. The raw data and its version history on the off-chain database are generated from their original version, corresponding to the *derivations* in Component 2. The sig field in the data objects on blockchain belongs to the executor who performs the *activities*, corresponding to the *agents* in Component 3. *Bundles* on blockchain are used to support the provenance of provenance. For users to decide whether they can place their trust in something, they may want to analyze its provenance and determine the agent to which its provenance is attributed and when it was generated. The t, sig fields as the provenance of provenance correspond to Component 4. The k field in the data object on blockchain marks different versions of the same data object as a keyword, corresponding to Component 5. And the *collection* in Component 6 is implemented by the inter-block index in Sect. 4.3.

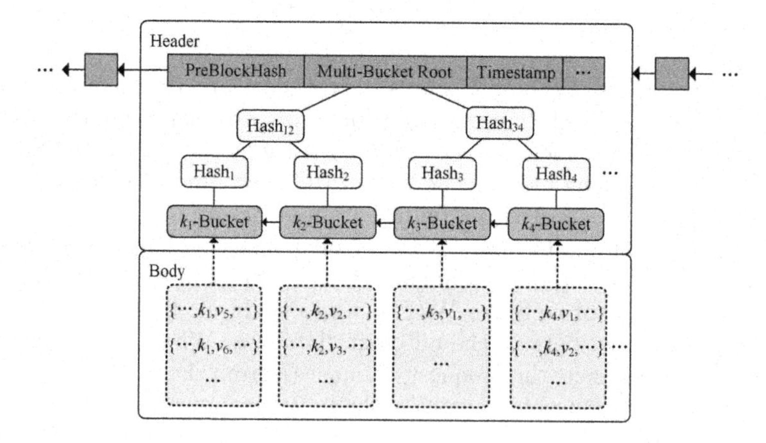

Fig. 4. Intra-block index.

4.3 Index Mechanisms

It is inefficient to process a request by scanning blocks one by one because the provenance information of the same data may be stored in different blocks, and the data in each block is out of order. Therefore, to improve the query efficiency of provenance information, we design a multi-bucket index as the intra-block index and a skip list index as the inter-block index.

Multi-bucket Index Blockchain data is stored as a Merkle tree in each block. The Merkle tree is characterized in that data can quickly verify whether it exists in the blockchain by using the hash value generated by the Merkle tree. When users want to access provenance information in the blockchain, all the data stored in the block by the Merkle tree need to be traversed for the blockchain nodes. Therefore, we propose a multi-bucket index, which not only realizes the characteristics of the Merkle tree. That is, a node can verify whether a block contains a certain transaction without downloading the whole block, but it also improves the query efficiency of provenance information. The multi-bucket index combines the characteristics of the Merkle tree, bucket structure, and sequence list, and its data structure is shown in Fig. 4. In a block, a multi-bucket structure is constructed for data objects with different keywords, and the data objects belonging to the same keyword are stored in the same bucket. And the buckets are sorted according to the value of keywords and linked into a sequence list. In addition, the Merkle tree structure is constructed according to each bucket.

Intra-block Index In queries among blocks on blockchain, if the users want to query all the provenance information corresponding to a certain keyword, they need to traverse all the blocks in blockchain, starting from the genesis block to the latest block. This query method is undoubtedly inefficient. To improve the query efficiency, we build a skip list index among blocks to open a shortcut for querying the provenance information. The keyword tags the data object on blockchain, so we use the keyword to concatenate the data on blockchain and construct the skip list index, as shown in Fig. 5. This index can significantly improve the query efficiency of provenance information. In addition, the index combines the provenance information of the same data into a collection to facilitate data provenance, which satisfies the requirements of Component 4 in the W3C PROV data model.

5 Verifiable Query Algorithm

In the blockchain system, maintaining a full copy of the entire blockchain might be too costly to an ordinary user, as it requires considerable storage, computing, and bandwidth resources. Therefore, many users participate in blockchain networks as light nodes. Light nodes only store block header information, and data query is carried out through full nodes storing all blockchain information.

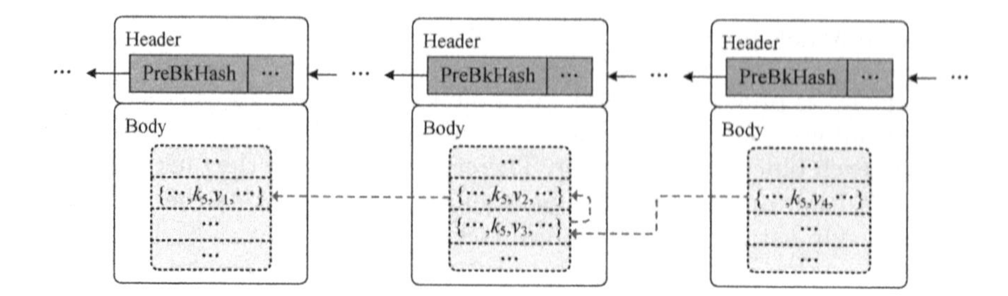

Fig. 5. Inter-block index.

Therefore, the query results obtained by light nodes need to be verified. Currently, the blockchain data verification method is through the Merkle tree. The user recalculates the hash value of Merkle root, calculates the hash of the leaf node corresponding to the query result and its sibling node after connecting them, then calculates the connection value of the parent node and the uncle node and calculates the hash, and so on until the root hash value is generated. Users can compare the computed hash value with the root hash stored in the block header to determine whether the data has been tampered with.

This verification method can only verify a data object at a time. In data provenance, the query results of provenance information usually include multiple data objects, which makes the verification efficiency low. To ensure the verification efficiency of provenance information, we propose a hash-linked list, as shown in Fig. 6. For a certain piece of raw data, we add a $PreHash$ field corresponding to the raw data in the blocks where there is provenance information of the raw data. Each $PreHash$ is generated from the previous $PreHash$ and the bucket hash in the current block. Therefore, using $PreHash$ can verify whether the provenance information from the current block and up to the genesis block has been tampered with.

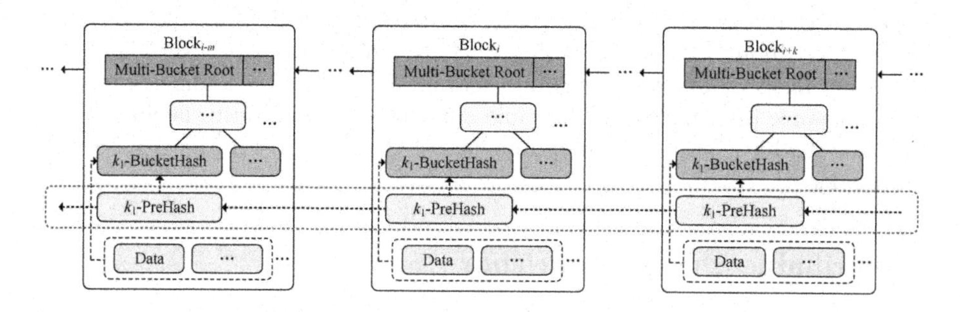

Fig. 6. Provenance information integration verification structure.

We propose a verifiable data query algorithm based on the query and verification indexes introduced above. The user (light node) sends the query request in the data query process. Full nodes query provenance information through query index according to keywords. At the same time, full nodes record the $PreHash$ corresponding to the last provenance information. After obtaining all provenance information, the query results with verification fields are returned to the user. A detailed description of the blockchain data query is shown in Algorithm 1.

Algorithm 1: Blockchain data query algorithm.

```
   Input: Blockchain data BC, keyword k.
   Output: Provenance information prov, verification fields V.
 1 prov = ∅;
 2 PreHash = ∅;
 3 while skiplist ∈ ∅ do
 4     new p = query(skiplist, k);
 5     if thenumberofp ≠ 1 then
 6         while bucket ∈ ∅ do
 7             p = query(bucket, k);
 8             prov = prov + p;
 9         end
10     else
11         prov = prov + p;
12     end
13     V = PreHash;
14 end
15 return prov, V;
```

6 Experimental Results and Analysis

The hardware environment for the experiments is a 2.6 GHz Intel Core i7 processor with 16 GB of RAM. The blockchain system we use is the BlockChainDemo system[3]. The experimental data are generated for simulation. Two samples are simulated in this paper. According to the characters of the proposed model, 1–10 provenance information is simulated for each raw data in sample 1 and sample 2. The first provenance information of each user is to add operation records, and then the provenance information will generate modification and deletion operations. The two types of information are generated at a 1: 1 in sample 1 and a 7: 3 in sample 2. The parameters of the experiment are shown in Table 1. The experiment is divided into query performance evaluation and result verification performance evaluation.

[3] https://github.com/zestaken/BlockChainDemo.

Table 1. Experimental parameters.

Data Items	Provenance Items (Sample 1)	Provenance Items (Sample 2)
2000	9030	11309
4000	21816	20326
6000	32430	30790
8000	41168	39971
10000	49047	52564

6.1 Query Performance Evaluation

The first part of the experiments is the query performance evaluation experiment, which involves four evaluation schemes: the first scheme uses the method in the current system(Origin Scheme); the second scheme uses the intra-block index (Intra-Block Scheme); the third scheme uses the inter-block index (Inter-Block Scheme); the fourth scheme uses both the intra-block and the inter-block index(Mix Scheme). And the experiment is carried out from blockchain construction time, data query time, and query traversal number.

The results from the first experiment (as shown in Fig. 7) reveal that the time curves of the Intra-Block schema and Origin schema for creating the blockchain basically overlap, indicating that the time for creating the blockchain is the same for both, which also proves that the use of inter-block tree structure has little impact on the time for building a blockchain. In contrast, the Inter-Block and Mix schemes require indexing between blocks, so the creation time is longer.

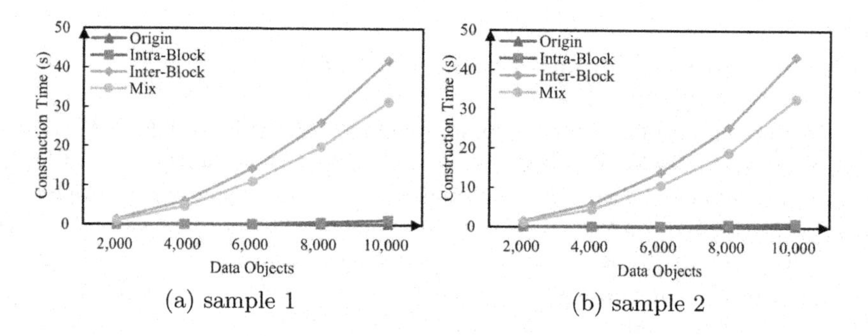

Fig. 7. Blockchain Construction Time.

The query time in the second experiment (as in Fig. 8) is the average time to query all the operation data corresponding to a certain raw data. By comparing the two sample experiments in the second group, we found that the query speed of all three Schemes using the optimized structure has improved. Among them, the Mix Scheme is the most efficient. The reason is that the Intra-Block schema is a subtle optimization of the internal structure of the block, and the speed is slightly improved. The Inter-Block schema adds a skip list index between blocks, and the query speed naturally increases substantially. The Mix schema combines the intra-block and inter-block indexes, so it is the fastest data query speed.

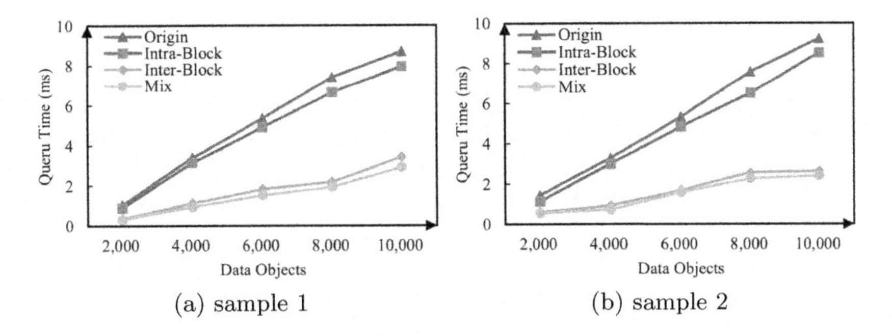

(a) sample 1 (b) sample 2

Fig. 8. Query Time.

Figure 9 depicts the results of the third experiment. The record data is the average number of records traversed when querying the data of all operations corresponding to a certain raw data. The experimental results show that the number of records traversed by the three optimization schemas is significantly reduced compared to the Origin scheme at query time, most notably by the Mix scheme. It indicates that using the intra-block and inter-block indexes can effectively reduce the number of traversed records, which is also the reason for faster queries after optimization.

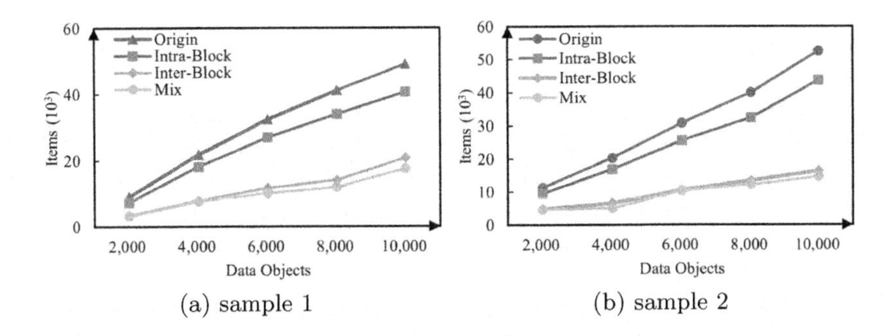

(a) sample 1 (b) sample 2

Fig. 9. Number of Records Traversed.

The above experiments show that using the architecture proposed in this paper, although the time to create the blockchain increases, the time to query the data is significantly reduced. After the on-chain data is uploaded to the blockchain, the data is not changed, and the users can use the indexes to query provenance information multiple times. Therefore, it is beneficial to build indexes. In addition, as the number of data and provenance information increases, the time to create the blockchain and the time to query data using the architecture grows linearly, which shows that the optimized method has high scalability.

6.2 Validation Performance Evaluation

The second part of the experts evaluates the performance of data promotion verification. The verification time is shown in Table 2. Compared with the original verification method, the verification method proposed in this paper is more effective. This is because the original method uses the Merkle structure to verify every traceability information, and the verification efficiency is not high. However, the method proposed in this paper only needs to hash the provenance information and compare it with the recorded information in the block header.

Table 2. Verification cost.

Provenance Items	Time-Original (ms)	Time-Optimized (ms)
2	19.9994	1.7111
4	39.9796	3.3732
6	60.0406	5.8493
8	79.9876	9.8161
10	97.8773	10.5993

7 Summary

Efficient and trusty data provenance is crucial for data mining. We propose an efficient blockchain data trusty provenance architecture based on the W3C PROV model to achieve this goal. For the lack of unified provenance specification in data provenance architecture based on blockchain, we propose integrating the W3C PROV data model into the blockchain system. We can support the standard W3C PROV model in the blockchain provenance process by integrating various components. To address the lack of indexes for querying provenance information, we design multi-bucket indexes with blocks and a skip list index to connect related provision information among blocks. To improve the verification efficiency of query results, we design hash-linked list indexes among blocks. Experiments prove the advantages of our architecture. In the future, we will continue to import the performance of our architecture and plan to construct more research on data promotion based on blockchain.

Acknowledgements. This research was partially supported by the National Key R&D Program of China (No. 2022YFB4500800), the National Natural Science Foundation of China (No. 62072089) and the Fundamental Research Funds for the Central Universities of China (Nos. N2116016, N2104001, N2019007).

References

1. Bose, R.J.C., Phokela, K.K., Kaulgud, V., Podder, S.: Blinker: a blockchain-enabled framework for software provenance. In: 2019 26th Asia-Pacific Software Engineering Conference (APSEC), pp. 1–8. IEEE (2019)
2. Chenli, C., Jung, T.: ProvNet: networked blockchain for decentralized secure provenance. In: Chen, Z., Cui, L., Palanisamy, B., Zhang, L.-J. (eds.) ICBC 2020. LNCS, vol. 12404, pp. 76–93. Springer, Cham (2020). https://doi.org/10.1007/978-3-030-59638-5_6
3. Dang, T.K., Anh, T.D.: A pragmatic blockchain based solution for managing provenance and characteristics in the open data context. In: Dang, T.K., Küng, J., Takizawa, M., Chung, T.M. (eds.) FDSE 2020. LNCS, vol. 12466, pp. 221–242. Springer, Cham (2020). https://doi.org/10.1007/978-3-030-63924-2_13
4. Dang, T.K., Duong, T.A.: An effective and elastic blockchain-based provenance preserving solution for the open data. Int. J. Web Inf. Syst. **17**(5), 480–515 (2021)
5. Demichev, A., Kryukov, A., Prikhod'ko, N.: Business process engineering for data storing and processing in a collaborative distributed environment based on provenance metadata, smart contracts and blockchain technology. J. Grid Comput. **19**(1), 1–30 (2021)
6. Dinh, T.T.A., Wang, J., Chen, G., Liu, R., Ooi, B.C., Tan, K.L.: Blockbench: a framework for analyzing private blockchains. In: Proceedings of the 2017 ACM International Conference on Management of Data, pp. 1085–1100 (2017)
7. Goebel, M., Gruenwald, L.: A survey of data mining and knowledge discovery software tools. ACM SIGKDD Explor. Newsl. **1**(1), 20–33 (1999)
8. Lautert, F., Pigatto, D.F., Gomes, L.: A fog architecture for privacy-preserving data provenance using blockchains. In: 2020 IEEE Symposium on Computers and Communications (ISCC), pp. 1–6. IEEE (2020)
9. Liu, D., Ni, J., Huang, C., Lin, X., Shen, X.S.: Secure and efficient distributed network provenance for IoT: a blockchain-based approach. IEEE Internet Things J. **7**(8), 7564–7574 (2020)
10. Qi, X., Zhang, Z., Jin, C., Zhou, A.: BFT-store: storage partition for permissioned blockchain via erasure coding. In: 2020 IEEE 36th International Conference on Data Engineering (ICDE), pp. 1926–1929. IEEE (2020)
11. Tang, W., Chenli, C., Jung, T.: PRV 2 chain: storage of tree-structured provenance records in blockchain with linkage privacy. In: 2021 IEEE International Conference on Blockchain and Cryptocurrency (ICBC), pp. 1–3. IEEE (2021)
12. Wang, H., Xu, C., Zhang, C., Xu, J.: vChain: a blockchain system ensuring query integrity. In: Proceedings of the 2020 ACM SIGMOD International Conference on Management of Data, pp. 2693–2696 (2020)
13. Wu, X., Zhu, X., Wu, G.Q., Ding, W.: Data mining with big data. IEEE Trans. Knowl. Data Eng. **26**(1), 97–107 (2013)
14. Xu, C., Zhang, C., Xu, J.: vChain: enabling verifiable Boolean range queries over blockchain databases. In: Proceedings of the 2019 International Conference on Management of Data, pp. 141–158 (2019)

15. Xu, Z., Han, S., Chen, L.: Cub, a consensus unit-based storage scheme for blockchain system. In: 2018 IEEE 34th International Conference on Data Engineering (ICDE), pp. 173–184. IEEE (2018)
16. Zhang, C., Xu, C., Wang, H., Xu, J., Choi, B.: Authenticated keyword search in scalable hybrid-storage blockchains. In: 2021 IEEE 37th International Conference on Data Engineering (ICDE), pp. 996–1007. IEEE (2021)
17. Zhang, C., Xu, C., Xu, J., Tang, Y., Choi, B.: Gemˆ2-tree: a gas-efficient structure for authenticated range queries in blockchain. In: 2019 IEEE 35th International Conference on Data Engineering (ICDE), pp. 842–853. IEEE (2019)
18. Zhu, Y., Zhang, Z., Jin, C., Zhou, A., Yan, Y.: SebDB: semantics empowered blockchain database. In: 2019 IEEE 35th International Conference on Data Engineering (ICDE), pp. 1820–1831. IEEE (2019)

Cardinality Estimation of Subgraph Search Queries with Direction Learner

Wenzhe Hou[1], Xiang Zhao[1(✉)], and Wei Wang[2]

[1] National University of Defense Technology, Changsha, China
{wzhou,xiangzhao}@nudt.edu.cn
[2] The Hong Kong University of Science and Technology (Guangzhou), Guangzhou, China
weiwcs@ust.hk

Abstract. In recent years, graph data has been widely used in various fields, which has made graph data management a hot topic of research. Subgraph searching is a crucial operation for graph data management, but its NP-complete nature makes exact matching algorithms time-consuming. To address this problem, new methods have begun to employ deep learning techniques to produce approximate results in less time. However, the distinctiveness of representations of different type of nodes are still limited, and edge information are not well used. To address these limitations, a novel graph neural network architecture, Directional Embedding Network (DENet), is proposed for cardinality estimation of subgraph search queries. It effectively captures node relationships to distinguish different nodes through direction discrepancy, and accurately represents edges with learnable structural information. Experimental results demonstrate that DENet could help improve the expressive power of existing approaches, which has a direct impact on producing more accurate cardinality estimation and optimize the subgraph search query procedure during graph data analysis.

Keywords: data management · subgraph matching · cardinality estimation · ai4db

1 Introduction

With the rapid increase of online information, data comes in complex and diverse forms. With the unique and powerful modeling ability for relational data, graph shows its natural advantages in representing the association relationships in the real world (such as social networks, molecular structures, and reference networks). Thus, graph database has become the forefront and hot spot in the field of data management.

Subgraph Search Query is important basic operation in graph database. Given a *data graph* and a *query graph*, the subgraph search query aims to return all subgraphs of data graph whose structure is the same with the query graph.

X. Yang et al. (Eds.): ADMA 2023, LNAI 14180, pp. 77–93, 2023.
https://doi.org/10.1007/978-3-031-46677-9_6

Subgraph search query plays a important role in graph analysis applications, typical including pattern mining, anomaly monitoring, knowledge modeling, etc. [12].

The key of subgraph search query is subgraph matching condition detection, which is a typical NP-complete [16,18] question and the execution cost of different subgraph search queries may vary greatly in practical applications. Therefore, A mature and sound graph data management system need cardinality estimators to estimate the cost of subgraph search queries and schedule multiple queries efficiently to get optimal performance.

However, it is challenging to accurately count all the matches of a query graph in the data graph. The existing methods are usually based on sampling to achieve cardinality estimation of subgraph search query [9,17,18]. But they may fail to give sound estimation result if few samples are successfully returned. As a result, they can hardly be used in large data graph or complex query graphs.

In recent years, the development of deep learning is in full swing, and deep learning methods achieve good performance. They utilize graph neural networks [3,4,25] to capture information on graphs and some even realize information passing between query and data graphs [13,22]. However, many of them are limited by the over-smooth issues [24,25]. After multiple neural network layers, the representations of different nodes become indistinguishable. Additionally, while existing methods can integrate edge labels into node representations [27], they are unable to generate representations with learnable structural information of edges. This means that their edge representations only captures features of the edge label, ignoring the high-level features of its connected nodes. However, an edge connecting two nodes is the key part inferring the matching conditions which is essential for subgraph search queries, and its representation is supposed to capture the structural information inferred in node representations.

Aiming at above challenges, this paper proposes a learnable Directional Embedding graph Network (DENet), which is a framework for graph data analysis aiming to optimize and enhance representations of both nodes and edges to give robust cardinality estimation of subgraph search queries.

Specifically, DENet is comprised of five key components, namely node representation, edge representation, directional learning, cross-graph learning, and estimation. These components enable DENet to capture the characteristics of nodes and edges in graphs more accurately and effectively, and to compute the final prediction using MLP. Compared with traditional methods, DENet can effectively distinguish different structures and produce a robust representation for the following cardinality estimation. Thereby, DENet can be extensively applied in various fields of graph data analysis, such as social network analysis, bioinformatics, and data mining.

This research takes an exploration on AI4DB [10] for the query cardinality estimator in the graph data management system. The main contributions can be summarized as follows:

- Propose a novel graph neural network architecture for the subgraph matching problem, which can be applied to existing neural subgraph isomorphism cardinality estimation methods and achieve performance improvements.
- Capture structural information of edges in graph data, which is crucial for condition detection in subgraph search.
- Enable the model to learn the similarities and difference between nodes by leveraging directional information in node representations, thereby enhancing the model's discriminative power for nodes.

The following sections of the article are arranged as follows: Sect. 2 defines the problem of cardinality estimation of subgraph search queries, and Sect. 3 outlines the existing methods of cardinality estimation and other related work. Section 4 introduces the technical details of DENet. Section 5 uses real datasets to evaluate different estimation methods in multiple metrics, while Sect. 6 is the summary of the research and the prospect of work.

2 Problem Definition

Subgraph search is a key technology for managing and analyzing graph data. Subgraph matching looks up important query graphs by catching subgraphs with specific rules and structures. This paper studies the cardinality estimation of the subgraph matching, which are defined as follows:

Definition 1 (Graph). *A graph is a data structure composed of nodes and edges representing relational data, which can be defined by a set of edges, nodes, labels, and a label mapping function. A graph is formally defined as a quaternion* $(V, E, L, f^{(L)})$*, where* V *denotes a set of nodes in the graph,* $E \subset V \times V$ *is a set of edges, each node* $v \in V$ *and* $e \in E$ *has an attribute value as a label.* L *denotes a set of all valid labels.* $f^{(L)} : V \cup E \to L$ *denotes a function mapping node or edge to a label.*

Definition 2 (Subgraph Search). *Given a data graph* $G = (V_G, E_G, L_G, f_G^{(L)})$ *and a query graph* $P = (V_P, E_P, L_P, f_P^{(L)})$*, let* $G_{sub} = (V_{sub}, E_{sub}, L_{sub}, f_{sub}^{(L)})$ *denote a subgraph of* G*, if there is an onto function* $f : V_P \to V_{sub}$*, have (1)* $\forall v \in V_P, f(v) \in G_{sub}$*, and (2)* $\forall (u, v) \in E_P, (f(u), f(v)) \in E_{sub} \wedge f_P^{(L)}(u, v) = f_{sub}^{(L)}(f(u), f(v))$*,*

The query graph P matches the subgraph of data graph G, and f is a subgraph matching function for P and G. The subgraph matching function f obtained under the above conditions is a subgraph homomorphic function (if f is a bijection function, f is a subgraph isomorphic function of query graph P to data graph G).

Definition 3 (Subgraph Matching Cardinality). *Given a data graph* G *and query graph* P*, the set of all subgraph matching functions of* P *to* G *is denoted as* $\mathcal{F}_G(P)$*, then the number of functions in* $\mathcal{F}_G(P)$*, denoted as* $|\mathcal{F}_G(P)|$*, is the subgraph matching cardinality of query graph* P *to data graph* G*.*

3 Related Work

3.1 Accurate Subgraph Searching

The accurate subgraph searching methods commonly enumerate the search space to return the matched subgraphs. To the best of our knowledge, Ullmann algorithm [21] is first subgraph searching algorithm, and it reduces the search space through a series of simple pruning rules. Based on this, VF2 algorithm [16] was proposed, which greatly reduced the search space by preselecting the candidate nodes considering in/out degrees. However, these only perform pruning for one node and can not handle large queries. In fact, the absence of candidates of one's neighborhoods may lead to fail matches of the central node. From this point of view, more and more algorithms maintain candidate sets [1,2,5,6] to filter search space for the whole query, which lead to significant efficiency improvement. Among which, GraphQL [6] is a powerful filter methods producing smallest search space. It maintains a candidateset for each query node, and iteratively remove unqualified nodes using the neighboring information.

3.2 Cardinality Estimations of Subgraph Search Query

For the cardinality estimation of subgraph search query, a series of methods [15,20] are given in resource description framework, but these methods have special requirements on the structure of data graph and query graph, and are usually used to find a single entity with multiple attributes for specialized problems. Park et al. tried to treat the nodes of graph data as tables in relational databases and query graphs as joins of multiple tables, thereby applying the cardinality estimation method Wander Join [9], which is originally used in relational database, to the cardinality estimation of subgraph matching [18]. The above heuristic methods are typically based on statistic information and may be hard for complex queries.

Deep learning methods to achieve estimation of subgraph matching. NSIC [13] encode and input both query graph and data graph into graph neural networks and reads out representations for them to produce a final estimation. During the process, it utilize attention to pass information between query and data graph, and DIAMNet [13] is proposed to achieve information passing in limited RAM. However, experiments show that DIAMNet has a high cost for memory during training, which makes it difficult to deal with large-scale query graphs. NeurSC [22] solve the problem by extracting substructures from data graph, avoiding the whole input of large data graphs. It also propose inter-graph learning to align node representations in query and data graph. LSS [27] only input the query graph into the estimator. To preserve the information from data graph, pretrains are carried out to obtain embeddings for certain labels.

However, the information passing between query and data graph make potential nodes to match similar with each other, this on the contrary limits the discriminative power of data graph nodes. If two nodes in data graph are potentially matched to the same query node, they tend to be similar in representation space under the message passing, but they may have different local structures in fact, which further leads to unexpected estimation result.

3.3 Graph Neural Networks

Graph neural networks map nodes to representations which are high-dimensional vectors. It utilize non-linear transformations to capture various features in the graph. The original Graph Neural Networks (GNNs) [19] captures graph features via random walks. Graph Convolution Network (GCN) [7] is a graph neural network that uses the convolution operation on the graph. GraphSAGE [4] samples and aggregate multi hop neighborhood information; Graph Isomorphism Network (GIN) [25] Improving the ability to distinguish between different nodes through a learnable injective function, thus is proved to have the same expressive power as WL test [23], and is widely used in multiple studies.

However, with the stacking of multilayer graph neural networks, the representations become hard to distinguish with each other, leading the bad performances.

4 DENet

In this study, The main task of this deep learning model framework is to implement cardinality estimation for subgraph matching queries using a neural network model. The framework consists of several key components: (1) Node Representation: It represents nodes as vectors that contain hidden state information. (2) Edge Representation: Existing methods have limited capability to represent edges and fail to effectively capture the distinctive features of edges in the graph. This framework captures local features of edges and represents them in a more expressive way. (3) Direction Learning: The framework adjusts the direction of representation vectors to make the representations of nodes or edges with similar features point in similar directions. (4) Cross-Graph Interaction: For related nodes between the query graph and the data graph, the framework employs contrastive learning to align their vector representations. (5) Estimation: The framework utilizes a Multi-Layer Perceptron (MLP) to provide the final prediction value.

Next, we will introduce the details of each major operation in the cardinality estimation processes.

4.1 Node Representation

As discussed in Sect. 3.3, graph representation learning is crucial for capturing local structural features of graphs. To achieve so, nodes in the graph are initialized as fixed-length vectors, known as embeddings. These embeddings are then fed into multi layer graph neural networks, which help aggregate neighborhoods of nodes and finally output representations with structural information.

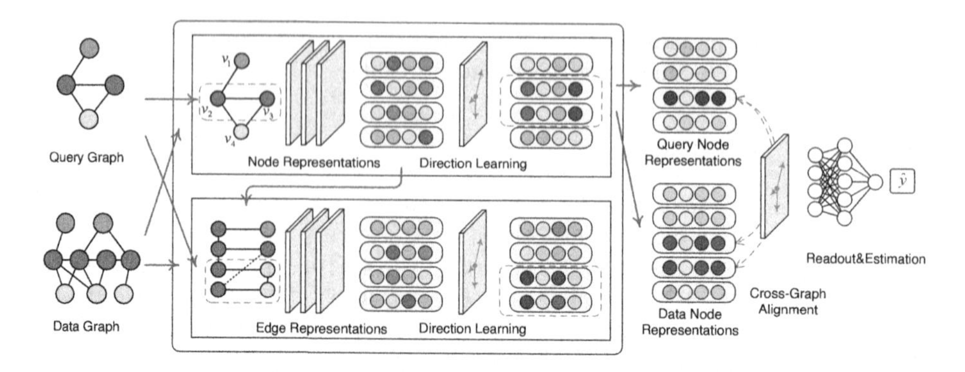

Fig. 1. Framework of DENet

Node Embedding encode label for each node to obtain an embedding.

For the given input graph, DENet captures node labels and initializes an embedding $\mathbf{h}^{(0)}$ for each node, which implies the attribute of the node itself. The components of vector $\mathbf{h}_u^{(0)}$ infer different aspects of features of the node u, and it will be updated and learned through subsequent training process. The initialization may be divided into two categories, *Single Node Embedding* and *Pre-trained Node Embedding*. The former is applied in methods that input both query and data graph into the model [13,14,22], while the latter only input query graphs and they capture data graph information via pretrained embeddings [27]. We will focus on the former in the following.

In graphs, labels can generally be divided into two types, namely category type and numerical type. DENet adopts different embedding methods for them.

For categorical data, we use dict searching to return the feature vector. Taking the embedding process of node labels as an example, DENet obtains the corresponding feature vectors according to different labels, and updates these vectors through learning. It maintains a learnable lookup matrix, where L_V is the set of labels that the nodes may get. After inputting the edge-quintuple, we map the label of the node to a single integer, and query the corresponding column vector in W_R according to the integer:

$$\mathbf{h}_u^0 = W_R^{(vertex)}[l_u], \tag{1}$$

where, l_u is the label of the node, $W_R^{(vertex)}[\cdot]$ is an index operation that returns a row in the vector corresponding to a specific label.

For numerical data, its own size is meaningful, and there is no need to do category mapping. In order to adapt it to the model input, linear mapping can be directly adopted to convert it into a vector in length k:

$$\mathbf{h}_u^0 = l_u \cdot \mathbf{w}_T, \tag{2}$$

where $\mathbf{w}_T \in \mathbb{R}^k$ is a learnable conversion vector of nodes, and l_u is the label value of nodes.

Representation Learning inputs embeddings and update the representation vectors via multi layer graph neural networks.

Various architectures of GNNs [4,8,25] are widely used for graph structural feature extraction, among which, GIN has been proved to have the same expressive power as WL test [23], which is widely used in neural subgraph searching cardinality estimation [13,22,27]. The k-th layer of GIN is illustrated in Eq. (3).

$$\mathbf{h}_u^{(k)} = \sigma\left(\texttt{MLP}^{(k)}\left(\left(1+\epsilon^{(k)}\right)\cdot \mathbf{h}_u^{(k-1)} + \sum_{v\in\mathcal{N}(u)} \mathbf{h}_v^{(k-1)}\right)\right), \tag{3}$$

where ϵ is a learnable scalar, and σ is an activation function.

DENet also applies GIN for local structure capturing:

$$\mathbf{h}_u^{(k)} = \texttt{GIN}^{(k)}\left(\mathbf{h}_{\mathbf{u}}^{(\mathbf{k-1})}, A\right), \tag{4}$$

where A is the adjacency matrix implying the structure.

4.2 Edge Representation

As far as our knowledge goes, the existing methods of graph representation learning involve incorporating edge labels as a component of the data that is fed into graph neural networks. These methods may view edges with differing labels as constituting different views of the graph, or they may directly aggregate edge information into connected nodes [27]. However, the implicit structural information of the edges themselves is yet to be learned.

In the process of subgraph matching, edge information contains the basic constraints to determine whether the query graph can match the data graph. DENet is capable of learning a representation vector for each edge in the output of any layer of the graph neural network, which is updated in learning procedure. This representation vector contains the label attributes of the edge, and also captures the structural information that the edge represents in the graph.

To obtain the label attributes of the edges themselves, the model employs the same approach as that used for processing the node labels, as described in the Sect. 4.1 with the following equation:

$$\mathbf{x}_{u,v}^{(l)} = W_R^{(edge)}[l_{u,v}], \tag{5}$$

where $W_R^{(edge)}$ is a learnable label embedding matrix, and the label representation is to select a column vector corresponding to $l_{u,v}$.

To capture the structural information of the edge, we use the nodes connected by it to provide local structural information of the graph. For an undirected graph, we can treat these two nodes as two elements in a set without an order, and use Deep Set [26], which is a widely used deep learning framework defined on unordered sets, to merge their representations:

$$\mathbf{h}_{u,v}^{\prime(s)} = \psi\left(\phi\left(\mathbf{h}_u^{(l)}\right) + \phi\left(\mathbf{h}_v^{(l)}\right)\right), \tag{6}$$

where ψ and ϕ are two MLPs with activation functions for output layers. The output is in the same shape as input.

For directed graphs, the roles of the two nodes connected by an edge is different, so they cannot be regarded as two elements of a set, therefore, we concatenate the representation vectors of the two in the order of starting point followed by ending point, then feed them into a multilayer perceptron and output a vector of length k:

$$\mathbf{h}'^{(s)}_{u,v} = \rho([\phi\left(\mathbf{h}^{(l)}_u\right); \phi\left(\mathbf{h}^{(l)}_v\right)]) \tag{7}$$

Here, ρ is defined as a multilayer perceptron with an input of length $2k$, which outputs a k-dimensional vector.

4.3 Direction Learning

Over-smoothing is a common issue faced by many graph neural networks in existing graph representation learning frameworks. It refers to the phenomenon that, during the optimization of node embeddings, the values of different nodes gradually converge to the average due to the way the embeddings propagate to neighboring nodes, which is through weighted averaging or sampling instead of direct copying. This averaging effect leads to the feature similarity between adjacent nodes increasing over time, resulting in the loss of local structural information of the graph, and ultimately weakening the model's representation capability.

It is fortunate that according to Fig. 1, after being processed by the graph neural network, the representation vectors tend to concentrate in similar directions, indicating that there is still much unused space in the representation space. This presents an opportunity to further improve graph neural networks by developing techniques that encourage the exploration of this unused space while simultaneously avoiding over-smoothing. By doing so, we can potentially improve the discriminating power and generalization capabilities of graph neural networks, leading to better performance on various graph-related tasks.

Therefore, in this section, we will adopt the Directional Embedding method to handle representation vectors. Considering the intrinsic characteristics of graphs, different nodes may possess distinct labels. We aim to ensure that nodes with the same label are closer in direction in the representation space, while increasing directional discrepancies between nodes with different labels.

For the input graph of a graph neural network layer, We first sample k positive and k negative example pairs based on their labels, each positive pair of nodes $(u^+, v^+) \in S$ has the same label: $L(u^+) = L(v^+)$, whereas negative pair $(u^-, v^-) \in N$ is the opposite: $L(u^-) \neq L(v^-)$.

Similarly, k positive and k negative pairs for edges are sampled. The set of positive pairs and negative pairs are denoted by P_E and N_E respectively. A positive pair of edges (u_1^+, v_1^+) and (u_2^+, v_2^+) have the same labels of corresponding endpoints $L(u_1^+) = L(u_2^+)$, $L(v_1^+) = L(v_2^+)$, and the labels of two edge themselves are the same, too: $L(u_1^+, v_1^+) = L(u_2^+, v_2^+)$. If any of the above conditions are not satisfied, an edge pair will be considered as a negative pair.

Next, node and edge representations will be updated with cosine embedding loss, which aligns nodes with the same label to similar directions and ensures nodes with different labels to have distinct directions in their representation space. Thus we propose the loss function to achieve so: $\mathcal{L}_{dir} = \mathcal{L}_{dir}^{(p)} + \mathcal{L}_{dir}^{(n)}$, where

$$\mathcal{L}_{dir}^{(p)} = \frac{1}{k} \sum_{(u,v) \in P_V} (1 - \cos(\mathbf{h}_u, \mathbf{h}_v)) + \frac{1}{k} \sum_{(e_1,e_2) \in P_V} (1 - \cos(\mathbf{h}_{e_1}, \mathbf{h}_{e_2})); \quad (8)$$

$$\mathcal{L}_{dir}^{(n)} = \frac{1}{k} \sum_{(u,v) \in N_V} \max(0, (\cos(\mathbf{h}_u, \mathbf{h}_v) - \theta) + \frac{1}{k} \sum_{(e_1,e_2) \in N_V} \max(0, (\cos(\mathbf{h}_{e_1}, \mathbf{h}_{e_2}) - \theta). \quad (9)$$

Here, θ is a threshold, the smaller it is, the more different the distinct directions are.

4.4 Cross-Graph Learning

State of the art method of neural cardinality estimation of subgraph search queries which is known as NeurSC [22] inputs both query and data graph into learning models. It also utilizes inter-graph learning to pass information between query nodes and their qualified candidate nodes. The candidate nodes are promising nodes in data graph that potentially match the query nodes and are selected by GraphQL filter [6], whose simplified procedures are listed in Algorithm 1.

Next, an inter-graph $G^{(inter)}$ is constructed where the nodes are the union of query nodes and all candidate nodes:

$$V^{(inter)} = V_P \cup \left(\bigcup_{u \in V_P} C_u \right), \quad (10)$$

and the candidate nodes are connected to their corresponding query node:

$$E^{(inter)} = \bigcup_{u \in V_P} \{(u, v) \mid v \in C_u\}. \quad (11)$$

Then the inter-graph is fed into graph neural networks to learn cross-graph information between query node and its candidates. During inter-graph learning, the query nodes and their candidate nodes get more similar in representation space due to the homogeneity assumption of GNNs.

NeurSC makes the distance between query node and its candidate nodes closer to align representation space. However, this would decrease the flexibility of the node representations. Please consider the following example:

Algorithm 1: Candidate Filtering

```
Input  : A query graph P = (V_P, E_P, L_P)
         A data graph G = (V_G, E_G, L_G)
Output : Candidate sets C_1, C_2, ..., C_|V_P| for every query node
1  for node u in V_P do
       // Set each candidate set to graph nodes with same label as the
          query node.
2      C_u := {m ∈ V_G | L_P(u) = L_G(m)};
3  repeat
4      for query node u in V_P do
           // Remove graph nodes without expected neighbors in
              candidate sets.
5          for graph node m in C_u do
6              tag := false;
7              for node v in N(,u) do
8                  if {n ∈ C_v | (m,n) ∈ E} == ∅ then
9                      tag = true;
10             if tag == true then
11                 C_u = C_u − {m};
12 until all candidate sets are not updated;
```

Example 1. Two nodes in a data graph are both candidate nodes of a same query graph node, however, their neighbors are quite different, in such case inter-graph learning would still bring their representations closer. Such alignment method leads to reduced distinguish ability of the model for different nodes.

From this point of view, we propose direction-based cross graph alignments, in which the directions of representation vectors of query node and its candidates are tend to be similar, while the directions tend to be different on the contrary.

To achieve so, positive and negative pair sets are sampled. The former (denoted as P_C) contains k node pairs (u, v) where v is a candidate node of u. While the latter (denoted as N_C) contains k node pairs (u, v) where u and v are in candidate sets of different query nodes. Then we compute the cross graph learning loss for aligning: $\mathcal{L}_{cross} = \mathcal{L}_{cross}^{(p)} + \mathcal{L}_{cross}^{(n)}$, where

$$\mathcal{L}_{cross}^{(p)} = \frac{1}{k} \sum_{(u,v) \in P_I} (1 - \cos(\mathbf{h}_u, \mathbf{h}_v)); \tag{12}$$

$$\mathcal{L}_{cross}^{(n)} = \frac{1}{k} \sum_{(u,v) \in N_I} \max(0, (\cos(\mathbf{h}_u, \mathbf{h}_v) - \theta)). \tag{13}$$

4.5 Readout and Estimation

After representation, each node in query and data graphs obtains a representation implying the node attribute and local structural features, and the representation vectors are finally read out as two representations for the whole query and data graph, denoted as $\mathbf{h}_P$ and $\mathbf{h}_G$ respectively. The readout operation is as:

$$\mathtt{READOUT}(\mathbf{H}) = \sigma\left(\sum_u \mathtt{MLP}(\sigma(\mathbf{h}_u))\right). \tag{14}$$

For the representation of the query graph, the cardinality estimation result is obtained via the fully connected neural network after concatenating representations of query and data graph:

$$\hat{y}_{P,G} = \mathtt{MLP}\left(\mathbf{h}_P || \mathbf{h}_G\right). \tag{15}$$

DENet contains deep learning modules such as GNN and MLP, and the model parameters are updated supervised by optimizers once we get the loss value. Combined with aforementioned losses in Sects. 4.3 and 4.4, The total loss is:

$$\mathcal{L} = \sum\left(\hat{y}_{P,G} - y_{P,G}\right)^2 + \mathcal{L}_{dir} + \mathcal{L}_{cross}, \tag{16}$$

where $y_{P,G}$ is the true cardinality of query P and data graph G.

5 Experiments

In order to in DENet's ability to estimate the cardinalities of subgraph matching, we introduce our experiments.

5.1 Experimental Setups

In the experiments, we first generate query sets on three widely used real-world data graphs, namely Yeast, Wiki and Wordnet. We sample subqeuries as query graphs from the data graphs, and compute the true cardinalities via GraphQL [6], then the query graphs are divided in two groups, known as training set and test set. Among them, we get 80% for training and 20% for testing following the settings in [22,27].

Datasets. The datasets used in the experiments was four labeled graph data commonly used in subgraph matching research [13,18,22,27], namely, Yeast, Wiki and Wordnet. The statistical information of the dataset is shown in Table 1, where $|V|$ represents the number of nodes in the data graph, and $|E|$ represents the number of edges in the data graph. d_{avg} denotes the average degrees of nodes in the data graph. $|L_v|$ represents numbers of distinct nodes labels. $|V_Q|$ infers the numbers of node in sampled queries, while $|P|$ indicates the number of query graphs extracted from the dataset.

- **Yeast** is used to study protein-protein interactions, where each node represents a protein structure, and if two types of proteins can interact with each other, there will be an edge connecting them.

- **Wiki** is a hyperlinked dataset that uses a graph to represent the hyperlink relationships between web pages with the topic squirrel. Each node represents a Wikipedia web page, and its label discretely infers the page's visit count. Each edge represents a jumping link between web pages.
- **WordNet** models semantic relationships between words in the form of a knowledge graph, where nodes represent individual words and edges represent the relationships between different words. As the nodes themselves do not have labels, we assign random labels to them.

Table 1. Statistics of datasets

Dataset	$\|V\|$	$\|E\|$	d_{avg}	$\|L_v\|$	$\|V_P\|$	$\|P\|$
Yeast	3,112	12,519	8.046	71	{5,10,20,40}	1,200
Wiki	5,201	198,353	14	76.275	{5,10,20}	900
Wordnet	40,559	71,925	16	3.547	{5,10,20}	900

Evaluation Metrics. In order to evaluate the cardinality estimation results of subgraph matching, the experiment uses $q-$error as a metric. $q-$error is a commonly used evaluation for cardinality estimation [18,22,27], which is defined as:

$$q\text{-error}(\hat{y}, y) = \max(\frac{\hat{y}}{y}, \frac{y}{\hat{y}}). \tag{17}$$

5.2 Baselines

We compare with two methods with best performance in [18], known as **Wander Join (WJ)** and **Join Sampling with Upper Bounds (JSUB)**. For deep learning-based methods, DIAMNet [13], NeurSC [22] and LSS [27] are taken into account. All the parameters are keep the default for the compared methods.

5.3 Experimental Analysis

This section will shows the results and analysis of the experiments.

Accuracy. Figure 2 illustrates the comparisons for accuracy. The results are shown in the form of boxplot, where the upper and lower bound of the box represent the 75% percentile and 25% percentile respectively, while the whisker shows the max errors. On most cases, DENet overperform others with smaller q-errors, especially in 20-node queries on Yeast and 20-node queries on Wordnet as shown in Fig. 2.

The reason is that our proposed DENet utilizes directional learning to distinguish the directions of different types of nodes, this enhance the ability of nodes to keep their characteristics and may have a larger representation space to adjust its feature vectors. What's more, cross-graph learning makes the matched nodes have feature vectors pointing to the same direction, which can be significant instructions of matching detection. Thus it improves the accuracy for cardinality estimation of subgraph search queries. In Fig. 2, DENet has much better performance on large query sets. Because complex queries with more nodes are more likely to face up with the over-smoothing issue [11,24], which is relieved via directional learning. Another key point is that DENet explicitly represents the edges, which infers the structure of the graphs.

Ablation studies in Sect. 5.3 further improve the effects of the mentioned techniques.

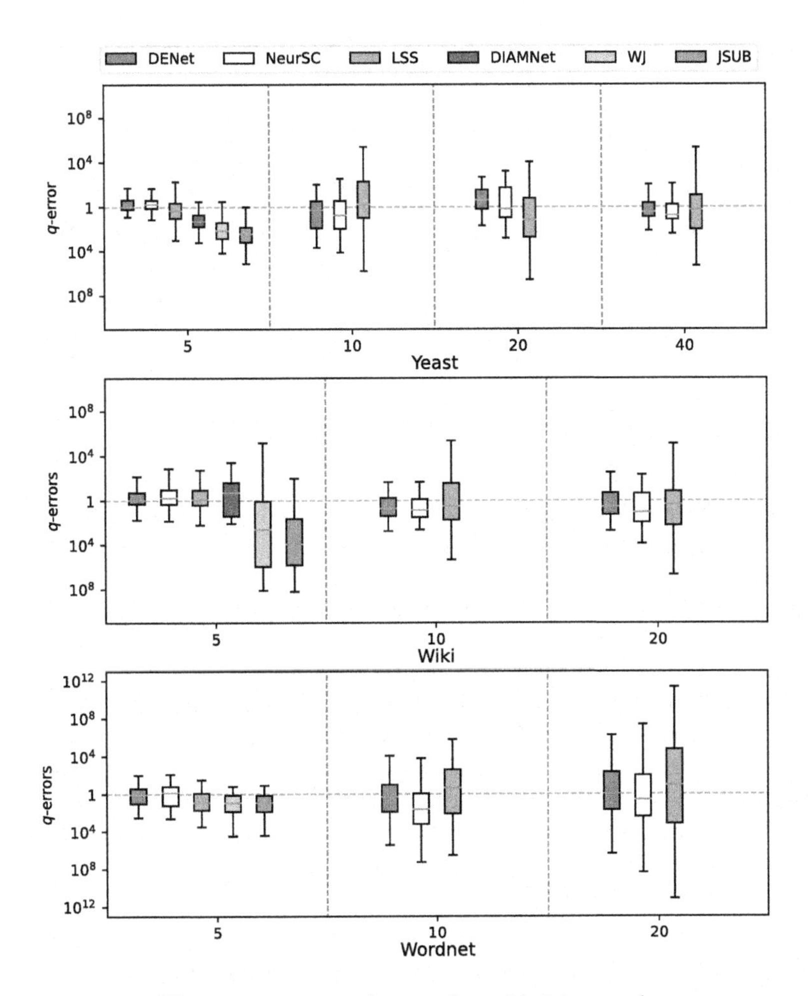

Fig. 2. q-errors on three real-world data graphs

Please notice that, compared with deep learning-based methods, sampling methods have few successful samples once the query size is larger than 10, and tend to return 0 for all the queries. So we only reports 5-node queries for them. Besides, DIAMNet suffers from heavy RAM cost, and may be easily out of memory on large queries, so we only report in small data graphs.

Efficiency. Efficiency evaluation is based on Yeast dataset. We report the training time in one epoch of DENet, NeurSC and LSS. And the average estimation cost.

In the training process of the deep learning method, the learning models are trained using the 80% of 1, 200 generated queries in size 5, 10, 20 and 40 (300 queries for each size). Each model is trained for 100 epochs (LSS contains 50 train epochs and 50 active learning epochs). Then we summarize the average cost for a single epoch. We use the rest 20% queries to evaluate estimation cost. After obtaining all the estimations, we calculate the mean cost for each size.

Table 2. Elapsed time for training and estimating on Yeast

Method	WJ	JSUB	NeurSC	LSS	DENet
Training (/s)	–	–	105.3	18.19	107.9
Estimating on $N = 5$ (/ms)	10.53	32.72	74.75	10.25	75.31
Estimating on $N = 10$ (/ms)	21.63	56.88	79.73	16.11	81.82
Estimating on $N = 20$ (/ms)	38.51	79.36	84.73	27.31	86.39
Estimating on $N = 40$ (/ms)	–	–	89.63	45.72	40.17

The evaluation results are shown in Table 2. On the one hand, Learning-based models (i.e., DENet, NeurSC, LSS) need to be trained before estimating the cardinality of subgraph search queries. Among them, DENet and NeurSC input both query graph and data graph to learn the cross-graph information, and execute the candidate filtering, which lead to extra cost, so they require more time than LSS. DENet further perform directional learning and cross-graph learning, but the total cost for each epoch is still similar with NeurSC which is satisfying for users. On the other hand, a trained model produce estimation in short time on large queries, as shown in Table 2, the cost of JSUB significantly grows with the query size get larger, however, DENet and NeurSC have stable time cost, which mainly from candidate filtering. Overall, despite of the additional components, DENet can return an relatively accurate estimation cardinality in comparable time as existing methods.

Ablation Studies. In order to study the effect of proposed techniques of DENet, we perform ablation studies, where three variants of DENet are taken into account. DENet-D is the variant which do not adjust the representation using directional learning, DENet-C performs directional learning with out cross-graph information, while DENet-E drops the edge representation parts of DENet.

Table 3. Estimation Errors of Ablation Study

Method	Yeast		Wordnet		Wiki	
	Mean	Median	Mean	Median	Mean	Median
DENet-D	785.2	8.632	4.2×10^8	195.6	1.1×10^9	18.25
DENet-C	1,061	8.384	3.7×10^8	165.2	9.4×10^8	15.84
DENet-E	426.8	7.141	$\mathbf{2.8} \times \mathbf{10^8}$	185.2	2.4×10^8	**11.37**
DENet	**119.0**	**6.812**	3.0×10^8	**143.7**	$\mathbf{2.1} \times \mathbf{10^8}$	12.58

The experimental is based on 20-size queries in three data graphs, the results are the mean and median of q-error as shown in Table 3. Among the results, DENet perform the best over all, meaning that each part has positive effects on DENet. Especially, DENet-D and DENet-C significantly underperforms original DENet. This illustrates that the directional learning and cross-graph learning are two key parts for the improvement of the performance. Besides, in most cases, DENet surpass DENet-E (if not, the gap is small). Thus edge representation also help improve the estimation ability of DENet.

6 Conclusion

Cardinality estimation of subgraph search queries plays an important role in graph data management and analysis. DENet method uses deep learning to capture both edge's and node's labels and structural information where the former are typically ignored by existing methods. What's more, DENet novelly adjust the feature vectors based on its directions to align potentially matched nodes, which improves the ability of estimating the cardinality of subgraoh search queries.

DENet improves the overall cardinality estimation accuracy compared with other estimation methods. Experiments show that DENet can give sound estimation results in a comparable time as other deep learning based methods, and is suitable for multiple tasks in graph analysis.

Acknowledgment. This work was supported by NSFC under grants Nos. 62272469, 71971212 and U19B2024.

References

1. Bi, F., Chang, L., Lin, X., Qin, L., Zhang, W.: Efficient subgraph matching by postponing cartesian products. In: SIGMOD, pp. 1199–1214 (2016)
2. Carletti, V., Foggia, P., Saggese, A., Vento, M.: Challenging the time complexity of exact subgraph isomorphism for huge and dense graphs with VF3. TPAMI **40**(4), 804–818 (2018)
3. Gardner, M., Dorling, S.: Artificial neural networks (the multilayer perceptron) - a review of applications in the atmospheric sciences. Atmos. Environ. **32**, 2627–2636 (1998)
4. Hamilton, W.L., Ying, Z., Leskovec, J.: Inductive representation learning on large graphs. In: NIPS, pp. 1024–1034 (2017)
5. Han, M., Kim, H., Gu, G., Park, K., Han, W.S.: Efficient subgraph matching: harmonizing dynamic programming, adaptive matching order, and failing set together. In: SIGMOD, pp. 1429–1446 (2019)
6. He, H., Singh, A.K.: Graphs-at-a-time: query language and access methods for graph databases. In: SIGMOD, pp. 405–418 (2008)
7. Kipf, T., Welling, M.: Semi-supervised classification with graph convolutional networks. ArXiv abs/1609.02907 (2017)
8. Kipf, T.N., Welling, M.: Semi-supervised classification with graph convolutional networks. In: 5th International Conference on Learning Representations, ICLR 2017, Toulon, France, 24–26 April 2017, Conference Track Proceedings (2017). https://openreview.net/forum?id=SJU4ayYgl
9. Li, F., Wu, B., Yi, K., Zhao, Z.: Wander join and XDB: online aggregation via random walks. ACM Trans. Database Syst. **44**(1), 1–41 (2019)
10. Li, G., et al.: A survey of machine learning based database techniques. Chinese J. Comput. **43**(11), 2019–2049 (2020)
11. Li, Q., Han, Z., Wu, X.M.: Deeper insights into graph convolutional networks for semi-supervised learning. In: Proceedings of the Thirty-Second AAAI Conference on Artificial Intelligence and Thirtieth Innovative Applications of Artificial Intelligence Conference and Eighth AAAI Symposium on Educational Advances in Artificial Intelligence, AAAI 2018/IAAI 2018/EAAI 2018 (2018)
12. Li, R., Hong, L.: Method for subgraph matching with inclusion degree. J. Softw. **29**, 1792–1812 (2018)
13. Liu, X., Pan, H., He, M., Song, Y., Jiang, X., Shang, L.: Neural subgraph isomorphism counting. In: KDD 2020, pp. 1959–1969 (2020)
14. Liu, X., Song, Y.: Graph convolutional networks with dual message passing for subgraph isomorphism counting and matching. In: AAAI, vol. 36, pp. 7594–7602 (2022)
15. Neumann, T., Moerkotte, G.: Characteristic sets: accurate cardinality estimation for RDF queries with multiple joins. In: Proceedings of the 2011 IEEE 27th International Conference on Data Engineering, ICDE 2011, pp. 984–994 (2011)
16. Cordella, L.P., Foggia, P., Sansone, C., Vento, M.: A (sub)graph isomorphism algorithm for matching large graphs. IEEE Trans. Pattern Anal. Mach. Intell. **26**(10), 1367–1372 (2004)
17. Paradies, M., Vasilyeva, E., Mocan, A., Lehner, W.: Robust cardinality estimation for subgraph isomorphism queries on property graphs. In: Wang, F., Luo, G., Weng, C., Khan, A., Mitra, P., Yu, C. (eds.) Big-O(Q)/DMAH 2015. LNCS, vol. 9579, pp. 184–198. Springer, Cham (2016). https://doi.org/10.1007/978-3-319-41576-5_14

18. Park, Y., Ko, S., Bhowmick, S.S., Kim, K., Hong, K., Han, W.S.: G-care: a framework for performance benchmarking of cardinality estimation techniques for subgraph matching. In: Proceedings of the 2020 ACM SIGMOD International Conference on Management of Data, SIGMOD 2020, pp. 1099–1114 (2020)
19. Scarselli, F., Gori, M., Tsoi, A.C., Hagenbuchner, M., Monfardini, G.: The graph neural network model. IEEE Trans. Neural Netw. **20**, 61–80 (2009)
20. Stocker, M., Seaborne, A., Bernstein, A., Kiefer, C., Reynolds, D.: SPARQL basic graph pattern optimization using selectivity estimation. In: Proceedings of the 17th International Conference on World Wide Web, WWW 2008, pp. 595–604 (2008)
21. Ullmann, J.R.: An algorithm for subgraph isomorphism. J. ACM **23**(1), 31–42 (1976)
22. Wang, H., Hu, R., Zhang, Y., Qin, L., Wang, W., Zhang, W.: Neural subgraph counting with Wasserstein estimator. In: SIGMOD, pp. 160–175 (2022)
23. Weisfeiler, B., Leman, A.: A reduction of a graph to a canonical form and an algebra arising during this reduction. Nauchno-Technicheskaya Informatsia **9** (1968). (in Russian)
24. Wu, Z., Pan, S., Chen, F., Long, G., Zhang, C., Yu, P.S.: A comprehensive survey on graph neural networks. IEEE Trans. Neural Networks Learn. Syst. **32**(1), 4–24 (2021)
25. Xu, K., Hu, W., Leskovec, J., Jegelka, S.: How powerful are graph neural networks? In: ICLR. OpenReview.net (2019)
26. Zaheer, M., Kottur, S., Ravanbhakhsh, S., Póczos, B., Salakhutdinov, R., Smola, A.J.: Deep sets. In: NIPS, pp. 3394–3404. Curran Associates Inc., Red Hook (2017)
27. Zhao, K., Yu, J.X., Zhang, H., Li, Q., Rong, Y.: A learned sketch for subgraph counting. In: SIGMOD, pp. 2142–2155 (2021)

Approximate Continuous k Representative Skyline Queries over Memory Limitation-Based Streaming Data

Yunzhe An[1], Zhu Zhen[1], Shuangshuang Zhang[2], Rui Zhu[1](✉), and Chuanyu Zong[1]

[1] Shenyang Aerospace University, Shenyang, China
{anyunzhe,zhurui,zongcy}@sau.edu.cn, 10215887562@qq.com
[2] Shenyang Aircraft Corporation, Shenyang, China

Abstract. Continuous skyline queries over sliding windows is an important problem in the field of streaming data management. The query monitors the query window, and returns all skyline objects to the system whenever the window slides. However, this type of query is greatly influenced by the data set size, data dimensions, and distribution. In many cases, the scale of skyline objects may be large, leading that it is difficult for users to find suitable query results from a large number of skyline objects. In addition, the space cost may be very high. The state of the arts efforts cannot work under memory limited based environment. To solve the above problems, in this paper, we propose a novel framework named ρ-AKRS(short for $\rho-$Approximate K-Representative Skyline) to support k-representative skyline query under memory limited based streaming data. Unlike traditional k-representative skyline queries that retrieves query results based exact skyline objects, it selects k approximate skyline objects with high representative as query results. In order to support query processing, we propose a $\rho-$quad tree based index to support approximate skyline objects search, and propose a M-tree-based index to support k-representative skyline search. The comprehensive experiments on both real and synthetic data sets demonstrate the superiority of both efficiency and quality.

Keywords: Streaming Data · Continuous k-Representative Skyline Query · Dominance · Memory Limitation

1 Introduction

This paper studies the problem of continuous $\rho-$approximate representative skyline query over data stream, an important variant of skyline query over data stream.

Formally, let $o[1, \cdots, d]$ and $o'[1, \cdots, d]$ be two $d-$dimensional objects in the object set $\mathcal{D}$ in $[0, 1]^d$ space. o' is dominated by o if $\forall\ u \in [1, d]$, $o'[u] \leq o[u]$, and $\exists\ v \in [1, d]$,$o'[v] < o[v]$. Here, we use $o[u]$ to express the coordinate of o in

X. Yang et al. (Eds.): ADMA 2023, LNAI 14180, pp. 94–106, 2023.
https://doi.org/10.1007/978-3-031-46677-9_7

the dimension u. In addition, if an object $o \in \mathcal{D}$ is not dominated by any object in $\mathcal{D}$, o is regarded as a *skyline object*. All skyline objects in $\mathcal{D}$ form the skyline set $\mathcal{SK}$. In many cases, the scale of $\mathcal{SK}$, i.e., denoted as $|\mathcal{SK}|$ may be very large, leading that it is difficult for users to find valuable information from $\mathcal{SK}$ in real time. Therefore, $k-$representative skyline query is proposed. It selects k objects with the highest representative from $\mathcal{SK}$, returns them to the system.

In this paper, we use sliding window to model streaming data. Formally, let $\mathcal{S}$ be the set of streaming data in the query window. A continuous $\rho-$approximate representative skyline query(CARS for short) monitors objects in the window, and returns to k objects with the highest representative to the system whenever the window slides. It has many real applications.

For example, in a stock analysis system, stock information are updated quickly in real-time, and users are usually interested in stock information that are generated in the last few minutes. Assuming a user prefers to invest stocks with low-risk and low-cost, a skyline query could be submitted to monitor stock trading information that are generated in the last few minutes, and return important information to the system whenever some new trading information are generated/expired. If the number of skyline objects is large, the system can selects k trading information with the highest representative from skyline object set, returns them to the system.

However, in real applications, memory resource of streaming data system is usually limited, but the space cost of supporting CARS may be very high, that is, k representative skyline objects are selected based on maintaining ALL skyline objects. Since the skyline object scale is sensitive to data distribution, data scale and so on, the space cost of the corresponding algorithm may be very high. In many cases, if the memory resource of the system is limited, these algorithms cannot effectively work.

To address this issue, many scholars have proposed some approximate algorithms that reduce space costs by introducing user-defined error thresholds, which return approximate query results to users. Compared with exact algorithms, these approximate algorithms can discard part of skyline objects, and select skylines with high representative from the reminder skyline objects. In other words, these algorithms can select skylines with high representative based on partial skyline objects. However, these error thresholds are typically pre-set by users and cannot be flexibly changed in real time. If the error threshold is small, the size of the skyline set is still high. By contrast, the algorithm cannot return high quality result to the system.

In this paper, we propose a novel framework named ρ-AKRS(short for ρ-Approximate K-Representative Skyline) to support k-representative skyline query under memory limited based streaming data. We use a rho-quadtree-based index to maintain streaming data, where we uses it to support ρ-approximate skyline query. Based on the searching result, we further propose an M-Tree-based index to find approximate k-representative skyline objects. Above all, the contribution of this paper is as follows.

- (i)We propose a novel query named ρ-approximate continuous k-representative skyline query, where ρ, ranging from 0 to 1, is a error threshold. We use it for bounding the error ratio between exact and approximate results;
- (ii) We propose an M-Tree-based index to find approximate k-representative skyline objects. This method efficiently retrieves query result objects in the query window by efficiently grouping objects based on the distance relationship among approximate skyline objects. Finally, extensive experiments were conducted to evaluate the effectiveness of the proposed algorithm.

The rest of this paper is as follows. Section 2 reviews the related work and proposes the problem definition. Section 3 explains the framework ρ-AKRS. Section 4 evaluates the performance of ρ-AKRS. Section 5 concludes this paper.

2 Preliminary

In this section, we first discuss the related work. Next, we discuss the problem definition. Last of this section, we introduce a quad-tree based index named $\rho-$quadtree.

2.1 Related Works

In recent years, many scholars have studied the problem of skyline queries, and many efforts, such as BNL[1], DC[1], SFS[2], NN[4] and BBS[5], are proposed that can support skyline queries over various of data environments. In this section, we only skyline queries and its variants under data stream. In the following, we first discuss algorithms about skyline query algorithms over data stream. Next, we discuss efforts about k-representative skyline queries.

Skyline Query Over Data Stream. Due to the importance of skyline queries over data stream, many scholars have studied the problem of skyline queries over sliding windows. Among all efforts, Lin et al. [10] first proposed the algorithm named n-of-N, which can support continuous skyline queries over data stream. Compared with skyline queries algorithms, it supports query processing via considering dominance and arrived order relationships among objects. For example, if an object o is dominated by another object o' and o' arrives later than that of o, o cannot become skyline object before it expires from the window, and it could be regarded as a meaningless object. Accordingly, the algorithm removes meaningless objects and uses meaningful objects to support query processing. In other words, the algorithm only needs to maintain meaningful objects that are not dominated by other objects that arrive later than them. However, in the worst case, the number of meaningful objects may be linear to the data set scale. Tao et al.[6] proposed algorithms named LAZY and Eager to support skyline queries over data stream. The algorithm emphasizes that the query results will only change when a new data object arrives or a result object expires.

k-representative Objects Query Algorithms. Many scholars have conducted extensive research on variant algorithms of the skyline queries. Among all of them, k-representative object query is an important variant. Currently, the existing k-representative skyline algorithms can be divided into two categories: *non skyline*-based and *skyline*-based. For the former ones, their target is to select k objects with highest representative, but query result objects may not be skyline objects. For the later ones, query result objects must be skyline objects. Among all algorithms, Lin et al.[10] first proposed a new k-dominant skyline query algorithm, named RSP. Tao et al. proposed the algorithm named DRS, which returns the k points with the largest number of dominant data objects in the entire data set. Bai et al.[14] proposed the k-LDS algorithm, which returns k representative skyline points in the data stream environment. They first proposed an effective accurate algorithm, called prefix-based algorithm PBA, to solve the k-LDS problem over two-dimensional space. Based on it, they further design a greedy algorithm to answer the k-LDS problem in three-dimensional space.

Discussion. The DRS and RSP algorithms mentioned earlier are based on static data, which are not suitable for supporting continuous queries over data stream. Although the k-LDS algorithm can support continuous query over data stream, this algorithm requires maintaining all candidate objects generated by the data set, resulting in high space costs and making it unable to work over memory-limited environments. Therefore, it is essential to propose an error bounded-based approximate query algorithm over memory-limited-based data stream.

2.2 Problem Definition

A continuous k representative skyline query q, expressed by the tuple $q\langle n, s, k, \rho\rangle$, monitors the objects in the window W. Whenever the window slides, the query q first finds all skyline objects in the window, and then returns k-skyline with the highest representative to the system. Here, the representative of these k objects is evaluated via computing their combined dominance area.

Figure 1 shows the corresponding running example. Let W be the query window that contains the set of 12 objects. Skyline objects under the current window are $\{o_4, o_6, o_7, o_8, o_9, o_{12}\}$. Given the parameter $k = 2$, the 2 representative skyline objects are $\{o_6, o_7\}$, where the combined dominance area under these 2 objects is 0.44. When the window slides to W_1, object $\{o_1, o_2, o_3\}$ are expired from the current window, and object $\{o_{13}, o_{14}, o_{15}\}$ flow into the window. Skyline objects in W_1 are $\{o_4, o_7, o_6, o_8, o_{13}\}$, and the 2$-$representative skyline objects are $\{o_4, o_{13}\}$. In the following, we will formally explain the concept of $\rho-$approximate continuous k-skyline queries over memory limitation-based streaming data.

Definition 1. Approximate continuous k-skyline queries. *Approximate continuous k-skyline queries q, denoted as $\langle \rho, n, s, k, M\rangle$, monitors objects in the window. Whenever the window slides, q returns a set of $\rho-$approximate k represents skyline objects $\mathcal{KS} = \{ks_1, ks_2, \cdots, ks_r\}$ to the system,i.e., $(\rho, k)-$ARSS for short. Given the $(\rho, k)-$ARSS $\mathcal{RK}$, there exists element $mk \in \mathcal{RK}$ satisfying $\frac{DomSize\{ks_1, ks_2, \cdots, ks_r\}}{DomSize\{sk_1, sk_2, \cdots, sk_r\}} \geq \rho$ compared with the exact skyline set $\mathcal{SK} =$*

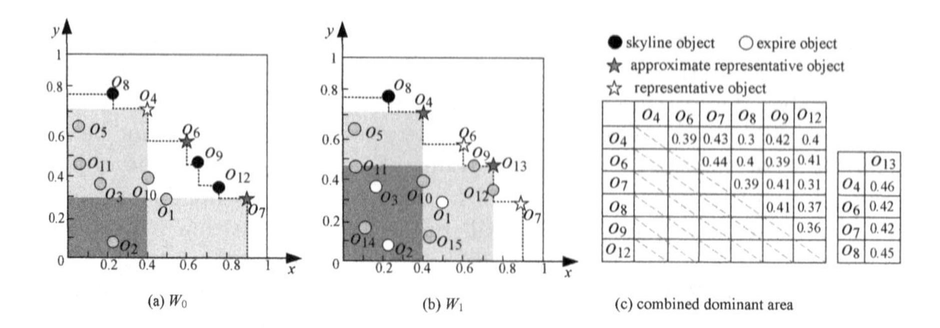

	o_4	o_6	o_7	o_8	o_9	o_{12}
o_4		0.39	0.43	0.3	0.42	0.4
o_6			0.44	0.4	0.39	0.41
o_7				0.39	0.41	0.31
o_8					0.41	0.37
o_9						0.36
o_{12}						

	o_{13}
o_4	0.46
o_6	0.42
o_7	0.42
o_8	0.45

Fig. 1. Running Example of Problem Definition($s = 3, \rho = 0.80$).

$\{sk_1, sk_2, \cdots, sk_s\}$ of the window. Here, $DomSize\{o_i, o_j\}$ is the combined dominant size of o_i and o_j. M refers to the maximal number of objects the system can maintain.

Back to the example in Fig. 1. Given the parameter $k = 2$ and $\rho = 0.80$, under the current window W_0, the exact query results are $\{o_6, o_7\}$, where the combined dominance area under these 2 objects is 0.44. Compared with these 2 objects, $\{o_4, o_7\}$ could be used as approximate query results, where the combined dominance area under these 2 objects is 0.43 and $0.43/0.44 < 0.8$. When the window slides from W_0 to W_1, 2−representative skyline objects are updated to $\{o_4, o_{13}\}$. In this case, objects $\{o_6, o_7\}$ could be used as approximate query results, where the combined dominance area under these 2 objects is 0.44 and $0.44/0.46 < 0.8$. Also, objects $\{o_8, o_{13}\}$ could be used as approximate query results. The corresponding combined dominance area under these 2 objects is 0.45.

An et al. [16] proposed a novel framework named $\rho-$SEAK(short for $\rho-$Self-adaptive Error-based Approximate Skyline) to support approximate continuous skyline queries over memory limitation-based streaming data. The key behind it is using a quad-tree T with height bounded by $\log \frac{1}{\rho}$,i.e., denoted as $\rho-$Quadtree, to maintain streaming data, and support approximate continuous skyline queries. It is based on an important observation. That is, let o and o' be two objects contained in the cube c with $|c|$ being ρ. If $T(o) \geq T(o')$, o must $\rho-$dominate o'. Accordingly, $\rho-$SEAK only needs to maintain one object in the same cube and delete the other one. Here, given any two objects o and o' in the window W, if they satisfy the following two conditions, we say o ρ-dominates o'. That are: (i)$T(o)$ is no smaller than $T(o')$ with $T(o)$ being the arrived order of o; and (ii) $\forall j \in [1, d], o[j] + \rho \geq o'[j]$.

2.3 The $\rho-$Selection Algorithm

Whenever the number of objects maintained by T achieves to M, $\rho-$SEAK enlarges ρ so as to delete more objects. Under this, $\rho-$SEAK combines some nodes maintained in T. After combination, some objects located in the same

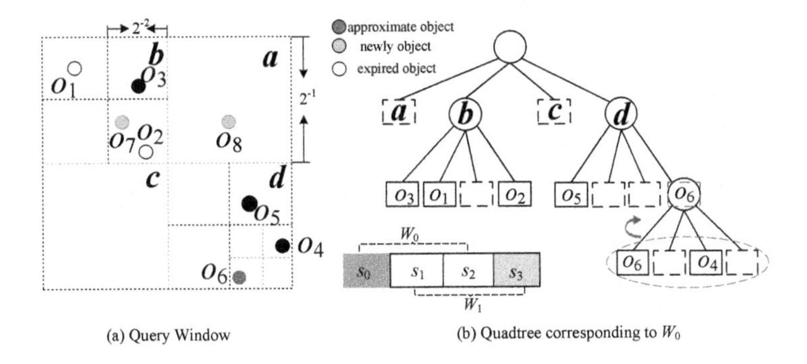

(a) Query Window (b) Quadtree corresponding to W_0

Fig. 2. The Example of the ρ-TREE($\rho = 2^{-2}$).

leaf nodes could be deleted. In other words, $\rho-$SEAK can find a suitable ρ via controlling the height of T. Clearly, if the number of objects maintained in T is small, we can set ρ to a small value so as to find high-quality query results. Under this case, $\rho-$SEAK enlarges the maximal allowed height of T. In other words, if a leaf node contains more than one object, we can further split the space occupied by the corresponding cube.

Take an example in Fig. 2, Let W_0 be the query window that contains the set of 6 objects. The object o_6 is ρ-dominate the object o_4 as (i) $T(o_4) < T(o_6)$, and (ii) $\forall i \in [1, d]$, $o_6[i] + 0.25 > o_9[i]$. In particular, suppose $\{e_1, e_2, \cdots, e_M\}$ represents the leaf nodes of a quad-tree, and $\{c_1, c_2, \cdots, c_M\}$ are cubes that serve as bounding regions for objects within the leaf nodes of the quad-tree. We iterate through the cubes in $\{c_1, c_2, \cdots, c_M\}$, and then determine the median value of the side lengths of these cubes. denoted as c_med. We use $c_{med} \cdot 2$ as the parameter ρ, which is 2^{-2}.

3 The Framework $\rho-$AKRS

This section proposes a novel framework named $\rho-$AKRS(short for $\rho-$Approximate K-Representative Skyline) to support ρ-CASQ over memory limitation-based streaming data. It mainly contains two parts: (i) using the framework $\rho-$SEAK to maintain $\rho-$approximate skyline objects; and (ii) maintaining $(\rho, k)-$ARSS based on $\rho-$approximate skyline objects. In this following, we first discuss the initialization algorithm, and then explain the incremental maintenance algorithm.

3.1 The Initialization Algorithm

In this section, we mainly discuss how to form the set $(\rho, k)-$ARSS based on objects in the query window. As shown in algorithm 1, we first form the index $\rho-$quadtree T as the manner discussed in []. Next, we form the set C_ρ based on objects maintained in T. Here, C_ρ refers to the set of objects that are not

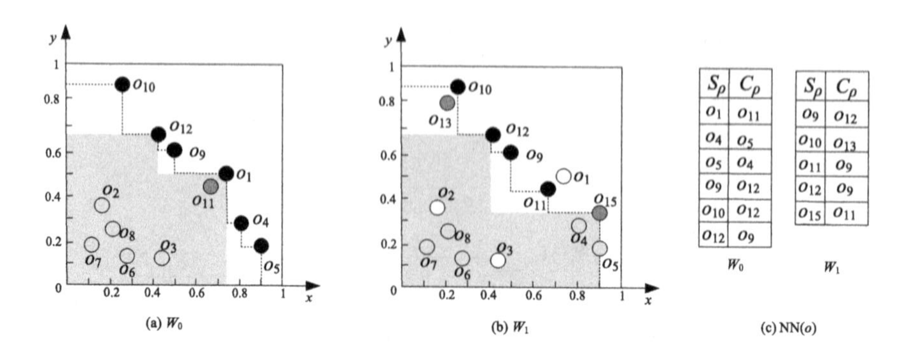

Fig. 3. Running Example of the incremental maintenance($s = 3, N = 12, k = 2$)

$\rho-$domianted by other objects in the query window. Based on C_ρ, we further form the $\rho-$approximate skyline set S_ρ, i.e., the set of $\rho-$skyline objects in the current window.

Once C_ρ is formed, we further form the set AS_ρ. Specially, we first form an M-Tree T_M based index to maintain objects in C_ρ. Next, for each object $o \in S_\rho$, we search on T_M for finding its nearest neighbor. After searching, we temporarily use C_ρ as AS_ρ, and repeatedly use the following operations to reduce the scale of AS_ρ. When the number of objects contained in AS_ρ is reduced to k, the algorithm is terminated.

Specially, we sort objects in AS_ρ based on the distances among them to their nearest neighbors. For simplicity, we regard the distance between each object o and its NN in AS_ρ as the score of o. Next, we scan objects in AS_ρ and delete some objects based on the current scanned object o. Let RNN(o) be the set of objects that regard o as their nearest neighbor, and θ_k be $k-$th highest score among all objects in AS_ρ. For each object $o' \in AS_\rho$, if the score of o' is smaller than θ_k, we remove it from AS_ρ. After scanning, if $|AS_\rho|$ is still large than k, we research objects' NN if its NN has been removed from AS_ρ, and then repeat the above operations. The algorithm is terminated when the number of objects in AS_ρ is reduced to k.

The entire process of index construction is depicted in Fig. 3 shows the sliding window W_0 with a window length of 12. After applying the $\rho-$SKYLINE algorithm described in Sect. 3.1, an approximate set satisfying the error threshold is obtained, which contains 6 objects $\{o_1, o_4, o_5, o_9, o_{10}, o_{12}\}$. When $\rho = 0.05$, $|C_{0.05}| = 7$ with $C_{0.05}$ being $\{o_1, o_4, o_5, o_9, o_{10}, o_{11}, o_{12}\}$. Find the nearest neighbor of each object in C_ρ, delete o_4, o_9. The 2-representative objects is $\{o_1, o_{12}\}$.

3.2 The Incremental Maintenance

When the query window slides from W_t to W_{t+1}, a new set of objects flows into the window, and an incremental maintenance algorithm should be executed at

Algorithm 1: The Initialization Algorithm

```
   Input: Stream Data S, the Maximal Memory M
   Output: (ρ,k)−ARSS AS_ρ
 1 T ←formQuadTree(S);
 2 S_ρ ←formRSkyline(T);
 3 if |S_ρ| ≤ k then
 4     AS_ρ ← S_ρ;
 5     return;
 6 T_M ←formMTree(C_ρ);
 7 AS_ρ ← S_ρ;
 8 while |AS_ρ| ≥ k do
 9     for i from 1 to |AS_ρ| do
10         let o be AS_ρ[i];
11         NN_o ← findNN(o, S_ρ, T_M);
12     for i from 1 to |AS_ρ| do
13         o ← AS_ρ[i];
14         for j from 1 to |RNN(o)| do
15             if the score of RNN(o)[j] is smaller than θ_k then
16                 AS_ρ ← AS_ρ− RNN(o)[j];
17 return AS_ρ;
```

this time. In this section, we will discuss how to support query processing under data stream.

As shown in Algorithm 2, we first insert newly arrived objects into the window and delete the expired objects from the window. Next, we update the $\rho-$quadtree T, meaningful object set C_ρ and the $\rho-$skyline set S_ρ as the manner discussed in []. Thirdly, we are going to update the set AS_ρ. Specially, we first check whether S_ρ has been updated. If the answer is no, the algorithm could be terminated. Otherwise, we have to update AS_ρ. Under this case, we first insert all newly generated(or arrived) skyline objects, i.e., denoted as S'_ρ, into AS_ρ. For each object $o \in S'_\rho$, we compute its score by searching its nearest neighbor. During the search, we should update the score of objects that have been contained in the set AS_ρ. From then on, we follow the logic discussed in Algorithm 1 to reduce the scale of AS_ρ. When $|AS_\rho|$ achieves k, the algorithm is terminated.

As shown in Fig. 3, when the query window slides from W_0 to W_1, $s = 3$, objects in Object o_1 expires from the window, and the approximate skyline set in the window is updated to $\{o_9, o_{10}, o_{11}, o_{12}, o_{15}\}$. $\{o_{15}\}$ are new approximate skyline objects that arrive. We repeat the above operations to handle objects $\{C_\rho\}$ and $\{S_\rho\}$, the $2-$representative objects updated to $\{o_{12}, o_{15}\}$.

Algorithm 2: The Incremental Maintenance Algorithm

Input: Stream Data S, the Maximal Memory M, Newly arrived object set S_{in}, Expired object set S_{out}

Output: $(\rho, k)-$ARSS AS_ρ

1 $W \leftarrow W \cup S_{in}$, $W \leftarrow W - S_{out}$;
2 $T \leftarrow$insertion(S_{in});
3 $T \leftarrow$deletion(S_{out});
4 $S'_\rho \leftarrow$ update(T, S_ρ);
5 $AS_\rho \leftarrow AS_\rho \cup S'_\rho$;
6 **for** *i from 1 to* $|S'_\rho|$ **do**
7 $\quad o.NN \leftarrow$ searchNN(o, AS_ρ),$o' \leftarrow o.NN$;
8 $\quad$ **if** *the score of* o' *is smaller than* $D(o, o')$ **then**
9 $\quad\quad o'.NN \leftarrow o$;
10 repeat lines 8-16 in the initialization algorithm to from AS_ρ;
11 return AS_ρ;

4 Performance Evaluation

4.1 Experiment Settings

Data sets. The experiment is based on 3 data sets, one real data set STOCK, and 2 synthetic data sets. STOCK contains trading records o 2,300 stocks. Each record is expressed by the tuple $\langle ID, time, Price, vol\rangle$. Objects in these two synthetic data sets follow normal distribution and random distribution respectively. Each data set contains a set of 1G objects.

Table 1. Parameter Settings.

Parameter	value
N	5MB, 10MB, **15MB**, 20MB,25MB
s	0.1%,0.5%,**1%**,5%, 10%($\times$ N)
ρ	0.001,0.005, **0.01**, 0.05, 0.1
k	5, **10**, 15, 20, 25

4.2 Experimental Evaluation

Parameter Settings, Competitors and Metrics. In our experiments, we evaluated the performance of algorithms under the following three parameters as shown in Table 1. In addition to the $\rho-$AKRS discussed in this paper, we use algorithms named RSP and k-LDS as competitors. Last of all, total running time, memory consumption, and accuracy are used as metrics of our paper. Here, the total running time records the time used to process all objects in the data set.

Memory size refers to the space cost for maintaining candidate objects. Accuracy evaluates the quality of the results.

In this section, we first compare the performance of $\rho-$AKRS with other algorithms. The first group of experiments tested the impact of window lengths on the algorithm performance, where the window size ranges from 5M to 25M. As shown in Fig. 4(a)-(b), when the window length increases, we have the following findings. First, the performance of $\rho-$AKRS is the best, and its running time is almost unaffected by the window length. In reason behind it is when newly arrived objects flow into the window, $\rho-$AKRS can prune most meaningless objects. Accordingly, the running cost could be reduced a lot. Another reason is $\rho-$AKRS only maintains a subset of skyline objects, which also could reduce the running cost.

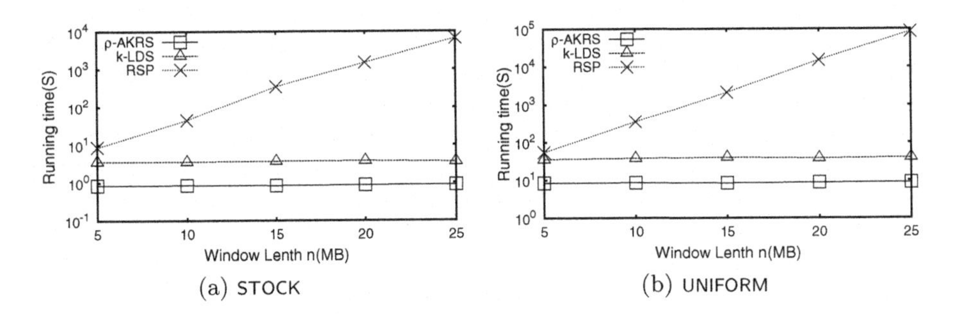

(a) STOCK (b) UNIFORM

Fig. 4. Running time comparison of different algorithms under different data sets.

Next, we evaluate the performance of algorithms under different s (data stream speed), that is, which ranges from 0.5M to 2.5M. The other parameters are set to default values. As shown in Fig. 5(a)-(b), the performance of $\rho-$AKRS is still the best of all. As s increases, the running time of these three algorithms increases, and RSP increases the most significantly because the algorithm needs to re-calculate the k most representative objects when the skyline objects change. $\rho-$AKRS is not sensitive to the parameter s. As the reason discussed before, $\rho-$AKRS only maintains a subset of skyline objects.

The third group of experiments evaluated the effect of parameter k on the algorithm performance. As shown in Fig. 6(a)-(b), the performance of $\rho-$AKRS is higher than that of RSP algorithm and k-LDS algorithm, the running time of RSP is increasing with the increasing of k, while the algorithm $\rho-$AKRS is not sensitive to the parameter k. The reason is main running cost is contributed by maintaining skyline objects in the window.

In this experiment, the algorithm presets the ρ in a range from 0.001 to 0.1, and default setting values were used for other parameters. As shown in Table 2. As the ρ increases, the running time of the algorithm decreases accordingly. The reason is that as the error threshold increases, the query window can delete more meaningless objects. However, the accuracy of the observed discovery algorithm decreases as the error threshold increases. The reason is that the

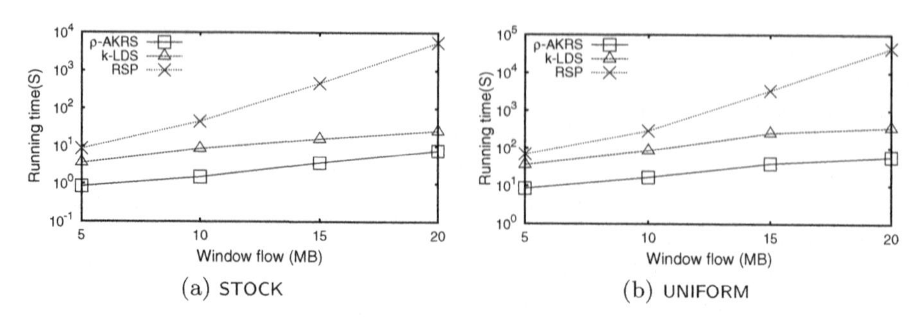

(a) STOCK (b) UNIFORM

Fig. 5. Running time comparison of different algorithms under different data sets.

large error threshold will make the system and the exact result object gap. As the error threshold increases, the spatial cost of the algorithm decreases accordingly. The reason is that larger error thresholds will cause more pruned objects and fewer objects in the candidate set. Morever, the algorithm only recalculates the approximate k representive objects when the window slides and its calculation frequency is relatively low, reducing overall computational cost by s, resulting in a lower overall calculation cost for the approximate k representive objects.

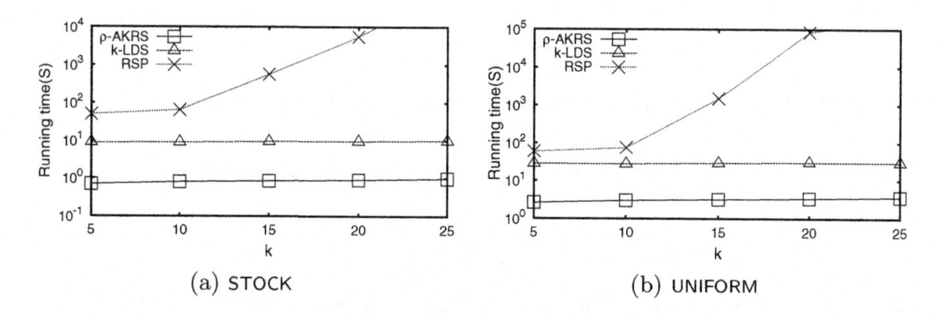

(a) STOCK (b) UNIFORM

Fig. 6. Running time comparison of different algorithms under different data sets.

Table 2. Effect of the parameter ρ

Dataset		The Parameter(ρ)									
		0.001	0.002	0.003	0.005	0.01	0.02	0.03	0.04	0.05	0.1
STOCK	Running Time	45.9	43.39	48.19	41.20	40.23	36.18	35.35	32.99	31.60	30.12
	Accuracy	0.99	0.99	0.99	0.98	0.96	0.95	0.93	0.92	0.90	0.89
	cost	0.31	0.31	0.30	0.29	0.27	0.27	0.26	0.24	0.23	0.22
UNIFORM	Running Time	41.23	39.36	38.25	36.83	35.38	34.29	33.37	32.29	31.66	28.68
	Accuracy	0.99	0.99	0.99	0.97	0.96	0.94	0.92	0.91	0.90	0.88
	cost	0.29	0.29	0.28	0.27	0.26	0.24	0.22	0.21	0.20	0.19

5 Conclusion

In this paper, we propose a novel framework named ρ−AKRS for supporting ρ-approximate continuous k representative skyline query over data stream. It uses ρ-tree and M-tree for indexing streaming data and achieves the goal of supporting ρ-approximate continuous k representative skyline query under a memory limitation-based data stream. We have conducted extensive experiments to evaluate the performance of ρ-AKRS on several data sets under different distributions. The results demonstrate the superior performance of ρ-AKRS.

Acknowledgements. This paper is partly supported by the National Key Research and Development Program of China(2020YFB1707901), the National Natural Science Foundation of Liao Ning(2022-MS-303, 2022-MS-302, and 2022-BS-218), the National Natural Science Foundation of China (62102271, 62072088, Nos. U22A2025, 62072088, 62232007, 61991404), and Ten Thousand Talent Program (No. ZX20200035).

References

1. Xuemin, L., Yidong, Y., Wei, W., Hongjun L.: Stabbing the sky: efficient skyline computation over sliding windows. In: ICDE, pp. 502–513 (2005)
2. Michael, D.M., Jignesh, M.P., William, I.G.: Efficient continuous skyline computation. In: ICED, p. 108 (2006)
3. Zhenjie, Z., Reynold, C., Dimitris, P.: Minimizing the communication cost for continuous skyline maintenance. In: SIGMOD, pp. 495–508 (2009)
4. Nikos, S., Gautam, D., Nick, K.: Categorical skylines for streaming data. In: SIGMOD, pp. 239–250 (2008)
5. Junchang, X., Guoren, W., Lei, C., Xiaoyi, Z., Zhenhua, W.: Continuously maintaining sliding window skylines in a sensor network. In: DASFAA, pp. 509–521 (2007)
6. Yufei, T., Papadias, D.: Maintaining sliding window skylines on data streams. In: IEEE Transactions on Knowledge and Data Engineering, pp. 377–391 (2006)
7. Xinjun, C., Bai, M., Dong, H., Wangguo, R.: An efficient processing algorithm for ρ-dominant skyline query. Chinese J. Comput. (2011)
8. Liang, S., Peng, Z., Yan, J.: Adaptive mining the approximate skyline over data stream. In: International Conference on Computational Science (3), pp. 742–745 (2007)
9. Xuemin, L., Yidong, Y.: Selecting stars: the k most representative skyline operator. In: ICDE, pp. 86–95 (2007)
10. Lin, X., Yuan, Y., Zhang, Q.: Selecting stars: the k most representative skyline operator. In: IEEE 23rd International Conference on Data Engineering. Istanbul, pp. 86–95 (2006)
11. Borzsonyi, S., Kossmann, D., Stocker, K.: The skyline operator. In: ICDE, pp. 421–430 (2001)
12. Donald, K., Frank, R., Steffen, R.: Shooting stars in the sky: an online algorithm for skyline queries. In: VLDB, pp. 275–286 (2002)
13. Dimitris, P., Yufei, T., Greg, F., Bernhard, S.: An optimal and progressive algorithm for skyline queries. In: SIGMOD, pp. 467–478 (2003)

14. Mei, Bai., Wang, G., Xin, J.: Discovering the k representative skyline over a sliding window. In: IEEE Transactions on Knowledge and Data Engineering, pp. 2041–2056 (2016)
15. Tianyi, L., Lu, C., Christian, S.: Evolutionary clustering of moving objects. In: ICDE, pp. 2399–2411 (2022)
16. Yunzhe, A., Zhu, Z., Rui, Z.: Approximate continuous skyline queries over memory limitation-based streaming data. In: APWebWAIM (2023)

AAP: Defending Against Website Fingerprinting Through Burst Obfuscation

Zhenyu Yang[1], Xi Xiao[1,4](✉), Bin Zhang[2], Guangwu Hu[3], Qing Li[2], and Qixu Liu[4]

[1] Shenzhen International Graduate School, Tsinghua University, Beijing, China
yangzy20@mails.tsinghua.edu.cn, xiaox@sz.tsinghua.edu.cn
[2] Peng Cheng Laboratory, Shenzhen, China
bin.zhang@pcl.ac.cn
[3] Shenzhen Institute of Information Technology, Shenzhen, China
hugw@sziit.edu.cn
[4] Institute of Information Engineering, Chinese Academy of Sciences, Beijing, China
liuqixu@iie.ac.cn

Abstract. Website fingerprinting enables eavesdroppers to identify the website a user is visiting by network surveillance, even if the traffic is protected by anonymous communication technologies such as Tor. To defend against website fingerprinting attacks, Tor provides a circuit padding framework as the official way to implement padding defenses. However, the circuit padding framework can not support additional delay, which makes most defense schemes unworkable. In this paper, we study the patterns of HTTP requests and responses generated during website loading and analyze how these high-level features correlate with the underlying features of network traffic. We find that the HTTP requests sent and responses received continuously in a short period of time, which we call HTTP burst, have a significant impact on network traffic. Then we propose a novel website fingerprinting defense algorithm, Advanced Adaptive Padding(AAP). The design principle of AAP is similar to Adaptive Padding, which works by obfuscating burst features. AAP does not delay application packets and is in line with the design philosophy of low latency networks such as Tor. Besides, AAP uses a more sensible traffic obfuscation strategy, which makes it more effective. Experiments show that AAP outperforms other zero-delay defenses with moderate bandwidth overhead.

Keywords: Website fingerprinting defense · Tor · Circuit padding framework · Traffic analysis

1 Introduction

Tor is an anonymous communication system based on the second-generation onion router [1]. In the Tor network, communication data is first encrypted at multiple layers and then forwarded by several proxies called onion routers. Onion

X. Yang et al. (Eds.): ADMA 2023, LNAI 14180, pp. 107–121, 2023.
https://doi.org/10.1007/978-3-031-46677-9_8

routers are randomly selected and no single node can know the IP addresses of both the source and the destination. The purpose of Tor is to protect people from third-party trackers, surveillance, and censorship. Due to its high security, easy deployment, and low latency, Tor has become the most popular anonymous communication system today. Tor has over 6,000 intermediate server nodes worldwide, and over 3 million people use Tor clients to communicate anonymously. However, recent studies [2] have shown that Tor is not resistant to website fingerprinting attacks. Using the Tor network to access the web is still at risk of privacy leakage. Website fingerprinting attacks have been proven to be applied in real scenarios with a high success rate [3].

Most existing defenses work in the network layer by traffic padding and traffic delaying, they incur too much bandwidth overhand or latency overhead that makes them difficult, even impossible, to deploy in real-world environments. To counter website fingerprinting attacks, Tor provides a circuit padding framework for implementing defenses. The framework does not provide any mechanism to delay actual user traffic deliberately, which makes delay-based defenses unworkable.

Tor developers prefer Adaptive Padding [4] style defenses, the most famous of which is WTF-PAD [5]. However, WTF-PAD fails to defend against advanced website fingerprinting attacks. In this paper, we analyze burst features in the network traffic and propose a new website fingerprinting defense scheme *Advanced Adaptive Padding*(AAP). AAP obfuscates traffic features by sending fake bursts in the gap between real bursts, changing the size and number of bursts. It only sends dummy packets without delaying user traffic. Besides, AAP has a finite state machine like WTF-PAD, which makes it can be deployed by circuit padding framework. We conduct comprehensive experiments to evaluate AAP. Experimental results show that AAP can achieve a good defensive effect with moderate overhead.

In summary, we make the following contributions:

- We analyze the distribution patterns of HTTP requests and responses during the loading of different websites and their correlation with network traffic. We find that the HTTP requests sent and responses received continuously in a short period of time, which we call HTTP burst, have a significant impact on network traffic.
- We propose a novel website fingerprinting defense scheme called AAP. AAP changes burst patterns of website traffic by sending dummy packets without delaying user traffic. Our experiments show that AAP can outperform other zero-delay defenses with moderate bandwidth overhead.
- We collect a new dataset in the live Tor network. The dataset contains in total of 20,000 instances with 100 monitored websites (each load 100 times) and 10,000 non-monitored websites (each loaded once).

The structure of this paper is as follows. We first introduce the background knowledge in Sect. 2, then we discuss related works in Sect. 3. We illustrate the motivation in Sect. 4 and details of AAP in Sect. 5. We introduce the dataset we collected in Sect. 6. Then we present the experiment settings and results in Sects. 7 and 8. Finally, we summarize our work in Section 9.

2 Background

In this section, we report preliminary notions related to Tor and website fingerprinting largely used in the paper, with the aim to make the paper self-contained.

2.1 Tor

Tor [1] is currently the most popular anonymous communication system, which consists of more than 6,000 volunteer nodes distributed around the world. The anonymity of Tor is guaranteed by multiple proxies. The proxy in Tor is called onion router (OR). A Tor circuit contains three ORs selected from volunteer nodes randomly, each of which knows only the IP addresses of the previous and next hops. These three ORs are the entry node, middle node, and exit node. Application traffic is transmitted through three ORs in turn, so that no one can obtain the identity of end users at the same time. The structure of the Tor circuit is illustrated in Fig. 1.

Tor sends data using 512-byte fixed-length cells. Each cell contains a header and a payload. There are two fields in the header, the first field is circuit ID, indicating which circuit the cell belongs to, and the second is the command field, indicating the type of the cell. Based on their command, cells are either control cells or relay cells. The control cell is responsible for circuit establishment and destruction, and each OR needs to resolve and perform related operations. The relay cell carries end-to-end stream data.

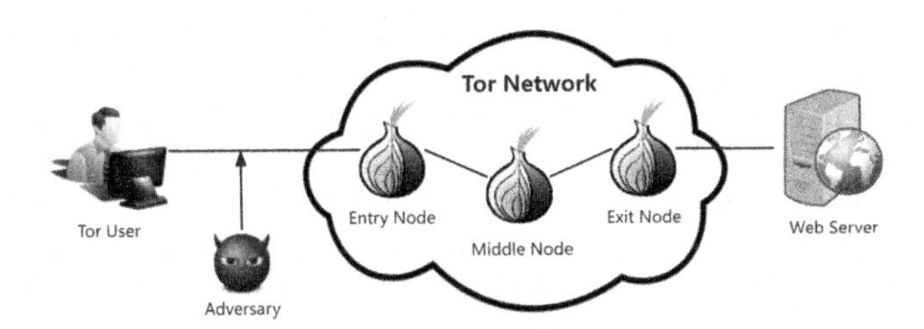

Fig. 1. Tor circuit and website fingerprinting attack model.

2.2 Website Fingerprinting

Website fingerprinting is a technique for traffic analysis. It can be seen as a classification problem that determines which website a user has visited by analyzing encrypted network traffic. Potential attackers include anyone that can sniff the communication between the user and the entry node of Tor, such as Internet Service Provider (ISP) and local network administrator because these individuals have access to the user's IP address and can observe the network traffic. Figure 1 shows the threat model of website fingerprinting. The attacker is passive, meaning that he only observes and records the traffic traces that pass through the network and does not have the ability to drop, delay, or modify packets in the traffic stream.

In order to conduct a website fingerprinting attack, the attacker needs to collect a dataset of website traffic first. Then he can use the dataset to train a machine-learning-based or deep-learning-based classifier. Finally, the attacker sniffs network traffic when victims visit websites and uses the trained classifier to determine which websites victims have visited.

3 Related Work

Website fingerprinting defenses defend against website fingerprinting attacks by changing traffic patterns observed by attackers. To ensure that website content is loaded correctly, the original application traffic cannot be modified or discarded. Common methods of website fingerprinting defense include sending dummy packets and delaying packet delivery.

Regularization defenses use fixed patterns for sending packets, they minimize information leakage and provide strong security guarantees. BuFLO family defenses [6,7] send packets with the same packet length and time interval. They send data even after the website has finished loading to mask the total load time. TAMARAW [8] performs best among them by using different sending rates at different directions. Lu et al. [9] introduce DynaFlow, which is a defense with dynamically-changing intervals and fixed burst patterns. GLOVE [10] and Supersequence [11] cluster websites according to their similarity, then calculate the super sequence of each class. The traces in the same class are morphed into the super sequence so they can not be distinguished. Walkie-Talkie [12] modifies the browser in half-duplex mode to produce easily moldable burst sequences, then it uses burst molding to change burst patterns. Regularization defenses typically have high latency and bandwidth overhead and are therefore not suitable for low overhead networks like Tor.

Distribution-based defenses work by sending dummy data to change the traffic features. Some function at the network layer and others function at the application layer. Application layer defenses include Decoy [13], HTTPOS [14], LLaMA [15] and ALPaCA [15]. While a website is loading, Decoy loads another page in the background to camouflage. HTTPOS uses a variety of strategies such as enabling HTTP pipeline and sending invalid requests to change traffic patterns. It was defeated by Cat et al. [16]. LLaMA is a client-side website fingerprinting defense, it works by adding extra delay to HTTP requests and sending redundant requests. ALPaCA is a server-side website fingerprinting defense, it works by morphing the size of HTTP objects in the web server, which makes it hard to deploy. Others work in the network layer. WTF-PAD [5] is an approach based on adaptive padding [4]. It decides whether to send dummy packets based on the difference between the sampled time interval and the real time interval. However, it fails to defend against deep learning-based website fingerprinting attacks [2]. FRONT [17] is a lightweight defense method proposed in recent years. It aims to obfuscate the feature-rich front part of the trace.

Multipath defenses establish multiple links to transmit a website through different links, ensuring that an attacker can only observe part of the traffic. Henri et al. [18] introduce a defense named HyWF that exploits multihoming. HyWF requires user devices to establish multiple different physical links to the entry node and assign network traffic to different physical links. Another multipath defense is TrafficSliver [19]. TrafficSliver requires users to establish multiple Tor connections simultaneously, each passing through a different entry node, so a malicious entry node can only get part of the traffic. Multipath defenses require extra infrastructure or modify Tor's protocol. Therefore, they are difficult to be implemented.

In recent years, deep learning has made rapid progress and achieved excellent results in the field of website fingerprinting. Many defenses use adversarial-based techniques to defend against deep learning-based website fingerprinting attacks. Hou et al. [20] propose WF-GAN to fight back against website fingerprinting attacks. The structure of WF-GAN is based on AdvGAN [21] with improvements. Mockingbird [22] fools deep learning-based classification models with adversarial examples. Both of them take the entire web traffic trace as input to produce perturbations which makes them impossible to deploy in practice. Other adversarial-based defenses [23] achieve good defensive effectiveness with a small amount of overhead. However, if the attacker knows the defense scheme used by the victim and trains attack models with the defended data, adversarial-based defenses become ineffective.

4 Motivation

There are two commonly used methods in designing website fingerprinting defense algorithms.

- Make different website traffic the same, including converting the traffic pattern from one website to another or regularizing traffic across multiple websites. [6,7,11]
- Randomize website traffic to hide features in network traffic that are valid for website fingerprinting. [5,17]

Since website content is constantly changing, more latency and bandwidth overheads need to be added to make all website traffic the same. Usually low overhead website fingerprinting defense algorithms are randomized algorithms. Therefore we choose to use randomization to implement website fingerprinting defense. The traffic generated in website loading is mainly generated by HTTP requests and responses, so we analyze the association between HTTP and website traffic, and then explore the website traffic characteristics.

Table 1 shows HTTP requests and responses generated when loading the homepage of *Google*. Some of these HTTP request and response transmission times are very close or even overlap. We define HTTP requests sent in a short time period as an HTTP burst. There are 6 HTTP bursts in this website, which can be denoted as [(1),(2–5),(6),(7–9),(10),(11)]. In each HTTP burst, the sum

Table 1. HTTP requests and responses recorded when visiting https://www.google.com/ Request time indicates the time from the headers sent to the last byte sent. Response time indicates the time from the headers received to the last byte received. Time is in seconds and size is in bytes.

ID	Request Time	Response Time	Request Size	Response Size
1	0.0–0.001	0.372–0.664	422	48362
2	1.268–1.278	1.663–1.667	693	6723
3	1.276–1.278	1.667–1.674	709	352
4	1.277–1.278	1.777–1.859	687	1414
5	1.276–1.278	1.858–1.859	824	352
6	2.163–2.164	3.012–3.014	645	2304
7	3.473–3.475	3.873–3.874	501	1524
8	3.457–3.460	3.756–3.875	506	66251
9	3.556–3.557	3.872–3.986	388	1376
10	4.064–4.065	4.556–4.557	842	352
11	5.977–6.055	6.493–6.494	769	37654

of the HTTP request size determines the amount of data sent, and the sum of the HTTP response size determines the total amount of data received. We count the total size of requests and responses in an HTTP burst. The size of each HTTP burst is shown on the left-hand side of Fig. 2. The right-hand side of Fig. 2 shows the trend of the traffic rate during the loading of *www.google.com*.

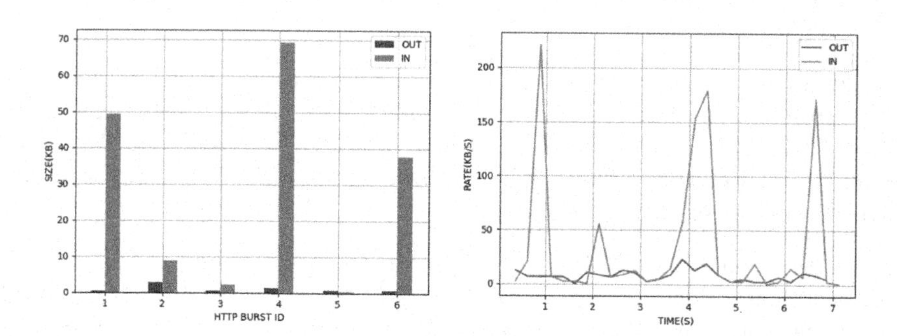

Fig. 2. The left graph shows the size of the six HTTP bursts generated when accessing *www.google.com*. The right graph shows the trend of network transfer rate when accessing *www.google.com*. It can be clearly seen that there is a correlation between the two graphs.

5 AAP

AP algorithm chooses an expected inter-packet interval(EIPI) after receiving a packet and decides whether to send a dummy packet based on EIPI. Tor developers are interested in deploying AP-style defenses because they have no latency overhead. We propose a new website fingerprinting defense, Advanced Adaptive Padding(AAP). AAP is an AP-style defense that tries to obfuscate burst patterns by sending fake bursts, it does not bring any latency overhead.

The state of AAP is shown in Fig. 3. AAP has three modes: real burst mode, fake burst mode, and gap mode.

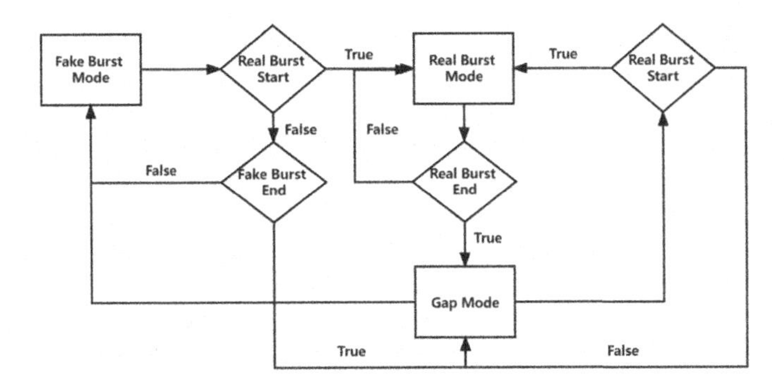

Fig. 3. Finite state machine to illustrate the AAP algorithm.

Real Burst Mode. In the real burst mode, the browser is performing network activities, such as sending HTTP requests or receiving HTTP responses. User traffic is being transmitted and the network utilization is high. Therefore, in this mode, AAP only forwards user traffic and does not send dummy packets to avoid network congestion which affects the real user traffic. AAP uses a sliding window to determine whether to be in the real burst mode. We denote the window size as W. If at least K application packets have been forwarded in the past W seconds, AAP is in the real burst mode. When the real traffic forwarding is completed, i.e., less than K packets have been sent in the past W seconds, the real burst mode ends and AAP switches to Gap mode.

Gap Mode. In the gap mode, the browser is idle, so no user traffic is currently being transmitted. AAP needs to decide when the next fake burst will be sent. For the randomness of the burst interval, AAP samples t from the distribution of real burst intervals and uses $g = t * Q$ as the time interval between two bursts. Q is a parameter greater than 0 and less than 1. Q is used to control the expectation of the time interval and has an impact on the bandwidth overhead. When AAP enters the gap mode, it immediately calculates g and starts timing. If a real burst starts before g expires, then AP transfers to the real burst mode, otherwise, it transfers to the fake burst mode.

Fake Burst Mode. In the fake mode, AAP generates a fake burst and sends it. AAP uses the feature distribution of real bursts to generate fake bursts, in order to prevent attackers from being able to distinguish the fake bursts. First, AAP randomly selects the fake burst length k from the length distribution of real bursts. Then, AAP selects $k-1$ time intervals as the intervals between packets in the fake burst. Due to significant differences in burst characteristics between incoming and outgoing directions, it is necessary to use different distributions for each direction. During the process of sending a fake burst, AAP always detects if a real burst is sent. If so, AAP stops sending fake bursts and switches to the real burst mode. In this case, the sent dummy packets and the application packets sent later together form a large burst, changing the length of the real burst. When the fake burst is sent, AAP switches to the gap mode.

The first few seconds of each trace leak the most useful features, because the first request sent by each website is a request for an HTML file which is relatively stable and sent separately. So AAP starts in the fake mode to obfuscate the feature-rich front part of a trace. In particular, the first fake burst does not terminate early, even if it encounters a real burst.

To control the bandwidth overhead, AAP uses a parameter N to limit the total number of dummy packets. To increase the intra-class variance, an intuitive idea is to use random N for different instances of the same website. But after trying different random strategies, we find that using a fixed N is more effective and stable. As the proportion of incoming and outgoing packets is an important feature, we use a random value P to determine the ratio of outgoing packets. P is sampled from a uniform distribution between 0 and 1 for each trace.

6 Dataset

In this section, we describe the dataset collection process and the data representation in our dataset.

6.1 Data Collection

We collect a new dataset between November and December 2022 to investigate the association of HTTP requests and responses with network traffic characteristics. During access to the website, we not only capture the generated packets on the network card but also record the HTTP requests sent and responses received.

For the data collection process, we used 10 virtual machines in a cloud environment, each virtual machine is provisioned with 2 CPUs and 4GB of RAM. We select 100 popular websites from Alexa[1] as the monitored set, each visited 100 times, and 10,000 other websites as the non-monitored set, each visited once. We use selenium to control Firefox for web access. We use mitmproxy [24] to log the information of HTTP requests and responses. At the same time, we use tcpdump to capture packets on the net card and save them in pcap files. Since

[1] www.alexa.com.

the packet payloads are encrypted and thus have no value for the adversary, we extract metadata from the traffic traces and discard pcap files for saving storage. Each website is given 180 s to load before the browser is killed, and the timeout is marked as invalid. Upon loading the page, it is left open for additional 10 s, after which the browser is closed and any profile information is removed. Besides, we remove pages with request failure rates greater than 50%.

6.2 Data Representation

We follow the approach proposed by Wang and Goldberg [25] to process data and extract Tor cells from the captured pcap file. We use a sequence of cells as a traffic trace, denoted as $T = [(t_1, d_1), (t_2, d_2)...(t_{|T|}, d_{|T|})]$ where $|T|$ is the total number of cells in the trace, t_i is the timestamp of the i-th cell, d_i shows the direction of the i-th cell. The incoming and outgoing cells are represented as -1 and +1, respectively.

6.3 Ethical Consideration

Since large-scale data collection may have some impact on the Tor network, we try to mitigate the adverse effects on the Tor network. We use scripts to directly drive website visiting, so none of those visits come from real users. We just keep the minimal information that is necessary for our experiment. We only visit one web page at a time, there is no parallel processing, so for Tor, it is just one more user visiting and no additional burden.

7 Experiment Settings

To evaluate the improvements in performance offered by AAP. We use CUMUL [26], k-FP [26], DF [2] as attackers to evaluate AAP. Because these methods are state-of-the-art for different times and all achieve a high level of accuracy. CUMUL derives features from the cumulative representation of the trace and uses LibSVM with a Radial Basis Function (RBF) for classification. k-FP uses random forests to extract a fingerprint for each trace and uses KNN for classification. DF is the current state-of-the-art website fingerprinting attack algorithm, which is based on deep learning. It uses packet direction sequence as input and uses CNN for feature extraction and uses fully-connected layers for classification.

We choose four defenses, TAMARAW [11], WTF-PAD [5], FRONT [17] and WF-GAN [20] as competitors to our defenses. TAMARAW is a regularization defense with high bandwidth overhead and latency overhead. TAMARAW has high security, so it is often used as a benchmark for comparison. WTF-PAD, FRONT and WF-GAN are lightweight obfuscation defenses, they all have no latency overhead without delaying user traffic. WTF-PAD is an improvement on AP, and it is the main subject of our comparison. FRONT obfuscates the front part of the trace, it is the state-of-the-art zero-delay defense. WF-GAN is an adversarial-based defense that has very low bandwidth overhead.

We use simulation experiments to verify the effectiveness of website fingerprinting defenses. For each trace, we generate post-defense traces based on different defense protocols. Then we use attack methods to validate their defensive effect on the defended dataset. To ensure the accuracy of the results, we perform 10-fold cross-validation on the dataset. We evalute them in both closed world scenario and open world scenario.

Closed World Scenario. In the closed world scenario, the victim is restricted to access only the websites in the monitored set, the attacker's goal is to identify which website the victim has visited. So it can be seen as a multi-classification problem. The closed world scenario is an ideal scenario for testing website fingerprint attacks and defenses.

Open World Scenario. In the open world scenario, the victim can access any of the websites on the Internet, the goal of the attacker is to determine whether the website visited by the victim is in the monitored set or not. So it can be seen as a binary classification problem. The open world scenario is a relatively realistic scenario.

Metrics. Website fingerprinting defenses should be evaluated in terms of both overhead and defense effect. Overhead includes latency overhead and bandwidth overhead. Latency overhead is the additional time needed to load the website as a percentage of the original loading time. Bandwidth overhead is the ratio of the additional traffic transferred to the original website traffic. Defense effect can be shown by attack effect. In the closed world scenario, the effect of website fingerprinting is usually judged using accuracy. Accuracy is the proportion of correctly classified traces to the total number of traces. In the open world scenario, we use TPR, FPR, and F1 to judge the effect of website fingerprinting attacks. TPR is the percentage of samples in the monitored traces that are correctly classified. FPR is the percentage of samples in the non-monitored traces that are wrongly classified. F1 is the reconciled average of precision and recall.

8 Experiment Results

In this section, we provide our experiment results and compare the performance of AAP with other defenses.

8.1 Overhead

Table 2 summarizes the bandwidth overhead and latency overhead for each defense on our data set. With the exception of TAMARAW, other defenses have no latency overhead due to the fact that they do not intentionally delay sending packets. TAMARAW uses a fixed time interval to send packets, so it delays the delivery of user traffic, causing it takes too long to load websites. Its bandwidth and latency overheads are 85.92% and 73.66%, respectively, which are too high to be deployed in Tor. WTF-PAD is relatively lightweight, and it results in 66.59% bandwidth overhead. FRONT uses parameters to control the number of injected packets, with the default parameters, it incurs 54.42% bandwidth overhead. WF-GAN uses Generative Adversarial Networks to generate small perturbation. It has the lowest bandwidth overhead 6.17%. We set $N = 4000, Q = 0.4$, which makes AAP have a relatively small overhead compared to FRONT. In this setting, AAP incurs 40.37% bandwidth overhead. The parameters we used are shown in Table 2.

Table 2. Parameters and overheads of different defenses. BO for bandwidth overhead and LO for latency overhead. Overheads are all percentages.

Defenses	Parameters	BO	LO
No defense	None	0	0
TAMARAW	$\rho_{in} = 40, \rho_{out} = 12, L = 100$	85.92	73.66
WTF-PAD	$nomal\_rcv$	66.59	0
FRONT	$N_s = N_c = 2500, W_{min} = 1, W_{max} = 14$	54.42	0
WF-GAN	$\alpha = 1, \beta = 3$	8.91	0
AAP	$N = 4000, Q = 0.4$	40.37	0

8.2 Closed World

The closed world setting is useful for illustrating the effectiveness of website fingerprinting attacks and defenses. Table 3 shows how well website fingerprinting attacks perform against our evaluated defenses in the closed world setting.

Table 3. Accuracy of website fingerprinting attacks against different defenses in the closed world setting.

	CUMUL	k-FP	DF
No defense	94.55%	93.18%	97.10%
TAMARAW	16.19%	9.35%	5.19%
WTF-PAD	75.49%	74.38%	86.57%
FRONT	31.23%	59.27%	41.07%
WF-GAN	83.75%	89.17%	92.62%
AAP	**31.24%**	**47.49%**	**36.35%**

All attacks achieve high accuracy of over 93% on the undefended dataset. DF is the strongest attack since its accuracy is 97.10% and k-FP is the weakest attack with 93.18% accuracy.

The performance of AAP is worse compared to TAMARAW. However, TAMARAW has the highest bandwidth overhead and latency overhead, which makes it impossible to be deployed in practice. The other three defense schemes have cheaper bandwidth overhead and no latency overhead, but they are less effective than AAP. FRONT is the most effective method of them and has a similar overhead to AAP, but it has a lower defensive success rate against k-FP than AAP by 11.78%.

We find that AAP is less effective against k-FP and more effective against CUMUL and DF. Since AAP does not delay user traffic delivery, it leaks some time features, which k-FP can use effectively. CUMUL and DF use cumulative packet length and packet direction as features, respectively, and thus cannot exploit these time features, resulting in their poor effectiveness in combating AAP.

8.3 Open World

Table 4 shows how well website fingerprinting attacks perform against our evaluated defenses in the open world scenario.

Table 4. Defense performances in the open world scenario. A low F1 score represents a better defense. TPR and FPR are in percentage.

Defenses	CUMUL			k-FP			DF		
	TPR	FPR	F1	TPR	FPR	F1	TPR	FPR	F1
No defense	57.40	31.82	0.61	86.81	5.68	0.90	94.26	13.59	0.91
TAMARAW	12.80	21.90	0.19	52.85	47.22	0.53	23.10	82.20	0.23
WTF-PAD	50.79	33.20	0.55	52.82	7.30	0.66	70.63	23.72	0.72
FRONT	58.52	45.29	0.57	36.92	8.50	0.51	70.39	45.49	0.64
WF-GAN	54.70	42.31	0.59	87.03	9.35	0.85	92.45	82.20	0.91
AAP	**52.34**	**46.31**	**0.53**	**52.00**	**28.02**	**0.52**	**61.83**	**16.00**	**0.57**

When no defense is implemented, DF is the strongest website fingerprinting attack with a 0.91 F1 score and 94.26% TPR. CUMUL performs worst, it only has 57.40% TPR and 0.61 F1 score.

In the open world scenario, AAP can reduce the F1 score from 0.91 to 0.57 when defending against DF. AAP is still worse compared to TAMARAW, but better than the other defense schemes.

8.4 Discussion

AAP is a distribution-based defense with randomness, which produces different results for the same input. So it does not provide a theoretically provable security guarantee. In terms of defensive effectiveness, AAP is inferior to defense algorithms that can provide theoretical security guarantees, such as TAMARAW. However, these defense algorithms have a heavy overhead, which is not tolerated by Tor. As a result, these methods exist only on paper and cannot be deployed on Tor.

AAP and WTF-PAD are based on AP, but AAP outperforms WTF-PAD for many reasons. First, WTF-PAD uses packets interval to determine the interval of bursts. This approach is problematic due to the presence of background noise in the network. AAP uses a sliding window to make the distribution of bursts more reasonable. Besides, AAP uses a more random strategy, such as controlling the proportion of packets going in and out of the direction which increases intra-class variation. Finally, AAP only sends fake bursts between real bursts, when a real burst starts, it stops sending fake bursts. This avoids network congestion and improves network quality.

FRONT only obfuscates the front part of the trace, it sends a lot of noise when the website starts to load, without taking into account the impact on network congestion. The latter part of trace still gives away information. AAP only sends fake bursts when the network is idle and covers a larger area, so it works better.

Adversarial-based defenses such as WF-GAN can spoof fixed attack models with low overheads. However, when training with defended data, adversarial defenses perform poorly.

9 Conclusion

In this paper, we propose a new concept, HTTP burst, and analyze its association with network traffic. We find that HTTP burst is a key feature affecting network traffic, and based on this, we propose a new website fingerprinting defense method AAP. AAP obfuscates burst patterns by sending fake bursts in the long gap between real bursts. We evaluate AAP in both closed world scenario and open world scenario. Experiment results show that AAP outperforms other zero-delay defenses with lower overhead. AAP does not delay user traffic and it only incurs moderate bandwidth overhead, which makes it highly available to be adopted by Tor.

Acknowledgment. This work was supported inpart by the National Natural Science Foundation of China (61972219), the Research and Development Program of Shenzhen (JCYJ20190813174403598), the Overseas Research Cooperation Fund of Tsinghua Shenzhen International Graduate School (HW2021013), the Youth Innovation Promotion Association CAS (2019163), the Guangdong Basic and Applied Basic Research Foundation (2022A1515010417), the Key Project of Shenzhen Municipality(JSGG20211029095545002), the Science and Technology Research Project of Henan Province(222102210096), the Key Laboratory of Network Assessment Technology, Institute of Information Engineering, Chinese Academy of Sciences, and Shenzhen Science and Technology Innovation Commission (Research Center for Computer Network (Shenzhen) Ministry of Education).

References

1. Syverson, P., Dingledine, R., Mathewson, N.: Tor: the secondgeneration onion router. In: Proceedings of the 13th USENIX Security Symposium, (San Diego, CA, USA), pp. 303–320, USENIX Association (2004)
2. Sirinam, P., Imani, M., Juarez, M., Wright, M.: Deep fingerprinting: undermining website fingerprinting defenses with deep learning. In: Proceedings of the 2018 ACM SIGSAC Conference on Computer and Communications Security, CCS, (Toronto, ON, Canada), pp. 1928–1943 (2018)
3. Cherubin, G., Jansen, R., Troncoso, C.: Online website fingerprinting: evaluating website fingerprinting attacks on tor in the real world. In: 31st USENIX Security Symposium (USENIX Security 22), pp. 753–770 (2022)
4. Shmatikov, V., Wang, M.-H.: Timing analysis in low-latency mix networks: attacks and defenses. In: Gollmann, D., Meier, J., Sabelfeld, A. (eds.) ESORICS 2006. LNCS, vol. 4189, pp. 18–33. Springer, Heidelberg (2006). https://doi.org/10.1007/11863908_2
5. Juarez, M., Imani, M., Perry, M., Diaz, C., Wright, M.: Toward an efficient website fingerprinting defense. In: Askoxylakis, I., Ioannidis, S., Katsikas, S., Meadows, C. (eds.) ESORICS 2016. LNCS, vol. 9878, pp. 27–46. Springer, Cham (2016). https://doi.org/10.1007/978-3-319-45744-4_2
6. Dyer, K.P., Coull, S.E., Ristenpart, T., Shrimpton, T.: Peek-a-boo, i still see you: why efficient traffic analysis countermeasures fail. In: 33rd IEEE Symposium on Security and Privacy, pp. 332–346 (2012)
7. Cai, X., Nithyanand, R., Johnson, R.: CS-BuFLO: a congestion sensitive website fingerprinting defense. In: Proceedings of the 13th Workshop on Privacy in the Electronic Society, WPES, (Scottsdale, AZ, USA), pp. 121–130. ACM (2014)
8. Cai, X., Nithyanand, R., Wang, T., Johnson, R., Goldberg, I.: A systematic approach to developing and evaluating website fingerprinting defenses. In: Proceedings of the 2014 ACM SIGSAC Conference on Computer and Communications Security, CCS, (Scottsdale, AZ, USA), pp. 227–238. ACM (2014)
9. Lu, D., Bhat, S., Kwon, A., Devadas, S.: DynaFlow: an efficient website fingerprinting defense based on dynamically-adjusting flows. In: Proceedings of the 2018 Workshop on Privacy in the Electronic Society, WPES, (Toronto, Canada), pp. 109–113. ACM (2018)
10. Nithyanand, R., Cai, X., Johnson, R.: Glove: a bespoke website fingerprinting defense. In: Proceedings of the 13th Workshop on Privacy in the Electronic Society, pp. 131–134 (2014)
11. Wang, T., Cai, X., Nithyanand, R., Johnson, R., Goldberg, I.: Effective attacks and provable defenses for website fingerprinting. In: Proceedings of the 23rd USENIX Security Symposium, (San Diego, CA, USA), pp. 143–157, USENIX Association (2014)
12. Wang, T., Goldberg, I.: Walkie-Talkie: an efficient defense against passive website fingerprinting attacks. In: Proceedings of the 26th USENIX Security Symposium, (Vancouver, BC), pp. 1375–1390, USENIX Association (2017)
13. Panchenko, A., Niessen, L., Zinnen, A., Engel, T.: Website fingerprinting in onion routing based anonymization networks. In: Proceedings of the 10th annual ACM workshop on Privacy in the electronic society, WPES, (Chicago, IL, USA), pp. 103–114 (2011)

14. Luo, X., Zhou, P., Chan, E.W., Lee, W., Chang, R.K., Perdisci, R., et al.: HTTPOS: sealing information leaks with browser-side obfuscation of encrypted flows. In: Proceedings of the Network and Distributed System Security Symposium, NDSS, (San Diego, CA, USA) (2011)
15. Cherubin, G., Hayes, J., Juárez, M.: Website fingerprinting defenses at the application layer. Proc. Priv. Enhan. Technol. **2017**(2), 186–203 (2017)
16. Cai, X., Zhang, X.C., Joshi, B., Johnson, R.: Touching from a distance: website fingerprinting attacks and defenses. In: Proceedings of the 2012 ACM conference on Computer and communications security, CCS, pp. 605–616. ACM (2012)
17. Gong, J., Wang, T.: Zero-delay lightweight defenses against website fingerprinting. In: Proceedings of the 29th USENIX Security Symposium, (Boston, MA, USA), pp. 717–734, USENIX Association (2020)
18. Henri, S., García, G., Serrano, P., Banchs, A., Thiran, P., et al.: Protecting against website fingerprinting with multihoming. Proc. Priv. Enhan. Technol. **2020**(2), 89–110 (2020)
19. De la Cadena, W., et al.: TrafficSliver: fighting website fingerprinting attacks with traffic splitting. In: Proceedings of the 2020 ACM SIGSAC Conference on Computer and Communications Security, CSS, (Virtual Event, USA), pp. 1971–1985. ACM (2020)
20. Hou, C., Gou, G., Shi, J., Fu, P., Xiong, G.: WF-GAN: fighting back against website fingerprinting attack using adversarial learning. In: IEEE Symposium on Computers and Communications, ISCC, (Rennes, France), pp. 1–7. IEEE (2020)
21. Xiao, C., Li, B., Zhu, J.-Y., He, W., Liu, M., Song, D.: Generating adversarial examples with adversarial networks. In: Proceedings of the 27th International Joint Conference on Artificial Intelligence, IJCAI, pp. 3905–3911. AAAI (2018)
22. Rahman, M.S., Imani, M., Mathews, N., Wright, M.: Mockingbird: defending against deep-learning-based website fingerprinting attacks with adversarial traces. IEEE Trans. Inf. Forensics Secur. **16**, 1594–1609 (2020)
23. Nasr, M., Bahramali, A., Houmansadr, A.: Defeating DNN-based traffic analysis systems in real-time with blind adversarial perturbations. In: Proceedings of the 30th USENIX Security Symposium, (Vancouver, B.C., Canada), pp. 2705–2722, USENIX Association (2021)
24. Cortesi, A., Hils, M., Kriechbaumer, T.: mitmproxy: a free and open source interactive HTTPS proxy (2010) [Version 9.0]
25. Wang, T., Goldberg, I.: Improved website fingerprinting on tor. In: Proceedings of the 12th ACM Workshop on Workshop on Privacy in the Electronic Society, WPES, (Berlin, Germany), pp. 201–212. ACM (2013)
26. Panchenko, A., et al.: Website fingerprinting at internet scale. In: Proceedings of Internet Society Symposium on Network and Distributed Systems Security, (San Diego, CA, USA). IEEE (2016)

ReviewLocator: Enhance User Review-Based Bug Localization with Bug Reports

Renjie Xiao[1], Xi Xiao[1](✉), Le Yu[2], Bin Zhang[3], Guangwu Hu[4], and Qing Li[3]

[1] Shenzhen International Graduate School, Tsinghua University, Shenzhen, Guangdong, China
xrj20@mails.tsinghua.edu.cn, xiaox@sz.tsinghua.edu.cn
[2] Hong Kong Polytechnic University, Hong Kong, China
[3] Peng Cheng Laboratory, Shenzhen, Guangdong, China
bin.zhang@pcl.ac.cn
[4] Shenzhen Institute of Information Technology, Shenzhen, Guangdong, China
hugw@sziit.edu.cn

Abstract. Improving and updating applications based on user reviews is crucial to the continuous development of modern mobile applications. However, software bug descriptions in user reviews are often written by non-professional users, and contain a lot of irrelevant text, making it challenging to conduct bug localization. The current software bug localization technologies based on user reviews are not able to address these challenges effectively, resulting in suboptimal results. To address this issue, we propose ReviewLocator, which focuses on key phrases and learning from historical bug reports. It first utilizes syntactic analysis or source file parsing to convert each user review or source file into phrase representations. Then it depends on Key Phrase-based Ranking using a newly proposed Bug Report-based Term Weight to map review phrase sets to source file phrase sets. In our experiments on eight applications from the Google Play Store, the results prove our proposal surpasses ChangeAdvisor and Where2Change with an absolute improvement of 0.076 and 0.055 in terms of MAP correspondingly.

Keywords: Bug Localization · Bug Report · User Review

1 Introduction

The 21st century has witnessed an explosive growth of mobile applications, with Android emerging as one of the most popular mobile operating systems. As of January 1st, 2023, there are over 2.5 billion active Android users in 190 countries around the world[1], and the Google Play Store has more than 2.6 million mobile applications[2]. DevOps (Development and Operations) [1] has become a

[1] https://www.businessofapps.com/data/android-statistics.
[2] https://www.appbrain.com/stats/number-of-android-apps.

X. Yang et al. (Eds.): ADMA 2023, LNAI 14180, pp. 122–136, 2023.
https://doi.org/10.1007/978-3-031-46677-9_9

common development mode for mobile applications. DevOps emphasizes continuous integration and user feedback analysis since user reviews play a significant role in fixing bugs, improving functions, optimizing performance, and driving innovation. For instance, a recent study [2] collected 4,193,549 user reviews of 623 mobile applications and discovered that 77% of user-proposed features were added to the next version of the application. However, developers of mobile applications receive hundreds of user reviews [3] from the application market every day, and manually processing each review can be time-consuming and costly. According to Reference [4], developers spend an average of 1.9 working days to locate the class that requires modification for software bugs mentioned in user reviews. Therefore, the automated processing of user reviews has become a crucial research area.

The current research on user reviews of applications encompasses various techniques such as classification [5–7], clustering [8,9], summary generation [10–12], and bug localization [3,4,13–16] and so on. Information Retrieval-based Bug Localization of User Reviews (UR-IRBL) is an important area in the automated processing of user reviews. This is because software testing before release may not be able to detect all software bugs, and developers need to continuously perform bug repairs and software updates based on user reviews.

Recent research has focused on two aspects of UR-IRBL. The first is how to obtain more accurate bug description information and source file representation information from user reviews and source files. The second is how to use a more accurate semantic matching scheme to find the buggy source codes with the extracted review information. Existing approaches for extracting information from user reviews typically rely on sentiment analysis, sentence component extraction, and clustering to represent user reviews as word sets [4]. Then use the Vector Space Model (VSM) or average word vectors to represent user reviews. For the representation of the source files, it is generally to use the identifiers of the source files, the comment information of commit history, and the resource information (string, configuration, permission, etc.) obtained by the static analysis of the APK. However, the current technologies contain two significant issues. Firstly, user reviews are written by non-professional users. These statements always contain too much irrelevant information and cannot describe software bugs explicitly. Secondly, these schemes require extensive expert knowledge and manual construction of word mappings, review patterns, or key API information for manual recognition.

To address these issues, we propose ReviewLocator. It extracts key semantic phrases from user reviews and source files through syntactic and code analyses. Then, it transfers the bug localization task to matching user review phrases to source file phrases. The contributions of our work are as follows:

- We propose and implement ReviewLocator which locates buggy source files with non-professional and vague bug descriptions in user reviews. It outperforms the baseline schemes ChangeAdvisor and Where2Change by 0.076 and 0.055 absolutely for MAP.

- We transfer bug localization task to matching user review phrases to source file phrases and propose a **Key Phrase-based Ranking (KPR)** algorithm that efficiently matches phrases to phrases.
- We propose **Bug Report-based Term Weight (BRTW)** using historical bug reports to focus on keywords that are more significant for bug localization in ReviewLocator.

The organization of the paper is as follows. Section 2 introduces related work on applications and research about user reviews. Section 3 describes the framework and details of ReviewLocator. Section 4 describes the dataset, metrics, research questions, and experimental results. Section 5 concludes this paper.

2 Related Work

Recent studies have focused on proposing different practical uses of user reviews or making changes to previous algorithms or pipelines to get better effectiveness. These studies can be broadly categorized into three categories – user review classification, key information extraction, and software bug localization.

2.1 User Review Classification

User review classification refers to categorizing user reviews into different types based on their content and context and then used for different feedback processing. By classifying user reviews, it becomes easier to process and analyze large volumes of feedback data and to prioritize the most pressing needs of users. Reference [17] used natural language analysis, text analysis, and sentiment analysis to automatically classify valid user feedback contained in app reviews. Reference [18] used Bayesian models, decision trees, and other techniques to classify app reviews into four types, namely bug reports, feature requests, user experience, and text ratings. In a similar vein, Reference [12] presented an automated method for classifying user reviews into bug reports and feature requests, which can serve as a processing basis for bug-tracking systems. In this work, text classification, natural language processing, and sentiment analysis are used in conjunction with several machine learning algorithms. Finally, Reference [19] proposed to use context-sensitive features of user reviews such as the total number of reviews, percentage of reviews per star, and review frequency to distinguish real user reviews from fake ones.

2.2 Key Information Extraction

Key information extraction refers to identifying and extracting the most relevant and significant information from a large number of user reviews. This step helps developers to focus on the most important issues that users are facing. Reference [20] proposed an algorithm to prioritize user reviews for developers to focus on in the next software version. Reference [21] adopted a keyword-based method to extract security-related sentences from user reviews using "misbehavior-aspect-opinion" triplets based on sentiment and sentence structure. Reference [9] proposed an adaptive online topic model for short texts and alleviate the sparsity problem, helping developers identify version-sensitive content in user reviews. Reference [22] introduced KEFE, a method for identifying key application features using application descriptions and user reviews. KEFE matches feature description phrases from application descriptions with related user reviews and uses a regression model to identify features with a significant impact on application ratings. Reference [23] proposed a method of generating software change request documents from user reviews about additions, customizations, deletions, bug localization, and improvements of software features.

2.3 Software Bug Localization

Software bug localization is to identify the software codes responsible for a bug based on user reviews. Reference [13] proposed a clustering method to identify topic words from user reviews and map them to related classes using Dice similarity. Reference [4] proposed the Where2Change method, which uses clustering to obtain word sets from user reviews and corrects the weights of different words using the mapping relationship between historical bug reports and source files. Where2Change also uses bug reports to enhance the word set representations of user reviews. Reference [3] proposed RISING, a bug localization method for user reviews that uses classification and clustering. RISING incorporates expert knowledge in the text preprocessing and clustering stages to ensure that the clustering results accurately reflect the semantic associations with the classes. Reference [14] proposed an approach that transfers the historical experience of an application from the same category to quickly discover software bugs in new applications. This is implemented by matching historical bug reports of applications in the same category with user reviews of the new application.

3 ReviewLocator Framework

3.1 Framework Overview

The overall framework of ReviewLocator is presented in Fig. 1. When dealing with a user review related to a software issue, ReviewLocator first needs to determine the corresponding version of the APK associated with that review. Since user reviews in application markets typically do not provide this information, ReviewLocator assumes the latest released version of the application software before the publish time of the review is the right version. Moreover, open-source software platforms such as F-droid[3], Github[4], APKPure[5], etc., make it possible to download different versions of APK.

[3] https://fdroid.org.
[4] https://github.com.
[5] https://apkpure.com.

Once the relevant APK is downloaded, ReviewLocator extracts key phrases from the user reviews that capture the essential semantics with sentence screening, syntax analysis tools, regular expressions, and general natural language processing tools. ReviewLocator then extracts multiple phrases as semantic representations from the APK with APK parsing tools, source code processing tools, and general natural language processing tools.

Additionally, ReviewLocator uses bug reports to calculate BRTW, which allows it to focus on important words when computing the matching score with different source files. Finally, ReviewLocator employs KPR to rank and recommend the source files to developers based on their relevance to the key phrases from user reviews.

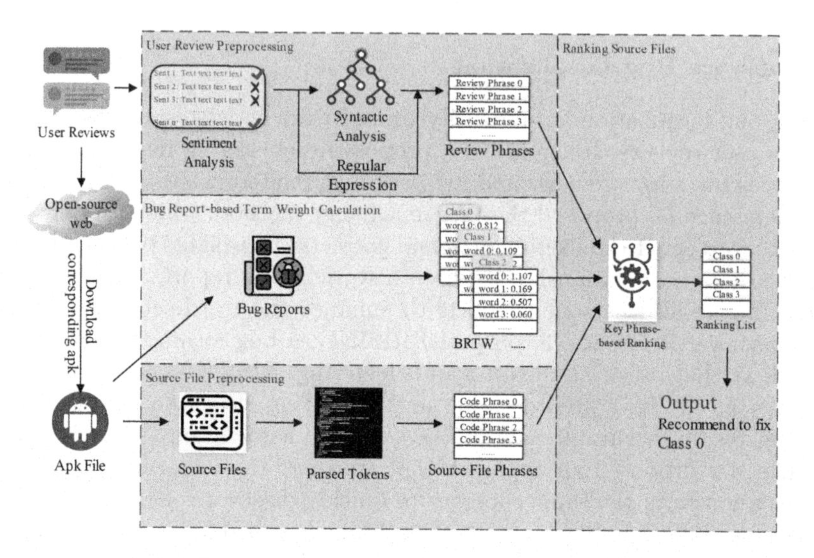

Fig. 1. Framework of ReviewLocator

3.2 User Review Preprocessing

User reviews are typically informal language inputs and may contain non-English characters or incorrect words. Therefore, we need some general preprocessing steps, including non-English character removal, stopwords removal, spelling correction, and word stemming. In user reviews, it's common to have multiple statements in a single review. ReviewLocator employs the approach provided by Reference [7] to filter out feature-request, information-seeking, and information-giving statements, which are irrelevant to software bugs. Furthermore, while filtering statements in user reviews, we also split compound statements. This is because different parts of a compound statement usually convey different meanings. For instance, the sentence "Sometimes crashes on my s3 but it is great otherwise" can be split into two parts using the word "but" as a delimiter: "Sometimes crashes on my s3" and "it is great otherwise." Compound statements are split according to coordinating conjunctions in the sentences.

Table 1. Dependencies from Stanford Parser

Tag	Meaning
nsubj	Noun subject
nsubjpass	Passive noun subject
dobj	Direct object
amod	Adjectival modifier
compound	Compound phrase

After completing the previous steps, statements that describe functional errors can be obtained. Software developers often use specific phrases to represent the semantics related to a class when naming identifiers in the source code. For instance, method names are usually verb phrases, and class names are typically noun phrases. Therefore, ReviewLocator extracts such phrases from user reviews. Firstly, syntactic parsing is performed on each sentence using the Stanford Parser[6]. The syntax tree and word dependency relationship of each sentence can be obtained from the syntactic parsing. Based on the word dependency relationship, phrases with relationship labels "nsubj", "nsubjpass", "obj", "amod", "compound" are selected. Table 1 provides the meanings of various dependency relationships and examples of phrases. Then, based on the syntax tree, rows marked as NP (Noun Phrases) are also selected.

We also take into account that user reviews may directly include descriptions that users see in the graphical interface of the application, which may directly match the strings in the source file. These descriptions are referred to as **Direct Reference (DR)**. For instance, in the sentence "but it just says 'unknown error' when I try to...", "unknown error" may be the prompt that the user sees in the graphical interface. Users often indicate Direct References with single or double quotes. Therefore, we use regular expressions to match all text enclosed in single or double quotes in user reviews as Direct References.

3.3 Source File Preprocessing

For APKs, ReviewLocator uses Jadx[7] and Javalang[8] to get class names, methods names, variable names, method calls, and strings for each class. These components are split into word lists by punctuation marks, spaces, and the camel case naming [24] rule. Especially, for some method names that only contain a verb, ReviewLocator adds an **Enhanced Semantic Phrase (ESP)** to the class, which consists of the method name and the class name. For instance, in the Cgeo application, the method name "create" in "cgeo.geocaching.ImagesActivity" can generate an ESP "create image activity" which effectively represents the activity related to creating images.

[6] https://nlp.stanford.edu/software/lex-parser.html.
[7] https://github.com/abdihaikal/pyjadx.
[8] https://github.com/c2nes/javalang.

3.4 Bug Report-Based Term Weight Calculation

User reviews of applications often describe function error information using non-professional language intangibly. Relying solely on user reviews may not lead to satisfactory bug localization results. A survey conducted by Reference [4] showed that 72.4% of developers still need to depend on historical bug reports to locate the source code classes related to user reviews. Therefore, detailed bug descriptions in historical bug reports can be used in conjunction with user reviews to achieve better software bug localization results. To this end, Reference [4] adjusted the weights of different words in TFIDF (Term Frequency - Inverse Document Frequency) vectors using bug reports and then merged the text of relevant bug reports and user reviews for bug localization. However, this method is time-consuming during the word weight iteration step and the simple merging of bug reports and user reviews introduces redundant information, leading to inaccurate bug localization. In practice, when using historical bug reports to assist in bug localization of user reviews, certain specific words are related to specific source files. For example, class "LocalStorage" in the application Cgeo is related to "backup". In this regard, ReviewLocator calculates the **Bug Report based Term Weight** (BRTW) using historical bug reports which means the word weights when matching different source files. The BRTW calculation process is shown in Equation (1):

$$BRTW(s_i, t_j) = \log\left(|G(s_i)| + 1\right) \cdot \frac{1}{|G(s_i)|} \sum_{b_k \in G(s_i)} tfidf(t_j, b_k), \tag{1}$$

where the variable s_i represents a source file, t_j represents a word, b_k represents a bug report, and $G(s_i)$ represents all bug reports related to s_i, $tfidf(t, d)$ are calculated according to Equation (2):

$$tfidf(t, d) = tf(t, d) \cdot idf(t), \tag{2}$$

$$tf(t, d) = \frac{\text{Number of } t \text{ in } d}{\text{Number of words in } d}, \tag{3}$$

$$idf(t) = \log \frac{\text{Number of docs}}{\text{Number of docs containing } t}, \tag{4}$$

where t is a word and d means a document. $tfidf(t, d)$ means the TFIDF value of term t in document d. The principle behind BRTW calculation is that if a particular word B frequently appears in the bug reports related to source file A, then it should be assigned a higher weight. However, when there are too few bug reports related to source file A, the calculated weight may not be reliable. Therefore, the number of bug reports needs to be considered to enhance the credibility of the weight. A higher value of $BRTW(s_i, t_j)$ indicates that t_j is more significant in calculating the similarity between user reviews and s_i. Theoretically, BRTW can be any non-negative number.

3.5 Ranking Source Files

To employ the key phrases extracted from user reviews and source codes for software bug localization, we introduce **Key Phrase-based Ranking (KPR)**, which is presented in Algorithm 1. KPR initially represents each key phrase in the user reviews as a vector. ReviewLocator uses the average word vector to represent key phrases.

Algorithm 1. Key Phrase-based Ranking

Input: General Parameters: RP: User review phrases, $SP = [SP^1, SP^2, \cdots, SP^k, \cdots, SP^K]$: Source file phrases, K is the number of source files, SP^k is the phrases of source file s_k.
Hyperparameters: h: Similarity threshold, α: Adjust the effect of BRTW;
Output: $classList$: A list of classes according to the suspicious score with user review;

1: $v_i = avgVector(RP_i)$;
2: $w_i = avgVector(SP_i)$;
3: $avgSim_i^j = cosineSimilarity(v_i, w_j)$;
4: $matchValue = \{\}$;
5: **for** k in $[1, K]$ **do**
6:   **for** i in $[1, len(RP)]$ **do**
7:     **for** j in $[1, len(SP^k)]$ **do**
8:       **if** $avgSim_i^j \geq h$ **then**
9:         $phraseWeight = 0$;
10:         **for** $term$ in RP_i **do**
11:           $phraseWeight = phraseWeight + BRTW[term]$;
12:         **end for**
13:         $matchValue[k] = matchValue.get(k, 0) + 1 + \alpha \cdot phraseWeight$;
14:       **end if**
15:     **end for**
16:   **end for**
17: **end for**
18: $classList \leftarrow$ Ranking the keys in $matchValue$ according to corresponding values;
19: **return** $classList$;

Algorithm 1 illustrates the KPR process. To begin, KPR vectorizes the input phrases (lines 1–2). Then, it computes the cosine similarity between different phrases (line 3). After that, KPR calculates the similarity between **Review Phrase and Source file Phrase pairs (RPSP)** from different reviews and source files. If the similarity is greater than the threshold, the RPSP is considered similar and used for ranking. The importance of an RPSP is obtained by summing up the BRTW of all words in the comment word group (line 11). The importance of a **Review and Source Pair (RSpair)** is obtained by summing up the weights of all RPSPs in the RSpair (line 13). Each related RPSP increases the suspiciousness of the source file (1 in line 13), and the importance of the RPSP is calculated according to the BRTW of the review phrases ($\alpha \cdot phraseWeight$ in line 13).

4 Experimental Settings

4.1 Dataset

To assess the efficacy of ReviewLocator, we performed experiments on a dataset collected by Reference [16]. The dataset comprises user reviews from the Google Play Store, with mappings between software bugs described in the reviews and the source files of the corresponding applications annotated by professional software engineering experts using bug reports and commit records by software developers. The dataset covers user reviews of eight distinct applications spanning diverse domains. Table 2 provides specific details of this dataset.

Table 2. User Review Dataset

Application	Number of User Reviews	Number of Source Files	Type
Cgeo	179	577	Map
SeriesGuide	221	402	Video
K-9 Mail	159	412	Email
OneBusAway	146	243	Traffic
AntennaPod	82	702	Music
Twidere	247	2004	Blog
Signal	204	707	Chat
WordPress	298	630	Media

4.2 Metrics

To evaluate the performance of ReviewLocator and other UR-IRBL schemes, we choose Top-k accuracy (Top-k), Mean Reciprocal Rank (MRR), and Mean Average Precision (MAP) as evaluation metrics. The three metrics are widely used for evaluation in retrieval systems and recommendation systems.

(1) Top-k: It is the proportion that UR-IRBL schemes successfully recommend the right source files at the top k results.
(2) MRR: It is the mean of the reciprocals of the best ranks. MRR is calculated as follows:

$$MRR = \frac{1}{N}\sum_{i=1}^{N}\frac{1}{Rank_i}, \tag{5}$$

where N means the number of user reviews and $Rank_i$ means the best rank of the ith user review with corresponding relevant source files.
(3) MAP: It is the mean of the average precisions (AP) of all user reviews. MAP is calculated as follows:

$$MAP = \frac{1}{N}\sum_{i=1}^{N}AP_i, \tag{6}$$

$$AP_i = \frac{1}{M_i}\sum_{j=1}^{K}\frac{T_{ij} \times Q_{ij}}{j}, \tag{7}$$

where N means the number of user reviews, M_i means the number of the relevant source files of the ith review, K means the number of all source files, T_{ij} means the number of relevant files in the top j files from the ranking list of the ith review, Q_{ij} is 1 if the jth file in the rank list of the ith review is the relevant file else 0.

4.3 Research Questions and Experiments

To validate the efficacy of ReviewLocator, we put forth several research questions and conduct corresponding experiments to answer each question.

(1) How does the performance of ReviewLocator compare to other UR-IRBL techniques?
Up to now, there are four typical benchmark techniques in UR-IRBL. In this study, ChangeAdvisor [13] and Where2Change [4], are selected. The other two are RISING [3] and ReviewSolver [16], which depend on expert knowledge in some procedures. So we do not choose the two techniques for comparison. ChangeAdvisor locates bugs by computing the Dice similarity between the sets of words in user reviews and source file classes, while Where2Change locates bugs by the weighted average vectors. The weights of words are calculated iteratively based on historical bug reports. ReviewLocator, ChangeAdvisor, and Where2Change were then applied to the dataset introduced in Sect. 4.1 for bug localization. The h parameter in ReviewLocator is set to 0.68, following the setting in Reference [25], and the α parameter is set to 34, obtained through linear search.

Table 3. Compare ReviewLocator with ChangeAdvisor [13] and Where2Change [4]. (RLoc: ReviewLocator, CA: ChangeAdvisor, W2C: Where2Change)

Application	Reviews	Maps	Top-1			MRR			MAP		
			RLoc	CA	W2C	RLoc	CA	W2C	RLoc	CA	W2C
Cgeo	179	1147	**35**	23	28	**0.302**	0.216	0.262	**0.178**	0.092	0.144
SeriesGuide	221	1545	**43**	6	12	**0.316**	0.098	0.141	**0.129**	0.037	0.057
K-9 Mail	159	591	**17**	7	12	**0.207**	0.099	0.161	**0.106**	0.042	0.071
OneBusAway	146	428	**17**	0	10	**0.356**	0.177	0.224	**0.147**	0.078	0.071
AntennaPod	82	422	**28**	7	11	**0.462**	0.178	0.290	**0.218**	0.060	0.129
Twidere	247	2874	**65**	22	18	**0.338**	0.177	0.143	**0.062**	0.028	0.025
Signal	204	1387	56	39	**60**	**0.429**	0.274	0.394	**0.129**	0.078	0.115
WordPress	298	3146	**115**	39	34	**0.499**	0.244	0.211	**0.139**	0.061	0.057
Sum	1536	11540	**376**	143	185	–	–	–	–	–	–
Average	–	–	–	–	–	**0.364**	0.183	0.228	**0.138**	0.060	0.083

Table 3 compares the experimental results of ReviewLocator (RLoc), ChangeAdvisor (CA), and Where2Change (W2C). Overall, RLoc outperforms CA and W2C with the highest Top-1, MRR, and MAP performance. When examining each technique's performance individually on each application, RLoc demonstrates superior performance in all applications except for Signal, where its localization performance slightly lags behind that of W2C.

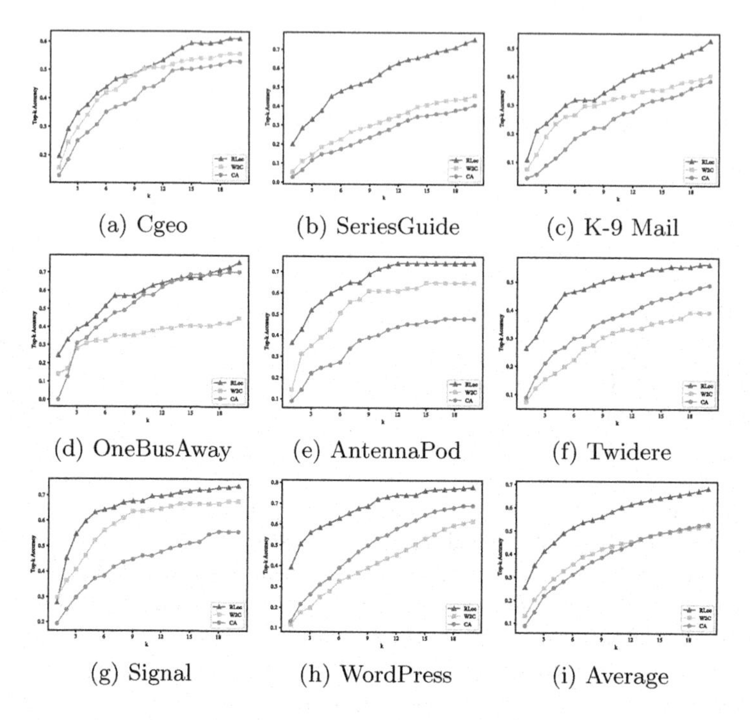

Fig. 2. Compare Top-k of ReviewLocator with Where2Change and ChangeAdvisor on Eight Applications (Figures (a-h) show the Top-k of each application.

Figure 2 illustrates the Top-k accuracy of ReviewLocator (RLoc), ChangeAdvisor (CA), and Where2Change (W2C) on eight software applications, with k ranging from 1 to 20. Overall, RLoc demonstrates a higher accuracy than both CA and W2C across all eight applications.

(2) How is the hyperparameter α in Algorithm 1 selected?
Algorithm 1 contains two hyperparameters, with h set to 0.68 based on the work of Reference [25]. (We did experiments to verify the effects of different h and found that $(0.60 - 0.76)$ is a suitable value range. The results would not change much. Due to the space limitation, we do not talk in detail in this paper.) To determine the optimal value of α, we test α with values from 0 to 50 using a step size of 2. The effectiveness of ReviewLocator is evaluated on the dataset introduced in Sect. 4.1.

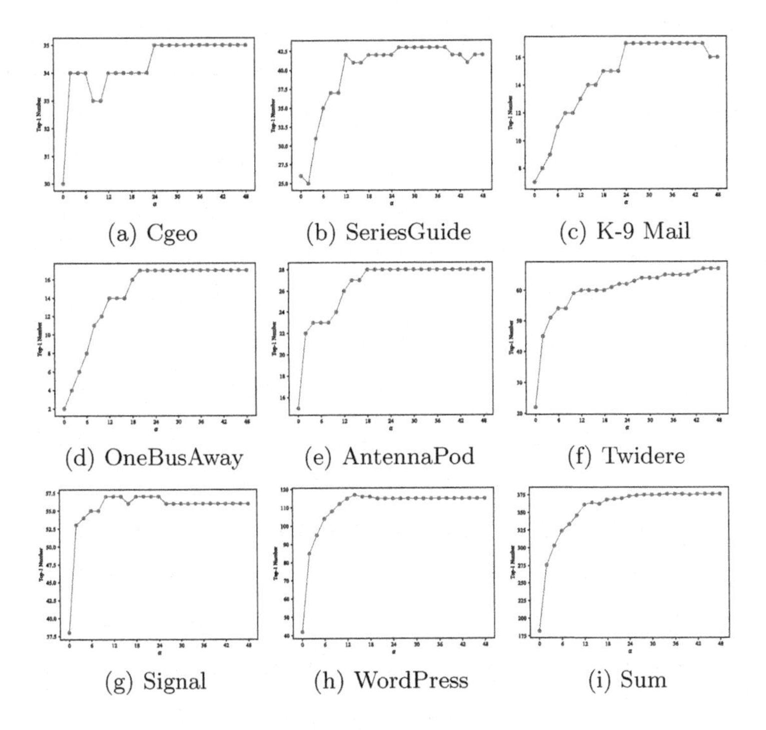

(a) Cgeo (b) SeriesGuide (c) K-9 Mail
(d) OneBusAway (e) AntennaPod (f) Twidere
(g) Signal (h) WordPress (i) Sum

Fig. 3. Results of ReviewLocator with Different α (The horizontal axis means different α and the vertical axis means corresponding Top-1 hit numbers.

Figure 3 depicts the Top-1 hit counts of ReviewLocator on eight applications at various values of α. Overall, the Top-1 hit counts of ReviewLocator initially increase and then keep almost unchanged as α increases across these applications. Therefore, we determined the optimal value of α by selecting the one that yielded the highest total Top-1 hit count across all eight applications ($\alpha = 34$). As illustrated in Fig. 3i, the number of user reviews that ReviewLocator can hit increases gradually as α increases, and it reaches its maximum value at $\alpha = 34$, after which it stabilizes. Consequently, we used $\alpha = 34$ as the optimal configuration for ReviewLocator in this study.

(3) What is the significance of BRTW in ReviewLocator?
To investigate the role of BRTW in ReviewLocator, we evaluate the bug localization performance of a variant of ReviewLocator called RLoc-BRTW, which does not use BRTW which means setting the hyperparameter α in Algorithm 1 as 0. The performance of ReviewLocator is compared with that of RLoc-BRTW to validate the importance of BRTW in ReviewLocator.
Table 4 displays the results of RLoc and its variants without using BRTW in locating reviews of eight applications. The results indicate that RLoc performs much better than RLoc without BRTW. The total Top-1 hit count of RLoc is 194 more than that of RLoc-BRTW, which accounts for approximately 51.6%. The overall average MRR and MAP of RLoc-BRTW are

Table 4. Compare RLoc with RLoc-BRTW. "RLoc" Means ReviewLocator, "RLoc-BRTW" Means ReviewLocator without BRTW.

Application	Top-1		MRR		MAP	
	RLoc	RLoc-BRTW	RLoc	RLoc-BRTW	RLoc	RLoc-BRTW
Cgeo	35	30	0.302	0.283	0.178	0.167
SeriesGuide	43	26	0.316	0.246	0.129	0.095
K-9 Mail	17	7	0.207	0.129	0.106	0.076
OneBusAway	17	2	0.356	0.179	0.147	0.082
AntennaPod	28	15	0.462	0.311	0.218	0.129
Twidere	65	22	0.338	0.172	0.062	0.034
Signal	56	38	0.429	0.312	0.129	0.094
WordPress	115	42	0.499	0.312	0.139	0.082
Sum(S)—Average(A)	376(S)	182(S)	0.364(A)	0.243(A)	0.138(A)	0.095(A)

relatively reduced by 33.2% and 31.5% compared to RLoc, respectively. Thus, it can be concluded that BRTW is the key factor in achieving good bug localization results in RLoc.

However, the calculation of BRTW depends on the historical bug reports, which may not be available for some applications to learn BRTW. For a brand-new application, ReviewLocator can only run its variant version RLoc-BRTW. In this case, the bug localization effect is equivalent to that of Where2Change.

5 Conclusion

User review is a crucial factor in the development and enhancement of applications. One key application of user review is identifying and locating software bugs. However, user reviews often contain too much irrelevant information, and the description of software bugs is vague. To address these issues, we propose ReviewLocator. It extracts key phrases from user reviews and matches user reviews to source files using a novel ranking method namely KPR with these key phrases. For the first time, we calculated BRTW with historical bug reports to achieve efficient ranking even if the reviews are inexplicable. Our experiments on user reviews from eight applications show that ReviewLocator exceeds previous baselines ChangeAdvisor and Where2Change by an absolute improvement of 0.076 and 0.055 in terms of MAP correspondingly. In the future, we consider utilizing API interface description information and developer comment information committed with source files to calculate keyword weights. This approach may overcome the cold start problem and further improve the performance of ReviewLocator.

Acknowledgment. This work was supported in part by the National Natural Science Foundation of China (61972219), the Research and Development Program of Shenzhen (JCYJ20190813174403598), the Overseas Research Cooperation Fund of Tsinghua Shenzhen International Graduate School (HW2021013), the Guangdong Basic and Applied Basic Research Foundation (2022A1515010417), the Key Project of Shenzhen Municipality (JSGG20211029095545002), the Science and Technology Research Project of Henan Province (222102210096), the Key Laboratory of Network Assessment Technology, Institute of Information Engineering, Chinese Academy of Sciences, and Shenzhen Science and Technology Innovation Commission (Research Center for Computer Network (Shenzhen) Ministry of Education).

References

1. Leite, L., Rocha, C., Kon, F., Milojicic, D., Meirelles, P.: A survey of devops concepts and challenges. ACM Comput. Surv. **52**(6) (2019)
2. Noei, E., Zhang, F., Zou, Y.: Too many user-reviews! what should app developers look at first? IEEE Trans. Software Eng. **47**(2), 367–378 (2021)
3. Zhou, Y., Su, Y., Chen, T., Huang, Z., Gall, H., Panichella, S.: User review-based change file localization for mobile applications. IEEE Trans. Softw. Eng. **47**(12) (2021)
4. Zhang, T., Chen, J., Zhan, X., Luo, X., Lo, D., Jiang, H.: Where2change: change request localization for app reviews. IEEE Trans. Softw. Eng. **47**(11), 2590–2616 (2021)
5. Ciurumelea, A., Schaufelbühl, A., Panichella, S., Gall, H.C.: Analyzing reviews and code of mobile apps for better release planning. In: 24th International Conference on Software Analysis. Evolution and Reengineering, pp. 91–102. IEEE, Klagenfurt (2017)
6. Kurtanović, Z., Maalej, W.: On user rationale in software engineering. Requir. Eng. **23**, 357–379 (2018)
7. Panichella, S., Di Sorbo, A., Guzman, E., Visaggio, C.A., Canfora, G., Gall, H.C.: How can i improve my app? classifying user reviews for software maintenance and evolution. In: Proceedings of the 2015 IEEE International Conference on Software Maintenance and Evolution, pp. 281–290. IEEE Computer Society, Bremen (2015)
8. Chen, R., Wang, Q., Xu, W.: Mining user requirements to facilitate mobile app quality upgrades with big data. Electron. Commer. Res. Appl. **38**(C), 100889 (2019)
9. Hadi, M.A., Fard, F.H.: Aobtm: adaptive online biterm topic modeling for version sensitive short-texts analysis. In: International Conference on Software Maintenance and Evolution, pp. 593–604. IEEE, Adelaide (2020)
10. Di Sorbo, A., Panichella, S., Alexandru, C.V., Visaggio, C.A., Canfora, G.: Surf: summarizer of user reviews feedback. In: Proceedings of the 39th International Conference on Software Engineering Companion, pp. 55–58. IEEE Press, Buenos Aires (2017)
11. Gao, C., et al.: Listening to users' voice: automatic summarization of helpful app reviews. IEEE Trans. Reliab. 1–13 (2022)
12. Phetrungnapha, K., Senivongse, T.: Classification of mobile application user reviews for generating tickets on issue tracking system. In: 12th International Conference on Information & Communication Technology and System, pp. 229–234. IEEE, Surabaya (2019)

13. Palomba, F., et al.: Recommending and localizing change requests for mobile apps based on user reviews. In: Proceedings of the 39th International Conference on Software Engineering, pp. 106–117. IEEE Press, Buenos Aires (2017)
14. Tang, X., Tian, H., Kong, P., Liu, K., Klein, J., Bissyande, T.F.: App review driven collaborative bug finding. arXiv preprint arXiv:2301.02818 (2023)
15. Yu, L., Chen, J., Zhou, H., Luo, X., Liu, K.: Localizing function errors in mobile apps with user reviews. In: 48th Annual IEEE/IFIP International Conference on Dependable Systems and Networks, pp. 418–429. IEEE, Luxembourg (2018)
16. Yu, L., et al.: Towards automatically localizing function errors in mobile apps with user reviews. IEEE Trans. Softw. Eng. **49**(4), 1464–1486 (2022)
17. Panichella, S., Di Sorbo, A., Guzman, E., Visaggio, C.A., Canfora, G., Gall, H.C.: Ardoc: app reviews development oriented classifier. In: Proceedings of the 2016 24th ACM SIGSOFT International Symposium on Foundations of Software Engineering, pp. 1023–1027. Association for Computing Machinery, Seattle (2016)
18. Maalej, W., Kurtanović, Z., Nabil, H., Stanik, C.: On the automatic classification of app reviews. Requir. Eng. **21**(3), 311–331 (2016)
19. Martens, D., Maalej, W.: Towards understanding and detecting fake reviews in app stores. Empir. Softw. Eng. **24**(6), 3316–3355 (2019)
20. Mahmud, O., Niloy, N.T., Rahman, M.A., Siddik, M.S.: Predicting an effective android application release based on user reviews and ratings. In: 7th International Conference on Smart Computing & Communications, 1–5. IEEE, Sarawak (2019)
21. Tao, C., Guo, H., Huang, Z.: Identifying security issues for mobile applications based on user review summarization. Inf. Softw. Technol. **122**, 106290 (2020)
22. Wu, H., Deng, W., Niu, X., Nie, C.: Identifying key features from app user reviews. In: Proceedings of the 43rd International Conference on Software Engineering, pp. 922–932. IEEE Press, Madrid (2021)
23. Nadeem, M., Shahzad, K., Majeed, N.: Extracting software change requests from mobile app reviews. In: 36th IEEE/ACM International Conference on Automated Software Engineering Workshops, pp. 198–203. IEEE, Melbourne (2021)
24. Binkley, D., Davis, M., Lawrie, D., Morrell, C.: To camelcase or under_score. In: Proceedings of the 17th International Conference on Program Comprehension, pp. 158–167. IEEE, Vancouver (2009)
25. Qu, Z., Rastogi, V., Zhang, X., Chen, Y., Zhu, T., Chen, Z.: Autocog: measuring the description-to-permission fidelity in android applications. In: Proceedings of the 2014 ACM SIGSAC Conference on Computer and Communications Security, pp. 1354–1365. Association for Computing Machinery, Scottsdale (2014)

Improving Adversarial Robustness via Channel and Depth Compatibility

Ruicheng Niu[1,2], Ziyuan Zhu[1,2](✉), Tao Leng[1,2], Chaofei Li[1,2], Yuxin Liu[1,2], and Dan Meng[1,2]

[1] Institute of Information Engineering, Chinese Academy of Sciences, Beijing 100085, China
zhuziyuan@iie.ac.cn
[2] School of Cyber Security, University of Chinese Academy of Sciences, Beijing 100049, China

Abstract. Several deep neural networks are vulnerable to adversarial samples that are imperceptible to humans. To address this challenge, a range of techniques have been proposed to design more robust model architectures. However, previous research has primarily focused on identifying atomic structures that are more resilient, while our work focuses on adapting the model in two spatial dimensions: width and depth. In this paper, we present a multi-objective neural architecture search (NAS) method that searches for optimal widths for different layers in spatial dimensions, referred to as DW-Net. We also propose a novel adversarial sample generation technique for one-shot that enhances search space diversity and promotes search efficiency. Our experimental results demonstrate that the proposed optimal neural architecture outperforms state-of-the-art NAS-based networks widely used in the literature in terms of adversarial accuracy, under different adversarial attacks and for different-sized tasks.

Keywords: Robust architecture · Neural architecture search · Deep learning

1 Introduction

Deep neural networks (DNNs) have shown great success in a wide range of complex tasks, including target recognition and natural language processing [5, 11,23]. However, recent studies have highlighted a significant weakness of DNNs - their vulnerability to adversarial samples [7,21]. This poses a critical challenge for the deployment of DNNs in safety-critical applications, such as autonomous driving [6].

Several methods have been proposed to defend against adversarial samples, including denoising [13], security distillation [18], adversarial training [16], and others [3,9]. Adversarial training(AT), in particular, is considered to be one of the safest and most effective defense methods. This approach involves optimizing the weights by including adversarial samples in the training set.However,

X. Yang et al. (Eds.): ADMA 2023, LNAI 14180, pp. 137–149, 2023.
https://doi.org/10.1007/978-3-031-46677-9_10

current techniques for defending against adversarial samples focus primarily on weight optimization, while neglecting the potential contribution of the network architecture to robustness. Recent studies have shown that the network topology is closely related to intrinsic robustness [2,16]. A well-designed network architecture can significantly enhance the model's resilience against adversarial attacks.Therefore, we believe that a proper network topology plays a critical role in achieving superior intrinsic robustness.

Prior research in the field of robust neural network architectures has relied on expert knowledge and collaboration to design networks capable of withstanding specific challenges. However, such an approach is not scalable due to its dependence on extensive trial and error by domain experts. As an alternative, Neural Architecture Search (NAS) [30] has been proposed as a means of generating resilient networks. Recent work has highlighted the importance of cell topology in the search for more reliable models [10,17]. The spatial structure of a neural network model has been shown to have a significant impact on its robustness, as demonstrated in recent research [4]. To delve deeper into the relationship between topology and robustness, this study aims to investigate the design of a more resilient neural network topology in terms of the spatial properties of the model architecture.

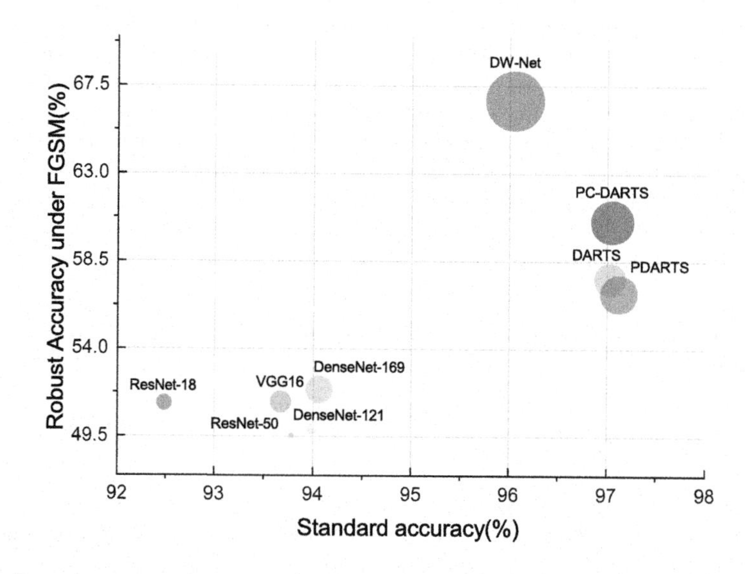

Fig. 1. Quantitative analysis of clean accuracy and adversarial accuracy on CIFAR10 data.

In this work, if the correlation between the spatial properties (width and depth) of a model and its robustness can be learned, then a more robust model topology can be designed. Our search strategy is motivated by the observation that different convolutional layer depths have different perceptual fields and feature information requirements and that a mismatch between these two spatial features can result in a massive amount of redundant data in the processed feature maps. In contrast to previous approaches, we emphasize exploring for the optimal depth and width of the model to enhance the model's robustness. Our proposed method significantly enhances the model's robustness, as seen in Fig. 1. Our main contributions are summarized as follows:

(1) Our proposed search strategy, which adjusts widths for different depth filters, reduces the amount of redundant information passed, rendering the model less prone to overfitting and enabling the search for a more robust network.

(2) We propose a new search objective based on Lipschitz's constant, which reduces the search error caused by insufficient model convergence during the one-shot search. And the robustness of different stages can be evaluated based on it.

(3) We propose a novel adversarial sample generation technique for a one-shot that dramatically reduces the number of parameters used in backpropagation, enhances the effectiveness of the search and diversifies the search space.

2 Related Work

Adversarial Attack. Adversarial samples are a well-known problem in the field of deep learning, where small perturbations are added to an input image, rendering it unrecognizable to humans but causing the model to misclassify it with high confidence. Adversarial attacks can take various forms, including LBGS [23], FGSMFGSM [8], PGD [23], and others [12,22]. In this paper, all experiments are conducted on white-box attacks, where the attacker has full access to the model's architecture and parameters.

Adversarial Defense. Over the years, numerous adversarial defense strategies [1] have been proposed in response to the increasing threat of adversarial attacks. Among these defense mechanisms, adversarial training [16,24] has been shown to be one of the most effective. Adversarial training involves solving a min-max optimization problem:

$$\min_{\theta} \max_{\boldsymbol{x}' \in \mathcal{B}_\epsilon(\boldsymbol{x})} \mathcal{L}\left(\mathcal{F}\left(\boldsymbol{x}', \theta\right), y\right) \tag{1}$$

$\mathcal{F}$ represents a DNN,x is a clean sample with label y, x' is the adversarial sample of x manufactured by the Lp−norm attack, $\mathcal{F}(x', \theta)$ is the output of the network, and $\mathcal{L}$ is the classification loss. Despite the effectiveness of current state-of-the-art adversarial training methods, such as TRADES [28], DART [25], and FAT [29], they do not take into account the impact of model architecture on inherent robustness. While other defense mechanisms such as defense distillation and gradient confusion have been proposed, they also neglect this important aspect.

Neural Architecture Search(NAS) and Robustness. NAS has emerged as a promising approach for automating the design of neural network topologies. Depending on the search algorithm, NAS can be classified into three categories: reinforcement learning-based algorithms like ENAS [19], evolutionary algorithms like NSGA-NET [15], and differentiable algorithms like DRATS [14]. Recent research has shown that the topology of neural networks is inherently robust, and NAS has been proposed as a tool for discovering more resilient architectures. RNAS [20], a DARTS-based algorithm, has been proposed for discovering more robust neural networks, while other studies [26] have focused on constructing lightweight robust neural networks through new atomic structure discovery. However, further research is needed to understand how network spatial properties such as width and depth impact the robustness of the architecture.

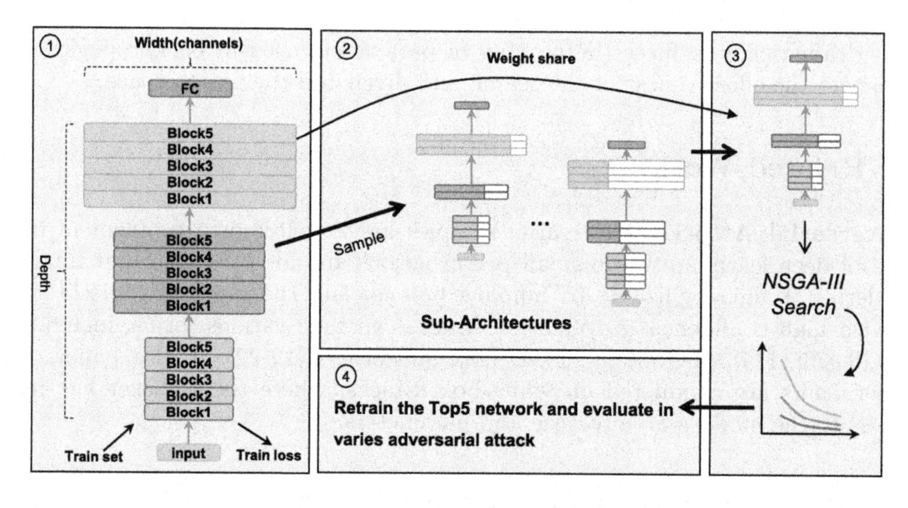

Fig. 2. The overall framework of DW-Net.

3 Dynamic Adversarial Multi-objective One-Shot NAS

Our methodology is comprised of three main steps. The first step involves training a supernet, which uses a single-path training strategy to eliminate coupling between each operation of the search space. In the second step, we conduct a robust architecture search by sampling subnets from the supernet and inheriting parameters from it to calculate the fitness. We then use the NSGA-III evolutionary algorithm to optimize the architecture parameters. Finally, in the third step, we retrain the Top-5 network and carry out adversarial attack validation. The entire system is illustrated in Fig. 2. It is important to note that all of the validation tests conducted in this study are done without any adversarial training. In the following sections, we will discuss the details of each step.

3.1 Supernet Training

In the pursuit of discovering robust network structures, the design of the search space plays a crucial role. The search space in this paper is referenced to ResNet, WRN, and NSGA-Net. Our proposed DW-Net explores how the intricate topology of the model influences its vulnerability to adversarial attacks.

Problem Definition. Our search space S can be represented by a directed graph, denoted as $F(S,W)$, W represents the parameters of the supernet. The sampled subnet s, is a subgraph $s \in S$. A single-path training approach is utilized to diminish the coupling of each operation in the supernet. $\Gamma(S)$ is a prior distribution of $s \in S$. Each iteration selections a path from $\Gamma(S)$ for training, denoted as $N(s, W(s))$. The supernet optimization can be expressed as follows:

$$W_s = \operatorname{argmin} E_{s \sim \Gamma(\mathrm{S})}\left[L_{\text{train}}\left(F(s, W(s))\right)\right] \tag{2}$$

3.2 Robustness Architecture Search

Initialization. In the evolutionary search, the neural network structure is encoded in an N-dimensional space, where N represents the number of choice blocks. We initialize a D-sized population, each representing a different neural network architecture, expressed as:

$$S_d = \{l_1, l_2, l_3, \ldots, l_N\} \tag{3}$$

l_n represents the state of the n-th selection block. Its value range is $[0, 1.5 \times C]$, and C represents the number of channels preset in the current layer. When the value of l_n is less than $0.15 \times C$, the Indentity operation is selected. When the value of l_n is in $[0.15 \times C, 1.5 \times C]$, the operation of the current layer is maintained and the number of channels in the current block is dynamically modified according to the value of l_n. Note that the first layer of each stage belongs to the downsampling layer, keep its channel number greater than $0.15 \times C$.

Objective Function. The adversarial accuracy is one of the search targets in the current method for finding robust neural network architectures. The sampled subnets would not, however, accurately reflect the robustness of the search space since the insufficient convergence of the search space(Overfitting is the primary cause of the adversarial samples). To diminish the search's impact of non-convergence. PGD_lip, FGSM_lip, and clean error are the search objectives in this resarch.The first two targets are Lipschitz constants about the subnet under FGSM and PGD attacks. The Lipschitz constants are calculated as follows:

$$\|f_s(x) - f_s(\boldsymbol{x}')\| \leq L \|x - x'\| \tag{4}$$

$f_s(x)$ represents the output of subnet,x represents the natural sample,and x' represents the adversarial sample. Recent research has demonstrated that robust models typically exhibit a small Lipschitz constant, which is a reliable indicator of a model's ability to withstand adversarial attacks. The Lipschitz constant is an effective measure of robustness since it captures changes in input and output, thereby mitigating the impact of incomplete supernet convergence on the search for robust models. The clean error measures the prediction error rate of the model on the original data, and the PGD_lip and FGSM_lip metrics assess the model's resilience against adversarial samples. By minimizing these three objectives, the challenge of balancing accuracy rate and robustness is alleviated, and the resulting architecture is capable of defending against a variety of attacks. It is express the multi-objective optimization as:

$$\begin{aligned} &\min \{fitness_1(s), fitness_2(s), fitness_3(s)\} \\ &\text{s.t. } s \in S \end{aligned} \tag{5}$$

Multi-objective Search. In this work, the NSGA-III technique was employed as the multi-objective search method. Compared to NSGA-II, NSGA-III exhibits superior convergence and diversity in multi-objective optimization problems with more than three objectives and is less likely to converge to a local optimum. A critical requirement for the search process is that the performance of the sampled subnets, which utilize inherited weights from the supernet without additional fine-tuning or training from scratch, should be highly predictive. It expresses as follows:

$$F_i(\mathrm{s}) = ACC_{\mathrm{val}}\left(N\left(\mathrm{s}, W_S(\mathrm{s})\right)\right) \tag{6}$$

Referring to the format of Eq. 6, effectively computing the remaining fitness functions reduces the time complexity of the search.

Generation of Adversarial Samples. To improve the efficiency of the search process, we introduce a new method for generating adversarial samples inspired by YOPO [27]. Traditional energy-based approaches are not suitable for this task due to the large number of required adversarial samples. YOPO assumes that the adversarial perturbation is only related to the weights of the first layer. To further reduce the number of forward and backward propagations, we incorporate a slack variable into the method:

$$\begin{aligned} p = \nabla_{s_{\tilde{W(s)}}} &\left(\ell\left(s_{\tilde{W(s)}}\left(f_0\left(x_i+\eta_i^{j,0}, W(s)_0\right)\right), y_i\right)\right) \cdot \\ &\nabla_{f_0}\left(s_{\tilde{W(s)}}\left(f_0\left(x_i+\eta_i^{j,0}, W(s)_0\right)\right)\right) \end{aligned} \tag{7}$$

f_0 denotes the first layer, $s_{\tilde{W(s)}}$ represents the remaining layers. Freeze p as a constant when generating adversarial samples to verify subnet robustness. According to this scheme, the processes for manufacturing adversarial samples via PGD are modified, expressed as follows:

$$\eta_i^{j,s+1} = \eta_i^{j,s} + \alpha_1 p \cdot \nabla_{\eta_i} f_0\left(x_i + \eta_i^{j,s}, W(s)_0\right), i = 1, \cdots, B \tag{8}$$

The modified PGD adversarial sample generation requires only one backpropagation on the entire subnet, and all the remaining iterations will be done on the first layer of the subnet. Our proposed method drastically reduces the time consumed by generating adversarial samples during the search. With the limited resource consumption, it can make the search space more diverse.

Retrain. After the search process was completed, the top 5 architectures were selected and subjected to a retraining phase using normal training procedures. Subsequently, their robustness was assessed in relation to the specified attack methodology, in order to demonstrate that the neural network topologies developed in this study are inherently robust without the need for adversarial training.

Table 1. Architecture of the entire search space

Input shape	Block	Channels	Repeat	stride
$32 \times 32 \times 3$	3×3 Conv	16	1	1
$32 \times 32 \times 16$	CB	160	9	2
$16 \times 16 \times 160$	CB	320	9	2
$8 \times 8 \times 320$	CB	640	9	2
$4 \times 4 \times 640$	GAP		1	
640	FC	10	1	

4 Experiments

4.1 Data Set and Experimental Implementation

The two datasets that are most frequently used to assess model robustness, CIFAR10, and CIFAR100, were employed for the experiments in this work. 60,000 images from CIFAR10 were all 32×32 pixels in resolution. There are 10,000 test photos and 50,000 training images altogether among these ten classes of images. Collected data from 100 distinct classes are included in CIFAR100. Each class has 600 32×32 RGB pictures, of which 500 are used for training and 100 are utilized for testing.

The search for the research's architecture is autonomous, and the top-5 model that is eventually discovered will be retrained and applied for validation. NSGA-III was used to perform the search with populations of 20 and 50 evolutions. The whole search space is represented in Table 1. The selection block is represented by CB. For adversarial robustness, we test against standard attacks like FGSM and PGD. For all these attacks, we use a perturbation value of $8/255(0.03)$. The step size is $2/255$ (0.007) with the attack iterations set as 10.

Table 2. Adversarial success and natural accuracy of FGSM on CIFAR10.

Model	Clean(%)	FGSM(%)
ResNet-18	92.48	51.22
ResNet-50	93.78	49.55
DenseNet-121	93.98	49.88
DenseNet-169	94.06	51.91
VGG16 BN	93.67	51.29
DARTS	97.03	57.61
PDARTS	97.12	56.87
PC-DARTS	97.05	60.55
DW-Net(ours)	96.05	**66.75**

4.2 Results on CIFAR10

Table 2 exhibits the adversarial accuracy of our proposed DW-Net against FGSM attacks. DW-Net achieves an impressive adversarial accuracy of 67%, which is 6.2% higher than the state-of-the-art NAS approach. Furthermore, compared to the hand-crafted model, DW-Net outperforms the adversarial accuracy by a significant margin of 14.84%. The unique feature of DW-Net lies in its ability to consider channel activation differently for adversarial and practical data. The network's depth and width of each layer are optimized dynamically based on the complexity of the task to reduce the redundancy of intermediate features. These design choices lead to a substantial improvement in the inherent robustness of the architecture, while minimizing the risk of model overfitting. Overall, our

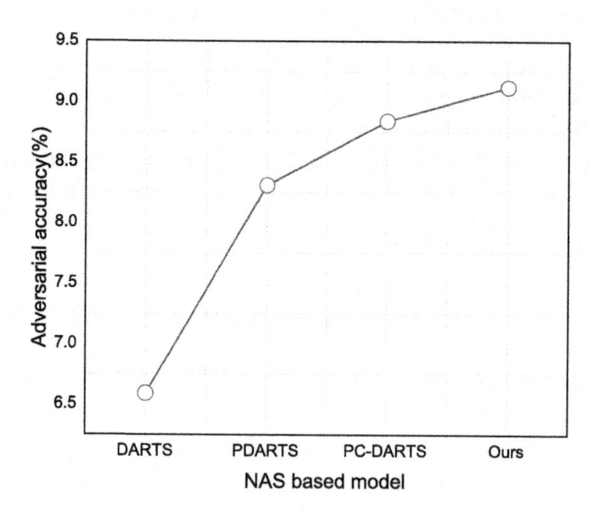

Fig. 3. Quantitative analysis of PGD adversarial accuracy on CIFAR10 data.

results demonstrate that DW-Net is a promising approach for enhancing the robustness of neural network models against adversarial attacks. Fig. 3 displays the adversarial accuracy of PGD on CIFAR10. The results indicate that DW-Net outperforms the NAS-based model. However, as the attack's complexity increases, hand-crafted architectures tend to be more robust than DW-Net. Our findings suggest that DW-Net demonstrates superior architectural robustness over other models when subjected to small-scale attacks such as FGSM, with a considerable margin.

4.3 Experimental Results on CIFAR100

Based on the results presented in Table 3, DW-Net shows superior performance compared to other models in terms of robustness against FGSM on more complex tasks, surpassing the NAS-based model by 4.57% and the hand-crafted model by 0.22%. The architecture discovered by DW-Net also shows improved robustness against PGD attacks, with a difference ranging from 0.3% to 1.1% compared to previous NAS techniques. Despite the increased complexity of the task, DW-Net still demonstrates good robustness against simple attacks like FGSM, and its adversarial accuracy against PGD is nearly equivalent to that of the hand-crafted network architecture.

Table 3. Adversarial success and natural accuracy on CIFAR100.

Model	Clean(%)	FGSM(%)	PGD(%)
ResNet-18	63.77	16.58	6.01
ResNet-50	72.69	17.94	5.46
DenseNet-121	78.21	22.54	6.55
DenseNet-169	82.14	22.23	7.17
VGG16 BN	71.95	16.89	4.17
DARTS	82.11	24.55	2.01
PDARTS	82.08	26.89	2.81
PC-DARTS	81.63	25.92	2.24
DW-Net(ours)	80.93	**27.11**	3.11

In conclusion, the proposed DW-Net architecture exhibits remarkable superiority over other models in terms of ensuring robustness against small-scale attacks as the task complexity increases. Additionally, DW-Net shows competitive adversarial accuracy to the hand-designed architecture against large-scale attacks in fine-grained tasks. These results suggest that the matching depth and width search method, which reduces the redundancy of intermediate features, significantly strengthens the architecture's inherent robustness. Overall, the findings of this study provide important insights for designing robust neural networks and advancing the field of deep learning.

4.4 Ablation Learning

Depth-Only Search. In this study, we conducted an experiment to evaluate the robustness of a model using depth search without scaling the width of each layer. Our findings, illustrated in Fig. 4, indicate that the success rate of recognition decreases for both clean data and adversarial samples when only depth search is performed. The analysis of experimental data revealed that the number of channels activated by the adversarial sample and the clean sample may be distributed differently, with the adversarial sample becoming uniform. This discrepancy in channel activation can be attributed to the high information redundancy of intermediate features due to the mismatched depths and widths of the model, leading to the risk of overfitting and vulnerability. To address this issue, we propose using DW-Net, which adapts the optimal width at various depths, to exclude redundant channel information from the model. As a result, the channel activation patterns of adversarial samples and natural data converge, reducing the model's overfitting and vulnerability. These findings suggest that DW-Net may be a promising approach to improving the robustness of deep neural networks.

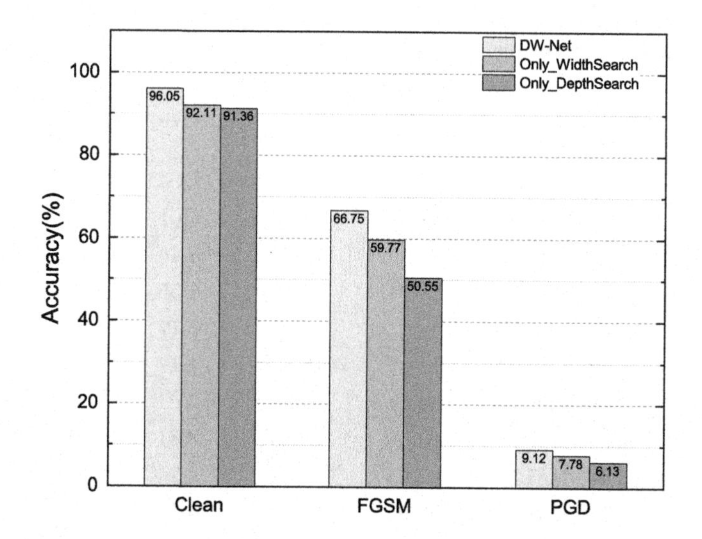

Fig. 4. Quantitative profiling of accuracy under different search conditions.

Width-Only Search. In our experiment, we conducted a width search by setting the depth of each stage to 5 while keeping other experimental settings unchanged. The results, as shown in Fig. 4, reveal a significant decrease in adversarial accuracy of over 15% under the FGSM attack and nearly 3% under the PGD attack. It is worth noting that solely adjusting the width of the model without considering depth does not effectively address the overfitting issue that arises from increasing model complexity. The discrepancy between the channel activation patterns of clean samples and adversarial samples is a key factor in understanding the impact of width search. Clean samples tend to activate only a small number of channels, whereas adversarial samples tend to activate a more uniform distribution of channels. By cropping out redundant channels, the width-only search limits the activation of excessive channels by adversarial samples, thus preventing the model from making incorrect predictions. However, it is important to note that scaling the width while keeping the depth constant may lead to model instability. Without appropriate adjustments, the model could become overwhelmed and fail to perform effectively. To address this challenge, we introduced DW-Net search, which simultaneously explores the model's two spatial dimensions (width and depth). By considering both dimensions, the search process allows the model to achieve improved resilience while maintaining stability.

5 Conclusion

This study proposes a robust multi-objective neural network structure search algorithm based on the depth and width of the model's two spatial dimensions. The main objective is to enhance the model's resilience by optimizing the association between the depth of the entire model and the width of the various layers. To mitigate the effect of partial supernet convergence on the model's robustness, DW-Net introduces the Lipschitz constant as a search target. Additionally, we present an adversarial sample generation strategy for NAS that enhances the diversity of the search space and reduces the temporal complexity of validating the subnets' robustness. The entire search process is one-shot, which significantly increases the search space's diversity. Our research shows that even in the absence of adversarial training, complex network architecture can afford robustness against attacks. Furthermore, the robustness of a neural network is dominated by the depth and width of its various stages. By selecting suitable depth and width parameters for each stage, a model's robustness can be significantly increased. In the future, we will investigate the robustness of different manual modules at different phases and how to merge different modules to increase the model's resilience.

Acknowledgment. This work is supported by Chinese Government Key R&D Project - High Security Isolated Execution Technology and Application based on Reconfigurable Technology in Specific Areas of Information Security

References

1. Athalye, A., Carlini, N., Wagner, D.: Obfuscated gradients give a false sense of security: circumventing defenses to adversarial examples. In: International Conference on Machine Learning, pp. 274–283. PMLR (2018)
2. Bai, Y., et al.: Improving adversarial robustness via channel-wise activation suppressing (2021)

3. Das, N., et al.: Compression to the rescue: Defending from adversarial attacks across modalities. In: ACM SIGKDD Conference on Knowledge Discovery and Data Mining (2018)
4. Devaguptapu, C., Agarwal, D., Mittal, G., Gopalani, P., Balasubramanian, V.N.: On adversarial robustness: a neural architecture search perspective http://arxiv.org/abs/2007.08428
5. Devlin, J., Chang, M.W., Lee, K., Toutanova, K.: Bert: pre-training of deep bidirectional transformers for language understanding (2018)
6. Eykholt, K., et al.: Robust physical-world attacks on deep learning visual classification. In: Proceedings of the IEEE Conference on Computer Vision and Pattern Recognition, pp. 1625–1634 (2018)
7. Goodfellow, I.J., Shlens, J., Szegedy, C.: Explaining and harnessing adversarial examples (2017)
8. Goodfellow, I.J., Shlens, J., Szegedy, C.: Explaining and harnessing adversarial examples (2014)
9. Gu, S., Rigazio, L.: Towards deep neural network architectures robust to adversarial examples. arXiv preprint arXiv:1412.5068 (2014)
10. Guo, M., Yang, Y., Xu, R., Liu, Z., Lin, D.: When NAS meets robustness: in search of robust architectures against adversarial attacks. In: 2020 IEEE/CVF Conference on Computer Vision and Pattern Recognition (CVPR), pp. 628–637 (2020). https://doi.org/10.1109/CVPR42600.2020.00071, ISSN: 2575-7075
11. He, K., Zhang, X., Ren, S., Sun, J.: Deep residual learning for image recognition (2016). http://arxiv.org/abs/1512.03385
12. Kurakin, A., Goodfellow, I., Bengio, S.: Adversarial machine learning at scale (2016)
13. Liao, F., Liang, M., Dong, Y., Pang, T., Hu, X., Zhu, J.: Defense against adversarial attacks using high-level representation guided denoiser. In: Proceedings of the IEEE Conference on Computer Vision and Pattern Recognition, pp. 1778–1787 (2018)
14. Liu, H., Simonyan, K., Yang, Y.: DARTS: differentiable architecture search (2018). http://arxiv.org/abs/1806.09055
15. Lu, Z., Whalen, I., Boddeti, V., Dhebar, Y., Deb, K., Goodman, E., Banzhaf, W.: NSGA-net: neural architecture search using multi-objective genetic algorithm. In: Proceedings of the Genetic and Evolutionary Computation Conference, pp. 419–427 (2019)
16. Madry, A., Makelov, A., Schmidt, L., Tsipras, D., Vladu, A.: Towards deep learning models resistant to adversarial attacks (2017)
17. Ning, X., Zhao, J., Li, W., Zhao, T., Yang, H., Wang, Y.: Multi-shot NAS for discovering adversarially robust convolutional neural architectures at targeted capacities. arXiv preprint arXiv:2012.11835 (2020)
18. Papernot, N., McDaniel, P., Wu, X., Jha, S., Swami, A.: Distillation as a defense to adversarial perturbations against deep neural networks. In: 2016 IEEE Symposium on Security and Privacy (SP), pp. 582–597. IEEE (2016)
19. Pham, H., Guan, M.Y., Zoph, B., Le, Q.V., Dean, J.: Efficient neural architecture search via parameter sharing (2018)
20. Qian, Y., Huang, S., Wang, Y., Li, S.: RNAS: robust network architecture search beyond DARTS (2021)
21. Szegedy, C., et al.: Intriguing properties of neural networks (2013)
22. Tramr, F., Kurakin, A., Papernot, N., Goodfellow, I., Boneh, D., McDaniel, P.: Ensemble adversarial training: attacks and defenses (2017). http://arxiv.org/abs/1705.07204, issue: arXiv:1705.07204

23. Wang, Y., Deng, X., Pu, S., Huang, Z.: Residual convolutional CTC networks for automatic speech recognition (2017)
24. Wang, Y., Ma, X., Bailey, J., Yi, J., Zhou, B., Gu, Q.: On the convergence and robustness of adversarial training (2021)
25. Wang, Y., Ma, X., Bailey, J., Yi, J., Zhou, B., Gu, Q.: On the convergence and robustness of adversarial training (2021). arXiv preprint arXiv:2112.08304
26. Xie, G., Wang, J., Yu, G., Zheng, F., Jin, Y.: Tiny adversarial mulit-objective oneshot neural architecture search (2023)
27. Zhang, D., Zhang, T., Lu, Y., Zhu, Z., Dong, B.: You only propagate once: accelerating adversarial training via maximal principle, vol. 32 (2019)
28. Zhang, H., Yu, Y., Jiao, J., Xing, E., El Ghaoui, L., Jordan, M.: Theoretically principled trade-off between robustness and accuracy. In: International Conference on Machine Learning, pp. 7472–7482. PMLR (2019)
29. Zhang, J., et al.: Attacks which do not kill training make adversarial learning stronger. In: International Conference on Machine Learning, pp. 11278–11287. PMLR (2020)
30. Zoph, B., Le, Q.V.: Neural architecture search with reinforcement learning (2016)

Boosting Adversarial Attacks with Improved Sign Method

Bowen Guo[1], Yuxin Yang[1], Qianmu Li[1](✉), Jun Hou[2], and Ya Rao[1]

[1] Nanjing University of Science and Technology, Nanjing, China
{newbg,yangyx,qianmu,raoya}@njust.edu.cn
[2] Nanjing Vocational University of Industry Technology, Nanjing, China

Abstract. Deep neural networks (DNNs) can be susceptible to adversarial examples, which involve adding imperceptible perturbations to benign images in order to deceive the models. Although many current adversarial attack methods achieve nearly 100% success rates in white-box attacks, their effectiveness often diminishes when targeting black-box models due to limited transferability. The transfer-based attack technique involves choosing a surrogate model that closely resembles the target model. By exploiting the shared decision boundaries among different models, this method generates highly transferable adversarial examples. Most attack methods change the gradient using the Sign Method (SM) and add small perturbation to the original image. Although SM is simple and effective, it only extracts the sign of the gradient unit, ignoring the size of the gradient value, and resulting in an inaccurate update direction in the iterative process. In our research, we introduce an innovative approach called the Improved Sign Method (ISM). This strategy involves assigning weights to gradients, aiming to mitigate this concern and thereby enhance the efficacy of black-box attacks in terms of transferability. The effectiveness of our proposed Improved Sign Method (ISM) is strongly validated through a series of comprehensive experiments. These results unequivocally demonstrate its capability to significantly enhance the transferability of adversarial examples within black-box attack scenarios. Furthermore, ISM seamlessly integrates with the Fast Gradient Sign Attack Method (FGSM) family, making it applicable across various scenarios. Notably, the computational overhead associated with this integration is practically negligible. Moreover, by incorporating with other advanced attack methods, the performance of black-box attack has been significantly enhanced.

Keywords: Black-box attack · Transferability · Adversarial example · Sign method

1 Introduction

In recent years, Deep neural networks (DNNs) [1–4] have achieved remarkable accomplishments, asserting dominance across various fields, such as object detection [5–7]. Nonetheless, recent research reveals the susceptibility of deep neural networks (DNNs) to adversarial examples [8–10,37]. These examples involve

X. Yang et al. (Eds.): ADMA 2023, LNAI 14180, pp. 150–164, 2023.
https://doi.org/10.1007/978-3-031-46677-9_11

imperceptible perturbations and pose significant concealed threats to applications demanding heightened security, like autonomous driving [11], face recognition [12], and voice assistants, among others. Particularly within the realm of autonomous driving, the real-time recognition of traffic signs holds critical importance. The potential impact of adversarial examples on such systems could result in catastrophic outcomes. This situation has propelled research into two primary directions. The first involves augmenting the transferability of black-box attacks, while the second focuses on enhancing the resilience of detection models. These two directions not only have the potential to mutually reinforce one another but also contribute to a symbiotic developmental trajectory.

Presently, a multitude of researchers have introduced various attack techniques for generating adversarial examples. These include methods like the fast gradient sign method [10], iterative fast gradient sign method [14], and C&W attack [15]. These strategies fall under the umbrella of white-box attacks [16], as they leverage intricate gradient information from the target models. Nonetheless, when applied to black-box models, these methods often exhibit lackluster performance and low transferability. This is particularly evident when confronting models that have undergone adversarial training [14,17] or have implemented alternative defense mechanisms.

In response to this issue, many researchers are committed to enhancing the transferability of adversarial attack under the black-box setting, mainly by improving gradient calculation (e.g. Momentum, Nesterov's accelerated gradient, Variance Tuning, etc.) [21,22,34,36] or adding diverse transformations on the input (e.g. random resizing and padding, scale, admix, etc.) [16,22–24,35]. Furthermore, engaging in simultaneous attacks on multiple models can lead to enhanced performance in adversarial attacks. Conventionally, the following transfer-based iterative attack methods, in the recent years, are all based on Sign Method (SM) [33] to boost the adversarial attack.

However, although SM is simple and effective, it also has some limitations. SM only extracts the sign information of each gradient unit and assigns [0, 1, -1] to each pixel. This approach overlooks the variations in gradient magnitudes and doesn't fully harness the potential of gradient information. As a result, it might veer away from the globally optimal attack region that effectively deceives both the surrogate and target models, consequently diminishing transferability.

As a solution to this issue, we introduce an enhanced technique known as the Improved Sign Method (ISM). This method leverages the complete gradient information from the surrogate model, preserving the strong performance observed in white-box attacks, while simultaneously enhancing the transferability of adversarial examples within black-box attacks. Notably, we have devised a weight assignment strategy within ISM. This strategy allocates specific weights based on the magnitude of gradient unit values, thereby ensuring a more precise adjustment of the update direction. In short, we merely manipulate the sign perturbation added on the image, so ISM can be generally integrated into the family of FGSM algorithms. In sum, our key contributions are as below:

- We find the limitation of the Sign method (SM) in the FGSM family, the most popular gradient attack algorithm at present. The insufficient use of gradient information leads to a coarse update step size, which makes the adversarial examples easily deviate from the global optimal attack region.
- We propose an Improved Sign Method (ISM) to reduce the negative influence of the above problems. By fully capitalizing on the pixel-level gradient values, we employ a heuristic approach to assign weights to each unit. This enhances the precision of update steps, yielding more effective results. Additionally, ISM can seamlessly integrate with advanced techniques within the FGSM family.
- We have conducted extensive experiments on the ImageNet [25] dataset, validating the effectiveness of our proposed approach. Notably, by incorporating advanced FGSM attack methods, the transferability of black-box attacks can be further improved, all while maintaining the exceptional performance observed in white-box scenarios.

2 Related Work

Given an image classifier denoted as f with parameters θ, an original image x, and its corresponding true label y_{true}, where $f(x)$ yields the classification outcome of x. In this context, we employ the cross-entropy function $J(x, y_{true})$ as the loss function. The resulting adversarial example x_{adv}, derived from the original image, resembles x but is misclassified by the network, with the aim of maximizing the loss function. To ensure that the generated adversarial examples remain imperceptible to the human eye, a small perturbation upper bound ε is typically set. The problem can be represented as follows:

$$f_{\theta}\left(x_{adv}\right) \neq y, \text{ s.t. } \left\|x_{adv} - x\right\|_{p} \leq \epsilon \tag{1}$$

where $\|.\|_p$ signifies the p-norm distance, where p can take values of 0, 2, or ∞. In particular, we underscore that $p = \infty$ corresponds to the convention employed in prior research.

2.1 Generating Adversarial Examples

Currently, researchers have put forth a multitude of attack methods for conducting adversarial attacks, which can be broadly categorized into two primary approaches. In the first approach, there is a persistent focus on refining the gradient calculation method to achieve a more precise update direction during the iterative process. The second approach centers around enhancing input diversity through the implementation of a range of input transformations. This serves to augment the transferability of adversarial attacks.

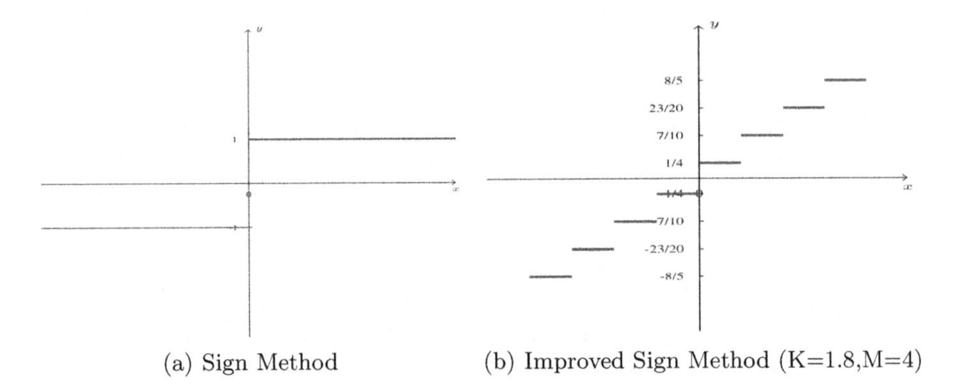

(a) Sign Method (b) Improved Sign Method (K=1.8,M=4)

Fig. 1. SM vs. ISM (K = 1.8 and M = 4 for example).

Momentum Iterative Fast Gradient Sign Method (MI-FGSM) [21] incorporates momentum into the existing iterative technique to achieve a more precise update direction, leading to improved transferability in black-box attacks. The summarized update formula is as follows:

$$g_{t+1}=\mu\cdot g_t+\frac{\nabla_x L\left(x_t^{adv},y\right)}{\left\|\nabla_x L\left(x_t^{adv},y\right)\right\|_1} \tag{2}$$

The Nesterov Iterative Fast Gradient Sign Method (NI-FGSM) [22] introduces Nesterov accelerated gradient into the iterative procedure, which improves forward-looking ability and enhances the accuracy of direction updates.

$$\begin{aligned} x_t^{nes}&=x_t^{adv}+\alpha\cdot\mu\cdot g_t\\ g_{t+1}&=\mu\cdot g_t+\frac{\nabla_x J\left(x_t^{nes},y^{\text{true}}\right)}{\left\|\nabla_x J\left(x_t^{nes},y^{\text{true}}\right)\right\|_1}\\ x_{t+1}^{adv}&=\operatorname{Clip}_x^{\epsilon}\left\{x_t^{adv}+\alpha\cdot\operatorname{sign}\left(g_{t+1}\right)\right\} \end{aligned} \tag{3}$$

where g_t represents the accumulated gradient in the t-th iteration, and μ denotes the decay factor.

The Variance Tuning Momentum-based Iterative Method (VMI-FGSM) [26] goes a step further by calculating the variance from the previous iteration to fine-tune the gradient value. This modification aims to yield a more substantial update direction and replaces the conventional gradient value update.

$$g_{t+1}=\mu\cdot g_t+\frac{\hat{g}_{t+1}+v_t}{\left\|\hat{g}_{t+1}+v_t\right\|_1} \tag{4}$$

Subsequently, the method estimates its value by selecting N data points from the surrounding vicinity of x, leading to the acquisition of $V(x)$.

$$V(x)=\frac{1}{N}\sum_{i=1}^{N}\nabla_{x^i}J\left(x^i,y;\theta\right)-\nabla_x J(x,y;\theta) \tag{5}$$

Here $x^i = x + r_i, r_i \sim U\left[-(\beta \cdot \epsilon)^d, (\beta \cdot \epsilon)^d\right]$, and $U\left[a^d, b^d\right]$ represents the uniform distribution in d-dimensional space.

Diverse Inputs Method (DIM) [16] utilizes diverse input methods, involving modifications to the original image in various ways, to generate adversarial examples. By introducing diversity into the attack process, the DIM method can increase the success rate of attacks and make it more challenging for the target model to defend against them.

$$x_{t+1}^{adv} = C\text{lip}\ p_{\boldsymbol{x},\epsilon}\left\{\boldsymbol{x}_t^{adv} - \alpha \cdot \operatorname{sign}\left(\nabla_{\boldsymbol{x}} J\left(D\left(\boldsymbol{x}_t^{adv}\right), y^{adv}\right)\right)\right\} \tag{6}$$

where $D(\boldsymbol{x})$ represents different transformations of the input image.

Transferable Input Method (TIM) [23] leverages pre-defined kernel functions to generate adversarial perturbations. These kernels can be tailored based on the characteristics and vulnerabilities of target models, enabling the creation of adversarial perturbations capable of transferring across different models.

$$\boldsymbol{x}^{adv} = \boldsymbol{x} - \epsilon \cdot \operatorname{sign}\left(\boldsymbol{W} * \nabla_{\boldsymbol{x}} J\left(\boldsymbol{x}_t^{adv}, y^{adv}\right)\right) \tag{7}$$

Scale-Invariant attack (SIM) [22] leverages images and their scaled versions with a factor $1/2^i$ exhibit similar losses, a form of data augmentation. It aids adaptability to varying image sizes during training, bolstering robustness for diverse resolutions in practical applications. Additionally, it augments generalization, aiding unseen data.

$$\begin{aligned} &\underset{x^{adv}}{\arg\max} \frac{1}{m} \sum_{i=0}^{m} J\left(S_i\left(x^{adv}\right), y^{\text{true}}\right), \\ &\text{s.t.}\ \left\|x^{adv} - x\right\|_{\infty} \leq \epsilon, \end{aligned} \tag{8}$$

2.2 Defending Against Adversarial Examples

Counterbalancing the focus on adversarial attacks, researchers have dedicated substantial attention to bolstering the robustness of Deep Neural Networks (DNNs). They have introduced an array of adversarial defense methods aimed at mitigating potential vulnerabilities. Among the various approaches, the most prevalent and efficacious technique is adversarial training [10,13,14]. This method enhances the neural network's robustness by incorporating adversarial examples during the model training process. However, Tramèr et al. [17] highlighted that conventional adversarial training still leaves networks vulnerable to adversarial attacks. In response, they introduced ensemble adversarial training, which involves augmenting the model training set with unique images generated by attacking multiple networks simultaneously. This innovative approach further fortifies the network's defense mechanisms. Tramèr et al. [17] emphasized that adversarially trained models remain susceptible to adversarial examples. To address this, they introduced ensemble adversarial training. This approach enriches the training data with perturbations transposed from other models, thereby further enhancing network robustness. Currently, adversarial training stands out as one of the premier techniques for fending off adversarial attacks. However, it often comes with a substantial computational cost, rendering it challenging to scale for larger datasets and intricate neural networks [13].

An alternative set of defense methods focuses on mitigating the impact of adversarial perturbations by altering the input data. One prevailing approach involves adjusting the input to counteract the adverse effects of adversarial examples. For instance, Guo et al. [18] have explored this defense strategy. They leverage a series of image transformations, such as JPEG compression and Total Variance Minimization, during the inference phase. These transformations are employed to alleviate the detrimental influence of adversarial perturbations on the input data. Xie et al. [19] applies R&P to the original images, creating modified versions that are then fed into the network. The aim is to reduce the impact of adversarial perturbations while retaining the essential visual information within the picture Bit-Red [28] is employed to assess whether an adversarial attack has been launched. This method involves reducing the bit depth of the input data as a means of detecting and mitigating adversarial attacks. Liao et al. [27] have suggested a novel approach involving training an exceptional representation denoiser named High-Order Gradient Descent (HGD). This denoiser effectively cleans up images, contributing to improved robustness. Liu et al. [29] have proposed a defense strategy that employs JPEG-based defensive compression. This technique is designed to rectify adversarial examples while preserving the classification performance of the original images.

3 Methodology

3.1 Motivation

The FGSM family of attack methods plays a pivotal role in facilitating transfer-based attacks for generating adversarial examples. In particular, these methods are all based on Sign Method(SM), which is simple but effective. It extracts the sign information of each gradient unit to decide the direction of update while the step size is often set to a fixed value, so that each pixel can modify as much information as possible. Besides, manipulating the gradients by SM for iterative attacks can quickly reach the boundary of ∞-ball with only a few iterations.

Indeed, focusing solely on the sign of gradient elements neglects the variations in gradient magnitudes. This approach inadvertently disregards crucial gradient information from the substitute model, leading to inefficiencies in gradient utilization. Despite the diverse gradient values across pixels, the conventional method allocates identical update steps to each pixel. Especially for target attacks, adversarial samples need to be within a specific range of the target label, and need more accurate directions and steps. Consequently, the resultant perturbation deviates from the target territory and decreases the transferability. Thus, we propose our ISM for making full use of the value of gradient unit to allocate more accurate update step for each pixel, so as to obtain more effective adversarial examples.

3.2 Improved Sign Method

To enhance the transferability of adversarial attacks more effectively, certain researchers have sought to optimize the gradient calculation methods, aiming to obtain more precise gradient values. Some avoid local overfitting by increasing the diversity of inputs, but few people point out that the shortcomings of SM and design better weight assignment strategy.

Hence, we introduce the Improved Sign Method (ISM), essentially a strategy for assigning weights, enabling comprehensive utilization of gradient unit values. Furthermore, our method exclusively operates on the gradient component, allowing seamless integration with a majority of FGSM family techniques, resulting in substantial improvements in black-box adversarial attack performance. For the sake of simplicity, let's consider our variant MI-FGISM (outlined in Algorithm 1) as an illustrative example to demonstrate the integration process.

Technically, the weight assignment strategy proposed by our method assigns different weight sizes according to the size of each pixel unit value, so that each pixel can heuristically obtain the appropriate step size. Especially for the target attack, it can reach the target territory more quickly and accurately. Our method can be summarized as follows:

Firstly, we calculate the gradient g_t of the substitute model at t-iteration with respect to the input, including the information of each pixel gradient unit.

$$\boldsymbol{g}_t = \nabla_{\boldsymbol{x}} \ \mathrm{J}\left(\boldsymbol{x}, y_{true}\right) \tag{9}$$

A total of M t_{th} segmentation points are computed from the gradient value g_t, with M denoting the number of segments. This process yields g^t (where t = 0, 1, 2, ..., $M-1$) representing the value for each segment point, ultimately forming a weight assignment function. Within this context, K signifies the maximum admissible weight for each pixel unit, and ω symbolizes the interval between segment points, calculated as $\omega = \frac{100}{M}$.

$$W_t = \begin{cases} \frac{\omega}{100}, & g^0 \leq \left|\boldsymbol{g}^{i,j}\right| \leq g^1 \\ \frac{(K+1)\omega}{100}, & g^1 < \left|\boldsymbol{g}_t^{i,j}\right| \leq g^2 \\ & \vdots \\ \frac{(Kt+1)\omega}{100}, & g^{t-1} < \left|\boldsymbol{g}_t^{i,j}\right| \leq g^t \\ & \vdots \\ K - \frac{K-1}{M}, & g^{M-2} < \left|\boldsymbol{g}_t^{i,j}\right| \leq g^{M-1} \end{cases} \tag{10}$$

Then, according to the weight assignment function we designed, each pixel point is assigned a different weight, and the weight of each pixel point is combined with a fixed step size to obtain a more precise step heuristically, which can improve the update speed and accuracy.

Finally, the formula need to be rewritten for generating adversarial samples,

$$\boldsymbol{x}_{t+1}^{adv} = \mathrm{clip}\,\boldsymbol{x}_{x,e}\left\{\boldsymbol{x}_t^{adv} - \alpha \cdot \mathrm{sign}\left(\boldsymbol{g}_t\right) \odot \boldsymbol{W}_t\right\} \tag{11}$$

where $\odot$ is Hadamard product.

3.3 Relationship with SM

In the conventional Sign Method (SM), each pixel point is assigned a weight of 1, resulting in a fixed step for updates. However, our proposed Improved Sign Method (ISM) employs a heuristic approach to assign varying weights to each pixel point based on the gradient unit magnitudes. This strategy leads to more precise update steps, enhancing the accuracy of each adjustment and subsequently further improving transferability within the black-box context.

When the number of segments M and the maximum acceptable weight K are all set to 1, our ISM degenerates to SM.

Algorithm 1. MI-FGISM

Require: We have the following components in our context: a benign example denoted as x along with its true label y; a classifier *f(.)* parameterized by θ; the loss function $J(x, \theta)$ for our surrogate models; the iteration count T; the decay factor μ; and the magnitude of perturbation denoted as ϵ.

Ensure: An adversarial example x^{adv}

1: $\alpha = \epsilon/T$
2: $g_0 = 0; x_0^{adv} = x$
3: **for** $t = 0 \to T-1$ **do**
4: Calculate the gradient $\hat{g}_{t+1} = \nabla_{x_t^{adv}} J\left(x_t^{adv}, y; \theta\right)$
5: calculate the p_{th} percentile
6:
$$\boldsymbol{W}_t = \begin{cases} \frac{\tau}{100}, & g_t^0 \leq \left|\boldsymbol{G}_t^{i,j}\right| \leq g_t^{\tau} \\ \frac{3\tau}{100}, & g_t^{\tau} < \left|\boldsymbol{G}_t^{i,j}\right| \leq g_t^{2\tau} \\ \vdots & \\ \frac{(2k+1)\tau}{100}, & g_t^{p-\tau} < \left|\boldsymbol{G}_t^{i,j}\right| \leq g_t^{p}, \\ \vdots & \\ \frac{(2K-1)\tau}{100}, & g_t^{100-\tau} < \left|\boldsymbol{G}_t^{i,j}\right| \leq g_t^{100}. \end{cases}$$
7: $\boldsymbol{x}_{t+1}^{adv} = \text{clip}_{x,\epsilon}\left\{\boldsymbol{x}_t^{adv} - \alpha \cdot \text{sign}\left(\boldsymbol{G}_t\right) \odot \boldsymbol{W}_t\right\}$
8: $x_{t+1}^{adv} = \text{clip}\left(x_{t+1}^{adv}, -1, 1\right)$
9: **end for**
10: $x^{adv} = x_{T-1}^{adv}$
11: **return** x^{adv}

3.4 Attacking an Ensemble of Models

Building upon the findings of Liu et al. [20], the utilization of ensemble attack methods, especially in the context of black-box models, can amplify the transferability of adversarial attacks. When adversarial examples lead multiple models to misclassify, it often becomes simpler to penetrate other detection models.

In line with this notion, we incorporate an ensemble attack approach to enhance the efficacy of adversarial examples. This method combines the logits from an ensemble of M models, as follows:

$$l(x) = \sum_{m=1}^{M} u_m l_m(x) \tag{12}$$

In this context, $l_m(\cdot)$ signifies the logits of the m-th model, and the parameters u_m represent the respective weights assigned, with the condition that $u_m > 0$ and $\sum_{m=1}^{M} u_m = 1$.

4 Experiment

To demonstrate the significance of our ISM, we carry out comprehensive experiments on the public ImageNet database.

Initially, we outline the experimental setup in Sect. 4.1. Following that, we conduct a comprehensive comparison of the performance of our methods against competitive baseline approaches. This assessment is carried out for both traditionally trained models (Sect. 4.2) and sophisticated defense models (Sect. 4.3). In the end, we performed ablation experiments on the number of segments and the maximum allowed weight to find the excellent hyper-parameters.

4.1 Experiment Setup

Dataset: From the ImageNet validation set, we randomly select 1000 clean pictures that are essentially all categorized correctly by testing models. Additionally, each image in this collection has a pre-determined target label that is often unique. All of these images have already been reduced to $299 \times 299 \times 3$.

Networks: To benchmark against state-of-the-art attack methodologies, we consider a range of seven distinct networks. This selection includes four traditionally trained networks, namely Inception-v3 (Inc-v3) [4], Inception-v4 (Inc-v4) [3], Resnet-v2-152 (Res-152) [1], and Inception-Resnet-v2 (IncRes-v2) [17]. Additionally, we encompass three adversarially trained networks: ens3-adv-Inception-v3 (Inc-v3ens3), ens4-adv-Inception-v3 (Inc-v3ens4), and ens-adv-Inception-ResNet-v2 (IncRes-v2ens). All of these aforementioned networks are publicly accessible for evaluation purposes.

Parameters: To facilitate a more convenient comparison of various methods, we adopt standardized parameters. Specifically, we set $\varepsilon = 16$ as the upper limit for perturbation, in line with prevailing research conventions. We establish $T = 16$ iterations, resulting in a single step of $\alpha = 1.0$ per iteration. In ensemble attacks across M networks, we maintain equal weights for each model's logits. Within the MI-FGSM approach, the decay factor is $\mu = 1.0$. For the DI-FGSM method [49], the probability of distinct transformations is set at $p = 0.5$. Regarding TI-FGSM, the Gaussian kernel dimensions are 5×5 for normally trained models, and 15×15 for defense models. In our ISM, we opt for $M = 4$ segments and a maximum acceptable weight of $K = 1.8$. This configuration has demonstrated remarkable performance in both normally trained and adversarially trained models, as evidenced by ablation experiments.

Baselines: We selected several advanced attack methods in FGSM family, such as MI-FGSM, NI-FGSM, DTS-FGSM, VMI-FGSM, which include improving the gradient calculation strategy and increasing the input diversity to increase the transferablility black-box attack. They all utilize SM to decide the direction of the update, and we each replace it with the ISM we proposed to further verify the feasibility of our approach.

4.2 Attacking a Single Network

Table 1. Adversarial attack success rates (%) against the seven baseline networks in a single-model setting. Four advanced FGSM-based/FGISM-based attack methods are used to construct adversarial examples for Inc-v3, Inc-v4, IncRes-v2, and Res-152 networks.

Model	Attack	Inc-v3	Inc-v4	IncRes-v2	Res-101	Inc-v3-ens3	Inc-v3-ens4	IncRes-v2-ens
Inc-v3	MI-FGSM	100%	44.6%	41.4%	35.3%	13.8%	11.6%	5.5%
	MI-FGISM	100%	49.2%	45.7%	39.8%	15.3%	14.9%	6.8%
Inc-v4	DTS-FGSM	87%	99.2%	82.2%	75.3%	69.8%	65.1%	55.9%
	DTS-FGISM	89.5	99.1%	86.6%	79.7%	74.2%	70.3%	63.7%
IncRes-v2	NI-FGSM	61.2%	53.6%	99.1%	45.1%	19.5%	14.6%	8.8%
	NI-FGISM	66.8%	57.2%	99.2%	49.6%	24.3%	18.2%	12.6%
Res-152	VMI-FGSM	74.3%	68.2%	69.6%	99.2%	45.2%	40.4%	30.1%
	VMI-FGISM	78.9%	72.6%	73.8%	99.3%	49.5%	45.1%	33.2%

We combine the weight assignment method with four relatively advanced adversarial attack methods, namely MI-FGSM, NI-FGSM, DTS-FGSM and VMI-FGSM. The methods after integration are recorded as MI-FGISM, NI-FGISM, DTC-FGISM, VMI-FGISM. Adversarial examples were crafted on four normally trained neural networks, followed by testing them on the aforementioned seven neural networks. The adversarial attack success rate, representing the likelihood of various models misclassifying when confronted with adversarial examples, is presented in Table 1. Rows correspond to the attack model, while columns signify the seven models under test. Evidently, upon integrating our weight assignment strategy with the four advanced attack methods, the success rate of black-box attacks significantly escalates, while retaining the strong white-box performance. We attacked the Inc-v3 model, for instance, to produce adversarial images, which can still achieve at a success rate of 100% for the white-box attack. When attacking the other three normally trained model, MI-FGISM increases the success rate by around 5% compared with MI-FGSM, while facing the adversarially trained model, the transferability of black-box has also been greatly increased. When generating adversarial examples with the Inc-v4 model, DTS-FGISM achieves the success rate of 74.2% for Inc-v3$_{ens3}$ and 63.7% for IncRes-v2$_{ens}$, which surpasses the baseline attacks DTS-FGSM for 4.4% and 7.8% respectively, which strongly supports the efficacy of our proposed weight assignment method.

4.3 Attacking an Ensemble of Networks

Table 2. Adversarial attack success rates (%) against the seven baseline networks in the ensemble-model setting. Four advanced FGSM-based/FGISM-based attack methods are utilized to generate adversarial examples for a subset of Inc-v3, Inc-v4, IncRes-v2, and Res-152 networks.

Model	Attack	Inc-v3	Inc-v4	IncRes-v2	Res-101	Inc-v3-ens3	Inc-v3-ens4	IncRes-v2-ens
Inc-v3	MI-FGSM	99.9%	98.6%	96.2%	99.8%	37.6%	34.9%	23.8%
	MI-FGISM	99.5%	99.2%	96.4%	99.9%	67.4%	59.2%	49.8%
Inc-v4	SDT-FGSM	99.9%	99.2%	96.9%	99.8%	90.1%	88.4%	85.3%
	SDT-FGISM	99.9	99.1%	97.2%	99.7%	92.4%	90.5%	89.7%
IncRes-v2	NI-FGSM	99.7%	98.9%	98.6%	99.8%	40.2%	32.4%	22.5%
	NI-FGISM	99.8%	99.5%	99.2%	98.6%	66.3%	62.4%	48.6%
Res-152	VMI-FGSM	99.3%	98.5%	96.0%	99.9%	66.2%	60.8%	45.1%
	VMI-FGISM	99.6%	99.3%	97.8%	99.8%	73.5%	68.1%	54.2%

It was suggested by Liu et al. [20] that simultaneously attacking several models might enhance the performance of the adversarial attack. Therefore, we attempt to employ the ensemble attack, which combines the logit outputs of many networks to evaluate the effectiveness of ISM. This occurs because it is more likely to migrate to other models if an example continues to be adversarial for several models. We specifically target an ensemble of models that have been trained normally, i.e. Equal ensemble weights are applied to Inc-v3, Inc-v4, IncRes-v2, and Res-152 using MI-FGISM, NI-FGISM, DTS-FGISM, and VMI-FGISM, respectively. The performance of the baselines on the models that have received adversarial training can be significantly enhanced by our attack methods(MI-FGISM, NI-FGISM), as shown in Table 2, by surpassing 20% and 25% for their baseline methods. While DTS-FGSM could obtain adversarial examples with sufficient level of transferability, our improved method DTS-FGISM greatly increases the performance of black-box attack. And VMI-FGISM defeats three adversarially trained models with success rates of 57% to 65%, exposing the vulnerability of defense strategies currently. In addition, our approaches could still achieve the success rate of 100% comparable to the baselines in the white-box attack (Figs. 2 and 3).

4.4 Ablation Studies

In order to investigate the effects of two hyper-parameters, K and M, in our proposed improved sign method, we conduct multiple ablation experiments. The adversarial examples are generated by attacking the Inc-v3 network, effectively deceiving the white-box model across various hyper-parameter settings.

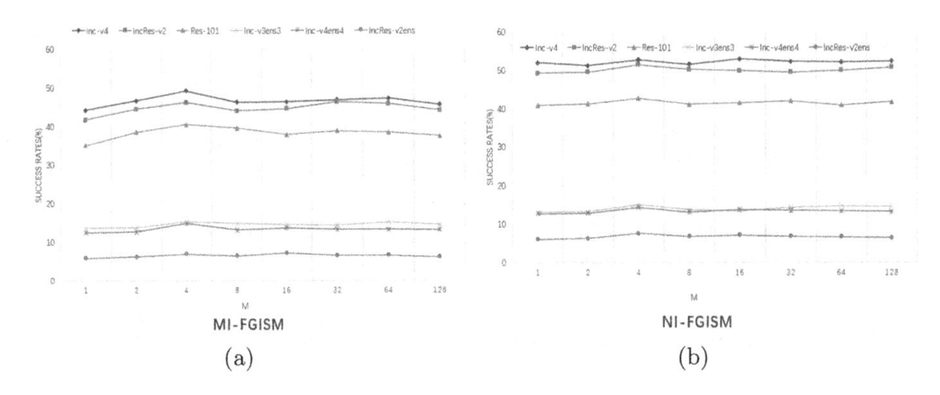

Fig. 2. Success rates (%) for the remaining six models using adversarial examples generated by the Inc-v3 network. The weight assignment methods of MI-FGISM and NI-FGISM are examined by varying the factor M for the number of segments.

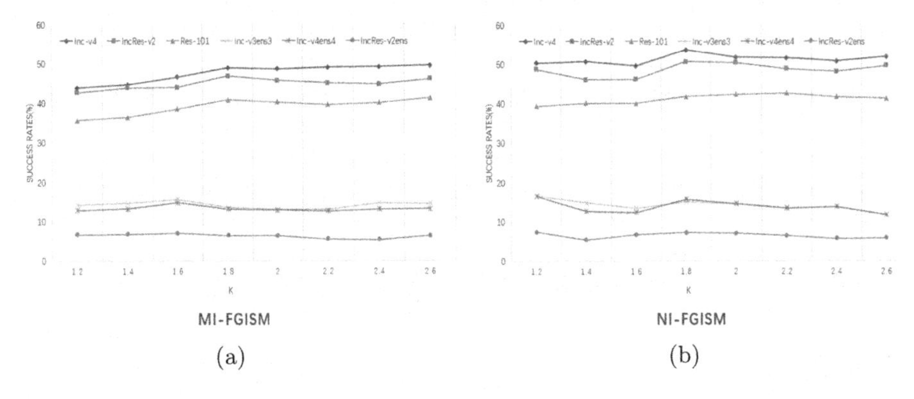

Fig. 3. Success rates (%) for the remaining six models using adversarial examples generated by the Inc-v3 network. The weight assignment methods of MI-FGISM and NI-FGISM are examined by varying the factor K for the maximum acceptable weight.

In Fig. 1, we study the effect of the maximum acceptable weight for the every pixel units, which is determined by the parameter K, and the number of segments M is fixed to 64. It is obvious that with the continuous increasement of K, the success rate of attacking the normally trained model and the adversarially trained model continues to improve, and reaches the peak when $K = 1.8$. However, if K continues to increase, the success rate of adversarial attack for adversarially trained models will decrease. In our study, to satisfy the compromise for performance on both normally trained models and adversarially trained models, we choose $K = 1.8$.

Regarding the number of segments in the weight assignment method, we also explore the difference of the number of segments M in the weight assignment method on tranferability of black-box attack. We can see that with the increasement of M, the performance of the attack normally trained model has been significantly improved, for example, the success rate of attacking Res152 in MI-FGISM increased from 74.3% to 78.9% and the effect of the adversarially trained model has also been improved, NI-FGISM is about 5% higher than the original method. At $M = 4$, the success rate peaks, surpassing MI-FGSM by a significant margin of 5.2% (rising from 44.2% to 49.4%). If M continues to increase, the curve will remain basically stable. Compared with forward propagation and backpropagation, the cost of percentage calculation required in ISM is negligible.

5 Conclusions

In this paper, we propose a weight assignment method named ISM to replace the traditional Sign Method (SM) in the current advanced FGSM family attack methods, assigning different weights to each pixel unit according to the value of the gradient, so as to heuristically obtain the step size and improve the attack ability of adversarial examples. Our method can be combined with most current attack methods that improve gradient calculation or make different transformations on the input, so as to further improve the performance of adversarial attack. Numerous experimental results highlight that ISM substantially enhances the success rates of black-box attacks while maintaining the excellence of white-box attacks. Moreover, even when confronting more sophisticated defense models, our approach enhances the fundamental efficacy of the original method. This observation underscores the limitations of current defense mechanisms and encourages us to explore more effective strategies for bolstering model robustness.

References

1. He, K., Zhang, X., Ren, S., Sun, J.: Deep residual learning for image recognition. In: Proceedings of the IEEE Conference on Computer Vision and Pattern Recognition, pp. 770–778 (2016)
2. Huang, G., Liu, Z., Van Der Maaten, L., Weinberger, K.Q.: Densely connected convolutional networks. In: Proceedings of the IEEE Conference on Computer Vision and Pattern Recognition, pp. 4700–4708 (2017)
3. Szegedy, C., Ioffe, S., Vanhoucke, V., Alemi, A.A.: Inception-v4, inception-ResNet and the impact of residual connections on learning. In: Thirty-First AAAI Conference on Artificial Intelligence (2017)
4. Szegedy, C., Vanhoucke, V., Ioffe, S., Shlens, J., Wojna, Z.: Rethinking the inception architecture for computer vision. In: Proceedings of the IEEE Conference on Computer Vision and Pattern Recognition, pp. 2818–2826 (2016)
5. Girshick, R.: Fast R-CNN. In: Proceedings of the IEEE International Conference on Computer Vision, pp. 1440–1448 (2015)
6. Ren, S., He, K., Girshick, R., Sun, J.: Faster R-CNN: towards real-time object detection with region proposal networks. In: Advances in Neural Information Processing Systems, vol. 28 (2015)
7. Zhang, Z., Qiao, S., Xie, C., Shen, W., Wang, B., Yuille, A.L.: Single-shot object detection with enriched semantics. In: Proceedings of the IEEE Conference on Computer Vision and Pattern Recognition, pp. 5813–5821 (2018)

8. Szegedy, C., et al.: Intriguing properties of neural networks. arXiv preprint arXiv:1312.6199 (2013)
9. Biggio, B., et al.: Evasion attacks against machine learning at test time. In: Blockeel, H., Kersting, K., Nijssen, S., Železný, F. (eds.) ECML PKDD 2013. LNCS (LNAI), vol. 8190, pp. 387–402. Springer, Heidelberg (2013). https://doi.org/10.1007/978-3-642-40994-3_25
10. Goodfellow, I.J., Shlens, J., Szegedy, C.: Explaining and harnessing adversarial examples. arXiv preprint arXiv:1412.6572 (2014)
11. Liu, A., et al.: Perceptual-sensitive GAN for generating adversarial patches. Proc. AAAI Conf. Artif. Intell. **33**(01), 1028–1035 (2019)
12. Guo, Y., Wei, X., Wang, G., Zhang, B.: Meaningful adversarial stickers for face recognition in physical world. arXiv preprint arXiv:2104.06728 (2021)
13. Kurakin, A., Goodfellow, I., Bengio, S.: Adversarial machine learning at scale. arXiv preprint arXiv:1611.01236 (2016)
14. Madry, A., Makelov, A., Schmidt, L., Tsipras, D., Vladu, A.: Towards deep learning models resistant to adversarial attacks. arXiv preprint arXiv:1706.06083 (2017)
15. Carlini, N., Wagner, D.: Towards evaluating the robustness of neural networks. In: IEEE Symposium on Security and Privacy (SP), pp. 39–57. IEEE (2017)
16. Xie, C., et al.: Improving transferability of adversarial examples with input diversity. In: Proceedings of the IEEE/CVF Conference on Computer Vision and Pattern Recognition, pp. 2730–2739 (2019)
17. Tramèr, F., Kurakin, A., Papernot, N., Goodfellow, I., Boneh, D., McDaniel, P.: Ensemble adversarial training: attacks and defenses. arXiv preprint arXiv:1705.07204 (2017)
18. Guo, C., Rana, M., Cisse, M., Van Der Maaten, L.: Countering adversarial images using input transformations. arXiv preprint arXiv:1711.00117 (2017)
19. Xie, C., Wang, J., Zhang, Z., Ren, Z., Yuille, A.: Mitigating adversarial effects through randomization. arXiv preprint arXiv:1711.01991 (2017)
20. Liu, Y., Chen, X., Liu, C., Song, D.: Delving into transferable adversarial examples and black-box attacks. arXiv preprint arXiv:1611.02770 (2016)
21. Dong, Y., et al.: Boosting adversarial attacks with momentum. In: Proceedings of the IEEE Conference on Computer Vision and Pattern Recognition, pp. 9185–9193 (2018)
22. Lin, J., Song, C., He, K., Wang, L., Hopcroft, J.E.: Nesterov accelerated gradient and scale invariance for adversarial attacks. arXiv preprint arXiv:1908.06281 (2019)
23. Dong, Y., Pang, T., Su, H., Zhu, J.: Evading defenses to transferable adversarial examples by translation-invariant attacks. In: Proceedings of the IEEE/CVF Conference on Computer Vision and Pattern Recognition, pp. 4312–4321 (2019)
24. Wang, X., He, X., Wang, J., He, K.: Admix: enhancing the transferability of adversarial attacks. In: Proceedings of the IEEE/CVF International Conference on Computer Vision, pp. 16158–16167 (2021)
25. Russakovsky, O., et al.: ImageNet large scale visual recognition challenge. Int. J. Comput. Vision **115**(3), 211–252 (2015)
26. Wang, X., He, K.: Enhancing the transferability of adversarial attacks through variance tuning. In: Proceedings of the IEEE/CVF Conference on Computer Vision and Pattern Recognition, pp. 1924–1933 (2021)
27. Liao, F., Liang, M., Dong, Y., Pang, T., Hu, X., Zhu, J.: Defense against adversarial attacks using high-level representation guided denoiser. In: Proceedings of the IEEE Conference on Computer Vision and Pattern Recognition, pp. 1778–1787 (2018)

28. Xu, W., Evans, D., Qi, Y.: Feature squeezing: Detecting adversarial examples in deep neural networks. arXiv preprint arXiv:1704.01155 (2017)
29. Liu, Z., et al.: Feature distillation: DNN-oriented jpeg compression against adversarial examples. In: 2019 IEEE/CVF Conference on Computer Vision and Pattern Recognition (CVPR), pp. 860–868. IEEE (2019)
30. Jia, X., Wei, X., Cao, X., Foroosh, H.: ComDefend: an efficient image compression model to defend adversarial examples. In: Proceedings of the IEEE/CVF Conference on Computer Vision and Pattern Recognition, pp. 6084–6092 (2019)
31. Cohen, J., Rosenfeld, E., Kolter, Z.: Certified adversarial robustness via randomized smoothing. In: International Conference on Machine Learning, pp. 1310–1320. PMLR (2019)
32. Salman, H., et al.: Provably robust deep learning via adversarially trained smoothed classifiers. In: Advances in Neural Information Processing Systems, vol. 32 (2019)
33. Gao, L., Zhang, Q., Zhu, X., Song, J., Shen, H.T.: Staircase sign method for boosting adversarial attacks. arXiv preprint arXiv:2104.09722 (2021)
34. Wang, X., Lin, J., Hu, H., Wang, J., He, K.: Boosting adversarial transferability through enhanced momentum. arXiv preprint arXiv:2103.10609 (2021)
35. Wang, G., Yan, H., Guo, Y., Wei, X.: Improving adversarial transferability with gradient refining. arXiv preprint arXiv:2105.04834 (2021)
36. Wang, G., Wei, X., Yan, H.: Improving adversarial transferability with spatial momentum. arXiv preprint arXiv:2203.13479 (2022)
37. Xiao, Z., et al.: Improving transferability of adversarial patches on face recognition with generative models. In: Proceedings of the IEEE/CVF Conference on Computer Vision and Pattern Recognition, pp. 11 845–11 854 (2021)

BRQG: A BART-Based Retouching Framework for Multi-hop Question Generation

TongXin Liao, Bin Xu(✉), YiKe Han, Shuai Li, and Shuo Zhang

SOE Lab, School of Computer Science and Engineering, Northeastern University, Shenyang, China
1309782308@qq.com

Abstract. BART is a powerful pre-trained model that has excelled in generative tasks such as text summarization, question answering, and machine translation. In previous studies, the BART model has often been used for multi-hop question generation(MQG) task, and it significantly improved the quality of generated questions compared to recurrent neural network-based models. However, due to the differences between downstream tasks and pre-training tasks, BART still generates some nonsensical and grammatically incorrect questions in multi-hop question generation tasks. These types of questions can have a negative impact on the user's reading experience. To address this challenge, we propose a BART-based retouching framework(BRQG), which builds upon BART. Specifically, BRQG uses BART-generated questions as a starting point, and introduces a Retouching Network module to reattend to the questions and context. The Retouching Gate layer then fuses this attention in an appropriate proportion to generate second-round questions that are more complete and readable. In addition, we propose a Entity Awareness Enhancement module, which construct graph structures from input documents to improve the correctness of entity generation. We conducted experiments on the HotpotQA dataset, and the results show that our model outperforms the currently proposed model on BLEU4, demonstrating the advantages and feasibility of BRQG in multi-hop question generation.

Keywords: BART · Multi-hop Question Generation · Retouching Framework

Funding provided by The Fundamental Research Funds for the Central Universities(N2116019), the National Natural Science Foundation of China (62137001,72271048), the Liaoning Natural Science Foundation (2022-MS-119), the Liaoning Province Discipline Inspection Supervision Big Data Key Laboratory (ZX20220460) and China University Industry-University-Research Innovation Fund(2022MU017).

X. Yang et al. (Eds.): ADMA 2023, LNAI 14180, pp. 165–179, 2023.
https://doi.org/10.1007/978-3-031-46677-9_12

Table 1. Example questions generated by the BART model based on the HotpotQA dataset. For the first example, the generated text is a subspan of context(yellow part), which is not a question. For the second example, the generated questions do not match the correct grammatical specification.

Example1 **Paragraph:** Frictional Games AB is an independent Swedish video game developer based in Helsingborg , Sweden , founded on 1 January 2007 by Thomas Grip and Jens Nilsson . The company specializes in the development of survival horror video games , and is best known for Frictional Games AB titles " " and " Soma " . Soma (stylized as SOMA) is a science fiction survival horror video game developed and published by Frictional Games for Microsoft Windows , OS X , Linux and PlayStation 4 . The game was released on 22 September 2015 . **Reference:** When was one of the science fiction survival horror video game released for who 's title Frictional Games AB was famous for ? **BART:** is a science fiction survival horror video game developed and published by Frictional Games for Microsoft Windows , OS X , Linux and PlayStation 4 , released on...
Example2 **Paragraph:**: Local H is an American rock band originally formed by guitarist and vocalist Scott Lucas , bassist Matt Garcia , drummer Joe Daniels , and lead guitarist John Sparkman in Zion , Illinois in 1987 . For Against is a United States post - punk / dream pop band from Lincoln , Nebraska . **Reference:** Are Local H and For Against both from the United States ? **BART:** Local H and For Against , are American bands ?

1 Introduction

Question Generation (QG) is a crucial aspect of text generation, producing fluent and natural questions from a given document. One of the primary applications of QG is in the educational field, where it is used to create reading exercises and assessments [1,2]. It can also complement question answering systems (QA) by improving their performance [3,4]. Moreover, QG can assist QA systems in generating training datasets. On the internet, there are numerous articles, news pieces, and literature that lack corresponding questions and must be manually marked, a labor-intensive and time-consuming process that requires skilled markers. A question generation system can generate matching questions automatically, thus saving significant labor and time costs.

Much of the previous work has focused on generating questions using SQuAD [5], a dataset contains simple and straightforward questions. Most questions can be answered using a few sentences, without the consideration of context over entire document, and more than 90% of the questions can be answered in a single sentence [6]. Recently, more and more researchers tend to use multi-hop datasets to generate more in-depth questions. Pre-trained language model (PLM) have shown strong performance on this task and are favored by an increasing number of researchers.

However, PLM may generate incomplete or syntactically incorrect sentences during training due to the inconsistency between the original and target domains and the complexity of the reasoning dataset. (In this paper, we focus on BART as the research target) Table 1 illustrates this situation, where BART generates meaningless output in Example1 by copying a subspan of the context without understanding its contextual meaning. In Example2, the generated question contains all the necessary components, but its question structure does not match the grammatical specification.

To address these challenges, we propose a BRQG framework, which improves the question generated by BART in two ways. First, we introduce a Retouching Network module, an additional GRU-based decoder that computes the attention to the context and to the first round of questions separately through the attention layer, fuses these two parts of attention through the Retouching Gate. This approach help us to control the model's allocation of attention to contextual and first-round generated questions, leading to higher quality second-round questions. In addition to our proposed approach, we also propose a Entity Awareness Enhancement module in combination with methods from previous research work. Specifically, we leverage techniques including the extraction of entities from input documents to construct graph structures and the utilization of graph attention network (GAT) to facilitate entity reasoning. By implementing these methods, we have effectively improved the quality of the questions as well as the accuracy of the entity predictions. We conducted experiments on HotpotQA, the results showed that our model outperformed the state-of-the-art model by 1.77 BLEU4 points.

2 Related Works

2.1 Question Generation

The task of Question Generation (QG) is to generate questions from given sentences or paragraphs. Early QG work typically used rule-based or manually crafted question templates to generate questions by transforming declarative sentences into interrogative ones [1,7]. This approach requires researchers to have a deep understanding of grammar and also requires a significant amount of manual effort. Its generality and scalability are relatively weak. With the development of deep learning and neural networks, neural question generation model using Sequence-to-Sequence combined with attention mechanism has been proposed [8]. Later, Zhou et al. [9] proposed a feature-rich encoder, which greatly improves the performance from the model by adding Answer Position Feature and Lexical Feature. Kim et al. [10] proposed to use a special token to mask the answer information from the input, which solved the problem that the generated question contains answer information. Zhao et al. [11] proposed a Max-out pointer mechanism to solve the problem of repeated word generation in the output sequence. Nema et al. [12] noticed that previous models are prone to generate incomplete questions, and proposed Refine Network to solve this problem.

2.2 Multi-hop Question Generation

Multi-hop question generation aims to generate deep questions that require reasoning before answering. Pan et al. [13] improve model reasoning capabilities by constructing semantic graphs and entity graphs, respectively. Su et al. [14] proposed the use of using QA model to enhance the pre-trained model BART to constrain the generated questions. Wang et al. [15] proposed an answer-aware initialization module that introduces document and answer information into the decoder, allowing the model to focus more on answer information. Fei et al. [16] found that there is no method to guarantee the complexity of the model generation questions, and thus proposed the use of controlled generation method to make more key entities appear in the questions. Although previous researchers have proposed powerful models based on BART, they have not solved the problems of generating incorrect entities and generating sentences of irrational length. Further improving the quality of its generated questions on a pre-trained language model with strong performance is the main goal of our current study.

3 Methodology

3.1 Task Define

Given a document D and an answer A, the goal is to find question $\overline{Q}$. Question Q and answer A are all composed of sequences of words, we define the MQG task as:

$$\bar{Q} = \underset{Q}{\text{argmax}} P(Q \mid D, A) \tag{1}$$

where each document $D = \{p_t\}_{t=1}^{m}$ consists of several paragraphs, and each paragraph $P = \{s_t\}_{t=1}^{n}$ has a corresponding title, consisting of several sentences. Some of these sentences are evidence sentences which can provide support for reasoning. Since answer A may not be a sub-span in document D, to answer question Q, it is necessary to reason about supporting sentences.

The structure of our proposed model is shown in Fig. 1.

3.2 BART-Based Question Generation Module

We chose BART as the basic question generator because of its excellent performance on MQG. BART consists of an encoder and a decoder, as shown in Fig. 2, we define the encoder part of BART as contextual feature extractor; the decoder part of BART as basic question generator.

The contextual feature extractor encodes the concatenation of the answer A and C and outputs the context embedding $\mathbf{C}^{bart} = \{c_1^{\text{bart}}, c_2^{\text{bart}}, \ldots, c_l^{\text{bart}}\} \in \mathbb{R}^{l \times d}$. Where the context C here is a concatenation of sentence and its corresponding paragraph title, and l denote the length of input, and d is the hidden dim of BART. While the basic question generator is responsible for generating basic question $\mathbf{Q}_1 = \{y_1, y_2, \ldots, y_{|y|}\}$ of length $|y|$ after receiving the encoder hidden states.

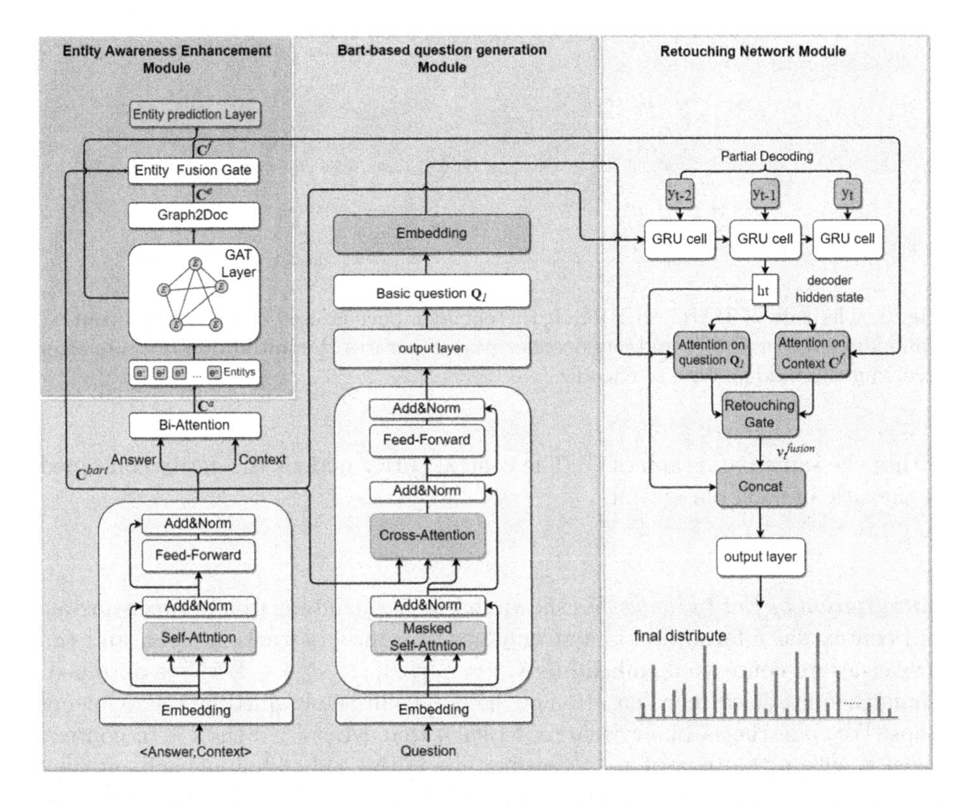

Fig. 1. BRQG Architecture. In the initial stage, BART generates the first round of question $\mathbf{Q}_1$ by context and computes the embedding representation to obtain $e^{\mathbf{Q}_1}$. The last layer of hidden state representation $\mathbf{C}^{bart}$ of the encoder part of BART is used as constructing the initial representation of entity nodes. After reasoning by GAT, the last layer of entity representation will flow back to the context representation to get $\mathbf{C}^e$ and the Entity Fusion Gate layer will fuse it with $\mathbf{C}^{bart}$ to get $\mathbf{C}^f$. In the second stage, the Retouching Network module computes the attention vector for $\mathbf{C}^f$ and $e^{\mathbf{Q}_1}$ respectively to generate the final question after fusion by the Attention Fusion Gate layer.

3.3 Entity Awareness Enhancement Module

The main goal of the multi-hop reasoning task is to link multiple pieces of information distributed in an article and do reasoning. In previous work, researchers have made the model's reasoning ability effectively improved by constructing entity graph from input documents. Reference to the work of Qiu et al. [17] we extracted entities from the context and connected them according to the rules: First, we establish sentence-level links for every pair of entities that co-occur within the same sentence of C. Second, we establish context-level links for every pair of entities that share identical mentioned text within C. Third, we establish paragraph-level links between a central entity node and other entities present

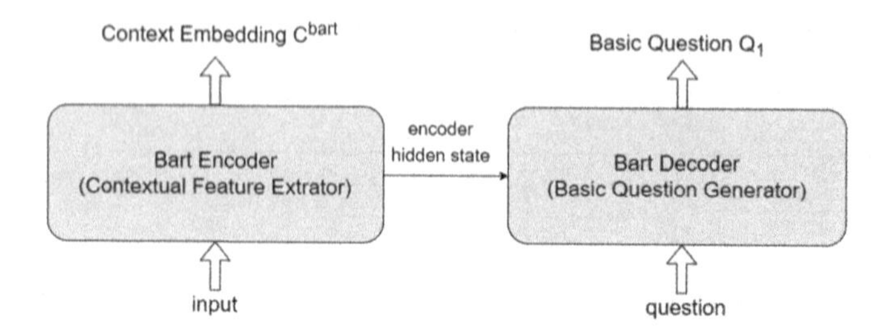

Fig. 2. The role of BART in BRQG, the encoder part is used to obtain the context embedding vector $\mathbf{C}^{bart}$, and the decoder part generates the initial question $\mathbf{Q}_1$ after receiving the hidden state of encoder.

within the same paragraph of C. The central entity node is the entity contained in the title of each paragraph.

Initialization of Entity Nodes. As shown in Fig. 1, the interaction between answer and contextual information is first enhanced by the bi-attention layer, and the answer-aware contextual embedding $\mathbf{C}^a = \{c_1^a, c_2^a, \ldots, c_m^a\} \in \mathbb{R}^{m \times d}$ is obtained. Then, the initialized representation of the node will be obtained by the following steps: We construct a binary matrix $\mathbf{M}$ such that $\mathbf{M}_{ij} = 1$ if the $i-th$ context token is within the span of $j-th$ entity. The initial embedding of each entity is obtained by calculating mean-max pooling for the span corresponding to the M matrix. And obtain the initial node embedding $E_0 = \{e_1^0, e_2^0, \ldots, e_n^0\}$.

GAT Reasoning Layer. Then we calculate the attention score α between two entity nodes as described in Velikovi et al. [18]:

$$\alpha_{i,j} = \frac{\exp\left(\text{LeakyReLU}\left(e_{i,j}\right)\right)}{\sum_{k \in \mathcal{N}_i} \exp\left(\text{LeakyReLU}\left(e_{i,k}\right)\right)} \tag{2}$$

$$e_{i,j} = \sigma\left(\mathbf{W}\left[h_i, h_j\right]\right) \tag{3}$$

$$h_i = \mathbf{W}^h e_i + b \tag{4}$$

where $\mathbf{W}$, $\mathbf{W}^h$ are trainable matrices, $\alpha_{i,j}$ is the attention score between two entities i and entity j , $\mathcal{N}_i$ is the neighbors of node i.

Then we aggregate the information of nodes through multi-head attention:

$$e_i^{t+1} = \|_{k=1}^{K} \sigma\Big(\sum_{j \in \mathcal{N}_i} \alpha_{ij}^k W^k h_j\Big) \tag{5}$$

where $||$ is concatenate operation, $\mathbf{W}^k$ is the parameter matrix shared for each head. Then we obtain the final node embedding $E_t = \{e_1^t, e_2^t, \ldots, e_n^t\}$.

Graph2Doc Layer. To facilitate the subsequent computation of the model, we refer to the work of Qiu et al. [17] to select the tokens of the involved entities by the matrix $\mathbf{M}$ computed above, and finally use LSTM to further make the entity information flow back into the context, and obtain the context embedding with entity information:

$$\mathbf{C}^{e(t)} = \mathrm{LSTM}([\mathbf{C}^{a(t-1)}, \mathbf{ME}^{(t)\top}]) \tag{6}$$

Entity Fusion Gate. After obtaining $\mathbf{C}^{e(t)}$ from Graph2Doc layer, we fuse it with the context embedding $\mathbf{C}^{bart}$ by Entity Fusion Gate to obtain the entity-fused context embedding $\mathbf{C}^{f}$:

$$\mathbf{C}^{f} = g^{e} \odot \mathbf{C}^{bart} + (1 - g^{e}) \odot \mathbf{C}^{e(t)} \tag{7}$$

$$g^{e} = \mathrm{sigmoid}(\mathrm{F}(\mathbf{C}^{\mathrm{bart}}) + \mathrm{F}(\mathbf{C}^{e(t)})) \tag{8}$$

where $\odot$ denotes element-wise multiplication. $\mathrm{F}(\cdot)$ is standard nonlinear transformation function (i.e., $\mathrm{F}(x) = \sigma(\mathbf{W}x + \mathbf{b})$, where σ indicates sigmoid function).

Entity Prediction Layer. To enhance the accuracy of the entity generation model within MQG, we introduced an entity prediction task, which involves predicting whether each entity should appear in the generated question. Consequently, we incorporated an additional feed-forward layer to perform binary classification on the entity nodes. Specifically, if the content of a node appears in the ground-truth question, we consider it as a positive ground-truth instance for training the key entity extraction task. The training process is optimized using the Binary Cross Entropy Loss (BCELoss).

3.4 Retouching Network Module

Inspired by Nema et al. [12] we propose Retouching Network, compared to the Refine Network proposed by them. The Retouching Network acts directly on the more powerful pre-trained model, and adds the Retouching Gate to control the reasonable distribution of attention ratio. It simulates the distribution of human attention to the original text when revising drafts of different quality.

Retouching network is responsible for modifying the questions generated by BART, which does the same work as decoder. It generates the second round of questions $\mathbf{Q}_2$ sequentially conditional on the entity-fused context embedding $\mathbf{C}^{f}$, the first round of questions $\mathbf{Q}_1$ and the previously decoded words y_{t-1}.

$$h_t = GRU\left(\left[\mathrm{emb}\left(y_{t-1}\right), v_{t-1}^{\mathrm{fusion}}\right], h_{t-1}\right) \tag{9}$$

where $v_{t-1}^{\mathrm{fusion}}$ is the attention vector obtained by fusing the context vector and the question vector at time $t-1$, which we will describe in detail in the following sections(Eq. 16.). h_{t-1} is the GRU hidden state at time step $t-1$ and $emb(\cdot)$ represents to obtain the embedding vector of word.

Attention Calculation. Here we use Luong attention mechanism [19] to calculate the attention score β^t of the t-th hidden state h_t and the entity-fused context embedding $\mathbf{C}^f$. Then we obtain the context attention vector c_t.

$$score(h_t, \mathbf{C}^f) = \mathbf{C}^f \mathbf{W}^f h_t \tag{10}$$

$$\beta^t = softmax(score(h_t, \mathbf{C}^f)) \tag{11}$$

$$c_t = C^f \cdot \beta^t \tag{12}$$

where $\mathbf{W}^f$ is a trainable parameter, β^t is the attention score of the t-th hidden state h_t to the entity-fused context embedding $\mathbf{C}^f$.

Since the main work of Retouching Network is to modify the BART-generated questions, we also need to calculate the question attention vector of the question $\mathbf{Q}_1$:

$$score(h_t, emb(\mathbf{Q}_1)) = emb(\mathbf{Q}_1)\mathbf{W}^e h_t \tag{13}$$

$$\gamma^t = softmax(score(h_t, emb(\mathbf{Q}_1))) \tag{14}$$

$$q_t = emb(\mathbf{Q}_1) \cdot \gamma^t \tag{15}$$

where $\mathbf{W}^e$ is a trainable parameter, γ^t is the attention score of the t-th hidden state h_t to embedding vector of question $\mathbf{Q}_1$.

Retouching Gate. After obtaining the context attention vectors c_t and question attention vector q_t, we propose to use a gate unit to control the fusion of these two vectors:

$$v_t^{\text{fusion}} = g_t^{\text{retouching}} \odot c_t + (1 - g_t^{\text{retouching}}) \odot q_t \tag{16}$$

$$g_t^{\text{retouching}} = \text{sigmoid}(\text{F}(c_t) + \text{F}(q_t) + \text{F}(h_t)) \tag{17}$$

where $\odot$ and $\text{F}(\cdot)$ are consistent with the usage in Eq. 8.

Output and Optimization. Eq. 18 shows the concatenation of the decoder hidden state h_t and the fusion attention vector v_t^{fusion}, which is used to generate the words for the time step t.

$$\widehat{h}_t = \tanh\left(W^p\left[h_t; v_t^{\text{fusion}}\right]\right) \tag{18}$$

Finally, y_t is obtained by selecting the highest probability word in p_g. Our training objective is to minimize the negative log-likelihood of the training data with respect to all the parameters:

$$p_g = \text{softmax}(\mathbf{W}^o[\widehat{h}_t]) \tag{19}$$

$$y_t = argmax\,(p_g) \tag{20}$$

$$L(\theta_b) = -\frac{1}{N}\sum_i^N \log P_{\text{generate}}\,(y \mid y_{<t}; \theta_b) \tag{21}$$

where $\mathbf{W}^o$ is a trainable parameter matrix with vocabulary size, θ_b is the parameters of the whole model.

4 Experiments and Results

4.1 Experimental Setup

To evaluate our model, we conduct experiments on the HotpotQA dataset, which contains 100k crowd-sourced questions that require reasoning over separate Wikipedia articles. Each question is paired with two supporting documents that contain the evidence necessary to infer the answer. We followed the spilt of 90440/6072 to divide the training set and test set respectively. And using BLEU1-4 [20], METEOR [21], and ROUGE-L [22] as evaluation metrics, where BLEU measures the average n-gram overlap of a set of reference sentences, METEOR uses the n-gram idea of BLEU for machine translation, while ROUGE-L is used for text summary evaluation, a measure of sentence fluency in text generation.

4.2 Implementation Details

We utilized the BART implementation provided by Huggingface for our experiments, employing both BART-large and BART-base models. It is worth noting that we employed a progressive training approach. Specifically, we initially trained the BART model independently. Subsequently, the pre-trained BART model with fixed parameters was utilized as the foundational model, which was further trained using the proposed BRQG framework. We use two layers of GRU as retouching network decoder. And we used AdamW optimizer with a learning rate of 3e-5 for BART and 4e-4 for BRQG. The training was carried out for 5 epochs for BART and 20 epochs for BRQG, with gradient accumulation steps set to 4. To handle input sequences, we set the maximum source sequence length to 512, while the hidden state size and embedding size were set to 384 and 300, respectively. For constructing graph structures, we set the layers of GAT to 2 and extracted 8 entities for each context. During inference, we adopted beam search with a beam size of 10 and set the maximum target length to 32. Finally, for this experiment, we employed two Tesla T4 GPUs.

4.3 Baselines

We compare our proposed model with the previous MQG models and strong conventional QG models.

NQG++ [9] It enhanced the model by using a feature-rich encoder.

S2sa-at-mp-gsa [11] The model follows the pointer-generator framework and proposes gated self-attention and maxout pointers to solve the problem that the model prefers to generate words that occur multiple times in context.

ASs2s-a [10] Replacing the target answer in the original passage with a special token allows the model to identify which interrogative word should be used and proposes a keyword-net to better utilize the information from the passage and the target answer.

Table 2. Our model scores on the HotpotQA dataset using BLEU, METEOR, and ROUGE-L as evaluation metrics.

Models	BLEU-1	BLEU-2	BLEU-3	BLEU-4	METEOR	ROUGLE
NQG++ [9]	32.97	21.11	15.41	11.81	18.19	33.48
S2sa-at-mp-gsa [11]	35.36	22.38	15.88	11.85	17.63	33.02
ASs2s-a [10]	37.67	23.79	17.21	12.59	17.45	33.21
SGGDQ(DP) [13]	40.55	27.21	20.13	15.53	20.15	36.94
ADDQG [15]	44.34	31.32	22.68	17.54	20.56	38.09
BART-base [23]	46.30	34.44	27.34	22.36	25.18	40.46
BART-large [23]	49.51	37.87	30.75	25.56	27.24	46.21
QA4QG(LARGE setting) [14]	49.55	37.91	30.79	25.70	27.44	46.48
BART-base + paragraph title	49.96	37.31	29.78	24.41	26.02	41.77
BART-large + paragraph title	51.24	39.59	32.40	27.18	**28.33**	47.41
BRQG(BASE setting)	**54.06**	40.74	32.83	27.19	26.78	48.04
BRQG(LARGE setting)	53.99	**40.86**	**33.03**	**27.45**	27.11	**48.13**

SGGDQ [13] The model constructs graphs at the semantic level for the input and encodes the semantic graphs using GGNN, fusing document-level and graph-level representations.

ADDQG [15] Proposed an answer-aware initialization module with a gated connection layer and designed with a semantic-rich fusion attention mechanism.

BART [23] A generative pre-trained language model that uses a combination of auto-regression and bi-directional architectures to generate high-quality text, achieving state-of-the-art results on multiple language generation tasks.

QA4QG [14] Add QA models to BART for generating higher quality questions using the constraints between questions and answers.

4.4 Main Results

As shown in Table 2, we have added paragraph title information, which gives a significant performance boost to the bart model. After using the BRQG framework, the performance further improved and achieved the best result in the BLEU4 metric, improving 1.44 BLEU points compared to the QA4QG model. Compared with BART-large, the BRQG based on BART-base has more obvious effect improvement.

4.5 Ablation Experiments

To evaluate the performance of each component in the model, we conducted ablation experiments for BRQG on base and large setting respectively. And show the result in Table 3.

BRQG w/o RNm: When removing **Retouching Network Module**, the performance of the large and base settings on BLEU-4 drops by 0.27 and 2.78

Table 3. Ablation experiment of BRQG.

Models	BLEU-4	METEOR	ROUGE-L	Models	BLEU-4	METEOR	ROUGE-L
BRQG(Base setting)	**27.45**	27.11	**48.13**	BRQG(Large setting)	**27.19**	**26.78**	**48.04**
w/o RNm	27.18	**28.33**	47.41	w/o RNm	24.41	26.02	41.77
w/o RGu	26.95	26.64	48.00	w/o RGu	26.05	26.12	47.04
w/o EFGu	26.87	26.55	47.93	w/o EFGu	26.74	26.77	47.71
w/o EAEm	26.65	26.94	47.85	w/o EAEm	26.42	26.69	47.66

points, respectively, demonstrating that Retouching Network is an important module in the BRQG, especially for BART-base, where Retouching Network has a great improvement on it.

BRQG w/o RGu: When removing **Retouching Gate Unit**, the performance of the model is greatly reduced, which is consistent with our intuition that different proportions of attention should be assigned to questions of different quality generated by BART.

BRQG w/o EFGu: When removing the **Entity Fusion Gate unit**, the performance of the model decreases because of the contextual information with entities and the contextual information are not effectively fused.

BRQG w/o EAEm: When removing **Entity Awareness Enhancement Module**, this means that GATs are no longer used to reason about entities, the model cannot enhance the entity information, and Retouching Network's awareness of the entity will be affected when generating the questions, thus the performance will be dropped.

4.6 Analysis of Question Quality

In this section, we evaluate the quality of questions generated by the BRQG framework in terms of the correctness of generated question words, the length distribution of generated questions, and the correctness of generated entities. To demonstrate the improvement of BRQG over BART, we compare the performance of BRQG with BART-base and BART-large, respectively.

Entity Prediction. As shown in Fig. 3, BART tends to generate more entities than BRQG, but many of these entities are incorrect. Here we calculate the accuracy of entities by $Entity_{correct}/Entity_{all}$, where $Entity_{correct}$ is the number of entities that appear in the reference in the question generated by the model, and $Entity_{all}$ is the total number of entities generated by the model. The accuracy of the entities generated by BRQG is 1.85% and 1.86% higher than that of BART, respectively, which indicates that BRQG has stronger entity awareness capability compared to BART.

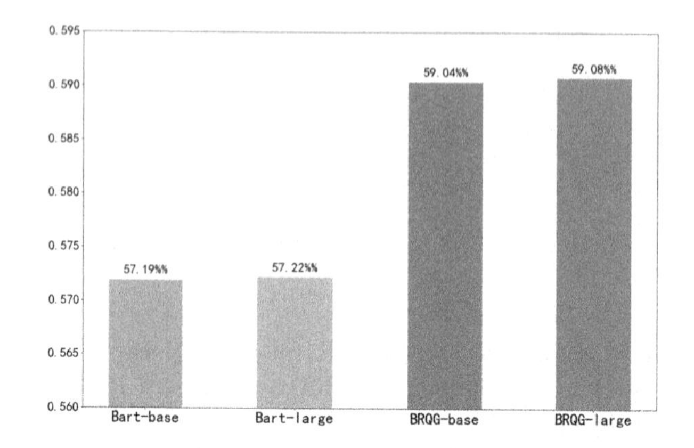

Fig. 3. Accuracy of entities generated by BRQG and BART.

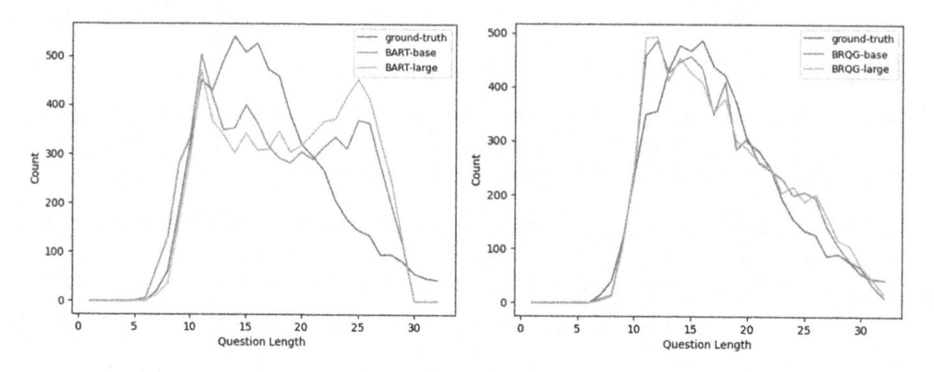

Fig. 4. Question length distribution of BART.

Fig. 5. Question length distribution of BRQG.

Question Length Distribution. We analyzed the length distribution of questions produced by both models and found that BART-generated questions significantly differ in length from the ground-truth questions, which matches our previous mention of the performance degradation of BART after task migration. However, when we use the BRQG framework, the length of the questions largely fit with groud-truth, thus proving that the BRQG framework can effectively modify the BART-generated questions (Fig. 4).

Generation of Question Words. In this section, we evaluate the number of correct question words generated by BART-base, BART-large, and BRQG. We consider the first word of each sentence as the question word and count it as correct if it matches the ground truth. Figure 6 shows that BART-base generated the least number of correct question words, while BRQG (Base setting) generated 1657 more correct question words, which is an increase of 203%. Compared to

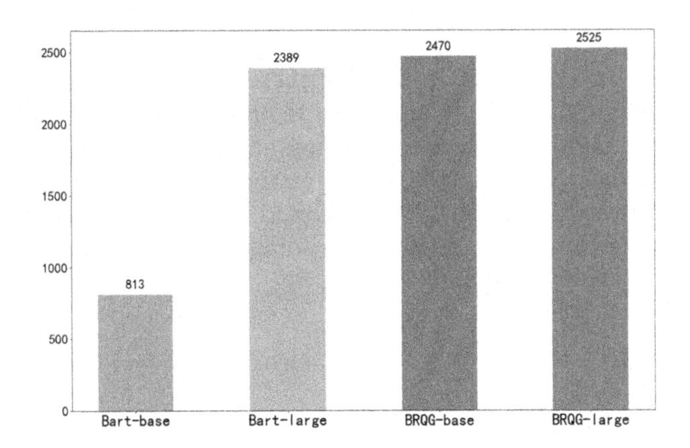

Fig. 6. Comparison of the number of correct question words.

Reference : Are Local H and For Against both from the United States ?

(a) BRQG without Retouching Gate

Are Local H and For Against both American bands ?

<s>	yes	</s>	<s>	Local	H	Local	H	is	an
American	rock	band	originally	formed	by	guitarist	and	vocalist	Scott
Lucas	,	bass	ist	Matt	Garcia	,	drummer	Joe	Daniels
,	and	lead	guitarist	John	Spark	man	in	Zion	,
Illinois	in	1987		For	Against	For	Against	is	a
United	States	post	-	punk	/	dream	pop	band	from
Lincoln	,	Nebraska	.	</s>					

(b) BRQG with Retouching Gate

Are Local H and For Against both bands from the United States ?

<s>	yes	</s>	<s>	Local	H	Local	H	is	an
American	rock	band	originally	formed	by	guitarist	and	vocalist	Scott
Lucas	,	bass	ist	Matt	Garcia	,	drummer	Joe	Daniels
,	and	lead	guitarist	John	Spark	man	in	Zion	,
Illinois	in	1987	.	For	Against	For	Against	is	a
United	States	post	-	punk	/	dream	pop	band	from
Lincoln	,	Nebraska	.	</s>					

Fig. 7. Compared to the removal of RetouchingGate, the addition of the Gate unit assigns more attention to contextually useful information and reduces the attention to characters such as "is", "." , "-" and other useless characters.

BART-large, BRQG (Base setting) generated an additional 136 correct question words, representing a 5.6% increase. Furthermore, BRQG (Base setting) outperformed BART-large in terms of the number of correct question words generated (Fig. 5).

4.7 Analysis of Retouching Gate

The purpose of this section is to assess the effectiveness of the Retouching Gate. To illustrate this, we present examples (a) and (b) in Fig. 7, which compare the attention of BRQG with Retouching Gate and BRQG without Retouching Gate on the given context. By examining these examples, it is apparent that without the gate unit, BRQG's attention is more scattered, resulting in less focus on relevant information and producing an erroneous entity "American". Conversely, after incorporating the gate, attention to the phrase "United States" considerably improved, while the attention paid to irrelevant characters such as "is", ".", and "-" decreased, ultimately leading to the creation of a question similar to the reference question.

5 Conclusion

In this paper, we propose the BRQG framework to enhance the quality of BART-generated questions. The framework comprises two additional modules, namely the Entity Awareness Enhancement Module and Retouching Network Module, integrated with the BART model. Our experiments show that the Retouching Network is highly effective in improving the quality of questions generated by BART, including aspects like the distribution of question length and the use of accurate question words. Additionally, the Entity Awareness Enhancement Module successfully enhances the correctness rate of entities generated by BRQG, when compared to BART. Finally, we demonstrate the effectiveness of Retouching Gate, which allocates attention effectively and helps the model focus on relevant information. In theory, our model is expected to perform well on generative language models such as BART and T5. However, for the current study, we only conducted experiments on the BART model. In future work, we plan to extend our experiments to include a wider range of models. We sincerely hope that our work will be useful for subsequent research on MQG.

References

1. Heilman, M., Smith, N.A.: Good question! statistical ranking for question generation. In: Human Language Technologies: The 2010 Annual Conference of the North American Chapter of the Association for Computational Linguistics, pp. 609–617 (2010)
2. Danon, G., Last, M.: A syntactic approach to domain-specific automatic question generation. arXiv preprint: arXiv:1712.09827 (2017)
3. Duan, N., Tang, D., Chen, P., Zhou, M.: Question generation for question answering. In: Proceedings of the 2017 Conference on Empirical Methods in Natural Language Processing, pp. 866–874 (2017)
4. Tang, D., Duan, N., Qin, T., Yan, Z., Zhou, M.: Question answering and question generation as dual tasks. arXiv preprint: arXiv:1706.02027 (2017)
5. Rajpurkar, P., Zhang, J., Lopyrev, K., Liang, P.: Squad: 100,000+ questions for machine comprehension of text. arXiv preprint: arXiv:1606.05250 (2016)
6. Min, S., Zhong, V., Socher, R., Xiong, C.: Efficient and robust question answering from minimal context over documents. arXiv preprint: arXiv:1805.08092 (2018)
7. KMazidi, K., Nielsen, R.: Linguistic considerations in automatic question generation. In: Proceedings of the 52nd Annual Meeting of the Association for Computational Linguistics (Volume 2: Short Papers), pp. 321–326 (2014)

8. Du, X., Shao, J., Cardie, C.: Learning to ask: neural question generation for reading comprehension. arXiv preprint: arXiv:1705.00106 (2017)
9. Zhou, Q., Yang, N., Wei, F., Tan, C., Bao, H., Zhou, M.: Neural question generation from text: a preliminary study. In: Huang, X., Jiang, J., Zhao, D., Feng, Y., Hong, Yu. (eds.) NLPCC 2017. LNCS (LNAI), vol. 10619, pp. 662–671. Springer, Cham (2018). https://doi.org/10.1007/978-3-319-73618-1_56
10. Kim, Y., Lee, H., Shin, J., Jung, K.: Improving neural question generation using answer separation. In: Proceedings of the AAAI Conference on Artificial Intelligence, vol. 33, pp. 6602–6609 (2019)
11. Zhao, Y., Ni, X., Ding, Y., Ke, Q.: Paragraph-level neural question generation with maxout pointer and gated self-attention networks. In: Proceedings of the 2018 Conference on Empirical Methods in Natural Language Processing, pp. 3901–3910 (2018)
12. Nema, P., Mohankumar, A.K., Khapra, M.M., Srinivasan, B.V., Ravindran, B.: Let's ask again: refine network for automatic question generation. arXiv preprint: arXiv:1909.05355 (2019)
13. Pan, L., Xie, Y., Feng, Y., Chua, T.S., Kan, M.Y.: Semantic graphs for generating deep questions. arXiv preprint: arXiv:2004.12704 (2020)
14. Su, D., Xu, P., Fung, P.: QA4QG: using question answering to constrain multi-hop question generation. In: ICASSP 2022–2022 IEEE International Conference on Acoustics, Speech and Signal Processing (ICASSP), pp. 8232–8236. IEEE (2022)
15. Wang, L., Xu, Z., Lin, Z., Zheng, H., Shen, Y.: Answer-driven deep question generation based on reinforcement learning. In: Proceedings of the 28th International Conference on Computational Linguistics, pp. 5159–5170 (2020)
16. Fei, Z., Zhang, Q., Gui, T., Liang, D., Wang, S., Wu, W., Huang, X.J.: CQG: a simple and effective controlled generation framework for multi-hop question generation. In: Proceedings of the 60th Annual Meeting of the Association for Computational Linguistics (Volume 1: Long Papers), pp. 6896–6906 (2022)
17. Qiu, L., et al.: Dynamically fused graph network for multi-hop reasoning. In: Proceedings of the 57th Annual Meeting of the Association for Computational Linguistics, pp. 6140–6150 (2019)
18. Velickovic, P., et al.: Graph attention networks. Stat **1050**(20), 10–48550 (2017)
19. Luong, M.T., Pham, H., Manning, C.D.: Effective approaches to attention-based neural machine translation. arXiv preprint: arXiv:1508.04025 (2015)
20. Papineni, K., Roukos, S., Ward, T., Zhu, W. J.: BLEU: a method for automatic evaluation of machine translation. In: Proceedings of the 40th Annual Meeting of the Association for Computational Linguistics, pp. 311–318 (2002)
21. Banerjee, S., Lavie, A.: METEOR: an automatic metric for MT evaluation with improved correlation with human judgments. In: Proceedings of the ACL Workshop on Intrinsic and Extrinsic Evaluation Measures for Machine Translation and/or Summarization, pp. 65–72 (2005)
22. Lin, C.-Y.: Rouge: a package for automatic evaluation of summaries. In: Text Summarization Branches Out, pp. 74–81 (2004)
23. Lewis, M., et al.: BART: denoising sequence-to-sequence pre-training for natural language generation, translation, and comprehension. arXiv preprint: arXiv:1910.13461 (2019)

MTSTI: A Multi-task Learning Framework for Spatiotemporal Imputation

Yakun Chen, Kaize Shi, Xianzhi Wang(✉), and Guandong Xu

University of Technology Sydney, Sydney, NSW 2007, Australia
yakun.chen@student.uts.edu.au,
{kaize.shi,xianzhi.wang,guandong.xu}@uts.edu.au

Abstract. Spatiotemporal data analysis is crucial for various fields of applications, such as transportation, healthcare, and meteorology. Spatiotemporal data collected in the real world often contain missing values due to sensor failures or transmission loss. Therefore, spatiotemporal imputation aims to fill in the missing values by leveraging the underlying spatial and temporal dependencies in the partially observed data. Previous models for spatiotemporal imputation focus solely on the imputation task as a preparatory step for solving the downstream tasks. Instead, we aim to use downstream tasks to reinforce spatiotemporal imputation and further propose a multi-task learning framework, MTSTI, for spatiotemporal imputation. Our proposed framework utilizes a graph neural network to learn spatiotemporal representations via message-passing. The multi-task learning structure, combining spatiotemporal imputation with the forecasting task, provides additional insights that enhance the model's performance and generality. Our empirical results demonstrate that our proposed framework outperforms state-of-the-art methods in the imputation task on various real-world datasets across different fields.

Keywords: Graph Neural Network · Multitask Learning · Spatiotemporal Imputation

1 Introduction

Multivariate time series analysis has found widespread application in numerous fields, such as economics, transportation, healthcare, and meteorology [5,17,30,35]. Many machine learning models can achieve competitive performance on complete datasets in various time-series analysis tasks, such as forecasting [13], classification [15], and anomaly detection [6]. However, multivariate time series datasets naturally contain missing values in the real world for various reasons, such as sensor failures and transmission loss. This may significantly reduce the quality of datasets and thereby impact the performance of pattern learning and inference on subsequent tasks.

X. Yang et al. (Eds.): ADMA 2023, LNAI 14180, pp. 180–194, 2023.
https://doi.org/10.1007/978-3-031-46677-9_13

Several efforts [20] have been recently made to address the spatiotemporal imputation problem. Typically, such studies leverage spatiotemporal dependencies learned from available observed data to reconstruct the entire sequence, and then existing models can be applied directly to accomplish the downstream tasks. Early research on spatiotemporal imputation relied on classical statistical or machine learning methods, typically including autoregressive moving average (ARMA) [2], expectation-maximization algorithm (EM) [26], and k-nearest neighbors (kNN) [14,23]. These models generally make strong assumptions, such as temporal smoothness and inter-series similarity about time series data, which may not hold in complex spatiotemporal data. Consequently, these models may not yield satisfactory performance in practice. As deep learning gains prevalence, researchers turn to exploring recurrent neural networks (RNN) as the backbone structure for spatiotemporal imputation. These models recursively update their hidden state to capture temporal dependencies with existing observations to impute missing values [7,8]. More recent research extends these studies by incorporating graph neural network (GNN) to learn the spatial dependencies between series and combining it with RNN-based methods [10,12]. All the above approaches for imputing missing data focus merely on the imputation tasks while ignoring the significance of downstream tasks in helping with the imputation tasks.

We propose a **M**ulti-**T**ask learning framework for **S**patio-**T**emporal **I**mputation (MTSTI), which jointly optimizes the imputation and downstream tasks to enhance the model's performance and generalization ability. In particular, we incorporate multivariate time series forecasting as an auxiliary task to aid the extraction of temporal features from time series data and apply these features to time series imputation. As such, our model can understand the input data more comprehensively, providing additional clues to reduce overfitting and increase the model's robustness and generalization ability. Specifically, MTSTI combines gated recurrent units (GRU) and graph neural networks (GNN) to comprehensively leverage features for the spatiotemporal imputation problem. Specifically, the GRU structure serves as the backbone to capture temporal dependencies and the GNN learns spatial dependencies among time series. The main contributions of this paper are summarized as follows:

- We propose MTSTI, a novel multi-task learning framework that combines graph neural network (GNN) and gated recurrent unit (GRU) for spatiotemporal imputation.
- We incorporate an auxiliary task of time series forecasting in its multi-task learning module, which effectively improves the model's generalization ability and performance in capturing spatiotemporal dependencies.
- We demonstrate the superiority of our proposed method in spatiotemporal imputation tasks through extensive experiments and analyses in comparison with state-of-the-art and competitive baselines on two real-world datasets.

The rest of this paper is organized as follows. Section 2 introduces the spatiotemporal imputation problem and presents a brief overview of graph definition. Section 3 presents the details of our proposed MTSTI framework, including the use of graph and recurrent neural networks and the introduction of a time series forecasting auxiliary task to enhance model performance. Section 4 reports our evaluation of our framework's performance on real-world spatiotemporal datasets. Section 5 reviews the related work on spatiotemporal imputation, and finally, Sect. 6 provides the concluding remarks.

2 Preliminaries

2.1 Spatiotemporal Imputation

Spatiotemporal imputation refers to the prediction of missing values in historical time-series records. We formalize spatiotemporal data as a sequence $\mathbf{X}_{1:T} = \{\mathbf{X}_1, \ldots, \mathbf{X}_t, \ldots, \mathbf{X}_T\} \in \mathbb{R}^{N \times T}$, where $\mathbf{X}_t = \{x_t^1, x_t^1, \ldots, x_t^N\} \in \mathbb{R}^N$ represents the values observed at time t by N observation nodes, e.g., air monitoring stations and traffic speed sensors. A binary mask matrix $\mathbf{M} \in \{0,1\}^{N \times T}$ is also an important corresponding input for spatiotemporal imputation task to denote the positions of missing values in $\mathbf{X}_{1:T}$. If $m_t^n = 0$, then x_t^n is missing, and $m_t^n = 1$ indicates otherwise. Besides, we use the imputation target (denoted by $\widehat{\mathbf{Y}} \in \mathbb{R}^{N \times T}$), i.e., the ground truth for the missing values to be predicted, for training and evaluating the imputation methods, and use the binary mask $\mathbf{M}$ to identify them among spatiotemporal data. Thus, the spatiotemporal imputation task is to The task of multivariate time series imputation is to find the closest approximations for the missing values in $\mathbf{X}_{1:T}$ and then generate a complete dataset.

2.2 Graph Definition

The observation nodes in spatiotemporal data can be represented as a graph $\mathcal{G} = (\mathcal{V}, \mathcal{E})$, where $\mathcal{V}$ and $\mathcal{E}$ denote the sets of nodes and edges, respectively. Each edge $e_{ij} \in \mathcal{E}$ corresponds to (v_i, v_j), representing the linkage between v_i and v_j. $\mathbf{A} \in \mathbb{R}^{N \times N}$ is the adjacency matrix used to indicate the relationships among all nodes in the graph, where N is the number of variables. If $e_{ij} \in \mathcal{E}$, then $a_{ij} \neq 0$; otherwise, $a_{ij} = 0$.

3 Methodology

Our proposed framework, **M**ulti-**T**ask **S**patio-**T**emporal **I**mputation (MTSTI), utilizes a multi-task learning approach to leverage spatiotemporal correlations and geographic relationships for imputation. MTSTI (illustrated in Fig. 1a) consists of a bidirectional pipeline, which processes the input sequence sequentially in both forward and backward directions. The final results for imputation and forecasting depend on the outputs of two MTSTI modules, whose learned representations are subsequently processed by two distinct last readout modules. In accordance with previous work [12], we incorporate a bidirectional extension to our framework, following the unidirectional model's detailed description.

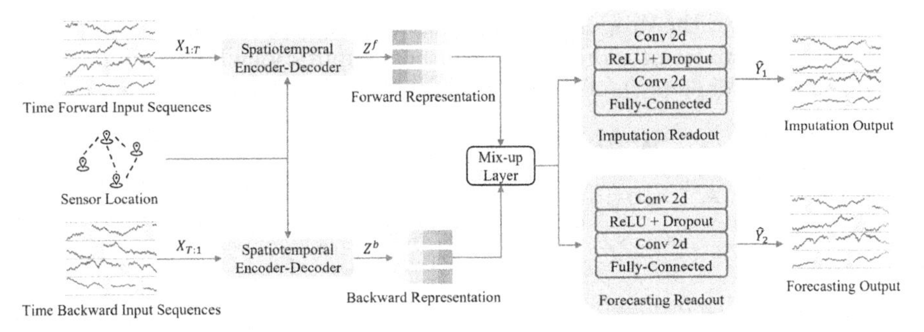

(a) Overall Architecture of MTSTI

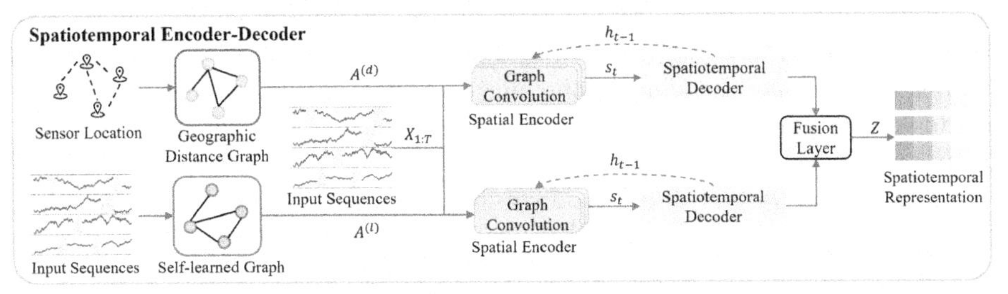

(b) Architecture of Spatiotemporal Encoder-Decoder

Fig. 1. Our MTSTI framework: (a) The overall structure of MTSTI, which consists of two Spatiotemporal Encoder-Decoder modules, a mix-up layer, and two readout modules; (b) The Spatiotemporal Encoder-Decoder module, which combines the Spatial Encoder, Spatiotemporal Decoder, and Fusion Layer.

3.1 Spatiotemporal Encoder-Decoder

The Spatiotemporal Encoder-Decoder structure (Fig. 1b) consists of two distinct blocks: the spatial encoder and the spatiotemporal decoder. The spatial encoder is responsible for combining the information derived from the input sequence along with that of its neighboring elements, thus producing a spatial representation. The spatiotemporal decoder utilizes the Gated Recurrent Unit (GRU) to facilitate spatiotemporal message passing.

Spatial Encoder. In the spatial encoder, we incorporate two graphs to represent spatial dependencies, namely the geographic distance graph and the self-learned graph. To obtain the adjacency matrix $\mathbf{A}^{(d)}$ for the geographic distance graph, we follow the previous study [12,30] and employ a thresholded Gaussian kernel on the distances between nodes. In contrast, the self-learned graph $\mathbf{A}^{(l)}$ is constructed based solely on the input sequences. It does not require external location information but can capture the implicit and sparse relationships between nodes. To obtain a sparse adjacency matrix for the self-learned graph, we adopt Gumble Softmax [25,33], commonly used in reinforcement learning-based methods, to design the dropout layer. By sparsifying the adjacency matrix, the subsequent information fusion process can be computationally streamlined.

Equation (1) depicts the Gumbel softmax dropout layer of our module:

$$\begin{aligned} \boldsymbol{\Theta} &= \mathbf{W}^{(l)}\mathbf{E}\mathbf{E}^T \\ \mathbf{A}^{(l)} &= \sigma((log(\theta_{ij}/(1-\theta_{ij})) + (g_{ij}^1 - g_{ij}^2))/s) \\ &\quad s.t. \quad g_{ij}^1, g_{ij}^2 \sim Gumbel(0,1) \end{aligned} \tag{1}$$

where $\mathbf{A}_{ij}^{(l)} = 1$ with probability θ_{ij} if there is an edge from node i to node j, and $\mathbf{A}_{ij}^{(l)} = 0$ otherwise. $\mathbf{E} \in \mathbb{R}^{N \times d}$ denotes the node embedding, $\mathbf{W}^{(l)}$ denotes the learnable model parameter, and $\boldsymbol{\Theta}$ denotes the probability matrix, with $\theta_{ij} \in \boldsymbol{\Theta}$ denoting the probability of keeping the edge from node i to node j. Note that the Gumbel softmax has the same probability distribution as the normal softmax, ensuring consistency between our graph convolution module and the probability matrix generation statistically.

Our proposed model employs the sensor location information and input sequences to generate the geographic distance and self-learned graphs, respectively. The following step involves combining input data with the neighborhood information to create an aggregated representation $\mathbf{s}_t$ at time t. This process is accomplished using a graph convolution module with D layers. The graph convolution module can be formulated as follows:

$$\begin{aligned} \mathbf{s}_t^{(0)} &= \mathcal{F}\left(\mathbf{x}_t \| \mathbf{m}_t \| \mathbf{h}_{t-1}\right) \\ \mathbf{s}_t^{(d)} &= \mathbf{A}\mathbf{s}_t^{(d-1)} \\ \mathbf{s}_t &= \mathcal{F}\left(\mathbf{s}_t^{(0)} \| \mathbf{s}_t^{(1)} \| \cdots \| \mathbf{s}_t^{(D)}\right) \end{aligned} \tag{2}$$

In Eq. (2), $\mathbf{x}_t$ and $\mathbf{m}_t$ denote the input sequences and the mask at time t. Our module receives the hidden representation $\mathbf{h}_{t-1}$ from the last time step. $\mathbf{A}$ is the adjacency matrix for geographic distance and self-learned graphs. $\mathbf{s}_t^{(d)}$ represents the aggregated representation in d-th layer. $\|$ and $\mathcal{F}(\cdot)$ represent concatenation operation and representation combination function, separately.

Spatiotemporal Decoder. After receiving the aggregated representation $\mathbf{s}_t$ at time t calculated by the spatial encoder, the spatiotemporal decoder integrates it with the hidden representation $\mathbf{h}_{t-1}$ at time $t-1$ to generate $\mathbf{h}_t$. The initial hidden representation $\mathbf{h}_0$ is randomly generated by our model. Consistent with prior research [19], our proposed model applies a GRU module to capture the temporal dependencies. Based on the above, the spatiotemporal decoder can be described as:

$$\begin{aligned} r_t &= \sigma\left(\mathbf{W}_r(\mathbf{s}_t \| \mathbf{m}_t \| \mathbf{h}_{t-1}) + b_r\right) \\ u_t &= \sigma\left(\mathbf{W}_u(\mathbf{s}_t \| \mathbf{m}_t \| \mathbf{h}_{t-1}) + b_u\right) \\ \mathbf{c}_t &= \tanh\left(\mathbf{W}_c(\mathbf{s}_t \| \mathbf{m}_t \| r_t \odot \mathbf{h}_{t-1}) + b_c\right) \\ \mathbf{h}_t &= \mathbf{c}_t \odot u_t + \mathbf{h}_{t-1} \odot (1-u_t) \end{aligned} \tag{3}$$

In Eq. (3), r_t, u_t, $\mathbf{c}_t$ and $\mathbf{h}_t$ are reset gate, update gate, memory cell, and hidden representation, respectively. $\sigma(\cdot)$ and $\tanh(\cdot)$ are two different activation functions. Upon completing calculations across all T time steps, $\mathbf{s}_t$ and $\mathbf{h}_t$ are merged to create the representation of the sequence.

3.2 Bidirectional Model

Expanding our model to handle both forward and backward dynamics is a natural extension and can be achieved by duplicating the architecture outlined in Sect. 3.1. After finishing all calculations from the unidirectional modules, we combine the forward and backward outputs to generate the final imputation and forecasting result (denoted as $\widehat{\mathbf{Y}}_1$ and $\widehat{\mathbf{Y}}_2$, respectively). Specifically, this combination can be formulated as follows:

$$\begin{aligned} \hat{\mathbf{z}}_t &= \mathcal{F}\left(\mathbf{s}_t \| \mathbf{h}_{t-1}\right) \\ \widehat{\mathbf{Y}}_1 &= \text{Readout}\left(\widehat{\mathbf{Z}}^f \| \widehat{\mathbf{Z}}^b \| \mathbf{M}\right) \\ \widehat{\mathbf{Y}}_2 &= \text{Readout}\left(\widehat{\mathbf{Z}}^f \| \widehat{\mathbf{Z}}^b\right) \end{aligned} \tag{4}$$

where $\mathbf{z}_t$ is the hidden representation for $\mathbf{x}_t$ at time t. $\widehat{\mathbf{Z}}^f$, $\mathbf{Z}^b$ are the forward and backward representations of the sequence, respectively. $\widehat{\mathbf{Y}}_1$ and $\widehat{\mathbf{Y}}_2$ are the final imputation and forecasting result respectively. $\mathcal{F}(\cdot)$ is the same representation combination function with the Eq. (2). Readout$(\cdot)$ is the readout function that maps the hidden representation to the final output.

3.3 Multi-task Learning Loss

We employ a multi-task learning strategy to optimize spatiotemporal imputation and forecasting tasks jointly. Specifically, the loss function is a combination of imputation and forecasting losses which are computed respectively for each learning task:

$$\begin{aligned} \mathcal{L}_1(\mathbf{Y_1}, \widehat{\mathbf{Y}}_\mathbf{1}, \overline{\mathbf{M}}) &= \sum_{n=1}^{N} \sum_{t=1}^{T} \frac{\langle \overline{m}_t^n, l(y_t^n, \widehat{y}_t^n) \rangle}{\langle \overline{m}_t^n, \overline{m}_t^n \rangle} \\ \mathcal{L}_2(\mathbf{Y_2}, \widehat{\mathbf{Y}}_\mathbf{2}) &= \sum_{n=1}^{N} \sum_{h=1}^{H} l(y_h^n, \widehat{y}_h^n) \\ \mathcal{L} &= \alpha \mathcal{L}_1 + (1-\alpha)\mathcal{L}_2 \end{aligned} \tag{5}$$

where $\mathcal{L}_1$ is the imputation task loss, $\mathcal{L}_2$ is the forecasting task loss, and $\mathcal{L}$ is the multi-task total loss. α is a weight hyperparameter to balance the loss between imputation and forecasting task. $\overline{\mathbf{M}}$ and $\overline{m}_t^n$ are the logical binary counterpart of $\mathbf{M}$ and m_t^n, respectively. $\widehat{\mathbf{Y}}$ and $\widehat{y}_t^n$ denote the imputed values for the missing input in $\mathbf{X}$; while, $\mathbf{Y}$ and y_t^n are the corresponding ground truth. $\langle \cdot, \cdot \rangle$ is the stand dot product. $l(\cdot, \cdot)$ represents an element-wise error function computed using L1 loss in the following experiments.

4 Experiments

We evaluate the proposed model through a series of experiments designed to answer the following research questions:

- **RQ1**: Does our model achieve superior imputation performance under various missing patterns in comparison with state-of-the-art baselines?
- **RQ2**: How does the imputation performance of our model vary across different parameter settings?
- **RQ3**: Does our model effectively capture the temporal and spatial dependencies present in partially observed spatiotemporal data?
- **RQ4**: Does the multitask learning design provide additional information that enhances the performance of both imputation and forecasting tasks?

4.1 Datasets

We evaluate the proposed model on two real-world datasets: AQI and METR-LA. These datasets contain spatiotemporal data related to air quality and traffic speed, respectively.

- **AQI**: The AQI dataset contains hourly sampled PM2.5 pollutants records from 437 stations in China, covering a total of 12 months.
- **METR-LA**: This dataset incorporates 207 sensor measurements within the Los Angles area, specifically recording traffic speeds at various highway locations every 5 min.

We have carefully selected these datasets to evaluate the effectiveness of our model in handling spatiotemporal imputation and forecasting tasks under diverse application backgrounds.

4.2 Baselines

To evaluate the performance of our proposed model, we compare it with a range of classic statistical methods (Mean), machine learning methods (kNN, MICE, and VAR), as well as state-of-the-art deep learning methods (rGAIN, BRITS, GRIN, and AGRN) for multivariate time series imputation. We briefly introduce each baseline method below:

- **Mean**: Filling the missing values directly with the variable-level historical average.
- **kNN** [14]: Apply the average value of the top-k nearest neighbor nodes to filling missing values. We use geographic distance to calculate and define the neighbor nodes.
- **MICE** [28]: Multiple imputation method by chain equations with up to 100 iterations.
- **VAR** [38]: Vector autoregressive method followed by a single-step prediction module.

- **rGAIN** [32]: Apply GAIN with a bidirectional recurrent encoder-decoder.
- **BRITS** [7]: A RNN-based imputation model with bidirectional structure.
- **GRIN** [12]: A bidirectional GRU-based method with pre-defined graph neural network for multivariate time series imputation.
- **AGRN** [10]: A model combined recurrent neural network and adaptive graph learning module.

4.3 Evaluation Metrics

Following previous work [10,12], we employ three established evaluation metrics to assess the effectiveness of our proposed model for spatiotemporal imputation. Specifically, we calculate the Mean Absolute Error (MAE), Mean Squared Error (MSE), and Mean Relative Error (MRE) over the imputation window.

4.4 Experiment Settings

Dataset Split. We divided the data into training, validation, and test sets following the prior study [12]. For the AQI dataset, we adopted the test set division proposed in [37], where data from March, June, September, and December were reserved for testing. As for the traffic dataset METR-LA, we performed a three-fold split, with 70% of the data reserved for training and the remaining 10% and 20% split chronologically for validation and testing, respectively.

Imputation Target. For the AQI dataset, we apply the same procedure as the previous work [31] to simulate the distribution of real missing data. For the METR-LA dataset, we simulate the presence of missing data by artificially masking 25% observations [12]. In addition to the manually masked missing values, every dataset has original missing data, i.e., 25.67% in the AQI dataset and 8.10% in the METR-LA dataset. We perform evaluations on the manually injected faults in the test set.

Hyperparameters. Table 1 presents the primary hyperparameters employed in the proposed MTSTI model. We conducted the training of the model on a single Nvidia Quadro RTX 6000, which has 24 GB of memory. To optimize the model parameters, we employed the Adam optimizer with a gradient clip of 10. Moreover, we adopted the cosine annealing learning rate decay strategy with the minimum learning rate of 0, where T_{max} represents the maximum number of epochs.

4.5 Results

Overall Performance (RQ1). We evaluate the spatiotemporal imputation performance of MTSTI in comparison with other baselines. The comparison results (Table 2) show our proposed model performs better than the baselines

Table 1. The hyperparameters of MTSTI for all datasets

Description	AQI	METR-LA
Batch size	32	64
Time length T	24	24
Horizon H	6	6
Max epochs	500	500
Learning rate	0.0003	0.0003
L2 regularization	10^{-6}	10^{-8}
Graph embedding size	128	128
Loss balance $1-\alpha$	0.0005	0.005

Table 2. Performance comparisons on AQI and METR-LA datasets. The best metrics are highlighted in bold, while the subsequent best metrics are underscored.

Method	AQI			METR-LA		
	MAE	MSE	MRE(%)	MAE	MSE	MRE(%)
Mean	39.60	3231.04	59.25	7.56	142.22	13.10
KNN	34.10	3471.14	51.02	7.88	129.29	13.65
MICE	26.98	1930.92	40.37	4.42	55.07	7.65
VAR	22.95	1402.84	33.99	2.69	21.10	4.66
rGAIN	21.78	1274.93	32.26	2.83	20.03	4.91
BRITS	20.21	1157.89	29.94	2.34	16.46	4.05
GRIN	14.73	775.91	21.82	1.91	10.41	3.30
AGRN	14.08	686.52	21.07	1.90	10.10	3.28
MTSTI	**13.92**	**651.14**	**20.83**	**1.81**	**9.32**	**3.13**

across different datasets. The statistical and classical machine learning methods exhibit poor performance on all datasets, as they fill missing values based on assumptions such as time series stability or seasonality, which fail to capture complex spatiotemporal correlations in real-world datasets. Among deep learning methods, both GRIN and AGRN utilize the graph and recurrent neural networks and perform better than other GAN-based or RNN-based methods (rGAIN and BRITS) owing to their ability to extract spatial correlations. Our proposed MTSTI performs even better than Graph-based methods. Furthermore, MTSTI outperforms other single-task learning models, indicating the effectiveness of integrating the imputation task and downstream forecasting task in enhancing the performance of Graph-based models on imputation tasks.

Sensitivity Study (RQ2). We have conducted sensitivity studies on the primary parameters of the multi-task learning module. Figure 2 shows part of representative results because of the space constraints. The observation on other

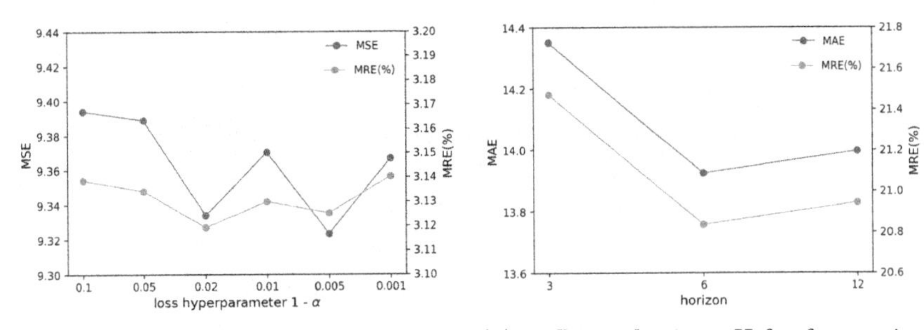

(a) **Different loss weight parameter** α for multi-task loss on METR-LA dataset.

(b) **Different horizon** H for forecasting module on AQI dataset.

Fig. 2. The sensitivity study of key parameters for multi-task learning structure.

datasets aligns with similar conclusions. The parameter α in Eq. (5) has shown to be a crucial factor in the joint task learning framework design while it balances the imputation loss $\mathcal{L}_1$ and forecasting loss $\mathcal{L}_2$ contributions in the final multi-task learning loss $\mathcal{L}$. Our experiments show that the choice of α significantly impacts the imputation and forecasting performance. For example, on the METR-LA dataset (results shown in Fig. 2a), the MSE and MRE remain relatively stable when $\alpha \in [0.9, 0.999]$ but ramp up drastically when α goes below or beyond this range. We believe that a larger weight should be assigned to the main imputation task since it is of higher importance, enabling the model to account for the main task's impact more. However, the model tends to overlook the auxiliary task and overfit the main task if setting a too small weight for the auxiliary task.

The forecasting horizon H is a critical parameter to consider when selecting forecasting as the auxiliary task. We tested our main imputation task results for $H \in 3, 6, 12$, following previous work on solving multivariate time series forecasting [29]. The experimental results on the AQI dataset (Fig. 2b) show our proposed model achieved the best performance when $H = 6$. While a small forecasting horizon may not provide sufficient spatiotemporal dependencies as a reference for the main imputation task, a large horizon may limit the effectiveness of feature extraction in the auxiliary forecasting task due to error accumulation. Thus, selecting an appropriate forecasting horizon is essential for achieving optimal results.

Ablation Study (RQ3 and RQ4). We conducted ablation studies to assess the effectiveness of each module in our proposed model. Specifically, we compared our full model against the following variants:

- *w/o distance graph*: The geographic distance graph is removed, and only the self-learned graph is used.
- *w/o learned graph*: The self-learned graph is removed, and only the geographic distance graph is used.

Table 3. Ablation results conducted on the air quality dataset.

Methods	MAE	MSE	MRE(%)
w/o distance graph	14.00	678.78	20.94
w/o learned graph	14.00	680.23	20.95
w/o bidirectional	16.88	890.40	25.26
w/o multitask learning	14.34	706.86	21.45
MTSTI	**13.95**	**651.14**	**20.88**

Table 4. Performance comparison for the auxiliary forecasting task on METR-LA dataset.

Methods	MAE	RMSE	MAPE(%)
w/o multi-task learning	2.91	5.83	7.67
MTSTI	**2.82**	**5.67**	**7.59**

- *w/o bidirectional branch*: The backward branch is excluded from the framework, utilizing solely the output of the forward branch as the final output.
- *w/o multitask learning module*: The auxiliary forecasting task is removed, and only the main imputation task is kept.

The variants *w/o distance graph* and *w/o learned graph* are utilized to assess the significance of spatial dependency learning, whereas the variant *w/o bidirectional structure* is employed to investigate whether the backward branch can provide additional information for temporal dependency learning. The variant *w/o multitask learning module* is used to evaluate the necessity of the auxiliary forecasting task. The evaluation is conducted on the AQI dataset, and the conclusion and results on the METR-LA dataset are the same. The performance of each module is measured using three evaluation metrics, with the results presented in Table 3.

The findings of the ablation study indicate that all modules significantly contribute to the performance of the proposed model. The distance and learned graphs play an important role in providing neighbor information to facilitate learning of the spatial dependencies among sensors. The bidirectional structure contributes the most to the overall performance of the model. The backward branch is necessary for understanding the temporal dependencies in spatiotemporal data. Furthermore, the inclusion of the auxiliary forecasting task in the multitask learning module improves the performance of the main imputation task and reduces the model's overfitting to the main task.

Influence on Forecasting Task (RQ4). We perform an investigation to determine whether the primary imputation task can offer external information to the auxiliary forecasting task. The outcomes in Table 4 demonstrate that the forecasting performance declines when we eliminate the main imputation task.

It suggests that both tasks can complement each other in capturing the spatiotemporal dependencies, leading to improved final results.

5 Related Work

5.1 Spatiotemporal Imputation

The spatiotemporal imputation problem can be solved along temporal or spatial dimensions. Early studies mostly use statistical and machine learning-based methods to reconstruct the missing values, such as local interpolation [1,18]. Some fill missing values based on the historical time series data using AutoRegressive Integrated Moving Average (ARIMA) [3], Multivariate Imputation by Chained Equations (MICE) [4] and the Expectation-Maximization (EM) algorithm [26]. Other studies focus on using spatial relationships or neighboring sequences to fill missing values with kNN [14] and Kriging [27] algorithms. Low-rank matrix factorization (MF) [9,24] is also a popular method for spatiotemporal imputation. For instance, TRMF [34] incorporates the structure of temporal dependencies into a temporal regularized matrix factorization framework.

Recently, deep learning methods were introduced to spatiotemporal imputation. The RNN-based method is first proposed for imputation in 2018 [8]. Later, BRITS [7] imputes the missing values on the hidden state through a bidirectional recurrent structure and considers the correlation between features. Until present, most deep learning methods have been using RNN as the backbone to capture the temporal dependencies for multivariate time series imputation. There are also a number of methods [36] using Generative Adversarial Networks (GAN) to generate missing data. For example, GAIN [32] imputes data conditioned on observation values by the generator and utilizes the discriminator to distinguish the observed and imputed parts; SSGAN [22] learns the data distribution with a semi-supervised classifier and the temporal reminder matrix for time series imputation. To learn the spatial and temporal dependencies simultaneously, GRIN [12] and AGRN [10] introduce graph neural network and combine it with a recurrent structure to exploit the inductive bias of historical spatial patterns for imputation. In summary, all the above methods are limited in focusing on the imputation task without obtaining insights from downstream tasks in the spatiotemporal field.

5.2 Multi-task Learning

Multi-task learning (MTL) has been proposed as a technique for optimizing multiple related tasks jointly, which generally enhances the generalization performance of a task by leveraging other related tasks. Although there has been no prior research on using MTL for spatiotemporal imputation, several multi-task learning models have been developed for other multivariate time series applications. For instance, recent studies [11,21] have employed a shared attention structure with multiple time series forecasting tasks to improve the overall model performance. Jawed et al. [16] have introduced a forecasting task as an auxiliary task to be jointly optimized with the main task of classification. All these previous studies work on other time series domains but provide valuable insights and motivated us to introduce an auxiliary task for imputation to enhance the performance and generalization of our model.

6 Conclusion

In this paper, we propose MTSTI, a multi-task learning framework for spatiotemporal imputation, which uses the forecasting task as the auxiliary task to improve the imputation performance. In particular, our proposed framework captures spatiotemporal dependencies by combining graph and recurrent neural networks. It achieves better imputation results than state-of-the-art baselines on two real-world datasets in the air quality and traffic domains. For future work, we will work towards devising a more general imputation framework that can handle more types of downstream tasks and exploring more diverse real-world datasets that reflect more complex problems in practice to inspire better designs.

References

1. Acuna, E., Rodriguez, C.: The treatment of missing values and its effect on classifier accuracy. In: Banks, D., McMorris, F.R., Arabie, P., Gaul, W. (eds.) Classification, Clustering, and Data Mining Applications. Studies in Classification, Data Analysis, and Knowledge Organisation, pp. 639–647. Springer, Berlin (2004). https://doi.org/10.1007/978-3-642-17103-1_60
2. Ansley, C.F., Kohn, R.: On the estimation of ARIMA models with missing values. In: Parzen, E. (ed.) Time Series Analysis of Irregularly Observed Data. Lecture Notes in Statistics, vol. 25, pp. 9–37. Springer, New York (1984). https://doi.org/10.1007/978-1-4684-9403-7_2
3. Arumugam, P., Saranya, R.: Outlier detection and missing value in seasonal ARIMA model using rainfall data. Mater. Today: Proc. **5**(1), 1791–1799 (2018)
4. Azur, M.J., Stuart, E.A., Frangakis, C., Leaf, P.J.: Multiple imputation by chained equations: what is it and how does it work? Int. J. Methods Psychiatr. Res. **20**(1), 40–49 (2011)
5. Bauer, P., Thorpe, A., Brunet, G.: The quiet revolution of numerical weather prediction. Nature **525**(7567), 47–55 (2015)
6. Blázquez-García, A., Conde, A., Mori, U., Lozano, J.A.: A review on outlier/anomaly detection in time series data. ACM Comput. Surv. (CSUR) **54**(3), 1–33 (2021)
7. Cao, W., Wang, D., Li, J., Zhou, H., Li, L., Li, Y.: BRITS: bidirectional recurrent imputation for time series. In: Advances in Neural Information Processing Systems, vol. 31 (2018)
8. Che, Z., Purushotham, S., Cho, K., Sontag, D., Liu, Y.: Recurrent neural networks for multivariate time series with missing values. Sci. Rep. **8**(1), 1–12 (2018)
9. Chen, X., Sun, L.: Bayesian temporal factorization for multidimensional time series prediction. IEEE Trans. Pattern Anal. Mach. Intell. **44**(9), 4659–4673 (2021)

10. Chen, Y., Li, Z., Yang, C., Wang, X., Long, G., Xu, G.: Adaptive graph recurrent network for multivariate time series imputation. In: Tanveer, M., Agarwal, S., Ozawa, S., Ekbal, A., Jatowt, A. (eds.) Neural Information Processing. Communications in Computer and Information Science, vol. 1792, pp. 64–73. Springer, Singapore (2022). https://doi.org/10.1007/978-981-99-1642-9_6
11. Chen, Z., Jiaze, E., Zhang, X., Sheng, H., Cheng, X.: Multi-task time series forecasting with shared attention. In: 2020 International Conference on Data Mining Workshops (ICDMW), pp. 917–925. IEEE (2020)
12. Cini, A., Marisca, I., Alippi, C.: Filling the gaps: multivariate time series imputation by graph neural networks. arXiv preprint: arXiv:2108.00298 (2021)
13. Han, Z., Zhao, J., Leung, H., Ma, K.F., Wang, W.: A review of deep learning models for time series prediction. IEEE Sens. J. **21**(6), 7833–7848 (2019)
14. Hastie, T., Tibshirani, R., Friedman, J.H., Friedman, J.H.: The Elements of Statistical Learning: Data Mining, Inference, and Prediction, vol. 2. Springer, Cham (2009)
15. Ismail Fawaz, H., Forestier, G., Weber, J., Idoumghar, L., Muller, P.A.: Deep learning for time series classification: a review. Data Min. Knowl. Disc. **33**(4), 917–963 (2019)
16. Jawed, S., Grabocka, J., Schmidt-Thieme, L.: Self-supervised learning for semi-supervised time series classification. In: Lauw, H.W., Wong, R.C.-W., Ntoulas, A., Lim, E.-P., Ng, S.-K., Pan, S.J. (eds.) PAKDD 2020. LNCS (LNAI), vol. 12084, pp. 499–511. Springer, Cham (2020). https://doi.org/10.1007/978-3-030-47426-3_39
17. Kaushik, S., et al.: Ai in healthcare: time-series forecasting using statistical, neural, and ensemble architectures. Front. Big Data **3**, 4 (2020)
18. Kreindler, D.M., Lumsden, C.J.: The effects of the irregular sample and missing data in time series analysis. In: Nonlinear Dynamics, Psychology, and Life Sciences (2006)
19. Li, Y., Yu, R., Shahabi, C., Liu, Y.: Diffusion convolutional recurrent neural network: data-driven traffic forecasting. arXiv preprint: arXiv:1707.01926 (2017)
20. Liu, Y., Yu, R., Zheng, S., Zhan, E., Yue, Y.: NAOMI: non-autoregressive multiresolution sequence imputation. In: Advances in Neural Information Processing Systems, vol. 32 (2019)
21. Ma, T., Tan, Y.: Multiple stock time series jointly forecasting with multi-task learning. In: 2020 International Joint Conference on Neural Networks (IJCNN), pp. 1–8. IEEE (2020)
22. Miao, X., Wu, Y., Wang, J., Gao, Y., Mao, X., Yin, J.: Generative semi-supervised learning for multivariate time series imputation. In: Proceedings of the AAAI Conference on Artificial Intelligence, vol. 35, pp. 8983–8991 (2021)
23. Oehmcke, S., Zielinski, O., Kramer, O.: kNN ensembles with penalized DTW for multivariate time series imputation. In: 2016 International Joint Conference on Neural Networks (IJCNN), pp. 2774–2781. IEEE (2016)
24. Salakhutdinov, R., Mnih, A.: Bayesian probabilistic matrix factorization using Markov chain monte Carlo. In: Proceedings of the 25th International Conference on Machine Learning, pp. 880–887 (2008)
25. Shang, C., Chen, J., Bi, J.: Discrete graph structure learning for forecasting multiple time series. arXiv preprint: arXiv:2101.06861 (2021)
26. Shumway, R.H., Stoffer, D.S.: An approach to time series smoothing and forecasting using the EM algorithm. J. Time Ser. Anal. **3**(4), 253–264 (1982)
27. Stein, M.L.: Interpolation of Spatial Data: Some Theory for Kriging. Springer, Cham (1999)

28. White, I.R., Royston, P., Wood, A.M.: Multiple imputation using chained equations: issues and guidance for practice. Stat. Med. **30**(4), 377–399 (2011)
29. Wu, Z., Pan, S., Long, G., Jiang, J., Chang, X., Zhang, C.: Connecting the dots: Multivariate time series forecasting with graph neural networks. In: Proceedings of the 26th ACM SIGKDD International Conference on Knowledge Discovery & Data Mining, pp. 753–763 (2020)
30. Wu, Z., Pan, S., Long, G., Jiang, J., Zhang, C.: Graph WaveNet for deep spatial-temporal graph modeling. arXiv preprint: arXiv:1906.00121 (2019)
31. Yi, X., Zheng, Y., Zhang, J., Li, T.: ST-MVL: filling missing values in geo-sensory time series data. In: Proceedings of the 25th International Joint Conference on Artificial Intelligence, IJCAI 2016, pp. 2704–2710 (2016)
32. Yoon, J., Jordon, J., Schaar, M.: Gain: missing data imputation using generative adversarial nets. In: International Conference on Machine Learning, pp. 5689–5698. PMLR (2018)
33. Yu, H., et al.: Regularized graph structure learning with semantic knowledge for multi-variates time-series forecasting. arXiv preprint: arXiv:2210.06126 (2022)
34. Yu, H.F., Rao, N., Dhillon, I.S.: Temporal regularized matrix factorization for high-dimensional time series prediction. In: Advances in Neural Information Processing Systems, vol. 29 (2016)
35. Yu, P., Yan, X.: Stock price prediction based on deep neural networks. Neural Comput. Appl. **32**, 1609–1628 (2020)
36. Zhang, Y., Zhou, B., Cai, X., Guo, W., Ding, X., Yuan, X.: Missing value imputation in multivariate time series with end-to-end generative adversarial networks. Inf. Sci. **551**, 67–82 (2021)
37. Zheng, Y., Capra, L., Wolfson, O., Yang, H.: Urban computing: concepts, methodologies, and applications. ACM Transa. Intell. Syst. Technol. (TIST) **5**(3), 1–55 (2014)
38. Zivot, E., Wang, J.: Vector autoregressive models for multivariate time series. In: Zivot, E., Wang, J. (eds.) Modeling Financial Time Series with S-PLUS®, pp. 385–429. Springer, New York (2006). https://doi.org/10.1007/978-0-387-32348-0_11

TIGAN: Trajectory Imputation via Generative Adversarial Network

Yichen Shi, Hongye Gao, and Weixiong Rao(✉)

Tongji University, Shanghai 201804, China
{2131494,2131493,wxrao}@tongji.edu.cn

Abstract. GPS trajectories are crucial for urban planning, traffic prediction, and location-based services. These applications often require dense trajectories, which is often not the case due to power limitations and privacy concerns. To this end, we propose a novel generative adversarial network-based model, namely TIGAN, for trajectory imputation. TIGAN inserts artificial GPS points between real ones, resulting in imputed trajectories that closely resemble those collected at much higher sampling rates. Unlike existing works, TIGAN does not require prior knowledge such as underlying road networks. Moreover, TIGAN incorporates transportation modes into trajectory imputation, leading to much better performance. Evaluation in two real-world datasets demonstrates the superior performance of TIGAN over state-of-the-art methods.

1 Introduction

Trajectory data, consisting of a sequence of spatio-temporal GPS points, are crucial for many applications such as urban computing, traffic management and location-based services [2,20]. However, due to power limitations and privacy concerns, trajectories are usually sparse with various spatio and temporal gaps between consecutive GPS points. Such sparse points severely degrade the accuracy of the applications that heavily rely on dense trajectories. To address the issue, some recent efforts are developed via the trajectory imputation technique. The main goal is to insert artificial points between each two consecutive trajectory points, such that the artificially inserted points are as accurate as if there were actual readings of trajectory data.

In order to impute dense trajectories from sparse ones, many existing imputation works require underlying road networks. However, considering that those sparse trajectories are generated by people who are walking randomly on campus and do not follow the constraint of road networks, the trajectory imputation methods, if exploiting road networks, may not work at all. Moreover, existing works do not take into account the transportation modes of trajectory data. If missing the transportation modes, the two sparse trajectories with different modes(e.g. driving and walking) but the same positions may lead to the exactly same dense trajectories generated by the trajectory imputation works.

To address the issue above, in this paper, we propose a Generative Adversarial Network (GAN)-based sparse trajectory imputation model, called TIGAN. More

X. Yang et al. (Eds.): ADMA 2023, LNAI 14180, pp. 195–209, 2023.
https://doi.org/10.1007/978-3-031-46677-9_14

specifically, the generator in TIGAN can imputes artifical GPS points into an input sparse trajectory. Meanwhile, the discriminator performs with two goals to determine whether or not a certain input trajectory is real, but also the transportation mode of this input trajectory is consistent with the original one.

In our generator, we utilize the seq2seq architecture, in which both the encoder and decoder employ an LSTM with time-interval attention. After using the encoder to obtain a vector representation of the trajectory, we concatenate it with a one-hot encoding representing the transportation mode and random noise sampled from a standard normal distribution. This combined input is then fed into the decoder to generate the imputed full trajectory. To capture the distinct movement patterns of various transportation modes, we have designed a classifier within the discriminator to identify the transportation mode to which the trajectory belongs after being imputed by the generator. Our contribution is summarized as follows.

- We propose a novel model based on generative adversarial network called TIGAN, to impute sparse trajectories with different transportation modes and irregular time intervals. To the best of our knowledge, TIGAN is the first model that introduces transportation mode feature to solve the trajectory imputation problem.
- We leverage generative adversarial network for trajectory imputation problem. In the generator, we adopt a LSTM with time interval attention as the encoder and sample multiple trajectories when imputing sparse trajectories to ensure that the model can cover as much of the sample space as possible.
- We conduct extensive experiments on the real-life mobility dataset. Results show that our TIGAN model significantly outperforms state-of-the-art baselines in terms of improving imputation accuracy by 5%∼11%.

The remainder of this paper is structured as follows. Section 2 provides a review of the related work. In Sect. 3, we present some of the definitions used in this paper. The details of TIGAN are presented in Sect. 4. Section 5 reports on the experiments and results obtained. Finally, Sect. 6 presents the conclusions and future work.

2 Related Work

Trajectory Imputation: Existing trajectory imputation methods can be divided into two categories, non-learning-based methods and learning-based methods. Non-learning-based methods typically employ mathematical functions for imputing trajectories, which do not necessitate extensive trajectory data but rely on a priori knowledge, such as historical data around missing points and road network information. Zheng *et al.* [18] employed travel patterns inferred from historical data and possible routes provided by the road network to impute sparse trajectories. M. Elshrif *et al.* [3] proposed TrImpute, a method that operates without the knowledge of the underlying road network, and instead, relies on the wisdom of nearby crowds to guide the imputation process.

In contrast, learning-based methods generally use machine learning algorithms [1,4] to learn patterns and regularities in trajectories, and then impute trajectories based on the learned patterns and regularities, which usually require a large amount of trajectory data for training. Xia *et al.* [14] introduced Attn-Move, an attention-based neural network model, for imputing users' missing locations at a fine-grained spatio-temporal level through an intra-trajectory mechanism. Sun *et al.* [12] proposed PeriodicMove, a graph neural network-based model, for imputing human mobility from lengthy and sparse individual trajectories, leveraging graph neural networks for complex model transitioning patterns. However, these two methods primarily focus on car trajectories by modeling trajectory data as transitions between grid cells, neglecting the differences arising from various transportation modes. Consequently, they are unable to handle fine-grained GPS points with small time intervals.

Generative Adversarial Network (GAN): GAN [5] consists of a generator network and a discriminator network, which competes with each other to improve the quality of generated data. The objective function is expressed as:

$$\min_{G} \max_{D} V(D,G) = \mathbb{E}_{x \sim p_{data}(x)}[\log D(x)] + \mathbb{E}_{z \sim p_z(z)}[\log(1 - D(G(z)))] \quad (1)$$

where $p_{data}(x)$ is the distribution of real data, $p_z(z)$ is the prior distribution of the generator's input noise.

GAN has demonstrated excellent performance in extracting features of different categories, with variants such as conditional GAN [11] and triple GAN [7] being used to generate or discriminate specific types of data. Moreover, GANs have been successfully applied to various types of data imputation tasks. GAIN [15], proposed by Yoon *et al.*, uses a masking vector and a two-stage training procedure to achieve state-of-the-art performance. E2GAN [10], proposed by Luo *et al.*, leverages a noised compressing and reconstructing strategy to ensure that the imputed values are reasonable. However, their focus is on time series data instead of trajectories.

3 Preliminaries

In this section, we introduce the definitions we use in this paper followed by the formal definition of the investigated problem.

Definition 1 (Trajectory). A *trajectory* can be represented as a set of geographic location points in a time series $(lon_i, lat_i, t_i)_{i=1}^{n}$, where lon_i, lat_i and t_i denote the longitude, latitude and timestamp of the i-th location point and n is the length of the trajectory, namely the total number of the location points. In practice, the location points are usually connected in chronological order to form a continuous trajectory line segment $T = <(lon_1, lat_1, t_1), (lon_2, lat_2, t_2), \ldots, (lon_n, lat_n, t_n)>$.

Definition 2 (Transportation Mode). Let $\mathcal{M} = \{m_1, m_2, \ldots, m_k\}$ be the set of all possible transportation modes, such as walking, cycling, driving, etc., where m_i represents the label of the i-th transportation mode. A *transportation mode* is a function $m : T \rightarrow \mathcal{M}$ that assigns a transportation mode label to a trajectory based on the means of transportation used during the trajectory. Now we give the following problem definition.

Problem Statement. In practice, *trajectory imputation* is often performed on a complete trajectory by randomly removing some of its values and then using the imputation algorithm to fill in the missing values. In this case, trajectory imputation can be defined as follows: Given a complete trajectory $T = \{(lon_1, lat_1, t_1), (lon_2, lat_2, t_2), \ldots, (lon_n, lat_n, t_n)\}$ and a trajectory sub-sequence with missing values $T_m = \{(lon_1, lat_1, t_1), (lon_2, lat_2, t_2), \ldots, (lon_m, lat_m, t_m)\}$, where $m < n$, the goal of trajectory imputation is to produce a trajectory sub-sequence T_i of the same length as T, which includes the known location points in T_m as well as the imputed points for the unknown locations.

4 TIGAN Framework

To address the problem described above, we propose a new trajectory imputation model called TIGAN, which is based on generative adversarial networks (GANs). TIGAN is designed to learn the unique characteristics of different transportation modes during the trajectory imputation process. The model's architecture is illustrated in Fig. 1. Similar to GAN, TIGAN consists of two components: a generator and a discriminator, which are trained in opposition to each other.

The generator of TIGAN utilizes an encoder-decoder architecture, where the encoder takes the trajectory as input and encodes it into a high-dimensional feature space that captures the underlying structure and patterns of the trajectory. This encoded representation is then used by the decoder to perform missing position imputation. Meanwhile, the discriminator not only distinguishes between real and imputed trajectories but also classifies the transportation mode of the imputed trajectory. This allows the imputed trajectories to closely fit the transportation modes.

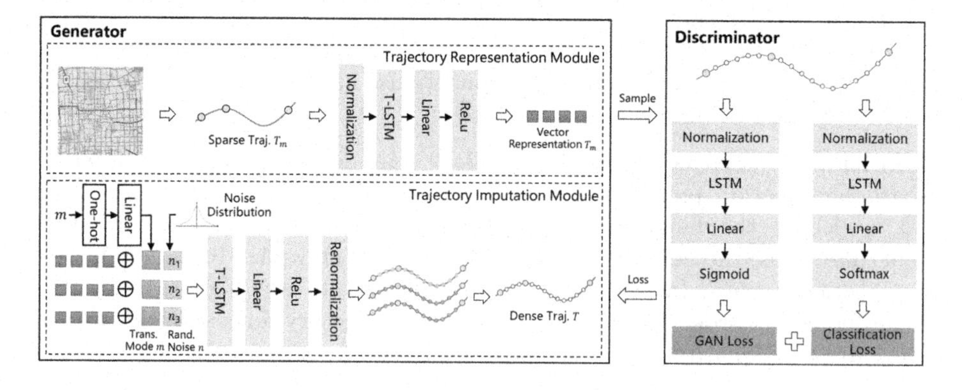

Fig. 1. Main architecture of the TIGAN.

4.1 Generator

The generator uses an encoder-decoder architecture to impute missing values in a partially observed trajectory T_m with corresponding transportation mode m, where T_m is obtained by masking the ground truth dense trajectory T with the binary mask array M:

$$T_m = T \odot M \tag{2}$$

The encoder maps T_m to a high-dimensional vector representation e:

$$e = Encoder(T_m) \tag{3}$$

The decoder concatenates e, its corresponding transportation mode m and randomly sampled noise n, and then maps them to the reconstructed dense trajectory $\hat{T}$:

$$d = Decoder(e, m, n) \tag{4}$$

Finally, we use the corresponding values of d to replace the missing positions of T_m to obtain the completed trajectory:

$$\hat{T} = T_m + d \odot \bar{\mathcal{M}} \tag{5}$$

where $\odot$ denotes element-wise multiplication, and $\bar{\mathcal{M}} = 1 - \mathcal{M}$.

In the following paragraph, we call the encoder *trajectory representation module* and the decoder *trajectory imputation module.*

Trajectory Representation Module. The Trajectory Representation Module models the trajectories with missing positions. However, the missing positions in the trajectory may not be uniform, such as the situation shown in Fig. 2 (a) and (b), where the red points represent the points to be imputed, and the blue points represent the observed points. Only two positions are missing in Fig. 2 (a) and (b), but their time slots are different.

Therefore, we have adopted the LSTM with time interval attention, proposed by Zhang et al. [16], whose structure is shown in Fig. 2.

As the forget gate determines how much information from the previous state is forgotten and the input gate selectively records the current input information into the cell state, so we assign weights to them respectively. Specifically, we design a linear layer with a tanh activation function, where the input is the time difference between the current state and the previous state:

$$\alpha_t = tanh(w^a \Delta t_{t-1,t} + b^a) \tag{6}$$

where w^a and b^a are both learnable parameters and $\Delta t_{t-1,t}$ is the timestamp difference between the current input x_t and the precious input x_{t-1}. Detailed

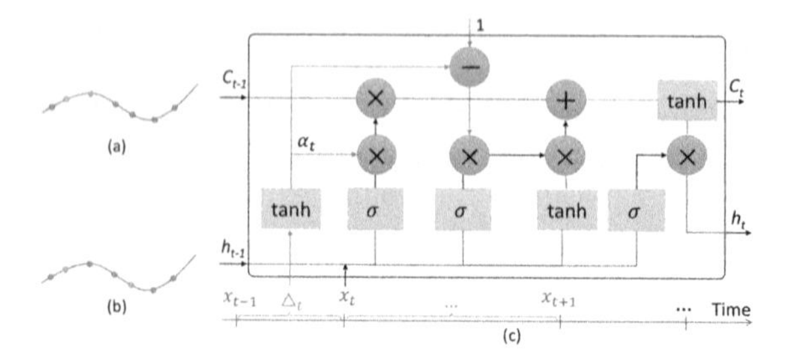

Fig. 2. LSTM Cell with time-interval attention.

mathematical expressions of the proposed LSTM with time interval attention are given below:

$$
\begin{aligned}
f_t &= \sigma(W_f \odot [h_{t-1}, x_t] + b_f) * \alpha_t \\
i_t &= \sigma(W_i \odot [h_{t-1}, x_t] + b_i) * (1 - \alpha_t) \\
\hat{C}_t &= tanh(W_C \odot [h_{t-1}, x_t] + b_C) \\
C_t &= f_t * C_{t-1} + i_t * \hat{C}_t \\
o_t &= \sigma(W_o \odot [h_{t-1}, x_t] + b_o) \\
h_t &= o_t * tanh(C_t)
\end{aligned}
\tag{7}
$$

where x_t represents the current input, h_{t-1} and h_t are previous and current hidden states, and C_{t-1} and C_t are previous and current cell memories. W_f, b_f, W_i, b_i, W_o, b_o, and W_c, b_c are the network parameters of the forget, input, output gates and the candidate memory, respectively. When the time interval becomes larger, the model should pay more attention to the current information and properly ignore the past information. At this time, the time interval attention first acts on the forgetting gate to discard the information from the past cell state by reducing the weight. Then, we use $1 - \alpha_t$ to appropriately increase the weight influence of the current input on the output.

Trajectory Imputation Module. After the trajectory representation module extracts the features of the sparse trajectory, the trajectory imputation module imputes it into a dense trajectory by imputing the missing locations. As shown in the lower left part of Fig. 1, we first concatenate the vectors representing the sparse trajectory, transportation mode, and noise together, and then use LSTM with time interval attention to output the dense trajectory.

The method described so far attempts to produce the "average" imputation in cases where there can be multiple outputs. To cover the sample space as much as possible, we generate multiple trajectory options by sampling the noise k times. Then we take the trajectory with the smallest distance loss from the real trajectory as the output of the generator. The distance loss is defined as

$$
\mathcal{L}_{distance} = \arg\min \sum_{i=0}^{n} Dist(l_i, \hat{l}_i^k) \tag{8}
$$

where l_i and $\hat{l}_i^k$ denotes the i-th location in the real and k-th generated trajectory, respectively. $Dist$ is the distance between two locations, which can be any type of distance, such as Euclidean Distance, Manhattan Distance and so on. In practice, we use the Haversine function to calculate the distance between two points:

$$Dist = 2r \arcsin\left(\sqrt{\sin^2\left(\frac{\phi_2 - \phi_1}{2}\right) + \cos(\phi_1)\cos(\phi_2)\sin^2\left(\frac{\lambda_2 - \lambda_1}{2}\right)}\right) \tag{9}$$

where r is the average radius of the Earth, ϕ_1 and ϕ_2 are the latitudes of points A and B, respectively, and λ_1 and λ_2 are their longitudes. All angles are expressed in radians.

In the inference stage, since there is no real trajectory, we input the generated multiple trajectories into the discriminator and select the one with the highest probability as the output trajectory.

4.2 Discriminator

As the dense trajectory is generated by the generator, the discriminator learns to distinguish the real dense trajectory from the generated dense trajectory. The structure of the discriminator is composed of an LSTM layer and a linear layer with sigmoid activation function, and its loss function is as follows:

$$D_{loss} = \frac{1}{n}\sum_{i=1}^{n} log(D(x_i)) + \frac{1}{n}\sum_{i=1}^{n} log(1 - D(G(x_i, z_i))) \tag{10}$$

where $D(x_i)$ is the output probability when the input trajectory is x_i and $G(x_i, z_i)$ is the generated trajectory via input of x_i and z_i.

Moreover, in order to enable the generator to learn the feature of transportation mode, we design a new classification task for the discriminator, which is to classify its transportation mode according to the input trajectory. The classifier is composed of an LSTM layer and a linear layer with a softmax activation function, and we adopted the multi-classification cross-entropy loss function as its loss function :

$$C_{loss} = -\frac{1}{n}\sum_{i=1}^{n}\sum_{j=1}^{k} y_j log(p(x_j)) \tag{11}$$

where k denotes the number of transportation modes, y_j is 1 when the sample belongs to transportation mode j, and $p(x_j)$ is the probability distribution given by the classifier.

4.3 Training

Pre-training. As one of the difficulties of GAN, training a GAN-based network is very difficult and cannot be directly trained like other networks. Especially in our proposed network, the imbalance of generator and discriminator and the fact that the trajectory imputation problem is fundamentally different from the generation problem make it not easy to obtain the promising performance when training the whole network. Therefore, we designed a pre-training process before the formal training to ensure that the generator and discriminator can maintain a balance during the training process and the result can meet the accuracy required by the trajectory imputation problem.

Specifically, we first train the generator separately for the trajectory imputation task, that is, input sparse trajectories and then output full trajectories via the generator, and we adopt MSE(Mean Squard Error, MSE) as its loss function:

$$MSE = \frac{1}{n}\sum_{i=1}^{n}(G(x_i) - y_i)^2 \tag{12}$$

Then, we train the discriminator to perform trajectory recognition and transportation mode classification tasks. Trajectory recognition task is designed as a binary classification task to distinguish whether the input trajectory conforms to the law of human movement. In order to obtain the training data, we construct the fake samples by randomly deleting several points in the real trajectory and using linear imputation method to impute the missing ones. Furthermore, we design a transportation mode classification task to pre-train the classifier.

Training. In the formal training phase, we train the discriminator and generator one by one. Firstly, we fix the generator to train the discriminator. Specifically, we input the training set into the generator to obtain a fake data set with the same sample size as the training set, and input both data set into the discriminator for training. Then, in the training process of the generator, we use the multi-sampling method mentioned above to enable the generator to generate multiple different outputs for the same input, so as to ensure that it can cover the sample space as much as possible.We calculate the distance error between each generated trajectory and the real trajectory, and select the trajectory with the smallest distance error and input it to the discriminator and classifier. Finally, the discriminator and classifier return the loss to the generator and then update the parameters of the generator via an optimizer:

$$G_{loss} = \frac{1}{n}\sum_{i=1}^{n} log(1 - D(G(x_i, z))) - w\frac{1}{n}\sum_{i=1}^{n}\sum_{j=0}^{k-1} y_j log(p(x_j)) \tag{13}$$

where, we set w as 0.5 in practice, and the later experiments will prove that 0.5 is the most suitable value.

The overall training process is shown in Algorithm 1.

5 Experiments

5.1 Dataset

We evaluate the performance of our framework on a large-scale real-world mobility dataset and a small-scale individual dataset, *GeoLife* [19] and *Jiading*. The

Algorithm 1: Training Algorithm for TIGAN

Input: Generator G_θ; Discriminator D_β; Classifier C_ϕ; Trajectory dataset $S = \{T_1, T_2, \cdots, T_n\}$; Training epochs t; Batch-size b; Discriminator training steps d; Generator training steps g

Output: Model Parameters θ, β, ϕ

```
1  Pre-train Gθ, Dβ, Cφ as mentioned in 4.3
2  for t-steps do
3      for d-steps do
4          Use the generator G to generate negative samples and combine with
           given positive samples
5          Train the discriminator D according to Eq. 10
6      end
7      for g-steps do
8          Shuffle the training set S into b batches
9          for b-batches do
10             Generate k negative samples with k different noises and output the
               one with the minimum Distance error
11             Input the negative samples to Dβ and Cφ and calculate the losses
               according to Eq. 13
12             Update Gθ parameters via the optimizer ADAM
13         end
14     end
15 end
```

GPS trajectory of the two dataset is represented by a sequence of time-stamped points, each of which contains the information of latitude, longitude, altitude, timestamp and transportation mode.

For pre-processing of *GeoLife*, we first extract the data with transportation mode in the range 115.41–117.51 in longitude and 39.44–41.06 in latitude and delete the trajectory data that does not meet to the characteristics of the transportation mode. We set time interval to 10 s and the duration to 10 min to get as much trajectory data as possible. The dataset is split into a training set for network training, a validation set for parameter tuning, and a testing set for evaluation, with a partition ratio of 7:2:1. For pre-processing of *Jiading*, we set the time interval to 15 s due to a sampling interval of 3 s and the duration to 5 min. We divided the dataset into 6:2:2 because the dataset contains less trajectory data. Table 1 shows the basic statistics of the pre-processed datasets.

5.2 Baselines

We compare the proposed methods with seven representative baselines.

- **Linear** [6]: This is a ruler-based method, which imputes the locations by assuming that users are moving straightly and uniformly.

Table 1. Statistics of the datasets.

	GeoLife	Jiading
Num. of Walking Trajectories	1618	796
Num. of Cycling Trajectories	1592	340
Num. of Car Trajectories	1490	185
Num. of Bus Trajectories	1558	–
Time Interval(s)	10	15
Num. of Points in a Trajectory	60	20

- **LSTM** [8]: This is a deep learning model which models the forward sequential transitions of mobility by recurrent neural network and uses the prediction for next time slot as imputation.
- **Bi-LSTM** [17]: Bi-LSTM extends LSTM by bidirectional RNN to consider the spatial-temporal constraints given by all observed locations.
- **GAN** [9]: Based on the general design of GAN, a Seq2Seq framework is selected as the generator, in which both the encoder and the decoder are composed of Bi-LSTM, and a Bi-LSTM is selected as the discriminator.
- **GAIN** [15]: GAIN is a GAN based imputation method that uses a hint vector to impute the missing values.
- **E2GAN** [10]: E2GAN is an end-to-end GAN based network that leverages GRU to compress and reconstruct the missing values.
- **TrImpute** [3]: TrImpute is an non-learning-based method which chooses candidate points from history points of all other trajectories. Due to the specific nature of TrImpute, it is only effective in locations with a high number of historical trajectories, so after comparing the above baseline methods, we compare it with TIGAN alone.

5.3 Experimental Settings

To evaluate performance, we masked certain time slots as ground truth to recover based on a missing rate of 20%, 40%, 60%, and 80%, where the average distance between adjacent points in trajectories with different missing rates is shown in Fig. 3. We utilized both uniform and non-uniform masking methods for missing data. For the uniform masking method, we first grouped every 5 location points as a set of data and then masked the 5th, 4th, 3rd, and 2nd location points based on the missing rate. For the non-uniform masking method, we grouped 10 location points as a set of data and ensured that the first location point was always known to limit the length of the continuous mask under any missing rate to no greater than 8. This was done to avoid too many unknown points being missing continuously, which would result in large imputation errors. All results of the non-uniform masking experiments were obtained by calculating the average of the results of 50 experiments.

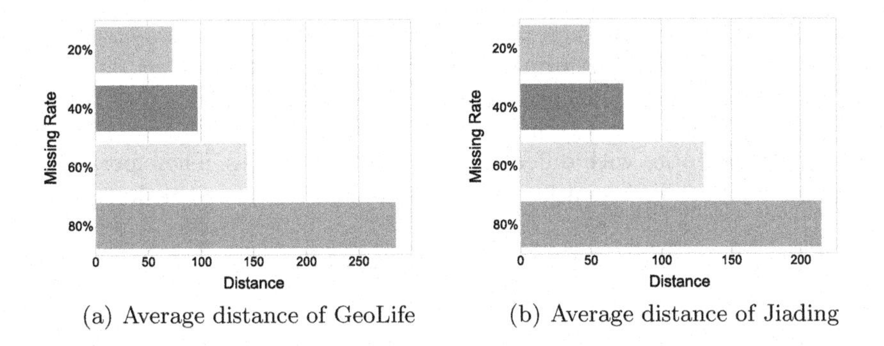

(a) Average distance of GeoLife (b) Average distance of Jiading

Fig. 3. The average distance between adjacent points in trajectories with different missing rates.

Due to the difficulty of achieving exact imputation in GPS coordinate points, we make use of the metric of *Distance*, which is the average geographical distance between the imputed points and the ground-truth. The smaller the *Distance* is, the better the performance will be.

5.4 Experiment Results

Overall Performance: We report the overall performance in Table 2. For experiment results on GeoLife dataset, we have the following finding.

The Linear method performed the worst among all baselines with a distance of more than 20 m in all missing rates in the GeoLife dataset, indicating that treating trajectories as linear motion alone cannot achieve favorable performance. Models based on RNN have achieved better results than the Linear method, especially the bi-directional mechanism of Bi-LSTM which enables it to take into account the context information of the trajectory. As the trajectory becomes sparser, the performance of Bi-LSTM is greater than that of LSTM. However, GAN-based methods generally outperform RNN-based methods, with the two state-of-the-art methods, GAIN and E2GAN, achieving good results. TIGAN achieved optimal performance under all missing rates, partially due to the different transportation modes of the trajectory data, making it difficult for other methods to impute the missing locations by a fixed pattern.

Furthermore, as the trajectories become sparser, the distance errors of GAIN and E2GAN increased significantly, with growth rates of 43% and 68%, respectively. In contrast, TIGAN only had a 30% increase in error rate, proving that TIGAN is effective in dealing with sparse trajectories.

For experiment results on Jiading dataset, we have the following finding. Unlike the experimental results on the GeoLife dataset, the difference between our approach and the baselines on the Jiading dataset was smaller. This was due to the dataset having fewer trajectories, resulting in the model not learning effective movement patterns. Additionally, the number of trajectories with different transportation modes was unbalanced, with the number of driving trajectories being much smaller than the number of walking ones. Despite this, our model still achieved good performance, reaching optimal performance at a 40% missing rate and the second-best performance at other missing rates.

In conclusion, TIGAN achieved preferable results compared to all learning-based baselines. This justifies our model's effectiveness in capturing the mobility patterns of trajectories with different transportation modes. Therefore, TIGAN is a powerful model for imputing sparse trajectories into dense trajectories.

Table 2. Overall performance comparison.

Model	GeoLife								Jiading							
	Uniform				Non-Uniform				Uniform				Non-Uniform			
	20%	40%	60%	80%	20%	40%	60%	80%	20%	40%	60%	80%	20%	40%	60%	80%
Linear	20.24	25.13	26.54	24.90	20.35	25.33	26.76	29.64	13.33	15.85	16.38	16.80	**12.83**	16.12	16.49	16.64
LSTM	17.76	19.44	22.41	23.18	18.02	18.89	22.72	23.90	14.78	16.67	17.05	16.74	14.90	16.34	16.23	16.67
Bi-LSTM	17.72	18.56	21.05	21.93	17.80	18.57	20.63	22.73	14.03	15.63	15.47	16.07	14.12	16.00	15.71	16.32
GAN	15.52	15.74	18.08	19.25	14.79	15.66	17.96	20.49	13.87	14.98	15.26	15.50	13.63	15.41	15.48	15.63
GAIN	**12.23**	14.87	17.93	20.57	**13.27**	16.31	19.00	21.40	13.59	14.06	13.82	15.36	13.76	14.18	**13.85**	**14.69**
E2GAN	12.85	15.42	16.80	18.42	13.79	16.77	17.63	19.73	13.61	14.14	14.30	15.03	13.21	14.25	14.71	15.33
TIGAN	12.66	**14.54**	**15.68**	**16.46**	13.45	**14.73**	**15.77**	**16.39**	**13.23**	**13.05**	**13.51**	**14.49**	12.95	**13.73**	14.17	15.02

For comparison with TrImpute, as TrImpute requires an extremely large number of historical track points to achieve good results, we take tracks from the data with longitude at [39.97, 40.02] and latitude at [116.29, 116.35]. TrImpute will impute an unknown number of points between two consecutive points, for example, even though we have only masked one point between two consecutive points, TrImpute may still add two or more points. Therefore, we choose the artificially imputed point that is closest in time to the real point as the output of TrImpute and the result is shown in Fig. 4(a), where we choose three combinations of hyper-parameters for TrImpute, whose performance is shown in [3] to be weak to strong. From the result we can find that TIGAN and the best TrImpute are on par, but TIGAN can be applied to any situation while TrImpute only works when there are a large amount history trajectories around, which proves that TIGAN is not only more accurate in imputation, but also more suitable for a wider range of situations.

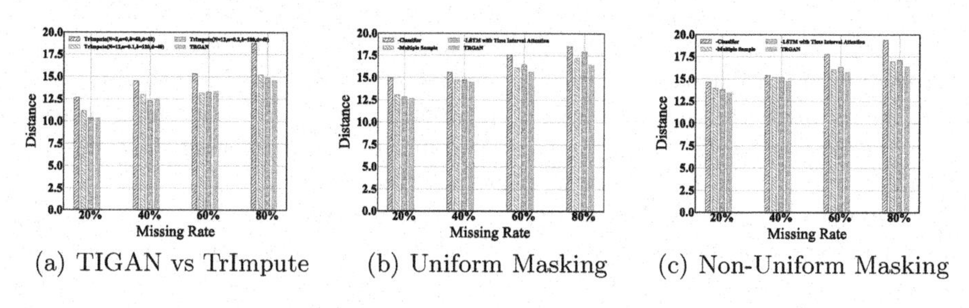

(a) TIGAN vs TrImpute (b) Uniform Masking (c) Non-Uniform Masking

Fig. 4. Figure (a) shows the performance between TIGAN and TrImpute with different hyper-parameters. Figures (b) and (c) show the effect of uniform and non-uniform masking approaches on GeoLife Dataset, respectively.

Ablation Study. We propose a number of components and training methods in our TIGAN model and, in order to verify their effectiveness, we create ablations by removing them one by one, i.e., removing the classifier and the transportation mode inputted to the generator, using vanilla LSTM to replace LSTM with time interval attention, and canceling multiple sampling of different noises in the generator to generate different trajectories. We report the results in Figs. 4(b) and 4(c).

As expected, TIGAN outperforms all the ablations, which indicates the importance of each component in improving imputation accuracy. Specifically, the performance drops the most significantly when we remove the classifier. It is worth noting that the LSTM with the time interval attention has a low positive impact on the model when the trajectory missing rate is low, and an increasing positive impact on the model when the trajectory missing rate increases, indicating its effectiveness in handling trajectories with irregular time intervals.

Sensitivity of Hyper-parameters. We investigate the sensitivity of two important hyper-parameters including the number of noises sampled and hidden size of LSTM with time interval attention.

Firstly, we observe the performance change at a missing rate of 80% by tuning the number of noises sampled in the range of {1, 2, 3, 4, 5}. From the result presented in Fig. 5(a), we can see that the model performance progressively improves as the sample size increases from 1 to 3, but afterwards it fluctuated up and down, with little change in overall performance.

Then, we tune the hidden size of LSTM with time interval attention in the range of {8, 16, 32, 64, 128, 256, 512}, which is also at a missing rate of 80%. As Fig. 5(b) shows, the performance of the model improves significantly when the hidden size increases from 8 to 64, but only marginally when it increases from 64 to 128, and decreases when it increases from 128 to 512.

Finally, we conducted experiments on the weight w of the G_{loss}, with values of 2, 1, 0.5, 0.25, 0.1. Figure 5(c) proves that the model performed the best when the weight w was set to 0.5.

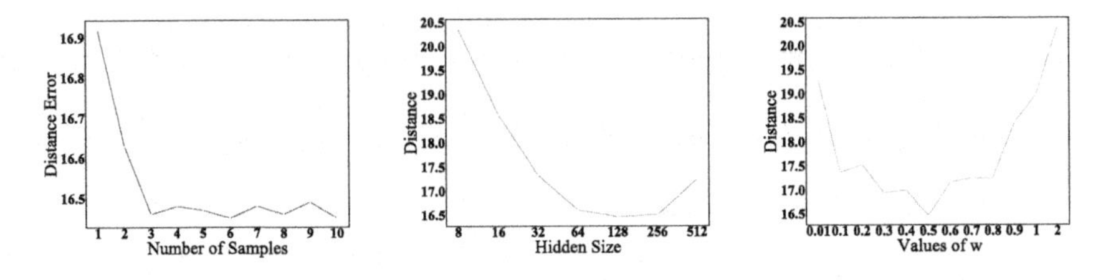

Fig. 5. Sensitivity Study. (a) Sample size (b) Hidden size (c) Weight w.

Robustness Analysis. We also conduct experiments to evaluate the robustness of TIGAN towards datasets of single transportation mode. We compare TIGAN with Linear, GAIN and E2GAN on GeoLife with 80% non-uniform missing rate, and the result is demonstrated in Table 3. From the result, we can see: TIGAN outperforms other state-of-the-art methods in all transportation modes except cycling. In fact, the difference between the models on the walking and cycling data set is not significant, which is easily influenced by the random missing positions. In contrast, TIGAN outperforms other methods by at least 23% on the driving and bus datasets, which proves its robustness.

Table 3. Robustness Analysis.

Mode	Model			
	Linear	GAIN	E2GAN	TIGAN
	Distance			
Walking	14.90	11.58	12.04	**10.79**
Cycling	19.58	**15.40**	16.49	16.20
Driving	65.80	33.89	31.55	**24.16**
Bus	55.50	29.09	25.23	**18.97**

6 Conclusion and Future Works

In this paper, we propose a GAN-based sparse trajectory imputation framework, namely TIGAN. We adopted the GAN approach, using adversarial training between the generator and discriminator to achieve more complete trajectory imputations. Additionally, to handle trajectories with different transportation modes, we designed a classifier to aid the generator in capturing these differences. Extensive experiment results on the real-life dataset GeoLife demonstrate the effectiveness of TIGAN compared with the state-of-the-art baselines.

As for future work, we plan to study a section of the trajectory containing different transportation modes, which is not only a much more complex issue but also allows our model to be applied to a wider and more complex range of situations [13]. Also, we will introduce road network information without affecting the different transportation modes, which will greatly improve the imputation ability of our model.

Acknowledge. This work is partially supported by National Key R&D Program of China (No. 2022YFE0208000, 2021YFE204500, 2021YFC3340601), National Natural Science Foundation of China (No. 61972286), the Shanghai Science and Technology Development Funds (No. 22410713200, 20ZR1460500), the Shanghai Municipal Science and Technology Major Project (2021SHZDZX0100), and Shanghai Key Lab of Vehicle Aerodynamics and Vehicle Thermal Management Systems, and the Fundamental Research Funds for the Central Universities.

References

1. Cao, W., Wang, D., Li, J., Zhou, H., Li, L., Li, Y.: BRITS: bidirectional recurrent imputation for time series. In: NeurIPS 2018, pp. 6776–6786 (2018)
2. Di, X., Xiao, Y., Zhu, C., Deng, Y., Zhao, Q., Rao, W.: Traffic congestion prediction by spatiotemporal propagation patterns. In: IEEE MDM 2019, pp. 298–303. IEEE (2019)
3. Elshrif, M.M., Isufaj, K., Mokbel, M.F.: Network-less trajectory imputation. In: Renz, M., Sarwat, M. (eds.) SIGSPATIAL 2022, pp. 1–10. ACM (2022)
4. Feng, J., et al.: DeepMove: predicting human mobility with attentional recurrent networks. In: WWW 2018, pp. 1459–1468. ACM (2018)
5. Goodfellow, I.J., et al.: Generative adversarial nets. In: NeurIPS 2014, pp. 2672–2680 (2014)
6. Hoteit, S., Secci, S., Sobolevsky, S., Ratti, C., Pujolle, G.: Estimating human trajectories and hotspots through mobile phone data. Comput. Netw. **64**, 296–307 (2014)
7. Li, C., Xu, T., Zhu, J., Zhang, B.: Triple generative adversarial nets. In: Guyon, I., et al. (eds.) NeurIPS 2017 (2017)
8. Liu, Q., Wu, S., Wang, L., Tan, T.: Predicting the next location: a recurrent model with spatial and temporal contexts. In: Schuurmans, D., Wellman, M.P. (eds.) AAAI 2016 (2016)
9. Luo, Y., Cai, X., Zhang, Y., Xu, J., Yuan, X.: Multivariate time series imputation with generative adversarial networks. In: Bengio, S., Wallach, H.M., Larochelle, H., Grauman, K., Cesa-Bianchi, N., Garnett, R. (eds.) NeurIPS 2018 (2018)
10. Luo, Y., Zhang, Y., Cai, X., Yuan, X.: E2gan: end-to-end generative adversarial network for multivariate time series imputation. In: IJCAI 2019 (2019)
11. Mirza, M., Osindero, S.: Conditional generative adversarial nets. CoRR, abs/1411.1784 (2014)
12. Ren, H., et al.: MTrajRec: map-constrained trajectory recovery via seq2seq multi-task learning. In: KDD 2021 (2021)
13. Tian, L., Zhao, K., Yin, J., Vo, H., Rao, W.: The levy flight of cities: analyzing social-economical trajectories with auto-embedding. CoRR, abs/2112.14594 (2021)
14. Xia, T., et al.: AttnMove: history enhanced trajectory recovery via attentional network. In: AAAI 2021 (2021)
15. Yoon, J., Jordon, J., van der Schaar, M.: GAIN: missing data imputation using generative adversarial nets. In: Dy, J.G., Krause, A. (eds.) ICML 2018 (2018)
16. Zhang, Y., Rao, W., Zhang, K., Chen, L.: Outdoor position recovery from heterogeneous telco cellular data. CoRR, abs/2108.10613 (2021)
17. Zhao, J., Xu, J., Zhou, R., Zhao, P., Liu, C., Zhu, F.: On prediction of user destination by sub-trajectory understanding: a deep learning based approach. In: CIKM 2018, pp. 1413–1422 (2018)
18. Zheng, K., Zheng, Y., Xie, X., Zhou, X.: Reducing uncertainty of low-sampling-rate trajectories. In: ICDE 2012, pp. 1144–1155. IEEE Computer Society (2012)
19. Zheng, Y., Xie, X., Ma, W.: GeoLife: a collaborative social networking service among user, location and trajectory. IEEE Data Eng. Bull. **33**(2), 32–39 (2010)
20. Zhu, F., et al.: City-scale localization with telco big data. In: ACM CIKM 2016, pp. 439–448. ACM (2016)

Boolean Spatial Temporal Text Keyword Skyline Query

Chenyang Li[1] and Leigang Dong[2](✉)

[1] Jilin Institute of Chemical Technology, Jilin City, Jilin, China
[2] Baicheng Normal University, Baicheng, Jilin, China
Lgdong010@163.com

Abstract. Spatial text keyword skyline query, as an efficient data retrieval method, mainly considers the spatial distance and textual relevance between query points and objects. However, the existing query algorithms do not consider spatial, temporal, numerical and keyword attributes simultaneously and cannot meet the user-specific preference requirements. Based on this, this paper proposes Boolean Spatial Temporal Text Keyword Skyline Query (BSTTKSQ), which is used to find those objects that contain query keywords and query times and are not dominated by other objects in terms of both spatial and numerical text attributes. Firstly, we design a spatial index NTIR-Tree that can store the above four attributes, then, we propose the pruning strategy for nodes and objects in NTIR-Tree and two pruning methods for candidate object dominance determination, namely, the sorting summation comparison method and the dimensional incremental solution method, and propose a complete skyline query algorithm based on the above pruning strategies and methods, and finally, the effectiveness of the algorithm was verified through experiments.

Keywords: Spatial text skyline query · Boolean query · Time · NTIR-Tree index

1 Introduction

With the rapid development of technology such as mobile Internet and mobile terminals, a huge amount of data information has been generated on the Internet. However, not all data information is valid for users, and usually users have specific query needs for data information on the Internet. How to query the information that satisfies users' preferences from the huge amount of data has gradually become the focus of scholars' research. Börzsönyi et al. [1] first applied skyline query to the database field in 2001, and it has been widely used in many fields such as multi-objective decision-making, environmental monitoring, market analysis and data mining because of its efficient data retrieval capability [2]. The result of a skyline query is a set of skyline objects, none of which can be dominated by any other object in the same dataset.

With the diversity of user query requirements, the existing skyline query algorithm can no longer meet the user's needs. For example, a tourist plans to book a hotel that can be open during the time period of 11:00–13:00 and has a close distance to a tourist

X. Yang et al. (Eds.): ADMA 2023, LNAI 14180, pp. 210–224, 2023.
https://doi.org/10.1007/978-3-031-46677-9_15

spot, low price and good service quality, and also requires that the hotel has Wi-Fi and air conditioning. The information of the four hotel is listed in Table 1, which contains the spatial distance from the hotel to the query point, the per capita consumption price of the hotel, the user rating, the keyword information and the opening hours. The initial skyline query only considered numerical attributes, and the query of the above example could only get the objects {a, c} with close distance, low price and good quality of service, and then the research introduced the keyword attribute into the skyline query and proposed the keyword skyline query, after adding the keyword attribute, the query with the keywords "Wi-Fi" and "air conditioning" gets the result set {a}. With the development and application of spatial database, scholars applied skyline query to spatial database and proposed spatial text skyline query, considering attributes such as spatial, numerical text and keywords. For the above example, if the result set obtained by using spatial text skyline query is {a}, it is obvious that a is invalid for the user's query, because the query does not consider the effect of temporal attributes and a only contains part of the query keywords, and usually, the user may prefer the object containing all the query keywords. Based on the shortcomings of existing algorithms, this paper proposes a Boolean spatial temporal text keyword skyline query, which yields objects that contain query time and all query keywords and are not dominated by other objects in terms of both spatial distance and numerical text attributes. The query using BSTTKSQ for the above example returns a result set of {b} that satisfies the user's preferences on all four attributes: spatial, temporal, numerical, and keyword.

Table 1. Hotel information.

Hotel name	Distance (Km)	Per capita price	User rating	Keywords information	Opening hours
a	3.6	15	8.0	parking, air conditioning	05:30–09:00
b	4.0	18	7.0	Wi-Fi, air conditioning	10:00–22:00
c	2.2	25	8.0	parking	22:00–03:00
d	4.5	20	7.0	Wi-Fi, air conditioning	11:00–14:00

In this paper, we design NTIR-Tree, a spatial index structure capable of storing four attributes of spatial, temporal, numerical text and keywords simultaneously, and propose an algorithm based on NTIR-Tree index that can effectively solve Boolean spatial temporal text keyword skyline query. The main contributions of this paper are as follows:

(1) The first Boolean spatial temporal text keyword skyline query is proposed.
(2) The NTIR-Tree spatial index structure is designed, and the pruning strategy of nodes and objects and the method of domination determination are proposed.

(3) The query algorithm STTK and the pruning algorithm domCalculation are proposed based on the pruning strategies and methods.
(4) Experiments are conducted on real and simulated datasets, and the experimental results verify the accuracy and effectiveness of the algorithms.

2 Related Work

Börzsönyi et al. first applied skyline queries to the database domain and proposed the classical BNL algorithm and D&C algorithm in 2001 [1], since then skyline queries have been widely studied by scholars, and the NN algorithm and BBS algorithm based on R-Tree indexing were proposed in the literature [3] and [4], respectively, early algorithms such as BNL, D&C, NN and BBS were studied based on numerical text attributes of objects.

With the development of spatial databases, scholars have introduced skyline queries into spatial databases. In literature [5], spatial skyline queries are proposed, algorithms B^2S^2 and VS^2 for static queries are proposed, and a dynamic query algorithm VCS^2, which uses the changing pattern of query points to avoid unnecessary computations and thus executes the algorithm efficiently. Considering spatial attributes, algorithms for solving the skyline query problem in Euclidean space and road network space are presented in the literature [6, 7]. The calculation of spatial distances is extended in literature [8], and spatial skyline queries based on Manhattan distances are proposed instead of Euclidean distances. The literature [9] applied K-domination to road network skyline query and proposed a K-dominated spatial skyline query method to handle multi-attribute data objects in road network environment. With further research, considering only spatial attributes does not satisfy the textuality demand of users in practical applications. The literature [10, 11] considers the spatial location of query points and keyword attributes, and the query results returned can better satisfy the textuality preference of users. The literature [12] combines R* trees and inverted files to index data objects, and the speed of inserting and querying data is significantly improved by this index structure. Literature [13] proposed a spatial skyline query based on weighted distance, assigning different weights according to different importance levels among interest points, using weighted Euclidean distance and proposing a MapReduce approach to obtain skyline point sets, reducing the number of dominant judgments to improve query efficiency.

As the research on skyline queries gradually extends to some other issues, including skyline queries in sensor networks [14], skyline queries in data flow environments [15], subspace skyline queries [16], Top-k skyline queries [17], and spatial text skyline queries based on attributes such as direction, social, time etc. In the mobile Internet environment, considering that some query results are different from the user's walking direction and thus cannot satisfy the user's directional preference, the literature [18] proposes a direction-based spatial skyline, which returns skyline objects in different directions, and proposes the concept of pseudo-skyline, where if there is no skyline object in a certain direction, a pseudo skyline object is replaced by a pseudo-skyline object. Considering the impact of user social interaction on queries, literature [19] combined keyword skyline queries with social networks to propose geosocial skyline queries, which mainly rely on users' spatial location information and social network information to perform queries. Considering the importance of temporal attributes in the query, literature [20] applied temporal information to spatial keyword query, and also considered location relevance, text relevance and temporal relevance between objects and query points, and defined two evaluation functions to meet different needs of users, which literature [21] extended to spatial text skyline query, and designed spatial text relevance and temporal text relevance calculation functions to optimize the query results.

In summary, existing query algorithms are not able to solve objects that contain query keywords and query time, as well as objects that are not dominated by other objects on both spatial distance and numerical text attributes. Inspired by the above literature, this paper proposes Boolean spatial temporal text keyword skyline queries to find excellent objects that satisfy user query preferences on all four attributes: spatial, temporal, numerical, and keyword.

3 Problem Definition

The objects in the dataset D are represented by o(l, t, nt, k), o.l denotes the spatial location of the object, represented by a set of coordinate pairs (x, y), o.t denotes the time period contained in the object, $o.t = [t_1, t_2] \cup [t_3, t_4] \cup \ldots \cup [t_i, t_{i+1}]$ and $t_i < t_{i+1}$, where t_i, $t_{i+1} \in [0, 24]$, t_i is the start timestamp and t_{i+1} is the end timestamp, o.nt denotes the set of numerical text attributes of the object, and the set $o.nt = \{nt_1, nt_2, \ldots, nt_j\}$, where $|o.nt_j|$ denotes the value of the j-th dimensional numerical text attribute of the object, and o.k denotes the set of keywords of the object, and the set $o.k = \{k_1, k_2, \ldots, k_m\}$. Boolean spatial temporal textual keyword skyline query is represented by q(l, t, nt, k, r), q.l denotes the spatial location of the query point, q.t denotes the query time period, q.nt denotes the set of numerical textual attributes of the query point, the set $q.nt = \{nt_1, nt_2, \ldots, nt_j\}$, q.k denotes the set of keywords of the query point, the set $q.k = \{k_1, k_2, \ldots, k_n\}$, and q.r denotes the query range, which is represented by a circle with radius r.

In this paper, the Euclidean distance $d(q, o_i)$ between the object o_i and the query point q is considered as a numerical text attribute, and it is set that the smaller the value of the numerical text attribute, the better the object performs. If a certain object performs better with larger values on numerical text attributes, the object o is preprocessed first: $o_i' = max_i - o_i$, where max_i denotes the maximum value on the i-th dimensional numerical text attribute and o_i denotes the value of the object o on the i-th dimensional numerical text attribute.

Definition 1. (Numerical Domination) Given any two objects o_i, o_j with n-dimensional numerical text attributes in a data set, if o_i is less than or equal to o_j in every dimensional numerical text attribute and less than o_j in at least one dimensional numerical text attribute, then o_i is said to numerically dominate o_j, denoted as $o_i \prec_{SN} o_j$.

Definition 2. (Boolean Spatial Temporal Text Keyword Domination) Given a query point q and any two objects o_i, o_j in spatial dataset D, if o_i, o_j contain both query keywords and query time, and $o_i \prec_{SN} o_j$, then it is called o_i Boolean spatial temporal text keyword domination o_j, denoted as $o_i \prec_{BSTTK} o_j$.

Definition 3. (Boolean Spatial Temporal Text Keyword Skyline) Given a query point q and a dataset D, a Boolean spatial temporal text keyword skyline is the set of objects that return those objects from D that cannot be dominated by any other object, denoted the set SP, if $o \in SP$ when and only when $\forall o' \in D$, $o' \nprec_{BSTTK} o$.

4 NTIR-Tree Index

Spatial indexes can efficiently obtain data objects and are very effective in performing query operations. R-Tree [22] is a classical spatial index data structure that divides objects based on their spatial information, and scholars have later extended on the basis of R-Tree to propose spatial indexes such as IR-Tree [23] and IR^2-Tree [24], but they all cannot store spatial, temporal, numerical text and keyword information of objects simultaneously. Therefore, this paper proposes an NTIR-Tree index that can simultaneously store the spatial, temporal, numerical text and keyword information of an object. The structure of NTIR-Tree index is shown in Fig. 1. Where N denotes node, O denotes object, the node containing the object is a leaf node, and other upper level nodes are non-leaf nodes.

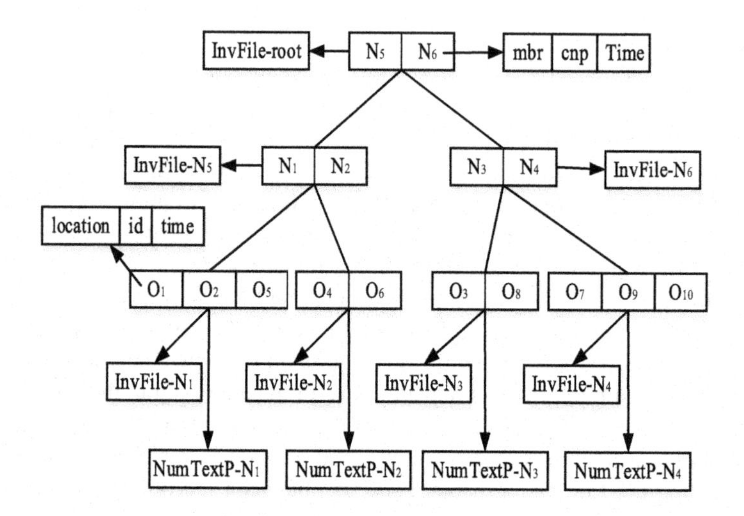

Fig. 1. NTIR-Tree index

NTIR-Tree index leaf node contains the following object information: location: indicates the object's spatial location information, id: indicates the object's identifier in the dataset, time: indicates the time period information contained in the object, InvFile: indicates a pointer to the inverted file of this node. The keywords in the inverted file are composed of the union of the keywords of all objects contained in the node. The objects o_1, o_2, o_4, o_5, o_6 contain the time period information as shown in Table 2. The inverted file of the leaf node is shown in Table 3. NumTextP: indicates the pointer to the numerical text information of the node, and the numerical text information of the node contains both numerical text information of all objects of the node. Example of numerical text information of the leaf node is shown in Table 4.

A non-leaf node contains the following node information: mbr: denotes the smallest bounding rectangle containing all child nodes of the node, cnp: denotes a pointer to a child node of the node, Time: represents the union of the time periods of all child nodes contained in this node, InvFile: indicates the pointer to the inverted file of this node. The keywords in the inverted file are composed of the union of the keywords of all the

Table 2. Time period information of objects

Objects	Time Period
O_1	06:00–08:00
O_2	09:00–12:00
O_4	10:00–17:00
O_5	08:00–20:00
O_6	18:00–24:00

Table 3. Inverted file of leaf nodes

InvFile-N_1	InvFile-N_2	InvFile-N_3	InvFile-N_4
k_1: O_1 、 O_2	k_1: O_4 、 O_6	k_1: O_3	k_1: O_9 、 O_{10}
k_2: O_5	k_2: O_6	k_2:	k_2: O_7
k_3: O_1 、 O_5	k_3:	k_3: O_8	k_3: O_7 、 O_{10}
k_4: O_2 、 O_5	k_4: O_4	k_4: O_3 、 O_8	k_4: O_7

Table 4. Numerical text information

NumTextP-N_1			NumTextP-N_2		
Object	Price per capita	User rating	Object	Price per capita	User rating
O_1	77	8.5	O_4	55	8.5
O_2	65	8.8	O_6	71	7.6
O_5	81	9.0			
NumTextP-N_3			NumTextP-N_4		
Object	Price per capita	User rating	Object	Price per capita	User rating
O_3	95	9.2	O_7	66	7.5
O_8	110	8.9	O_9	72	7.1
			O_{10}	73	8.2

child nodes contained in this node. The nodes N_1, N_2, and N_5 contain the time period information shown in Table 5. The inverted file of non-leaf nodes is shown in Table 6.

Table 5. Time period information of nodes

Nodes	Time Period
N_1	06:00–20:00
N_2	10:00–17:00, 18:00–24:00
N_5	06:00–24:00

Table 6. Inverted file for non-leaf nodes

InvFile-root	InvFile-N_5	InvFile-N_6
k_1: N_5 、 N_6	k_1: N_1 、 N_2	k_1: N_3 、 N_4
k_2: N_5 、 N_6	k_2: N_1 、 N_2	k_2: N_4
k_3: N_5 、 N_6	k_3: N_1	k_3: N_3 、 N_4
k_4: N_5 、 N_6	k_4: N_1 、 N_2	k_4: N_3 、 N_4

5 BSTTKSQ Algorithm Description

This section will introduce the query algorithm STTK and pruning algorithm domCalculation of BSTTKSQ. The BSTTKSQ algorithm consists of two main steps:

First, start from the root node of NTIR-Tree to traverse the index by depth-first search, and select the candidate set of objects containing query time and query keywords. When determining whether the node or object contains query keywords, first, determine whether the inverted file of the node contains query keywords, and if it does, take the intersection of all the keywords to be queried in the inverted file to get the node or object that contains all the query keywords. For example, the keywords contained in leaf node N_1 in Fig. 1 are shown in Table 3. Suppose the query keywords are k_1 and k_3, take the intersection of keywords k_1 and k_3 in the inverted file of node N_1 to get the candidate object o_1 that contains all the query keywords.

Second, the candidate set objects are subjected to the Boolean spatial temporal text keyword domination calculation to obtain the result set objects. In the NTIR-Tree based query algorithm and pruning algorithm, this paper uses a priority queue to maintain the candidate set C obtained in the first step and the result set R obtained in the second step, and the objects in the priority queue are sorted according to the increasing order of the Euclidean distance $d(q, o_i)$ between the query point q and the object o_i (the case of equal Euclidean distance between the query point and the object is not considered for the time being, and this case is finally processing). Since the numerical domination determination in the second query process is the most time-consuming and frequent, this paper proposes two pruning methods, the sorting summation comparison method and the dimensional incremental solution method, to improve the query efficiency.

The following will give the corresponding pruning strategy based on the above two-step solution process.

5.1 Pruning Strategy

In the first step of the query process, nodes and objects are used in the following pruning strategy:

Pruning strategy 1: When traversing the NTIR-Tree index, if the time period of the node or object does not contain the query time period, the node and the child nodes or objects of the node are pruned directly without the judgment of whether the keyword is included.

Pruning strategy 2: When traversing the NTIR-Tree index, if the keyword of the node or object does not contain the query keyword, the node and the child nodes or objects of the node are pruned directly.

Pruning strategy 3: When traversing the NTIR-Tree index, if both the keyword of the node or object contains the query keyword and the time period of the node or object contains the query time period, the child node of the node will be taken to continue the judgment until the leaf node is judged to select the candidate objects that meet the conditions, otherwise the node and the child node or object of the node will be pruned.

Theorem 1. [21] In the priority queue that is dequeued in the increasing order of $d(q, o_i)$, the first dequeued object must be the skyline object.

Theorem 2. In the priority queue dequeued in the increasing order of $d(q, o_i)$, if the dequeued object is o, and any object dequeued after o is o′, there must be $o' \not\prec_{BSTTK} o$.

Proof: By the properties of the priority queue it is known that $d(q, o) < d(q, o')$ and by Definition 2, $o' \not\prec_{BSTTK} o$.

Theorem 3. In the process of queuing out in the increasing order of $d(q, o_i)$, if the sum Sum(o) of the numerical text attributes of the candidate object o is less than or equal to the $Sum(p_i)$ of the numerical text attributes of each object p_i in the result set R, then o is a skyline object.

Proof: According to the nature of the priority queue it is known that $d(q, p_i) < d(q, o)$, and since $Sum(o) \le Sum(p_i)$, the value of p_i is greater than o at least on the one-dimensional numerical text attribute, and according to Definition 2 it is known that p_i and o cannot dominate each other, and o is a skyline object.

Theorem 4. In the process of queuing out in the increasing order of $d(q, o_i)$, the candidate object o and each object p_i in the result set are judged from the second-dimensional numerical text attribute, if the value of p_i on a certain attribute if the value is greater than o, then o and p_i cannot dominate each other, and o is a skyline object.

Proof: According to the nature of the priority queue it is known that $d(q, p_i) < d(q, o)$ because the value of p_i on a property is greater than o, according to Definition 2, it is known that o and p_i cannot dominate each other and o is a skyline object.

During the second query step, the candidate objects are used in the following pruning strategy:

The sorting summation comparison method is proposed based on Theorem 3 first, the candidate objects are sorted according to the increasing order of $d(q, o_i)$ using the priority queue, then, the sum of numerical text attributes of each object is calculated separately, and finally, all the candidate objects satisfying the conditions of Theorem 3 are put into the result set R.

The dimensional incremental solution method is proposed based on Theorem 4: the sorting and summation comparison method is used to filter the remaining objects for judgment, and all the candidate objects that satisfy the conditions of Theorem 4 are put into the result set R.

5.2 Algorithms

Algorithm 1. STTK query algorithm

```
    Input: query point q, NTIR-Tree index, spatial object point set O
    Output: result set R
1.  R = ∅ ; C = ∅ ;
2.  C ← NewPriorityQueue;
3.  R ← NewPriorityQueue;
4.  While not Stack.isEmpty() do
5.    N ← Stack.pop();
6.    If d(q,N) < q. r
7.      If q.t⊆N.t
8.        If q.k⊆N.k
9.          If N.isLeaf() then
10.           For each o in N do
11.             If q.t⊆o.t
12.               If q.k⊆o.k
13.                 C.Enqueue(o);
14.         Else
15.           Stack.push(N.ChildNode);
16. R=domCalculation(q,C);
17. return R;
```

Algorithm 1 is the specific process of BSTTKSQ. Lines 2–3 initialize the candidate set and result set priority queue, lines 4–5 maintain the index in the form of a stack, and lines 6–9 select the nodes containing query keywords and query times within the query range until traversing to the leaf nodes. Lines 10–13 traverse the leaf nodes, pick the objects containing query keywords and query times, and put the objects into the candidate set. Line 16 performs dominance calculation on the objects in the candidate set, and puts the objects that are not dominated into the result set queue.

Algorithm 2. domCalculation pruning algorithm

Input: candidate set C, query point q
Output: result set R
1. R ← getCFirst();
2. For each o in C from the queue do
3.   If Sum(o)≤ Sum(p_i)
4.   insert o into R
5.   Else
6.   insert o into C
7. For other o in C from the queue do
8.   If $|o.nt_j| < |p_i.nt_j|$, j from 2 to n, $p_i \in R$
9.   insert o into R
10 return R

Algorithm 2 is a pruning algorithm to determine the dominance relationship between objects in the candidate set. Line 1 puts the first object out of the queue in the candidate set into the result set. Lines 2–6 select skyline objects using the sorting summation comparison method, and lines 7–9 select skyline objects using the dimension incremental solution method.

Taking the numerical text information of the objects in Table 4 as an example, set {13, 3, 8, 11, 5, 10, 2, 6, 1, 15} as the spatial distances from objects o_1, o_2..., o_{10} to the query point, respectively. Suppose the objects containing the query time and keywords are {o_2, o_3, o_4, o_6, o_8, o_9}, firstly, the objects are sorted in increasing order of spatial distance to get {o_9, o_2, o_8, o_3, o_6, o_4}, and the user rating attributes are preprocessed to get the attribute values of the objects as {o_9(2.1), o_2(0.4), o_8(0.3), o_3(0), o_6(1.6), o_4(0.7)}, find the sum of the numerical text attributes {o_9(75.1), o_2(68.4), o_8(116.3), o_3(103), o_6(82.6), o_4(66.7)} for each object. Then, from the priority queue, we first get o_9 as the result set object, and according to the sorting summation comparison method, we select the candidate objects whose sum of numerical text attributes is less than or equal to all skyline objects in the result set in turn, and get o_2 and o_4, and then compare the remaining objects with the objects in the result set from the second dimensional numerical text attributes according to the dimensional increasing solution method, and get the objects o_8 and o_3 that are not dominated, so the final skyline objects are {o_2, o_3, o_4, o_8, o_9}.

6 Experimental Analysis

In this section, the algorithm is validated under real and simulated datasets. Since the existing algorithm cannot solve the BSTTKSQ problem, the improved INKS algorithm [25], which can handle the problem, is compared with the STTK algorithm. INKS is a keyword skyline query algorithm based on inverted index, and the improved INKS algorithm is referred to as IINKS. The IINKS algorithm first retrieves objects containing query keywords based on the inverted index, then retrieves objects containing query time based on the BNL algorithm and performs domination calculations on the candidate objects. The hardware device used in the experiment is 64-bit Windows 10 operating system, Intel Core i7 CPU @2.10 GHz processor, 8 G memory, Java language to implement the algorithm, and IntelliJ IDEA as the integrated development environment. The real dataset uses the open source dataset on yelp website, which includes 150,346 merchant information in 11 cities such as Cleveland and Toronto, etc. The latitude and longitude, classification and business hours of merchants in the dataset are used as spatial information, keywords and temporal information of the objects respectively, and the number of visits and star rating are used as numerical text information. The method in literature [1] is used to generate simulated positive correlation dataset and simulated anti-correlation dataset and generate corresponding numerical textual information. Experiments are conducted to test the effectiveness of the algorithm by comparing the IINKS algorithm with the STTK algorithm, and the average of 10 tests under the same environment is taken as the final result for each test.

6.1 Impact of the Number of Query Keywords

The impact of the change in the number of query keywords on the algorithm is verified under the real dataset by setting the spatial location of the query points, the query range and the query time period unchanged, the numerical text attribute as 3-dimensional, and the query keywords varying from 1 to 8. Figure 2(a) shows the effect of the change in the number of query keywords on the algorithm running time. The overall running time of both algorithms gradually decreases as the number of query keywords increases, which is because the more the number of query keywords contains fewer objects of query keywords, the less time it takes to dominate the decision. In general, the running time of the STTK algorithm is shorter than that of the IINKS algorithm because the STTK algorithm filters the objects that do not contain query keywords and query time based on the pruning strategy in the first step of traversing the NTIR-Tree, while some objects are filtered based on the two pruning strategies in the second step of dominance determination, which reduces the number of dominance determinations, while the IINKS algorithm only filters quickly based on the inverted index when retrieving objects that contain query keywords, and needs to compare them one by one when retrieving objects that contain query time and dominance calculation, which consumes a lot of time. Figure 2(b) shows the effect of the change in the number of query keywords on the pruning rate. The pruning rate is the ratio of the number of nodes or objects that are cropped to the number of summary points or objects. As the number of query keywords keeps increasing the pruning rate of both algorithms increases, because the more the number of query keywords contains fewer objects of query keywords, the higher the pruning rate is, and the STTK algorithm has a much higher pruning rate than the IINKS algorithm based on an effective pruning strategy.

6.2 Impact of Query Time Period Size

The impact of the change of query time period on the algorithm is verified under the real data set by setting the query keywords as 3, the numerical text attributes as 3-dimensional, the spatial location of the query points and the query range are unchanged, and the query time period keeps increasing. The effect of the change of query time

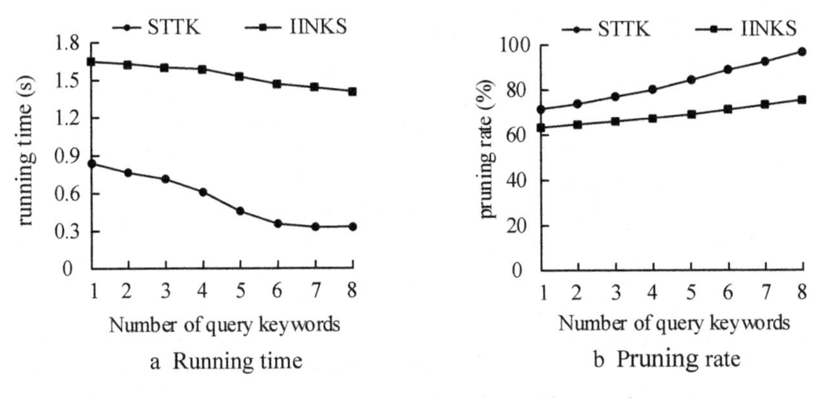

Fig. 2. Impact of the number of query keywords

period on the running time of the algorithm is shown in Fig. 3(a). The overall running time of both algorithms decreases slightly as the query time period keeps increasing. The effect of the change in the query time period on the running time of the algorithm is not obvious, and the trend of decreasing the running time of the STTK algorithm is more obvious than that of the IINKS algorithm, because the STTK algorithm filters more objects according to the pruning strategy and the number of dominant determinations of candidate objects is less. Figure 3(b) shows the effect of the change in query time period on the pruning rate. As the query time period keeps increasing the pruning rate of the IINKS algorithm remains basically the same because the IINKS algorithm does not have an effective pruning strategy and needs to judge all objects containing the query keywords, while the STTK algorithm applies a pruning strategy thus the pruning rate increases.

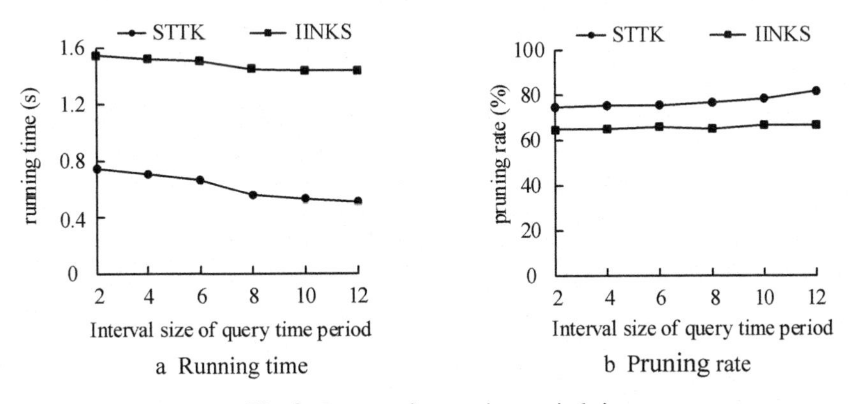

Fig. 3. Impact of query time period size

6.3 Impact of Numerical Text Attribute Dimensions

As the numerical text attribute dimension increases, its domination calculation process is the most frequent and time-consuming, so the algorithm as a whole is first tested under the real dataset, and then numerical domination is tested under the simulated dataset to further verify the effectiveness of the algorithm. The spatial location of query points, query range and query time period are set unchanged under the real dataset, the query keywords are three, and the numerical text attribute dimension varies from 1 to 8 dimensions. Set the objects in the simulated dataset to 100 000, and the numerical text attribute dimension varies from 1 to 8 dimensions. The effect of the change of numerical text attribute dimension on the algorithm running time is shown in Figs. 4(a), (b), and (c). With the increasing dimensionality of numerical text attributes, both algorithms gradually increase the running time on both real and simulated datasets. Because of the need to judge the candidate objects one by one, the running time of the IINKS algorithm tends to rise more obviously, while the STTK algorithm improves the query efficiency by two effective pruning strategies. Since the anti-correlation dataset contains more skyline points, the running time of the anti-correlation dataset is longer than that of the positive correlation dataset. Figure 4(d) shows the effect of the change of numerical text attribute dimensions on the pruning rate under the real dataset. The pruning rate of the IINKS algorithm is basically unchanged because the IINKS algorithm does not have an

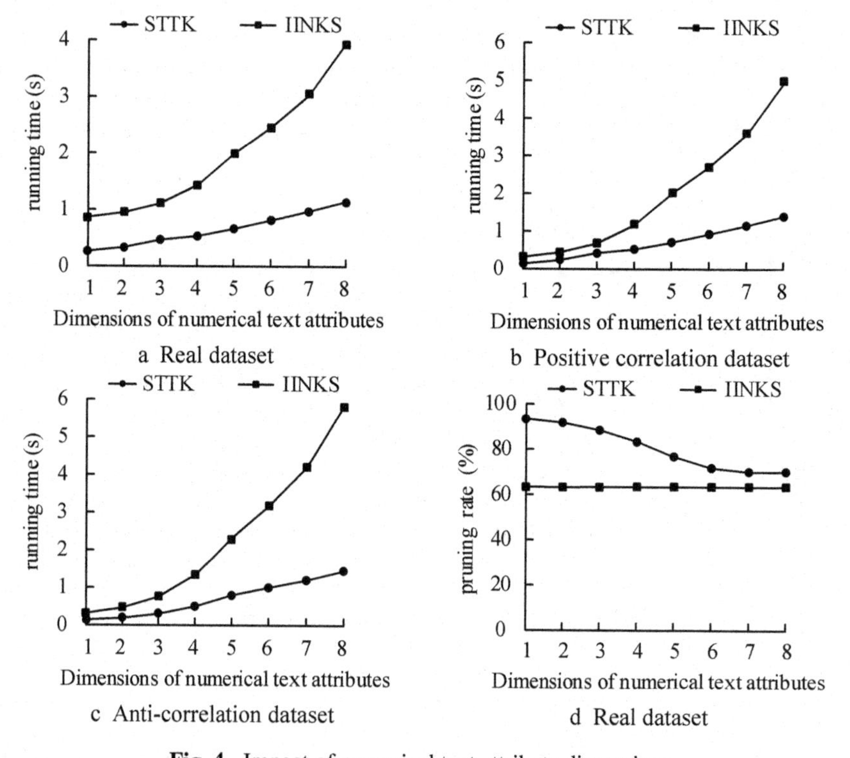

Fig. 4. Impact of numerical text attribute dimensions

effective pruning strategy. As the dimensionality of numerical text attributes increases and more and more objects are not dominated, the pruning rate of the STTK algorithm gradually decreases, but the overall STTK algorithm has a higher pruning rate than that of the IINKS algorithm.

7 Conclusions

To address the diversity of user query requirements, this paper proposes a Boolean spatial temporal text keyword skyline query, which returns those objects that contain query keywords and query times and are not dominated by other objects in terms of both spatial distance and numerical text attributes, and designs the NTIR-Tree, a spatial index that can store four attributes of spatial, time, numerical text and keywords simultaneously. Then, the node and object pruning strategies and query algorithms are proposed based on the NTIR-Tree index, and finally, tests are conducted on real and simulated datasets, and the experimental results show that the proposed algorithm can efficiently solve the Boolean spatial temporal text keyword skyline query problem. The subsequent work will considers the study of the solution to the BSTTKSQ problem in road networks.

Acknowledgements. This paper was supported by the Natural Science Foundation of Jilin Province (YDZJ202201ZYTS666) and the Scientific Research Project of Jilin Education Department (JJKH20210005KJ).

References

1. Borzsony, S., Kossmann, D., Stocker, K.: The skyline operator. In: Proceedings 17th International Conference on Data Engineering, pp. 421–430. IEEE, Heidelberg (2001)
2. Wang, Y., Shi, Z., Wang, J., Sun, L., Song, B.: Skyline preference query based on massive and incomplete dataset. IEEE Access **5**(99), 3183–3192 (2017)
3. Kossmann, D., Ramsak, F., Rost, S.: Shooting stars in the sky: an online algorithm for skyline queries. In: Proceedings of the 28th International Conference on Very Large Data Bases, pp. 275–286. Morgan Kaufmann, Hong Kong (2002)
4. Papadias, D., Tao, Y., Fu, G., Seeger, B.: An optimal and progressive algorithm for skyline queries. In: Proceedings of the 2003 ACM SIGMOD International Conference on Management of Data, pp. 467–478. Association for Computing Machinery, New York (2003)
5. Sharifzadeh, M., Shahabi, C.: The spatial skyline queries. In: Proceedings of the 32nd International Conference on Very Large Data Bases, pp. 751–762. VLDB Endowment, Seoul (2006)
6. Mao, R.: Spatial skyline query problem in Euclidean and road-network spaces. Simon Fraser University, Canada (2020)
7. Cai, Z., Cui, X., Su, X., Guo, L., Liu, Z., Ding, Z.: Continuous road network-based skyline query for moving objects. IEEE Trans. Intell. Transp. Syst. **22**(12), 7383–7394 (2020)
8. Son, W., Stehn, F., Knauer, C., Ahn, H.K.: Top-k manhattan spatial skyline queries. In: Pal, S.P., Sadakane, K. (eds.) Algorithms and Computation. WALCOM 2014, LNCS, vol 8344, pp. 22–33. Springer, Cham (2014). https://doi.org/10.1007/978-3-319-04657-0_5
9. Li, S., Dou, Y.N., He, X.H.: The method of the k-dominant space skyline query in road network. J. Comput. Res. Dev. **57**(1), 227–239 (2020)

10. Li, X.L., Qin, X.L., Wang, N.: Spatial keywords skyline query algorithm. J. Chin. Comput. Syst. **40**(10), 2175–2181 (2019)
11. Metre, K.V., Kharat, M.: Efficient processing of continuous spatial-textual queries over geo-textual data stream. Indonesian J. Electr. Eng. Comput. Sci. **25**(2), 1094–1102 (2022)
12. Bavirthi, S.S., Supreethi, K.P.: An approach for combining spatial and textual skyline querying using indexing mechanism. Turkish J. Comput. Math. Educ. **12**(11), 672–680 (2021)
13. Gavagsaz, E.: Weighted spatial skyline queries with distributed dominance tests. Clust. Comput. **25**(5), 3249–3264 (2022)
14. Dong, L., Liu, G., Cui, X., Li, T.: G-skyline query over data stream in wireless sensor network. Wireless Netw. **26**, 129–144 (2020). https://doi.org/10.1007/s11276-018-1784-2
15. Dzolkhifli, Z., Ibrahim, H., Hassin, M.H.B.M.: Review on skyline query processing techniques over data stream. In: 2021 International Conference on Software Engineering & Computer Systems and 4th International Conference on Computational Science and Information Management, pp. 443–446. IEEE, Pekan (2021)
16. Jiang, T., Zhang, B., Lin, D., Gao, Y., Li, Q.: Efficient column-oriented processing for mutual subspace skyline queries. Soft. Comput. **24**, 15427–15445 (2020)
17. Wei, L., Lin, Z.Y., Lai, Y.X.: DFTS:a Top-k skyline query for large datasets. Comput. Sci. **46**(5), 150–156 (2019)
18. Chen, Z., Guo, S., Liu, W.: Direction-based spatial-textual skyline. Int. J. Innov. Comput. Inf. Control **13**(6), 1813–1828 (2017)
19. Attique, M., Afzal, M., Ali, F., Mehmood, I., Ijaz, M.F., Cho, H.J.: Geo-social top-k and skyline keyword queries on road networks. Sensors **20**(3), 798 (2020)
20. Chen, Z., Zhao, T., Liu, W.: Time-aware collective spatial keyword query. Comput. Sci. Inf. Syst. **18**(3), 1077–1100 (2021)
21. Guo, S.S., Li, S., Yan, H.C.: Time-aware spatial-textual skyline query. Comput. Eng. Appl. **56**(24), 59–65 (2020)
22. Guttman, A.: R-trees: a dynamic index structure for spatial searching. In: Proceedings of the 1984 ACM SIGMOD International Conference on Management of Data, pp. 47–57. Association for Computing Machinery, New York (1984)
23. Cong, G., Jensen, C.S., Wu, D.: Efficient retrieval of the top-k most relevant spatial web objects. Proc. VLDB Endowment **2**(1), 337–348 (2009)
24. De Felipe, I., Hristidis, V., Rishe, N.: Keyword search on spatial databases. In: 2008 IEEE 24th International Conference on Data Engineering, pp. 656–665. IEEE, Cancun (2008)
25. Choi, H., Jung, H., Lee, K.Y., Chung, Y.D.: Skyline queries on keyword-matched data. Inf. Sci. **232**(5), 449–463 (2013)

Continuous Group Nearest Neighbor Query over Sliding Window

Rui Zhu[1,2](✉), Chunhong Li[1], Xiangpeng Meng[1], Chuanyu Zong[1], and Tao Qiu[1]

[1] Shenyang Aerospace University, Shenyang, China
{zhurui,zongcy,qiutao}@sau.edu.cn
[2] Avic Shenyang Aircraft Company Limited, Shenyang, China

Abstract. Group nearest neighbor query(GNN for short) is a classic problem in the spatial database field. Given a data point set D, a query point set Q, the goal of the Group nearest neighbor query(GNN) is to select an object point o in D to minimize the total distance between o and all query points in Q. In this paper, we study GNN in the streaming data environment, i.e., continuous group nearest neighbor search(CGNN for short) over sliding window. In this paper, we propose a continuous query processing framework named BGPT. The idea of the framework is to partition the window and prune some meaningless objects through the dominant relationship between partitions. In order to efficiently support CGNN, we propose a grid-based index to manage streaming data. At the same time, we propose a partition-based method that can use a small number of objects in the streaming data set to monitor query result object. The comprehensive experiments on both real and synthetic data sets demonstrate the superiority of both efficiency and quality.

Keywords: Streaming Data · Continuous Group Nearest Neighbor Query · Dominance · Partition

1 Introduction

Nearest neighbor search is a classic problem in the domain of data management [3]. It also has many variants. Among all of them, group nearest neighbor query(GNN for short) is an important one [8]. Let $\mathcal{D}$ be the set of $d-$dimensional objects, $q(\mathcal{P}, k)$ be a GNN search with $\mathcal{Q}$ being a set of m query points. The goal of GNN search is to find an object $o \in \mathcal{D}$ such that $\sum_{i=1}^{i=k} D(P[i], o)$ is minimal [6]. Here, $Q[i]$ refers to the $i-$th query point.

In this paper, we study the problem of continuous GNN(CGNN for short) search over data stream [2]. Without of generality, we use sliding window for modelling streaming data. Formally, a sliding window W, denoted by the tuple $\langle n, s\rangle$, contains a set of n objects in the window [1]. Whenever the window slides, s objects arrive in the window, and another set of s objects expire from the window. A CGNN, denoted as the tuple $\langle n, s, \mathcal{Q}\rangle$, monitors objects in the window W, which

X. Yang et al. (Eds.): ADMA 2023, LNAI 14180, pp. 225–236, 2023.
https://doi.org/10.1007/978-3-031-46677-9_16

returns one object with the smallest distance sum to the system whenever the window slides. It has many important real applications, such as online conference planning, wireless sensor networks, stock analysis, and so on.

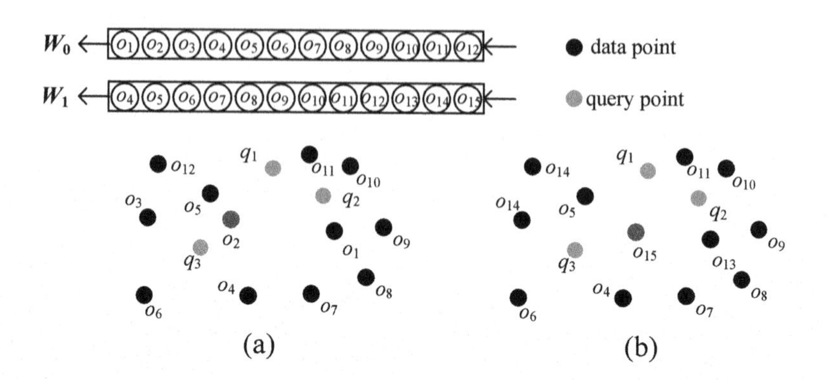

Fig. 1. The running example of CGNN

Take an example in Fig. 1. The window W_0 $\langle 12, 3\rangle$ consists of 12 objects, and the query $\langle 12, 3, Q, 3\rangle$ returns the object o_2 with the smallest distance sum to the system. When windows W_0 slides from W_0 into W_1, objects $\{o_1, o_2, o_3\}$ expires from the window, and objects $\{o_{13}, o_{14}, o_{15}\}$ flows into the window. At that moment, the query $\langle 12, 3, Q, 3\rangle$ returns the object o_{15} to the system.

Currently, some researchers have studied the problem of GNN search over static databases [12]. Their key idea is to use pruning strategies to eliminate data that can't become the nearest neighbor, thereby reducing the search space and speeding up query processing. However, these efforts do not consider how to effectively maintain newly arrived and expired objects, leading that they cannot meet the real-time requirement of users.

In this paper, we propose a novel framework named BGPT to support CGNN search over streaming data [5]. Firstly, we propose a grid-based index to maintain objects in the sliding window. Based on the index we propose, we further propose a group of algorithms that can monitor objects in the window, check which objects have chance to become query result objects and use dominance relationships to delete objects that cannot become query result objects so as to reduce cost both in computational and space. Overall, the contributions of this paper are as follows.

1) We propose a novel query named CGNN. Its target is to monitor objects in the window, returns an object o with minimal score to the system whenever the window slides. Here, the score of an object o, denoted as $F(o)$, equals to the distance sum among query points and o.
2) We propose a novel index named P-Grid to maintain streaming data. For one thing, it uses grid to maintain position relationships among objects. For another, it uses an efficient method named partition [10] to maintain arrived

order among objects. Our goal is to use both position and arrived order relationships among objects to reduce running costs.

3) We propose a group of algorithms to support CGNN search. The goal of these algorithms is to form domination relationship among objects for pruning, and achieve the goal of using a small number of objects to support query processing.

The rest of this paper is as follows. Section 2 reviews the related work and proposes the problem definition. Section 3 explains the framework BGPT. Section 4 evaluates the performance of CGNN. Section 5 concludes this paper.

2 Preliminary

In this section, we first review some important existing results about various types of queries over d-dimensional objects including continuous top-k query and group nearest neighbor query. Then, we introduce the problem definition.

2.1 Related Works

Continuous top-k Query is an important query in the domain of streaming data management, which retrieves k objects with the highest scores whenever the window slides. Based on whether approximate query results are allowed, existing algorithms could be divided into two types: exact algorithms and approximate algorithms. For the former one, Yang at al. [11] proposed a MinTopk algorithm. Its main idea is to maintain a relatively small set of objects for the current window such that the top-k results can be retried from this set as many as possible. For the latter one, zhu et al. [13] proposed an efficient framework, named PABF to support approximate top-k query over a sliding window. It consists of two steps. First, when the window slides, PABF scans the newly arrived objects in batch and utilizes a pruning value to filter out objects whose scores are lower than the pruning value. Then, adopts the idea of merge sort to merge the remaining objects to the candidate set [7]. Accordingly, PABF can support the approximate query by maintaining a few moving objects.

Group Nearest Neighbor Query was proposed by Papadias et al. [9] in 2004. This problem is an important variant of the NN search. Papadias et al. first proposes a R-Tree [4] based index to manage objects. Then, they propose two algorithms named threshold and SPM to support query processing. The key behind these two algorithms is using triangle inequality to compute the lower bound of the total distance sum from the object or node to the query points. If the lower bound is larger than or equal to the score of objects we have accessed so far, then the object or node can be safely pruned. Xu et al. [10] proposed two algorithms, named MTO and TOO in 2010 respectively, which are proposed to support group visible nearest neighbor query. The key behind the algorithms MTO and TOO are by defining an invisible region within the minimum bounding rectangle (MBR) of the query set. This region is used to efficiently eliminate irrelevant data and obstacles. Additionally, the algorithms optimize the traversal of the obstacle R*-tree by performing it only once, further improving efficiency in solving the problem.

2.2 Problem Definition

In this section, we first introduce the concept of the sliding window. It is expressed as the tuple $\langle n, s \rangle$, where n refers to the number of objects contained in the current window, and s refers to the number of objects flowing into(or expiring from) the window each time the window slides. In the following, we will explain the problem definition.

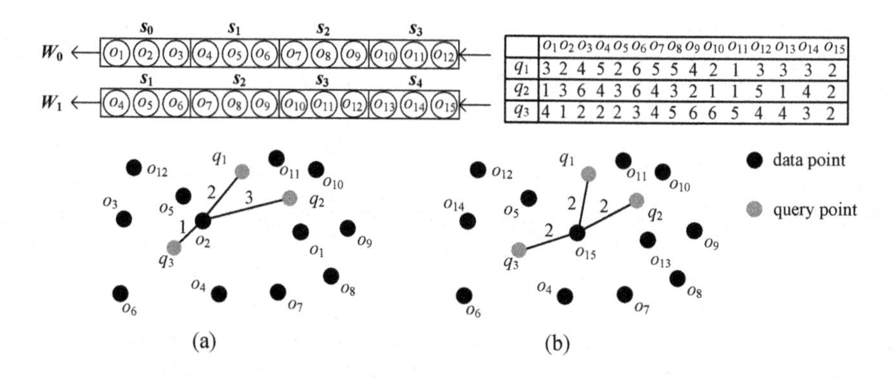

	o_1	o_2	o_3	o_4	o_5	o_6	o_7	o_8	o_9	o_{10}	o_{11}	o_{12}	o_{13}	o_{14}	o_{15}
q_1	3	2	4	5	2	6	5	5	4	2	1	3	3	3	2
q_2	1	3	6	4	3	6	4	3	2	1	1	5	1	4	2
q_3	4	1	2	2	2	3	4	5	6	6	5	4	4	3	2

Fig. 2. Sample problem definition diagram

Definition 1: Continuous Group Nearest Neighbor Query. Let $q\langle \mathcal{Q} \rangle$ be a continuous group nearest neighbor query(CGNN for short), $\mathcal{Q}\{q_1, q_2, \cdots, q_m\}$ be the set of query points with scale being m. It monitors objects in the window and returns one object o with the smallest score to the system. Here, the score of o, denoted as $F(o)$, refers to the distance sum among query points in $\mathcal{Q}$ and o.

$$F(o) = \sum_{i=1}^{m} dist(o, q_i) \tag{1}$$

Backing to the example in Fig. 1. Let q be a CGNN with s being 3 and $\mathcal{Q}$ being the query point set that contains 3 query points $\{q_1, q_2, q_3\}$. In Fig. 1(a). The current window W_0 consists of 12 objects, When windows W_0 slides from W_0 into W_1, objects $\{o_1, o_2, o_3\}$ expires from the window, and $\{o_{13}, o_{14}, o_{15}\}$ flows into the window. We can calculate the distance among all these 3 query points and objects in the window. Because the distance sum among $\{q_1, q_2, q_3\}$ to o_2 is minimal, i.e., equaling to 1 + 2 + 3=6, we return o_2 to the system. In Fig. 2(b), when the window slides in W_1, the query result object is updated to o_{15}, i.e., equaling 2 + 2 + 2=6.

3 Initial Algorithm

In this section, we propose a novel framework named BGPT(short for Bound Grid Partition algorithm) for supporting CGNN over streaming data. In the following, we first explain the structure of our method, and then explain our proposed algorithms.

3.1 The Basic Idea

In this section, we will introduce the initialization algorithm which consists of two main steps, namely The Index Construction, and Candidate Set Maintenance.
The Index Construction. First of all, we form a grid G, and then insert objects in the current window into G. After insertion, for each cell $c \in G$, we further partition objects in it into a group of partitions, i.e., $P(c,m)=\{c_1, c_2, c_3, \cdots, c_{m-1}\}$ following the logic discussed in [10]. After partitioning, the following two conditions should be satisfied: (i) for any two partitions c_i and c_j, $c_i \cap c_j = \emptyset$; (ii) given two objects $o \in c_i$ and $o' \in c_{i+1}$, $o.t$ should be larger than $o'.t$.

Taking the example in Fig. 2. First, we form a grid G, and then insert the objects in the window into G according to the position relationship between the objects and the index grid. For example, the objects o_{11}, o_7 and o_6 are inserted in the c_7, then we further partition the c_7 into $P(c_7, 3)=\{c_1, c_2, c_3\}$. Here, c_1 contains the object o_{11}, c_2 contains the object o_7, and c_3 contains the object o_6. Here the arriving order of o_{11} in c_1 is later than o_7 and o_6 in c_2 and c_3.

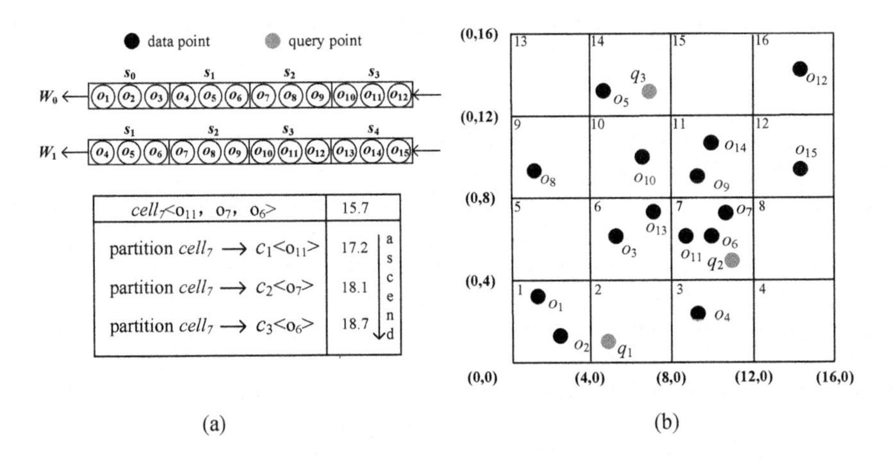

Fig. 3. Example diagram of initialization algorithm

Candidate Set Maintenance. After the index is formed, we are going to evaluate which objects have chance to become query result objects. Specially, we compute the score(or score lower-bound) of each object. After calculation, we further evaluate whether o is dominated by the others. If the answer is yes, we delete it [7]. Otherwise, we regard it as a candidate object. As the logic of the algorithm is simple, we only discuss how to efficiently calculate the score of each object.

Algorithm 1: The Initialization Algorithm

```
Input: The Query Point Set {q1, q2, ......qm}, The Data Object Set O, s
Output: The Object With The Minimal F(oi)
1  P-Grid(c) ← formGridIndex;
2  P-Grid(c) ← insert the object;
3  for i from 0 to c do
4      DLB(c) ←Sum of LB EuclideanDistance(c,Q);
5  for i from 0 to c do
6      DLB(c) ←Sum of LB EuclideanDistance(ci,Q);
7      DLB(c') ← Sum of LB EuclideanDistance(c-ci,Q);
8      if DLB(c) ≥ DLB(c') and TLB(c) ≤ DLB(c') then
9          continue;
10     else
11         C ← insert the smallest DLB ci;
12         for j from 0 to ci do
13             {c1, c2, ··· , cm−1} ← Partition ci;
14             DLB(c) ←Sum of LB EuclideanDistance(cj,Q);
15             DLB(c') ← Sum of LB EuclideanDistance(ci-cj,Q);
16             if DUB(c) ≥ DLB(c') and TUB(c) ≤ DLB(c') then
17                 delete cj;
18 return The Object With The Minimal F(oi);
```

We first sort non-empty cells based on the score lower-bound of objects contained in them. Let c_1 be the cell with the lowest score lower-bound. We access each object contained in c_1, and then compute its score. After computation, we initialize $\mathcal{C}\{c_1, c_2, \cdots, c_{m-1}\}$. Here, c_i is used for maintaining the object with the minimal score in the partition P_i. specially, for each object $o \in c_1$, if $c_i = \infty$ or $c_i > F(o)$, c_i is set to $F(o)$. Otherwise, we ignore o.

We then access other cells based on their corresponding score lower-bound. For the cell c_i, if existing an element $c \in \mathcal{C}$ satisfying: (i) DLB(c_i) is larger than m_j; (ii) objects in c_i arrive later than the ones contained in the partition P_j, we could avoiding access objects contained in c_i as objects contained in it cannot become query result objects. Otherwise, we should further access objects contained in c_i.

$$DLB(c) = \sum_{j=1}^{m} LB(c, q_j) \tag{2}$$

Backing the example in Fig. 3. we compute the score lower-bound of each cell, and the calculation result is in Fig. 4. According to sorting non-empty cells

	$cell_7$	$cell_6$	$cell_2$	$cell_3$	$cell_{11}$	$cell_{10}$	$cell_5$	$cell_1$	$cell_9$	$cell_{12}$	$cell_{14}$
q_1	4.2	3	0	3	7.6	7	3.2	1	7.1	9.9	11
q_2	0	3	3.2	1	3	4.2	7.1	7.1	7.6	3.2	7.6
q_3	3.2	3	7.1	7.1	1.4	1	4.2	7.6	3.2	5	0
sum	7.4	9	10.3	11.1	12	12.2	14.5	15.7	17.9	18.1	18.6

Fig. 4. Example diagram of the distance lower-bound

based on the score lower-bound of objects contained in them, we can get the DLB of $cell_7$ as the smallest, After computation, we initialize $C\{c_1, c_2, c_3\}$. Here, c_1 is used for maintaining the object o_{11} with a minimal score of 17.2 in the partition s_3, c_2 is used for maintaining the object o_7 with a minimal score of 18.1 in the partition s_2, and c_3 is used for maintaining the object o_6 with a minimal score of 18.7 in the partition s_1 respectively.

3.2 The Incremental Maintenance Algorithm

In this section, we explain the incremental maintenance algorithm. When a newly arrived object o flow into the window, we first insert it into the cell c it is located. Then, we evaluate whether we should further compute the score of o as follows. If c is empty before o is inserted into c. We first compute DLB(c) as the manner discussed before. Otherwise, we compare DLB(c) with θ_{m-1}. If $\theta_{m-1} \geq$ DLB(c), we need not to further compute score of o. Otherwise, we compute score of o. After computation, if $F(o) < \theta_{m-1}$, we update θ_{m-1} to $F(o)$.

After updating θ_{m-1}, we further check whether o can dominate objects in the other partitions. Specially, we search the min-heap H for finding objects with scores no smaller than o. If the searching result set is empty, it means no objects in H is dominated by o. Otherwise, the corresponding objects could be deleted. When the partition P_m before the first partition of the window, if o is still contained in H, it means some objects in P_m also may have chance to become query result objects. In this case, we should further access objects in P_m with arrived order larger than o.

In this section, we can further use our proposed index to reduce the scanning cost. Specially, for each object o' we have to access, we first find the cell c that contains o'. If DLB(c) is larger than θ_{m+1}, we need not to further access o'. Otherwise, we should further access o, calculate the real score of o'.

Take an example in Fig. 5, we can see that the candidate result set C under the initialization window W_0 is stored in the form of a tuple, the θ_{m-1} is 17.2 at this moment. When the new object o_{13}, o_{14} and o_{15} flow into the window as shown in W_1. Then the object o_{13} and o_{14} are inserted into the $cell_6$ and $cell_{11}$ respectively, because $cell_6$ and $cell_{11}$ are not empty before o_{13} and o_{14} are inserted. According to the lower-bound score in Fig. 4, we can know the DLB($cell_6$) and DLB($cell_{11}$)

are 9 and 12, which are less than 17.2, so we partition this cell into c_4 and c_5, computing its score are 16.7 and 19.3. Because the c_4 can dominate the c_1, c_2, and c_3, delete this cell in the candidate set $\mathcal{C}$. The object o_{15} is inserted in the empty $cell_{12}$, calculating the DLB as 18.1, which is not less than 17.2. Then partition the cell into c_6 and insert it into $\mathcal{C}$, computing its score as 20.3.

Algorithm 2: The Incremental Maintenance Algorithm

```
   Input: The initial set C, the newly arrived object o
   Output: Updated set C
 1 θ_{m-1} ← C;
 2 Grid(c_i)← according to the position of o;
 3 if c_i is empty then
 4 |   DLB(c) ←Sum of LB EuclideanDistance(c_i,Q);
 5 |   P-Grid(c_i) ← insert the newly arrived object o;
 6 else
 7 |   P-Grid(c_i) ← insert the newly arrived object o;
 8 |   if θ_{m-1} ≥ DLB(c_i) then
 9 |   |   C ← insert c_i;
10 |   else
11 |   |   F(o) ← Sum of EuclideanDistance(o,Q);
12 |   |   if θ_{m-1} ≥ F(o) then
13 |   |   |   θ_{m-1} ← F(o);
14 if P_m is the first partition and o is existing then
15 |   rescan P_m;
16 |   C ← insert the object arrived order larger than o;
17 for j from 0 to C do
18 |   F(o) ← Sum of EuclideanDistance(o,Q);
19 |   θ_{m-1} ← C_j;
20 |   if C_j ≥ DLB(c') then
21 |   |   delete C_j;
22 |   else
23 |   |   continue;
24 return C;
```

4 The Experiment

In this section, we conduct extensive experiments to demonstrate the efficiency of the GNN. The experiments have been conducted using a Microsoft Windows 10 PC (Intel Core2 6226R CPU, 256 GB memory), and all the algorithms are implemented with C++. The experiments are based on both real datasets and synthetic datasets. In the following, we first explain the datasets used in our experiments and the settings of our experiments and then report our findings.

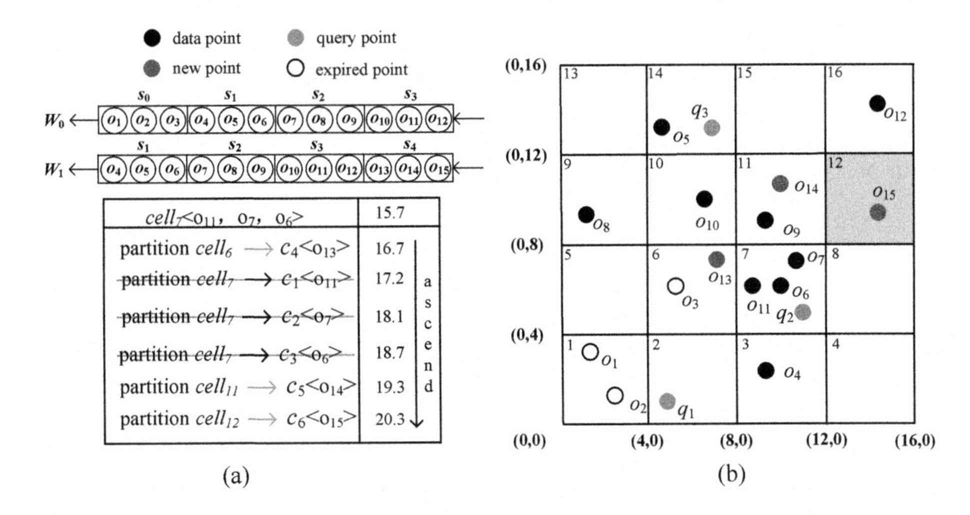

Fig. 5. Example diagram of updating algorithm

4.1 Experiment Settings

Data Set. A total of three data sets were used in our experiment, one real data set named DIDI, and two synthetic data sets named SYN-R and SYN-U. DIDI contains the 1 GB location information of Shanghai/Beijing in the last two years. We cleaned the data with the original size of 30 GB, deleted missing values, duplicates, non-numeric types, and unnecessary data columns, and finally obtained data containing three attributes (taxi Id, location longitude, location latitude) for each record. In the synthetic data set SYN-R, objects are normally distributed. In the synthetic data set SYN-U, objects obey uniform distribution.

Parameters. In our experiment, we evaluate the performance of different algorithms under three parameters, including N, Q, and s. Here, N refers to the window size, Q refers to the number of query points, and s refers to the flow rate of the sliding window. The parameter settings are listed in Table 3 with the default values bolded.

Table 1. Parameter Settings

Parameter	value
N	200 KB, 400 KB, **600 KB**, 800 KB,1 MB
Q	5, **10**, 15, 20, 25 (× 100)
s	2%, 4%, **6%**, 8%, 10% (× N)

Performance Metrics. In continuous group nearest neighbor queries, we evaluated the difference between the initialization time, which is the time to build the initial window and the query, and the update time, which is all the time spent updating the candidate set and the query. Take these two times as the main performance indicators of our experiment.

Competitors. In addition to the algorithm proposed by us, we also implement multiple query method MQM and single point method SPM. MQM algorithm uses the main idea of threshold algorithm, that is, to conduct incremental nearest neighbor query for each point in the query set sum and combine its results. SPM algorithm is to calculate the centroid Q of the query set q and use q to trim data objects or nodes when traversing the R tree. Both algorithms are implemented in a static environment. For better comparison, we implement them in a streaming data environment.

4.2 Performance Comparison

Initialization Running Time Comparison. In the first set of experiments, we compared the initialization times of the algorithm framework BGPT with two comparison algorithms under different sliding window sizes N, and the other parameters were set to default values. From Fig. 4(a)–(c), we can observe that the initialization time of our algorithm is minimal. This is because we calculate the boundary value from the query point to the grid, while the two comparison algorithms calculate the distance from the query point to multiple areas. By calculating the boundary value of the grid, we can avoid repeated calculations in the same grid object, thus saving a lot of time.

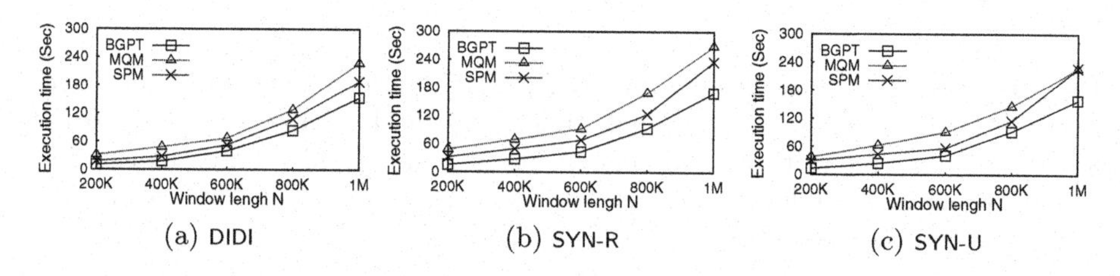

(a) DIDI (b) SYN-R (c) SYN-U

Fig. 6. Initialization cost comparison under different N

Updating Running Time Comparison. Next, we compare the performance of BGPT with its competitors when supporting data streams.

In the second set of experiments, we compared the update time of the algorithm framework BGPT with the two comparison algorithms under different numbers of query points Q, and the other parameters were set to default values. From Fig. 5(a)–(c), we can observe that the query time increases with the increase in the number of query points, and the update time of our algorithm is close to that of the comparison algorithm SPM because our algorithm reduces the time spent on new objects by maintaining candidate sets and, although SPM is close to our time, the pruning time is not as fast as the time we spent processing the sum of candidate sets.

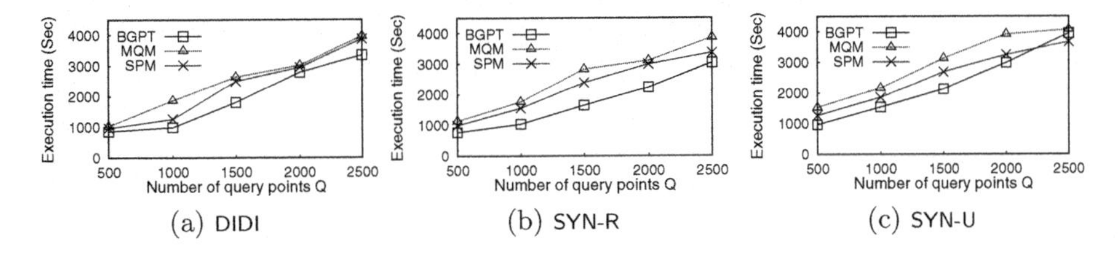

Fig. 7. Updating time comparison under different Q

In the third set of experiments, we compared the update time of the algorithm framework BGPT with the two comparison algorithms at different window flow rates s, and the other parameters were set to default values. We can observe from Fig. 6(a)–(c) that our algorithm is still optimal. Although our algorithm can prune a lot of meaningless objects by maintaining candidate sets and updating, with the increase of s, the partition frequency we process will become faster, so the time difference between our algorithm and other algorithms becomes smaller. However, in daily life, most applications do not have excessive data flow rates, and our algorithm is effective in the general case.

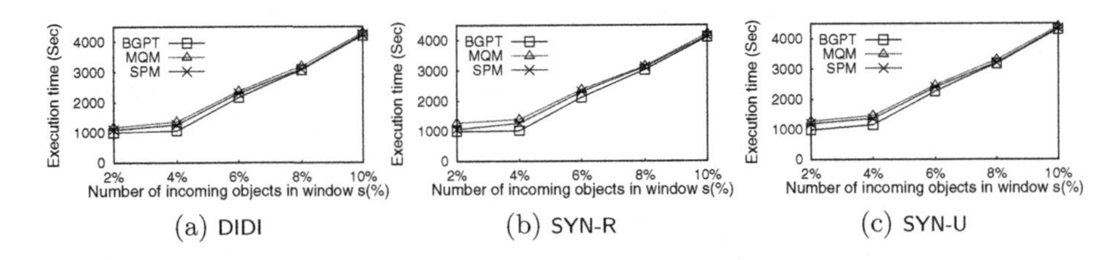

Fig. 8. Updating time comparison under different s

To sum up, BGPT is both stable and efficient. It requires the lowest running time to support continuous group nearest neighbor search compared with SPM and MQM.

5 Conclusion

In this paper, we propose a novel framework BGPT to support continuous group nearest neighbor query on sliding Windows. By calculating the distance lower bound among query points and objects, the framework uses a group of thresholds to prune objects that cannot become query result in a batch way. We conducted experiments to evaluate the performance of the framework BGPT on multiple data sets with different distributions. The results show that BGPT has superior performance.

Acknowledgements. This paper is partly supported by the National Key Research and Development Program of China(2020YFB1707901), the National Natural Science Foundation of Liao Ning(2022-MS-303, 2022-MS-302, and 2022-BS-218).

References

1. Baig, F., Teng, D., Kong, J., Wang, F.: Spear: dynamic spatio-temporal query processing over high velocity data streams. In: 2021 IEEE 37th International Conference on Data Engineering (ICDE), pp. 2279–2284. IEEE (2021)
2. Koç, C.K.: Analysis of sliding window techniques for exponentiation. Comput. Math. Appl. **30**(10), 17–24 (1995)
3. Kollios, G., Gunopulos, D., Tsotras, V.J.: Nearest neighbor queries in a mobile environment. In: Böhlen, M.H., Jensen, C.S., Scholl, M.O. (eds.) STDBM 1999. LNCS, vol. 1678, pp. 119–134. Springer, Heidelberg (1999). https://doi.org/10.1007/3-540-48344-6_7
4. Kothuri, R.K.V., Ravada, S., Abugov, D.: Quadtree and r-tree indexes in oracle spatial: a comparison using gis data. In: Proceedings of the 2002 ACM SIGMOD International Conference on Management of Data, pp. 546–557 (2002)
5. Li, T., Chen, L., Jensen, C.S., Pedersen, T.B.: Trace: real-time compression of streaming trajectories in road networks. Proc. VLDB Endowment **14**(7), 1175–1187 (2021)
6. Li, T., Chen, L., Jensen, C.S., Pedersen, T.B., Gao, Y., Hu, J.: Evolutionary clustering of moving objects. In: 2022 IEEE 38th International Conference on Data Engineering (ICDE), pp. 2399–2411. IEEE (2022)
7. Li, T., Huang, R., Chen, L., Jensen, C.S., Pedersen, T.B.: Compression of uncertain trajectories in road networks. Proc. VLDB Endowment **13**(7), 1050–1063 (2020)
8. Moutafis, P., García-García, F., Mavrommatis, G., Vassilakopoulos, M., Corral, A., Iribarne, L.: Algorithms for processing the group k nearest-neighbor query on distributed frameworks. Distrib. Parallel Databases **39**, 733–784 (2021)
9. Papadias, D., Shen, Q., Tao, Y., Mouratidis, K.: Group nearest neighbor queries. In: Proceedings. 20th International Conference on Data Engineering, pp. 301–312. IEEE (2004)
10. Xu, H., Li, Z., Lu, Y., Deng, K., Zhou, X.: Group visible nearest neighbor queries in spatial databases. In: Chen, L., Tang, C., Yang, J., Gao, Y. (eds.) Web-Age Information Management, pp. 333–344. Springer, Berlin Heidelberg, Berlin, Heidelberg (2010)
11. Yang, D., Shastri, A., Rundensteiner, E.A., Ward, M.O.: An optimal strategy for monitoring top-k queries in streaming windows. In: Proceedings of the 14th International Conference on Extending Database Technology, pp. 57–68 (2011)
12. Yiying, J., Liping, Z., Feihu, J., Xiaohong, H.: Groups nearest neighbor query of mixed data in spatial database. J. Front. Comput. Sci. Technol. **16**(2), 348 (2022)
13. Zhu, R., Wang, B., Luo, S.Y., Yang, X.C., Wang, G.R.: Approximate continuous top-k query over sliding window. J. Comput. Sci. Technol. **32**(1), 93–109 (2017)

Exploration of Stochastic Selection of Splitting Attributes as a Source of Inducing Diversity

Md. Nasim Adnan(✉)

Department of Computer Science and Engineering, Jashore University of Science and Technology, Jashore7408, Bangladesh
nasim.adnan@just.edu.bd

Abstract. One of the most important requirements for a decision forest to secure better ensemble accuracy is generating simultaneously accurate as well as diverse decision trees as base classifiers. Most of the decision forest algorithms in literature exploit either or both of the two major sources of diversity: subspacing and sampling for inducing diversity among the decision trees. Recently, a decision forest algorithm named "Stochastic Forest" introduces stochastic selection of splitting attributes in the process of inducing decision trees and reports promising results. In this paper, we explore the worthiness of "stochastic selection of splitting attributes" as a source of inducing diversity in comparison with the existing two major sources of diversity. We carry out experiments on twenty-five popular data sets that are publicly available from the UCI Machine Learning Repository. The experimental analysis demonstrates the worthiness of "stochastic selection of splitting attributes" as an effective source of diversity.

Keywords: Decision Tree · Decision Forest · Diversity · Stochastic Process

1 Introduction

The use of ensembles in the domain of "classification" is an active area of research [1–5]. An ensemble of classifiers is found to be effective when it is generated from unstable classifiers such as decision trees [6]. Concept Learning System (CLS) [7] is regarded as the pioneering effort for inducing top-down decision trees. In CLS, the induction of a decision tree starts by selecting a non-class attribute A_i to split the training data set $\boldsymbol{D}$ into a disjoint set of horizontal partitions [6,8] in such a way that each horizontal partition contains the same value (or range of values) for the non-class attribute. The purpose of this splitting is to generate more homogeneous distribution of class values in the succeeding partitions than the distribution in $\boldsymbol{D}$. The homogeneity of class distribution in the succeeding

X. Yang et al. (Eds.): ADMA 2023, LNAI 14180, pp. 237–249, 2023.
https://doi.org/10.1007/978-3-031-46677-9_17

partitions is measured for all non-class attributes and the attribute that delivers more homogeneous class distributions in the succeeding partitions than others and than the preceding partition is selected as the splitting attribute. The process of selecting the splitting attribute continues recursively for each succeeding partition until either every partition obtains "absolutely homogeneous class distribution" or there is no scope of obtaining more homogeneous class distributions in the succeeding partitions than the preceding partition. By "absolutely homogeneous class distribution" we mean the presence of a distinct class value for all records in a partition. Popular top-down decision tree induction algorithms such as CART [9] and C4.5 [10,11] follow similar workflow as CLS [7].

If the scope of obtaining more homogeneous class distributions in the succeeding partitions than the preceding partition remains achievable until each partition obtains "absolutely homogeneous class distribution", the generated decision tree will be tightly coupled with the examples (records) of the training data set and should be able to correctly classify every record from the training data set. However, if there are noise(s) or insufficient records in the training data set, such tight coupling with the training data set may backfire on a testing data set whose records can different. Hence, relaxation of tight coupling through the selection of appropriate subset of attributes and/or records may improve the generalization ability of decision tree.

Decision trees are regarded as an unstable classifier since minor differentiations (through the selection of different subset of attributes and/or records) of the training data set can cause significant differences in the structure of ensuing decision trees. A decision forest (in short, forest) is an ensemble of decision trees (in short, trees) where an individual tree acts as a base classifier [6]. A forest is said to impart better generalization ability and is more robust to noise(s) as the trees in a forest are less tightly coupled with the training data set through the selection of appropriate subset of attributes and/or records [5].

We understand that trees generated from the same training data set will be identical. However, it is easy to understand that trees to be accumulated in a forest should be as diverse (in terms of classification errors) as possible; otherwise if they were similar, there could be marginal or no improvement from their ensuing forest [12]. For example, a forest of three identical trees with 95% individual accuracy may perform worse than a forest of three trees being least correlated in terms of classification errors with 67% individual accuracy [13].

If the training data set is differentiated from its original composition through the selection of appropriate subset of attributes and/or records then individual accuracy is bound to decrease due to loss of information. On the other hand, due to their unstable nature, trees become structurally different when they are generated from differently differentiated training data sets. We know, structurally different trees can increase uncorrelated classification errors [14] and hence can increase the ensemble accuracy of the forest [5,15]. Therefore, it is required to differentiate the training data set differently for each tree to make them diverse in terms of classification errors. A number of forest algorithms have been proposed in literature that differentiates the training data set differently for each tree by exploiting the two sources of diversity: subspacing and sampling (through the selection of appropriate subset of attributes and/or records). We present three most prominent of the forest algorithms as follows.

In Bagging [16], a bootstrap sample $\boldsymbol{D}_i$ is generated by selecting a subset of records from the original training data set $\boldsymbol{D}$ in the following way: a $\boldsymbol{D}_i$ contains the same number of records as in $\boldsymbol{D}$ and each record of a $\boldsymbol{D}_i$ is selected from $\boldsymbol{D}$ at random. As a result, some records from $\boldsymbol{D}$ may be selected multiple times or once or may not be selected at all in a $\boldsymbol{D}_i$. According to probabilistic deduction, a subset of around 63.2% original records from $\boldsymbol{D}$ is selected in a $\boldsymbol{D}_i$ [17]. A tree induction algorithm such as CART [9] is applied on a predefined number ($|\boldsymbol{T}|$) of bootstrap samples ($\boldsymbol{D}_1, \boldsymbol{D}_2, ..., \boldsymbol{D}_{|T|}$) to generate the forest.

In Random Subspace algorithm [18], a subset of attributes (subspace) $\boldsymbol{f}$ is generated from the entire attribute space $\boldsymbol{m}$ in the following way: an attribute in $\boldsymbol{f}$ is selected from $\boldsymbol{m}$ at random and can not be selected more than once. Attributes in $\boldsymbol{f}$ can be selected either at "node level" or at "tree level". When selected at "node level", attributes in $\boldsymbol{f}$ can be different for different nodes in the tree; however when selected at "tree level", attributes in $\boldsymbol{f}$ remains the same for each node of the tree. The best non-class attribute (in terms of generating more homogeneous class distributions in the succeeding partitions) from $\boldsymbol{f}$ is selected as the splitting attribute. Instead of using bootstrap samples or any other types of samples, the Random Subspace algorithm uses the original training data set for generating the trees.

The Random Forest algorithm [19] is regarded as a combination of random subspacing of attributes (in short, subspacing) and bootstrap sampling. In the simplest form of Random Forest, the attributes in subspaces ($\boldsymbol{f}$s) are randomly selected at "node level" and $|\boldsymbol{f}| = int(\log_2 |\boldsymbol{m}|) + 1$ [19]. Since its inception in 2001, Random Forest [19] attains considerable interest from the research community and thereby numerous variants have been proposed in recent years [8,20–25].

All the forest algorithms discussed so far exploit either subspacing or bootstrap sampling or both as source(s) of diversity. As mentioned earlier, differentiating the training data set away from its original composition through subspacing or bootstrap sampling or both will certainly cause loss of information. Now, if we want to induce more diversity among the trees by exploiting the two available sources of diversity, it will certainly cause more loss of information. Hence, we need to explore a different source of diversity which will not cause loss of information any further. In doing so, the following opportunity can be explored.

As described earlier, the induction of a decision tree starts by selecting the best non-class attribute (in terms of generating more homogeneous class distributions in the succeeding partitions) to split the training data set into a disjoint set of horizontal partitions [6,8]. Here we see that the best non-class attribute is deterministically selected as the splitting attribute. This deterministic selection of the best non-class attribute as the splitting attribute eventually actuates a tree to be individually highly accurate. As the process of selecting the splitting attribute in no way causes loss of information, here the opportunity of injecting some randomness to induce diversity while protecting individual accuracy remains to be explored thoroughly.

Recently, a decision forest algorithm named "Stochastic Forest" [26] introduces stochastic selection of splitting attributes in the process of inducing decision trees. According to [26], at first a set of non-class attributes ($\boldsymbol{F}$) is constructed whose members have classification capacity more than or equal to the average classification capacity of all non-class attributes. Then, from $\boldsymbol{F}$ the splitting attribute is selected at random, with the probability of being selected as the increasing function of the non-class attributes' classification capacities. The increasing function is realized in such a way that a non-class attribute from $\boldsymbol{F}$ is selected as the splitting attribute with a probability $p(A_i) = \frac{CC(A_i)}{\sum_{i=1}^{|F|} CC(A_i)}$ (where $CC(A_i)$ is the classification capacity of a non-class attribute A_i and $|\boldsymbol{F}|$ is the number of all non-class attributes in $\boldsymbol{F}$. Hence, according to [26], the probability of being selected as the splitting attributes for those attributes that have classification capacity lower than the average classification capacity of all non-class attributes is zero. It is important to note that classification capacity can be measured using Information Gain [10,11] or Gini Index [9].

In this way, the introduction of some randomness through stochastic selection of the splitting attribute over deterministic selection (selection of the best splitting attribute) can exploit the unstable behaviour of tree and contribute to induce diversity while preventing loss of information. Besides, stochastic selection of the splitting attribute over random selection will protect individual accuracy as attributes with higher classification capacities will get more preference to be selected as the splitting attributes. Moreover, since only attributes with classification capacity more than or equal to the average classification capacity of all non-class attributes are considered to be the splitting attribute, individual accuracy will be protected further.

In [26], the authors apply the above mentioned stochastic selection of splitting attributes on the original training data set to generate "Stochastic Forest" and report promising results that focuses mainly on increasing different dimensions of classification accuracy. However, no concrete clarification was provided to explain the reason behind such improvement. In order to excavate the underlying reason behind such improvement as well as to confirm the above mentioned assumptions, we use a simplified version that selects the splitting attribute from the set of all non-class attributes at random, with the probability of being selected as the increasing function of the non-class attributes' classification capacities. The increasing function is realized through the Roulette Wheel technique [27] where a non-class attribute is selected as the splitting attribute with a probability $p(A_i) = \frac{CC(A_i)}{\sum_{i=1}^{|m|} CC(A_i)}$ (where $CC(A_i)$ is the classification capacity of a non-class attribute A_i and $|\boldsymbol{m}|$ is the number of all non-class attributes in the training data set $\boldsymbol{D}$. We choose this Simplified Version of "stochastic selection of splitting attributes" (SV3SA) over the one proposed in [26] to conduct an impartial evaluation of SV3SA as a source of inducing diversity while protecting individual accuracy in comparison with the existing two major sources of diversity: subspacing and bootstrap sampling. Hence, there should be no superficial component in SV3SA that forcefully contributes to increase individual accuracy or the like to impart advantage over the existing two major sources of diversity.

The rest of the paper is organized as follows. In Sect. 2, we discuss the experimental analysis in detail. Finally, we provide the concluding assertions in Sect. 3.

2 Experimental Analysis

We carry out experiments on twenty-five (25) popular data sets that are publicly available from the UCI Machine Learning Repository [28]. Data sets used in experiments are described in Table 1. The main goal of this paper is to exhibit the effectiveness of SV3SA as a source of inducing diversity while protecting individual accuracy when applied in the framework of an existing forest algorithm. Hence, we choose three most prominent forest algorithms described in Sect. 1 i.e. Bagging (BG) [16], Random Subspace (RS) [18] and Random Forest (RF) [19] as the benchmark. For consistency with Random Forest (RF) [19], CART [9] is used as the tree induction algorithm and hence classification capacity is measured using Gini Index [9]. Also, the attributes in subspaces ($\boldsymbol{f}$) are randomly selected at "node level" and $|\boldsymbol{f}| = int(\log_2 |\boldsymbol{m}|) + 1$ [19].

We now apply SV3SA on each of Bagging (BG) [16], Random Subspace (RS) [18] and Random Forest (RF) [19] and call them "Stochastic Bagging (SBG)", "Stochastic Random Subspace (SRS)" and "Stochastic Random Forest (SRF)", respectively. Improving Ensemble Accuracy (EA) is the primary performance objective for a forest algorithm [4,5,15,29]. In [26], "Stochastic Forest" was generated using C4.5 [10,11] tree induction algorithm and hence Gain Ratio [10,11] was used as the measure of classification capacity. Moreover, in [26] the authors apply the stochastic selection of splitting attributes on a set of non-class attributes whose members have classification capacity more than or equal to the average classification capacity of all non-class attributes to boost up individual accuracy. On the other hand, in SV3SA there is no such superficial component to boost up any parameter value (e.g. individual accuracy) to impart comparison advantage over the existing two major sources of diversity. Therefore, we exclude "Stochastic Forest" [26] from this comparison spectrum.

We now evaluate SV3SA in terms of improving EA. In doing so, we compare the EA (in percentage) of BG vs. SBG, RS vs. SRS and RF vs. SRF in Table 2 for all data sets described in Table 1. One hundred (100) trees are generated for each forest as the number is considered to be large enough to ensure convergence of the ensemble effect [2,29] and majority voting is used to determine the EA for each forest [1,5]. All results reported in this paper are obtained using 10-fold Cross Validation (10-CV) [4,15,29,30] and best results are stressed through **bold-face**
From Table 2, we see that SV3SA delivers consistent improvement over existing deterministic selection (selection of the best splitting attribute) in terms of EA in the framework of all three benchmark forest algorithms. This indicates that SV3SA can collaborate effectively with the two major sources of diversity: subspacing and bootstrap sampling used in the framework of benchmark forest algorithms. In support of this observation, we see that SBG (bootstrap sampling+SV3SA) performs better than all of BG (bootstrap sampling), RS

Table 1. Description of the Data Sets

Data Set Name (DS)	Non-class Attributes	Records	Distinct Class Values
Abalone (AB)	08	4177	28
Balance Scale (BS)	04	625	3
Chess (CHS)	36	3196	2
Credit Approval (CA)	15	653	2
Ecoli (EC)	07	336	8
Hayes-Roth (HR)	04	132	3
Hepatitis (HEP)	19	80	2
Image Segmentation (IS)	19	2310	7
Iris (IRS)	04	150	3
Libras Movement (LM)	90	360	15
Liver Disorder (LD)	06	345	2
Lung Cancer (LC)	56	27	2
Musk (Version 2) (MV2)	166	476	2
Nursery (NUR)	08	12960	5
Pen-Based Recognition of Handwritten Digits (PD)	16	10992	10
Pima Indians Diabetes (PID)	08	768	2
Soybean (SOY)	35	47	4
Statlog Heart (SH)	13	270	2
Statlog Vehicle (SV)	18	846	4
Thyroid Disease (TD)	05	215	3
Tic-Tac-Toe (TTT)	09	958	2
Wine (WNE)	13	178	3
Wine Quality (WQ)	11	6497	7
Yeast (YST)	08	1484	10
Zoo (ZOO)	16	101	7

(subspacing) and RF (bootstrap sampling+subspacing) in terms of EA (SBG: **82.9%**, BG: 80.2%, RS: 79.2%, RF: 81.4%). This result conforms with the anticipation made by the authors in [26]. Similarly, SRS (subspacing+SV3SA) performs better than all of BG (bootstrap sampling), RS (subspacing) and RF (bootstrap sampling+subspacing) in terms of EA (SRS: **81.5%**, BG: 80.2%, RS: 79.2%, RF: 81.4%). Overall, SRF (bootstrap sampling+subspacing+SV3SA) emerges to be the best among all in terms of EA (BG: 80.2%, SBG: 82.9%, RS: 79.2%, SRS: 81.5%, RF: 81.4%, SRF: **83.4%**).

We already know that the trees in a forest are required to be simultaneously accurate as well as diverse to achieve better EA [18]. Hence, in order to explain the reason behind such improvements from SV3SA, we first compute individual accuracy (in percentage) of each tree in a forest and then determine Average Individual Accuracy (AIA) for the forest [2,5,15,29]. Hence, a higher AIA value indicates better individual accuracy of the trees in a forest. Kappa is popularly adopted as a measure of diversity among the trees in a forest [2,5]. When Kappa (K) of each tree in a forest is computed, we determine Average Individual Kappa (AIK) for the forest [2,5,15,29]. Unlike AIA, a lower AIK value indicates a higher diversity among the trees in a forest [2,5,15,29]. In Tables 3 and 4, we present

Table 2. Ensemble Accuracy (EA) in percentage

	BG vs. SBG		RS vs. SRS		RF vs. SRF	
DS	BG	SBG	RS	SRS	RF	SRF
AB	25.1	26.2	25.0	26.7	25.0	25.5
BS	77.5	82.6	72.2	78.4	80.5	85.4
CHS	97.9	99.3	95.1	96.7	95.2	97.4
CA	86.4	86.5	85.9	87.2	86.1	87.3
EC	83.1	83.8	83.1	84.4	85.0	84.7
HR	71.3	75.7	45.4	50.8	69.5	76.6
HEP	85.0	87.5	86.3	86.3	86.3	87.5
IS	92.6	97.1	97.6	97.1	97.1	97.1
IRS	95.3	95.3	94.0	94.7	96.0	96.0
LM	76.9	78.9	78.9	76.1	76.1	78.1
LD	68.7	70.4	69.8	71.0	71.5	71.9
LC	63.9	63.9	68.9	68.9	68.9	73.9
MV2	85.0	91.1	79.5	91.1	89.0	89.8
NUR	97.6	98.0	94.9	97.1	95.1	96.9
PD	97.6	98.8	98.7	98.9	98.6	98.7
PID	75.6	75.6	76.2	76.6	75.9	76.7
SOY	85.0	99.1	97.5	98.2	99.1	100.0
SH	81.1	83.7	83.7	83.3	83.0	83.3
SV	73.6	75.3	73.5	75.4	74.1	74.6
TD	93.6	95.4	93.1	94.5	94.6	95.5
TTT	91.8	92.1	80.7	89.5	84.5	88.5
WNE	96.5	98.4	97.2	99.0	97.2	99.0
WQ	57.6	64.0	53.0	63.3	54.4	64.3
YST	60.5	61.8	58.6	61.8	59.5	61.7
ZOO	87.0	91.0	90.0	92.0	92.0	94.0
Avg.	80.2	**82.9**	79.2	**81.5**	81.4	**83.4**

AIA (in percentage) and AIK of BG vs. SBG, RS vs. SRS and RF vs. SRF for all data sets described in Table 1.

From Table 4, we see that forests using SV3SA (i.e. SBG, SRS and SRF) are consistently and significantly more diverse than their respective benchmark forest algorithms (i.e. BG, RS and RF). This is due to the fact that when applied in the framework of the benchmark forest algorithms, SV3SA is able to collaborate effectively with their respective source(s) of diversity. As expected, with three sources of diversity (bootstrap sampling + subspacing + SV3SA), SRF is found to be the most diverse (**0.53**) among all. On the other hand, from Table 3, we see a decrease in AIAs for forests using SV3SA compared to their respec-

Table 3. Average Individual Accuracy (AIA) in percentage

	BG vs. SBG		RS vs. SRS		RF vs. SRF	
DS	BG	SBG	RS	SRS	RF	SRF
AB	21.5	18.3	24.1	22.2	24.1	18.4
BS	64.1	64.8	67.4	68.3	64.9	65.5
CHS	97.8	91.9	65.3	79.9	67.9	84.2
CA	85.2	76.3	80.9	77.7	74.0	74.1
EC	78.7	71.9	80.0	65.7	77.1	70.0
HR	55.9	55.3	48.3	51.4	55.4	54.3
HEP	81.4	76.7	81.4	80.9	79.1	77.6
IS	92.8	89.3	93.3	89.6	92.0	88.3
IRS	93.8	91.8	94.1	92.9	93.3	91.6
LM	53.7	43.3	58.0	48.8	49.8	42.9
LD	61.0	58.3	62.6	60.4	60.1	58.8
LC	66.3	49.9	60.9	50.2	58.9	50.3
MV2	75.3	72.5	79.5	75.7	73.3	72.1
NUR	95.8	94.1	70.4	88.1	70.2	87.1
PD	94.5	89.6	93.6	90.3	92.5	88.8
PID	69.8	67.3	71.8	69.7	70.0	67.4
SOY	74.8	71.5	74.7	72.4	66.7	65.9
SH	74.7	72.9	76.9	74.4	74.1	72.3
SV	68.5	63.6	68.6	65.2	66.7	63.0
TD	91.4	89.4	92.0	91.0	91.1	89.6
TTT	79.0	74.6	40.5	71.8	54.3	70.6
WNE	89.5	85.6	90.4	88.1	88.9	85.2
WQ	52.3	44.7	43.9	49.3	44.6	45.2
YST	47.9	46.4	52.1	50.1	47.9	45.7
ZOO	89.6	82.6	88.1	83.6	85.9	81.2
Avg.	**74.2**	69.7	**70.4**	70.3	**68.9**	68.4

tive benchmark forest algorithms. This conforms with our earlier discussion that deterministic selection of the best non-class attribute as the splitting attribute contributes to tight coupling with the training data set and actuates trees to be individually highly accurate. We now present this increase-decrease phenomenon more precisely in Fig. 1. For the purpose of better presentation, in line with AIA values, AIK values are also shown in percentage in Fig. 1.

From Fig. 1, we see that AIA:AIK values from SBG, SRS and SRF consistently remain higher compared to their respective benchmark forest algorithms (BG vs SBG: 1.16 vs. 1.25, RS vs. SRS: 1.09 vs. 1.21, RF vs. SRF: 1.22 vs. 1.29). This indicates that there are more individual accuracy (AIA in percent-

Table 4. Average Individual Kappa (AIK)

	BG vs. SBG		RS vs. SRS		RF vs. SRF	
DS	BG	SBG	RS	SRS	RF	SRF
AB	0.33	0.23	0.58	0.35	0.58	0.22
BS	0.46	0.41	0.61	0.53	0.43	0.42
CHS	0.98	0.86	0.47	0.65	0.49	0.70
CA	0.92	0.70	0.75	0.68	0.64	0.63
EC	0.79	0.69	0.83	0.64	0.76	0.66
HR	0.32	0.35	0.53	0.42	0.32	0.33
HEP	0.25	0.17	0.23	0.18	0.17	0.17
IS	0.86	0.88	0.90	0.88	0.90	0.86
IRS	0.92	0.90	0.95	0.92	0.92	0.88
LM	0.55	0.42	0.59	0.49	0.50	0.42
LD	0.36	0.28	0.49	0.41	0.31	0.28
LC	0.13	0.01	0.11	0.01	0.10	0.02
MV2	0.59	0.46	1.00	0.54	0.49	0.46
NUR	0.95	0.92	0.64	0.83	0.63	0.82
PD	0.95	0.89	0.93	0.90	0.92	0.88
PID	0.50	0.46	0.58	0.50	0.49	0.44
SOY	0.74	0.60	0.62	0.60	0.55	0.53
SH	0.62	0.53	0.68	0.59	0.59	0.53
SV	0.66	0.58	0.67	0.61	0.62	0.57
TD	0.87	0.81	0.88	0.84	0.84	0.80
TTT	0.55	0.44	0.28	0.39	0.31	0.34
WNE	0.84	0.77	0.85	0.82	0.83	0.77
WQ	0.51	0.31	0.46	0.39	0.41	0.30
YST	0.47	0.49	0.60	0.59	0.47	0.47
ZOO	0.89	0.77	0.85	0.78	0.81	0.75
Avg.	0.64	**0.56**	0.64	**0.58**	0.56	**0.53**

age) per unit of diversity (AIK in percentage) for SBG, SRS and SRF compared to their respective benchmark forest algorithms. Therefore, it is reasonable that as a source of diversity SV3SA can prevent loss of information while inducing diversity. AIA:AIK values also reveal how SV3SA collaborates with the other two major sources of diversity. For example, when SV3SA is applied on RS (subspacing) to form SRS (subspacing + SV3SA), AIA:AIK goes up from 1.09 to 1.21. This indicates that SV3SA collaborates better in the framework of RS. Similarly, when SV3SA is applied on RF (bootstrap sampling + subspacing) to form SRF (bootstrap sampling + subspacing + SV3SA), AIA:AIK experiences an increase from 1.22 to 1.29. This indicates that with two sources of diversity (bootstrap

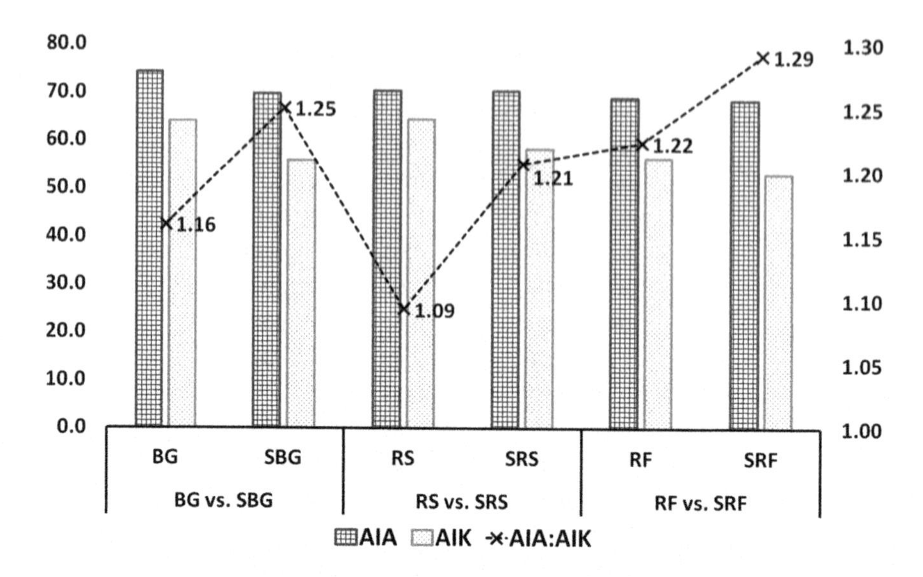

Fig. 1. Increase-Decrease phenomenon of AIA and AIK

sampling + subspacing) in place, RF already enjoys a favourable AIA:AIK value (compared to other two benchmark algorithms). In spite of that, when SV3SA is included in the spectrum, it renders considerable improvement. Furthermore, it is important to note that there is a strong correlation between the values of EA and AIA:AIK. From Table 2 and Fig. 1 we compile the following values: BG (EA: 80.2%, AIA:AIK value: 1.16), SBG (EA: 82.9%, AIA:AIK value: 1.25), RS (EA: 79.2%, AIA:AIK value: 1.09), SRS (EA: 81.5%, AIA:AIK value: 1.21), RF (EA: 81.4%, AIA:AIK value: 1.22) and SRF (EA: 83.4%, AIA:AIK value: 1.29). From the compiled values, we see that RS has the lowest AIA:AIK value (1.09) and the lowest EA (79.2%). Similarly, SRF has the highest AIA:AIK value (1.29) and the highest EA (83.4%). Equivalent level of strong correlation prevails for rest of the forest algorithms. Therefore, we comprehend that "the higher the AIA:AIK value for a forest algorithm, the higher the EA".

We now analyze the impact of SV3SA on tree structure. In doing so, we report Average Tree Depth (ATD) of BG vs. SBG, RS vs. SRS and RF vs. SRF in Table 5 for all data sets described in Table 1.

From Table 5, we see that trees generated from SV3SA reach comparatively more depth. The reason behind this is that it may require more splitting events to achieve the state where "every partition obtains absolutely homogeneous class distribution or there is no scope of obtaining more homogeneous class distributions in the succeeding partitions than the preceding partition" when selecting the splitting attributes stochastically instead of selecting the best non-class attributes as the splitting attributes.

Table 5. Average Tree Depth (ATD)

	BG vs. SBG		RS vs. SRS		RF vs. SRF	
DS	BG	SBG	RS	SRS	RF	SRF
AB	20.19	26.72	15.52	22.48	15.60	25.44
BS	3.02	3.05	3.00	3.00	3.01	3.02
CHS	10.90	16.07	6.89	15.94	7.50	17.42
CA	3.35	6.69	4.73	6.75	4.52	6.53
EC	7.49	8.23	8.22	7.50	7.94	8.60
HR	2.38	2.54	2.72	2.88	2.59	2.67
HEP	2.62	3.57	4.09	4.78	3.46	4.18
IS	13.48	14.45	13.68	15.08	13.18	14.97
IRS	3.51	4.15	3.99	4.55	3.81	4.48
LM	10.67	12.92	11.17	13.15	11.01	12.85
LD	9.11	11.85	10.12	11.84	10.11	11.95
LC	1.48	2.05	2.66	2.01	2.43	2.21
MV2	8.80	14.22	10.10	15.41	10.34	14.23
NUR	7.14	7.46	6.09	7.00	6.22	7.49
PD	12.88	17.68	14.13	17.99	13.83	18.05
PID	11.60	14.11	12.97	13.76	12.31	14.11
SOY	1.05	2.27	2.04	2.59	1.80	2.28
SH	6.35	8.16	7.53	8.40	7.17	8.42
SV	12.64	16.32	16.32	16.20	14.37	16.17
TD	4.06	5.13	4.88	5.68	4.36	5.38
TTT	5.97	6.12	3.29	6.00	4.59	6.28
WNE	3.30	4.81	4.13	5.08	3.87	5.03
WQ	19.47	30.78	9.91	28.41	13.35	29.42
YST	18.90	22.34	14.79	21.01	17.21	22.22
ZOO	5.81	5.31	5.47	5.48	5.29	5.35
Avg.	**8.25**	10.68	**7.94**	10.52	**7.99**	10.75

3 Conclusion

In this paper, we explore the potential of a Simplified Version of "stochastic selection of splitting attributes" (SV3SA) as a source of inducing diversity in comparison with the existing two major sources of diversity: subspacing and bootstrap sampling. In doing so, we apply SV3SA in the framework of three most prominent forest algorithms namely Bagging [16], Random Subspace [18] and Random Forest [19] (used as the benchmark). From experimental analysis, we find that SV3SA delivers consistent improvement in terms of Ensemble Accuracy in the framework of all the benchmark forest algorithms. We also find

SV3SA to be an effective source of inducing diversity while protecting individual accuracy. Through stochastic selection of the splitting attributes over deterministic selection (selection of the best splitting attributes), SV3SA can help relaxing the tight coupling with the training data set and hence has a potential to reduce the overfitting problem both in tree level and forest level. This can be a future research direction of this paper.

References

1. Polikar, R.: Ensemble based systems in decision making. IEEE Circ. Syst. Mag. **6**(3), 21–44 (2006)
2. Amasyali, M.F., Ersoy, O.K.: Classifier ensembles with the extended space forest. IEEE Trans. Knowl. Data Eng. **26**(3), 549–562 (2014)
3. Rokach, L.: Decision forest: twenty years of research. Inf. Fusion **27**, 111–125 (2016)
4. Adnan, M.N., Islam, M.Z.: Forest CERN: a new decision forest building technique. In: Lecture Notes in Computer Science (including subseries Lecture Notes in Artificial Intelligence and Lecture Notes in Bioinformatics), vol. 9651, pp. 304–315 (2016)
5. Adnan, M.N.: Decision tree and decision forest algorithms: on improving accuracy, efficiency and knowledge discovery. PhD thesis, School of Computing and Mathematics, Charles Sturt University, Bathurst, Australia (2017)
6. Tan, P.N., Steinbach, M., Kumar, V.: Introduction to data mining, vol. 12. Pearson Education (2011)
7. Simon, H.A., Hunt, E.B., Marin, J., Stone, P.: Experiments in Induction, vol. 80. Academic Press, New York (1967)
8. Adnan, M.N., Islam, M.Z.: Effects of dynamic subspacing in random forest. In: Lecture Notes in Computer Science (including subseries Lecture Notes in Artificial Intelligence and Lecture Notes in Bioinformatics), vol. 10604 LNAI, pp. 303–312 (2017)
9. Breiman, L., Friedman, J.H., Olshen, R.A., Stone, C.J.: Classification and Regression Trees. U.S.A, Wadsworth International Group, CA (2017)
10. Quinlan, J.R.: C4.5 - Programs for Machine Learning. Morgan Kaufmann Publishers, San Mateo, U.S.A (1993)
11. Quinlan, J.R.: Improved use of continuous attributes in C4.5. J. Artif. Intell. Res. **4**, 77–90 (1996)
12. Shipp, C.A., Kuncheva, L.I.: Relationships between combination methods and measures of diversity in combining classifiers. Inf. Fusion **3**(2), 135–148 (2002)
13. Zhang, Y., Burer, S., Nick Street, W.: Ensemble pruning via semi-definite programming. J. Mach. Learn. Res. **7**, 1315–1338 (2006)
14. Kuncheva, L.I.: Using diversity measures for generating error-correcting output codes in classifier ensembles. Pattern Recogn. Lett. **26**(1), 83–90 (2005)
15. Adnan, M.N., Islam, M.Z.: Forest PA: constructing a decision forest by penalizing attributes used in previous trees. Expert Syst. Appl. **89**, 389–403 (2017)
16. Breiman, L.: Bagging predictors. Mach. Learn. **24**(2), 123–140 (1996)
17. Han, J., Kamber, M., Pei, J.: Data Mining: Concepts and Techniques. Morgan Kaufmann Publishers, Burlington (2012)
18. Tin Kam Ho: The random subspace method for constructing decision forests. IEEE Trans. Pattern Anal. Mach. Intell. **20**(8), 832–844 (1998)
19. Breiman, L.: Random forests. Mach. Learn. **45**(1), 5–32 (2001)

20. Adnan, M.N.: On dynamic selection of subspace for random forest. In: Lecture Notes in Computer Science (including subseries Lecture Notes in Artificial Intelligence and Lecture Notes in Bioinformatics) **8933**, 370–379 (2014)
21. Adnan, M.N., Islam, M.Z.: A comprehensive method for attribute space extension for random forest. In: 2014 17th International Conference on Computer and Information Technology, ICCIT 2014, pp. 25–29 (2003)
22. Adnan, M.N., Islam, M.Z.: ComboSplit: combining various splitting criteria for building a single decision tree. In: International Conference on Artificial Intelligence and Pattern Recognition, AIPR 2014, Held at the 3rd World Congress on Computing and Information Technology, WCIT, pp. 1–8 (2014)
23. Adnan, M.N., Islam, M.Z.: Complement random forest. In: Conferences in Research and Practice in Information Technology Series, vol. 168, pp. 89–97 (2015)
24. Adnan, Md.N., Islam, Md.Z.: Improving the random forest algorithm by randomly varying the size of the bootstrap samples for low dimensional data sets. In: 23rd European Symposium on Artificial Neural Networks, Computational Intelligence and Machine Learning, ESANN 2015 - Proceedings, pp. 391–396 (2015)
25. Adnan, M.N.: On reducing the bias of random forest. In: Chen, W., Yao, L., Cai, T., Pan, S., Shen, T., Li, X., editors, Advanced Data Mining and Applications, pp. 187–195. Springer Nature Switzerland, Cham (2022). https://doi.org/10.1007/978-3-031-22137-8_14
26. Tsipouras, M.G., Tsouros, D.C., Smyrlis, P.N., Giannakeas, N., Tzallas, A.T.: Random forests with stochastic induction of decision trees. In: 2018 IEEE 30th International Conference on Tools with Artificial Intelligence (ICTAI), pp. 527–531 (2018)
27. Liu, Y., Xindong, W., Shen, Y.: Automatic clustering using genetic algorithms. Appl. Math. Comput. **218**(4), 1267–1279 (2011)
28. Lichman, M.: UCI machine learning repository [http://archive.ics.uci.edu/ml]. http://archive.ics.uci.edu/ml/datasets.html (2013)
29. Adnan, M.N., Leung Ip, R.H., Bewong, M., Islam, M.Z.: BDF: a new decision forest algorithm. Inf. Sci. **569**, 687–705 (2021)
30. Arlot, S., Celisse, A.: A survey of cross-validation procedures for model selection. Statist. Surv. **4**, 40–79 (2010)

Machine Unlearning Methodology Based on Stochastic Teacher Network

Xulong Zhang[1], Jianzong Wang[1(✉)], Ning Cheng[1], Yifu Sun[1,2], Chuanyao Zhang[1,3], and Jing Xiao[1]

[1] Ping An Technology (Shenzhen) Co., Ltd., Shenzhen, China
jzwang@188.com
[2] School of Computer Science, Fudan University, Shanghai, China
[3] University of Science and Technology of China, Hefei, People's Republic of China

Abstract. The rise of the phenomenon of the "right to be forgotten" has prompted research on machine unlearning, which grants data owners the right to actively withdraw data that has been used for model training and requires the elimination of the contribution of that data to the model. A simple method to achieve this is to use the remaining data to retrain the model, but this is not acceptable for other data owners who continue to participate in training. Existing machine unlearning methods have been found to be ineffective in quickly removing knowledge from deep learning models. This paper proposes using a stochastic network as a teacher to expedite the mitigation of the influence caused by forgotten data on the model. We performed experiments on three datasets, and the findings demonstrate that our approach can efficiently mitigate the influence of target data on the model within a single epoch. This allows for a one-time erasure and reconstruction of the model, and the reconstructed model achieves the same performance as the retrained model.

Keywords: Machine Unlearning · Stochastic Network · Knowledge Distillation

1 Introduction

Regarding user privacy, recent laws such as the General Data Protection Regulation (GDPR) [22] and the California Consumer Privacy Act (CCPA) [11] have bestowed users with the privilege of exercising their right to be forgotten. However, deleting user data does not guarantee that the model does not retain any information about that data. This can result in the model making predictions that reveal customer privacy. From a system security perspective, the intentional or unintentional use of low-quality or outdated data during the training phase can exert a detrimental influence on the model. Previous studies [3,6,7,17,20,26] have shown that the model may be vulnerable to data poisoning attacks from malicious clients. Therefore, there is a need to remove specific data contributions from a trained model to enhance the security and dependability of the model

X. Yang et al. (Eds.): ADMA 2023, LNAI 14180, pp. 250–261, 2023.
https://doi.org/10.1007/978-3-031-46677-9_18

system. However, the relationship between data and specific parameters of the model is not clear in the current field of deep learning [18,24], making it difficult to modify model parameters directly to effectively mitigate the influence of target data on the model and minimize its impact.

One simple method to forget target data is to delete it from the training dataset and perform model retraining using the remaining data. However, this approach is time-intensive and not feasible for other data owners who continue to participate in training. Machine unlearning [4,10,13,21] provides a way to forget target data by effectively mitigating the influence of the forgotten data on the model and reconstructing the model on the remaining data sets. The objective is to achieve the same effectiveness as a retrained model but with a faster and more efficient process.

The main difficulties of machine unlearning come from three aspects [24]: the interpretability of the deep learning model [28], the Randomness of model training process and the Increment of model training process. In [13], the author proposes to use the contrastive labels to help the model eliminate the impact of target data. In [16], the author proposes to rebuild the model by historical parameters. In [2,5] proposed a machine unlearning method for simple machine learning models. But existing machine unlearning methods are difficult to quickly remove knowledge from complex deep learning models.

This paper introduces a novel approach of machine unlearning which can realize fast forgetting of target data and fast reconstruction of model. When a model completely forgets the target data, its performance on the target data should be consistent with that of the model completely untrained by the target data. Therefore, we use a stochastic network that has not been trained by the target data as the teacher network [8,9]. By fitting the output probability distribution generated by the stochastic network for the target data, we make the trained model to forget the target data. Then, we leverage the original trained model as a mentor network to quickly reconstruct the model on the remaining data set. We have conducted extensive experiments on 3 pulic datasets, and the results of our experiments demonstrate that our method can quickly eliminate the impact of target data on the model through only one epoch, achieve the one-time erasure and reconstruction of the model on target data. The reconstructed model has achieved comparable performance to the retrained model.

2 Related Work

2.1 Machine Unlearning

With the development of artificial intelligence, the model is increasingly dependent on training data, and there is an increasing number of model parameters and an expanding scale of training data. Simultaneously, the problem of data ownership is more strict, and the owner of the data possesses the right to ask for the model to remove the contribution of personal data from its records. From the perspective of security, the system may be attacked by data poisoning, and malicious customers may provide incorrect data to impact the model's performance.

The system should possess the capability to mitigate the influence of malicious data on the model through deletion. From the perspective of system availability, The system should allow the customer to recall the wrong data entered unexpectedly. For example, in the recommendation system, the user may enter a certain product by mistake. If the influence of the unexpected data cannot be eliminated, it may lead to frequent recommendation of the wrong product for the user. The objective of machine unlearning is to effectively mitigate the impact of target data on the model.

In [4], the author divides the training data into smaller data sets by slicing the data, and reduces the computational overhead of the unlearning phase through the restriction of data point impact on the training process. In [1], the author proposed to randomly merge the training sample data into a smaller number of samples to train weak learners, which used random projection to model nonlinear relationships, and proposed to use random linear code to speed up the machine unlearning process. In [16], the author proposes a machine unlearning method for distributed computing, which helps to accelerate the effect of deleting client data on the global model by saving the historical update data of different clients. In [23], the author proposes a channel pruning method to eliminate information about specific categories in the model. By quantifying the relationship between the activation of different channels and specific categories in the neural network, the author identifies the corresponding relationship between different channels and specific categories. Then, specific categories are forgotten by pruning the parameters of specific channels. In [13], the author uses contrastive labels to help the model eliminate the effect of the requested forgotten data on the model. The contrastive label is an error label generated automatically utilizing the output of the original model as a basis for the request for forgotten data. In [24], the author effectively mitigates the influence of client data by directly subtracting the parameters of historical updates from it. However, this approach may lead to model skewness, and to address this issue, the author uses knowledge distillation to reconstruct the model.

2.2 Knowledge Distillation

The concept of knowledge distillation [12], involves extracting the knowledge encapsulated in a larger teacher model and impart it to a smaller student model. This is a knowledge transfer process. It obtains more output distribution information about the teaching network by introducing soft labels. In [19], to facilitate the training of a more compact student network, the author utilizes the feature maps from the intermediate layer of the teacher network to guide the corresponding student network's layer. In [27], this paper proposes a method to accomplish the transfer of knowledge from the teacher to the student through the attachment map of the network middle layer. In this work, we employ knowledge distillation as a technique to help the original training model achieve fast data information erasure and rapid model reconstruction.

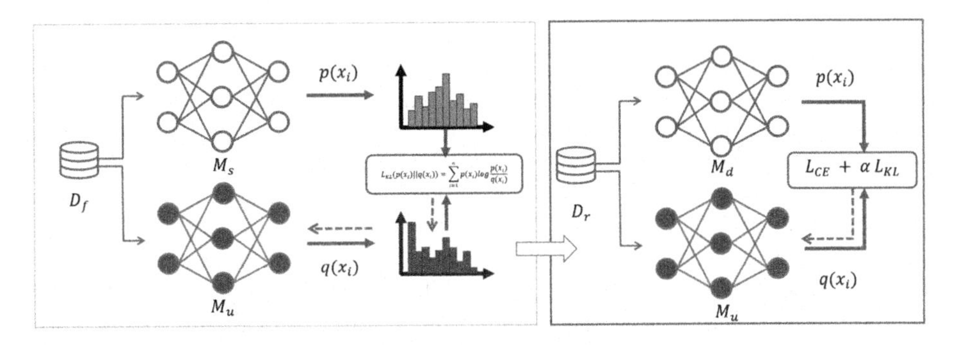

Fig. 1. Our algorithm is divided into two stages. The left represents the stage of knowledge erasure, and the right represents the stage of model reconstruction.

3 Methodology

3.1 Problem Definition

We use $D = (x_i, y_i)_{i=0}^{n}$ to represent the original dataset participating in model training, where n denotes the total number of training samples, and we use M_d to represent the model trained using the original dataset D. D_f indicates the target data requested to be forgotten by the model M_d. D_r represents the data remaining after the target data D_f is deleted from the original dataset D, where $D = D_f \cup D_r$. The objective is to effectively erase the influence caused by the D_f on the model M_d to obtain the unlearning model M_u. One approach to achieve complete eradication of the data from the model is utilizing the remaining data D_r to retrain a model M_r. The model retrained under using the remaining data D_r ensures that it does not contain any information about the target data D_f. However, the retraining cost required is unacceptable. We define a ideal machine unlearning process $L(\cdot)$ as follows:

$$I(M_r) = I(L(M_d, D, D_f)), \tag{1}$$

where $I(\cdot)$ represents a model evaluation function.

3.2 Our Approach

We introduce a machine unlearning methodology grounded in stochastic network to help the model eliminate the relevant impact of the target data and can quickly reconstruct the model, so that it can still maintain good performance on the remaining data sets. As shown in Fig. 1. Our algorithm comprises two stages. The left represents the stage of knowledge erasure, and the right represents the stage of model reconstruction.

In the stage of knowledge erasure, the goal is to erasure the relevant impact of forgotten data D_f on the model M_d. If the model completely eliminates the impact of the target data D_f requested to be recalled, the classification results of the model M_u for this data D_f should be consistent with the results of random selection. When the classification results are greater or less than the results of random selection, we believe that the model still retains the relevant prior knowledge of D_f. We introduce a stochastic initialization model M_s, which does not contain any relevant knowledge of D_f. We hope M_u to retain the knowledge related to data D_r as much as possible while deleting the relevant knowledge of D_f. Since the teacher network is a stochastic initialization network and has not undergone any data training, its classification results of data D_f are random. The results of this "stupid" teacher network for target data D_f are exactly what we hope M_d to achieve through machine unlearning. In other words, we hope that machine unlearning algorithms can help M_d forget all relevant knowledge of D_f. We use the method of knowledge distillation and use the parameters of M_d to initialize the model M_u. M_s will be used as a teacher's network and M_u will be used as a student's network. Through knowledge distillation, we can help the student network M_u forget the relevant knowledge of the target data D_f.

We use the parameters of M_d to initialize the model M_u, and make the probability distribution of M_u for the dataset D_f consistent with the probability distribution of the stochastic initialization model M_s for the target data D_f.

As shown on the right of Fig. 1, the model M_s and model M_u are composed of convolution neural network with the same structure. We use the parameters of the model M_d trained with complete data to initialization model M_u and random initialization the parameters of model M_s. The goal of knowledge erasure is to use M_s as teacher network help model M_u delete the impact of request forgotten data D_f on the model M_d. For a sample $x_i \in D_f$, passing x_i through model M_s and M_u respectively to get probability distributions $p(x_i)$ and $q(x_i)$. We use Eq. (2) to obtain the soft label of the model for the samples in the dataset D_f, where the temperature parameter τ used in the softmax function can help to obtain more information about the distribution, z_i represents the probability values of x_i for different categories, and n denotes the total number of categories of datasets D.

$$p(x_i) = \frac{exp(z_i/\tau)}{\sum_{j=0}^{n} exp(z_j/\tau)} \tag{2}$$

We use Eq. (3) to calculate the disparity in probability distribution between model M_u and M_s with respect to D_f by Kullback-Leibler Divergence (KLD) and update the parameters of M_u to forget about the knowledge about D_f in the model M_d.

$$L_{KL}(p(x_i)||q(x_i)) = \sum_{i=1}^{n} p(x_i) \log \frac{p(x_i)}{q(x_i)} \tag{3}$$

In the rapid model reconstruction stage, due to the modification of model M_u parameters, can result in a deterioration of the performance of the model M_u on data D_r while forgetting the relevant knowledge of the target data D_f. We use the remaining data set D_r and the original model M_d to help the model M_u rebuild quickly to restore the performance of the model M_u. As shown on the right of Fig. 1, the original model M_d is regarded as a teacher network, and the model M_u is regarded as a student network. For a sample $x_i \in D_r$, we use Eq. (3) to calculate the KLD of the probability distribution output $p(x_i)$ and $q(x_i)$ by M_d and M_u on dataset D_r. At the same time, the Cross Entropy (4) is used as the loss function of M_u on D_r.

$$L_{CE} = -\sum_{i=0}^{n} y_i log(q(x_i)) \tag{4}$$

The total loss is shown in Eq. (5), where α is the hyper parameter. We utilize the total loss to aid the model M_u fine tune parameters to quickly restore the performance on D_r. The knowledge erasure and model reconstruction are shown in Algorithm 1.

$$L = L_{CE} + \alpha L_{KL} \tag{5}$$

Algorithm 1. Knowledge Erasure and Model Reconstruction

Input: Request forgotten dataD_f, Remaining data D_r, Original modelM_d, Stochastic initialization network M_s.

Output: Model unlearning relevant knowledge of forgotten data:M_u

```
1  Knowledge Erasure(D_f, M_s, M_d):
2      Random initialization model M_s parameters
3      Using M_D parameters to initialization model M_u
4      for batch x ∈ D_f do
5          Compute L_KL loss by Eq.(3)
6          Update M_u parameters with gradient descent
7      return M_u // scratch model
8  Model Reconstruction(D_r, M_u, M_d):
9      for batch x ∈ D_r do
10         Compute L loss by Eq.5
11         Update model M_u parameters according to L
12     return M_u
```

4 Experiments and Results

4.1 Datasets

We conduct experiments on three popular benchmarks: CIFAR-10 [14], MNIST [15] and Fashion-MNIST [25].

The CIFAR-10 dataset comprises 60,000 color images, each with a size of 32×32 pixels. The dataset consists of 10 classes, with each class containing 6,000 images.

The MNIST and Fashion-MNIST datasets consist of 70,000 28×28 images each, respectively, with 10 classes and 7,000 images per class. The MNIST dataset consists of 60,000 images for training and 10,000 images for testing, while the Fashion-MNIST dataset has the same split of training and testing images.

4.2 Training Details

Our goal is to verify that our method can effectively erase information from the model related to the target data D_f and quickly reconstruct the model instead of obtaining the highest classification accuracy on a specific dataset. Therefore, we use a simple model architecture, which includes two convolution layers and a full connection layer. Firstly, we train an original model on the complete dataset D, and then utilize our algorithm to make the model *Original* unlearn the relevant information of the requested data D_f to obtain model *Scratch*. Then we use the remaining datasets D_r to help the model *Scratch* rebuild quickly to restore the model performance. We studied two cases of removing 10% and 20% data, and The reconstruction time and forgetting effect are compared with the method of retraining the model.

4.3 Evaluation Results

To showcase the efficacy of our approach in quickly erasing relevant information of requested forgotten data, we compared our model with the Original model, the Scratch model, and the Retrained model. In Table 1, we present a detailed analysis of the erasure effect of our method when deleting a certain category. The second to tenth rows of Table 1 represent the classification accuracy of each category in the remaining data, "Remaining data" represents the average classification accuracy of the remaining categories, and "cat" represents the target data requested to be forgotten. We list the accuracy of the original model, the Scratch model, the Retrained model, and the unlearning model obtained by our method in detail in Table 1. The table shows that our method can quickly erase the relevant knowledge of the target data "cat" in the knowledge erasure stage. After the knowledge erasure stage, the classification accuracy of the Original model for the "cat" category decreased from 49.70% to 8.10%, while the average classification accuracy of the remaining data decreased from 76.67% to 72.40%, and the classification accuracy only decreased by 4.27%. This indicates that our method can accurately erase specific knowledge in the model. After the model reconstruction stage, our model for the "cat" category was restored to 11.40%, which is equivalent to the result of random classification for the CIFAR-10 dataset containing 10 categories. Moreover, the average classification accuracy of the remaining dataset increased from 72.40% to 80.11%.

Retrained results were acquired by training the model again using the remaining dataset. Since the "cat" class did not participate in the training during the retraining process, all "cat" instances in the testing process were incorrectly predicted by the Retrained model, which we believe is due to the inductive bias

Table 1. The accuracy of each class in CIFAR-10. The requested forgotten data D_f accounts for 10% of the total data D.

Categories	Original	Scratch	Retrained	Our
Plane	68.80%	75.80%	81.30%	79.70%
Car	85.30%	91.50%	84.20%	83.80%
Bird	65.20%	54.80%	56.00%	67.80%
Deer	69.00%	64.90%	70.20%	77.80%
Dog	69.00%	45.60%	83.90%	78.10%
Frog	85.00%	74.50%	86.80%	85.00%
Horse	79.10%	80.40%	76.00%	80.00%
Ship	89.10%	90.10%	85.80%	85.50%
Truch	79.50%	74.00%	80.30%	83.80%
Remaining data	76.67%	72.40%	78.28%	**80.11%**
Cat(Forgotten)	49.70%	8.10%	0.00%	**11.40%**

of the model itself. The average classification accuracy of the Retrained model in the remaining dataset is 78.28%. Compared with the Retrained model, our method quickly erases the relevant knowledge of the target data set and achieves better results on the remaining data.

We show the average classification accuracy of our method for the remaining data sets after information erasure and model reconstruction in Tables 2 and 3.

Table 2. Average test accuracy for remained dataset D_r of all *Dataset*, The requested forgotten data D_f accounts for 10% of the total data D. Blue text indicates training epochs

Baselines	CIFAR-10	MNIST	FashionMNIST
Original	76.67%	99.25%	91.98%
Scratch	72.40%	97.48%	89.21%
Retrained	78.28% (20)	99.19%(10)	92.88% (20)
Ours	**80.11%** (1)	**99.21%** (1)	**94.13%** (1)

Table 2 illustrates the outcomes of the request to forget 10% of the data (deleting one category of data), and Table 3 displays the outcomes of the request to forget 20% of the data (deleting two categories of data). It can be seen from the Table 2 and Table 3 that after the knowledge erasure stage of the original model, the accuracy of classification for the remaining data has decreased. After the model reconstruction, the accuracy of the remaining data have been restored. On three data sets, we request to forget one or two types of data, and our model has achieved better performance than the Retrained model. The blue

Table 3. Average test accuracy for remained dataset D_r of all *Dataset*, The requested forgotten data D_f accounts for 20% of the total data D. Blue text indicates training epochs

Baselines	CIFAR-10	MNIST	FashionMNIST
Original	75.11%	99.34%	91.01%
Scratch	60.35%	91.85%	83.66%
Retrained	79.82% (20)	99.41% (10)	93.26% (20)
Ours	**81.45%** (1)	**99.42%** (1)	**93.76%**(1)

font represents the epochs of the model training. We can see that our model only needs one epoch to quickly recover performance, and retraining the model requires more epochs. We have studied the time required for retrained model and our method in detail in Fig. 2.

4.4 The Speed of Machine Unlearning

Training speed is an important factor in machine unlearning. It is essential to quickly erase the knowledge of requested forgotten data and regain the model's performance on the remaining data. In order to showcase the efficacy of our proposed method, we conducted experiments on a dataset with two requested forgotten categories. In Fig. 2 (a)-(c), we present the classification accuracy of each category after the knowledge erasure and model reconstruction stages. We trained our model using one epoch of knowledge erasure to erase the information of requested forgotten categories, and after one epoch of model reconstruction, our model quickly recovered the performance on the remaining data. Moreover, the classification accuracy of the requested forgotten data remained at random classification level, indicating that our model could erase the knowledge of requested forgotten data effectively.

In Fig. 2 (d)-(f), we compare the training speed of our method with the retrained model. The plots show the changes in the accuracy of the remaining data with the training epochs. Our model outperforms the Retrained model in terms of training speed and accuracy. After one epoch, our model quickly recovered the performance of the Scratch model, whereas retraining the model requires more time. Additionally, our model achieved better results than the Retrained model. Based on these experimental findings, it can be concluded that our method is indeed effective in quickly erasing the knowledge of requested forgotten data and restoring the model's performance on the remaining data.

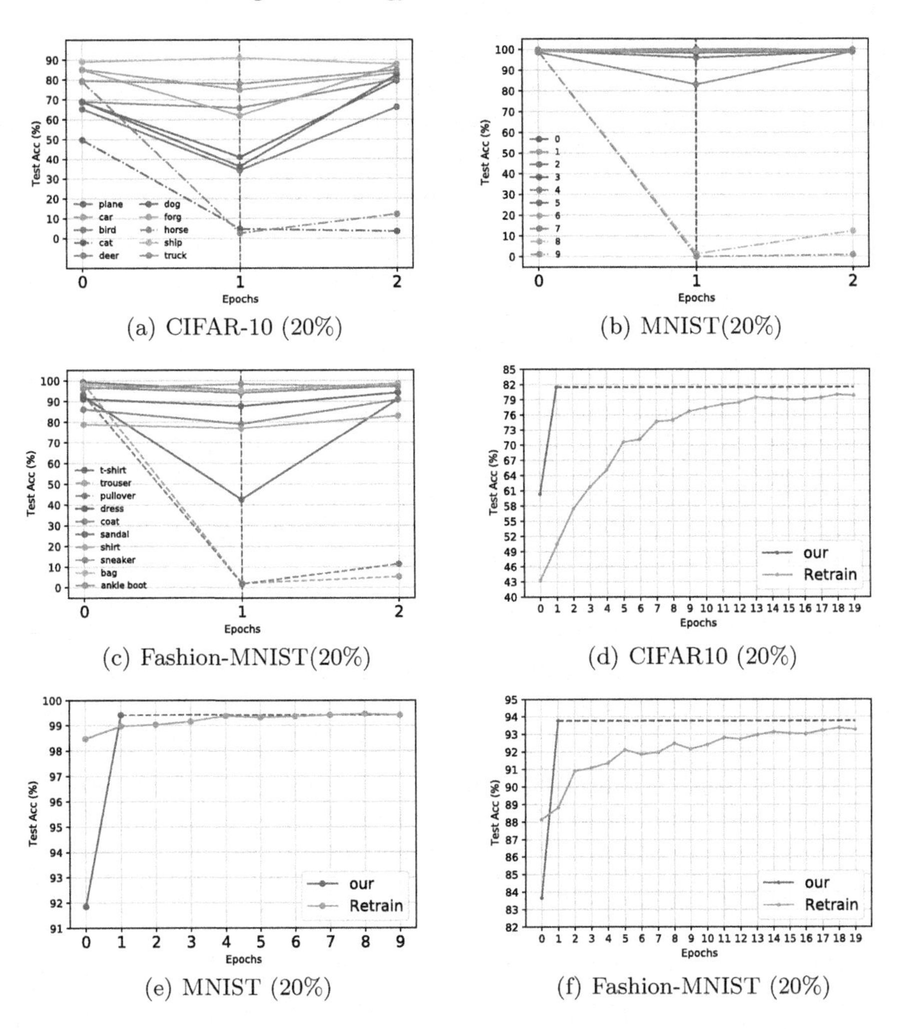

(a) CIFAR-10 (20%) (b) MNIST(20%) (c) Fashion-MNIST(20%) (d) CIFAR10 (20%) (e) MNIST (20%) (f) Fashion-MNIST (20%)

Fig. 2. (a),(b),(c) represent the change of classification accuracy of each category during information erasure and model reconstruction with the training epoch. (d),(e),(f) represent the process of model reconstruction and model retraining.

5 Conclusion

This paper presents a novel approach to quickly erase the impact of target data information on the model. Our method is based on an intuitive understanding that when a model completely forgets the target data, its performance on the target data should be consistent with that of the model completely untrained by the target data. We propose to use a stochastic initialized network as a teacher network to help the original model forget the target data information. Then, we use the remaining data to help model quickly rebuild. We conduct experiments on three benchmarks, and study the accuracy of different categories in the datasets during the knowledge erasure process. The results validate the efficacy of our method for erasing target data information. Our approach enables rapid mitigation of the influence exerted by the target data on the model through only one epoch and achieve the one-time erasure and reconstruction of the model on target data.

Acknowledgments. This paper is supported by the Key Research and Development Program of Guangdong Province under grant No.2021B0101400003.

References

1. Aldaghri, N., Mahdavifar, H., Beirami, A.: Coded machine unlearning. IEEE Access **9**, 88137–88150 (2021)
2. Baumhauer, T., Schöttle, P., Zeppelzauer, M.: Machine unlearning: linear filtration for logit-based classifiers. Mach. Learn. **111**(9), 3203–3226 (2022)
3. Biggio, B., Nelson, B., Laskov, P.: Poisoning attacks against support vector machines. arXiv:1206.6389 (2012)
4. Bourtoule, L., et al.: Machine unlearning. In: 2021 IEEE Symposium on Security and Privacy, pp. 141–159. IEEE (2021)
5. Brophy, J., Lowd, D.: Machine unlearning for random forests. In: International Conference on Machine Learning, pp. 1092–1104. PMLR (2021)
6. Cao, Z., Wang, J., Si, S., Huang, Z., Xiao, J.: Machine unlearning method based on projection residual. In: 9th IEEE International Conference on Data Science and Advanced Analytics, pp. 1–8. IEEE (2022)
7. Chen, M., Zhang, Z., Wang, T., Backes, M., Humbert, M., Zhang, Y.: When machine unlearning jeopardizes privacy. In: Proceedings of the 2021 ACM SIGSAC Conference on Computer and Communications Security, pp. 896–911 (2021)
8. Cho, J.H., Hariharan, B.: On the efficacy of knowledge distillation. In: Proceedings of the IEEE/CVF International Conference on Computer Vision, pp. 4794–4802 (2019)
9. Gou, J., Yu, B., Maybank, S.J., Tao, D.: Knowledge distillation: a survey. Int. J. Comput. Vis. **129**(6), 1789–1819 (2021)
10. Gupta, V., Jung, C., Neel, S., Roth, A., Sharifi-Malvajerdi, S., Waites, C.: Adaptive machine unlearning. Adv. Neural. Inf. Process. Syst. **34**, 16319–16330 (2021)
11. Harding, E.L., Vanto, J.J., Clark, R., Hannah Ji, L., Ainsworth, S.C.: Understanding the scope and impact of the California consumer privacy act of 2018. J. Data Protect. Priv. **2**(3), 234–253 (2019)
12. Hinton, G., Vinyals, O., Dean, J.: Distilling the knowledge in a neural network. arXiv:1503.02531 (2015)
13. Kim, J., Woo, S.S.: Efficient two-stage model retraining for machine unlearning. In: Proceedings of the IEEE/CVF Conference on Computer Vision and Pattern Recognition, pp. 4361–4369 (2022)
14. Krizhevsky, A., Hinton, G., et al.: Learning multiple layers of features from tiny images (2009)
15. LeCun, Y.: The mnist database of handwritten digits. http://yann.lecun.com/exdb/mnist/ (1998)
16. Liu, G., Ma, X., Yang, Y., Wang, C., Liu, J.: FedEraser: enabling efficient client-level data removal from federated learning models. In: 2021 IEEE/ACM 29th International Symposium on Quality of Service, pp. 1–10. IEEE (2021)

17. Marchant, N.G., Rubinstein, B.I., Alfeld, S.: Hard to forget: poisoning attacks on certified machine unlearning. In: Proceedings of the AAAI Conference on Artificial Intelligence, vol. 36, pp. 7691–7700 (2022)
18. Nguyen, T.T., Huynh, T.T., Nguyen, P.L., Liew, A.W.C., Yin, H., Nguyen, Q.V.H.: A survey of machine unlearning. arXiv:2209.02299 (2022)
19. Romero, A., Ballas, N., Kahou, S.E., Chassang, A., Gatta, C., Bengio, Y.: FitNets: hints for thin deep nets. arXiv:1412.6550 (2014)
20. Rong, D., Ye, S., Zhao, R., Yuen, H.N., Chen, J., He, Q.: FedRecAttack: model poisoning attack to federated recommendation. In: 38th IEEE International Conference on Data Engineering, pp. 2643–2655. IEEE (2022)
21. Sekhari, A., Acharya, J., Kamath, G., Suresh, A.T.: Remember what you want to forget: algorithms for machine unlearning. Adv. Neural. Inf. Process. Syst. **34**, 18075–18086 (2021)
22. Voigt, P., von dem Bussche, A.: The EU General Data Protection Regulation (GDPR). Springer, Cham (2017). https://doi.org/10.1007/978-3-319-57959-7
23. Wang, J., Guo, S., Xie, X., Qi, H.: Federated unlearning via class-discriminative pruning. In: Proceedings of the ACM Web Conference 2022, pp. 622–632 (2022)
24. Wu, C., Zhu, S., Mitra, P.: Federated unlearning with knowledge distillation. arXiv:2201.09441 (2022)
25. Xiao, H., Rasul, K., Vollgraf, R.: Fashion-MNIST: a novel image dataset for benchmarking machine learning algorithms. arXiv:1708.07747 (2017)
26. Yang, C., Wu, Q., Li, H., Chen, Y.: Generative poisoning attack method against neural networks. arXiv:1703.01340 (2017)
27. Zagoruyko, S., Komodakis, N.: Paying more attention to attention: Improving the performance of convolutional neural networks via attention transfer. In: 5th International Conference on Learning Representations (2017)
28. Zhang, Q.S., Zhu, S.C.: Visual interpretability for deep learning: a survey. Front. Inf. Technol. Electron. Eng. **19**(1), 27–39 (2018)

CNMBI: Determining the Number of Clusters Using Center Pairwise Matching and Boundary Filtering

Ruilin Zhang[1], Haiyang Zheng[1], and Hongpeng Wang[1,2](✉)

[1] Harbin Institute of Technology, Shenzhen, China
wanghp@hit.edu.cn
[2] Peng Cheng Laboratory, Shenzhen, China

Abstract. One of the main challenges in data mining is choosing the optimal number of clusters without prior information. Notably, existing methods are usually in the philosophy of cluster validation and hence have underlying assumptions on data distribution, which prevents their application to complex data such as large-scale images and high-dimensional data from the real world. In this regard, we propose an approach named CNMBI. Leveraging the distribution information inherent in the data space, we map the target task as a dynamic comparison process between cluster centers regarding positional behavior, without relying on the complete clustering results and designing the complex validity index as before. Bipartite graph theory is then employed to efficiently model this process. Additionally, we find that different samples have different confidence levels and thereby actively remove low-confidence ones, which is, for the first time to our knowledge, considered in cluster number determination. CNMBI is robust and allows for more flexibility in the dimension and shape of the target data (e.g., CIFAR-10 and STL-10). Extensive comparisof-the-art competitors on various challenging datasets demonstrate the superiority of our method.

Keywords: Number of clusters · Cluster center · Complex data · Pairwise matching · Boundary filtering

1 Introduction

The automatic determination of parameters has become one of the main considerations for the popularity of a learning solution in machine learning or pattern recognition [7]. Clustering, as an essential analysis tool, endeavors to discover underlying structure from unlabeled data, ensuring that samples[1] in the same group have high homogeneity and that different groups have the maximum difference [9]. It should be noted that, typically, most clustering approaches suffer from the limitation that the number of clusters has to be fed by a human user,

[1] Data points, objects, and samples are used exchangeably in this paper.

X. Yang et al. (Eds.): ADMA 2023, LNAI 14180, pp. 262–277, 2023.
https://doi.org/10.1007/978-3-031-46677-9_19

which is fundamental to obtaining a good data partition. However, the number of clusters in practice is usually unknown, as Salvador stated in [14], which often results in suboptimal clustering results. Determining the number of clusters automatically can enhance the potential of clustering methods, while also being instructive for other learning tasks, like counting the amount of farmland in remote sensing images and setting the type of anchors in object detection [10]. Although several solutions [1,4,16,19] have recently been proposed to estimate the number of clusters, they either perform unstably or are challenging to employ in practice when handling complex data like large-scale images and high-dimensional data from the real world. Technically, existing methods are generally based on the clustering validation philosophy, which uses the quality of clustering results to reveal the number of clusters, i.e., evaluating a clustering validity index (named CVI) over the target data and optimizing it as a function of the number of clusters. With a clear semantic and mathematical background, this paradigm has gained significant attention [8,13], mainly covering the design of cluster validity index (CVI) [13,18,19] and the embedding of clustering schemes [4,10,16].

Nevertheless, the unsupervised context limits existing methods to design CVI based solely on the subjective morphological characteristics of final clusters, such as compactness and separation, which inevitably imposes potential assumptions on the target data distribution. Moreover, the clustering results of each round during iterative optimization are sometimes not discriminative and instructive, as in the case of complex data, due to the inherent limitations of the chosen clustering scheme. Commonly used K-means, for instance, favors spherical or well-separated clusters, so the consistency of the optimal number of clusters, the best clustering result, and the best score, i.e., the philosophy underlying such methods, could be invalidated by some challenging data. Additionally, we observe that existing work generally focuses on the methodological level while ignoring the significance of the sample to the target task in practice.

With this in mind, we bridge these bottlenecks by proposing a novel algorithm (CNMBI). Our work demonstrates the following contributions:

1) We propose a scheme for determining the number of clusters based on center pairwise matching. Instead of previous approaches resorting to CVI scoring and complete clustering results, we use inherent distribution information in the data space to model cluster number determination as a dynamic contrastive process of cluster centers regarding positional behavior.
2) We develop a boundary filtering method to remove low-confidence samples that interfere with the cluster number determination task, allowing our method to have greater flexibility and robustness in terms of data shape and dimension. To the best of our knowledge, this aspect has not been considered.
3) Our method outperforms the state-of-the-art competitors on diverse benchmark datasets (e.g., STL-10, MNIST, and CIFAR-10), and we are the first to report the results of a cluster number determination method on the challenging dataset such as STL-10 and CIFAR-10. The in-depth discussions of boundary information, and robustness jointly demonstrate the superiority of CNMBI.

2 Related Work

For the goal of determining the number of clusters, the philosophy behind most methods is that the actual number of clusters, the best clustering result, and the best clustering validity score are consistent. In this context, designing an effective validity index is essential because the optimal number of clusters needs to be reflected by evaluating the clustering results. Conventionally, most CVIs take distance information of resulting clusters as the main focus, considering the intra-cluster distance (compactness), inter-cluster distance (separation), or the statistical variants of both to describe clustering results, such as DBI [5], DUNN [6], CH [3]. Combined with partition-based clustering, these indexes can suggest the desired number of clusters under the spherical or well-separated data. Additionally, some work attempts to construct various curves formed by intra-cluster variance, using characteristics such as "maximum curvature" [21], "knee of the curve" [14], "elbow of the curve", "max magnitude" [17], or "sharp jump" [15] to visually locate the number of clusters.

Recently, in addition to the work above relying on distance information, several efforts have leveraged density information to define CVI or update clustering schemes. For instance, [16] uses cluster density to evaluate the clustering results. Likewise, LCCV [4] selects a group of highly representative density cores by density information to simplify the silhouette coefficient and employs agglomerative hierarchical clustering as the assignment scheme. Besides, there are some promising density clustering methods that work with CVI, one of the typical paradigms is density peaks clustering (DPC) [11]. As in the case of [18], which simplifies the adapted silhouette coefficient by considering density centers from DPC as representative. Another example is DPC-AHS [19], which takes the skew of distribution within the cluster as the validity index by the advantages of density clustering for arbitrary-shaped data. Despite their simplicity and ease of implementation, these methods are not practical as they usually perform unstably on some complex data such as large-scale images and high-dimensional data from the real world, as discussed in the introduction. In contrast, our method addresses the above limitations.

3 Proposed Method: CNMBI

3.1 Center-Pairwise Matching

Cluster number determination is essentially a dynamic optimization process, so one potential solution is to improve identification performance by leveraging valuable pattern information. However, in such an unsupervised context, significant external information, such as data labels, is unavailable. It should be noted that the distribution information from the data space is also instructive, such as the cluster center generated by density or geometric information, which is closely related to the number of clusters and, more importantly, is the most representative in the data space.

With this in mind, we find a fundamental insight that the distance between the density centers (using density information) and the mean centers (relying geometric information) is closest when and only when the number of clusters is actual, which is enlightening for cluster number determination. As shown in Fig. 1, the distances between the mean centers and the density centers are small and the positions almost overlap in the case of the number of clusters $k = 3$. On the contrary, there is a significant distance between the two in cases smaller than the actual number of clusters (Fig. 1(a), k=2) or larger (Fig. 1(c), k=5). To further discuss the principle behind this observation above, we have some formal definitions as follows.

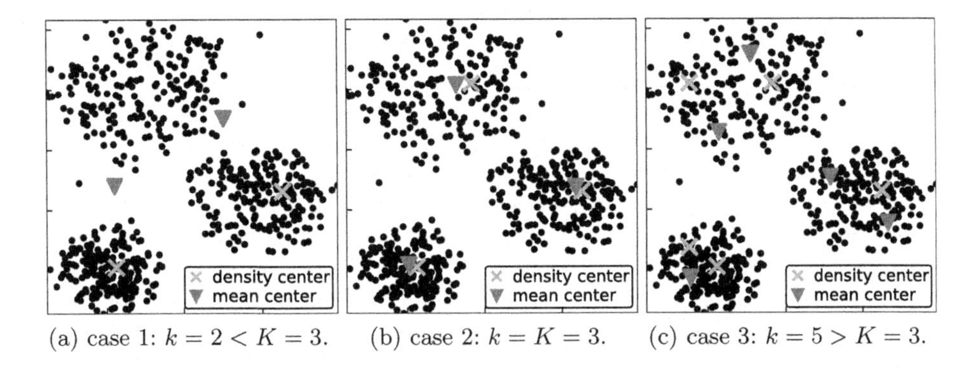

(a) case 1: $k = 2 < K = 3$. (b) case 2: $k = K = 3$. (c) case 3: $k = 5 > K = 3$.

Fig. 1. Density center versus mean center for different cluster number settings.

Given a dataset $X \in \mathbb{R}^d$ containing K actual clusters of n samples. Let k denote the number of clusters currently under consideration, the mean center is obtained by

$$S_{\mu}^{(k)} = \left\{ \frac{\sum\limits_{x \in C_{\mu}^{(i)}} x}{\left| C_{\mu}^{(i)} \right|} \middle| 1 < i \leq k \right\} \tag{1}$$

where $S_{\mu}^{(k)}$ represents a set containing k mean centers and $C_{\mu}^{(i)}$ denotes a cluster generated by partitioning-based clustering (K-means is used here), i.e. $X = \{C_{\mu}^{(1)}, ..., C_{\mu}^{(k)}\}$. In particular, Eq. 1 illustrates that the mean centers are geometric centroids of the resulting clusters and their positions are unstable. As shown in Fig. 1, in the case of k=2 or $k = 5$, the mean centers are generally distributed at the edges or outside of the actual clusters in X, due to cluster splitting or cluster merging behaviors.

In addition, the density centers are described as those objects surrounded by neighbors with lower density and have a relatively large distance from any data points with higher density, as discussed in DPC. To this aim, the two quantities: local density ρ and δ distance are used to measure the possibility of each object becoming a cluster center from the density view.

$$\rho_i = \sum_{x_j \in X \setminus \{x_i\}} \exp(-(\frac{d_{ij}}{dc})^2) \tag{2}$$

$$\delta_i = \begin{cases} \min\{dist(x_i, x_j)| \rho_i < \rho_j, \ \ x_i \in X\} & if\ \exists x_j \in X, \ \ \rho_i < \rho_j \\ \max\{dist(x_i, x_j)| x_j \in X\} & otherwise \end{cases} \tag{3}$$

where dc acts as the sampling radius of density calculation, it is determined by sorting n^2 distances in X from ascending order and selecting a number ranked at 2%. With the product $\gamma_i = \rho_i \times \delta_i$, the greater the score γ_i, the more likely the x_i is considered as a cluster center.

Definition 1. ***Density Center Set*** *Given the current cluster number k, the density center set $S_\gamma^{(k)}$, including k centers determined by DPC, is given by*

$$S_\gamma^{(k)} = \{y_i \,|\gamma_i \geq \bar{\gamma}_k \ , \ \ 1 \leq i \leq n\} \tag{4}$$

where $\bar{\gamma}_k$ is the k-th largest gamma score by sorting objects in γ descending order. Note that the cardinality of both sets $S_\gamma^{(k)}$ and $S_\mu^{(k)}$ is k. According to the above formula, we can also conclude that the density center tends to be the most core region of each actual cluster in X, and the position is stable because the density information directly determines it.

Inspired by above observation, we draw a promising insight favoring the determination of the number of clusters: "the difference between the mean center and the density center in terms of location distribution is minimal when and only when the actual number of clusters is given, quantified as the sum of the k pairwise distances formed is minimal compared to the other cases".

To apply this insight, we propose to model the relationship between the density centers and mean centers using graph theory to quantify the actual distance between them. The solution regarding the optimal number of clusters can be transformed into the searching the "Center Pairwise Matching" below.

We formally treat Mean Center Set S_μ (following Eq. 1) and Density Center Set S_γ (following Eq. 4) as two graph node sets, and the elements of which are regarded as individual nodes. After that, Center Bipartite Graph is defined as.

Definition 2. ***Center Bipartite Graph*** *Let $S_\gamma^{(k)}$ and $S_\mu^{(k)}$ be the set of mean center and density center at number of clusters k, respectively. We then define the center bipartite graph $G_{\mu,\gamma}^{(k)} = (V^{(k)}, E^{(k)})$ such that it should obey the properties:*

(1) $V^{(k)} = \{S_\mu^{(k)} \cup S_\gamma^{(k)}\}$ and $S_\mu^{(k)} \cap S_\gamma^{(k)} = \emptyset$; (2) for each node $x_i \in S_\mu^{(k)}$(*or* $x_j \in S_\gamma^{(k)}$), there must be k edges between it and another node subset; (3) for each edge $e(x_i, x_j) \in E^{(k)}$, that means $x_j \in S_\mu^{(k)}$ as well as $x_j \in S_\gamma^{(k)}$.

Definition 3. ***Center Complete Matching*** *Let CM represent a complete matching on $G_{\mu,\gamma}^{(k)} = (V^{(k)}, E^{(k)})$, a subset of size k of the edge set E^k, which must satisfy the following conditions:*

(1) For any two edges $e(x_i, x_j)$, $e(x_p, x_q) \in CM$, $x_i \cap x_p = \emptyset$ and $x_j \cap x_q = \emptyset$ should be hold; (2) $|CM| = \left|S_\gamma^{(k)}\right| = \left|S_\mu^{(k)}\right|$. Note that for the graph $G_{\mu,\gamma}^{(k)}$, there are $k!$ complete matching cases: $CM \in P^{(k)} = \{M_1, M_2, ..., M_{k!}\}$.

Definition 4. ***Center Similarity Matrix*** *For any complete matching $PM \in P^{(k)}$ of the center bipartite graph $G_{\mu,\gamma}^{(k)}$, the corresponding "center similarity matrix" with sparse property, denoted by CD, is defined as:*

$$CD_{[p,q]}^{(k)} = \begin{cases} \|x_p - x_q\|_2^2, & if\ e(x_p, x_q) \in CM \\ 0 & otherwise \end{cases} \tag{5}$$

Definition 5. ***Center Pairwise Loss*** *Given a matching scheme CM, the expectation of center similarity matrix CD, called "center pairwise loss", is used to quantify the center pairwise relationships in this matching CM*

$$L(CM) = \frac{1}{k} I_{1*k}(CD^{(k)}) * I_{1*k}^T \tag{6}$$

where $I_{1*k} = (1, 1, ..., 1)_{1*k}^T$ is an auxiliary matrix for vectorization.

Definition 6. ***Center Pairwise Matching*** *Given the number of clusters k, there exist Ki complete matches between the mean center and the density center. We use the "center pairwise matching" as the matching scheme under the current number of clusters, expressed as follows.*

$$CPM^{(k)} = \left\{ \bigcup_{e(x_p,x_q) \in PM} e(x_p, x_q) \middle| \arg \min_{PM \in P^{(k)}} (L(PM)) \right\} \tag{7}$$

It is worth mentioning that the loss value of the center pairwise matching is considered as the unique feature value of the current number of clusters during cluster number optimization. To this end, the optimal number of clusters should be projected as the global minimum center loss, thus we define

$$K^* = \arg_k \min(L(CPM^{(k)})) \tag{8}$$

3.2 Low Confidence Sample Filtering

Inspired by self-paced learning in which samples have different confidence levels for a given learning task, we note that some extreme points in the data space, especially at the edges of clusters, between clusters, or away from clusters, may disturb the cluster number identification. These points do not reveal the core structure of the clusters as well as tend to incur cluster merging, weaken the independence of clusters, eventually breaking the consistency between "the actual number of clusters," "the best clustering result," and "the best score," which is the main factor that current work is hard to deal with complex data. It is noteworthy that the existing methods are not aware of this fact. To this aim, we propose a boundary filtering method based on space vector decomposition, aiming to enhance the robustness of our method against complex data.

We find that the boundary object and its neighbors have strong unbalance in the projection subspace, forming the typical skew distribution in the direction of the base vector. By contrast, the neighbors of the core object show significant symmetry in the direction of the basis vector [20]. With this in mind, we first treat the close relationship between target object and its neighbors as independent space vectors, described as follows.

Let x_i be any object in data set $X = \{x_1, x_2, ..., x_n\} \in \mathbb{R}^{m\times n}$. According to the space vector decomposition theorem, there exists a unique ordered sequence of real number $\{\lambda_1, \lambda_2, ..., \lambda_m\}$ such that $x_i = \lambda_1 e_1 + \lambda_2 e_2 + ... + \lambda_m e_m$, where $\{e_1, e_2, ..., e_m\}$ denotes the basis vectors for the data space $\mathbb{R}^m$. For $\forall\, x_i, x_j \in X$, the neighborhood vector formed by the two in $\mathbb{R}^m$ is given by

$$h_{i,j} = (\lambda_{i,1} - \lambda_{j,1}, ..., \ \lambda_{i,d} - \lambda_{j,d}, ..., \lambda_{i,m} - \lambda_{j,m})(e_1, ..., e_m)^T \ 1 \le i, j \le n \quad (9)$$

where $\lambda_{i,d} - \lambda_{j,d}$ denotes the projection of $h_{i,j}$in the direction of basis vector e_d.

Notably, the neighbors of x_i refers to objects within the open neighborhood of x_i with the radius as dc, i.e.,$N_{dc}(x_i)$. As the typical operation to describe spatial position relations, the local representative vector $\mathcal{H}_i$ is customized as the vector addition regarding x_i and its neighbors.

$$\mathcal{H}_i = \sum_{x_j \in N_{dc}(x_i)} (h_{i,j}) {=} (\sum_{x_j} (\lambda_{i,1} - \lambda_{j,1}), ..., \sum_{x_j} (\lambda_{i,m} - \lambda_{j,m}))(e_1, ..., e_m)^T \quad (10)$$

where $\sum (\lambda_{i,d} - \lambda_{j,d})$ symbolizes the projection of the neighborhood vectors of x_i in the direction of the basis vector e_d. In this regard, we make the following derivation about projection of the neighborhood vector onto the feature space:

Theorem 1. *Given a one-dimensional feature space $v_d (1 \le d \le m)$ and the neighborhood vector $h_{i,j}$, then the projection of the neighborhood vector $h_{i,j}$ onto it is $e_d (e_d^T e_d)^{-1} e_d^T h_{ij}$, where e_d is the basis vector of v_d.*

Proof. Assuming $\mathcal{S}$ is a subspace of $\mathbb{R}^m$ ($\mathcal{S} \in \mathbb{R}^p, s.t. p \le m$) and $\{e_1', e_2', e_3', e_4', ..., e_p'\}$ is a basis, then for any object $a \in \mathcal{S}$, there is only $a = y_1 e_1' + y_2 e_2' + y_3 e_3' + y_4 e_4' + ... + y_p e_p'$. Let matrix $A = [e_1', e_2', e_3', e_4', ..., e_p']_{m\times p}$, then $a = Ay, y \in \mathbb{R}^p$.

According to the above statement, let $Proj_s h_{i,j}$ denote the projection of neighborhood vector $h_{i,j}$ (composed of $x_i \in \mathbb{R}^m$ and its neighbor $x_j \in \mathbb{R}^m$) on subspace $\mathcal{S}$, then $Proj_s h_{i,j} \in \mathcal{S}$ with

$$Proj_s h_{i,j} = Ay, y \in R^p \quad (11)$$

Based on the fundamental theory of linear algebra, the neighborhood vector $h_{i,j}$ is such that $h_{i,j} = Proj_{s^\perp} h_{i,j} + Proj_s h_{i,j}$, where $Proj_{s^\perp} h_{i,j}$ denote the projection of $h_{i,j}$ on the orthogonal complement of subspace S, which is equal to the null space of A^T. We have:

$$Proj_{s^\perp} h_{i,j} = h_{i,j} - Proj_s h_{i,j} \in Null(A^T) \quad (12)$$

Left-multiplying both sides by A^T and simplifying, we obtain $A^T h_{i,j} - A^T Ay = 0$, then $y = \left(A^T A\right)^{-1} A^T h_{i,j}$. Substituting the expression for y in the $Proj_s h_{i,j}$, we have $Proj_s h_{i,j} = Ay = A(A^T A)^{-1} A^T h_{i,j}$. Now, Eq. 11 can be rewritten as

$$Proj_s h_{i,j} = Ay = A(A^T A)^{-1} A^T (x_i - x_j) \quad (13)$$

Without loss of generality, we assume that the subspace $\mathcal{S}$ is instantiated as a one-dimensional space s_d, thus the matrix A could be transformed into a column vector e_d, i.e. the basis vector of this one-dimensional space. Consequently, projection is formalized as $e_d(e_d^T e_d)^{-1} e_d^T h_{ij}$. The theorem is proved. □

With the help of orthogonality, we utilize a standard orthogonal basis $\{e_d\}_{1 \le d \le m}$,(s.t.$\langle e_i, e_j{=}0\rangle \cap e_i^{\mathrm{T}} e_i{=}1$),to instantiate matrix A in Eq. 13. After that, the projections of the neighborhood vector $h_{i,j}$ composed of x_i and x_j on the directions of the m orthogonal basis vectors can be expressed as:

$$Proj_{v_d} h_{i,j} = (e_d(e_d^T e_d)^{-1} e_d^T h_{ij}) = (e_d e_d^T x_i - e_d e_d^T x_j) = (x_{id} - x_{jd}), 1 \le d \le m \quad (14)$$

Thus, local representative vector from Eq. 10 is formalized as follows:

$$\begin{aligned} \mathcal{H}_i &= (\sum_j Proj_{v_1} h_{i,j}, \sum_j Proj_{v_2} h_{i,j}, ..., \sum_j Proj_{v_m} h_{i,j})(e_1, e_2, e_3, ..., e_m)^T \\ &= (\sum_j e_1 e_1^T x_i - e_1 e_1^T x_j, \sum_j e_2 e_2^T x_i - e_2 e_2^T x_j, ..., \sum_j e_m e_m^T x_i - e_m e_m^T x_j) \begin{bmatrix} 1 & 0 & ... & 0 \\ 0 & 1 & 0 & ... \\ ... & 0 & 1 & 0 \\ 0 & ... & 0 & 1 \end{bmatrix} \\ &= (\sum_j x_{i1} - x_{j1}, \sum_j x_{i2} - x_{j2}, ..., \sum_j x_{im} - x_{jm}) \end{aligned} \quad (15)$$

Mathematically, the length of the local representative vector $\mathcal{H}$ formed by the boundary object is significantly longer than that of the core object. in this case, the P-norm is preferred here.

Definition 7. ***Boundary degree*** *Considering the computational cost, the Boundary degree φ is defined by Eq. 16 with p=1.*

$$\varphi_i = ||\mathcal{H}_i||_p = (\sum_d (|\sum_j x_{id} - x_{jd}|)^p)^{\frac{1}{p}} \quad s.t.\ 1 \le d \le m,\ x_j \in N_{dc}(x_i) \quad (16)$$

where the role of boundary degree φ is to determine whether x_i is a boundary point; a larger value suggests that x_i should be removed first.

After that, the dataset X can be evolved into a core subset:

$$X' = \{x_i \in X : \varphi_i \leq A_\varphi[\lfloor n * \lambda \rfloor]\} \tag{17}$$

where λ denotes the proportion of low confidence sample in X and A is a descending list of φ. λ is empirically set to 10%. Subsequently, the density centers and mean centers are derived from X' by CNMBI, which finally predicts the number of clusters by the proposed center pairwise matching. As summarized in Algorithm 1.

Algorithm 1: CNMBI

Input: Input data: X
Output: the optimal number K^*
1 Calculate Boundary degree of each sample in X, φ_i using Eq. 16 ;
2 Remove the low-confidence samples to form the core set X' of X using Eq. 17;
3 **for** $k = \hat{k}_{\min} : \hat{k}_{\max}$ **do**
4 | Find the density center set $S_\gamma^{(k)}$ using Eq. 4
5 | Find the mean center set $S_\mu^{(k)}$ using Eq. 1
6 | Construct the Center Bipartite Graph $G_{\mu,\gamma}^{(k)}$ using Definition 2
7 | Search the Center Complete Matching CM and calculate the Center Pairwise Loss $L(CM)$ using Definition 3 and Eq. 7
8 | Determine the Pairwise Matching $CPM^{(k)}$ under the k using Eq. 7
9 **end**
10 $K^* = \arg\min(L(CPM^{(k)})), k \in [\hat{k}_{\min}, \hat{k}_{\max}]$
Result: K^*

---

4 Experiments

Data Sets. To demonstrate the effectiveness of the proposed CNMBI, a variety of data scenarios are covered, the information of which is presented in Table 1. The 8 synthetic datasets are generated from Shape sets of the Clustering basic benchmark[2], and 3 high-dimensional datasets are obtained from the UCI [2]. The image datasets (CIFAR-10[3], STL-10[4], MNIST[5], Pendigits, Optical Recognition, and Pointing[6]) are considered. **Evaluation Metrics.** We record the optimal number of clusters (NC) and calculate the accuracy metric (ACC: i.e., the proportion of successful estimates in 50 runs). **Experimental Settings.** We set the

[2] http://cs.joensuu.fi/sipu/datasets/.
[3] http://www.cs.toronto.edu/kriz/cifar.html.
[4] https://cs.stanford.edu/acoates/stl10/.
[5] http://yann.lecun.com/exdb/mnist/.
[6] http://www-prima.inrialpes.fr/Pointing04.

number of clusters k to the acknowledged range of 2 to $\sqrt{n}$, performing 50 runs. We compare the proposed CNMBI with 4 state-of-the-art approaches, LCCV [4], VCIM [1], DPC-AHS [19], CNAK [13] and well-known validity index SC [12].

4.1 Quantitative Study

Table 2 summarizes the experimental results of the quantitative study, from which we can draw some conclusions. First, our algorithm correctly identified 13 out of 17 datasets in 50 runs, 8 more than the second place, achieving top-1 performance. Second, our method is suitable for data containing arbitrary-shaped clusters. For instance, the clusters in datasets "S2" (Fig. 2(a)), "A1" (Fig. 2(b)), "Flame" (Fig. 2(d)), and "Asymmetric" (Fig. 2(f)) almost overlap visually. Whereas datasets "Unbalanced" (Fig. 2(c)) and "Jain" (Fig. 2(e)) are examples of multi-density, showing manifold and unbalanced, respectively. As a result, only our method accurately identified all of them. Besides, Table 2 also reports the more challenging datasets from real world. For example, REUTERS-10K is a typical text type; Covid-19 records biochemical indicators of 3 types of COVID-19 with significant dimensional-difference; RNA-seq records a surprising 20,000 dimensions of gene transcription data. The six large-scale image datasets further validate the effectiveness of our algorithm CNMBI, especially for complex CIFAR-10 and STL-10, where only CNMBI achieves the correct cluster number estimation. The above satisfactory performance is mainly thanks to the

Table 1. Basic information about the studied data sets.

Name	Size	Dimension	Class	Description
Unbalanced	1500	2	3	Multi-density
Flame	240	2	2	Overlapping
Jain	373	2	2	Crescent, Multi-density
S2	5000	2	15	Sub-cluster,Overlapping
A1	3000	2	20	Small cluster
Asymmetric	1000	2	5	Unbalanced, noisy
Heterogeneous	400	2	3	Heterogeneous geometric
Multi-objective	1000	2	4	elongated, circle-like, multi-objective
REUTERS-10K	10000	12000	4	Word, News, Text,
Covid-19	171	53	3	Covid-19
RNA-seq	801	20531	5	Gene expression, Nonlinear
MNIST	10000	28*28	10	OCR, high-dimensional
Pendigits	10992	4*4	10	Handwritten Digits
Optical Recognition	5620	64	10	OCR, Handwritten Digits
Pointing Data	1395	384*284*3	15	Posture recognition
STL-10	13000	3*96*96	10	Complex Contextual
CIFAR-10	10000	3*32*32	10	Complex Contextual, Large-Scale

joint modeling of the center pairwise matching scheme and the boundary filtering principle, the former ensuring the flexibility of our method in terms of data dimension and shape, while the latter enhancing the robustness of CNMBI when handling complex data. It is also worth mentioning that the scores under the ACC metric indicate that our method is more stable than other methods over multiple runs. In particular, Fig. 3 visualizes the optimization process of part of the data set.

4.2 Ablation Study

Table 3 reports the experimental results, from which we can draw some conclusions: 1) Method-1 removes the boundary filtering strategy compared to our method, resulting in a significant decrease in NC score from a perfect score of 1 to 0.2 on the Pointing Dataset, which demonstrates that this strategy plays an important role in improving robustness; 2) The results of Method-5 indicate that our method has good scalability, i.e., cluster number identification can also be achieved by comparing the mean centers generated twice, although it is not as stable as our original method, as in the case of NC score.

Table 2. Cluster number prediction on challenging datasets over 50 runs.

Dataset	SC		DPC-AHS		VCIM		LCCV		CNAK		CNMBI	
	NC	ACC	NC	ACC	NC	ACC	NC	ACC	NC	ACC	NC	ACC
Unbalanced	2	0	2	0	2	0	2	0	**3**	0.9	**3**	**1**
Flame	4	0	**2**	**1**	4	0	4	0	1	0	**2**	**1**
Jain	**2**	0.5	**2**	**1**	3	0	7	0	3	0	**2**	**1**
S2	**15**	0.4	13	0	2	0	23	0	**15**	**1**	**15**	**1**
A1	**20**	0.1	**20**	**1**	2	0	4	0	**20**	**1**	**20**	**1**
Asymmetric	2	0	**5**	**1**	2	0	**5**	**1**	**5**	0.5	**5**	**1**
Heterogeneous	2	0	2	0	5	0	4	0	2	0	**3**	**1**
Multi-objective	12	0	2	0	3	0	**4**	**1**	3	0	**4**	**1**
REUTERS-10K	78	0	**4**	**1**	2	0	92	0	1	0	**4**	0.1
Covid-19	2	0	**3**	**1**	**3**	**1**	2	0	**3**	**1**	**3**	**1**
RNA-seq	**5**	0.6	**5**	**1**	**5**	**1**	**5**	**1**	**5**	0.6	**5**	**1**
MNIST	**10**	0.1	8	0	3	0	**10**	**1**	**10**	0.1	**10**	**1**
Pendigits	15	0	13	0	2	0	13	0	11	0	**10**	0.8
Optical	**10**	0.7	5	0	7	0	**10**	**1**	**10**	**1**	**10**	**1**
Pointing Data	14	0	25	0	3	0	24	0	9	0	**15**	**1**
STL-10	20	0	1	0	-	-	-	-	1	0	**10**	0.2
CIFAR-10	1	0	1	0	-	-	56	0	**10**	0.1	**10**	0.3

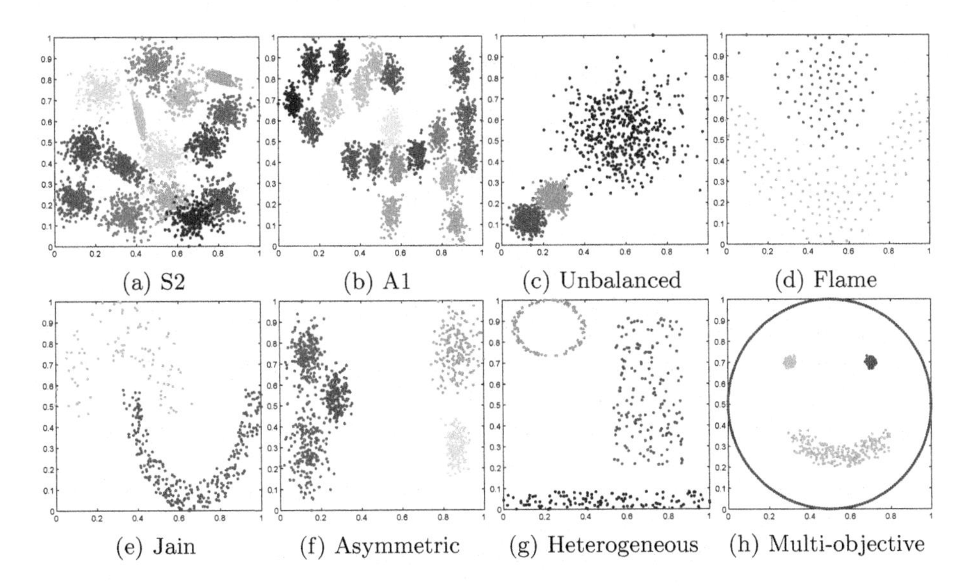

(a) S2 (b) A1 (c) Unbalanced (d) Flame

(e) Jain (f) Asymmetric (g) Heterogeneous (h) Multi-objective

Fig. 2. Synthetic dataset containing arbitrary-shaped clusters.

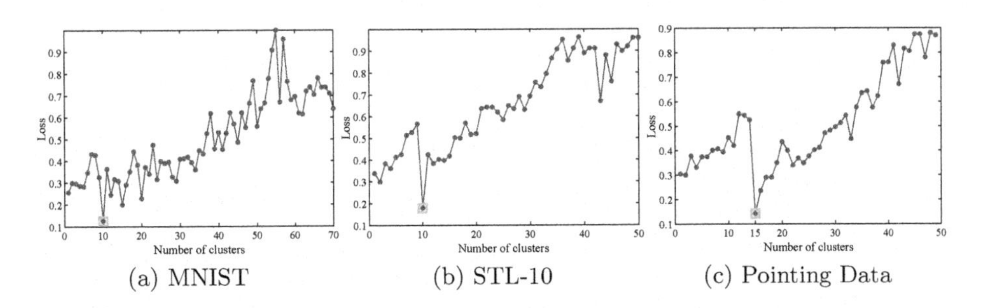

(a) MNIST (b) STL-10 (c) Pointing Data

Fig. 3. Plots of k and center pairwise loss on partial image data sets.

Table 3. Ablation study of CNMBI and its degradation methods. S_μ, S_γ and φ denote the Mean center, Denisty center, and boundary filtering, respectively.

Methods	S_μ	S_γ	φ	MNIST	Pointing Data
Ours	✓	✓	✓	NC: 10 ACC: 1	NC: 15 ACC: 1
Method-1	✓	✓	×	NC: 10 ACC: 0.6	NC: 15 ACC: 0.2
Method-2	✓	×	×	NC: 1 ACC: 0	NC: 1 ACC: 0
Method-3	×	✓	×	NC: 10 ACC: 0.3	NC: 15 ACC: 0.1
Method-4	✓	×	✓	NC: 1 ACC: 0	NC: 1 ACC: 0
Method-5	×	✓	✓	NC: 10 ACC: 0.5	NC: 15 ACC: 0.4

4.3 Parameter Sensitivity

CNMBI is parameter-free since the parameters, cutoff distance dc in density center and proportion λ in boundary filtering, are dynamically adaptive. Here we test CNMBI in three typical scenarios, each with four datasets, as reported in Fig. 4. For scenario "Noise," different levels of noise interference are added. The "Density" consists of 8 clusters with unbalanced density. For "Number," the four data sets contained have different numbers of clusters, 5,10,20,40.

The results are concentrated in Fig. 4(m). Concretely, as given by the red line in Fig. 4, it can be found that from the first to the fourth dataset, CNMBI behaves satisfactorily without error despite the increase in the number of clusters. The Noise-40 and Noise-50 dataset (Fig. 4(g),(h)) cannot be visually separated due to intense noise disturbance, CNMBI can still determine the desired number of clusters 2 (see the blue-green line segment in Fig. 4). Moreover, the density imbalance gradually increases from Density-1 to Density-4, and the green line segment in Fig. 4 shows that CNMBI is robust to density changes. Generally, benefiting from the consideration of active boundary filtering and simple but effective cluster number features generated by center matching, CNMBI has better robustness against data with complex distribution than other methods.

4.4 Boundary Pattern Recognition

Our method can extract valuable boundary pattern information while determining the number of clusters. Although some points cause undesirable results in cluster number determination, in practice, these objects often possess the characteristics of two or more groups simultaneously and have great potential value, such as populations carrying viruses but not manifested in epidemiology and scribbled characters in optical character recognition.

Figure 5(a) shows the boundary samples we extracted from the handwriting dataset (MNIST), from which it can be seen that the characters are illegible, with blurred, hollow, continuous strokes that are distinctly different from normal fonts. The pointing dataset records 23 types of head posture images from 15 volunteers. As shown in Fig. 5(b), the identified images have exaggerated movements and complex expressions from Pointing Data, which is beneficial for determining the amplitude threshold of each action in the pose recognition. Moreover, our proposed boundary filtering method is informative for other learning tasks.

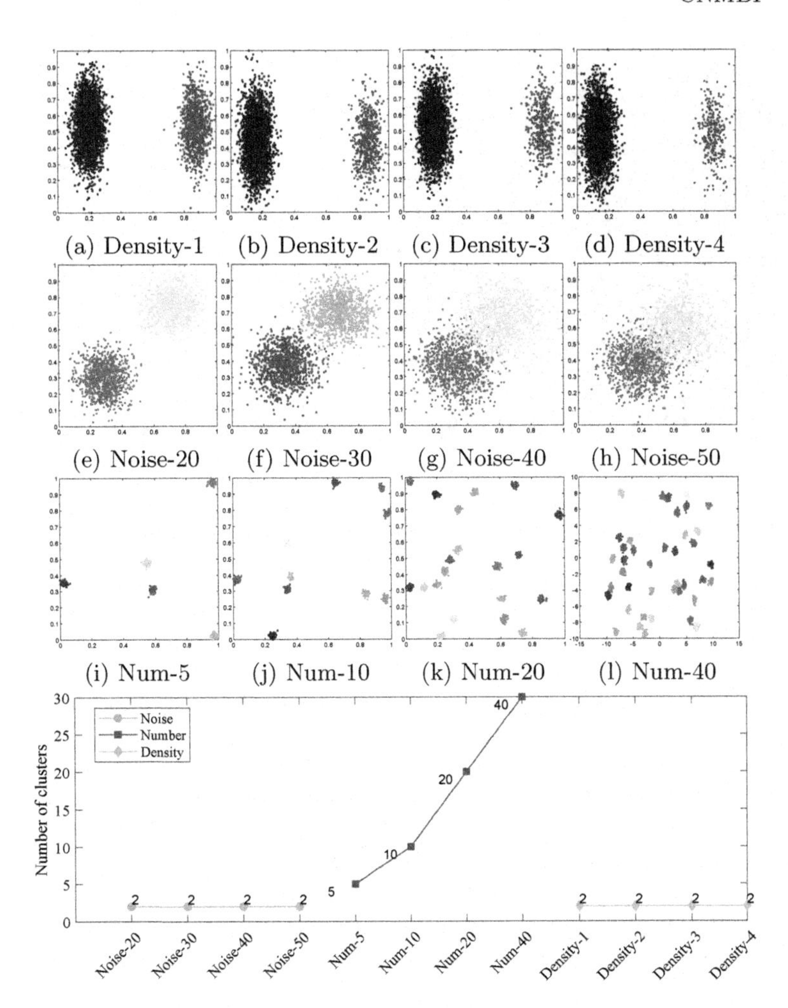

Fig. 4. Three extreme scenarios in Robustness

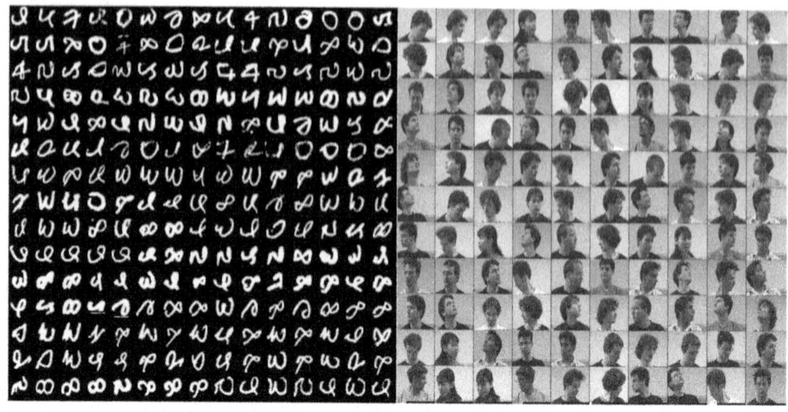

(a) Illegible handwritten characters. (b) Complex gestures.

Fig. 5. Boundary pattern information extraction.

5 Conclusion

In this paper, we propose a novel method, CNMBI, to identify the number of clusters, which is capable of being applied to some challenging data such as large-scale images and high-dimensional real-world data. We are the first to report the results on STL-10 and CIFAR-10. We do not rely on the clustering results of the entire data but instead utilize a few but highly representative cluster centers. This purposely straightforward contrast allows our method to be more scalable and flexible in dimension and shape, most importantly, sufficient for discovering actual cluster numbers through the positional behavior of the cluster center. Additionally, the principle of filtering low-confidence samples further enhances the robustness of our method, which is the first attempt in this field.

Acknowledgements. This work was supported in part by the Shenzhen Fundamental Research Fund (JCYJ20210324132212030) and the Guangdong Provincial Key Laboratory of Novel Security Intelligence Technologies (2022B1212010005).

References

1. Abdalameer, A.K., Alswaitti, M., Alsudani, A.A., Isa, N.A.M.: A new validity clustering index-based on finding new centroid positions using the mean of clustered data to determine the optimum number of clusters. Expert Syst. Appl. **191**, 116329 (2022)
2. Bache, K., Lichman, M.: UCI machine learning repository (2013). https://doi.org/10.1145/2063576.2063689
3. Calinski, T., Harabasz, J.: A dendrite method for cluster analysis. Commun. Stat. **3**, 1–27 (1974). https://doi.org/10.1080/03610917408548446
4. Cheng, D., Zhu, Q., Huang, J., Wu, Q., Yang, L.: A novel cluster validity index based on local cores. IEEE Trans. Neural Netw. Learn. Syst. **30**(4), 985–999 (2019). https://doi.org/10.1109/TNNLS.2018.2853710

5. Davies, D.L., Bouldin, D.W.: A cluster separation measure. IEEE Trans. Pattern Anal. Mach. Intell. **PAMI-1**(2), 224–227 (1979)
6. Dunn, J.C.: A fuzzy relative of the isodata process and its use in detecting compact well-separated clusters. J. Cybern. **3**(3), 32–57 (1973)
7. Li, Y., Hu, P., Liu, Z., Peng, D., Zhou, J.T., Peng, X.: Contrastive clustering. Proc. AAAI Conf. Artif. Intell. **35**(10), 8547–8555 (2021). https://doi.org/10.1609/aaai.v35i10.17037
8. Nguyen, S.D., Nguyen, V.S.T., Pham, N.T.: Determination of the optimal number of clusters: a fuzzy-set based method. IEEE Trans. Fuzzy Syst. **30**(9), 3514–3526 (2022). https://doi.org/10.1109/TFUZZ.2021.3118113
9. Qiu, T., Li, Y.: Fast LDP-MST: an efficient density-peak-based clustering method for large-size datasets. IEEE Transactions on Knowledge and Data Engineering **35**(5), 4767–4780 (2022)
10. Rasool, Z., Zhou, R., Chen, L., Liu, C., Xu, J.: Index-based solutions for efficient density peak clustering. IEEE Trans. Knowl. Data Eng. **34**(5), 2212–2226 (2022). https://doi.org/10.1109/TKDE.2020.3004221
11. Rodriguez, A., Laio, A.: Clustering by fast search and find of density peaks. Science **344**(6191), 1492–1496 (2014). https://doi.org/10.1126/science.1242072
12. Rousseeuw, P.J.: Silhouettes: a graphical aid to the interpretation and validation of cluster analysis. J. Comput. Appl. Math. **20**, 53–65 (1987). https://doi.org/10.1016/0377-0427(87)90125-7
13. Saha, J., Mukherjee, J.: CNAK: cluster number assisted k-means. Pattern Recogn. **110**, 107625 (2021)
14. Salvador, S., Chan, P.: Determining the number of clusters/segments in hierarchical clustering/segmentation algorithms. In: 16th IEEE International Conference on Tools with Artificial Intelligence, pp. 576–584 (2004)
15. Sugar, C.A., James, G.M.: Finding the number of clusters in a dataset: an information-theoretic approach. J. Am. Stat. Assoc. **98**, 750–763 (2003)
16. Tavakkol, B., Choi, J., Jeong, M.K., Albin, S.L.: Object-based cluster validation with densities. Pattern Recogn. **121**, 108223 (2022)
17. Tibshirani, R., Walther, G., Hastie, T.: Estimating the number of clusters in a data set via the gap statistic. J. Roy. Stat. Soc. B **63**(2), 411–423 (2001). https://doi.org/10.1111/1467-9868.00293
18. Xu, X., Ding, S., Wang, L., Wang, Y.: A robust density peaks clustering algorithm with density-sensitive similarity. Knowl.-Based Syst. **200**, 106028 (2020)
19. Zhang, R., Miao, Z., Tian, Y., Wang, H.: A novel density peaks clustering algorithm based on hopkins statistic. Expert Syst. Appl. **201**, 116892 (2022)
20. Zhang, R., Zheng, H.: Density clustering based on the border-peeling using space vector decomposition. Acta Automatica Sinica **49**(6), 1–19 (2023)
21. Zhang, Y., Mańdziuk, J., Quek, C.H., Goh, B.W.: Curvature-based method for determining the number of clusters. Inf. Sci. **415–416**, 414–428 (2017)

TS-MVP: Time-Series Representation Learning by Multi-view Prototypical Contrastive Learning

Bo Zhong, Pengfei Wang, Jinwei Pan, and Xiaoling Wang(✉)

School of Computer Science and Technology, East China Normal University, Shanghai, China
{bzhong,pfwang,jwpan}@stu.ecnu.edu.cn, xlwang@cs.ecnu.edu.cn

Abstract. IoT and wearable devices generate large amounts of time series data daily, providing opportunities for the development of human-computer interaction and digital services through learning powerful representations from these rich data. While Masked Autoencoders (MAE) have been used for time series representation learning, contrastive learning has superior performance. However, existing contrastive learning methods often utilize perturbation operations that may disrupt the local and global structure of time series data, and they do not explicitly model the relationship between downstream classification tasks. In this paper, we propose a framework based on multi-view prototypical contrastive learning for learning multivariate time-series representations from unlabeled data. Our approach involves transforming the original data into time-based and feature-based views using innovative masking technology based on state transfer probabilities and then embedding them using an encoder along with the original data. Moreover, a novel prototype contrastive module is designed that learns similar outputs from different views using clustered soft labels generated by the original data and prototypes, which helps the model develop fine-grained representations that can be effectively integrated into classification tasks. We conducted experiments on four real-world time series datasets, and the results demonstrate that our proposed TS-MVP framework outperforms previous time series representation learning methods when training a linear classifier on top of the learned features.

Keywords: Time series · Contrastive learning · Representation learning

1 Introduction

As Big Data evolves, large volumes of data from multiple sensors and wearable devices can be collected in real-time. Utilizing and mining these data is extremely beneficial to the medical [4,13,27], industrial [17,22], and manufacturing sectors [18]. For example, it will be possible to improve human-computer interaction

X. Yang et al. (Eds.): ADMA 2023, LNAI 14180, pp. 278–292, 2023.
https://doi.org/10.1007/978-3-031-46677-9_20

as well as the warning system for hazards. Time-series modeling has seen vast improvements in recent years, owing to the design of powerful deep learning architectures but also to the ever-increasing volume of available training data. However, it presents unique challenges compared to other data types, such as images or natural language text. On the one hand, the process of understanding and manually annotating multivariate time series data is more complex and challenging than images due to its inherent multidimensionality and temporal nature [7,8]. On the other hand, time series data typically exhibits lower semantic density compared to natural language text, where words within sentences are highly relevant to each other in terms of meaning. Extracting meaningful semantic information from time series data becomes further complicated by the fact that it can only be obtained at the segment-level, such as identifying trends or patterns, rather than relying on isolated data points.

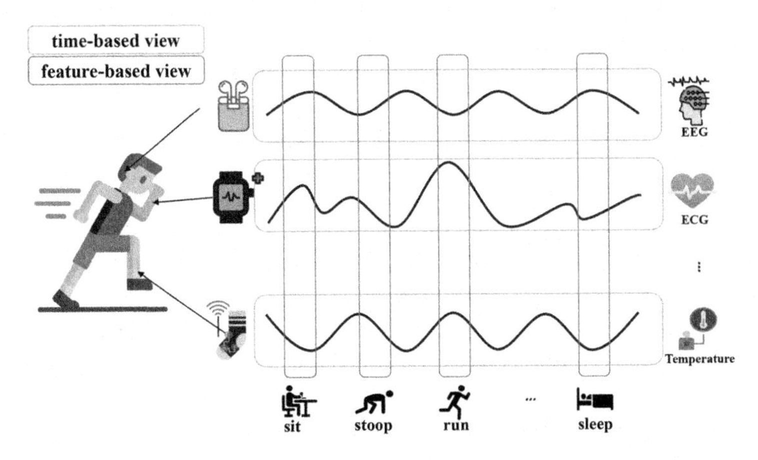

Fig. 1. An illustration of two kinds of views. In the time-based view, each series represents a record from the same sensor at different moments. This view captures dynamic patterns or changes over time, such as temperature change patterns or heart rate variations. It provides insights into the temporal dynamics of the data. In the feature-based view, each temporal record represents data from different sensors at the same moment. This view captures the global attributes or characteristics of the data, such as movement type or pose recognition. It provides insights into the overall properties of the data at a particular moment.

Despite the field's impressive evolution, dealing with large volumes of data often precludes access to labels and highlights the need for effective self-supervised learning strategies, which has been phenomenally successful in natural language processing [16,19,23] and computer vision [5,7,10,20] powering robust representation. Inspired by these work, two types of multivariate time series self-supervised representation learning methods based on reconstruction [21,26] and contrastive learning [9,25,28] have been proposed.

The reconstruction method for the masked prediction task is a relatively simple approach. It is motivated by the idea that a model can learn to represent data by learning to reconstruct it from a subset of the data itself.

The underlying assumption is that the data contains useful information that can be captured by the model through the reconstruction process. However, since multivariate time series are numerical series that naturally have interpolation and fitting capabilities, this easy pretext task leads to low generalization. It is worth noting that for certain classification tasks, it is only necessary for the model to discriminate between multivariate time series. Therefore, reconstruction is not required. Contrastive learning has demonstrated powerful self-supervised representation learning capabilities in computer vision as it can learn the essential representation of data from different augmented data. Thus, a growing body of work has attempted to incorporate contrastive learning into self-supervised time series representation learning.

The effectiveness of contrastive learning heavily depends on the way to generate positive samples. Previous work [25] has shown that data augmentation approaches in images, such as rotation, flipping etc., cannot be directly applied to time series. Existing work has developed data augmentation methods for time series in terms of strength, time and frequency, time and space, etc., but they all include perturbation operations. For multivariate time series, a small perturbation in both the time domain and frequency can have a massive impact on the semantic information of the series. This form of data augmentation does not suffice in capturing the rich local and global series structure. It can even interfere with the classification task. Furthermore, the assumptions based on predictive consistency or time-frequency consistency [28] are difficult to combine with downstream tasks such as classification.

To tackle the aforementioned issues, we propose the TS-MVP model, which leverages two natural perspectives for multivariate time series: the time-based view and the feature-based view. As shown in Fig. 1, the former represents data collected by the same sensor at different times, capturing changing patterns in sensor attributes, while the latter represents data collected from various sensors at the same moment, aggregating local attributes from different sensors and reflecting the more general semantics of global properties. Our proposed data augmentation method effectively utilizes both views, enabling the model to learn from diverse perspectives and develop robust representations. We also introduce a prototype contrastive module that utilizes clustered soft labels to learn similar outputs from both views, allowing the model to develop fine-grained representations that can be better integrated into classification tasks. Additionally, we propose an instance-level contrastive module that utilizes data augmentation for the two views and utilizes the InfoNCE [16] loss function to maximize similarity within instances and minimize similarity between different instances. Our self-supervised learning approach enhances the model's ability to capture meaningful patterns and relationships within the data. In summary, the contributions of our work are the TS-MVP model, the proposed data augmentation method, the prototype contrastive module, and the instance-level contrastive module, which improve the model's ability to learn from both perspectives and capture meaningful patterns in the data.

- **Data augmentation from both time-based and feature-based views**: We propose a novel data augmentation approach that masks the original multivariate time series data using state transition probabilities, capturing both time and feature-based patterns. This results in augmented data from both views, enriching the diversity of the training data and allowing the model to learn from multiple perspectives.
- **Prototype contrastive module for denoising and robust representation learning**: We introduce a prototype contrastive module that implicitly denoises the representations at the representation level. By clustering the masked input representations under both time-based and feature-based views to match the unmasked input representations, the model learns invariant-based representations that are more robust and aligned with classification tasks. This pre-training step enhances the feature extractor's ability to capture high semantic-level representations.
- **Effective and practical approach validated through comparison and ablation experiments**: We conducted comparison and ablation experiments on four public datasets to evaluate the effectiveness and practicality of our proposed approach. The results demonstrate the effectiveness of our data augmentation method and multi-view prototypical contrastive learning in improving the model's performance in classification tasks.

2 Related Work

2.1 Self-supervised Learning for Time Series

Reconstruction and contrastive learning are two types of multivariate time series self-supervised representation learning. Inspired by pre-training work in natural language processing, TST [26] was borrowed from the language model Transformer [23] to learn a better representation of multivariate time series. But reconstruction may not be suitable for classification. Based on the contrastive learning work proposed in computer vision, such as CPC [16], SimCLR [5], BYOL [10] and MoCo [6], existing work has applied contrastive learning to the pre-training of time series to learn the contextual representation of time series in a self-supervised manner. CLOCS [11] performed contrastive learning of physiological signals based on temporal and spatial invariance. In this method, similar temporal and spatial representations of signals for the same patient were encouraged. Mixing-up [24] used data augmentation in the form of mixture generation by predicting the mixture components as a soft target for contrastive learning. TS-TCC [9] proposed strong and weak data augmentation for time series and constructed a complex cross-prediction task based on such data augmentation. In order to construct a general time series representation learning framework, TS2Vec [25] attempted to learn scale-invariant representations of time series under different data augmentation. TF-C [28] assumed that the same samples could be closely combined in the time-frequency space by a time-based representation and a frequency-based representation. These methods introduced some perturbations and affine transformations into data augmentation. However, unlike images, small perturbations can have a massive impact on the semantic information in the time series. To generate different augmented samples that assist the model in learning advanced semantic representations, we innovate data augmentation from two views, time and feature. It ensures that the pre-training task is challenging while preserving the semantic information of the original multivariate time series.

2.2 Consistent Representation Learning

Consistent learning of Kullback-Leibler (KL) loss is widely used in computer vision to assist in representation learning. To strengthen the consistency of the representations, RELIC [15] added regularization to the embedding of different data augmentation. Most of the work is based on consistent representation learning techniques such as clustering using pseudo-cluster labels to learn visual representations of samples. SWAV [3] learned prototype clusters by improving the consistency between clusters of different views. MSN [2] combined with a random masking mechanism rely on view-invariant representation learning to match randomly masked images with the original image based on invariance and mask denoising, thus improving the model's performance on few-sample image classification. The common approach for time series analysis is to use forecasting techniques, in which the model is trained to predict future time steps based on past observations. However, existing time series representation learning approaches rarely perform view-consistently or with strong constraints. To address this issue, we propose consistent representation learning, where view-consistent inductive preference methods are used to force the model to learn similar outputs for different views of the same sample. This approach can reduce the hypothesis space size and improve the model's robustness to some extent.

3 Proposed Method

In this section, we will describe our proposed model **TS-MVP** in detail. Given a collection of time series data, denoted as $\mathcal{X} = \{x^1, x^2, \cdots, x^N\}$, containing N instances, our objective is to leverage the local and global characteristics of $x^i \in R^{T \times F}$, where T represents the length of time and F denotes the feature dimension. As shown in Fig. 2, **TS-MVP** consists of four parts. First, the original data are augmented with views from different perspectives Sect. 3.1. They are used as input to the encoder Sect. 3.2 together with the original data. Second, the prototype contrastive module Sect. 3.3 learns similar outputs from different views based on clustered soft labels from the original data, which in turn will help the model to develop fine-grained representations that can be more effectively integrated into classification tasks. Last, to further learn the distinguished representations, we introduce an instance contrastive module Sect. 3.4 that built upon the prototype contrastive module. It attempts to maximise the similarity between views for the same instance and minimise the similarity between views for different instances. Finally, we present our overall training objective Sect. 3.5. Next, we will introduce each module in each subsection.

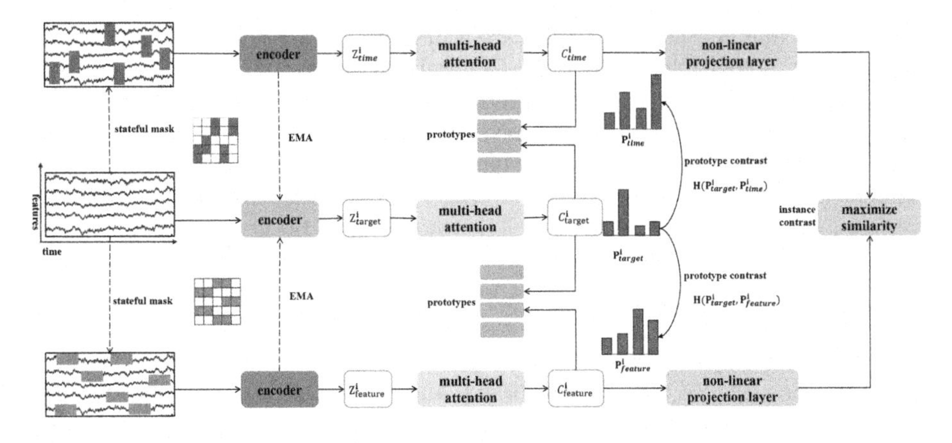

Fig. 2. The overall framework of the **TS-MVP** model.

3.1 Multi-view-based Time Series State Masking

Contrastive learning aims to capture the invariance of representations under different transformations of the same instance by encouraging the similar representations between positive sample pairs. The algorithm's performance depends on data augmentation to construct powerful pairs of positive and negative samples. Specifically, for each input sample $x^i \in R^{T \times F}$, we denote its time view augmentation as x^i_{time} and its feature view augmentation as x^i_{feature}. For data augmentation, we employ masking mechanisms more suitable for time series. Make a time binary mask $M^i_{\text{time}} \in R^{T \times F}$ and a feature binary mask $M^i_{\text{feature}} \in R^{T \times F}$ for each instance separately (Fig. 3).

$$x^i_{\text{time}} = M^i_{\text{time}} \odot x^i \tag{1}$$

$$x^i_{\text{feature}} = M^i_{\text{feature}} \odot x^i \tag{2}$$

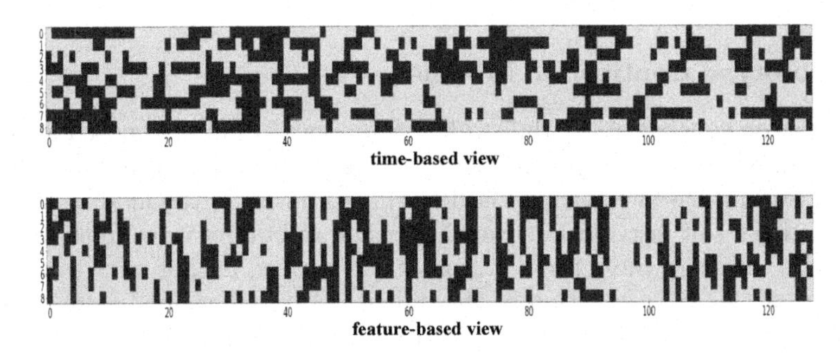

Fig. 3. An illustration of the time and feature binary masks for a sample $x^i \in R^{128 \times 9}$.

We generate a masking segment $L \in R^F$ with a masking rate r_{time} for each time point in x^i during each training epoch for time view augmentation. And for feature view augmentation, we generate a masking segment $L \in R^T$ with masking rate r_{feature} for each feature in x^i. Generally, we set $r_{\text{feature}} \neq r_{\text{time}}$. To avoid all masking elements are randomly distributed, we adopt geometric distribution sampling [26] to generate masking matrices with state transfer probabilities. As a result of the set state transfer probabilities r_{time} and r_{feature}, each masked segment (sequence of 0s) follows an average l_m geometric distribution, and the average length of the unmasked segment (sequence of 1s) is $l_u = \frac{1-r}{r} l_m$.

3.2 Feature Extractor

A feature extractor extracts the raw data x^i and its corresponding time-based augmentations x^i_{time} and feature-based augmentations x^i_{feature} into a high dimensional representation z^i_{target}, z^i_{time}, and z^i_{feature}. For the purpose of comparison, we employ the same feature extractor as the TS-TCC [9].

$$z^i_{\text{time}} = F_A\left(x^i_{\text{time}}\right) \tag{3}$$

$$z^i_{\text{feature}} = F_A\left(x^i_{\text{feature}}\right) \tag{4}$$

$$z^i_{\text{target}} = \tilde{F_A}\left(x^i\right) \tag{5}$$

It is significant to note that the time-based augmentation and the feature-based augmentation share a common feature extractor F_A. Original data are passed through momentum updated feature extractor $\tilde{F_A}$, which is copied in depth from feature extractor F_A. F_A and $\tilde{F_A}$ have the same structure. During the training process, however, the weights θ_{F_A} are updated in real-time by gradients and $\theta_{\tilde{F_A}}$ are updated iteratively by an exponential moving average (EMA) based on the updated weights of F_A:

$$\theta_{\tilde{F_A}} \leftarrow m\theta_{\tilde{F_A}} + (1-m)\theta_{F_A} \tag{6}$$

where $m \in [0, 1)$ is the momentum coefficient.

3.3 Prototype Contrastive Module

The prototype contrastive module, a key component of our proposed approach, combines invariance-based pre-training and mask-denoising techniques. Similar to the clustering-based self-supervised learning approach, the module is trained by computing a soft distribution over the anchor and target view prototypes. The objective is to assign the representation of the masked anchor view to the same prototype as the representation of the unmasked target view. To optimize this criterion, we utilize a standard cross-entropy loss, which allows us to effectively train the module and learn robust representations that capture the underlying patterns in the data.

An autoregressive model is used to map the time view augmentation representation z^i_{time} to c^i_{time}, the feature view augmentation representation z^i_{feature} to c^i_{feature}, and the original data representation z^i to c^i. Due to the efficiency of

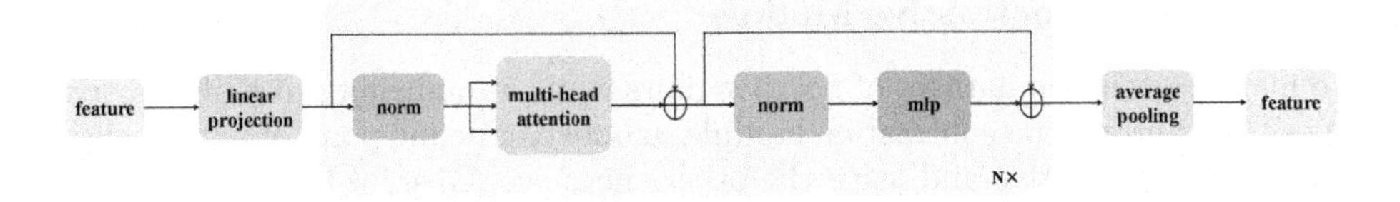

Fig. 4. Transformer architecture used in the clustering contrast module.

the Transformer in modelling across a wide range of domains, we also use Transformer Encoder as an autoregressive model. The model framework is shown in Fig. 4. Unlike the previous approach, we do not add [CLS] to the output to consider it a representative context vector of all the outputs. We use the mean of all output vectors at layer N to avoid the loss of temporal information. Example of mapping z^i to c_{target}^i:

$$\widetilde{\Phi} = MHA\left(\text{Norm}\left(W\left(z^i\right)\right)\right) \tag{7}$$

$$\Phi = MLP(\text{Norm}(\widetilde{\Phi})) + \widetilde{\Phi} \tag{8}$$

$$c_{\text{target}}^i = \text{Average}\left(\Phi_N\right) \tag{9}$$

The prototype network $q \in R^{K \times d}$ is designed to force the feature extractor to learn fine-grained category features in the distribution of data, where K represents the number of prototypes that can be learned, and d represents the dimension of each prototype. We calculate the similarity between the prototype mapping and the time view augmentation mapping, the feature view augmentation mapping, and the original data representation mapping as in the following equations:

$$p_{\text{time}}^i = \text{softmax}\left(\frac{c_{\text{time}}^i \cdot q}{\tau}\right) \tag{10}$$

$$p_{\text{feature}}^i = \text{softmax}\left(\frac{c_{\text{feature}}^i \cdot q}{\tau}\right) \tag{11}$$

$$p_{\text{target}}^i = \text{softmax}\left(\frac{c_{\text{target}}^i \cdot q}{\tau}\right) \tag{12}$$

where τ denotes the temperature coefficient. To avoid the appearance of trivial solutions and uniform distributions, we introduced target sharpening and regularization from the MSN [2]. We measure the difference in distribution between the original data, the feature view, and the time view using cross-entropy. Due to these two differences, the encoder is penalized.

$$L_{PC} = \frac{1}{2N}\sum_{i=1}^{N}\left(H\left(p_{\text{target}}^i, p_{\text{time}}^i\right) + H\left(p_{\text{target}}^i, p_{\text{feature}}^i\right)\right) - \lambda H\left(\overline{p^i}\right) \tag{13}$$

In this case, $\overline{p^i} = \frac{1}{2N}\sum_{i=1}^{N}\left(p_{\text{time}}^i + p_{\text{feature}}^i\right)$ refers to mean entropy maximisation (ME-MAX). We encourage the model to use all prototypes by setting $\lambda > 0$.

3.4 Instance Contrastive Module

To maximize the similarities of positive pairs while minimizing those of negative ones. After prototype contrastive module, model can define pairs of within-class samples to be positive and leave the others negative. Given a batch of N input samples, there will be N time view mappings and N feature view mappings, respectively. The two view mappings from the same sample are considered positive samples, and the remaining $(2N-2)$ mappings within the same batch are considered negative samples. The loss function of a positive sample pair (a,b) is defined as follows:

$$L_{IC} = -\sum_{i=1}^{N} \log \frac{\exp\left(\operatorname{sim}\left(c_a^i, c_b^i\right)/\tau\right)}{\sum_{m=1}^{2N} I_{[m \neq i]} \exp\left(\operatorname{sim}\left(c_a^i, c_b^m\right)/\tau\right)} \tag{14}$$

where sim is cosine similarity function, $I_{[m \neq i]} \in \{0,1\}$ is an indicator function, evaluating to 1 if $m \neq i$, and τ is a temperature parameter.

3.5 Objective Function

In summary, the overall self-supervised loss is composed of both cluster-level loss and instance-level loss, which are simultaneously optimized, as follows:

$$L = \lambda_1 \cdot L_{PC} + \lambda_2 \cdot L_{IC} \tag{15}$$

where λ_1 and λ_2 are hyperparameters used to balance the losses of the two components. We will discuss in the experiment section.

4 Experiments

4.1 Experimental Setup

Datasets. We evaluate our proposed model on four real-world datasets, i.e., HAR [1], Gesture [28], Uwave [14] and WISDM [12]. HAR was collected from 30 subjects with the accelerometer, gyroscope, and body 3-axis acceleration components in 9 channels at 50 HZ and 20 HZ, converted into 128 time-step segments. There are a total of six types. Gesture and Uwave come from the UCR database and use linear acceleration in three coordinate directions, 3 channels in total, at a frequency of 100 HZ. These segments are processed as segments of 206 and 315-time steps. A total of 8 hand gestures are included. WISDM acquired 3-axis accelerometer measurements from the subject on 3 channels at a frequency of 20 HZ, which were then processed into 128 segments of time. The statistics of the processed datasets are summarized in Table 1.

Implementation Details. In this work, we built our model using PyTorch 1.11 and trained it on a NVIDIA GeForce RTX 2080 Ti GPU. The key hyperparameter settings for each dataset are shown in Table 2.

Table 1. Description of datasets

Dataset	Train	Val	Test	Length	Channel	Class
HAR	5881	2947	2947	128	9	6
Gesture	320	120	120	206	3	8
Uwave	2686	896	896	315	3	8
WISDM	3579	1193	1193	128	3	6

Table 2. The key hyperparameter on four datasets.

Hyperparameter symbols	*HAR*	*Gesture*	*Uwave*	*WISDM*
r_{feature}	0.5	0.2	0.2	0.5
r_{time}	0.4	0.4	0.6	0.4
l_m	3	3	3	3
K	40	15	15	45
m	0.915	0.965	0.975	0.92
λ_1	0.6	1.3	0.7	0.6
λ_2	0.8	1.5	0.6	0.8

Baseline Methods. To evaluate the effectiveness of the model we proposed, we compare it with the following baseline methods. **Random Initialization** is a method for training a linear classifier based on a randomly initialized encoder. **Supervised** is to train both the encoder and linear classifier simultaneously. **TS-TCC** [9] is constructed as a complex cross-prediction task for time series with strong and weak data augmentation. **Ts2Vec** [25] is a hierarchical contrastive learning approach that enables scale-invariant representations of time series to be learned. **Mixing-up** [24] uses mixture generation to augment data by predicting components as soft targets for contrastive learning. **SimCLR** [5] learns the representation by maximizing the similarity of views from the same sample and minimizing the similarity of views from different samples. **TF-C** [28] expects the same sample to be tightly coupled in time-frequency space through a time-based representation and a frequency-based representation. TF-C primarily investigates time-series domain adaptivity. For a fair comparison, we use its proposed time- and frequency-domain data augmentation as an alternative to our data augmentation approach.

Evaluation Metrics. To evaluate the performance of all models, we apply *Accuracy*(ACC), *Macro-averaged F1-score* (MF1) and report the mean and standard deviation.

4.2 Comparison with Baseline Methods

Using Linear Classifier. A linear classifier is trained based on a frozen self-supervised pre-trained encoder model. In Table 3, we present the linear evaluation results of our approach in comparison with those of the baseline methods. As a result, our proposed **TS-MVP** outperforms all of the baseline methods. Demonstrate that our **TS-MVP** model has strong representation learning abilities. Accordingly, our clustering consistency method performs better than TS2Vec and TS-TCC, indicating it is more effective for multivariate time series classification. Further, it has the potential to play a significant role in the real-time warning function of wearable devices. However, our results are better than TF-C, showing that the time-feature domain is superior to the time-frequency domain for modelling multivariate time series.

Table 3. Performance comparison on four datasets using linear classifier evaluation experiment.

Models	*HAR*		*Gesture*		*Uwave*		*WISDM*	
	ACC	MF1	ACC	MF1	ACC	MF1	ACC	MF1
Random Init	78.54 ± 3.39	76.59 ± 3.84	68.49 ± 0.37	64.35 ± 1.02	90.66 ± 3.67	90.84 ± 3.53	77.31 ± 5.20	69.67 ± 6.97
TS-TCC	91.04 ± 0.78	90.62 ± 0.78	74.48 ± 0.37	72.13 ± 0.41	95.43 ± 0.84	95.09 ± 1.09	84.22 ± 0.71	73.03 ± 2.56
TS2Vec	91.04 ± 0.28	90.96 ± 0.29	75.39 ± 0.34	72.78 ± 1.02	95.57 ± 0.37	95.52 ± 0.37	89.08 ± 0.48	84.78 ± 0.61
Mixing-up	91.24 ± 1.04	91.10 ± 1.24	66.39 ± 0.40	60.06 ± 0.16	95.54 ± 0.33	95.49 ± 0.33	88.60 ± 0.14	83.28 ± 0.05
SimCLR	90.77 ± 0.89	90.39 ± 0.91	70.05 ± 1.61	66.57 ± 1.19	95.68 ± 0.28	96.35 ± 1.33	83.56 ± 0.65	73.15 ± 2.05
TF-C	89.33 ± 0.52	88.74 ± 0.54	72.40 ± 2.66	71.60 ± 2.89	96.43 ± 0.81	96.46 ± 0.72	89.53 ± 0.54	85.29 ± 0.90
TS-MVP (ours)	**92.62 ± 0.35**	**92.36 ± 0.42**	**76.28 ± 0.74**	**74.31 ± 0.78**	**96.54 ± 0.24**	**96.51 ± 0.23**	**90.40 ± 0.32**	**85.93 ± 0.29**

Fine-Tuning. In fine-tuning, we used labeled samples to fine-tune the pre-trained encoder. In this study, we compared our proposed model with supervised learning and TS-TCC to verify its effectiveness. The results are shown in Table 4. As expected, performance can be improved by self-supervised learning.

Table 4. Performance comparison on four datasets with a fine-tuned encoder.

Models	*HAR*		*Gesture*		*Uwave*		*WISDM*	
	ACC	MF1	ACC	MF1	ACC	MF1	ACC	MF1
Supervised	92.70 ± 0.72	92.39 ± 0.80	76.30 ± 2.57	74.85 ± 2.57	97.39 ± 0.10	97.33 ± 0.10	96.24 ± 0.01	94.00 ± 0.14
TS-TCC	93.78 ± 0.38	93.58 ± 0.45	**77.34 ± 1.10**	**75.52 ± 0.96**	97.51 ± 0.37	97.41 ± 0.40	96.29 ± 0.98	94.58 ± 1.28
TS-MVP (ours)	**94.01 ± 0.37**	**93.84 ± 0.31**	76.30 ± 0.74	75.25 ± 0.65	**97.54 ± 0.09**	**97.49 ± 0.10**	**96.71 ± 0.17**	**95.03 ± 0.35**

4.3 Ablation Study

In order to verify the effect of each module, we design five models for comparison, as shown in Table 5. To begin with, we train the prototype contrastive module (PC only) without the instance contrast module. Similarly, we train only the instance contrast module (IC only). For time-view mask only and feature-view mask only, we set up $r_{\text{feature}} = 0$ and $r_{\text{time}} = 0$ separately for time-view and feature-view masks only. In addition, we use the Bernoulli distribution mask instead of the state transition probabilities mask. The Bernoulli distribution can be used to generate random masks for time series data, resulting in a completely random masking pattern. The proposed prototype contrastive module enhances the model's performance by allowing features to be discriminated more effectively. Besides, the most efficient performance is obtained by making data augmentation with our proposed augmentation method that uses both views jointly.

Table 5. TS-MVP ablation study performed with linear classifier evaluation.

Component	*HAR*		*Gesture*		*Uwave*		*WISDM*	
	ACC	MF1	ACC	MF1	ACC	MF1	ACC	MF1
PC only	91.12 ± 1.09	90.71 ± 1.14	76.04 ± 1.60	73.61 ± 1.72	95.69 ± 0.14	95.62 ± 0.16	87.59 ± 0.56	81.68 ± 0.35
IC only	90.85 ± 0.68	90.40 ± 0.71	66.93 ± 1.48	64.45 ± 1.78	95.54 ± 0.42	95.55 ± 0.40	90.53 ± 0.23	87.05 ± 0.19
TS-MVP	92.62 ± 0.35	92.36 ± 0.42	76.28 ± 0.74	74.31 ± 0.78	96.54 ± 0.24	96.51 ± 0.23	90.40 ± 0.32	85.93 ± 0.29
Time-view mask only	85.03 ± 1.76	85.02 ± 1.62	69.79 ± 1.48	67.09 ± 1.82	96.17 ± 0.37	96.14 ± 0.38	88.30 ± 0.85	83.65 ± 1.18
Feature-view mask only	86.25 ± 3.06	85.63 ± 2.93	71.35 ± 0.37	68.16 ± 1.22	95.57 ± 0.65	95.53 ± 0.65	89.40 ± 0.96	84.77 ± 1.18
Bernoulli distribution mask	91.54 ± 1.22	91.19 ± 1.59	75.52 ± 0.97	72.61 ± 0.81	95.80 ± 0.59	95.61 ± 0.80	89.72 ± 0.72	84.72 ± 1.19

4.4 Sensitivity Analysis

Figure 5 shows the effect of the number of prototypes K on different datasets. Choosing the correct number of prototypes can improve performance. However, too large and too few prototypes can negatively affect the model's performance. The selection of the optimal number of prototypes based on domain knowledge is worth considering in practical applications. Figure 6 shows the model's performance on the HAR dataset as λ_1, λ_2, r_{time}, and r_{feature} vary over the range. We find that our model is more sensitive to r_{time}. This may be due to its higher redundancy in the time dimension for multivariate time series, and thus r_{time} is more likely to affect the challenge of the self-supervised task.

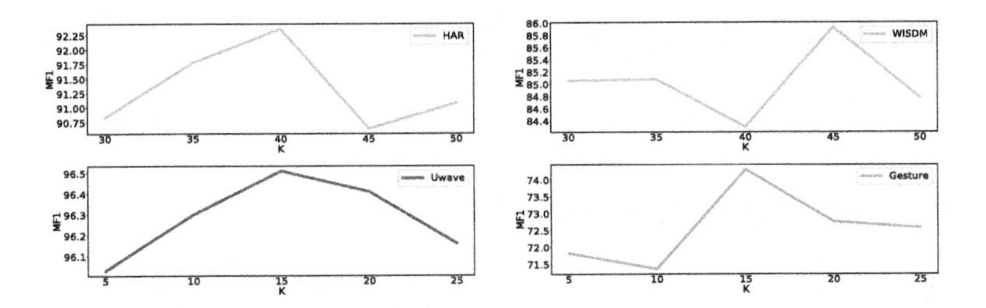

Fig. 5. Analysis the number of prototypes on four datasets.

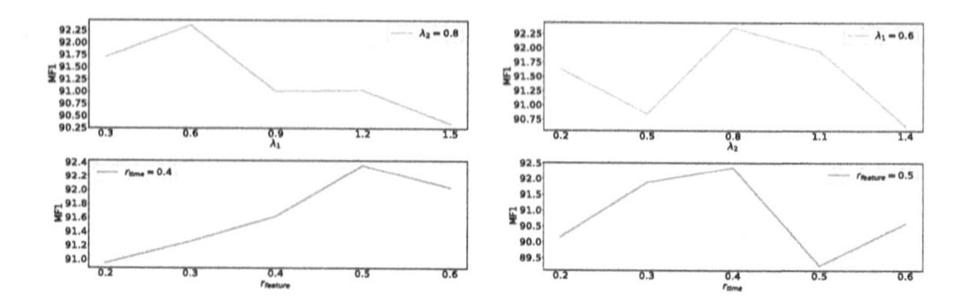

Fig. 6. Sensitivity analysis experiments on HAR dataset.

4.5 Case Study

To illustrate the information preserved after self-supervised learning, we used t-SNE to visualize the embeddings in the HAR dataset. Figure 7 demonstrates the performance of the feature extractor for random initialization, pre-training using TS-TCC and pre-training using our proposed TS-MVP method. The visualizations show that the learned representations are capable of distinguishing between different classes in the latent space. Furthermore, the embeddings obtained from TS-MVP demonstrate enhanced differentiation within the same class, with a greater number of smaller clusters compared to TS-TCC. This suggests that the proposed Multi-View Prototypical Contrastive Learning approach is effective in mining fine-grained label information and can better capture the subtle differences between subclasses within the same category.

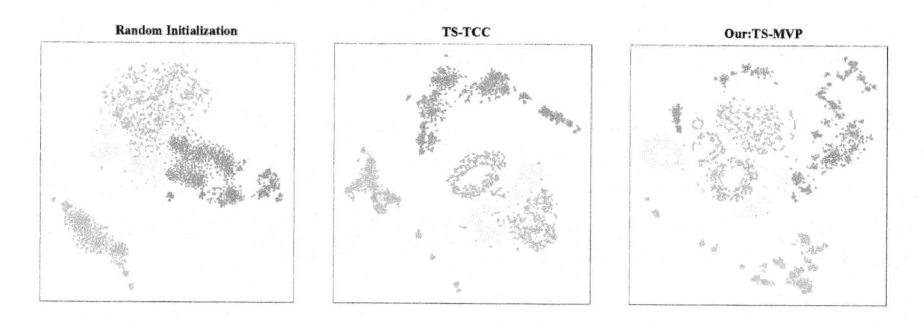

Fig. 7. t-SNE analysis on HAR dataset. Different colors represent different classes.

5 Conclusion and Future Work

This paper proposes a Multi-View Prototypical Contrastive Learning method for Time-Series representation learning (TS-MVP). TS-MVP utilizes a novel mask-denoising technique to improve the data from time-based and feature-based views. The prototype contrastive module clusters the masked input representations under both views to align with the unmasked input representations, resulting in invariance-based pre-training and meaningful representations that capture shared semantics. The cluster-level and instance-level contrastive losses optimize the model's ability to capture local and global patterns in multivariate time series data. Our proposed model outperforms all the baselines in experiments on four real-world datasets, demonstrating its effectiveness in learning meaningful representations of time-series data. In future work, we will test the performance of our model on datasets from a wider range of domains and investigate the optimal number of prototype settings through adaptive learning.

Acknowledgements. We thank editors and reviewers for their suggestions and comments. This work was supported by National Key R&D Program of China (No. 2021YFC3340700), NSFC grants (No. 62136002 and No. 61972155), and Shanghai Trusted Industry Internet Software Collaborative Innovation Center.

References

1. Anguita, D., Ghio, A., Oneto, L., Parra Perez, X., Reyes Ortiz, J.L.: A public domain dataset for human activity recognition using smartphones. In: Proceedings of the 21th International European Symposium on Artificial Neural Networks, Computational Intelligence and Machine Learning, pp. 437–442 (2013)
2. Assran, M., et al.: Masked Siamese networks for label-efficient learning. In: Avidan, S., Brostow, G., Cissé, M., Farinella, G.M., Hassner, T. (eds.) European Conference on Computer Vision, vol. 13691, pp. 456–473. Springer, Cham (2022). https://doi.org/10.1007/978-3-031-19821-2_26
3. Caron, M., Misra, I., Mairal, J., Goyal, P., Bojanowski, P., Joulin, A.: Unsupervised learning of visual features by contrasting cluster assignments. Adv. Neural. Inf. Process. Syst. **33**, 9912–9924 (2020)
4. Cerqueira, V., Torgo, L., Soares, C.: Early anomaly detection in time series: a hierarchical approach for predicting critical health episodes. Mach. Learn. 1–22 (2023)
5. Chen, T., Kornblith, S., Norouzi, M., Hinton, G.: A simple framework for contrastive learning of visual representations. In: International Conference on Machine Learning, pp. 1597–1607. PMLR (2020)
6. Chen, X., Fan, H., Girshick, R., He, K.: Improved baselines with momentum contrastive learning. arXiv preprint arXiv:2003.04297 (2020)
7. Chen, X., He, K.: Exploring simple Siamese representation learning. In: Proceedings of the IEEE/CVF Conference on Computer Vision and Pattern Recognition, pp. 15750–15758 (2021)
8. Cheng, Y., et al.: CUTS: neural causal discovery from irregular time-series data. arXiv preprint arXiv:2302.07458 (2023)
9. Eldele, E., et al.: Time-series representation learning via temporal and contextual contrasting. arXiv preprint arXiv:2106.14112 (2021)
10. Grill, J.B., et al.: Bootstrap your own latent-a new approach to self-supervised learning. Adv. Neural. Inf. Process. Syst. **33**, 21271–21284 (2020)
11. Kiyasseh, D., Zhu, T., Clifton, D.A.: CLOCS: contrastive learning of cardiac signals across space, time, and patients. In: International Conference on Machine Learning, pp. 5606–5615. PMLR (2021)
12. Kwapisz, J.R., Weiss, G.M., Moore, S.A.: Activity recognition using cell phone accelerometers. ACM SIGKDD Explor. Newsl. **12**(2), 74–82 (2011)

13. Li, H., Yu, S., Principe, J.: Causal recurrent variational autoencoder for medical time series generation. arXiv preprint arXiv:2301.06574 (2023)
14. Liu, J., Zhong, L., Wickramasuriya, J., Vasudevan, V.: uWave: accelerometer-based personalized gesture recognition and its applications. Pervasive Mob. Comput. **5**(6), 657–675 (2009)
15. Mitrovic, J., McWilliams, B., Walker, J., Buesing, L., Blundell, C.: Representation learning via invariant causal mechanisms. arXiv preprint arXiv:2010.07922 (2020)
16. van den Oord, A., Li, Y., Vinyals, O.: Representation learning with contrastive predictive coding. arXiv preprint arXiv:1807.03748 (2018)
17. Ravuri, S., et al.: Skilful precipitation nowcasting using deep generative models of radar. Nature **597**(7878), 672–677 (2021)
18. Rebjock, Q., Kurt, B., Januschowski, T., Callot, L.: Online false discovery rate control for anomaly detection in time series. Adv. Neural. Inf. Process. Syst. **34**, 26487–26498 (2021)
19. Seong, H.S., Moon, W., Lee, S., Heo, J.P.: Leveraging hidden positives for unsupervised semantic segmentation. arXiv preprint arXiv:2303.15014 (2023)
20. Shao, R., Wu, T., Liu, Z.: Detecting and grounding multi-modal media manipulation. arXiv preprint arXiv:2304.02556 (2023)
21. Shao, Z., Zhang, Z., Wang, F., Xu, Y.: Pre-training enhanced spatial-temporal graph neural network for multivariate time series forecasting. In: Proceedings of the 28th ACM SIGKDD Conference on Knowledge Discovery and Data Mining, pp. 1567–1577 (2022)
22. Su, B., Wen, J.R.: Temporal alignment prediction for supervised representation learning and few-shot sequence classification. In: International Conference on Learning Representations (2022)
23. Vaswani, A., et al.: Attention is all you need. In: Advances in Neural Information Processing Systems, vol. 30 (2017)
24. Wickstrøm, K., Kampffmeyer, M., Mikalsen, K.Ø., Jenssen, R.: Mixing up contrastive learning: self-supervised representation learning for time series. Pattern Recogn. Lett. **155**, 54–61 (2022)
25. Yue, Z., et al.: Ts2vec: towards universal representation of time series. In: Proceedings of the AAAI Conference on Artificial Intelligence, vol. 36, pp. 8980–8987 (2022)
26. Zerveas, G., Jayaraman, S., Patel, D., Bhamidipaty, A., Eickhoff, C.: A transformer-based framework for multivariate time series representation learning. In: Proceedings of the 27th ACM SIGKDD Conference on Knowledge Discovery & Data Mining, pp. 2114–2124 (2021)
27. Zhang, X., Zeman, M., Tsiligkaridis, T., Zitnik, M.: Graph-guided network for irregularly sampled multivariate time series. arXiv preprint arXiv:2110.05357 (2021)
28. Zhang, X., Zhao, Z., Tsiligkaridis, T., Zitnik, M.: Self-supervised contrastive pre-training for time series via time-frequency consistency. arXiv preprint arXiv:2206.08496 (2022)

An Efficient Feature Selection Method for High Dimensional Data Based on Improved BOA in AIoT

Weifeng Sun(✉), Hao Xu, Bo Liu, and Bowei Zhang

School of Software, Dalian University of Technology, Dalian 116620, China
wfsun@dlut.edu.cn, 1074145767@qq.com

Abstract. Feature selection can eliminate irrelevant redundant information in high-dimensional data, reduce dimensions, improve classification performance, and relieve the computational burden of machine learning. In this paper, we propose a feature selection method based on improved binary butterfly op-timization algorithm (BCSBOA). BCSBOA uses Tent chaotic mapping to improve the population initialization method, which makes the distribution of population individuals more uniform and improves population diversity. The mutation perturbation strategy is adopted to improve the ability of the algorithm to escape from local optimum. Through the adaptive weight strategy, the optimization ability of the global and local search stages is further balanced. In addition, BCSBOA maps the solution to the binary space through the Sigmoid function, proposes a new fitness function, jointly optimizes the classification accuracy and the number of features, and finally improves the feature selection effect. Simulation results of the algorithm performance over 10 UCI data sets show that, the classification accuracy and data-feature subset size have been improved. These findings also support the feasibility of our algorithm in improving machine learning performance.

Keywords: Butterfly Optimization Algorithm · Artificial Intelligence of Things · Chaotic Mapping · Mutation Perturbation Strategy

1 Introduction

With the advent of the era of big data and the popularization of AIoT, various database systems are widely used, and a large amount of data is generated every day in the fields of industry, medical care, finance, and education. The continuous increase of data scale will lead to a great reduction in the efficiency of various models and software. However, large data sets are flooded with a large amount of irrelevant data, which has a negative impact on the efficiency and accuracy of machine learning, the phenomenon also known as the curse of dimensionality. In order to solve the problem of information redundancy in high-dimensional data sets in industrial applications, feature selection has gradually become a research hotspot. As an important link in data mining, it can eliminate redundant data, reduce dimensions, and play an important role in improving mining capabilities and reducing computing burdens [1].

X. Yang et al. (Eds.): ADMA 2023, LNAI 14180, pp. 293–308, 2023.
https://doi.org/10.1007/978-3-031-46677-9_21

Feature selection is essentially to filter out the optimal feature subset from a high-dimensional data set. For each feature, there are only two states: selected and unselected. Researchers transformed the feature selection problem into a binary solution space optimization problem, and explored better feature selection effects through the combination of feature selection methods and swarm intelligence. When the RSO algorithm is applied to feature selection, the convergence speed of the algorithm is improved to a certain extent, and the feature dimension is reduced [2]. However, the test data set used in the experiment is very small, which cannot explain the universality of the algorithm. Through the combination with algorithms such as PSO [3], SSA [4] and CSO [5], the classification accuracy, F1, accuracy and other performance indicators of feature selection can be improved to a certain extent, but the performance improvement of swarm intelligence is not enough. Big. In addition, Rim et al. proposed a random walk binary gray wolf optimization algorithm to improve searchability, improve prediction accuracy, and reduce computational complexity by randomly walking in the search space [6]. Ivana et al. achieved feature selection for the COVID-19 dataset by mixing the ant lion algorithm and the Firefly algorithm [7]. In the above researches, the improvement of swarm intelligence algorithm itself is less, the importance of initial population to the optimization results of the algorithm is not taken into account, the search performance of the algorithm and the ability to escape from the local optimal is not improved, resulting in a large difference between the iterative results and the expected. In addition, the above research work only uses fixed indicators to evaluate the feature selection results, and does not construct a suitable fitness function. Therefore, it is necessary to continue to study swarm intelligence that are more suitable for feature selection, and further improve the effect of feature selection on different types of data sets.

This paper improves the butterfly optimization algorithm and proposes the BCSBOA for feature selection. The main contributions are divided into two aspects:

1. Overcome the shortcomings of BOA itself and improve the ability to solve the optimal value. Aiming at the problem that the initial population is too random and the individual distribution is sparse, the Tent chaotic map is introduced to enrich the diversity of the population. The adaptive weight strategy is introduced to balance the global and local search capabilities, and the mutation disturbance strategy is introduced to prevent the algorithm from falling into local optimum.
2. Convert the feature selection into a binary solution space problem, continue to introduce the Sigmoid function to discretize the position information on the basis of improvement, construct a fitness function with a weight coefficient, improve the classification accuracy, reduce the dimension of the feature subset, and obtain better feature selection results.

2 Related Work

Feature selection research based on swarm intelligence is a research direction that has developed rapidly in the field of data mining in recent years, which helps to solve practical problems in financial risk control, medical diagnosis, AIoT and other fields.

More and more research work has begun to combine different swarm intelligence to improve the accuracy and precision of feature selection and classification. Lee et al.

proposed a bearing fault diagnosis model based on machine learning, combined with GWO and HBO for feature selection, using MRA and FFT extracts features from raw signals measured by rotating machinery, using SVM and LDA as classifiers, which effectively improves the efficiency of bearing fault diagnosis [8]. Nath et al. designed a method to select a subset of features from a pool of physiological signals based on the QA algorithm [9]. Embed the extracted feature variables into the binary model, use the Pearson correlation coefficient to calculate the deviation of the feature variables, resample the underlying solution, and return the clique with the lowest energy as the optimal solution. Zeggari et al. proposed a feature selection technique for logo recognition for a large number of images in a trademark registration system [10]. Based on a genetic algorithm and feature weighting process, the method combines different types of features and applies them to the evolution process of the four markers. Kumar et al. proposed a hybrid feature selection method based on particle swarm optimization and image classification, and compared with supervised and unsupervised feature selection techniques, the image classification results obtained are even better [11].

In terms of swarm intelligence improvement, Huang et al. proposed a multi-objective particle swarm feature selection algorithm based on a cyclic penalty factor, using the PBI decomposition method to cyclically correct the penalty in the region, provide directional guidance for the algorithm, and balance convergence and distribution [12]. Wang et al. proposed a GA-based feature selection algorithm, using population information entropy to measure population diversity, and adding a consensus mechanism after genetic mutation, and using it as an alternative operator to speed up the convergence speed [13]. Ewees et al. proposed a locust optimization algorithm fused with a crossover operator to solve feature selection [14]. The crossover operator is used to increase the population of SSA and use it as a local search method to improve the solution effect of GOA. Apriyadi et al. mainly studied the feature selection technology based on particle swarm and genetic algorithm to predict students' learning performance, used SVR to predict student performance modeling, and used 10-fold crossover method to evaluate the student performance prediction model [15].

In addition to the above research work, some novel swarm intelligence has also been applied to the feature selection problem. For the SSO algorithm, Slezkin et al. proposed a hybrid method based on algebraic operations and MERGE operations to improve, using Friedman's rank variance two-way analysis to evaluate the binarization method, and using fuzzy rules and KNN classifiers to evaluate Feature subset effects [16]. Jameel et al. proposed a feature selection method based on the mongoose algorithm, and the MAD was introduced into the fitness function for evaluation [17]. Experimental results also show that the algorithm has higher feature selection accuracy.

In order to improve the existing problems in the standard BOA, this paper proposes the BCSBOA and applies it to feature selection.

3 Improved Butterfly Optimization Algorithm

3.1 Problem Description

The BOA is inspired by the foraging behavior of butterflies. The core idea is to update and search the position according to the fitness value of individual butterflies. All individuals in the population can release fragrances of different concentrations and be felt by other individuals. The fragrance formula of an individual is as follows:

$$f = cI^{\alpha} \tag{1}$$

f is the individual fragrance, c represents the perception mode, I represents the stimulus intensity, and α is the power exponent with a value ranging from 0 to 1.

The BOA mainly includes two key processes, the global search stage and the local search stage. In each iteration, a random number r from 0 to 1 is generated, and compared with the switch probability ρ. If $r < p$, the global search is performed, otherwise a local search is performed. The specific position update formula is as follows:

$$x_i^{t+1} = \begin{cases} x_i^t + \left(r^2 * g^* - x_i^t\right) * f_i, & r < p \\ x_i^t + \left(r^2 * x_k^t - x_j^t\right) * f_i, & r \geq p \end{cases} \tag{2}$$

Among them, x_i^t represents the solution vector of the i-th butterfly in the t-th iteration, g^* represents the optimal solution, f_i represents the fragrance of the i-th butterfly, x_k^t and x_j^t denote the kth and jth butterflies randomly selected from the solution space.

The standard BOA mainly has the following three problems.

P1: The algorithm uses random vectors and random number generators, which have certain randomness, which will lead to uneven distribution of individuals in the search area and poor solutions.
P2: During the global search, the current individual moves to the global optimal position, only relying on the information of the current butterfly x_i^t and the global optimal individual g^*, once the current individual feels a stronger fragrance, it will quickly move to g^* close, which will make the algorithm fall into local extremum faster. In the local search stage, $r^2 * x_k^t - x_j^t$ represents the distance between two random solution vectors, the search range will stay around the individual, and it is easy to miss the better solution or the optimal solution.
P3: The global search and local search of the algorithm are only controlled by the parameter p, and the search cannot be dynamically adjusted during the population iteration process, which will easily cause the algorithm to spend too much time at a certain stage and reduce the convergence speed of the algorithm.

3.2 Proposed Method (BCSBOA)

In order to solve the above problems, this paper proposes an improved butterfly optimization algorithm based on chaotic maps and search strategies, namely BCSBOA.

A good initial population will affect the process of the algorithm to find the global optimum, accelerate the convergence of the population, and improve the accuracy of the final solution. Since the initialization of the population depends on random factors, the distribution of individuals is unstable. This will cause the algorithm to fall into local optimum prematurely and deviate from the optimal solution. In order to avoid the interference of random factors on the stability of the algorithm, a chaotic map is added in the population initialization stage to replace the pseudo-random number generator and map the generated chaotic sequence to the search space.

In order to solve P1, BCSBOA uses Tent mapping to increase population diversity and improve algorithm solution accuracy. Compared with Logistic, Sine, Henon and other methods, Tent mapping is more controllable, the chaotic sequence is more representative, and the initial population generated based on Tent is more uniform. The mathematical description of Tent mapping is as follows:

$$x_{n+1} = f(x_n) = \begin{cases} x_n/\alpha, & x_n \in [0, \alpha) \\ (1 - x_n)/(1 - \alpha), & x_n \in [\alpha, 1] \end{cases} \tag{3}$$

The value of α ranges from 0 to 1. When x_n and α are the same, it will evolve into a periodic system and will no longer have the characteristics of chaotic maps.

Based on the above initial population, the algorithm needs to find the optimal individual through global and local searches in the iterative process. The BOA is easy to fall into local optimum, because the leader butterfly does not play a leading role in finding the optimal value. In order to allow the individual to jump out of the current area during the optimization process and check whether there are better solutions in other positions, it is necessary to update the individual to a new area, so as to achieve a better global optimization effect. In order to solve P2, BCSBOA introduces the Cauchy mutation perturbation strategy in the search phase, and its mathematical formula is as follows:

$$F_{cauchy}(x) = \frac{b}{\pi(x-a)^2+b^2} \tag{4}$$

Based on the above initial population, the algorithm continuously searches for the optimal individual in the population through global and local searches in the iterative process. In order to solve P2 and prevent the butterfly optimization algorithm from falling into local optimum and deviating from the optimal value during the search process, BCSBOA introduces the Cauchy mutation strategy. In the process of finding the optimal value, update the individual to a new position, jump out of the previous area, and check whether there is a better solution in other positions, so as to achieve a better global optimization effect. Its mathematical formula is as follows:

$$P_{newbest} = P_{best} + F_{cauchy}(P_{best}) \tag{5}$$

$P_{newbest}$ represents the new optimal individual position, and P_{best} represents the particle position of the old optimal individual position.

During the whole search process, the global search and local search will be switched. In order to solve P3, balancing the ability of the algorithm in global search and local refinement, BCSBOA introduced adaptive strategies to improve the way of position update. In the early search phase, the adaptive weight value is large, which is used to enhance the global search ability of the algorithm in the early stage, so that individuals can search each area of the population in a more comprehensive way. When the individual is close to the local optimal position, the weight value gradually decreases to ensure that the algorithm can perform a more detailed search. In this way, the search efficiency and precision are improved, the population convergence speed is accelerated, and the ability of global exploration and local refinement is further balanced. The calculation method of the adaptive weight is as follows:

$$\omega = \omega_{max} - \frac{(\omega_{max} - \omega_{min})}{iter_{max}} * iter * exp_{factor} \tag{6}$$

Among them, ω is the specific weight coefficient value, ω_{max} and ω_{min} represent the maximum and minimum weight respectively, $iter_{max}$ is the maximum number of iterations, $iter$ is the current number of iterations, and exp_{factor} is used to control the change slope of the weight coefficient. When exp_{factor} is 1, the weight coefficient curve becomes a straight line. Based on the improved weight coefficient, the global position update formula is as follows:

$$x_i^{t+1} = \omega * x_i^t + \left(r^2 * g^* - x_i^t\right) * f_i \tag{7}$$

The improved local position update formula is as follows:

$$x_i^{t+1} = \omega * x_i^t + \left(r^2 * x_k^t - x_j^t\right) * f_i \tag{8}$$

In order to convert the feature selection problem into a binary solution space optimization problem, BCSBOA uses the Sigmoid discretization method to convert the continuous values in the butterfly position vector into discrete values and map them to the two-dimensional solution space. The Sigmoid conversion function is as follows:

$$S(x_i) = \frac{1}{1+e^{-x_i}} \tag{9}$$

Assuming that the position of a butterfly individual in the population before conversion is $X_i = \{x_1, x_2, \ldots, x_n,\}$, the corresponding binary vector is $X_{i'} = \left\{x_{binary}^1, x_{binary}^2, \ldots, x_{binary}^n,\right\}$, the value of x_{binary}^i is as follows, and $rand$ is the random function value, which is used to adjust the specific binary value of x_{binary}^i.

$$x_{binary}^i = \begin{cases} 1, rand \le S(x_i) \\ 0, rand > S(x_i) \end{cases} \tag{10}$$

The fitness function of feature selection is mainly used to measure the importance of each feature and evaluate the pros and cons of feature subset. There are two main goals of feature selection, one is to improve the accuracy of classification, the other is to reduce the data dimension. In order to balance these two objectives, the fitness function adopted in BCSBOA is as follows:

$$Fitness = \alpha * (1 - Accuary) + \beta * \frac{|S|}{N} \tag{11}$$

Fitness is the fitness value, and *Accuary* is the classification accuracy. N represents the feature dimension of the data set, and $|S|$ represents the feature subset dimension. α

and β are the weight coefficients of classification accuracy and feature subset dimension respectively, and satisfy $\alpha+\beta=1$. The constraints of the fitness function are as follows:

$$\begin{aligned} & min \qquad Fitness \\ & \text{s.t. } C_1 : x_i \in \{0,1\}, i = 1,2,\ldots,N \\ & C_2 : 1 < |S| < N \\ & C_3 : 0 \leq \text{ Accuracy } \leq 1 \end{aligned} \tag{12}$$

C_1 is the constraint on the value range of x_i being 0 or 1, C_2 is the constraint on the number of feature selections, and C_3 is the constraint on the classification accuracy.

The fitness function of BCSBOA will not only improve the classification accuracy, but also reduce the complex problem of subset dimension, and transform it into a single-objective optimization function. It can also flexibly adjust the weight size to adapt to different feature selection goals.

4 Evaluation

4.1 Test Functions and Data Sets

Table 1. Test functions.

Function		Feature	Domain	Best
F_1	Sphere	unimodal	[−100,100]	0
F_2	Schwefel 1.2	unimodal	[−100,100]	0
F_3	Schwefel 2.21	unimodal	[−100,100]	0
F_4	Schwefel 2.22	unimodal	[−10,10]	0
F_5	Sum of different power	unimodal	[−10,10]	0
F_6	Sum square	unimodal	[−5.12,5.12]	0
F_7	Rastrigin	multimodal	[−5.12,5.12]	0
F_8	Ackley	multimodal	[−32,32]	0
F_9	Griewank	multimodal	[−600,600]	0
F_{10}	Schaffer	multimodal	[−100,100]	0
F_{11}	Apline	multimodal	[−10,10]	0
F_{12}	Salomon	multimodal	[−100,100]	0

This paper verifies the optimization ability of BCSBOA through 12 test functions shown in Table 1. $F_1 \sim F_6$ are unimodal functions, which are used to test the refinement ability and convergence speed of the algorithm. $F_7 \sim F_{12}$ are multimodal functions, which are used to test the ability of the algorithm to escape from local optimum.

In addition, in order to verify the feature selection effect of the algorithm, 10 UCI datasets from different application fields, such as medicine, finance, chemistry, and politics, with different numbers of features and different numbers of samples were selected. The specific experimental data sets are shown in Table 2.

Table 2. Experimental data sets.

Number	Data set	Number of features	Number of samples	Field
D1	Breastcancer	9	699	Medical
D2	Congress	16	435	Politics
D3	Heart-StatLog	13	270	Medical
D4	Hepatitis	19	142	Medical
D5	German	24	1000	Finance
D6	Lymphography	18	148	Medical
D7	Tic-tac-toe	9	958	Entertainment
D8	Spect	22	267	Medical
D9	Vote	16	435	Socialize
D10	Wine	13	178	Chemistry

4.2 Experimental Design

In order to verify the effectiveness of the improved strategy, the BCSBOA is compared with the standard BOA on 12 test functions. In order to verify the feature selection effect, the BCSBOA is compared with BOA, BPSO [18], BGWO [19] and BGA [20]. The population size of all algorithms is 30, the maximum number of iterations is 500, each data set is run independently 30 times, and the final classification accuracy and feature subset dimension are recorded. Choose KNN classifier, where K = 5. The algorithm parameters that need to be set separately are shown in Table 3. The processor of the experimental equipment is Intel(R) Core (TM) i5-8250U, the main frequency is 1.6 GHZ, the RAM is 16.00 GB, and the operating system is 64-bit Windows10.

Table 3. The main parameters of each algorithm.

Algorithm	Main parameters
BCSBOA	$p = 0.8, c = 0.01, a = 0.1, \alpha = 0.99, \beta = 0.01$
BPSO	$C_1 = 2.0, C_2 = 2.0$
BGWO	$a, A, C \in [0, 1]$
BGA	$Mutation\ rate = 0.1, Crossover\ rate = 0.5$
BOA	$p = 0.8, c = 0.01, a = 0.1$

4.3 Experimental Results

The data in the table of experimental results are in the form of Scientific notation, for example, ‘2.32E-02’ is 0.0232, and ‘0.00E + 00’ is 0.

Table 4 shows the optimization results of the two algorithms on the test functions. It can be seen that the BCSBOA can correctly solve the theoretical optimal value on all functions, while the BOA only has stable performance on the unimodal function. The main reason is that the multimodal function has a higher dimension and is more difficult to solve, so the performance on the multimodal function will be reduced.

Table 4. Experimental results on 12 test functions.

Function	Algorithm	Best	Worst	Mean	Sd
F_1	BCSBOA	**0.00E + 00**	**0.00E + 00**	**0.00E + 00**	**0.00E + 00**
	BOA	1.74E-13	1.98E-12	1.76E-12	1.53E-13
F_2	BCSBOA	**0.00E + 00**	**0.00E + 00**	**0.00E + 00**	**0.00E + 00**
	BOA	1.97E-12	2.41E-09	1.33E-10	4.29E-10
F_3	BCSBOA	**0.00E + 00**	**0.00E + 00**	**0.00E + 00**	**0.00E + 00**
	BOA	2.86E-17	7.44E-15	1.98E-15	5.39E-16
F_4	BCSBOA	**0.00E + 00**	**0.00E + 00**	**0.00E + 00**	**0.00E + 00**
	BOA	2.79E-11	3.74E + 08	5.40E + 03	3.64E + 05
F_5	BCSBOA	**0.00E + 00**	**0.00E + 00**	**0.00E + 00**	**0.00E + 00**
	BOA	3.46E-12	1.82E-10	6.23E-12	2.62E-11
F_6	BCSBOA	**0.00E + 00**	**0.00E + 00**	**0.00E + 00**	**0.00E + 00**
	BOA	1.77E-11	2.45E-11	1.96E-11	7.82E-14
F_7	BCSBOA	**0.00E + 00**	**0.00E + 00**	**0.00E + 00**	**0.00E + 00**
	BOA	3.51E-16	5.28E + 01	1.07E + 01	3.31E + 01
F_8	BCSBOA	**0.00E + 00**	**0.00E + 00**	**0.00E + 00**	**0.00E + 00**
	BOA	3.51E-08	1.24E-06	8.69E-08	3.26E-07
F_9	BCSBOA	**0.00E + 00**	**0.00E + 00**	**0.00E + 00**	**0.00E + 00**
	BOA	2.37E-10	8.23E-10	4.28E-10	7.02E-10
F_{10}	BCSBOA	**0.00E + 00**	**0.00E + 00**	**0.00E + 00**	**0.00E + 00**
	BOA	3.61E-07	2.47E-03	1.39E-04	4.32E-03
F_{11}	BCSBOA	**0.00E + 00**	**0.00E + 00**	**0.00E + 00**	**0.00E + 00**
	BOA	2.88E-09	8.13E-09	6.42E-09	5.61E-09
F_{12}	BCSBOA	**0.00E + 00**	**1.29E-105**	**5.56E-107**	**4.29E-106**
	BOA	1.52E-12	2.32E-02	4.51E-03	3.39E-02

Table 5 shows the classification accuracy comparison results of 5 algorithms including BCSBOA, BPSO, and BGWO after 30 independent experiments on 10 data sets, including the best, worst, mean and standard deviation of classification accuracy. The optimal value of each experimental index is highlighted by bold.

It can be seen from Table 5 that the BCSBOA can obtain the optimal classification accuracy rate on 8 data sets such as Breastcancer and Congress, and can obtain the best average value on 8 data sets such as Heart-StatLog and Hepatitis. On the poor side, the results of the BCSBOA on all data sets are relatively small, and it always maintains a relatively stable classification accuracy range.

Table 6 shows the feature subset dimension results of each algorithm after 30 independent experiments on the data set, which mainly records the optimal value, worst value, average value and standard deviation of this indicator. It can be seen from Table 6 that BCSBOA can obtain the feature subset with the smallest dimension on eight datasets including Congress, Breastcancer, and Wine, and can obtain the best average feature subset dimension on eight datasets including Lymphography.

4.4 Analysis

It can be seen from Table 4 that BCSBOA can find the theoretical optimal value on the complex functions of each dimension, correctly converge to the value 0, and the convergence effect is better than that of BOA. This is mainly because the BCSBOA introduces the search range and granularity of the inertial weight adaptive adjustment algorithm in different evolutionary periods, which makes the algorithm expand the search range in the early stage, speed up the convergence speed, and carry out fine mining in the later stage to improve the optimization accuracy. The Cauchy mutation perturbation strategy enhances the jumping of butterfly individuals in the entire search space. When the individual X_i is the current local optimal g^*, the position is updated through the perturbation formula to generate new individual information, which is beneficial to the algorithm in high-dimensional situations. Jump out of the local extremum, get rid of the function peak trap, and better find the global optimal solution. When performing local walks, the disturbance factor plays a role in maintaining the diversity of the butterfly population. This diversity makes it easier for the algorithm to jump out of the current optimum, making the exploration of individual butterflies more directional, thereby speeding up the convergence speed of the algorithm. It is able to find the theoretical optimum within few evolutionary generations.

On the Salomon function, BCSBOA can find the theoretical optimal value 0, but the average value of the iterative process is 5.56E-107, and the standard deviation is 4.29E-106, indicating that BCSBOA is not stable enough on this function, mainly because of the Salomon function There are many local peaks, and the algorithm may make mistakes when continuously getting rid of the local optimum, causing the final solution to deviate from the target value. As the dimension of the test function increases, the difference between the optimal value obtained by the BOA and the theoretical optimal value becomes larger and larger, and the performance in terms of stability and running time also becomes worse. The performance of BCSBOA with Tent mapping is obviously better than that of BOA in high-dimensional functions. This shows that the Tent map, as a two-dimensional map, can better map the population individual sequence to the search space and speed up the search compared to the random initial method. The final convergence result of the BCSBOA also reflects the significant effect of the adaptive weight strategy and the mutation perturbation strategy on improving the global and local search capabilities of the algorithm.

Table 5. Classification accuracy on data sets(%).

Data set	Algorithm	Best	Worst	Mean	Sd
Breastcancer	BCSBOA	**97.36**	**96.52**	**97.12**	0.09
	BPSO	95.38	94.21	95.15	0.11
	BGWO	96.22	95.41	95.72	**0.07**
	BGA	97.17	95.22	96.74	0.15
	BOA	95.11	94.05	94.27	0.11
Congress	BCSBOA	**98.11**	**97.42**	**97.73**	0.12
	BPSO	97.56	97.13	97.39	**0.08**
	BGWO	97.33	95.21	96.48	0.83
	BGA	97.23	96.52	96.98	0.14
	BOA	96.91	96.15	96.23	0.14
Heart-StatLog	BCSBOA	**94.29**	**90.37**	**91.84**	1.59
	BPSO	86.93	85.41	85.81	0.17
	BGWO	89.61	84.39	88.62	0.46
	BGA	85.33	83.27	84.88	**0.09**
	BOA	85.77	82.14	84.09	2.73
Hepatitis	BCSBOA	**82.33**	**77.83**	**79.44**	2.16
	BPSO	76.85	74.21	75.16	1.28
	BGWO	81.03	74.87	75.23	1.96
	BGA	75.42	73.39	73.94	**0.87**
	BOA	76.29	74.05	74.91	1.44
German	BCSBOA	**80.14**	76.39	78.87	1.33
	BPSO	79.11	**78.89**	**79.35**	**0.34**
	BGWO	77.92	76.78	77.24	0.67
	BGA	78.45	76.47	78.25	0.46
	BOA	77.53	76.24	76.71	0.83
Lymphography	BCSBOA	92.15	86.69	87.62	1.47
	BPSO	91.74	**90.22**	**90.83**	**0.29**
	BGWO	87.46	81.21	82.99	5.12
	BGA	**92.41**	89.64	90.75	0.98
	BOA	89.69	82.31	84.27	3.26
Tic-tac-toe	BCSBOA	**87.62**	**86.73**	**86.81**	0.73
	BPSO	85.34	85.12	85.28	**0.11**

(continued)

Table 5. (*continued*)

Data set	Algorithm	Best	Worst	Mean	Sd
	BGWO	86.17	85.44	85.79	0.42
	BGA	85.21	84.73	85.02	0.22
	BOA	84.63	81.68	83.17	0.94
Spect	BCSBOA	**81.59**	75.53	76.16	1.52
	BPSO	81.42	**76.57**	77.78	1.63
	BGWO	79.25	76.42	**77.92**	**1.04**
	BGA	78.56	74.91	76.03	1.29
	BOA	78.88	74.04	75.72	1.71
Vote	BCSBOA	**96.62**	**95.89**	**96.43**	**0.46**
	BPSO	95.34	94.52	94.71	0.53
	BGWO	94.93	92.11	94.43	0.82
	BGA	95.33	94.11	94.67	**0.46**
	BOA	93.76	91.58	92.95	0.91
Wine	BCSBOA	99.41	**99.17**	**99.26**	**0.04**
	BPSO	98.84	98.31	98.56	0.08
	BGWO	98.31	97.14	98.08	0.67
	BGA	**99.44**	98.86	99.13	0.22
	BOA	98.73	97.42	97.95	0.59

The BCSBOA better solves the shortcomings of the BOA in solving the extreme value optimization problem of high-dimensional complex functions, such as low optimization accuracy and easy to fall into local extremum. The improvement of the BOA itself is effective.

This paper mainly evaluates the effect of feature selection through two indicators: classification accuracy and feature subset dimension. From the experimental results in Table 5 and Table 6, it can be found that compared with the comparison algorithm, the classification accuracy obtained by the BCSBOA is significantly higher, and the dimension of the feature subset is also significantly reduced. Feature selection on the dataset works better. When the dimension exceeds 30, the parameters in the BCSBOA can be adjusted appropriately for a specific data set to obtain better results. On the German data set, the feature subset dimension of the BCSBOA is 10, which is lower than that of BPSO, mainly because the number of samples in the original data set is large, which leads to poor iterative search performance of the algorithm, while BPSO can use the collaborative search method, which is better Guide each round of search, so that the dimension of the final feature subset is relatively better. The average value of BCSBOA on Heart-StatLog, Vote and other data sets is the highest, but the performance of the two aspects of the optimal dimension and the worst dimension is relatively general,

Table 6. Dimension of feature subset on data sets.

Data set	Algorithm	Best	Worst	Mean	Sd
Breastcancer	BCSBOA	**4**	**6**	**4.73**	1.02
	BPSO	6	**6**	6.00	**0.00**
	BGWO	6	8	7.13	2.24
	BGA	**4**	7	5.44	0.79
	BOA	**4**	8	5.98	1.64
Congress	BCSBOA	**4**	**8**	**5.18**	1.22
	BPSO	5	**8**	5.89	**0.69**
	BGWO	6	9	7.21	0.83
	BGA	**4**	**8**	6.64	1.43
	BOA	**4**	10	7.43	2.27
Heart-StatLog	BCSBOA	**4**	7	**4.79**	0.81
	BPSO	5	**6**	5.32	**0.53**
	BGWO	**4**	8	5.69	0.88
	BGA	**4**	8	5.17	1.04
	BOA	5	9	5.91	1.94
Hepatitis	BCSBOA	**5**	**11**	**6.90**	**0.97**
	BPSO	**5**	15	7.31	1.88
	BGWO	8	16	9.14	1.73
	BGA	6	13	8.75	2.91
	BOA	7	15	9.82	2.74
German	BCSBOA	10	17	11.73	2.18
	BPSO	**7**	**15**	**10.64**	1.62
	BGWO	12	19	14.92	2.41
	BGA	12	17	13.75	**1.52**
	BOA	11	17	13.38	2.26
Lymphography	BCSBOA	**4**	**7**	**5.91**	**0.92**
	BPSO	**4**	8	6.72	1.15
	BGWO	**4**	8	6.43	1.03
	BGA	**4**	11	6.96	1.76
	BOA	6	10	7.25	1.44
Tic-tac-toe	BCSBOA	**6**	**6**	**6.00**	**0.00**
	BPSO	7	9	8.41	0.56

(continued)

Table 6. (*continued*)

Data set	Algorithm	Best	Worst	Mean	Sd
	BGWO	**6**	8	7.19	0.33
	BGA	9	9	9.00	**0.00**
	BOA	7	9	7.53	0.49
Spect	BCSBOA	6	13	8.57	1.92
	BPSO	5	11	7.21	1.53
	BGWO	5	18	9.15	3.68
	BGA	8	14	10.37	1.39
	BOA	8	13	10.04	2.41
Vote	BCSBOA	3	9	4.91	1.58
	BPSO	5	7	5.53	0.67
	BGWO	5	12	7.07	1.69
	BGA	3	11	6.14	2.64
	BOA	6	11	7.15	1.83
Wine	BCSBOA	3	8	6.43	1.79
	BPSO	5	8	7.53	0.96
	BGWO	4	11	8.17	1.84
	BGA	3	10	7.56	2.04
	BOA	4	10	8.24	2.56

mainly because the weight β in the fitness function in the experiment is set smaller, the performance in terms of the number of subset features will be slightly weaker, and a more ideal result can be obtained by adjusting the size of the two weight coefficients.

Comparing and analyzing the experimental results of the BCSBOA and the BOA separately, it can be clearly seen that in all the experimental data sets in this paper, the classification accuracy and feature subset dimension of BCSBOA are significantly better than the standard BOA. The main reason is that the improved initial population based on the Tent chaotic map has laid a good population foundation for the subsequent optimal search, accelerated algorithm convergence, adaptive strategy and mutation perturbation strategy, and further improved the optimization performance of the algorithm. to avoid falling into local optimum. The comparison results fully verify the effectiveness of the improved strategy for the standard BOA.

Due to the introduction of multiple improved methods, the running time of the BCSBOA has increased compared with BOA, but the overall complexity of the algorithm has not changed. Under the premise that the population size is N and the dimension is n, the time complexity of the two algorithms is $O(n + f(n))$. With the maturity of cloud computing technology and the continuous increase of computing power, it is still feasible to spend a small amount of computing power to improve the overall performance of the algorithm.

To sum up, the BCSBOA can improve the classification accuracy, reduce the feature dimension, and have good algorithm stability when solving the feature selection problem. Compared with other traditional methods, it can achieve better feature selection effect.

5 Conclusion

Feature selection is an effective measure against the curse of dimensionality. It can reduce the use of features in the modeling process, thereby reducing the overall training time, effectively preventing overfitting, and improving the generalization ability of the model. Aiming at the feature information screening problem, this paper proposes a feature selection method based on the improved BOA(BCSBOA). The initial population is optimized through the Tent chaotic map, the adaptive weight strategy and the mutation perturbation strategy are introduced to improve the optimization ability of the algorithm, and the Sigmoid function is used to convert the feature selection into a binary space solution problem The BCSBOA effectively avoids the problem of classification performance degradation due to the extra consumption of a large amount of computing resources.The experimental results on the UCI test data sets show that compared with BOA, BPSO, BGWO and other algorithms, BCSBOA can effectively improve the classification accuracy, reduce the dimension of feature subsets, and obtain better feature selection results.

The BCSBOA still has shortcomings in terms of operating efficiency. In the future, it will continue to optimize the initial population through methods such as reverse learning and Levy flight. By comparing the distribution of the population, an improved method with better effect will be selected. At the same time, it is considered to introduce the Sine Cosin Algorithm in the search process to reduce the optimization error and improve the quality of the final solution.

Acknowledgement. This work is supported by National Key R\&D Program of China (2018YFB1700100), CERNET Innovation Project (NGII20190801) and the Fundamental Research Funds for the Central Universities under Grants (DUT21LAB115).

References

1. Khaire, U.M., Dhanalakshmi, R.: Stability of feature selection algorithm: a review. J. King Saud Univ.-Comput. Inf. Sci. **34**(4), 1060–1073 (2022)
2. Rahab, H., et al.: A modified binary rat swarm optimization algorithm for feature selection in Arabic sentiment analysis. Arab. J. Sci. Eng. **48**, 1–28 (2022). https://doi.org/10.1007/s13369-022-07466-1
3. Liu, Yi, et al.: An Interpretable Feature Selection Based on Particle Swarm Optimization. IEICE TRANSACTIONS on Information and Systems. **105**(8), 1495–1500 (2022)
4. Zivkovic, M., et al.: Novel improved salp swarm algorithm: an application for feature selection. Sensors **22**(5), 1711 (2022)

5. Hamdi, M., et al.: Chicken swarm-based feature subset selection with optimal machine learning enabled data mining approach. Appl. Sci. **12**(13), 6787 (2022)
6. Barioul, R., et al.: Random walk binary grey wolf optimization for feature selection in sEMG based hand gesture recognition. In: 2022 IEEE 9th International Conference on Computational Intelligence and Virtual Environments for Measurement Systems and Applications (CIVEMSA), pp. 1–6 (2022)
7. Strumberger, I., et al.: Feature selection by hybrid binary ant lion optimizer with covid-19 dataset. In: 2021 29th Telecommunications Forum (TELFOR), pp. 1–4 (2021)
8. Lee, C-Y., Truong-An, L., Yu-Ting, L.: A feature selection approach hybrid grey wolf and heap-based optimizer applied in bearing fault diagnosis. IEEE Access **10**, 56691–56705 (2022)
9. Nath, R.K., Himanshu, T., Travis S.H.: Quantum annealing for automated feature selection in stress detection. In: 2021 IEEE Computer Society Annual Symposium on VLSI (ISVLSI), pp. 453–457 (2021)
10. Zeggari, A., Zianou, A.S., Fella, H.: Re-weighting features selection based on wrapper filter and genetic algorithms for figurative images recognition. In: 2021 5th International Conference on Pattern Recognition and Image Analysis (IPRIA), pp. 1–6 (2021)
11. Siddamallappa, K.U., Nisarg, G.: Design feature selection and classifiers for hybrid feature selection-particle swarm optimization (HFS-PSO). In: 2022 4th International Conference on Inventive Research in Computing Applications (ICIRCA), pp. 822–826 (2022)
12. Huang, G., Fei, H., Qing-Hua, L.: An improved feature selection algorithm with cyclic penalty boundary interaction based on MOPSO. In: 2022 5th International Conference on Pattern Recognition and Artificial Intelligence (PRAI), pp. 164–168 (2022)
13. Wang, Jiachen.: An Improved Genetic Algorithm for Web Phishing Detection Feature Selection. In: 2022 Asia Conference on Algorithms, Computing and Machine Learning (CACML), pp. 130–134 (2022)
14. Ewees, A.A., Gaheen, M.A., Yaseen, Z.M., et al.: Grasshopper optimization algorithm with crossover operators for feature selection and solving engineering problems. IEEE Access **10**(6), 23304–23320 (2022)
15. Apriyadi, M.R., Dian, P.R.: Implementation of feature selection based on particle swarm optimization and genetic algorithm on support vector regression algorithm to predict student performance. In: 2022 International Conference on Informatics, Multimedia, Cyber and Information System (ICIMCIS), pp. 395–400 (2022)
16. Slezkin, A.O., Ilya A.H., Alexander, A.S.: Binarization of the Swallow swarm optimization for feature selection. Program. Comput. Softw. **47**, 374–388 (2021)
17. Jameel, N., Hasanen, S.A.: A proposed intelligent features selection method using meerkat clan algorithm. J. Phys. Conf. Ser. **1804**(1), 012061 (2021)
18. Huang, C.L., Wang, C.J.: A GA-based feature selection and parameters optimizationfor support vector machines. Expert Syst. Appl. **31**(2), 231–240 (2006)
19. Dara, S., Banka, H.: A binary PSO feature selection algorithm for gene expression data. In: International Conference on Advances in Communication and Computing Technologies (ICACACT 2014), pp. 1–6 (2014)
20. Emary, E., Zawbaa, H.M., Hassanien, A.E.: Binary grey wolf optimization approaches for feature selection. Neurocomputing **172**(4), 371–381 (2016)

DACI: An Index Structure Supporting Attributed Community Queries

Zirui Zhang, Jiayi Li, and Xiaolin Qin(✉)

Nanjing University of Aeronautics and Astronautics, Nanjing, China
qinxcs@nuaa.edu.cn

Abstract. With the development of social networks, the problem of community query on attributed graphs has received increasing attention. Attributed communities models based on attributed *k*-cores, which are *k*-cores with vertices sharing keywords, have achieved great performance in finding high quality communities. However, existing solutions primarily rely on online pruning, resulting in low efficiency. We propose a dynamic three-layer index, called DACI, and a query algorithm based on it to efficiently find attributed *k*-cores. In the first layer, a hash table indexes attribute information. The second layer stores the coreness information in attributed *k*-cores while the third layer uses coreness to index corresponding vertices in attributed *k*-cores. Additionally, DACI dynamically stores the attributed *k*-cores calculated during the query process to improve the efficiency of subsequent queries. We conduct experiments and analyze results on four real datasets. The results indicate that the proposed solution has better query efficiency than existing solutions.

Keywords: Graph Mining · Attributed Graph · Community Search

1 Introduction

With the advancement of graph data processing technology, graph model has gained significant attention. Graph model can reveal the interrelation between multiple targets, making it widely applicable in various fields, such as social networks, collaborative networks, biological networks, and others [8, 17]. As one of the important application areas of graph model, community query has been widely researched.

In recent years, the study of communities has expanded to attributed graphs. Attributed graphs are graphs associated with keywords or text strings, often represented as vertices with keywords or edges with keywords [2]. Communities on attributed graphs are referred to as attributed communities, which require not only structural cohesiveness, meaning that community members are closely connected in structure, but also attributed cohesiveness, meaning that members share commonalities in keywords. With the development of Internet technology, an increasing number of attributed graphs have emerged in our daily lives. Therefore, identifying attributed communities has practical applications in many fields. For instance, in social networks like Facebook, vertices represent

X. Yang et al. (Eds.): ADMA 2023, LNAI 14180, pp. 309–323, 2023.
https://doi.org/10.1007/978-3-031-46677-9_22

users, edges represent friendships between users, and keywords of each vertex represent the user's preferences. Attributed communities in this network reveal user groups with similar preferences and shared friends, and users' posts within the groups are preferentially pushed to each other [3]. Figure 1(a) provides an instance of an attributed graph. Each vertex represents a user in a musical social network, each edge represents a friendship, and the keywords of each vertex indicate the user's preferred music genres. In this example, Tom, John, Alex, and Nick form an attributed community because they share common preferences and are friends with each other. To simplify presentation, the attributed graph with the simplified names of vertices and keywords is shown in Fig. 1(b).

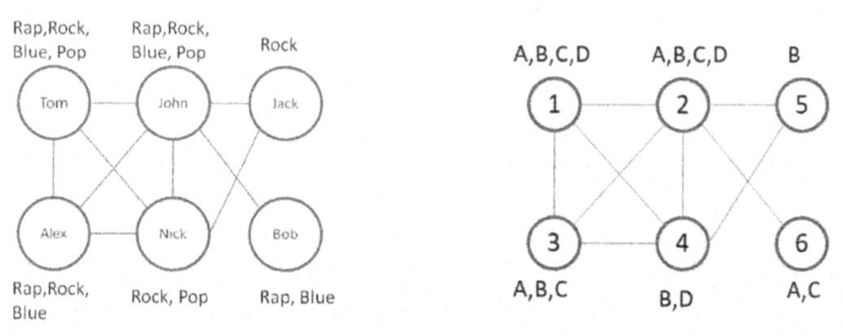

(a) An example of attributed graph (b) Simplified attributed graph

Fig. 1. Attributed graph

Existing researches on attributed community are mainly divided into community detection (CD) and community search (CS). Community detection, also known as clustering, aims to divide the whole graph into a number of closely connected clusters. In attributed graphs, clustering must satisfy not only the structural cohesiveness, but also the attribute cohesiveness [4–6, 16, 26]. However, community detection algorithm requires clustering the entire graph to identify all the communities, which may not be practical in many cases. Often, people only focus on the communities that meet specific conditions, and calculating all the communities may result in significant waste of resource. In contrast, attributed community search involves searching for qualified communities online using local information based on given structure and attribute requirements, thereby precisely obtaining desired results [2, 3, 8–11, 13–15, 23, 25].

Attributed community search has been studied deeply due to its practical applications. Among these, the research on attributed community based on common keywords has shown promising results. Existing approaches use a prune-based approach to filter out unqualified vertices to obtain attributed communities faster in remaining subgraph [2, 10]. However, this does not perform well in dense graphs with broad keyword coverage, as it becomes challenging to filter out vertices that are not in communities. For example, if the attributed graph is tightly connected with every vertex and a particular keyword covers 90% of all vertices, then when we query the attributed community with that keyword, only 10% of the vertices can be filtered out. In addition, in the case of high query frequency, this method based on online search can produce large repetitive calculations. For example, when querying attributed communities with common keyword set

S and S' that $S \subseteq S'$, the communities with S will be repeat calculation because they certainly include the communities with S'.

We propose a solution to the problems mentioned above by proposing a three-layer index called the Dynamic Attributed Core Index (DACI). The index stores attributed k-cores which are k-cores with vertices sharing keywords. DACI consists of three layers: the attribute index, the coreness index and the vertex index. The attribute index locates keyword sets to the pointers to the coreness index by hash. The coreness index stores coreness information of attributed k-cores, while the vertex index indexes the vertices in the attributed k-core by coreness. By searching DACI, we can quickly find the attributed k-core with a specific common keyword set. Furthermore, as queries proceed, DACI gradually records the attributed k-cores that have been computed, making subsequent queries faster.

We have made the following contributions in this paper:

1. We analyze the shortcomings of existing attributed community query methods and propose DACI index. We also provide the algorithm details for construction and dynamic update algorithm of DACI.
2. Based on DACI index, an attributed community query algorithm is proposed.
3. We have carried out sufficient experiments to verify the effectiveness of our index-based query method.

2 Related Work

Community Detection (CD). Community detection identifies all communities by grouping similar vertices into the same cluster. A survey of community detection on attributed graphs can be found in [24]. Reference [5] proposes a method called CODICIL, which expands the original graph by creating new edges based on content similarity, and then uses effective graph sampling to improve clustering efficiency. Reference [6] proposes a probabilistic model-based clustering method for attributed graphs, which combines structure and attribute information in a natural way rather than using artificially defined distance measures. Reference [4] improves the approach in [6], and proposes an incremental method to update edge weights, so as to improve the efficiency and stability of clustering. Reference [26] proposes an algorithm for community detection on multi-valued attributed graph. In order to prevent the clustering quality from being reduced duc to irrclcvant attributes, they only cluster in the subspace. Reference [16] proposes a local clustering method with conductance instead of topological constraints. However, as community detection seeks out all communities, it does not work well in large graphs, nor does it work well in most real-world situations.

Community Search (CS). Community search aims to capture communities "online" based on query requests. Based on the k-core model, reference [8] proposes Co-location community which is a k-core but requires the vertices in the community to be contained in a circle with a specific radius. For a given query vertex v, query keyword set S and integer k, reference [2] searches for k-cores that contains vertex v and shares the most keywords with the query keyword set S. Reference [11] supplements the method in [2] and proposes an index update algorithm under the condition of keyword update and edge update. Reference [21] extends the model of [2] to the edge attributed graph, but they don't propose a more efficient method. Reference [7] proposes EACS problem which aims to find the communities with the most similar edge attributes. Reference [13] proposes VAC model, but it may exclude vertices that should be in the community. References [14, 15, 22] solve the attribute-centered community search problem, but do not consider the query vertex. Reference [9] proposes (k, d)-truss, in which the distance from each vertex to the query vertex is no more than d. However, their problem is NP hard and the approximation ratio of their algorithm is not guaranteed. Reference [10] extends the triangle k-truss model defined in [12] to attributed graphs to ensure the stability of attributed communities. Reference [23] solves the community search problem over large semantic-based attribute graphs. Reference [25] proposes an efficient index to compute triangle k-truss communities, but attributes are not involved.

3 Problem Statement

Given an attributed graph $G = (V_G, E_G, A_G, \Sigma_G)$, V_G is vertex set of G, E_G is edge set such that $E_G \subseteq V_G \times V_G$, Σ_G is the universal attribute set, A_G is a set of vertex keywords and $\forall\, v \in V_G$, the keyword set of v is $A_G(v)$ such that $A_G(v) \subseteq \Sigma_G$. For a vertex v in V_G, $Nbr_G(v) = \{u \in V_G \mid (u, v) \in E_G\}$ says the neighborhood of v, $deg_G(v)$ says the vertex degree of v. The G in these subscripts can be omitted when the context is clear. For ease of notation, we say that vertex v contains S if $S \subseteq A(v)$.

A community is usually defined as a dense subgraph that satisfies structural cohesiveness. The minimum vertex degree is one of the most common structural cohesiveness constraints, which leads to the definition of k-core.

DEFINITION 1 (k-CORE [1]). *Given a graph G and an integer k, the k-core of G, denoted by* C_k, *is the largest subgraph of G, such that* $\forall\, v \in V_{C_k}$, $deg_{C_k}(v) \geq k$.

DEFINITION 2 (CORENESS). *Given a graph G and a vertex v, the coreness of v is the maximum k of the k-core containing v, denoted by* $core_G(v)$.

Coreness can be obtained by the k-core decomposition algorithm, which removes vertices with minimum degree iteratively to get coreness of all the vertices [1].

DEFINITION 3 (ATTRIBUTED K- CORE). *Given an attributed graph* $G = (V, E, A, \Sigma)$, *the attributed k-core of G is defined as a k-core* C_k, *in which there is at least a non-empty keyword set* S' *such that* $\forall\, u, v \in V_{C_k}$, $A(u) \cap A(v) \supseteq A'$. *We call the k-core with common keyword set* A' *as the attributed k-core generated by* A', *denoted by* $C_k(A')$.

The definition of Attributed Community Query (ACQ) problem is given as follow.

PROBLEM 1 (ACQ). *Given an attributed graph* $G = (V, E, A, \Sigma)$, *an integer k, a vertex* $v \in V$, *and a keyword set* $S \subseteq \Sigma$, *ACQ returns a connected attributed k-core* $C_k()$ *such that* $v \in V_{C_k()}$ *and the size of* $S' \cap S$ *is maximal.*

EXAMPLE 1. *Consider the attributed graph in Fig. 1(b), given a query* $\{v = 1, k = 2, S = \{A, B, D\}\}$, *ACQ returns two results respectively* $C_2(\{AB\}) = \{1, 2, 3\}$ *and* $C_2(\{BD\}) = \{1, 2, 4\}$. *For query* $\{v = 1, k = 2, S = \{A, B, C, D\}$, *it only returns a result* $C_2(\{ABC\}) = \{1, 2, 3\}$.

ACQ can find communities with high quality as they consider the query vertex and ensure that all members of communities have common keywords. Although there have been many studies on attributed communities, they either focus on a specific graph or fail to guarantee that community members share common keywords, so most of them are not applicable to the ACQ problem. However, existing approaches [2, 10, 11] to solve the ACQ problem may not perform optimally in attributed graphs with wide keyword coverage because their pruning based online search method will not be able to filter out a large number of vertices in such graphs, resulting in expensive k-core decomposition on the subgraphs. Additionally, when faced with a high query frequency, these methods may result in repeated calculations, reducing overall query efficiency.

Therefore, the problem to be solved in this paper is to propose an efficient solution based on index, which can effectively deal with the ACQ problem in the case of wide keyword coverage and high frequency queries.

4 Dynamic Attributed Core Index

We propose a dynamic three-layer index called Dynamic Attributed Core Index (DACI), which efficiently stores and accesses the computed attributed k-cores. The structure of DACI is depicted in Fig. 2. DACI is designed as a dynamic index, where it initially stores attributed k-cores generated by all singleton keyword sets. As the query proceeds, DACI dynamically stores the attributed k-cores calculated during the query process.

We introduce the structure of DACI in Sect. 4.1, and the algorithms for DACI initial construction and dynamic update in Sect. 4.2.

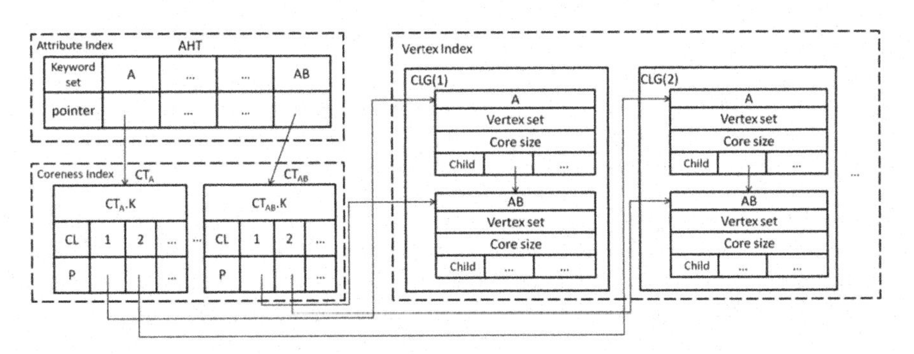

Fig. 2. The structure of DACI

4.1 Index Structure

The first layer of DACI is the attribute index, which employs a hash table, denoted as *AHT*, to map the keyword sets of attributed k-cores to pointers which point to the second layer. In turn, the second layer is the coreness index, which builds a coreness table for each keyword set in *AHT*. Specially, given a keyword set S, we denote the coreness table corresponding to S as CT_S. CT_S consists of three parts. The first part is the minimal

k value that can be queried, denoted as $CT_S.K$, which records the minimal k value for which $C_k(S)$ has been stored in DACI. The second part is an ascending array that records all the coreness existing in $C_k(S)$, denoted as $CT_S.CL$. The third part allocates a pointer list for each coreness x in $CT_S.CL$, where the pointers direct to the nodes in the third layer having the smallest size keyword sets that contains S. We denote the list of pointers for coreness x in CT_S as $CT_S.P(x)$.

The third layer is the vertex index, which indexes the vertices of attributed k-cores by coreness. It consists of multiple directed graphs, each identified by an integer x, denoted as $CLG(x)$, that indexes the vertices with coreness x in attributed k-cores. Given an attributed k-core $C_k(S)$, for each coreness x the vertices in $C_k(S)$ have, $CLG(x)$ creates a node, denoted as $N(S, x)$. $N(S, x)$ consists of four parts: a keyword set, a vertex set, a core size and a child pointer list. The keyword set of $N(S, x)$ is S, and the vertex set, denoted as $VS_{N(S,x)}$, stores the vertices with coreness x in $C_k(S)$ but not in $C_k(S')$ for any $S \subseteq S'$, and if there is no such vertex, remove $N(S, x)$. A vertex v is said to be contained in $N(S, x)$ if v is in $VS_{N(S,x)}$. The core size records the size of $V_{C_x(S)}$, denoted as $CS_{N(S,x)}$. For any two keyword sets S and S', if $S \subseteq S'$ and there is no node $N(W, x)$ such that $S \subseteq W \subseteq S'$, add a pointer to $N(S', x)$ into the child pointer list of $N(S, x)$. For each node, its child nodes are sorted by their core sizes in ascending order.

For $N(S', x')$ and its descendants, the vertices in their vertex sets must be vertices in $C_x(S)$, where $x' \geq x$ and $S \subseteq$. The correctness of the vertex index is guaranteed by the properties as follows.

PROPERTY 1 (STRUCTURAL NESTING). *Given a graph G and two integers k_1, k_2 such that $k_1 < k_2$, if C_{k_2} exists, then C_{k_1} must exist and $V_{C_{k_2}} \subseteq V_{C_{k_1}}$.*

PROPERTY 2 (ANTI-MONOTONICITY). *Given attributed graph G, an integer k and a keyword set S, if $C_k(S)$ exists, then for any keyword S' such that $S' \subseteq S$, $C_k(S')$ must also exist and $V_{C_k(S)} \subseteq V_{C_k()}$.*

Property 1 and property 2 can be derived by the definitions of k-core and attributed k-core respectively. With the definition of the vertex index, we can find that the vertex index obeys the following property.

PROPERTY 3. *Given two integers k, k', two keyword sets S, S' and $k < k\prime$, $S' \subseteq S$, if a vertex v is contained in N(S, k'), then v must not be contained in N(S', k') and N(S, k).*

Given a keyword set S and an integer k, we can get $C_k(S)$ by DACI if $C_k(S)$ is stored in DACI. Specifically, find the coreness table CT_S by AHT. If $CT_s.K$ is greater than k, $C_k(S)$ is not stored in DACI. Otherwise for each coreness x such that $x \geq k$, traverse the nodes which pointers in $CT_L.P(x)$ pointes to, as well as their descendants, and add vertices in their vertex sets to the result set.

Space Cost. The space cost of DACI in initial state is $O(|\Sigma| * |V|)$ because it stores attributed k-cores generated by all the singleton keyword sets. As the index updates, the space cost of DACI eventually becomes $O\left(2^{|\Sigma|} * |V|\right)$.

4.2 Index Construction and Dynamic Update Algorithms

DACI stores the attributed k-core generated by each singleton keyword set at initial state. Algorithm 1 presents the construction algorithm of DACI. The sets of vertices containing each keyword are first obtained by traversing the keyword sets of all vertices (line 1). Next, for each singleton keyword set L, store L in AHT and generateCT_L, where the $CT_L.K$ is 1 (lines 3–5). Then, perform k-core decomposition on the subgraph induced by of vertices containing L to obtain the coreness of each vertex. (lines 6–7). Then, for each coreness x in descending order, build node $N(L, x)$. All the vertices with coreness x in the induced subgraph are stored in$VS_{N(L,x)}$, and the size of $V_{C_x(L)}$ is recorded by $CS_{N(L,x)}$ (lines 9–11). Finally, store x in$CT_L.CL$, and a pointer to $N(L, x)$ is added to $CT_L.P(x)$ (lines 12–14).

Algorithm 1 BUILDDACI

Input: Attributed graph $G(V,E,A,\Sigma)$

Output: the initial state of DACI

1: Visit the attributed set of each vertex to obtain vertices containing each keyword.
2: **for** each $l \in \Sigma$
3: $L \leftarrow \{l\}$;
4: AHT.add(L); //store L into hash table;
5: $CT_L.K \leftarrow 1$;
6: Generate induced subgraph G_L by vertices containing L;
7: Perform k-core decomposition on G_L to get coreness of each vertex in G_L;
8: **for** each coreness x in G_L in descending order
9: Build node $N(L, x)$;
10: Add all vertices whose coreness is x into $VS_{N(L,x)}$;
11: $CS_{N(L,x)} \leftarrow |V_{C_x(L)}|$;
12: $CT_L.CL$.add(x); // add x into $CT_L.CL$
13: generate a pointer p to $N(L,x)$;
14: $CT_L.P(x)$.add(p); // add p into $CT_L.P(x)$

Algorithm 2 presents the dynamic update algorithm for DACI. Given an input attributed k-core $C_k(S)$, if S has not been stored in AHT, add S into the AHT and generate CT_S(line 2). For each coreness x in$C_k(S)$, node $N(S, x)$ is built and a pointer to $N(S, x)$ is added into $CT_S.P(x)$, then the function ADDintoCLG is called to insert $N(S, x)$ into the appropriate position, adjust $CLG(x)$ to maintain property 3, and adjust pointer lists in the coreness index if a node is removed(lines 3–7). Finally, set $CT_S.K$ be k (line 8). If S is already in AHT and $CT_S.K$ is larger than k, then for each coreness x less than $CT_S.K$, perform the same steps as above(lines 10–14). Finally, let $CT_S.K$ be k (line 15).

Algorithm 2 UPDATEDACI

Input: Attributed k-core $C_k(S)$
Output: DACI after update
1: **if** AHT[hash(S)] $\neq S$ // if S is not in hash table
2: Add S into AHT and generate CT_S;
3: **for** each coreness x in $C_k(S)$
4: CL_S.add(x);
5: Build $N(S, x)$;
6: Add a pointer to $N(S, x)$ into $CT_S.P(x)$
7: ADDintoCLG($N(S, x)$); //add $N(S, x)$ into $CLG(x)$ and adjust index
8: $CT_S.K \leftarrow k$;
9: **else if** $CT_S.K > k$
10: **for** each coreness x in $C_k(S)$ & $x \geq k_S$
11: $CT_S.CL$.add(x);
12: Build $N(S, x)$;
13: Add a pointer to $N(S, x)$ into $CT_S.P(x)$;
14: ADDintoCLG($N(S, x)$);
15: $CT_S.K \leftarrow k$;

Example 2. *Consider the attributed graph in Fig. 1(b). After dynamic update by adding the attributed k-cores $C_2(\{AB\})$, $C_1(\{AC\})$, $C_2(\{BD\})$, and $C_1(\{ABCD\})$ from the initial state, DACI is shown in Fig. 3. Note that because $V_{N(\{A\},2)}$ is empty, $N(\{A\}, 2)$ has been removed and the pointers in $CT_A.P(2)$ point to $N(\{AB\}, 2)$ and $N(\{AC\}, 2)$ respectively.*

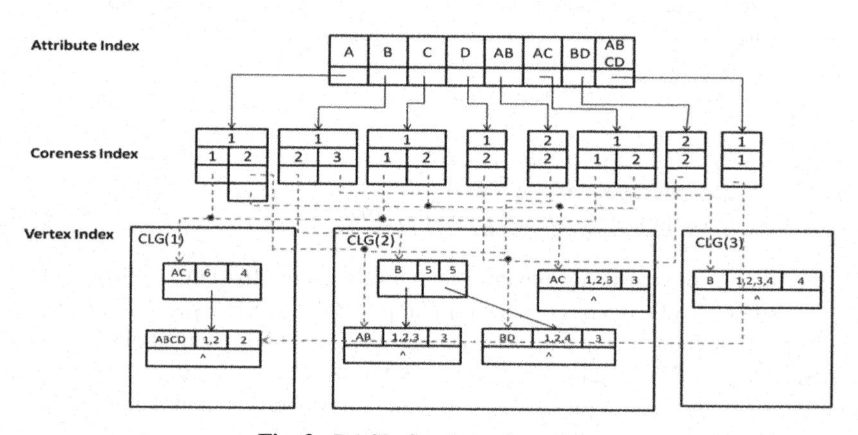

Fig. 3. DACI after dynamic update

5 Query Algorithm

The query algorithm based DACI is called DACI-Dec algorithm. It utilizes a frequent pattern mining algorithm to generate candidate keyword sets, which are used to filter out the subsets of the query keyword set that do not meet the query conditions in advance.

The algorithm then verifies the candidate sets in order of descending size. The procedure for DACI-Dec to generate candidate sets is based on the following lemma.

LEMMA 1 (NEIGHBORHOOD CONSTRAINT). *For a given attributed graph G and an attributed k-core $C_k(S)$ of G, for $\forall v \in V_{C_k(S)}$, there are at least k vertices $u \in Nbr_G(v)$ such that $A(u) \supseteq S$ and $core_G(u) \geq k$.*

Proof: By the definition of attributed *k*-core, every vertex *v* in $C_k(S)$ satisfies $deg(v) \geq k$ and $A(v) \supseteq S$, so for $\forall v \in V_{C_k(S)}$, there are at least *k* vertices $u \in Nbr_{C_k(S)}(v)$ satisfying $A(u) \supseteq S$ and $core_G(u) \geq k$. Since $C_k(S) \subseteq G$, $Nbr_{C_k(S)}(v) \subseteq Nbr_G(v)$, we can know that $u \in Nbr_G(v)$. Therefore the lemma 1 is proved.

Given a query $q = \{v_q, k, S_q\}$, lemma 1 shows that only keyword sets that are contained in the keyword sets of at least *k* vertices in neighborhood of v_q are likely to generate the attributed *k*-cores containingv_q. Frequent pattern mining algorithms can effectively solve this problem. We use the FP-growth algorithm [19] for candidate keyword set generation. The candidate keyword sets are divided into $\Psi_1, \Psi_2, \ldots, \Psi_h$, where Ψ_i ($1 \leq i \leq h$ and *h* is at most $|S_q|$) stores keyword sets with size *i*. After the candidate keyword sets are generated, they are verified in the order from largest to smallest. Algorithm 3 shows the pseudo-code of DACI-Dec.

Algorithm 3 DACI-Dec

Input: Attributed graph $G(V,E,A,\Sigma)$, query $q=\{v_q, k, S_q\}$
Output: sets of attributed community

```
1:  Ω_1, Ω_2, ... Ω_h ← ∅;
2:  Generate candidate keyword sets Ψ_1, Ψ_2, ..., Ψ_h by Fp-growth.
3:  for i=h to 1
4:     for each S ∈ Ψ_i
5:       if AHT[hash(S)]=S & CT_S.K<k //if S is in DACI
6:          C_k(S)←SearchDACI(S,k);
7:       else
8:          Vex←∅;
9:          for each l ∈S
10:            L←{l};V'←∅;
11:            V' ← GetVertexInCLG(L,S,V',k);
12:            Vex← V' ∩ Vex;
13:         Generate induced subgraph SubG by Vex;
14:         C_k(S)←KCOREDECOMPOSE (SubG);
15:         UPDATEDACI(C_k(S));
16:      if v_q ∈ C_k(S)
17:         R←the connected component contained v_q of C_k(S);
18:         Ω_i.add(R);  // add R into result set
19:    if Ω_i ≠ ∅
20:       break;
21: return ∅
```

Algorithm 3 initializes each result set Ω_i as an empty set and use FP-growth algorithm to generate sets of candidate keyword sets Ψ_1, Ψ_2, ..., Ψ_h (lines 1–2) at first. After that, set i equal to h and traverse Ψ_i in the order of i from largest to smallest. For each keyword set $S\in$, the algorithm first checks if $C_k(S)$ is in DACI. If so, the SearchDACI function is called to obtain $C_k(S)$ (lines 5–6). Otherwise, the algorithm generates a keyword set L for each keyword in S and calls GetVertexInCLG to obtain the vertex set V' such that $V_{C_k(S)}\subseteq V'\subseteq V_{C_k(L)}$. The intersection of each such V' yields a smaller candidate set *Vex* such that $V_{C_k(S)} \subseteq Vex$ and $S \subseteq A(v)$ for each vertex v in *Vex* (lines 8–12). Then, k-core decomposition is performed on the subgraph induced by *Vex* to obtain$C_k(S)$, and $C_k(S)$ is to be stored into DACI (lines 13–15). If the query vertex v_q is in$C_k(S)$, the connected component of $C_k(S)$ containing v_q is added to the Ω_i (lines 16–18). If Ω_i is not the empty set after all keyword sets in Ψ_i have been verified, then Ω_i is returned. Otherwise, proceed to the next loop to verify Ψ_{i-1} (lines 19–21). If all Ω_i are empty sets, the empty set is returned.

The pseudo-code of the GetVertexInCLG algorithm is shown in Algorithm 4. GetVertexInCLG(L,S,V,k) works recursively to obtain a vertex set V such that $V_{C_k(S)} \subseteq V \subseteq V_{C_k(L)}$. First obtain CT_L by AHT(1 line). For the first coreness x such that $x \geq k$ in $CT_L.CL$, iterate over the nodes that the pointers in $CT_L.P(x)$ point to and their descendants, and add their vertices to V (lines 2–4). During the traversal, find the node $N(W, x)$ such that $W \subseteq S$ and $CS_{N(W,x)}$ is minimum. If such $N(W, x)$ exists, then GetVertexInCLG(W,S,V,$x+1$) is called recursively and the coreness after x in $CT_L.CL$ is not considered again(lines 5–7). Otherwise repeat the above steps for the next coreness of x in $CT_L.CL$.

Algorithm 4 GetVertexInCLG

Input: current attribute set L, query keyword set S, candidate vertex set V, integer k
Output: candidate vertex set V
1: Get CT_S by AHT;
2: **for** each coreness x in $CT_L.CL$ & $x\geq k$
3: Visit all nodes pointed by pointers in $CT_L.P(x)$ and their descendents;
4: Add vertices in their vertex sets to V;
5: **if** can find $N(W, x)$ with minimal $CS_{N(W,x)}$ & $W\subseteq S$;
6: $V \leftarrow$GetVertexInCLG(W,S,V,x+1);
7: **break** ;
8: **return** V;

6 Experiments

We present the experiments in this section. The setup is shown in Sect. 6.1 and the experimental results are discussed in Sect. 6.2.

6.1 Setup

We consider four real-world datasets. For the dataset Steam [20], a vertex represents a game, and an edge represents two games sold in a bundle. The keywords of each vertex are the genres of the game. The dataset Movie is a network taken from Wikipedia, where each vertex represents an actor or director or movie, an edge represents two vertices appearing in the same Wikipedia pages and the keywords of a vertex is the frequent words in its description. For dataset Deezer_HU, each vertex represents a user, and the edge represents the friendship between users and keywords of a vertex are music types the user prefers. For dataset DBLP, each vertex represents an author in the database domain, an edge represents collaboration between authors, and the keywords of a vertex are the research directions of the author. Table 1 presents the information of each dataset.

Table 1. Information of each dataset.

Dataset	Vertices	Edge	$\lvert\Sigma\rvert$	$\hat{d}$
Steam	2797	24837	21	17.7
Movies	34283	142427	38	8.3
Deezer_HU	47538	222887	17	9.3
DBLP	228047	2828854	72	24.8

We conduct experiments to compare the index performance of DACI, CL-Tree [11], MTindex [10], as well as the query performance of query algorithms based on these index and a pruning-based algorithm BackTrack [23]. For the index performance, we measure the size and construction time of the three indexes. For the query performance, we use the algorithm that only uses the initial state of DACI and does not update DACI dynamically in the process of DACI-Dec as the baseline, and conduct two sets of experiments to compare the query efficiency of baseline, DACI-Dec, CLTree-Dec, MTindex-Dec and BackTrack for different k and different sizes of query keyword set S. In the first set of experiments, for each dataset, 300 queries are randomly generated, where S is set to the entire keyword set of the query vertex, and k is set to 4, 5, 6, 7 and 8 respectively. In the second set of experiments, for each dataset, 300 queries are randomly generated, where the k is set to 5, and the size of S is set to 1, 3, 5, 7 and 9 respectively. At the beginning of each set of experiments, we use the initial state of DACI and the queries are continuous. All algorithms are implemented in C++ and compiled by Visual Studio 2019. All experiments are conducted on a Windows machine with an Intel i7-7700 HQ processor at 2.8 GHz frequency and 40 gigabytes of memory.

6.2 Results

Index Construction Time. We compare the construction time of three indexes on each dataset, and the results are shown in Fig. 4(a). As can be observed from the figure, on the four datasets, the construction time of DACI is longer than that of CL-Trees and shorter than that of MTindex. This is because DACI need to calculate the attributed k-cores under each singleton keyword set, while CL-Tree only calculates the k-core on the original graph. On the other hand, MTindex calculates connected triangle k-trusses under each singleton keyword set, which takes longer time.

Index Size. We compare the index size of the three indexes on each dataset, where the size of DACI is divided into the initial size and the size after 300 random queries by DACI-Dec (denoted as Update-Dec), and the results are presented in Fig. 4(b). Figure 4(b) shows that for these four datasets, the initial size of DACI is smaller than that of CL-Tree and MTindex. This is due to the fact that CL-Tree stores all vertices under each keyword, while DACI records only the vertices in attributed k-cores generated by singleton keyword sets. Moreover, MTindex stores the triangle connected k-trusses generated by singleton keyword sets and builds a summary graph, which results in additional space storage. Additionally, after 300 queries, the size of DACI is still small, which is attributed to property 3.

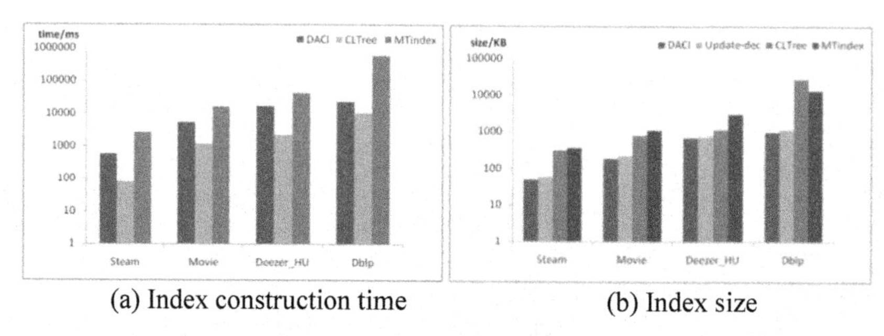

(a) Index construction time (b) Index size

Fig. 4. Results of Index Performance

Effect of Different k. Figures 5(a)–(d) present the query time of each query algorithm for different k. As k increases, the size of the subgraph obtained by index will become smaller, resulting in a decrease in the query time of all query algorithms. DACI-Dec consistently outperforms baseline on all datasets. CLTree-Dec only performs better than DACI-Dec on Steam because the dataset is small enough that a lightweight index like CL-Tree can prune vertices efficiently, while searching DACI is not faster than computing k-cores in this case. Additionally, it is observed that CLTree-Dec performs worse on Deezer_HU, because each keyword in Deezer_HU covers more vertices, and these vertices cannot form the attributed k-cores. Therefore, CLTree-Dec generates more candidate keyword sets and most of them cannot meet the query requirements, resulting in a high overhead of verification. However, MTindex-Dec is more efficient on Deezer_HU because there are fewer triangle connected k-trusses with k greater than 3 in Deezer_HU, leading to early end of queries. Moreover, the efficiency of BackTrack is relatively low on all the datasets.

Effect of Different Sizes of Query Keyword Set S. Figures 6(a)–(d) show the query time for different sizes of query keyword set S. Note that Movie has no vertex with keyword set size more than 7, Steam has no vertex with keyword set size more than 5, so the experiments on them only take the right $|S|$. As $|S|$ increases, the number and size of candidate keyword sets will increase, resulting in the query time longer. Results show that DACI-Dec consistently outperforms the baseline on all datasets, and outperforms

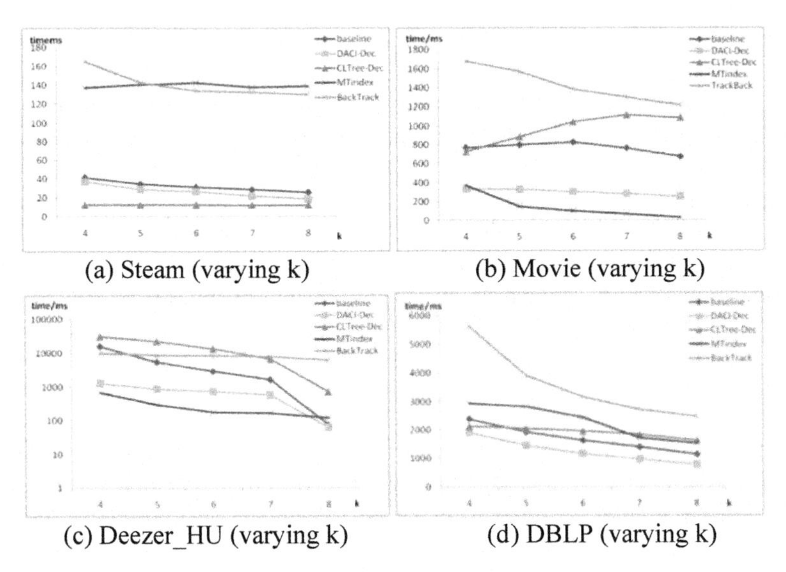

(a) Steam (varying k) (b) Movie (varying k)

(c) Deezer_HU (varying k) (d) DBLP (varying k)

Fig. 5. Results of query time for different k

CLTree-Dec on datasets other than Steam. The query speed of DACI-Dec is worse than that of MTindex-Dec on Movie and Deezer_HU, and better on Steam and DBLP. Moreover, BackTrack is less efficient on all datasets and is greatly effect by $|S|$.

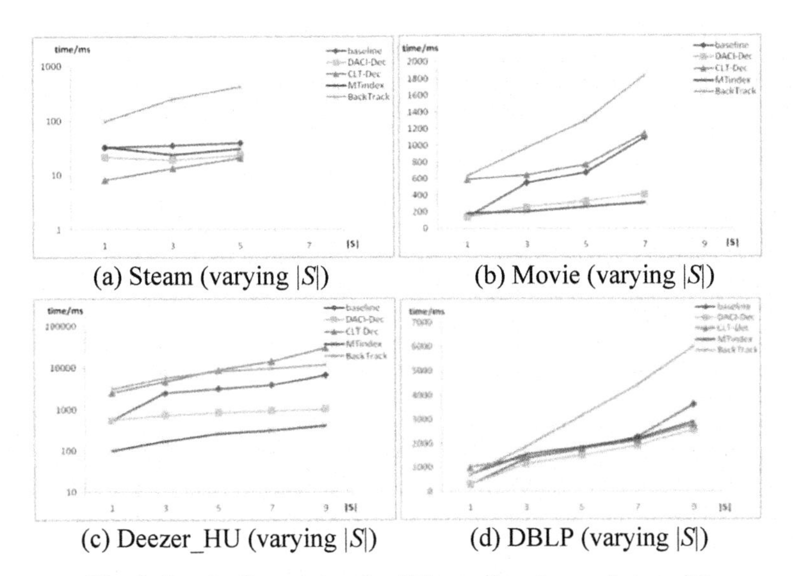

(a) Steam (varying $|S|$) (b) Movie (varying $|S|$)

(c) Deezer_HU (varying $|S|$) (d) DBLP (varying $|S|$)

Fig. 6. Result of query time for different size of query keyword S

Overall Evaluation. DACI-Dec achieves the highest comprehensive performance on all datasets. The baseline consistently outperforms CLTree-Dec except on Steam, indicating that DACI index is more query-friendly than CL-Tree even when only using initial state. However, the query efficiency of DACI-Dec is always better than that of the baseline, indicating that it is worthwhile to update DACI dynamically. Since MTindex indexes connected triangle k-trusses, which takes longer time to compute than k-cores, DACI-Dec has a faster query speed than MTindex-Dec on dense graphs. BackTrack is mainly designed for semantic-based attributed graphs. However, in the ACQ problem, attributes are all related by default, so the pruning effect of using semantics is greatly reduced.

7 Conclusion

The existing solutions for the ACQ problem are inefficient on graphs with high keyword coverage, and are prone to produce a large number of repeated computations when the query frequency is high. In order to tackle the challenges, we propose a three-layer dynamic index DACI, and develop a query algorithm based on it. Our experiments show that DACI requires less storage space than existing indexes, and our query algorithm outperforms others in terms of query efficiency.

References

1. Batagelj, V., Zaversnik, M.: An O(m) algorithm for cores decomposition of networks. arXiv (2003)
2. Fang, Y., Cheng, R., Luo, S., Hu, J.: Effective community search for large attributed graphs. Proc. VLDB Endow. **9**(12), 1233–1244 (2016)
3. Cui, W., Xiao, Y., Wang, H., Wang, W.: Local search of communities in large graphs. In: Proceedings of ACMSIGMOD International Conference on Management of Data, pp. 991–1002 (2014)
4. Cheng, H., et al.: Clustering large attributed information networks: an efficient incremental computing approach. Data Min. Knowl. Disc. **25**(3), 450–477 (2012). https://doi.org/10.1007/s10618-012-0263-0
5. Ruan, Y., Fuhry, D., Parthasarathy. S.: Efficient community detection in large networks using content and links. In: Proceedings of the 22th International Conference on World Wide Web, pp. 1089–1098 (2013)
6. Xu, Z., Ke, Y., Wang, Y., Heng, H., Cheng, J.: A Model-based approach to attributed graph clustering. In: Proceedings of the 2012 ACM SIGMOD International Conference on Management of Data, pp. 505–516 (2012)
7. Li, L., Zhao, Y., Luo, S., Wang, G., Wang, Z.: Efficient community search in edge-attributed graphs. IEEE Trans. Knowl. and Data Eng. **99**, 1–16 (2023)
8. Luo, J., Cao, X., Xie, X., Qu, Q.: Best co-located community search in attributed networks. In: Proceedings of the 28th ACM International Conference on information and Knowledge Management, Beijin, pp. 2453–2456, (2019)
9. Huang, X., Lakshmanan, L.: Attribute driven community search. Proc. VLDB Endow. **10**(9), 949–960 (2017)
10. Zhu, Y., et al.: When structure meets keywords: cohesive attributed community search. In: Proceedings of the 29th ACM International Conference on Information & Knowledge Management, Ireland, pp. 1913–1922, (2020)

11. Fang, Y., Cheng, R., Chen, Y., Luo, S., Jiafeng, H.: Effective and efficient attributed community search. VLDB J. **26**(6), 803–828 (2017). https://doi.org/10.1007/s00778-017-0482-5
12. Akbas, E., Zhao, P.: Truss-based community search: a truss equivalence based indexing approach. Proc. VLDB Endow. **10**(11), 1298–1309 (2017)
13. Liu, Q., Zhu, Y., Zhao, M., Huang, X., Xu, J., Gao, Y.: VAC: vertex-centric attributed community search. In: Proceedings of IEEE 36th International Conference on Data Engineering, pp. 937–948 (2020)
14. Zhu, Y., et al.: Querying cohesive subgraphs by keywords. In: Proceedings of IEEE 34th International Conference on Data Engineering, pp.1324–1327 (2018)
15. Zhang, Z., Huang, X., Xu, J., Choi, B., Shang, Z.: Keyword-Centric community search. In: Proceedings of IEEE 35th International Conference on Data Engineering (2019)
16. Yudong, N., Li, Y., Fan, J., Bao, Z.: Local clustering over labeled graphs: an index-free approach. In: Proceedings of IEEE 38th International Conference on Data Engineering, pp. 2805–2817 (2022)
17. Lappas, T., Liu, K., Terzi, E.: Finding a team of experts in social networks. In: Proceedings of the 15th ACM SIGKDD International Conference on Knowledge Discovery and Data Mining, pp. 467–476 (2009)
18. Li, R., et al.: Skyline community search in multi-valued networks. In: Proceedings of the 2018 ACM SIGMOD International Conference on Management of Data, pp. 457–472 (2018)
19. Han, J., Pei, J., Yin, Y., Mao, R.: Mining frequent patterns without candidate generation: a frequent-pattern tree approach. Data Min. Knowl. Disc. **8**(1), 53–87 (2004). https://doi.org/10.1023/B:DAMI.0000005258.31418.83
20. Pathak, A., Gupta, K., McAuley, J.: Generating and personalizing bundle recommendations on steam. In: ACM SIGIR forum 51, pp. 1073–1076 (2017)
21. Kang, Q., Kang, Y., Kong, H.: Edge-Attributed community search for large graphs. In: Proceedings of the 2nd International Conference on Big Data Research, pp. 114–118 (2018)
22. Chen, L., Liu, C., Liao, K., Li, J., Zhou, R.: Contextual community search over large social networks. In: Proceedings of IEEE 35th International Conference on Data Engineering (2019)
23. Lin, P., Siyang, Y., Zhou, X., Peng, P., Li, K., Liao, X.: Community search over large semantic-based attribute graphs. World Wide Web **25**(2), 927–948 (2022). https://doi.org/10.1007/s11280-021-00942-y
24. Bothorel, C., Cruz, J.D., Magnani, M., Micenkova, B.: Clustering attributed graphs: models, measures and methods. Network Sci. **3**(03), 408–444 (2015)
25. Xu, T., Lu, Z., Zhu, Y.: Efficient triangle-connected truss community search in dynamic graphs. Proc. VLDB Endow. **16**(3), 519–531 (2022)
26. Huang, X., Cheng, H., Yu, J.X.: Dense community detection in multi-valued attributed networks. Inf. Sci. **314**, 77–99 (2015)

Persistent Community Search Over Temporal Bipartite Graphs

Mo Li, Zhiran Xie, and Linlin Ding(✉)

School of Information, Liaoning University, Liaoning, China
dinglinlin@lnu.edu.cn

Abstract. Bipartite graphs are commonly used to model the relationship between two types of entities, and community search over them has recently gained much attention. However, in real-world scenarios, edges often contain temporal information, such as the timestamp of a user visiting a location at time instant t. Unfortunately, most previous studies have focused on finding communities without temporal information. In this paper, we study the problem of persistent community search over temporal bipartite graphs. We propose a novel persistent community model called (θ, τ)- persistent (α,β)-core to capture the persistence of a community, where α and β are the lower bound degree of each layer in the projected graph of the persistent community, and θ and τ represent the minimum and maximum temporal duration for the entire (α, β)-core. We have proved the NP-hardness of the problem. To solve this problem, we propose a PCSearch algorithm using the branch-reduce-and-bound strategy divided into four sub-algorithms. Finally, we conduct extensive experiments over several real-world temporal bipartite networks to demonstrate the efficiency and effectiveness of our proposed algorithm.

Keywords: Community search · Temporal bipartite graph

1 Introduction

Bipartite networks are ubiquitous in many real-world applications, since they can reflect the relationship between two different types of entities, such as people-location networks [3], and gene co-expression networks [13], to name just a few. It is widely recognized that practical networks inherently exist community structures. To capture these structures, a diverse range of dense subgraph models, including the (α, β)-core [5], bitruss [15], and biclique [7], have been devised. These models are designed for bipartite graphs and are exceptionally skilled at measuring the cohesiveness of community structures. Based on these models, community search over bipartite graphs aims to identify densely connected subgraphs containing the query node that satisfies specific query conditions.

Majority of the existing works focus on community search over static bipartite graphs. A (α, β)-core constraints the degree of the nodes in upper layer greater than α, whereas the degree of the nodes in lower layer greater than *beta*.

X. Yang et al. (Eds.): ADMA 2023, LNAI 14180, pp. 324–339, 2023.
https://doi.org/10.1007/978-3-031-46677-9_23

Figure 1 illustrates that the nodes enclosed by a red dashed box form a $(2, 3)$-core. Specifically, each node in the upper layer has a degree greater than 2, while each node in the lower layer has a degree greater than 3. Liu et al. [5] proposed a novel index-based method to solve the (α, β)-core community search over static bipartite graphs, which pre-computes all the (α, β)-cores with every possible α and β. Zhang et al. [14] focused on the community search problem over vertex-weighted bipartite graphs. Wang et al. [9] devised an efficient algorithm to solve the community search problem over the edge-weighted bipartite graphs.

It is important to note that the previous research efforts focus on assessing community structures based on the aggregated structural cohesiveness at a given time instant. However, this approach overlooks the crucial temporal information that each edge carries, which is a ubiquitous characteristic of real-world networks. For instance, consider a scenario where a user u visits location l_1 at time instant t_1 and subsequently visits location l_2 at time instant t_2. To accurately identify the community containing user u and recommend friends, it is imperative to consider the temporal information carried by each edge. Thus, incorporating temporal information in community search can yield more precise and reliable outcomes in real-world applications. Unfortunately, previous community search algorithms mostly ignored the temporal information of the graph, so they could not find time-related clusters. Li et al. [4] designed a persistent community model over temporal graphs, However, this model does not contain the target q, and cannot fully adapt to persistent community search over temporal bipartite graphs. To fill this research gap, we propose a novel persistent community model $(\theta,\ \tau)$-persistent $(\alpha,\ \beta)$-core over temporal bipartite graphs, where α and β are the lower bound degree of each layer in projected graph of the persistent community, θ and τ represent the minimum and maximum temporal duration for the entire $(\alpha,\ \beta)$-core. Additionally, the persistent community in the projected graph of the temporal bipartite graph in every $[t, t+\theta]$ is still a (α, β)-core.

For example, let $\theta = 60$ seconds, $\tau = 300$ s, α=10 and $\beta = 2$. Suppose we discover a (60,600)-persistent (10,2)-core through online market search. This means that within a 10-minute interval, there are 10 buyers and 2 sellers conducting transactions every minute. This time period can be considered a peak trading period, indicating the need to increase advertising efforts.

Applications. Finding the $(\theta,\ \tau)$-persistent $(\alpha,\ \beta)$-core that contains the target q has many real-world applications, such as fraud detection and personalized recommendation. Due to the space limitation, we just introduce the applications in the field of fraud detection. We can conduct a community search over temporal bipartite graphs to detect fraudulent behaviors in online marketplaces. In this scenario, the bipartite graph represents the interactions between buyers and sellers over time. The edges between buyers and sellers capture temporal information, such as the time of purchase and the transaction amount. By analyzing the evolving structure of the bipartite graph, persistent communities of buyers and sellers can be identified, representing groups of users that consistently engage in fraudulent activities over an extended period.

To the best of our knowledge, this is the first work to focus on persistent community search over temporal bipartite graphs. Although existing algorithms designed for community search over static bipartite graphs can be adapted to handle temporal graphs, such an approach is inefficient as it requires re-computation of communities at every time instant.

On the other hand, the k-core persistent community search algorithm proposed in [4] imposes significant limitations when applied to temporal bipartite graphs. The k-core persistent community search algorithm can only impose a degree constraint on a single degree, rather than constraining both degrees in a temporal bipartite graph. Furthermore, the main idea of this algorithm is to perform degree filtering on each vertex, without fully utilizing the characteristics of temporal bipartite graphs. Therefore, in this work, we propose a novel algorithm, PCSearch, specifically designed for efficient and effective persistent community search over temporal bipartite graphs. Our algorithm consists of several key steps. Firstly, we prune redundant vertices and subgraphs in the temporal bipartite network to reduce the search space. Next, we decompose the reduced temporal bipartite graph into smaller components using a branch-reduce-and-bound approach. Finally, we synthesize the persistent community (PC) from the temporal information. We validate the effectiveness and efficiency of our proposed algorithm through comprehensive experiments.

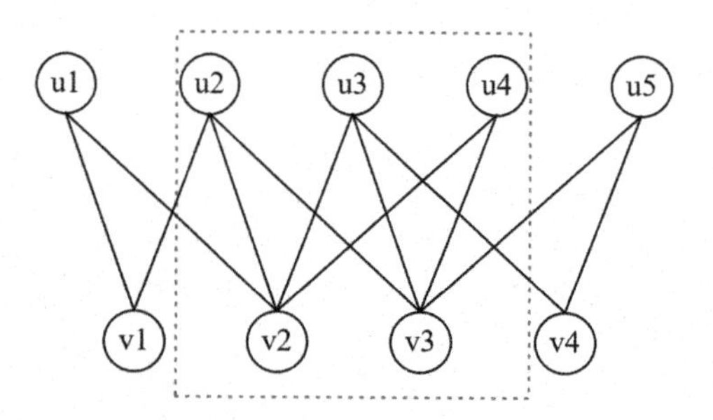

Fig. 1. A static bipartite graph and a (2, 3)-core

To sum up, our main contributions can be summarized as follows:

- We propose a (θ, τ)-persistent (α, β)-core community search problem over temporal bipartite graphs, and prove it is NP-hardness.
- We devise an efficient branch-reduce-and-bound algorithm PCSearch to solve the persistent community search over temporal bipartite graphs.
- Extensive experiments on several real datasets have been conducted to evaluate the effectiveness and efciency of our proposed algorithms.

2 Problem Definition

A temporal bipartite graph G can be defined as $G(V = (U, L), E)$, where $U(G)$ denotes the set of vertices in the upper layer, $L(G)$ denotes the set of vertices

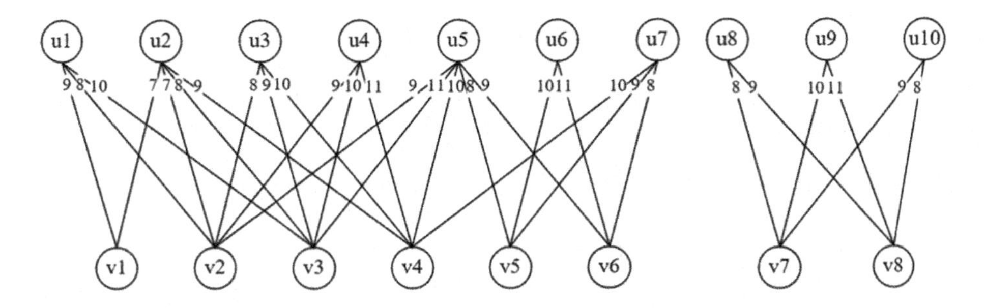

Fig. 2. A temporal bipartite network

in the lower layer, $U(G) \cap L(G) = \emptyset$, $V(G) = U(G) \cup L(G)$ denotes the vertex set and $E(G) \subseteq U(G) \times L(G)$ denotes the set of temporal edges. A temporal edge between two vertices u and v in G is denoted as $(u,\ v,\ t)$ or $(v,\ u,\ t)$, where u, v are vertices in V, and t is the interaction time between u and v. Let t_s and t_e be the time of the first and the last occurred temporal edges respectively. The interaction time t of any temporal edge $(u,\ v,\ t)$ in G must fall into the interval $[t_s,\ t_e]$. The length of the interval $[t_s,\ t_e]$ is equal to $[t_e - t_s]$. The temporal neighborhood of a node u in $G(V = (U, L), E)$ is defined by $N_u(G) = (u, v, t)|(u, v, t) \in E$. The degree of a node u in G, denoted by $\deg(u, G)$, is the number of neighbors in G. The projected graph denoted by $\mathcal{G}$ over the time interval $[t_s, t_e]$ is defined as $\mathcal{G} = (\mathcal{V}, \mathcal{E}, [t_s, t_e])$, where $\mathcal{V} = V$ and $\mathcal{E} = (u, v)|(u, v, t) \in E, t \in [t_s, t_e]$.

Definition 1. (**(α, β)-core**) Given a bipartite graph G and degree constraints α and β, a subgraph $R_(\alpha, \beta)$ is the (α, β)-**core** if it satisfies the following properties:

1) **Connectivity property.** R is a connected subgraph;
2) **Cohesiveness property.** Each vertex $u \in U(R)$ satisfies $\deg(u, R) \geq \alpha$ and each vertex $v \in L(R)$ satisfies $\deg(v, R) \geq \beta$;

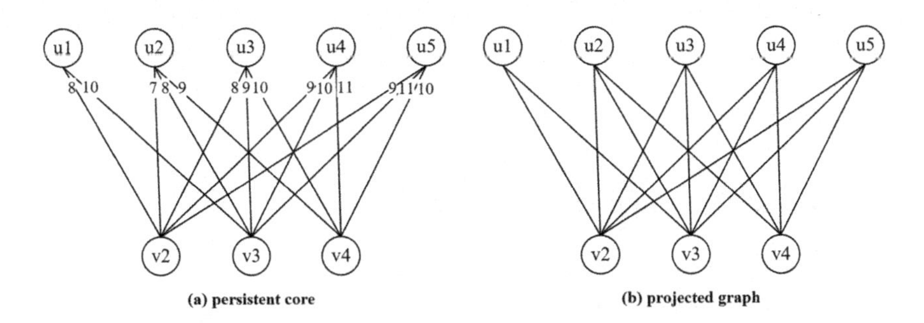

Fig. 3. A (3, 4)-persistent(2, 3)-core including vertex u_1 and the projected graph

Definition 2. (Maximal (θ, α, β)-persistent-core interval) Given a temporal bipartite graph G and the parameters θ, α and β, an interval I $[t_s, t_e]$ with $t_e - t_s \geq \theta$ is called Maximal (θ, α, β)-persistent-core interval if I satisfies the following properties:

1) **Persistent core property.** For any $t \in [t_s, t_e - \theta]$, the projected graph of G on the interval $[t, t + \theta]$ is a (α, β)-core.
2) **Maximality property.** There is no super-interval $I' \supseteq I$ and that satisfies the persistent property;

Definition 3. ((θ, τ)-persistent (α, β)-core), given a temporal bipartite graph G with parameters θ, τ, α and β, temporal bipartite subgraph $C \in G$ is a (θ, τ)-persistent(α, β)-core) if satisfies the following properties:

1) **Boundary property.** The interval of subgraph I_c is a Maximal (θ, α, β)-persistent-core interval, and $\theta \leq$ the length of $I_c \leq \tau$.
2) **Maximality property.** There is no subgraph $C' \supseteq C$ and that satisfies the boundary property;

Problem Statement. Given a temporal bipartite network W, parameters $\theta, \tau, \alpha, \beta$ and a query vertex q, the persistent community search problem aims to find all (θ, τ)-persistent (α, β)-cores including vertex q in W.

Hardness Result. Since the problem of finding the maximum persistent k-core in a temporal graph is NP-hard, our problem is also NP-hard.

Theorem 1. *The problem of enumerating all (θ, τ)-persistent(α, β)-core including vertex q in the temporal bipartite network is NP-hard.*

Proof. It has been shown in [4] that finding persistent k-core in a temporal graph is NP-hard, and our problem finding all (θ, τ)-persistent (α, β)-cores including vertex q in the temporal bipartite network is also NP-hard.

Example 1. Consider the temporal bipartite network W in Fig. 2. Let $\theta = 3$, $\tau = 4$, $\alpha = 2$, $\beta = 3$ and vertex is u_1. Fig 3(a) shows the (3, 4)-persistent(2, 3)-core including vertex u_1. And Maximal (3, 2, 3)-persistent-core interval I is [7,11], the length of $I \geq \theta$, and $\leq \tau$. In interval I, any projected graph with 3-length interval is (2, 3)-core.

3 PCSearch Algorithm

In the study of temporal bipartite graphs, we found some characteristics of temporal bipartite graph, which are now listed below.

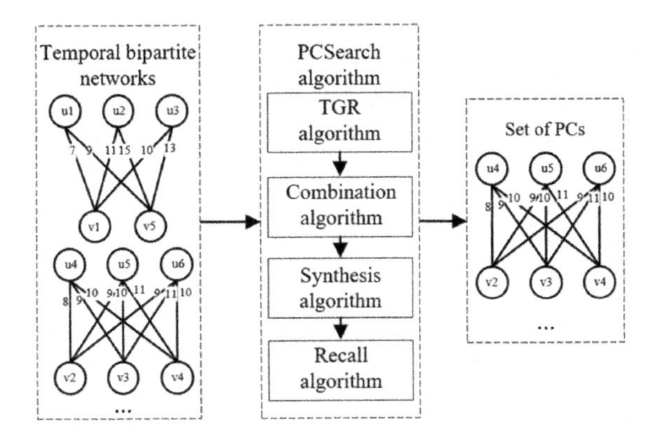

Fig. 4. Solution Overview

Observation 1. In the bipartite graph, for a (α, β)-core, the number of upper vertices of must be $\geq \beta$, and the number of lower vertices must be $\geq \alpha$.

Solution Overview. According to Definition 2, we have the following lemma.

Lemma 1. For the Maximal(θ, α, β)-persistent-core interval I, if t_1 and t_2 belong to this interval I and $0 < t_2 - t_1 < \theta$, then the projected graph in the interval $[t_2, t_1 + \theta]$ must be a (α, β)-core.

Proof. According to Definition 2, the intervals $[t_1, t_1+\theta]$ and $[t_2, t_2+\theta]$ are (α, β)-core. And according to Definition refdef2 shows that assuming $t_1 < t_3 < t_2$, then the projected graph on the interval $[t_3, t_3+\theta]$ must be a (α, β)-core, which means that the projected graph in the interval $[t_2, t_1 + \theta]$ must be a (α, β)-core.

As shown in Fig. 4, we propose a PCSearch algorithm to solve the persistent (α, β)-core community search problem, and it can be divided into four parts. Specifically, we first propose the TGR algorithm that remove irrelevant vertices from the temporal bipartite graph without losing precision according to Definition 1. Then, according to Definition 3, we can find (α, β)-core on each $[t, t + \theta]$ interval in the reduced temporal bipartite graph by combination algorithm. After that, according to Lemma 1, we combine these θ-persistent (α, β)-cores chronologically to form a subgraph by temporal information. When synthesizing a temporal bipartite subgraph, it is necessary to pay attention to the vertices in queue P. If target $q \notin P$, it stops running.

The key idea behind the PCSearch algorithm is the branch-reduce-and-bound strategy, where the branch phrase includes Combination algorithm, the reduce phrase is to reduce the search space by using TGR algorithm, and the bound phrase is to design some pruning rules to prematurely terminate the branch phrase by using Synthesis and Recall algorithm. Based on this on the above steps algorithmic framework, we will introduce our technique in the following sections.

3.1 Temporal Bipartite Graph Reduction

According to Theorem 1, the problem of enumerating all (θ, τ)-persistent(α, β)-core including vertex q in the temporal bipartite network is NP-hard, so the direct search method is not desirable. In order to solve this problem, we propose the TGR algorithm, which can efficiently cut out those irrelevant edges and narrow the search range.

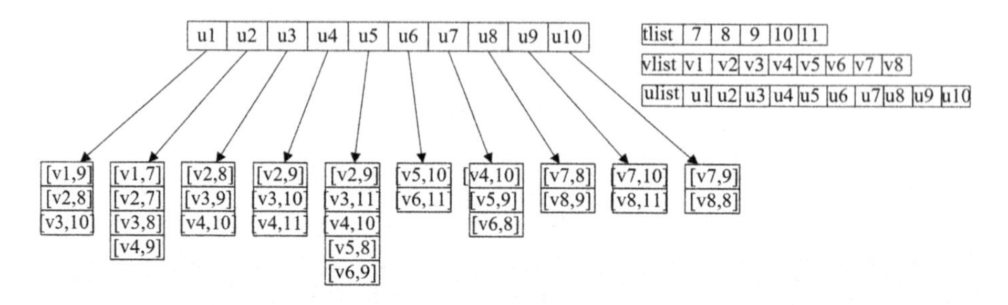

Fig. 5. An index of the upper vertex

In order to remove extraneous vertices, we propose the following rules.

Rule 1: For the degree information of a single vertex, a preliminary degree calculation is performed according to the given parameter α, β, and filter out the vertices that do not meet the degree requirements.

Rule 2: Since in the problem, our goal is to find the (θ, τ)-persistent(α, β)-core including vertex q, so remove all subgraphs that do not contain vertex q.

Through the above rules, we propose the TGR algorithm, which can reduce the irrelevant vertices of the temporal bipartite graph.

To improve the efficiency of the TGR algorithm, we propose the establishment of an index. The first layer consists of the upper or lower vertices, and the second layer contains their corresponding neighboring vertices and temporal information. Additionally, there are three sets: one to store upper vertex information, one to store lower vertex information, and one to store temporal information. Consider the temporal bipartite network W in Fig. 1. Construct an index of the upper vertex as the first layer, as shown in Fig. 5.

Example 2. Consider the index of the upper layer vertex in Fig. 5. Let $\alpha = 2$, $\beta = 3$ and vertex is u_1. Firstly, remove subgraphs that do not include vertex u_1, such as subgraphs composed of u_7, u_8, and u_9 and their neighbor. secondly, vertices that do not meet the degree of vertices are removed, such as vertex v_1.

Algorithm 1. TGR(W, q, α, β)

Input: W, q, α, and β
Output: list

```
1: list ← All-Components(G); // compute the connected components
2: for all G ∈ list do
3:    if q ∩ G = ∅ then
4:       remove G from list;
5:       continue;
6:    flag = false;
7:    while flag = false do
8:       if G = ∅ then
9:          remove G from list;
10:         break;
11:      u, α_min ← the minimum vertex degree in U(G);
12:      if α_min < α then
13:         remove u from G;
14:      else
15:         v, β_min ← the minimum vertex degree in L(G);
16:         if β_min < β then
17:            remove v from G;
18:         else
19:            flag = true;
20: return list;
```

3.2 Combination Algorithm

After the previous TGR algorithm reduction, we get that the temporal bipartite graphs in the list both contain the vertex u_1 and satisfy the (α, β)-cores. In this section, we introduce an Combination algorithm.

By Definition 3 and Observation 1 above, we know that the projected graph is the (α, β)-core in each θ interval. So we propose a combination algorithm based on the permutation and combination idea.

The Key Idea of the Algorithm: Since Algorithm 2 is based on the idea of permutation combinations, it is necessary to pass some necessary parameters for recursion. Among them, IPos is a fixed bit, IProc represents the starting subscript, ITol represents the termination subscript, and the des array stores the subscript of the widget, and the data array stores the subscript of graph G. In recursion, to prevent out-of-bounds access, IProc and ITol need to be compared (Lines 2-3). For components composed of β upper vertices, firstly, arrange the temporal information in this component in descending order (Lines 4-7), and then delete the projected graph of the θ-interval of each temporal information (Lines 8-13). If the number of upper vertices in projected graph satisfies β after reduction, then this component is put into the P (Lines 14-20).

Algorithm 2. Combination (IPos, IProc, ITol, des, data, $G, q, \alpha, \beta, \theta$)

Input: IPos, IProc, ITol, des, data, $G, q, \alpha, \beta, \theta$
Output: P
1: list $\leftarrow \emptyset$; $T \leftarrow \emptyset$;
2: **if** IProc > ITol **then**
3:     **return;**
4: **if** IPos = β **then**
5:     **for** i $\in$ des **do**
6:         T $\leftarrow$ G.index[i].temp;
7:     sort(T);// Sort in descending order
8:     **while** $|T| \neq 0$ **do**
9:         t $\leftarrow$ T.pop(); $G_c \leftarrow \emptyset$;
10:         **for** i $\in$ des **do**
11:             **if** t+$\theta \geq$ G.index[i].temp $\geq$ t **then**
12:                 G_c.push(G.index[i]);
13:         TGR(G_c, α, β);//TGR algorithm that does not consider vertices
14:         **if** $|G_c| = \beta$ **then**
15:             list.push(G_c);
16:     **return;**
17: **else**
18:     Combination (IPos, IProc+1, ITol,des, data, $G, q, \alpha, \beta, \theta$);
19:     des[iPos]=data[iProc];
20:     Combination (IPos+1,IProc+1,ITol,des, data, $G, q, \alpha, \beta, \theta$);
21: **return** P;

Example 3. Consider the Index of the upper layer vertex in Fig 6. Let $\alpha = 2$, $\beta = 3$ and vertex is u_1. After the Algorithm 2, some components in the P are shown in Fig. 6.

Through the Algorithm 2, the reduced temporal bipartite graph is decomposed into components with the length of time interval $\leq \theta$, and the upper vertex degree of each component is $\geq \alpha$ and the lower vertex degree is $= \beta$.

3.3 Recall Algorithm

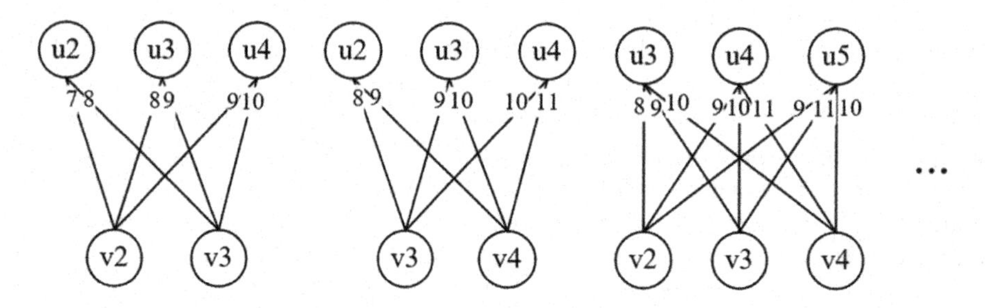

Fig. 6. Components in the P

Algorithm 3. Recall$(P, G, q, \alpha, \beta, \theta)$

```
Input: P, G, q, α, β, θ
Output: C
 1: CM ← All-Components(G); // compute the connected components
 2: for all C ∈ CM do
 3:   if q ∩ C = ∅ then
 4:     continue;
 5:   for edge ∈ C do
 6:     if edge. t +θ < C.t_e then
 7:       num_1, num_2 ← countMax(C, e);
 8:       if num_1 < α or num_2 < β then
 9:         remove edge from C;
10:     if edge. t −θ > C.t_s then
11:       num_1, num_2 ← countMin(C, e);
12:       if num_1 < α or num_2 < β then
13:         remove edge from C;
14: return C;
```

Because in the later Algorithm 4, the obtained subgraph only considers the component synthesis of different time periods, and does not consider the timestamp of a single vertex, which causes some vertices that do not meet the requirements of Definition 2, so we propose Algorithm 3 to prune the composite component operation for the subgraph that does not meet Definition 2.

The Key Idea of the Algorithm: First, skip the connected subgraphs of the input graph G that do not contain q (Lines 1-4). Then, examine the subgraphs that contain q based on temporal information. For edges with temporal information t less than $C.t_e - \theta$, check if there are more than α edges greater than t at the upper vertex and more than β edges greater than t at the lower vertex. Remove the edge from C if the requirements are not met (Lines 5-9). For edges where the temporal information t is greater than $C.t_s + \theta$, check if there are more than α edges less than t at the upper vertex and more than β edges less than t at the lower vertex. Similarly, remove the edge from C if the requirements are not met (Lines 10-13). Finally, return the subgraph C (Line 14).

3.4 Synthesis Algorithm

Through the previous Algorithm 3, we get the queue P where the components are stored. In this subsection, we will use Algorithm 4 on components in P to synthesize components into subgraphs.

Algorithm 4. Synthesis $(P, q, \alpha, \beta, \theta, \tau)$

Input: $P, q, \alpha, \beta, \theta, \tau$
Output: Q
1: $Q \leftarrow \emptyset$;
2: sort(P);
3: **while** $|P| \neq 0$ **do**
4: $R \leftarrow \emptyset$;
5: **if** $P \cap q = \emptyset$ **then**
6: **break;**
7: **for** $C \in P$ **do**
8: **if** $|R| = 0$ **then**
9: R.push(C); remove C from P; continue;
10: $C' = R$.pop();
11: **if** $C.t_s = C'.t_s$ and $C.t_e = C'.t_e$ **then**
12: R.push(C, C'); remove C from P;
13: **else if** $C.t_s \geq C'.t_s$ and $C.t_s < C'.t_e$ and $C.t_e \leq C'.t_s + \theta$ **then**
14: R.push(C, C'); remove C from P;
15: **else**
16: break;
17: $C \leftarrow \emptyset$; $C_s \leftarrow \emptyset$;
18: $C \leftarrow R$;
19: **while** $C \neq C_s$ **do**
20: $C_s = C$;
21: $C \leftarrow$ Recall($P, G, q, \alpha, \beta, \theta$);
22: $C \leftarrow$TGR(C, q, α, β);//TGR algorithm with a return value of C
23: **if** $\theta < C.t_e - C.t_s \leq \tau$ and $C \cap q \neq \emptyset$ **then**
24: Q.push(C);
25: **return** Q;

According to the previous Lemma 1, knowing the temporal information $t_1, t_2 (0 < t_2 - t_1 < \theta)$ in the I of a (θ, τ)-persistent (α, β)-core, then the projected graph on the interval $[t_2, t_1 + \theta]$ must be a (α, β)-core. Then our Algorithm 4 came into being. Firstly, sort P according to t_s in ascending order, compare t_e if t_s are equal (Line 1), then check P to whether it contains q(Lines 2-5), then synthesize t_s and t_e the same, and expand the temporal information of the subgraph C (Lines 7-13). Of course, because we only consider the component synthesis of different intervals and do not consider the vertices in the component, we have to check the subgraph C and delete the unqualified vertices (Lines 14-18), and because deleting the vertices will cause a chain reaction, we use Algorithm 1 to delete the vertices whose degrees are not satisfied(Line 19).

Example 4. Consider the P in Fig 6. Let $\alpha = 2$, $\beta = 3$, $\theta = 3$, $\tau = 4$, and vertex is u_1. Before the Algorithm 3, subgraph C are shown in Fig. 7. The red line is the temporal edge$[u_6, v_6, 11]$ that does not meet Definition 3 in subgraph C, and needs to be deleted by Algorithm 3, after reducing the temporal edge$[u_6, v_6, 11]$, and the other edges of subgraph C do not meet Definition 3, such as $[u_5, v_6, 9]$, etc., then Algorithm 1 needs to be used for deletion.

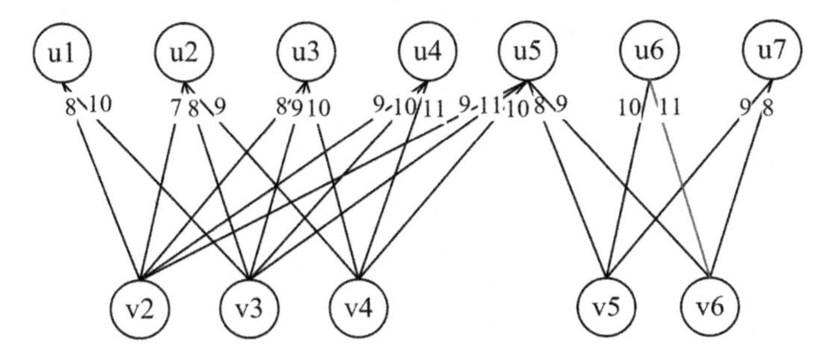

Fig. 7. Subgraph C before the Algorithm 3

3.5 The Pseudo-code of PCSearch Algorithm

Algorithm 5. PCSearch $(W, q, \alpha, \beta, \theta, \tau)$

Input: $W, q, \alpha, \beta, \theta, \tau$
Output: S
1: list$\leftarrow$ TGR(W, q, α, β); $S \leftarrow \emptyset$;
2: **for** $G \in$ list **do**
3: des.resize(β);
4: $P \leftarrow \emptyset$;
5: data $\leftarrow$ $cursor(G)$//Put the foot-tags into the array
6: $P \leftarrow$Combination (0, 0, G, des, data, $G, q, \alpha, \beta, \theta$)
7: $Q \leftarrow$Synthesis $(P, q, \alpha, \beta, \theta, \tau)$;
8: $S \cap Q$;
9: **return** S;

After introducing the three previous sub-algorithms, we can now present our Algorithm 5, which enumerates all the (θ, τ)-persistent (α, β)-cores in the temporal bipartite network that contain the vertex q.

4 Experiment

In this section, we first evaluate the efficiency of Algorithm 1 for temporal bipartite networks of different sizes. Finally, we compare the efficiency of Algorithm 5 with other algorithms.

We implement the following algorithm: i) **PCSearch algorithm**: the algorithm presented in this article; ii) **Index algorithm** [9]: it proposes an algorithm, which is suitable for enumerating the maximum (θ, τ)- persistent (α,β)-core that contains target q after our modification; iii) **Basic algorithm:** the temporal bipartite networks are screened one by one according to the temporal information, and finally the subgraphs are selected; As a reference for PCSearch algorithm. All algorithms are implemented in C++. All experiments are conducted on a computer with a Intel i7-12700 h 2.3 GHz CPU and 32 GB memory running Windows 11.

Table 1. Summary of Datasets.

Dataset	$\|U\|$	$\|L\|$	$\|E\|$
mrv	175	1146	6676
nsq	200	2311	7745
bsl	409	1841	8809
qta	316	2781	9857
bta	343	3530	10561
bbn	400	5288	11162
bka	258	5009	12212
nbg	324	4133	13336
mmr	281	4056	14353

Datasets. We use 9 real-world temporal bipartite networks in the experiments. The detailed statistics of our datasets are summarized in Table 1. All the datasets are downloaded from KONECT[1].

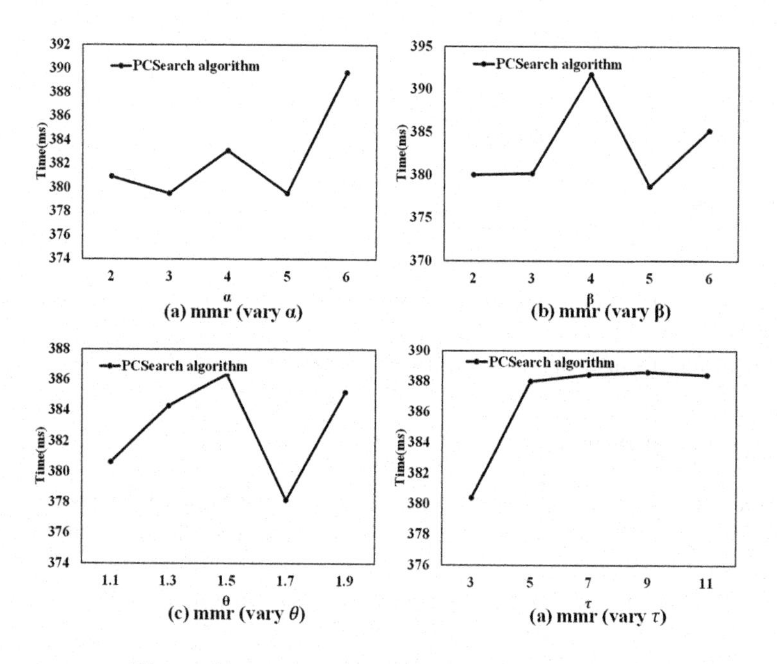

Fig. 8. Running time with different parameters

Exp 1. In this experiment, we observed the runtime of the PCSearch algorithm by varying parameter sizes. The mmr dataset was used, as shown in Fig. 8(a) (b).

[1] http://www.konect.cc/networks/.

As the degree increased, the algorithm's runtime also increased due to the need to delete more nodes. Figure 8(c) demonstrated that within a certain range, increasing the θ value also increased the algorithm's runtime. However, at $\beta = 5$ and $\theta = 1.7$, there was a sudden decrease in runtime. This occurred because the TGR algorithm phase deleted all nodes and subgraphs that didn't meet the degree requirement, significantly reducing the runtime. Figure 8(d) indicated that as the τ value increased, the algorithm's runtime leveled off since the subsequent searched subgraphs were identical.

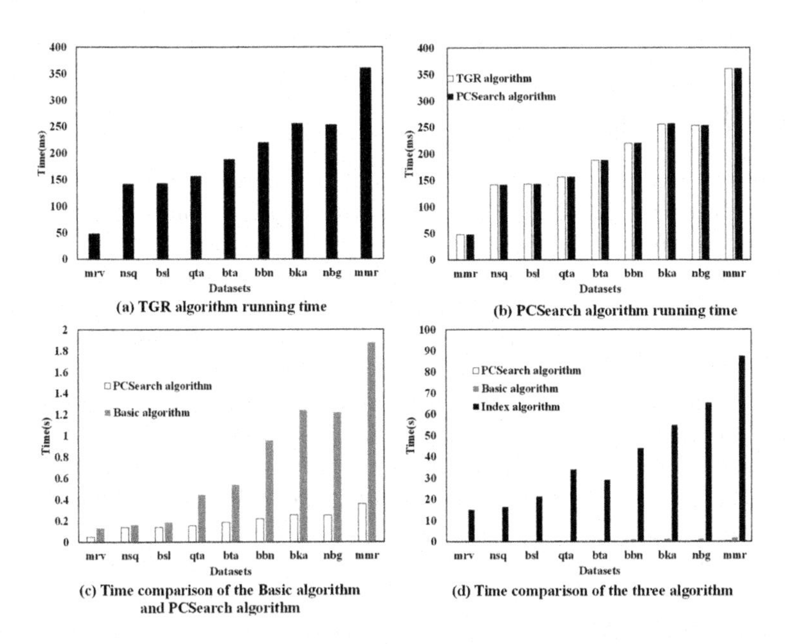

Fig. 9. Running time

Exp 2. In this experiment, the running time of TGR algorithm in temporal bipartite networks of different sizes is evaluated. As can be seen from Fig. 9(a), as the edges of the input temporal bipartite network increase, the running time of the algorithm also increases. The running time of TGR algorithm between *mrv* and *nsq* datasets increases abruptly because most of the subgraphs in the *mrv* dataset do not satisfy Rule 1.

Exp 3. In this experiment, we evaluate the effectiveness of TGR algorithm, because our PCSearch algorithm contains TGR algorithm, we run TGR algorithm and PCSearch algorithm on temporal bipartite networks of different sizes, as can be seen from Fig. 9(b), through their run time comparison, we can see that the run time of TGR algorithm and PCSearch algorithm are almost equal, which means that TGR algorithm can greatly reduce useless vertices and subgraphs.

Exp 4. In this experiment, we compare the running time of three algorithms in temporal bipartite networks of different sizes. As can be seen from Fig. 9(d), the running time of the Index algorithm is several orders of magnitude higher than that of the algorithm we propose, because the Index algorithm needs to establish an index for the entire temporal bipartite network, and then search for subgraphs that meet the criteria at index, and reduce and filter the subgraphs; For the Basic algorithm, as can be seen from Fig. 9(c), for datasets with less temporal information, the efficiency is higher than that of PCSearch algorithm, but if the temporal information is increased, the running time is greatly increased; This is because the Basic algorithm needs to sort and filterer for temporal information, and the running time and the amount of temporal information are positively correlated.

5 Related Works

In previous studies of Community search, different models of community cohesion have been proposed. Yao et al. [12] studied the identification of similar-bicliques in bipartite graph, looking for similar-bicliques by proposing the concept of similarity. Wang et al. [9] proposed to store (α, β)-cores that have been found in advance by indexing; This approach is very efficient in finding a specified (α, β)-core, but maintenance is more expensive for updating a small portion of the edge. Yang et al. [10] enumerate the (p, q)-clique model in the bipartite graph, and the proposed algorithm can be extended to the uncertain bipartite graph. Taher Alzahrani et al. [1] look for maximal bicliques by vertex similarity and propose a method that detects an order-limited number of overlapping maximal bicliques covering the network. In [5], Liu et al. first proposed storing bicore using an indexing approach and also introduced an indexing maintenance method. Yang et al. proposed a model of Bi-triangle to the bipartite graph [11]. Chen first proposed the problem of identifying (α, β)-core minimization in bipartite graphs in [2]. Lyu et al. [6] studied finding the biclique with the largest number of sides in a bipartite graph and solving the problem by decomposing the problem. Wang et al. [8] study the bitruss decomposition problem which aims to find all the k-bitrusses for $k \geq 0$. And proposed a novel online indexing algorithm, which reduces the time complexity of the existing algorithm. Doron et al. [7] study the problem of finding maximum edge bicliques in convex bipartite graphs. Li et al. [4]proposed a PC model in the temporal graph, which not only considers the cohesion in the time series interval, but also considers the relationship between each temporal interval, and proposes a novel branching and bond algorithm. But the authors only considered long-term cohesion, while sustained communities for short periods of time were not taken into account, and there was no query for subgraphs containing target q.

6 Conclusion

In this paper, we study the problem of searching for maximum (θ, τ)- persistent (α,β)-core that contains target q on a temporal bipartite graph. In order to solve this problem, we design a branch-reduce-bound algorithm, PCSearch.

Specifically, we first reduce the temporal bipartite network without losing accuracy, and then decompose the reduced temporal bipartite graph into small components C, and finally synthesize all the components C into (θ, τ)- persistent (α,β)-core that contains target q according to the temporal information. We conducted extensive experiments on real-world graphs, and the results proved the effectiveness of our proposed method.

Acknowledgement. This study was funded by the National Natural Science Foundation of China (Nos. 62072220), the Natural Science Foundation of Liaoning Province (Nos. 2022-KF-13-06, 2022-BS-111), the National Key Research and Development Program of China (No. 2022YFC3004603), and the Prosper Liaoning Talent Project (XLYC2203003).

References

1. Alzahrani, T., Horadam, K.: Finding maximal bicliques in bipartite networks using node similarity. Appli. Netw. Sci. **4**(1), 1–25 (2019)
2. Chen, C., Zhu, Q., Wu, Y., Sun, R., Wang, X., Liu, X.: Efficient critical relationships identification in bipartite networks. World Wide Web **25**(2), 741–761 (2021)
3. Kim, J., Guo, T., Feng, K., Cong, G., Khan, A., Choudhury, F.M.: Densely connected user community and location cluster search in location-based social networks. In: SIGMOD, pp. 2199–2209 (2020)
4. Li, R.H., Su, J., Qin, L., Yu, J.X., Dai, Q.: Persistent community search in temporal networks. In: ICDE, pp. 797–808 (2018)
5. Liu, B., Yuan, L., Lin, X., Qin, L., Zhang, W., Zhou, J.: Efficient (α, β)-core computation: an index-based approach. In: The World Wide Web Conference, pp. 1130–1141 (2019)
6. Lyu, B., Qin, L., Lin, X., Zhang, Y., Qian, Z., Zhou, J.: Maximum and top-k diversified biclique search at scale. VLDB J. **31**(6), 1365–1389 (2022)
7. Nussbaum, D., Pu, S., Sack, J.R., Uno, T., Zarrabi-Zadeh, H.: Finding maximum edge bicliques in convex bipartite graphs. Algorithmica **64**, 311–325 (2012)
8. Wang, K., Lin, X., Qin, L., Zhang, W., Zhang, Y.: Towards efficient solutions of bitruss decomposition for large-scale bipartite graphs. VLDB J. **31**(2), 203–226 (2022)
9. Wang, K., Zhang, W., Lin, X., Zhang, Y., Qin, L., Zhang, Y.: Efficient and effective community search on large-scale bipartite graphs. In: ICDE, pp. 85–96 (2021)
10. Yang, J., Peng, Y., Ouyang, D., Zhang, W., Lin, X., Zhao, X.: (p, q)-biclique counting and enumeration for large sparse bipartite graphs. VLDB J., 1–25 (2023)
11. Yang, Y., Fang, Y., Orlowska, M.E., Zhang, W., Lin, X.: Efficient bi-triangle counting for large bipartite networks. PVLDB **14**(6), 984–996 (2021)
12. Yao, K., Chang, L., Yu, J.X.: Identifying similar-bicliques in bipartite graphs. PVLDB **15**(11), 3085–3097 (2022)
13. Zhang, Y., Phillips, C.A., Rogers, G.L., Baker, E.J., Chesler, E.J., Langston, M.A.: On finding bicliques in bipartite graphs: a novel algorithm and its application to the integration of diverse biological data types. BMC Bioinform. **15**, 1–18 (2014)
14. Zhang, Y., Wang, K., Zhang, W., Lin, X., Zhang, Y.: Pareto-optimal community search on large bipartite graphs. In: CIKM, pp. 2647–2656 (2021)
15. Zhou, A., Wang, Y., Chen, L.: Butterfly counting and bitruss decomposition on uncertain bipartite graphs. VLDB J, 1–24 (2023)

Applications (Including Industry Track Papers)

D-Score: A White-Box Diagnosis Score for CNNs Based on Mutation Operators

Xin Zhang[1(✉)], Yuqi Song[1], Xiaofeng Wang[1], and Fei Zuo[2]

[1] University of South Carolina, 29210 Columbia, SC, USA
{xz8,yuqis}@email.sc.edu, wangxi@cec.sc.edu
[2] University of Central Oklahoma, 73034 Edmond, OK, USA
fzuo@uco.edu

Abstract. Convolutional neural networks (CNNs) have been widely applied in many safety-critical domains, such as autonomous driving and medical diagnosis. However, concerns have been raised with respect to the trustworthiness of these models: The standard testing method evaluates the performance of a model on a test set, while low-quality and insufficient test sets can lead to unreliable evaluation results, which can have unforeseeable consequences. Therefore, how to comprehensively evaluate CNNs and, based on the evaluation results, how to enhance their robustness are the key problems to be urgently addressed. Prior work has used mutation tests to evaluate the test sets of CNNs. However, the evaluation scores are black boxes and not explicit enough for what is being tested. In this paper, we propose a white-box diagnostic approach that uses mutation operators and image transformation to calculate the feature and attention distribution of the model and further present a diagnosis score, namely D-Score, to reflect the model's robustness and fitness to a dataset. We also propose a D-Score based data augmentation method to enhance the CNN's performance to translations and rescalings. Comprehensive experiments on two widely used datasets and three commonly adopted CNNs demonstrate the effectiveness of our approach.

Keywords: CNNs · computer vision · data augmentation

1 Introduction

Recently, convolutional neural networks (CNNs) have been increasingly used in safety-critical applications, including medical diagnostics [16,19] and autonomous vehicles [7,8]. Despite their impressive success, CNNs still face challenges related to robustness and accuracy. For instance, crashes caused by autonomous cars from Tesla and Google have led to significant losses [4]. Therefore, as with traditional software, testing is essential for CNN-based systems, which can effectively identify issues and improve the system's trustworthiness and robustness [29].

The most common method for evaluating CNNs is to assess its performance on selected evaluation metrics using a test set [17,24,29]. However, this method

X. Yang et al. (Eds.): ADMA 2023, LNAI 14180, pp. 343–358, 2023.
https://doi.org/10.1007/978-3-031-46677-9_24

heavily relies on the quality of the test set. In other words, if most instances in the test set have features similar to those in the training set, the model's testing results will likely be good. Conversely, due to reasons such as biased training data, overfitting, and underfitting, the trained model may show unexpected or incorrect behaviors on a test set that contains many corner cases [21]. Thus, when it comes to safety-critical areas, testing a trained model on an unevaluated test set and making decisions based on the test results can lead to catastrophic consequences.

To comprehensively evaluate CNNs, researchers have proposed several approaches, which can be divided into two categories. The first category involves introducing the traditional software engineering testing method, mutation testing, to CNNs [10,11,22]. This approach applies carefully designed mutation operators [17] to the CNN model to generate multiple variants. The higher the number of differences between the predictions of the variant models and the original model, the higher the quality of the test set. However, the score itself remains a black box, and the reasons behind the low quality of the test set are still unknown. Additionally, effective methods for selecting and combining mutation operators to detect test set quality remain unexplored [20]. The second category of approaches is based on neuron coverage [4,21,28]. These methods use gradient ascent to solve a joint optimization problem that maximizes both neuron coverage and the number of potentially erroneous behaviors, and eventually generate a set of test inputs [21]. However, as noted in [9], higher neuron coverage can lead to fewer defects detected, less natural inputs, and more biased prediction preferences. Therefore, developing effective methods for providing white-box scores for CNNs and proposing methods for enhancing these scores is critical for improving robustness and accuracy of CNNs.

This paper addresses the problem of how to diagnose CNNs using a white-box approach. First, we divide the neurons of each convolutional layer into several regions and delete the neurons of each region, enabling us to study the overall feature distribution of the test set due to the spatial character of CNNs. We then analyze the attention of the CNN model towards different regions by applying well-designed image transformations such as padding for given directions. Based on the overall feature distribution and attention distribution, we introduce the concept of D-Score for CNNs that reflects their robustness and fitness, where "robustness" of a CNN is the ability of the CNN in recognizing objects at any location of an image (e.g., translation invariance) and "fitness" of a CNN on a dataset means how well the attention of the CNN meets the feature distribution of the dataset. Based on the D-Score, we propose a scoring-guided data augmentation strategy to enhance CNN's robustness. It is known that CNNs are generally not robust enough to image transformation (e.g., translations and/or rescalings of the input image may drastically change the prediction of a CNN [1,30]). Although data augmentation is consistently considered as an effective strategy to address this issue [23], our experiments show that randomly and blindly selecting data augmentation techniques with little knowledge on the dataset will significantly limit such effectiveness. Instead, our data augmentation strategy fully utilizes D-Score, resulting in more targeted selection and design of augmentation techniques as well as their execution probabilities. This approach can effectively adjust the original feature distribution of the dataset, making the trained model more robust and reducing blind spots.

The main contributions are summarized as follows:

- By analyzing the impact of mutation operators on the accuracy of CNNs through deleting neurons in different regions and applying image transformations, we develop an approach to calculate the overall feature distribution of a dataset, as well as the attention distribution of models. This allows us to white-box diagnose CNNs and introduce a new concept of D-Score for CNN diagnosis, providing valuable insights into their performance.
- In order to showcase the efficiency of our D-Score and enhance the robustness of CNNs, we introduce a score-guided data augmentation approach that tackles the problem of CNNs' sensitiveness to shifts and rescalings.
- Our scoring and data augmentation method has been rigorously tested on two widely used datasets and three commonly adopted CNNs for these datasets, with comprehensive experiments confirming its effectiveness.

2 Mutation Testing for Deep Learning

The growing popularity of deep learning has raised concerns about the robustness and reliability of deep neural networks (DNNs), leading to a rise in research interest in mutation testing for deep learning [10,11,17,20]. However, unlike traditional software, where the decision logic is coded by developers, the behavior of deep learning systems is determined by the structures of DNNs and the parameters in the network [25] which is hard to foresee. Moreover, due to the randomness of the training process, it is common to observe different decisions when a DNN is retrained on the same dataset, even without any mutation operations. This makes it difficult to apply the mutation-killing metric of traditional software to learning systems directly [13].

In order to make the application of mutation testing in deep learning systems feasible, several different approaches have been proposed. MuNN [22] proposes five mutation operators, including replacements or deletions of neurons, activation functions, and parameters on trained CNNs. DeepMutation [17] focuses on mutation at both source-level and model-level, generating two types of operators to mutate the training data, model structures during and after training, and model parameters. DeepMutation++ [10] expands the prior work from CNN to RNN. DeepCrime [11] defines mutation operators based on studies of real faults in learning systems. Furthermore, the mutation-killing metric in deep learning systems is discussed in detail in [13]. As pointed out in [20], however, the use of these methods for evaluating a test set can only reveal the number of variant models that can be discovered by this test set. These scores do not provide clear insight into the trustworthiness of models. Therefore, these scores remain black boxes and are not sufficiently explicit in terms of what exactly is being tested.

3 Characters of CNNs

Spatial Characteristics. The spatial characteristics of CNNs mainly originate from the way convolution is calculated. CNNs rely on convolution kernels, which are typically small in spatial dimensionality but spread throughout the entirety of the input following a left-to-right, top-to-bottom order, to convolve the input matrices [18]. When the data hits a convolutional layer, the layer convolves each filter across the spatial dimensionality of the input to produce a 2D activation map. As a result, the relative positions of the features in the input are maintained after being convolved [26].

Translation Variance. CNNs are often assumed to be invariant to small image transformations in theory [5,6]. However, recent studies have shown that this is not always the case. Small translations or rescalings of the input image can significantly alter the network's prediction [1,12]. This issue can arise due to the concentration of features in the dataset [1] or the insufficient network architecture [30]. If a CNN is robust, it should be able to recognize the object wherever it appears in the image. In this sense, the lack of translation invariance will be considered as a manifestation of the model's poor robustness, which indicates that the current model may not accurately recognize certain corner cases.

4 Proposed Method

4.1 D-Score

Models and data are two critical components of deep learning. On the one hand, the data effectively guides the model to learn. On the other hand, the model can accurately predict the data. To ensure the impartiality and representativeness of the proposed D-Score, let us start our scoring method with the interplay between the data and the model. Specifically, our method takes the following two aspects into consideration:

- Keep the dataset fixed and evaluate it through changing the model by utilizing mutation operators to get the feature distribution of datasets.
- Keep the model fixed and evaluate it through changing the dataset by using image transformations to obtain the attention distribution of the CNN .

The overall pipeline of our scoring approach is shown in Fig. 1.

Mutation Operators. There exist different mutation operators, such as deleting neurons, deleting layers, adding layers, and changing activation functions, to name a few [17,20]. Since our goal is to first analyze the feature distribution of the dataset based on the spatial characteristics of the CNN, we only adopt the mutation operators for deleting neurons [22] to generate variants. It is worth mentioning that in the existing approaches, the neurons to be deleted are usually chosen randomly since they focus more on identifying variants. However, random deletion cannot meet our requirements for the purpose of analyzing the feature

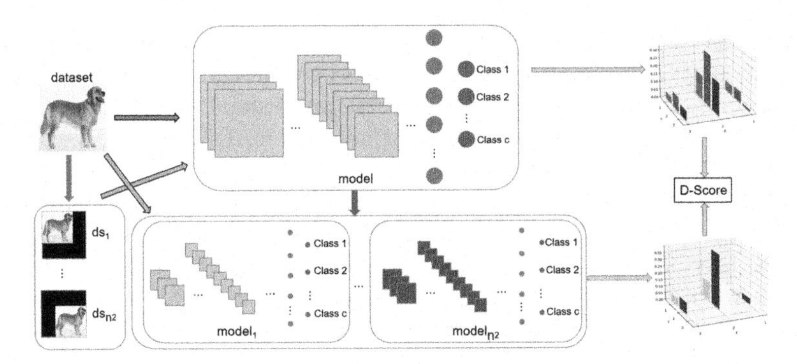

Fig. 1. The pipeline of D-Score calculation. Given a well-trained CNN (labeled as **model**), we generate n^2 model variants and n^2 new test sets using mutation operators and image transformations, respectively, and obtain $\mathbf{model}_1, \ldots, \mathbf{model}_{n^2}$ and $ds_1, \ldots, ds_{n^2}$. Next, we obtain the distribution of features and attention through the accuracy of n^2 model variants on the original test set and the accuracy of the original model on n^2 new test sets, respectively, based on which we calculate D-Score of **model**.

distribution of the dataset. Thus, we propose a region-based method to delete the neurons: First, divide the neurons of each convolutional layer into $n \times n$ equal-sized rectangular regions, where n is a hyperparameter, and index them in order from the upper-left corner to the lower-right corner before performing the deletion operation; Then, delete the corresponding regions of neurons with the same index in all convolutional layers from the well-trained CNN model to form n^2 variants, as shown in Fig. 2. This approach can prevent introducing the surrounding features into the deleted region (e.g., due to operations such as pooling) for a more accurate feature distribution of the dataset.

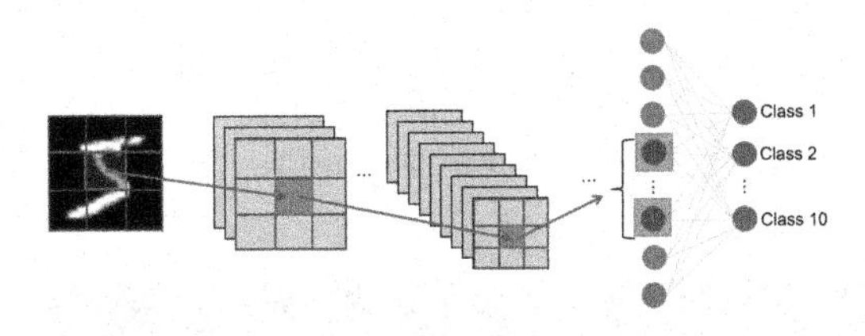

Fig. 2. Using deleting mutation operator to generate variants. In this case, we divide each convolutional layer into 3×3 regions, and then apply deleting operator on the region with index 5. The red masks in the convolutional layers indicate that the corresponding neurons in the 5th region will be deleted, which will result in the loss of input information at the fully connected layer. By "deleting a region of neurons", it means that this deletion operation is applied to the same region of every convolutional layer, which will result in the loss of features from certain regions of the input images. (Color figure online)

Feature Distribution. We take the original well-trained model as the baseline and calculate the accuracy difference between the variants and the baseline. Clearly, the larger the difference is, the worse performance the variant has, and the more important the related region of neurons is. Due to the spatial characteristics of the CNN, deleting the ith region of neurons implies loss of the information of the input image at a corresponding region when predicting. In this sense, the difference in accuracy of the ith variant model from the baseline can be used to represent the feature quantity of the dataset in that specific region. So we normalize the accuracy differences of the n^2 variants from the baseline to obtain the feature distribution of the dataset over regions:

$$\widetilde{f_i} = \frac{\max(f_b - f_i, 0)}{\sum_{i=1}^{n^2} \max(f_b - f_i, 0)}, \tag{1}$$

where $\widetilde{f_i}$ and f_i stands for the value of the i-th region after and before normalization, respectively, and f_b represents the accuracy of baseline model.

Image Transformation. To evaluate a model, we propose an image "translation" method to modify the dataset, which is different from the traditional translation that discards features outside the field of view (e.g., two sub-figures at the right side of the top row in Fig. 3). In our approach, we first divide the original image into $n \times n$ equal-sized regions and index them in order from the upper-left corner to the lower-right corner. Then we push the original image towards the target region by padding 0 around the original image. To be more specific, given the original image size $k \times l$ and a hyperparameter t, we fill $\frac{l}{t}$, $\frac{2l}{t}$, $\frac{2k}{t}$ and $\frac{k}{t}$ blank pixels in the top, bottom, left, and right directions of the original image, respectively, and then resize the newly generated image back to its original size so that it fits the dimensions of the trained model. The purpose of introducing t is to prevent the effective area of the image from being too small after transformation; otherwise, it may lead to low prediction accuracy. For example, the bottom row in Fig. 3 shows the translated images with different values of t, given $n = 4$ (16 regions in total) and the target region 7. With image transformation, we obtain n^2 new test sets.

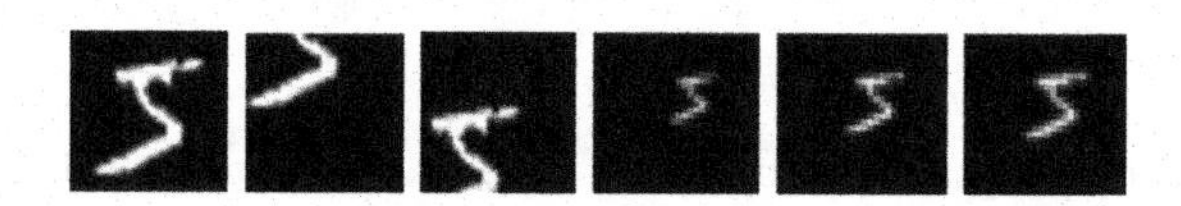

Fig. 3. Image transformation. The first one is the original image, followed by two images after traditional translation operations, which demonstrate a significant loss of information. The last three show the images after applying our image transformation method with $t = 2, 3, 4$, respectively.

Attention Distribution. We use the original trained model to predict the results on the newly generated n^2 test sets and then use the model's accuracy on the ith test set, as the indicator of the model's attention to the ith region of an image. Intuitively, the better the model performs on the ith test set, the more attention it pays to the ith region. To make it consistent with the feature distribution of datasets, we normalize the attention to obtain the attention distribution:

$$\widetilde{a}_i = \frac{a_i}{\sum_{i=1}^{n^2} a_i}, \tag{2}$$

where a_i and $\widetilde{a}_i$ stands for the accuracy before and after normalization.

D-Score Calculation. The similarity between the feature distribution and the attention distribution is crucial for successful learning. When these two distributions are close, namely that the model can focus exactly on the feature-dense regions of the dataset, the model's accuracy will be excellent on this dataset. Otherwise, it means that the model cannot capture the features of the dataset well. Therefore, $\frac{1}{n^2}\sqrt{\sum_{i=1}^{n^2}(\widetilde{f}_i - \widetilde{a}_i)^2}$ can, to some extent, reflect the accuracy degradation. Keeping this in mind, we define the fitness of CNN on a dataset as

$$v_{\text{fitness}} = \hat{a} - \frac{1}{n^2}\sqrt{\sum_{i=1}^{n^2}(\widetilde{f}_i - \widetilde{a}_i)^2}, \tag{3}$$

where $\hat{a}$ is the accuracy of the original trained model on the original test set.

To define the robustness index, let $a_{\text{avg}} = \frac{1}{n^2}$, which represents the average attention distribution of the model over n^2 regions, i.e., the model pays equal attention to each region of an image. So the difference between $\widetilde{a}_i$ and a_{avg} represents the unbalanced attention of the model on the ith region. Similarly, let $f_{\text{avg}} = \frac{1}{n^2}$ be the average feature distribution of the dataset. We can define the robustness index as

$$\begin{aligned} v_{\text{robust}} = \frac{1}{n^2}\Bigg(&\sqrt{\sum_{i=1}^{n^2}(\widetilde{f}_i - f_{\text{avg}})^2} + \sqrt{\sum_{i=1}^{n^2}(\widetilde{a}_i - a_{\text{avg}})^2} \\ &+ \sqrt{\sum_{i=1}^{n^2}(a_i - \hat{a})^2}\Bigg). \end{aligned} \tag{4}$$

Obviously, if the first two terms in the right side of Eq. (4) are large, it indicates that the model's attention is concentrated on specific regions, reflecting the model's sensitivity to translation and scaling. The third term, $\sqrt{\sum_{i=1}^{n^2}(a_i - \hat{a})^2)}$, measures the difference of accuracy when images appear in different regions from the accuracy of the original model on the original dataset. A large value of this term means that the model has a poor ability in handling corner cases.

Notice that v_{robust} is bounded. Since $\widetilde{f}_i \in [0,1]$, we have

$$\frac{1}{n^2}\sqrt{\sum_{i=1}^{n^2}(\widetilde{f}_i - f_{\text{avg}})^2} \leq \frac{1}{n^2}\sqrt{(1-\frac{1}{n^2})^2 + (n^2-1)(\frac{1}{n^2})^2}$$
$$= \frac{\sqrt{n^2-1}}{n^3}.$$

The maximum is achieved when $\widetilde{f}_i = 1$ for a specific i and the other $\widetilde{f}_i = 0$. Similarly, we have

$$\frac{1}{n^2}\sqrt{\sum_{i=1}^{n^2}(\widetilde{a}_i - a_{\text{avg}})^2} \leq \frac{\sqrt{n^2-1}}{n^3}.$$

Let c denote the number of classes in the dataset. The worst accuracy of the original model and its variants is $\frac{1}{c}$ in a probabilistic sense, namely picking classes randomly. So the smallest value for a_i and $\hat{a}$ is $\frac{1}{c}$. Then

$$\frac{1}{n^2}\sqrt{\sum_{i=1}^{n^2}(a_i - \hat{a})^2} \leq \frac{1}{n^2}\sqrt{n^2(1-\frac{1}{c})^2} = \frac{1}{n}\frac{c-1}{c}.$$

Therefore, with the inequalities above, we have $v_{\text{robust}} \leq g(n)$ where

$$g(n) = \frac{2\sqrt{n^2-1}}{n^3} + \frac{1}{n}\frac{c-1}{c}. \tag{5}$$

With v_{fitness} and v_{robust}, we can define D-Score as

$$\text{D-Score} = v_{\text{fitness}} - v_{\text{robust}}. \tag{6}$$

Notice that a small v_{robust} and a large v_{fitness} are expected, which will result in a large D-Score, meaning that the model achieves high robustness and fitness. The efficiency of D-Score will be demonstrated in the next section.

4.2 Score-Guided Augmentation

To demonstrate the effectiveness of D-Score in reflecting the robustness of CNNs, we propose a score-guided method to address the problem of CNNs' insensitivity to translations and rescalings (i.e., when an image is translated or rescaled, the performance of CNN decreases) [1,12,30]. To address this issue, a common solution is to locate and adjust objects to be detected in input images before performing predictions, such as adding spatial transformations [12] before CNNs. Essentially, this solution aims to change the feature distribution of datasets, but requires training or other geometric methods to gain prior knowledge. Here we remove this requirement of prior knowledge using D-Score.

Our method primarily utilizes the image transformation approach described in Sect. 4.1, which involves adding empty values around the image and resizing it. It is critical in this method to determine the execution probability p, image size after padding d, and the number of empty values to be padded in the four directions (left, right, up, down), denoted as d_l, d_r, d_u, and d_d. Our design is inspired by the idea that a model with lower robustness needs to apply this data augmentation technique with higher probability and a broader range to enhance its robustness. Since v_{robust} is an effective measure of the model's robustness, we utilize it to directly determine both p and d. In our method, the probability of execution and the size after padding are determined by the model's robustness score, namely that

$$p = \frac{v_{\text{robust}}}{g(n)}, \quad d = (1 + p) \times \hat{d}, \tag{7}$$

where $g(n)$ is defined in (5), $\hat{d}$ stands for the original size. Notice that $p \in [0, 1]$ since $g(n)$ is an upper bound on v_{robust}.

It is worth mentioning that $p = 1$ means $v_{\text{robust}} = g(n)$, which implies the worst robustness. It is corresponding to the case where all features of the dataset are concentrated in one region ($\widetilde{f}_{i^*} = 1$ for a specific i^* and $\widetilde{f}_i = 0$ for other is). Similarly, the attention of the model focuses on one specific region ($\widetilde{a}_{j^*} = 1$ for a specific j^* and $\widetilde{a}_j = 0$ for other js). Therefore, data augmentation is imperative in this case. We calculate d_l and d_r as follows:

$$d_l = \text{random}(0, p \times \hat{d}), \quad d_r = p \times \hat{d} - d_l, , \tag{8}$$

where random means the uniform distribution over $[0, p \times \hat{d}]$. Similarly, we can obtain d_u and d_d.

5 Experiments

The experiments are conducted based upon two datasets: MNIST [3] and CIFAR-10 [14]. For MNIST, we consider two widely used CNN models proposed in [15,27]. For CIFAR-10, we use the CNN model in [2]. The structures of these CNN models are summarized in Table 1. As suggested in [17], We follow the instructions described in [2,15,27] to train these three models. After training, the MNIST model A (MMA) achieves an accuracy of 98.56% and an average loss of 0.0413 on the test set, while the MNIST model B (MMB) achieves an accuracy of 99.08% and an average loss of 0.0149%, representing the state-of-the-art performance. For the CIFAR-10 model (CM), its accuracy on the training set can reach 98.01%, while only 79.66% on the test set with an average loss of 0.7049. The performance of these three models is nearly identical to [17].

5.1 Methods for Comparison

Though there exist methods that can score a test set [10,17,22], they cannot be applied to evaluate CNNs. To the best of our knowledge, there are few methods that can be directly compared to D-Score for evaluating CNNs. To demonstrate the effectiveness of our scoring-guided augmentation method, we compare it with

Table 1. The structures of our selected CNN models, which are widely adopted for MNIST and CIFAR-10 in the prior work. We use these three models as baselines and apply mutation operators of deleting to generate variants.

MNIST model A [15]	MNIST model B [27]	CIFAR-10 model [2]
Conv(6,5,5)+ReLU()	Conv(32,3,3)+ReLU()	Conv(64,3,3)+ReLU()
MaxPooling(2,2)	Conv(32,3,3)+ReLU()	Conv(64,3,3)+ReLU()
Conv(16,5,5)+ReLU()	MaxPooling(2,2)	MaxPooling(2,2)
MaxPooling(2,2)	Conv(64,3,3)+ReLU()	Conv(128,3,3)+ReLU()
Flatten()	Conv(64,3,3)+ReLU()	Conv(128,3,3)+ReLU()
FC(120)+ReLU()	MaxPooling(2,2)	MaxPooling(2,2)
FC(84)+ReLU()	Flatten()	Flatten()
FC(10)+Softmax()	FC(200)+ReLU()	FC(256)+ReLU()
	FC(10)+Softmax()	FC(256)+ReLU()
		FC(10)+Softmax()

several other commonly used data augmentation methods listed in Table 2. It is worth pointing out that some augmentation operations can result in a change of meaning for the MNIST data due to its specificity. For instance, vertical flipping can cause a digit 6 to become a digit 9. Hence, we only apply Random Padding + Resize (RPR) to the MNIST.

Table 2. The data augmentation methods adopted for comparisons.

Dataset	Method	Parameters
CIFAR-10	Random Horizontal Flip (RHF)	prob = 0.5
	Random Vertical Flip (RVF)	prob = 0.5
	Random Rotation (RR)	degree in (0,180)
	Random Hor + Ver Flip (RHV)	prob = 0.5
	Random Padding + Resize (RPR)	randomly
MNIST	Random Padding + Resize (RPR)	randomly

5.2 Experimental Results

Due to the space limitation, this subsection only presents the experimental results for $n = 3$ as an example.

Feature Distributions. We conduct the deleting mutation operator on MMA, MMB, and CM by dividing each model into $n \times n$ regions and then deleting the corresponding region's neurons one by one. We summarize the performance of

Table 3. The performance of variants generated by deleting mutation operators when $n = 3$. We bold the best-performing variant and underline the worst one in each group.

Model	MMA			MMB			CM		
Accuracy of Ori Model	98.56%			99.08%			79.66%		
Accuracy of Variants	**98.01%**	86.52%	97.35%	**99.01%**	95.64%	98.59%	77.35%	75.76%	**77.75%**
	93.79%	<u>75.89%</u>	96.54%	97.92%	<u>79.43%</u>	95.42%	76.09%	<u>71.04%</u>	76.15%
	97.51%	91.26%	97.80%	98.93%	88.16%	98.60%	77.08%	75.56%	77.60%

Table 4. The performance of MMA, MMB, and CM on newly generated test sets through image transformations when $n = 3$. We bold the best-performing variant and underline the worst one in each group.

Model	MMA			MMB			CM		
Accuracy of Ori Test set	98.56%			99.08%			79.66%		
Accuracy on New Test sets	21.65%	36.56%	18.66%	21.03%	41.85%	29.63%	42.31%	48.61%	<u>40.90%</u>
	45.23%	**94.36%**	43.54%	56.20%	**96.39%**	62.80%	51.22%	**62.40%**	50.88%
	28.51%	39.65%	<u>11.03%</u>	30.01%	42.18%	<u>10.03%</u>	44.29%	55.11%	45.28%

each variant in Table 3. Notice that the smaller the value of n is, the less computation is required, but the coarser the partition of the region is. The second row in Table 3 shows the accuracy of the original models on the original datasets. The third row in Table 3 lists the accuracy of the model variants. Regarding MMA and MMB, the performance gap reaches approximately 20% over regions. For the CM model, the difference is about 5%. Moreover, it is observed that the variants resulting from deleting the central region typically cause the most significant performance degradation, while the performance reduction of those variants deleting edge neurons is relatively minor. We then convert the performance differences to feature distributions using (1) and plot them in Fig. 4. It shows that the features in both MNIST and CIFAR are relatively concentrated in the central region, which aligns with our understanding of these two datasets, demonstrating the effectiveness of our approach to obtain feature distribution.

Another observation is that in Table 3 deleting neurons in the upper left region of the MMA model results in a 0.55% decrease in accuracy, while for MMB, the decrease is only 0.07%. We can take advantage of this result and refine the structure of CNNs. For instance, we can reduce **one-ninth** of the parameters used in all convolutional layers in this case for computational efficiency without sacrificing accuracy.

Attention Distributions. We generate new test sets by performing image transformation presented in Subsect. 4.1 with $n = 3$ and the hyperparameter $t = 5$. Table 4 shows the accuracy of the well-trained models MMA, MMB, and CM on the generated new test sets. We then transform these accuracy values into the models' attention allocations for each region based on (2) and present the attention distribution in Fig. 5. Table 4 indicates that all three models have a similar pattern: they exhibit higher accuracy for targets located in the central region of the image, and lower accuracy for targets located on the edges, particularly at four corners. It suggests that the model tends to pay more attention to the central region, especially for the MMA and MMB models, where the max-

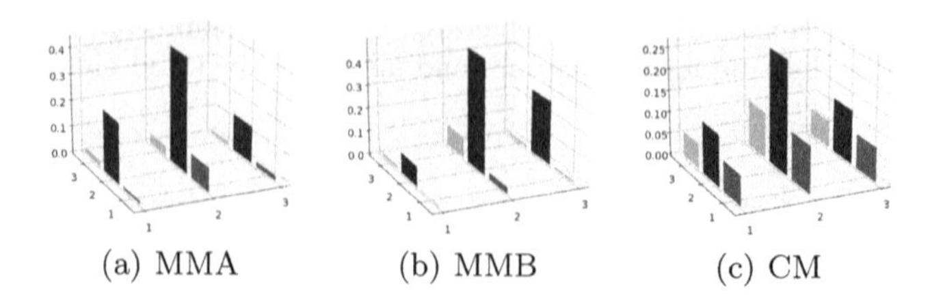

(a) MMA (b) MMB (c) CM

Fig. 4. The feature distribution calculated by deleting mutation operation on MMA, MMB, and CM when $n = 3$.

imum difference even reaches 72.71% and 86.33%, respectively. In comparison, the difference in the CM model is around 20%. This observation implies that the CM model exhibits stronger robustness with respect to the location of the object in an image. Similarly, the existence of the model's shift-invariance has also been demonstrated in this group of experiments.

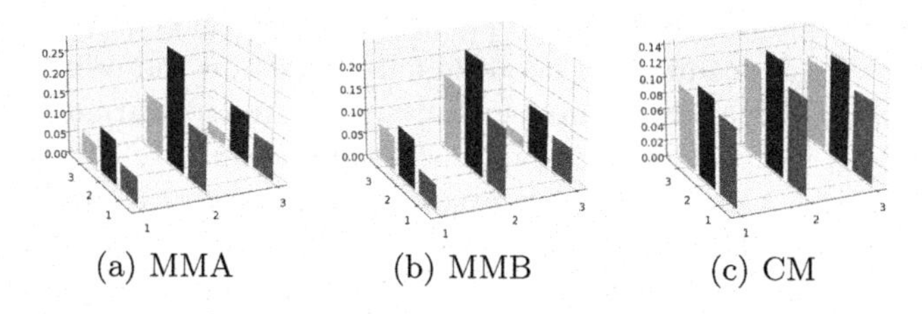

(a) MMA (b) MMB (c) CM

Fig. 5. The attention distribution calculated by feeding transformed test sets to MMA, MMB, and CM when $n = 3$.

Table 5. The D-Scores for MMA, MMB and CM. We set $\alpha = -1$ and $\beta = 1$ to equally evaluate the robust and accurate parts of these models. The highest score is indicated in bold, and the second best is underlined.

	v_{robust} (↓ better)	v_{fitness} (↑ better)	D-Score(↑ better)
MMA	0.2837	**0.9581**	<u>0.6744</u>
MMB	<u>0.2758</u>	<u>0.9527</u>	**0.6769**
CM	**0.1290**	0.7813	0.6523

Scoring. Combining the feature distribution and the attention distribution, we calculate D-Scores for the three models based on (6), as shown in Table 5. For robustness, the CM model performs the best with a score of 0.1290 (the lower the better), while MMA and MMB have scores of 0.2837 and 0.2758, respectively. This aligns with our understanding of the two datasets and the results

Table 6. The D-Score for CM with different data augmentation methods. We bold the best score and underline the second-best one.

Methods	a_i			$\widetilde{f}_i$			v_{robust}(↓ better)	v_{fitness} (↑ better)	D-Score (↑ better)
w/o Augmentation Loss:0.5694 <u>Acc:79.66%</u>	42.31%	48.61%	40.90%	7.095%	11.98%	5.866%	0.1290	0.7813	0.6523
	51.22%	62.40%	50.88%	10.96%	26.47%	10.78%			
	44.29%	55.11%	45.28%	7.924%	12.59%	6.327%			
RHF Loss:0.5694 **Acc:81.22%**	45.32%	52.10%	45.11%	3.543%	10.41%	6.011%	0.1221	**0.7934**	<u>0.6713</u>
	56.63%	66.98%	57.91%	6.634%	25.37%	8.257%			
	49.10%	58.23%	50.17%	9.870%	20.40%	9.491%			
RVF Loss:0.7573 Acc:73.70%	40.99%	47.32%	36.68%	2.678%	12.51%	4.407%	0.1265	0.7144	0.5878
	50.43%	58.73%	45.98%	10.64%	30.10%	9.152%			
	44.33%	49.17%	40.50%	8.576%	18.20%	3.723%			
RR Loss:0.9375 Acc:67.10%	32.10%	38.17%	27.98%	5.390%	7.338%	7.009%	0.1302	0.6576	0.5274
	41.23%	45.96%	34.47%	10.96%	22.60%	15.67%			
	35.60%	37.88%	26.35%	12.65%	11.36%	7.010%			
RHV Loss:0.7038 Acc:75.66%	38.01%	45.11%	39.81%	4.801%	12.03%	8.647%	<u>0.1219</u>	0.7458	0.6240
	47.30%	57.63%	47.83%	7.408%	20.65%	10.53%			
	43.26%	49.71%	39.99%	11.69%	15.02%	9.215%			
Ours (p=0.26) Loss:0.6274 Acc:79.44%	49.98%	61.12%	50.01%	7.117%	11.84%	7.056%	**0.0865**	<u>0.7839</u>	**0.6974**
	66.32%	76.42%	64.33%	10.17%	21.95%	9.448%			
	57.04%	66.08%	56.97%	8.177%	13.51%	10.72%			

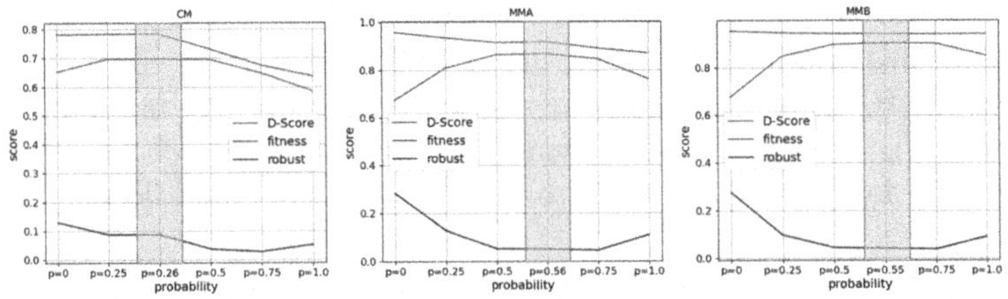

Fig. 6. The D-Scores of CM, MMA, and MMB after applying our data augmentation method with different probabilities. It is evident that the highest D-Score is obtained by using the execution probability calculated based on $p = \frac{v_{\text{robust}}}{g(n)}$, achieving a good balance between robustness and fitness. The shaded area indicates the range where the transitions of D-Scores from increasing to decreasing, which is consistent in all three models and two datasets, demonstrating the effectiveness of our score-guiding method and the rationality of D-Score in assessing CNN fitness and robustness.

in our previous experiments: the MNIST dataset primarily consists of digits in the center of the image, which leads to models trained on this dataset being insensitive to features located on the edges of the image and unable to recognize corner cases. Although the CIFAR dataset suffers from the same problem, it is less severe due to the more complex nature of the images, where the features are distributed in multiple locations besides the center of the image. Therefore, from the robustness perspective, the CM model demonstrates a more consistent ability to recognize features in all regions compared to the other two models. However, in terms of fitness, after calculating the difference in feature and attention distribution, MMB and MMA still outperform CM due to CM's lower accuracy on the original test set. Then we count both robustness and fitness equally to obtain the final D-Score of the model by setting $|\alpha| = \beta = 1$. Based on our evaluation, we believe that the MMB model performs the best overall, followed by MMA.

Score-Guided Augmentation. Both MNIST and CIFAR-10 contain 10 classes, which means $c = 10$. According to (5), the value of the boundary function $g(n)$ is around 0.5 where $n = 3$. We apply our data augmentation method to CIFAR-10 and compare it with other approaches, including RHF, RVF, RR, RHV and the case without augmentation. The results are summarized in Table 6. Our method achieves the best results in D-Score and v_{robust}, and the second-best result in v_{fitness}. Compared to the method without any data augmentation, our method significantly improves the robustness of the model while maintaining a similar level of accuracy (only 0.22% difference in the first column). The robustness can be clearly observed in the "a_i" column, where the model accuracy for objects appearing in different regions significantly increases (e.g., the accuracy for objects in the lower right corner increased from 45.28% to 56.97%. We notice that the RHF method achieves the best accuracy score (81.22%) and the highest fitness score (0.7934) in the test set, which shows the consistency between accuracy and our proposed fitness. From the "a_i" column, we can see that using our scheme the model's accuracy scores for various regions are significantly improved. So are the feature distributions of the dataset (features are more equally distributed), as shown in the "$\widetilde{f}_i$" column. For example, the top left corner of RHF only has 3.543% of the features, while the middle region has 25.37%, which is unbalanced. In contrast, through our method, the feature content in the top left corner increases to 7.117%, while the middle area decreases to 21.95%.

Another experiment is to use our augmentation method with different p on MNIST and CIFAR-10 to show the impact of the execution probability on the scores. The results are shown in Fig. 6. Clearly, using our proposed p in (7), based on the robustness score, yields the highest D-Score in all cases. Moreover, this calculated probability falls exactly in the interval where the D-Score transitions from increasing to decreasing in all cases, as indicated by the masked areas.

6 Conclusions

This paper studies how to effectively evaluate robustness of CNNs and their fitness to datasets, rather than just rely on their scores on an unevaluated test set. We propose a white-box method for this purpose, which analyzes the feature distribution of the dataset and the attention distribution of the model, using mutation operators and well-designed image transformations, respectively. With these distributions, we introduce D-Score to reflect the model's robustness and fitness. To demonstrate that our score can effectively represent the robustness of CNNs, we propose a score-guided data augmentation method to address the issue of CNN's lack of translation invariance. We validate our approach on two widely used datasets and three widely adopted models.

Acknowledgement. The authors gratefully acknowledge the partial financial support of the ASPIRE program at University of South Carolina.

References

1. Azulay, A., Weiss, Y.: Why do deep convolutional networks generalize so poorly to small image transformations? arXiv preprint arXiv:1805.12177 (2018)
2. Carlini, N., Wagner, D.: Towards evaluating the robustness of neural networks. In: 2017 IEEE Symposium on Security and Privacy (sp), pp. 39–57. IEEE (2017)
3. Deng, L.: The mnist database of handwritten digit images for machine learning research [best of the web]. IEEE Signal Process. Mag. **29**(6), 141–142 (2012)
4. Feng, Y., Shi, Q., Gao, X., Wan, J., Fang, C., Chen, Z.: Deepgini: prioritizing massive tests to enhance the robustness of deep neural networks. In: Proceedings of the 29th ACM SIGSOFT International Symposium on Software Testing and Analysis, pp. 177–188 (2020)
5. Fukushima, K.: Neocognitron: a self-organizing neural network model for a mechanism of pattern recognition unaffected by shift in position. Biol. Cybern. **36**(4), 193–202 (1980)
6. Fukushima, K., Miyake, S.: Neocognitron: a new algorithm for pattern recognition tolerant of deformations and shifts in position. Pattern Recogn. **15**(6), 455–469 (1982)
7. Grigorescu, S., Trasnea, B., Cocias, T., Macesanu, G.: A survey of deep learning techniques for autonomous driving. J. Field Robot. **37**(3), 362–386 (2020)
8. Gupta, A., Anpalagan, A., Guan, L., Khwaja, A.S.: Deep learning for object detection and scene perception in self-driving cars: survey, challenges, and open issues. Array **10**, 100057 (2021)
9. Harel-Canada, F., Wang, L., Gulzar, M.A., Gu, Q., Kim, M.: Is neuron coverage a meaningful measure for testing deep neural networks? In: Proceedings of the 28th ACM Joint Meeting on European Software Engineering Conference and Symposium on the Foundations of Software Engineering, pp. 851–862 (2020)
10. Hu, Q., Ma, L., Xie, X., Yu, B., Liu, Y., Zhao, J.: Deepmutation++: a mutation testing framework for deep learning systems. In: 2019 34th IEEE/ACM International Conference on Automated Software Engineering (ASE), pp. 1158–1161. IEEE (2019)
11. Humbatova, N., Jahangirova, G., Tonella, P.: Deepcrime: mutation testing of deep learning systems based on real faults. In: Proceedings of the 30th ACM SIGSOFT International Symposium on Software Testing and Analysis, pp. 67–78 (2021)
12. Jaderberg, M., Simonyan, K., Zisserman, A., et al.: Spatial transformer networks. In: Advances in Neural Information Processing Systems 28 (2015)
13. Jahangirova, G., Tonella, P.: An empirical evaluation of mutation operators for deep learning systems. In: 2020 IEEE 13th International Conference on Software Testing, Validation and Verification (ICST), pp. 74–84. IEEE (2020)
14. Krizhevsky, A., Sutskever, I., Hinton, G.E.: Imagenet classification with deep convolutional neural networks. Commun. ACM **60**(6), 84–90 (2017)
15. LeCun, Y., Bottou, L., Bengio, Y., Haffner, P.: Gradient-based learning applied to document recognition. Proc. IEEE **86**(11), 2278–2324 (1998)
16. Litjens, G., et al.: A survey on deep learning in medical image analysis. Med. Image Anal. **42**, 60–88 (2017)
17. Ma, L., et al.: Deepmutation: mutation testing of deep learning systems. In: 2018 IEEE 29th International Symposium on Software Reliability Engineering (ISSRE), pp. 100–111. IEEE (2018)
18. O'Shea, K., Nash, R.: An introduction to convolutional neural networks. arXiv preprint arXiv:1511.08458 (2015)

19. Ozturk, T., Talo, M., Yildirim, E.A., Baloglu, U.B., Yildirim, O., Acharya, U.R.: Automated detection of covid-19 cases using deep neural networks with x-ray images. Comput. Biol. Med. **121**, 103792 (2020)
20. Panichella, A., Liem, C.C.: What are we really testing in mutation testing for machine learning? a critical reflection. In: 2021 IEEE/ACM 43rd International Conference on Software Engineering: New Ideas and Emerging Results (ICSE-NIER), pp. 66–70. IEEE (2021)
21. Pei, K., Cao, Y., Yang, J., Jana, S.: Deepxplore: automated whitebox testing of deep learning systems. In: Proceedings of the 26th Symposium on Operating Systems Principles, pp. 1–18 (2017)
22. Shen, W., Wan, J., Chen, Z.: Munn: mutation analysis of neural networks. In: 2018 IEEE International Conference on Software Quality, Reliability and Security Companion (QRS-C), pp. 108–115. IEEE (2018)
23. Shorten, C., Khoshgoftaar, T.M.: A survey on image data augmentation for deep learning. J. Big Data **6**(1), 1–48 (2019)
24. Sun, Y., Huang, X., Kroening, D., Sharp, J., Hill, M., Ashmore, R.: Testing deep neural networks. arXiv preprint arXiv:1803.04792 (2018)
25. Wang, J., Dong, G., Sun, J., Wang, X., Zhang, P.: Adversarial sample detection for deep neural network through model mutation testing. In: 2019 IEEE/ACM 41st International Conference on Software Engineering (ICSE), pp. 1245–1256. IEEE (2019)
26. Wu, J.: Introduction to convolutional neural networks. In: National Key Lab for Novel Software Technology, vol. 5(23), p. 495. Nanjing University, China (2017)
27. Xiao, C., Li, B., Zhu, J.Y., He, W., Liu, M., Song, D.: Generating adversarial examples with adversarial networks. arXiv preprint arXiv:1801.02610 (2018)
28. Yu, J., Fu, Y., Zheng, Y., Wang, Z., Ye, X.: Test4deep: an effective white-box testing for deep neural networks. In: 2019 IEEE International Conference on Computational Science and Engineering (CSE) and IEEE International Conference on Embedded and Ubiquitous Computing (EUC), pp. 16–23. IEEE (2019)
29. Zhang, J.M., Harman, M., Ma, L., Liu, Y.: Machine learning testing: survey, landscapes and horizons. IEEE Trans. Softw. Eng. (2020)
30. Zhang, R.: Making convolutional networks shift-invariant again. In: International Conference on Machine Learning, pp. 7324–7334. PMLR (2019)

CTKM: Crypto-Based User Clustering on Web Transaction Data

Jiangfeng Li[1], Hao Luo[1], Qinpei Zhao[1](✉), Yang Shi[1], Chenxi Zhang[1], Ming Li[2], and Xuefeng Li[1]

[1] Tongji University, Shanghai 201804, China
{lijf,qinpeizhao,yangshi,xzhang2000,lixuefeng}@tongji.edu.cn, 2131484@tongji.edu.cn

[2] Shanghai Dianji University, Shanghai 201306, China
mingli@sdju.edu.cn

Abstract. User transaction data are rich, valuable, but sensitive. With the huge amounts of transaction data, data mining algorithms can make many applications practical, such as customer-behavior analysis, marketing, and forensics. The value behind the transaction data analysis on the other hand raises the risk of data leak. In this paper, we introduce a *C*rypto-based *KM*eans clustering algorithm (*CTKM*) on the *T*ransaction data of web users for user clustering and data protection as well. Considering the categoricalness of user transaction data, a taxonomy-based distance has been employed, which is applicable to the data encryption process also. In order to obtain efficient computations on the distance, a *distance batch computing*(*DBC*) protocol is designed and deployed in a two-server platform. We theoretically estimate both the computation and communication costs of the algorithm. Experimental results on a real data set demonstrate its practical value on web user clustering.

Keywords: transaction data · web user clustering · kmeans · data encryption

1 Introduction

More and more people prefer online shopping with the prosperity of e-commerce, which produces enormous transaction records everyday. These transaction records are data with privacy and valuable information. The data contains not only personal information like customers' interests but also latent information which is critical to data owners such as the tendency of a group of users and the sales volume of a certain product. Analysis of such valuable data attracts much attention. Among the analysis methods, user clustering based on transaction data is an important one, which has applications such as user recommendations and preprocessing before further analysis. An amount of effective user clustering algorithms [1–3] are proposed to mine the value of the transaction data.

Most user clustering algorithms have two natural assumptions. The first one is that the algorithms are employed in a credible environment by credible people

X. Yang et al. (Eds.): ADMA 2023, LNAI 14180, pp. 359–373, 2023.
https://doi.org/10.1007/978-3-031-46677-9_25

who analyze the data for good purposes, where the leak of data will not happen. The second one is that users know and permit their transaction records to be employed in such analysis. However, both assumptions usually do not hold in reality. The most famous example is the Cambridge Analytical scandal. The company analyzed behaviors of 50 million Facebook users illegally without informing related users to affect the American presidential election.

One solution to the conflicts is to make the user clustering applicable on encrypted data. Homomorphic encryption (HE) is a form of encryption that allows computations to be carried out on ciphertexts, thus generating an encrypted result which, when decrypted, matches the result of operations performed on the plaintext [4]. Fully homomorphic encryption (FHE) schemes allow arbitrary computations on encrypted data. Wu et al. proposed a privacy-preserving K-means clustering algorithm based on seal (an FHE library) [5]. However, efficiency is an obstacle to making the FHE practical. Partially homomorphic encryption (PHE) schemes, which offer the ability to carry out a certain type of operations on ciphertexts, are more practical from the efficiency point of view. For example, a privacy-preserving user profile matching algorithm is proposed in [6] based on ELGamal, a multiplicative homomorphic encryption scheme.

In the context of the existing HE schemes, more and more studies on the data mining methods for privacy-preserving are conducted [7,8]. Most of the work focuses on supervised learning [9–11]. For supervised learning, there are the training step and the testing step. Thereby, the classification on the ciphertexts can take the use of the trained model on plaintexts. On the other hand, unsupervised learning algorithms bring huge computations on the algorithm itself and the encryption step. The unsupervised learning on encrypted data has more challenges because of the high computation and communication cost, which brings challenges to the transaction data clustering as well.

In this paper, we study transaction-based user clustering in the context of data encryption. We mainly focus on the following scenario: due to some limitations like computation resources or knowledge of data analysis, data owners seek help from others. The original data and any information referring to the original data are not revealed. Under such requirements, we propose the CTKM (Crypto-based Transaction KMeans) for user clustering on the transaction data. The algorithm overcomes two challenges which are the definition of the distance on two web users and the computation burden during the whole analysis. A web user with categorical transaction data can be transformed into a vector with a category tree. Meanwhile, a more efficient and effective DBC protocol based on partially homomorphic encryption is introduced, which helps to calculate the distances between the encrypted data. The theoretical analysis of the cost is given, then we conduct experiments to verify the effectiveness and efficiency of the algorithm on real-life data.

The rest of the paper is organized as follows. Related works are discussed in Sect. 2. Section 3 explains the critical algorithms and protocols of the proposed CTKM scheme. In Sect. 4, we give the experiment results and discuss the validity and costs of the CTKM scheme. Section 5 concludes the paper.

2 Related Work

Currently, there have been work on the crypto-kmeans clustering in two categories, which are the multi-party or not.

The first mutual privacy preserving and collusion-resistant kmeans clustering scheme is introduced in [12]. The scheme enables the mutual privacy protection in clustering not only via keeping individuals' information private, but also by restraining the leakage of any cluster center information to the participants. However, the client side undertakes more computation than the server side. In 2018, a try on the kmeans clustering algorithm in the context of the FHE scheme is conducted in [13]. Even the implementation of the kmeans is improved to reduce 5% of running time, the algorithm is still impractical for real applications.

For multi-party crypto-kmeans, the communication cost also matters. The paper [14] is to reduce communication cost by sacrificing a little privacy. The kmeans clustering is for different sites containing different attributes for a common set of entities. Each site learns the cluster of each entity, but learns nothing about the attributes at other sites. An approach that enables the service provider to compare encrypted distances with the trapdoor information provided by the data owner is proposed [15]. In this approach, each iteration needs to take the cluster centroid for calculating the trapdoor information, which brings more cost on the communication between the server and client.

In crypto-clustering algorithms, the distance calculation is the most tricky one. In order to calculate the distances in the context of PHE, protocols on comparing two values are needed. Due to high complexity, comparison protocols with encrypted inputs have been a bottleneck in the design of the crypto-clustering algorithm. A range test protocol could be used for the following scenario: one party holds a ciphertext and needs to decide whether the corresponding plaintext's value lies in a certain interval range. However, existing range test [16,17] protocols are not suitable for our scheme because of inefficiency, and the protocol only needs to test the symbol rather than the range of certain ciphertexts in the proposed scheme.

Recently, two other types of clustering algorithms are introduced on clustering user transaction data. A *PurTreeClust* algorithm [1] focuses on the structure of the data, which organizes the products in trees. The *PurTreeClust* algorithm based on the tree structure is both effective and efficient for large transaction data. However, the *Jaccard Distance* in the *PurTreeClust* assumes the product tree has the same height which leads to limitations. An improvement on the distance metric is introduced in [18]. A *category tree* distance in introduced in [19,20], which makes use of the product taxonomy information to convert the user transaction data to vectors. Then, the similarity between web users can be evaluated by the vectors from their transaction data.

As shown in Table 1, the existing crypto-kmeans are compared from different aspects. The kmeans clustering is commonly for numerical data processing. None of the compared kmeans can handle categorical transaction data directly while protecting the information in the clustering.

Table 1. A comparison on different existing crypto-kmeans algorithms from clustering and security, where N is the data size, D is the dimension, k is the number of clusters and t is the running time.

paper	cate-gorical	multi-party	security tools	time	space	communication	user info.	cluster centroids	inter-mediate
[12]	✗	✗	Pailliar homomorphic encryption	✗	✗	✗	✓	✓	✓
[13]	✗	✗	TFHE library	N=400,D=2 k=3,t=0.85months	✗	✗	✓	✓	N/A
[21]	✗	✓	ElGamal's encryption	N=10000, k=6 D=3, t=1500s	✗	$O((k-1)(5+D/2)N)$	✓	✓	✓
CTKM	✓	✓	Pailliar homomorphic encryption	N=30000, D=22 k=100, t=196556s	✗	$O(NkD)+O(Nk)$	✓	✓	✓

3 Transaction Based User Clustering in Partially Homomorphic Encryption Context

Analysis of web users' purchase behaviors and consumption patterns by clustering users' transaction data has attracted widespread attention. We introduce the detailed design on transaction based user clustering in the context of PHE scheme. A suitable distance on two users is chosen for the CTKM algorithm based on the users' interests. Furthermore, we present a *DBC* protocol to calculate the distance in the kmeans. The CTKM scheme for web users is deployed on two servers to reduce the computation load.

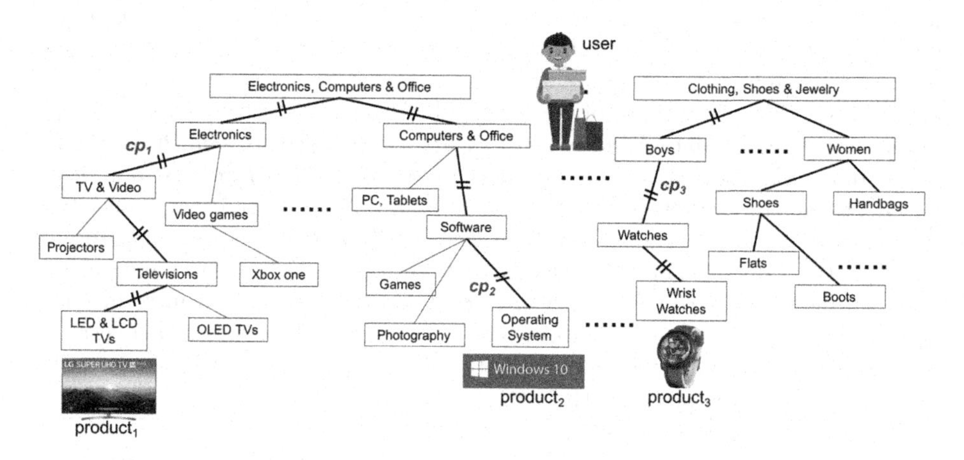

Fig. 1. An e-commerce transaction data is usually composed of users and products. A user can be represented by a set of products and a product is represented by its category information.

3.1 Distance on Two Users

Distance metric plays an important role in the cluster analysis. In clustering algorithms, the distance employed decides the characteristics of the algorithm. For a transaction data, products and users are always involved in the data. A user can be represented by the products he/she is interested in. It is quite natural to consider the similarity by means of fixed-length boolean attributes on the product. For example, take a complete product set as $P = \{p_1, ..., p_{10}\}$, and a transaction data as $T_1 = \{p_1, p_5\}$, then the data can be transformed to $T_1 = [1, 0, 0, 0, 1, 0, 0, 0, 0, 0]$. In the T_1, position i of the boolean vector represents product p_i. A value 1 corresponds to the presence of p_i in the transaction, while 0 corresponds to the absence. With such a representation, commonly used distances such as Euclidean distance can be applied directly. However, when there are huge numbers of products, i.e., the size of P is large, the vectors could be in high dimensions and be very sparse.

An e-commerce transaction data is composed of products and users (see Fig. 1). A user can be represented by the products he/she is interested in. A product on the other hand can be represented by its category information. For example, the product "LG TV" can use a category path "Electronics,Computers & Office→ Electronics→TV & Viddo→ Television → LED & LCD TVs" to represent it. Each user can be represented by several products, which can be in trees based on the category information.

In the context of Paillier encryption, only the addition and multiplication (plaintext with ciphertext) operations are supported. Therefore, the distance definition in the kmeans algorithm is quite critical. A *category tree* distance [19,20] is adopted in this paper, which makes use of the product taxonomy information to convert the user transaction data to vectors. Then, the similarity between web users can be evaluated by the vectors from their transaction data. The benefits are that the *category tree* distance is simple and satisfies the metric properties mathematically.

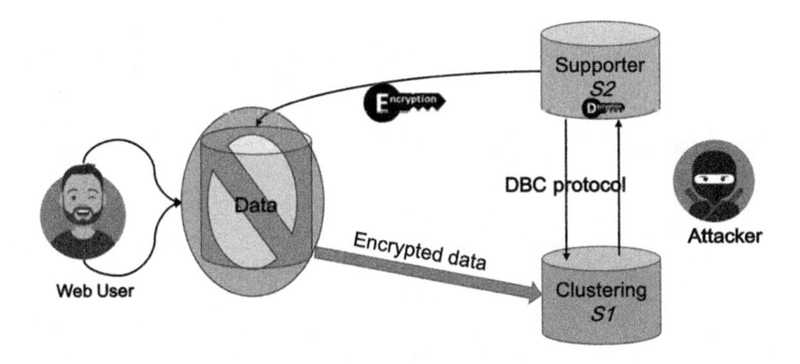

Fig. 2. Architecture of the proposed CTKM scheme.

3.2 The CTKM Scheme

In the proposed CTKM scheme, clustering server S_1 and supporting server S_2 are involved (see Fig. 2). In such an architecture, the client-side (i.e., data owner) needs no communication with the servers after the encrypted data has been uploaded, which enables a device with low computation capability (e.g. mobile-phone) be the client-side.

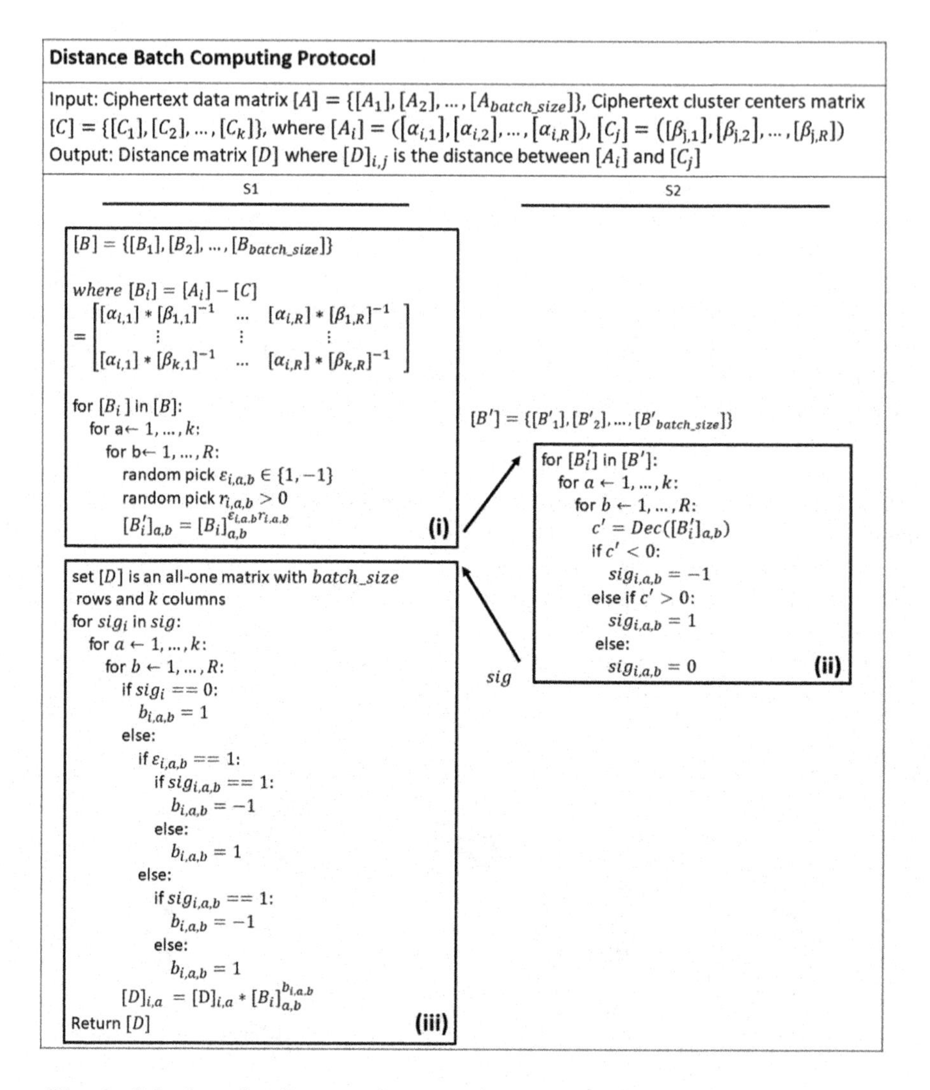

Fig. 3. Calculate the distances between data and cluster centroids in batches.

Building Blocks. Although there have been existing secure comparison protocols [16,17], these protocols have very high computation cost. In order to adapt to the specific scenario in our paper, we introduce a *DBC* protocol. Suppose m is the plaintext and $[m]$ is the corresponding ciphertext generated by a HE scheme with respect to addition of the ciphertext and scalar multiplication of the plaintext. Let K_{Enc} and K_{Dec} be the public key and private key of the encryption scheme, respectively. The protocol is to compare two values α and β based on their ciphertexts $[\alpha]$ and $[\beta]$. The protocol returns 0 if $\alpha \geq \beta$, and returns 1 otherwise. The goal is that, only clustering server S_1 obtains the comparison's result, plaintexts (e.g., α, β) and the result are kept secret to the supporting server S_2. To improve the computational efficiency, we proposed an approach for calculating the distances between data and cluster centroids in batches[1]. As shown in Fig. 3, the approach consists of three steps as follows.

(i) For each $i \in \{1, \dots, batch_size\}$, S_1 generates the intermediate values $[B_i]$ for encrypted data vectors $[A_i] = ([\alpha_{i,1}], \dots, [\alpha_{i,R}])$; and then each element $[B_i]_{a,b}$ in $[B] = [B_1], \dots, [B_{batch_size}]$ is randomized by multiplying $\varepsilon_{i,a,b}$ and $r_{i,a,b}$, where $r_{i,a,b}$ is a random positive floating point number and $\varepsilon_{i,a,b} \in_\$ \{-1, 1\}$, where $\in_\$$ stands for select an element uniquely randomly from a set (in the left of the symbol). Note that all the operations are conducted on ciphertexts using homomorphic properties. The obfuscated intermediate values $[B'] = [B'_1], \dots, [B'_{batch_size}]$ is sent to S_2 after then, where the corresponding plaintext $B_{i,a,b} = \varepsilon_{i,a,b} r_{i,a,b} (\alpha_{i,b} - \beta_{a,b})$.

(ii) The supporting server S_2 decrypts all the elements in $[B']$, compares them with zero and feeds back the results to S_1. Because $[B']$ is obfuscated, S_2 knows nothing about A and C. Moreover, clustering server S_1 only knows whether a dimension $\alpha_{i,b}$ is larger than the corresponding dimension $\beta_{a,b}$, the exact value of $\alpha_{i,b}$ and $\beta_{a,b}$ are not leaked.

(iii) S_1 computes the distances between data and cluster centroids in ciphertexts using the feedback from S_2.

We also give the *Argmin* protocol based on the introduced comparison protocol. The protocol is to let the S_1 find the minimum plaintext among the t ciphertexts $[a_1], [a_2], ..., [a_t]$ and S_2 owns the K_{Dec}. During the process, neither S_1 nor S_2 knows the corresponding plaintexts. We construct binary tress to iteratively compare the ciphertexts. The minimum value in the current iteration will be the input of the next iteration. Finally, the S_1 obtains the minimum value $[a_{min}]$. The protocol is described in Algorithm 1. After $\lceil log_2 t \rceil \cdot \lceil t/2 \rceil$ iterations, S_1 gets the minimum value among t ciphertexts.

In this scheme, the distance computation protocol (DCP) is used to calculate the proposed distance between two ciphertext vectors without compromising corresponding plaintexts. Given two vectors of ciphertexts $[x] = ([x_1], [x_2], ..., [x_D])$ and $[y] = ([y_1], [y_2], ..., [y_D])$, S_1 holds the vectors $[x]$ and $[y]$, and S_2 holds decryption key K_{Dec}, the *DCP* is described in Algorithm 2.

[1] We use $[X]$ to represent the ciphertext of the plaintext X.

Algorithm 1. *Argmin* protocol

1: to servers S_1 and S_2:
2: $num \leftarrow t$
3: **for all** $i = 1; i <= \lceil log_2 t \rceil; i++$ **do**
4: **for all** $j = 1; i <= \lfloor \frac{num}{2} \rfloor; j++$ **do**
5: **if** $i == 1$ **then**
6: $[a_{2j-1}] \leftarrow Compare([a_{2j-1}, a_{2j}])$
7: $[a_{2j}] \leftarrow 0$
8: **else**
9: $[a_{2i(j-1)+1}] \leftarrow Compare([a_{2i(j-1)+1}, a_{2ij-1}])$
10: $[a_{2ij-1}] \leftarrow 0$
11: **end if**
12: $num \leftarrow \lceil \frac{num}{2} \rceil$
13: **end for**
14: **end for**
15: $[a_{min}] \leftarrow [a_1]$

Algorithm 2. *DCP* protocol

1: to servers S_1 and S_2:
2: **for all** $i = 1; i <= D$ **do**
3: $sig \leftarrow Compare([x_i, y_i])$
4: **if** $sig == true$ **then**
5: $b_i = -1$ // $x_i < y_i$;
6: **else**
7: $b_i = 1$
8: **end if**
9: compute $[d_{xy}] = \prod_i ([x_i] * [y_i]^{-1})^{b_i}$
10: **end for**

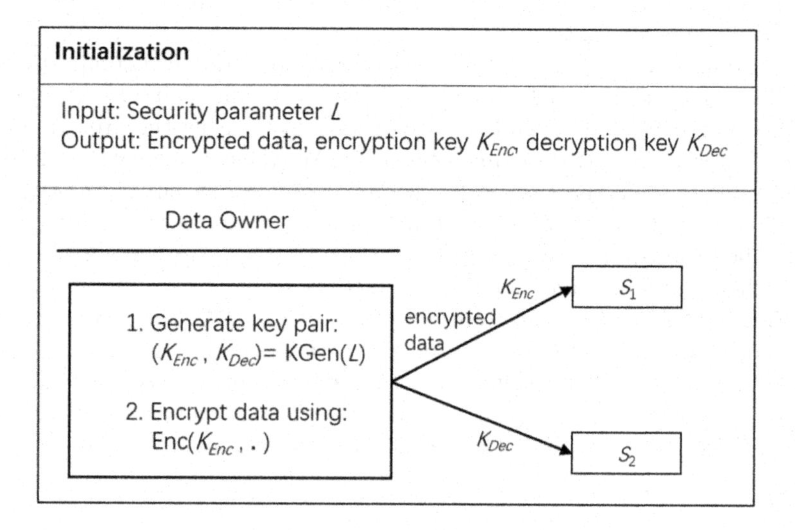

Fig. 4. The initialization step in the CTKM.

Main Scheme. Following the original kmeans clustering algorithm, the proposed Crypto-kmeans (CTKM) also contains the initialization, partitioning and centroids updating. Each iteration is composed of partitioning and centroids updating. The algorithm is stopped by a certain number of iterations.

In the initialization step (see Fig. 4), the public and private keys are firstly generated based on the given key size on the client side (i.e., Data Owner). Then the client side sends the encrypted plaintexts to the S_1 and the initial centroids $C = (c_1, c_2, ..., c_k)$ are selected randomly from the encrypted data on the S_1.

After the initialization, the CTKM algorithm iterates certain numbers. In each iteration, the distance between each point to all the centroids have to be calculated, which is to obtain the centroid that is the closest to the point. Based on the introduced *DBC* protocol and *DCP* protocol, we can calculate the encrypted distances between N points and kth clusters. After the distance calculation, the partitioning is the following step, which is to assign the points to their nearest centroids. The *Argmin* protocol is for the assignment. When S_1 gets the N nearest distances, which are encrypted, the assignment of the N points can be completed. During the assignment, a partition $Pa = (p_1, p_2, ..., p_N)$ is obtained, each of which represents the centroid index of the current point. For example, $p_i = k$ indicates that the ith point is assigned to the k cluster. The update on the centroids is the final step in each iteration. It is to take the average on each cluster according to the new assignment of each point. As shown in Algorithm 3, the input is the partitioning Pa and ciphertexts $[a]$. Since the PHE scheme supports only the addition and multiplication, we replace the division step by multiplying the inverse. Until then, one iteration of the CTKM is done and the centroids of clusters are updated.

The CTKM can reach a convergence after a certain number of iterations. The S_1 sends the cluster centroids with encryption to the client side, and the client obtains the final clustering result by the K_{Dec}.

Algorithm 3. centroids updating

Require: P, $[a]$
Ensure: C is updated.
1: **for all** $j = 1; j < k$ **do**
2: $sum = (0, 0, ..., 0)_D$, $|c_j| = 0$.
3: **for all** $i = 1; i <= N$ **do**
4: **if** $p_i == j$ **then**
5: $sum = sum \bigoplus [a_{p_i}]$.
6: $|c_j| = |c_j| + 1$.
7: **end if**
8: $c_j = \frac{1}{|c_j|} * sum$.
9: **end for**
10: **end for**
11: return $[C]$.

3.3 Complexity Analysis

We give an analysis on the time complexity of the CTKM. In the initialization step, the main computation comes from the encryption, which is $O(N * D)$ for a D-dimensional data set with size N. The initialization on the centroids is $O(k)$, where k is the number of clusters. It is mentioned that the initialization is done only once.

For each iteration, the distance computation takes the main cost. There are $N * k$ times of distance calculations for N points and k clusters. Each distance calculation calls the comparison protocol, which takes D times of comparisons. During the partitioning, one run of *Argmin* protocol is needed for each point to reach its nearest centroid. One run of *Argmin* protocol needs $k - 1$ times of comparisons. Therefore, the assignment takes $O(N*(k-1))$. When updating the centroids, N times of additions and k times of multiplications are needed. We list the details on the time complexity in Table 2. Totally, the time complexity of the CTKM is $O(N * k * D)$. Besides the time cost, the communication cost is also a burden to the whole scheme. In the initialization part, the client sends the encrypted data to S_1, therefore, the communication cost depends on the data size. The communication cost for each iteration is listed in Table 2. Since $N * k * D$ ciphertexts are delivered in a package, the cost is $O(N * k * D)$ for the distance calculation part and $O(N * k)$ for the partitioning part. As for the centroids updating, there is no cost. During the decryption, only k centroids are transferred, the cost is k ciphertexts.

Table 2. Time and communication costs of one iteration in the CTKM algorithm.

cost	dist. calculation	partitioning	centroid updates
time	Mul $O(N * k * D)$	Mul $O(N * (k - 1))$	Mul $O(k)$
	Dec $O(N * k * D)$	Dec $O(N * (k - 1))$	Add $O(N)$
communication	$O(N * k * D)$	$O(N * k)$	N/A

4 Evaluation

The experiments are conducted on a real web user data. The data set is from the Yelp Challenge[2] 2018, which includes user information for example user id, commercial store information with business id and categories, category dictionary[3] of Yelp and comments information which consists of ids of users who give comments, target stores and images. There are totally 1.2 million users in the dataset. In order to verify the performance of the CTKM algorithm, we select a

[2] https://www.yelp.com/dataset/challenge.

[3] https://www.yelp.com/developers/documentation/v3/all_categories.

dataset of 3,000 users (*yelp3k*) manually from the original Yelp data. The 3,000 users are selected according to their consuming products and they are distributed into four groups (i.e., $k = 4$) with similar purchasing interests (according to the product categories). The cluster sizes are [1000, 600, 600, 800] and the dimension of the data is 22. Another dataset (*yelp30k*) with 30,000 users is randomly selected from the Yelp data, which is to test the proposed distance and the performance of the CTKM algorithm. Since the *yelp30k* is randomly picked, the number of clusters is unclear. In order to test the performance of the clustering algorithm, the number of clusters of it in the experiment is set a large one, which is 100.

Table 3. Distribution of the Jaccard similarity values on the dataset *yelp30k*. It is clear that the majority of the users are recognized as different (i.e., = 0.0) by the Jaccard similarity.

Ranges	# of user pairs	proportion
= 0.0	440927700	98.0%
(0.0, 0.02]	3756311	0.8%
(0.02, 0.04]	2735026	0.6%
(0.04, 0.06]	1372083	0.3%
(0.06, 1.0]	1193800	0.31%

We tested the Jaccard distance for web users on the dataset *yelp30k*, as it is quite often used on transaction data, Given the 30,000 users, which produces 449985,000 pairs of users, the Jaccard similarity for each pair of users is distributed into seven ranges (see in Table 3). At first, we took 0.2 as the length of each range. However, it is found that most of the pairs are located in [0.00, 0.20], which takes 98.8%. More specifically, there are totally 98.0% of the user pairs are *0*s on the similarity. The results indicate that the Jaccard similarity is difficult to find the similarities between users, which will eventually cause a failure on the clustering. Therefore, we can conclude from the Table 3 that the Jaccard similarity is not suitable for the web user dataset in this paper.

4.1 Experiment Setup

The experiments are conducted on two computers $S1$ and $S2$. The $S1$ runs Ubuntu 16.04 operating system with 96 GB of RAM and two Intel(R) Xeon(R) E5-2640 CPUs(8 cores, 16 processors, 2.60 GHz). The $S2$ runs Ubuntu 16.04 with 64 GB memory and two Intel(R) Xeon(R) E5-2650 CPUs(12 cores, 24 processors, 2.20 GHz). In the experiments, we use 30 processors of $S1$ and 20 processors of $S2$ for distance calculating. The scheme is implemented in Python 3.6 with g++7 as the compiler. Since the original Paillier cryptosystem only supports elements of Z_n, in our implementation, the Paillier encryption scheme is implemented using the opensource PHE library[4], which is capable of processing floating-point numbers with a special encoding scheme given in Appendix II in [22].

4.2 The Validity on the CTKM Algorithm

We first check the validity of the clustering algorithm in the context of PHE scheme by comparing the clustering results on the ciphertexts and plaintexts. The validity is from internal and external [23]. The (*MSE*), which is commonly used as the result in the kmeans is evaluated. The external validity index is the Adjusted Rand Index (*ARI*). External validity index is usually to compare one clustering partition results with the ground truth partition results, which can be used for determining the number of clusters. For ARI, the higher the value, the higher matching of two partitions. As for MSE, the values are monotonously decreasing. The algorithms are tested on different settings of k (from 2 to 15). The results of the CTKM algorithm are compared with those of the kmeans on the plaintexts.

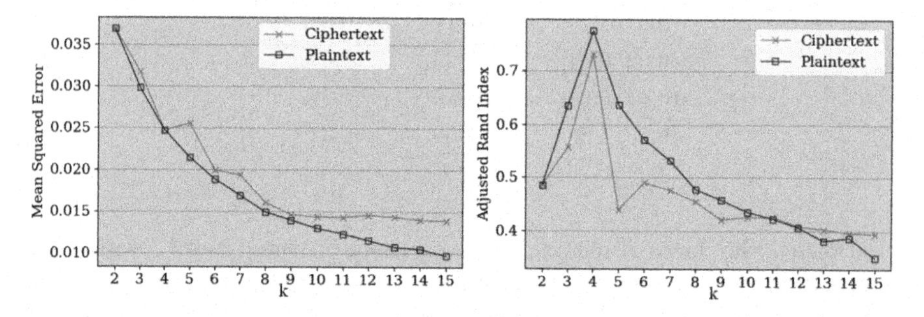

Fig. 5. The clustering results (*MSE* and *Adjusted Rand Index*) on different number of clusters (k) on ciphertexts and plaintexts. The dataset is *yelp3k*.eps

As shown in Fig. 5, the MSE values on different number of clusters from two clustering results on the *yelp3k* dataset have similar trends, where there are knee points at $k = 4$. Since the *yelp3k* dataset has the ground truth labels, the ARI values at $k = 4$ match to the ground truth that four clusters exist in the dataset. The MSE values and ARI values from two clusterings are at the same scale level. The results indicate that the CTKM algorithm works on the ciphertexts as the kmeans on the plaintexts.

4.3 The Costs in Computation and Communication

The efficiency of the CTKM is quite critical for a real application. We take a multi-core computer to run the experiment. For the clustering algorithm, the costs on each step in the algorithm can be seen from Table 4, together with the

[4] https://github.com/n1analytics/python-paillier.

whole running time. For the *yelp3k* dataset, the total running time is around 18.6 min. The distance calculations in the clustering algorithm take the most of time and communication costs. However, compared to the more secure Range Test protocol, the proposed *DBC* protocol already improves the efficiency of the distance calculation. The data size and the number of clusters affect the performance of the CTKM algorithm. Based on the results on the data *yelp30k*, the total running time, which is 196556 s (around 2.3 d), is around 176 times of that on the *yelp3k*, whereas the data size is 10 times of the *yelp3k* and the dimension is 25 times. Therefore, the running time is quite reasonable. On the other hand, with the increase of the data size and number of clusters, the communication cost is tremendously increased also. Although the running time is much higher than the kmeans on the plaintexts, it is already quite practical on the ciphertexts in a real application.

Table 4. The running cost (time and communication) on dataset *yelp3k* and *yelp30k* for each steps of the whole processing.

Steps	S_1 time (s)	S_2 time (s)	Comm	Delay (s)
yelp3k ($N = 3k, k = 4, D = 22$)				
key = 1024, iteration = 5, total time = 1117 s				
Initialization	58.7	-	27.1 MB	0.002
Dist. calculation	421.7	358.8	899.8 MB	24.4
Partitioning	36.6	20.3	45.8 MB	2.12
centroid updates	208.3	-	-	-
yelp30k ($N = 30k, k = 100, D = 22$)				
key = 1024, iteration = 5, total time = 196556 s				
Initialization	585.9	-	27.1 MB	0.99
Dist calculation	60660.4	88051.6	222.7 GB	16055.7
Partitioning	28333.0	7408.7	16.7 GB	2107.7
centroid updates	405.7	-	-	-

5 Conclusion

As an important analysis method, user clustering on transaction data has attracted many attentions. However, most existing work ignore the privacy and valuable information protection during the analysis. In this paper, we focus on the problem of web user clustering while protecting the user data. The users are transformed into vectors based on the *category tree* of the transactions. Then, the *category tree* distance employed in the clustering algorithm, which is also adaptive into the encryption, can be obtained. A *DBC* protocol is introduced for a more efficient distance calculation in the clustering algorithm. In order to show that the proposed CTKM is correct, the comparison of the clustering results on the ciphertexts and plaintexts is conducted. Meanwhile, the running and communication costs are also listed, indicating that it is quite practical for real applications.

Acknowledgements. This work is partially supported by National Key Research and Development Program of China (2021YFC3340601), National Natural Science Foundation of China (Grant No. 61972286, 62172301 and 61772371), the Science and Technology Program of Shanghai, China (Grant No. 20ZR1460500, 22511104300, 21ZR1423800), the Shanghai Municipal Science and Technology Major Project (2021SHZDZX0100) and the Fundamental Research Funds for the Central Universities.

References

1. Chen, X., Fang, Y., Yang, M., Nie, F., Zhao, Z., Huang, J.Z.: PurTreeClust: a clustering algorithm for customer segmentation from massive customer transaction data. IEEE Trans. Knowl. Data Eng. **30**(3), 559–572 (2017)
2. Guidotti, R., Monreale, A., Nanni, M., Giannotti, F., Pedreschi, D.: Clustering individual transactional data for masses of users. In: Proceedings of the 23rd ACM SIGKDD International Conference on Knowledge Discovery and Data Mining, pp. 195–204 (2017)
3. Carnein, M., Trautmann, H.: Customer segmentation based on transactional data using stream clustering. In: Yang, Q., Zhou, Z.-H., Gong, Z., Zhang, M.-L., Huang, S.-J. (eds.) PAKDD 2019. LNCS (LNAI), vol. 11439, pp. 280–292. Springer, Cham (2019). https://doi.org/10.1007/978-3-030-16148-4_22
4. Acar, A., Aksu, H., Uluagac, A.S., Conti, M.: A survey on homomorphic encryption schemes: theory and implementation. ACM Comput. Surv. (Csur) **51**(4), 1–35 (2018)
5. Wu, W., Liu, J., Wang, H., Hao, J., Xian, M.: Secure and efficient outsourced k-means clustering using fully homomorphic encryption with ciphertext packing technique. IEEE Trans. Knowl. Data Eng. **33**(10), 3424–3437 (2020)
6. Yi, X., Bertino, E., Rao, F.Y., Lam, K.Y., Nepal, S., Bouguettaya, A.: Privacy-preserving user profile matching in social networks. IEEE Trans. Knowl. Data Eng. **32**(8), 1572–1585 (2019)
7. Mendes, R., Vilela, J.P.: Privacy-preserving data mining: methods, metrics, and applications. IEEE Access **5**, 10562–10582 (2017)
8. Iezzi, M.: Practical privacy-preserving data science with homomorphic encryption: an overview. In: 2020 IEEE International Conference on Big Data (Big Data), pp. 3979–3988. IEEE (2020)
9. Zhang, L., Liu, Y., Wang, R., Fu, X., Lin, Q.: Efficient privacy-preserving classification construction model with differential privacy technology. J. Syst. Eng. Electron. **28**(1), 170–178 (2017)
10. Sun, X., Zhang, P., Liu, J.K., Yu, J., Xie, W.: Private machine learning classification based on fully homomorphic encryption. IEEE Trans. Emerg. Top. Comput. **8**(2), 352–364 (2018)
11. Vaidya, J., Shafiq, B., Fan, W., Mehmood, D., Lorenzi, D.: A random decision tree framework for privacy-preserving data mining. IEEE Trans. Dependable Secure Comput. **11**(5), 399–411 (2013)
12. Xing, K., Hu, C., Yu, J., Cheng, X., Zhang, F.: Mutual privacy preserving k-means clustering in social participatory sensing. IEEE Trans. Industr. Inf. **13**(4), 2066–2076 (2017)

13. Jäschke, A., Armknecht, F.: Unsupervised machine learning on encrypted data. In: International Conference on Selected Areas in Cryptography, pp. 453–478. Springer (2018). https://doi.org/10.1007/978-3-030-10970-7_21
14. Vaidya, J., Clifton, C.: Privacy-preserving k-means clustering over vertically partitioned data. In: Proceedings of the Ninth ACM SIGKDD International Conference on Knowledge Discovery and Data Mining, pp. 206–215 (2003)
15. Liu, D., Bertino, E., Yi, X.: Privacy of outsourced k-means clustering. In: Proceedings of the 9th ACM Symposium on Information, Computer and Communications Security, pp. 123–134 (2014)
16. Peng, Y., Li, H., Cui, J., Zhu, Y., Peng, C.: An efficient range query model over encrypted outsourced data using secure KD tree. In: 2016 International Conference on Networking and Network Applications (NaNA), pp. 250–253. IEEE (2016)
17. Guo, Y., Xie, H., Wang, M., Jia, X.: Privacy-preserving multi-range queries for secure data outsourcing services. IEEE Trans. Cloud Comput. **11**(3), 2431–2444 (2022)
18. Zhang, Y., Zhang, Y., Zhao, Q., Rao, W.: Automatic user categorization through large transaction data. In: 2019 IEEE International Conference on Multimedia and Expo (ICME), pp. 278–283. IEEE (2019)
19. Zhang, Y., Zhao, Q., Shi, Y., Li, J., Rao, W.: Category tree distance: a taxonomy-based transaction distance for web user analysis. Data Min. Knowl. Disc. **37**(1), 39–66 (2023)
20. Zhao, Q., et al.: TaxoVec: taxonomy based representation for web user profiling. In: Proceedings of the 2021 International Conference on Multimodal Interaction, pp. 548–556 (2021)
21. Yi, X., Zhang, Y.: Equally contributory privacy-preserving k-means clustering over vertically partitioned data. Inf. Syst. **38**(1), 97–107 (2013)
22. Hardy, S., et al.: Private federated learning on vertically partitioned data via entity resolution and additively homomorphic encryption. arXiv preprint arXiv:1711.10677 (2017)
23. Zhao, Q., Fränti, P.: WB-index: a sum-of-squares based index for cluster validity. Data Knowl. Eng. **92**, 77–89 (2014)

Calibrating Popularity Bias Based on Quality for Recommendation Fairness

Zhengyi Guo, Yanmin Zhu(✉), Zhaobo Wang, and Mengyuan Jing

Department of Computer Science and Engineering, Shanghai Jiao Tong University, Shanghai, China
yzhu@sjtu.edu.cn

Abstract. Recommendation fairness has become a significant topic recently. One obstacle to it is popularity bias. Previous works reduce the bias by eliminating popularity difference entirely. However, items with higher quality should be assigned with higher popularity. Completely excluding popularity will cripple recommender on recommendation accuracy and unfairly treat items with high quality. For this reason, we argue that the popularity of items should not be avoided at all costs but instead calibrated based on their quality, which is a depiction of items' natural properties. To this end, we propose the Quality Recommendation (QR) framework. Specifically, we separate item embedding into quality embedding and non-popularity ID embedding. The former efficiently encodes quality related features and the latter eliminates stereotype of specific items' popularity. To evaluate model vulnerability to popularity-quality discrepancy, we propose a novel evaluation method. Its core idea is simulate this discrepancy at the training stage. Experiments on datasets show the effective fairness and recommendation performance of our proposed methods.

Keywords: Recommender System · Fairness · Popularity Bias

1 Introduction

Recommender system provides personalized service for users to seek information, playing an increasingly important role in a wide range of online applications. Moreover, sellers or information providers rely on these systems to expose their items by recommending them to users. It is important for recommenders to ensure a fair recommendation rank mechanism for these suppliers. Unfairly under-ranking items provided by specific sellers could limit their motivation for the service. One vital obstacle to the recommendation fairness is popularity bias, i.e., recommenders typically tend to recommend popular items while neglecting less popular items [1,7]. This results in high-quality items with low popularity being unfairly under-ranked.

Previous popularity bias related works claim that popularity bias leads to Matthew's effect (i.e. popular items get more popularity), and long-tail distribution of popularity (i.e. most popularity is centralized to few items). They try to

X. Yang et al. (Eds.): ADMA 2023, LNAI 14180, pp. 374–388, 2023.
https://doi.org/10.1007/978-3-031-46677-9_26

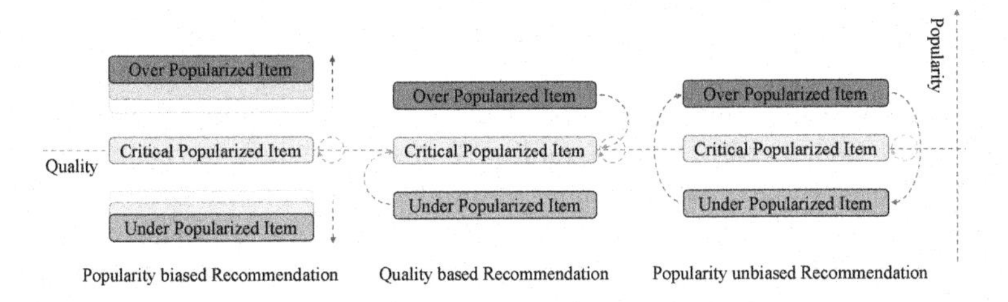

Fig. 1. Difference between popularity biased recommendation, quality recommendation and unbiased recommendation.

flatten the long-tail distribution (e.g., assign same exposure for different items) to solve the popularity bias [1,7]. Despite the influence of popularity bias on the long-tail distribution, the main and underlying reason behind this distribution is the mechanism by which users choosing items. To verify this claim, we conduct a theoretical analysis. It shows that in the full information setting (i.e. all users have the knowledge of every item and how much value or happiness these items can bring to them) the popularity will still be subject to the long-tail distribution. Further details can be found in Sect. 3. As such, flattening the long-tail distribution through methods such as assigning the same exposure to different items can deviate from the original intention of recommender systems to provide effective recommendations to users. Simply eliminating the long-tailed popularity distribution is harsh for recommendation fairness.

Instead of concentrating on the long-tail distribution of items' popularity as previous works, we argue the quality of items should be treated as a vital criterion for recommendation fairness. In the real scenario, high-quality items naturally deserve more exposure and attention for users' willingness to consume them. Assigning the same popularity weight to high-quality items and low-quality items is inherently unfair. For example, recommending a low-quality restaurant over a high-quality one solely for the purpose of flattening the long-tail distribution may lead to a poor user experience. This unfairness can discourage suppliers from improving their offerings. Additionally, if suppliers can generate sales by relying on the exposure provided by the recommender system and user misjudgment, they may have an incentive to lower the quality of their products in order to improve their profits. For such reasons, we use **quality**, a depiction of items' intrinsic properties (e.g. durability, functionality and price of a product) to describe how popular an item should be.

Compared to popularity-biased recommendation, quality-based recommendation can provide a more transparent and fair relationship between items' quality and their popularity. This can incentivize suppliers to improve the quality of their items and offer better services to users. As depicted in Fig. 1, biased recommendation ignore and exacerbate the popularity-quality discrepancy, while our quality recommendation aim to calibrate the discrepancy. However, it is important to note that unbiased recommendation may not always be the best solution, as they can also overcorrect the discrepancy. Instead, consistency between popularity and quality should be achieved to ensure that the recommender system is fair and effective.

Based on the above discussion, we propose that the popularity should not be eliminated but rather calibrated to quality to ensure fairness in recommender systems. The discrepancy between items' popularity and quality could be the main factor making popularity bias harmful to recommendation fairness. The aim of this paper is to eliminate the impact of this discrepancy in recommender systems, reducing the risk of high-quality items being under-ranked and low-quality items being over-ranked. We promote this by building an efficient quality encoder and eliminating stereotyped memory of item popularity. Specifically, we disentangle the item embedding into quality embedding and supplement ID embedding. The quality embedding is encoded from items' digitized properties and reviews that directly reflect items' natural attributes. The supplement ID embedding is inferred from item ID and eliminated popularity related information by an adversarial neural network to avoid the stereotype of item popularity. To evaluate the ability of recommenders on recommending based on quality, we simulate the popularity-quality discrepancy and check whether recommenders fairly recommend items with different discrepancies. Experiments conducted explore which model is more robust to items' popularity-quality discrepancy.

In summary, the main contributions of this paper are as follows:

1. We give a theoretical analysis of popularity distribution in the full information setting. This shows the long-tail distribution of popularity is inevitable. Hence, we need to focus more on recommendation based on quality. (presented in Sect. 3)
2. We propose a novel evaluation method for evaluating recommender's robustness to popularity-quality discrepancy. (presented in Sect. 6.2)
3. We propose a framework for recommending items by their quality. Experimental results show that our framework achieves a balanced performance in accuracy and fairness. (presented in Sect. 5 and Sect. 6.3)

2 Related Work

2.1 Popularity Bias in Recommendation

Popularity bias or popularity effect refers to the phenomenon that recommenders typically recommend popular items while ignoring unpopular items [1,4,7,10]. This concept and its influence on recommendations has been studied in [4,6,10,14]. Jannach et al. [6] empirically showed that different recommendation algorithms have different vulnerabilities to popularity bias.

Researchers have explored many approaches to counter popularity bias in recommender systems. Some studies offer a trade-off between recommendation accuracy and the coverage of long-tailed items. Liang et al. [7] propose Inverse Propensity Weighting (IPW). The core idea of this approach is lowering weight of popular items' interactions in the training loss. Abdollahpouri et al. [1] propose a regularization-based approach that aims to improve the rank of long-tail items. However, as discussed in Sect. 1, recommending long-tailed items to users forcefully deviates from their preferences.

Other works claim that the regular recommendation dataset is biased. As a result, they conduct experiments on an unbiased dataset to evaluate the capability of models to provide recommendations free from popularity bias. The unbiased dataset is sampled from the original dataset and ensure that item popularity is distributed uniformly. Bonner et al. [3] propose CauseE. This method trains unbiased embeddings on a small unbiased dataset. The unbiased embedding is more noisy because of insufficient training on the small dataset. Wei et al. [15] propose a model-agnostic framework (MACR) that performs a counterfactual inference to estimate the direct effect from item properties to the ranking score, which is removed to eliminate the popularity bias. Zheng et al. [21] propose a general framework for disentangling user interest and conformity for recommendation (DICE) with the causal embedding. Ren et al. [11] introduce a framework to mitigate popularity bias from a gradient perspective. However, as illustrated in Sect. 3, the popularity in the original dataset and the ideal case both follow a long-tailed distribution. Therefore, there is a gap between their experiments and the actual situation.

Zhu et al. [22] introduce the problem of popularity-opportunity bias, which takes users' preference into consideration compared to conventional popularity bias. Unlike their criterion, which aims to maintain the same true positive rate across items with varying popularity, our criterion ensures that the popularity of item i is greater than that of item j once it is established that the quality of i is better than that of j. We impose fewer restrictions on the recommender system, which is more aligned with users' preferences. Additionally, our criterion encourages suppliers to offer high-quality items instead of relying solely on the exposure from the recommender system and users' misjudgment.

Several recent works [16,18,19] claim to have eradicated the negative impact of popularity bias while retaining its positive effects. However, conventional recommendation experiments conducted by these studies are failed to demonstrate the successful elimination of the negative effects. To address this dilemma, we introduce a novel evaluation method. Similar to our work, Zhao et al. [19] suggest that quality is a beneficial aspect of popularity bias that should be preserved. However, their definition of quality only refers to the stable aspect of popularity bias. This definition is incomplete as the stable aspect of popularity bias can still include the harmful discrepancy between quality and popularity. In contrast to Zhao et al., we propose that the model can evaluate the quality of items based on their inherent attributes, such as the content of articles and reviews of products.

2.2 Recommendation Fairness

Recommendation fairness studies whether the recommender system treats different groups of items differently. These groups are often determined by sensitive attributes (e.g., gender, race). Zehlike et al. [17] propose goals for fair ranking. Zhu et al. [23] study statistical parity based fairness in recommenders, and argue that items should be shown at the same rate across groups. Singh and

Joachims [13] take a full-ranking view of fairness, but are able to apply this to recommender systems through a post-processing algorithm for model predictions. Beutel et al. [2] propose a set of novel metrics for measuring the fairness of a recommender system based on pairwise comparisons. Additionally, they show that this pairwise fairness metric directly corresponds to ranking performance and analyze its relation with pointwise fairness metrics. Given that our quality recommendation problem can be interpreted as treating items fairly despite differences in their popularity-quality discrepancy, we can leverage relevant prior research on fairness metrics to guide our evaluation. However, unlike sensitive attributes which are clearly identifiable to recommender systems, the discrepancy between popularity and quality is anonymous. Therefore, previous recommendation fairness methods cannot be applied to address our specific problem.

3 Theoretical Analysis for Popularity in Full Information Setting

In this section, we will show that popularity in the full information setting will be subject to long-tail distribution by a probabilistic analysis. This indicates that long-tail distribution is mainly caused by the mechanism of user choosing item. Flattening the long-tail distribution of popularity deviates from recommender's target that offer users with high-quality recommendation and is harsh for recommendation fairness.

We firstly give the definition of the full information setting: all users have the knowledge of every item and how much value or happiness these items can bring to them. In this case, the degree of information transmission is maximized and there is no need for an information retrieval system. To analyze the popularity in this setting, we first model the interaction mechanism between users and items following the assumptions in Probabilistic Matrix Factorization proposed by [9]:

$$\begin{aligned} \boldsymbol{\theta}_u &\sim \mathcal{N}(\mathbf{0}, \sigma_\theta^2 \boldsymbol{I}_k), \\ \boldsymbol{\beta}_i &\sim \mathcal{N}(\mathbf{0}, \sigma_\beta^2 \boldsymbol{I}_k), \\ y(u,i) &\sim \mathcal{N}(\boldsymbol{\theta}_u^\top \boldsymbol{\beta}_i, \sigma_y^2), \end{aligned} \tag{1}$$

where $\boldsymbol{\theta}_u$ and $\boldsymbol{\beta}_i$ represent user u's preferences and item i's attributes, k is the dimension of $\boldsymbol{\theta}_u$ and $\boldsymbol{\beta}_i$, $\mathbf{I}_k$ is k-order unit matrix and $\mathcal{N}$ represents normal distribution. Unlike previous works that define $y(u,i)$ as the feedback from user u to item i, we let $y(u,i)$ denote the utility that item i can bring to user u. The utility is a concept used to model worth or value in economics.

We assume the number of items that user u can consume is subject to the geometric distribution, i.e., $P(m_u = k) = p^{k-1}(1-p)$, where m_u is the quantity of user u's affordable items. p can be considered as the user retention rate, which is the probability that a user continues to interact with the recommender after accepting one item recommended by it.

For a user u, given the number of his affordable items m_u, he will choose m_u items to maximize his utility. In this situation, user u will greedily choose

m_u items with the greatest utility $y(u,i)$. The set of items chosen by user u, $\mathcal{I}_u$ can be described as: $\mathcal{I}_u = \underset{\mathcal{I}_u \subseteq \mathcal{I}, |\mathcal{I}_u|=m_u}{\arg\max} \sum_{i \in I_u} y(u,i)$, where $\mathcal{I}$ is the set of all items.

Obviously, the popularity of item i is the number of times that item i will appear in collections $\mathcal{I}_u$ for all users. The popularity of item i can be described as $pop_i = |\{\mathcal{I}_u | i \in \mathcal{I}_u, u \in \mathcal{U}\}|$, where $|\cdot|$ denotes the cardinal number of a set, and $\mathcal{U}$ is the set of all users.

The probability that an item i is in $\mathcal{I}_u$ can be expressed by the event that the number of the set that item j in $\mathcal{I}$ subject to $y(u,j) > y(u,i)$ is less than m_u, i.e.,

$$\begin{aligned} P(i \in \mathcal{I}_u) &= P\left(|\{j | j \in \mathcal{I}, y(u,j) > y(u,i)\}| < m_u\right) \\ &= \sum_{m=1}^{m_u} C_N^{N-m} P\left(y(u,i) > y(u,j)\right)^{N-m} \cdot \\ &\quad [1 - P\left(y(u,i) > y(u,j)\right)]^m, \end{aligned} \tag{2}$$

where N is the number of all items. C_N^{N-m} is the combination number.

$y(u,i) - y(u,j)$ can be considered as a random variable. According to properties of the Gaussian distribution, $y(u,i) - y(u,j)$ still follows Gaussian distribution:

$$y(u,i) - y(u,j) \mid \theta_u, \beta_i, \beta_j \sim \mathcal{N}\left(\theta_u^\top \beta_i - \theta_u^\top \beta_j, 2\sigma_y^2\right). \tag{3}$$

$$y(u,i) - y(u,j) \sim \mathcal{N}\left(\theta_u^\top \beta_i - \theta_u^\top \beta_j, 2(\sigma_y^2 + \sigma_\beta^2 \sigma_\theta^4)\right)$$

Now, we add randomness to β_j, and eliminate it from equation (3). $y(u,i) - y(u,j)$ still follows Gaussian distribution. According to the properties of conditional expectation, the expectation of $y(u,i) - y(u,j)$ is $\theta_u^\top \beta_i$ and variance is $\sigma_\beta^2 \theta_u^\top \theta_u + 2\sigma_y^2$, i.e.,

$$y(u,i) - y(u,j) \mid \theta_u, \beta_i \sim \mathcal{N}\left(\theta_u^\top \beta_i, \sigma_\beta^2 \theta_u^\top \theta_u + 2\sigma_y^2\right). \tag{4}$$

We can obtain $P(i \in \mathcal{I}_u \mid \theta_u, \beta_i)$ by integrating Eqs. (2) and (4). θ_u and m_u can be similarly eliminated from $P(i \in \mathcal{I}_u)$ through following Eqs. (5), (6).

$$P(i \in \mathcal{I}_u \mid \theta_u, \beta_i) = \sum_{m_u} P(i \in \mathcal{I}_u \mid \theta_u, \beta_i, m_u) P(m_u) \tag{5}$$

$$P(i \in \mathcal{I}_u \mid \beta_i) = \int P(i \in \mathcal{I}_u \mid \beta_i, \theta_u) P(\theta_u) d\theta_u \tag{6}$$

The popularity pop_i, follows a binomial distribution with the probability of $P(i \in \mathcal{I}_u)$ times M, which is the number of users. The expectation of popularity is $\mathbb{E}(pop_i) = M \cdot P(i \in \mathcal{I}_u)$. Finally, we can get the popularity expectation of different $\boldsymbol{\beta}_i$ by eliminating other variables.

To simplify and intuitively illustrate this question, we set number of items N to 1000, retention rate p to 0.9, σ_θ, σ_β, σ_y in Eq. (1) to 1, k in Eq. (1) to 2, i.e., $\boldsymbol{\theta}_u = (\theta_{u,1}, \theta_{u,2}), \boldsymbol{\beta}_i = (\beta_{i,1}, \beta_{i,2})$.

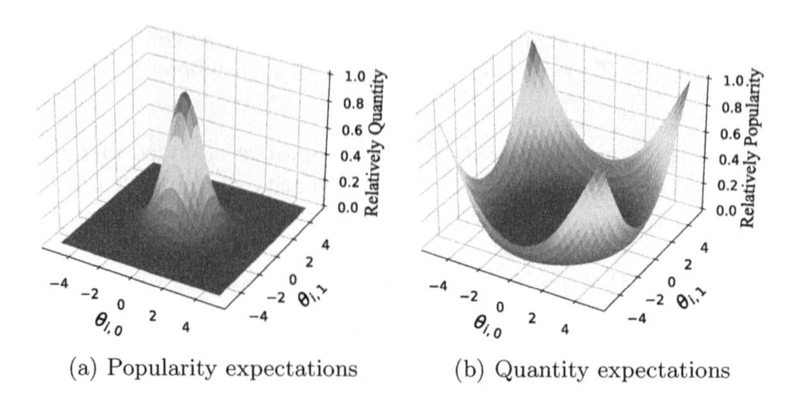

(a) Popularity expectations (b) Quantity expectations

Fig. 2. Results of popularity probabilistic distribution analysis.

Through calculation, we obtain the quantity expectations and popularity expectations of different $\boldsymbol{\beta}_i$, shown in Fig. 2. We can find that every dimension in Fig. 2(a) and 2(b) is equivalent, so our setting on k does not result in loss of generality. Combining two figures, we can find that a small number of items have a large amount of popularity, while a large number of items have a small amount of popularity. Thus, the popularity is subject to long-tail distribution, same to the popularity distribution in the real-word dataset.

In general, the assumptions we make about a user's utility regarding an item are not necessary conditions for the long-tail distribution of popularity. Intuitively, the greed characteristic of the users who tend to maximize their utility and the fact that they can only afford a small fraction of the available items are sufficient to cause the long-tail distribution of popularity in the full information setting.

Since popularity in the full information setting is also subject to long-tail distribution, long-tail distribution of popularity is unavoidable. Previous works' focus on popularity's long-tail distribution let them miss the true question and the recommender's main target. Some previous works, which use a test dataset with uniformly sampled popularity distribution to evaluate the resistance of recommenders to popularity bias [7,15,21], deviate from the true popularity distribution

4 Problem Definition

A wide variety of fairness concerns about recommender systems have been proposed e.g., fair ranking across items with different sensitive attributes. Inspired by Beutel et al. [2], who focused on the under-ranking of items from specific supplier groups with sensitive attributes (e.g., gender, race), we primarily focus on the risk of high-quality items being under-recommended or low-quality items being over-recommended in this work. For example, if a media network under-ranked high-quality videos, that could make up-loaders of these videos feel unfair and thus limit their engagement on the service. We will work to eliminate the impact of popularity-quality discrepancy for reducing the risk that items' recommendation weight is inconsistent with quality.

Except for fairness concerns need to be taken into consideration, providing users with satisfactory recommendation service is the main task of recommender systems. Suppose we have a recommender system with a user set $\mathcal{U} = \{u_1, u_2, \ldots u_M\}$ and an item set $\mathcal{I} = \{i_1, i_2, \ldots i_N\}$, where M, N is the number of users and items. Let $\mathcal{D}$ denote the historical interaction data, notated as a list of records, i.e., $\mathcal{D} = \{y(u,i)|u \in \mathcal{U}, i \in \mathcal{I}\}$, where $y(u,i)$ denotes whether the user u has interacted the item i. In the user side, the task of a recommender system can be stated as follows: learning a recommendation model from $\mathcal{D}$ so that it can capture user preference and make a high-quality recommendation.

5 Methodology

The structure of the **Quality Recommendation Framework** is shown in Fig. 3. Specifically, our QR framework separates item representation into quality representation $\boldsymbol{\beta}_{i,q}$ (Sect. 5.1) and supplement ID representation $\boldsymbol{\beta}_{i,id}$ (Sect. 5.2). Aggregation of both representations is the final embedding of each item, i.e., $\boldsymbol{\beta}_i = \boldsymbol{\beta}_{i,q} + \boldsymbol{\beta}_{i,id}$.

Since user is not the point of this paper, we simply infer user embedding by user ids. The click scoring model computes ranking score $\hat{y}$ based on the item embedding $\boldsymbol{\beta}_i$ and the user embedding $\boldsymbol{\theta}_u$ by inner dot product function, i.e., $\hat{y}_{u,i} = <\boldsymbol{\theta}_u \cdot \boldsymbol{\beta}_i>$.

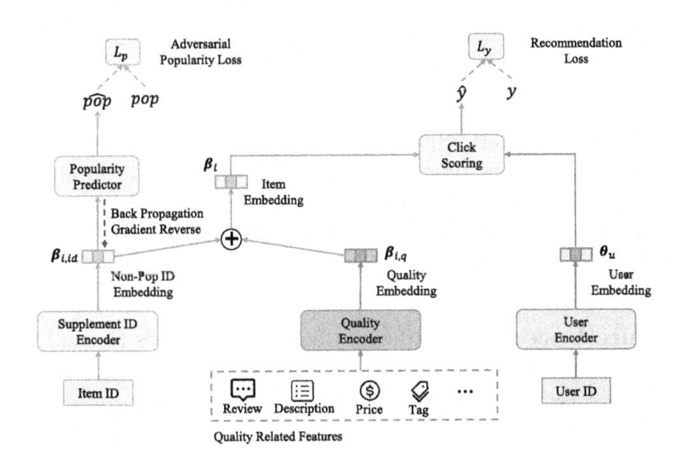

Fig. 3. The architecture of our QR approach.

5.1 Quality Embedding

We model quality representation by digitized item natural attributes (e.g. price, brand) and descriptive features (e.g. reviews, description) that reflect item intrinsic features. The implementation of quality embedding model includes encoders for review, textual features, discrete features and continuous features of items.

Review Encoder. To process and summarise the semantic information of each review, we first convert each review into a embedding vector by using the pre-trained BERT model, which is a recent and widely used language modeling approach. We froze the parameters of this language model, which reduces the risk of that semantic information in the reviews' embedding being twisted by the recommendation task. Then we use an attention pooling network to select reviews that are representative to the item i and then aggregate the representation of informative reviews. A two-layer network is applied to compute the attention score $a_{i,l}$. The input contains the semantic vector $R_{i,l}$ converted by the pre-trained BERT model from the lth review of item i, and the review's properties $P_{i,l}$ (e.g. time, length, helpfulness, etc.). Formally, the attention network is defined as:

$$a_{i,l}^{*} = h^{T}\,\mathrm{ReLU}\left(W_R R_{i,l} + W_P P_{i,l} + b_1\right) + b_2, \tag{7}$$

where $W_R \in \mathbb{R}^{t\times k_R}, W_P \in \mathbb{R}^{t\times k_P}, b_1 \in \mathbb{R}^t, h \in \mathbb{R}^t, b_2 \in \mathbb{R}^1$ are model parameters. K_R, K_P respectively denotes the size of the semantic vector and review property vector, and t denotes the hidden layer size of the attention network. ReLU is a nonlinear activation function.

The final weight of reviews are obtained by normalizing the above attention scores using the softmax function, i.e., $a_{i,l} = \frac{\exp\left(a_{i,l}^{*}\right)}{\sum_{l=1}^{L}\exp\left(a_{i,l}^{*}\right)}$. $a_{i,l}$ can be interpreted as the contribution of the lth review to the feature embedding $\boldsymbol{\beta}_i$ of item i.

After we obtain the attention weight of each review, the review feature vector of item i is calculated as the following weighted sum, $R_i = \sum_{l=1}^{L} a_{i,l} R_{i,l}$.

The output of the attention-based pooling layer is a k_R dimensional vector, which compresses all reviews of item i in the embedding space by distinguishing their contributions. Then we use a MLP(Multi-Layer Perceptron) to transform semantic vector R_i into interaction domain, i.e., $\tilde{R_i} = \mathrm{MLP}_R(R_i)$.

Textual Feature Encoder. Similarly, we process item i's textual features (e.g. description) by the pre-trained language model to get semantic vector D_i of these features. After that transform these vectors into interaction domain by MLP net, i.e., $\tilde{D_i} = \mathrm{MLP}_D(D_i)$.

Discrete Feature Encoder. For items' discrete features (e.g. tags, brand), we use an embedding lookup matrix to transform these into vectors and sum up to get items' discrete feature representation. $\tilde{T_i} = \sum_{t\in T_i}(\boldsymbol{e}_t)$, where t denotes a feature in discrete feature set T_i of item i and the $\boldsymbol{e}_t$ is the embedding of discrete feature t.

Continuous Feature Encoder. For items' continuous features (e.g. price, age) P_i, we use another MLP Network for encoding this, i.e., $\tilde{P}_i = \text{MLP}_P(P_i)$, where P_i denotes the vector of item i continuous feature vector.

Then $\tilde{R}_i, \tilde{D}_i, \tilde{T}_i, \tilde{P}_i$ are all sent to a sum pooling layer for computing the quality representation of item i: $\boldsymbol{\beta}_{i,q} = \tilde{R}_i + \tilde{D}_i + \tilde{T}_i + \tilde{P}_i$.

5.2 Supplement Id Embedding Without Popularity Information

Different from other item features, item ID is randomly assigned to an item and its value is not directly related to the natural attributes of the item. But ID is a unique identifier of an item to distinguish it from others. Since all items probably have unique and undigitized natural attributes, introducing ID embedding as a supplement is necessary. However, the existence of ID embedding makes the model attribute popularity only to the popular items themselves, rather than to its digitized natural properties. This lets the recommender system create a stereotyped impression that specific items are popular, increasing the risk of inconsistency between items' quality and recommendation weight. Our method to the above problems is to retain the ID embedding but eliminate popularity relate information from it.

For this reason, we introduce adversarial learning for eliminating popularity information from item ID embeddings. Specifically, we use a MLP predictor to predict item popularity according to the ID embedding as $\hat{pop}_i = \text{MLP}_{pop}(\boldsymbol{\beta}_{i,id})$.

During the back propagation, we take the gradient from the popularity predictor and **reverse** its sign, i.e., multiply it by -1, before passing it to the ID embedding layer. This step reverse the target of ID embedding, making the ID embedding model eliminating the popularity related information of item ID embedding to **avoid** the predictor from inferring popularity from it.

5.3 Model Training

Finally, we introduce how to train our QR model. The ranking score $\hat{y}_{ui}$ is mainly supervised to recover the historical interactions by a recommendation loss, i.e., $L_y = \text{BCELoss}(y, \hat{y})$, where BCE is Binary Cross Entropy loss.

To stabilize the change of popularity prediction loss in the whole training process, we choose the SmoothL1Loss as the loss function of popularity prediction, $L_p = \text{SmoothL1Loss}(pop, \hat{pop})$. SmoothL1Loss is formulated as follows:

$$\text{SmoothL1Loss}(x) = \begin{cases} 0.5x^2, & \text{if } |x| < 1, \\ |x|, & \text{otherwise} . \end{cases} \tag{8}$$

The total loss is formulated as $L = L_y + \lambda_p L_p + \lambda_\Omega \Omega(\Theta)$, where $\Omega(\Theta)$ is the regularization term, and λ_p and λ_Ω are coefficients that control the importance of their corresponding losses.

6 Experiment

6.1 Experimental Setup

Datasets. Since our method needs a variety of item information, we choose Amazon product review datasets as the source of the experimental dataset. We select three different domain datasets from it: Toys & Games, Grocery & Gourmet Food and CDs & Vinyl.

Baselines. We compare our methods with the following baselines including one classical baseline (BPR-MF [12]), three textual recommendation approaches (DeepCoNN [20], NARRE [5], DAML [8]), and five debiasing approaches (IPS [7], CausE [3], DICE [21], TIDE [22], MPBGP [11]).

6.2 Evaluation

Performance Evaluation. For recommend performance evaluation, we adopt three widely-used evaluation metrics: Hit Ratio at rank K (HR@K), Recall at rank K (Recall@K) and Normalized Discounted Cumulative Gain at rank K (NDCG@K). By default, we set K=10 in our experiments.

Fairness Evaluation. Since previous evaluation methods related to popularity bias are not reliable (as illustrated in Sect. 3), we need a new method for testing the model's robustness under popularity-quality discrepancy.

However, there is no definite and authoritative standard for quantitatively measuring the item quality nor its consistent popularity. This prevents us from further measuring the popularity-quality discrepancy and its impact on recommendations. On the condition that popularity distribution in reality and ideal case (as shown in Sect. 3) both subject to long-tail distribution, sub-optimally and practically, we can assume the popularity in the original dataset is consistent with quality, and disrupt the popularity in the training dataset to simulate popularity-quality discrepancy. Next, we examine the model's vulnerability to this manual discrepancy, in order to evaluate the recommendation vulnerability in reality.

Firstly, we simulate popularity-quality discrepancy in the following two steps:

1. Items are randomly separated into 3 groups: over, critical and under popularized groups. These simulates that these items get high, fair and low popularity compared to their quality for recommender system past unfair rank and recommendation.
2. Items in over, critical, and under popularized groups assigned will get the same $R\%$ of their interaction records in valid and test datasets but will get $(T - b)\%$, $T\%$ and $(T + b)\%$ records in train dataset, as shown in Fig. 4.

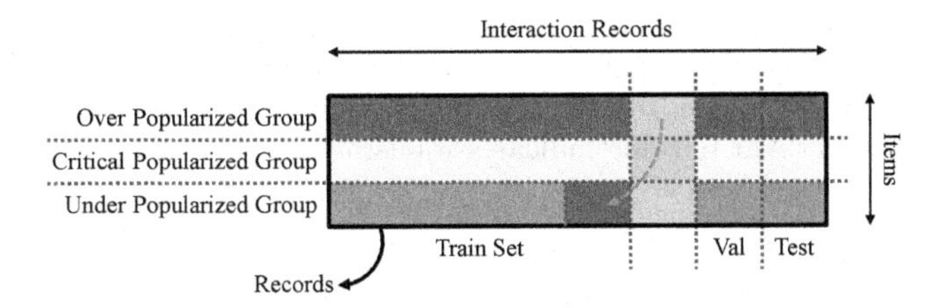

Fig. 4. Construction of the manual popularity-quality discrepancy. Red, yellow, blue parts represent interaction data of over-popularized, critical-popularized, under-popularized group. Grey parts represent the aborted data. (Color figure online)

From the global perspective, we can find that we separate the dataset into $T\%$ train part, $R\%$ valid, test parts, and $b\%$ abandoned part. b can also be taken as the degree of popularity-quality discrepancy. As different groups have the same quality expectation, if the recommender system pays great attention to over-popularized group while neglecting under-popularized group, it is vulnerable to popularity-quality discrepancy and therefore offers low fairness in recommendations.

For the evaluation of fair treatment, we introduce pairwise fairness proposed by [2]. A recommender is considered to obey pairwise fairness, if the likelihood of a clicked item being ranked above another relevant unclicked item is the same across different groups, i.e.,

$$\begin{aligned} &P\left(\hat{y}(u,i) > \hat{y}(u,j) \mid y(u,i) > y(u,j), i \in I_0, j \in I\right) \\ &\quad = P\left(\hat{y}(u,i) > \hat{y}(u,j) \mid y(u,i) > y(u,j), i \in I_1, j \in I\right), \end{aligned} \tag{9}$$

where I_0, I_1 are different subgroups of all items set I.

In practice, we use the variance of pairwise accuracy as the evaluation metric of fairness treatment across different popularized groups and the robustness to popularity-quality discrepancy. Low pairwise accuracy variance denotes more of pairwise fairness and low rank-quality inconsistency risk. Parameters of the manually simulated discrepancy, T, R and b are set as 75%, 10%, 5% respectively.

6.3 Performance Comparison

The overall performance is illustrated in Table 1, and the best results are highlighted in boldface. From the table, several observations can be made.

Previous debiasing models (IPS, DICE, TIDE, MPBPGP) based on MF have better performance on the discrepancy robustness metric, but at the expense of recommendation accuracy. This is because these de-bias works entirely exclude popularity and quality out of recommendation. Therefore, different popularized groups are treated equally. This makes models perform well on recommendation fairness. However, items in the same group with different quality are also treated equally, and this makes these models have a poor recommendation performance.

Table 1. The performance evaluation of the compared methods. Avg. and Var.R means the average pairwise accuracy and variance of pairwise accuracy relative to the BPR-MF model. Higher NDCG, Recall, HR, and Avg. values indicate better recommendation accuracy, while a low Var.R value indicates robustness and fairness in accounting for popularity-quality discrepancy.

	Toys & Games					Grocery & Gourmet Food					CDs & Vinyl				
Methods	Ndcg	Recall	Hr	Pairwise Acc.		Ndcg	Recall	Hr	Pairwise Acc.		Ndcg	Recall	Hr	Pairwise Acc.	
				Avg	Var.R				Avg	Var.R				Avg	Var.R
BPR-MF	0.148	0.252	0.322	0.599	1.000	0.198	0.286	0.360	0.643	1.000	0.141	0.257	0.335	0.57	1.000
IPS	0.134	0.194	0.233	0.612	**0.008**	0.181	0.267	0.336	0.636	**0.043**	0.129	0.232	0.323	0.562	**0.034**
CausE	0.101	0.175	0.230	0.566	1.291	0.116	0.176	0.230	0.556	0.635	0.107	0.157	0.237	0.546	0.122
DICE	0.185	0.249	0.314	0.555	0.036	0.172	0.240	0.307	0.578	0.246	0.139	0.279	0.319	0.616	0.143
TIDE	0.153	0.328	0.364	0.620	0.735	0.237	0.304	0.391	0.603	0.877	0.887	0.390	0.435	0.673	0.841
MPBGP	0.140	0.265	0.311	0.599	0.130	0.244	0.399	0.382	0.621	0.702	0.140	0.249	0.355	0.62	0.092
DeepConn	0.175	0.306	0.376	0.699	0.176	0.225	0.315	0.390	0.651	0.952	0.173	0.368	0.409	0.738	0.836
NARRE	0.190	0.244	0.393	0.697	0.806	0.234	0.376	0.454	0.711	1.562	0.187	0.411	0.452	0.745	0.769
DAML	0.227	0.423	0.471	0.756	0.432	0.193	0.376	0.455	0.741	1.485	0.212	0.401	0.479	0.75	0.945
QR	**0.248**	**0.462**	**0.537**	**0.784**	0.042	**0.264**	**0.465**	**0.549**	**0.791**	0.345	**0.219**	**0.423**	**0.533**	**0.753**	0.284

Methods considering reviews (DeepCoNN, NARRE, DAML) generally perform better on recommendation accuracy and bias robustness than the basic MF model. As reviews can help the recommender system to establish the relationship of interaction with the natural properties of items. However, as popularity stereotype remains, the fairness performance is unstable.

Our QR model achieves competitive performance on both recommendation performance and popularity-quality discrepancy robustness. This demonstrates the effectiveness of the quality representation and the supplement ID representation with popularity adversarial learning. The former can learn better representations through a large number of explicit and implicit natural attributes of items. The latter can make recommender better handle the different popularity distributions of training and testing stage. They both drive the model to think about why an item is popular instead of remembering it.

6.4 Ablation and Hyperparameter Study

We compare the recommendation pairwise accuracy and variance of QR and its variants, and the results are illustrated in Fig. 4. We have several observations from this plot. First, applying the adversarial popularity learning to QR helps improve bias robustness at the expense of a small drop in recommendation accuracy. This is because adversarial learning can encourage ID embedding to minimize the information of popularity. Second, the single item quality embedding is more performs better than the ID embedding, as the input of the quality embedding contains natural property information about the item. Third, the single ID embedding model with adversarial learning effectively removes the popular information, but model accuracy is greatly compromised.

We conduct experiments of the hyperparameter λ_P. Since λ_P only works to ID embedding, we explore the it's influence on fairness and performance of ID embedding. The result is shown in Fig. 5. We can find that a higher λ_P will encourage the ID embedding to be more fair at the expanse of performance (Fig. 6).

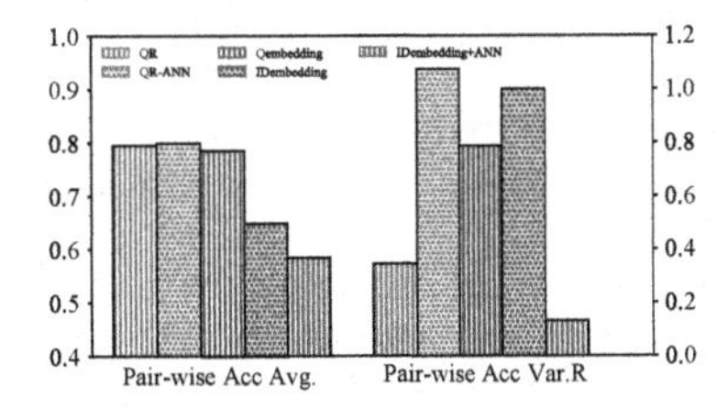

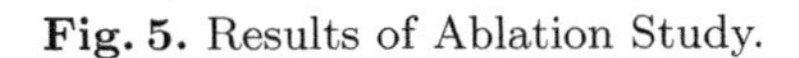

Fig. 5. Results of Ablation Study.

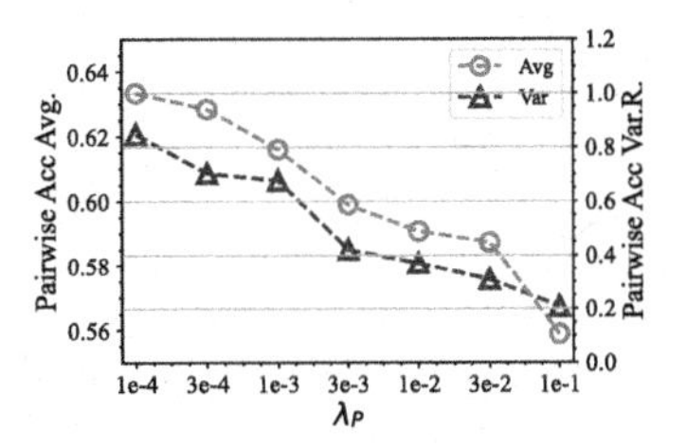

Fig. 6. Results of Hyperparameter Study.

7 Conclusion

In this work, we argue that the popularity of items should not be avoided at all costs but instead calibrated to their quality. To this end, we propose the Quality Recommendation (QR) framework to improve robustness to popularity-quality discrepancy. To evaluate models' robustness this discrepancy, we simulate this discrepancy on the training stage. Experiments on datasets show the effective fairness and recommendation performance of our proposed methods.

References

1. Abdollahpouri, H., Burke, R., Mobasher, B.: Controlling popularity bias in learning-to-rank recommendation. In: Proceedings of the Eleventh ACM Conference on Recommender Systems, pp. 42–46. RecSys '17, New York, NY, USA (2017)
2. Beutel, A., et al.: Fairness in recommendation ranking through pairwise comparisons. In: Proceedings of the 25th ACM SIGKDD International Conference on Knowledge Discovery & Data Mining, pp. 2212–2220. KDD '19, New York, NY, USA (2019)
3. Bonner, S., Vasile, F.: Causal embeddings for recommendation. In: Proceedings of the 12th ACM Conference on Recommender Systems, pp. 104–112. RecSys '18, New York, NY, USA (2018)
4. Celma, O., Cano, P.: From hits to niches? or how popular artists can bias music recommendation and discovery. In: Proceedings of the 2nd KDD Workshop on Large-Scale Recommender Systems and the Netflix Prize Competition. NETFLIX '08, New York, NY, USA (2008)
5. Chen, C., Zhang, M., Liu, Y., Ma, S.: Neural attentional rating regression with review-level explanations. In: Proceedings of the 2018 World Wide Web Conference, pp. 1583–1592. WWW '18, Republic and Canton of Geneva, CHE (2018)
6. Jannach, D., Lerche, L., Kamehkhosh, I., Jugovac, M.: What recommenders recommend: an analysis of recommendation biases and possible countermeasures. User Model. User-Adap. Inter. **25**(5), 427–491 (2015)
7. Liang, D., Charlin, L., Blei, D.M.: Causal inference for recommendation. In: Causation: Foundation to Application, Workshop at UAI. AUAI (2016)

8. Liu, D., Li, J., Du, B., Chang, J., Gao, R.: Daml: dual attention mutual learning between ratings and reviews for item recommendation. In: Proceedings of the 25th ACM SIGKDD International Conference on Knowledge Discovery & Data Mining, pp. 344–352. KDD '19, New York, NY, USA (2019)
9. Mnih, A., Salakhutdinov, R.R.: Probabilistic matrix factorization. In: Advances in Neural Information Processing Systems 20 (2007)
10. Park, Y.J., Tuzhilin, A.: The long tail of recommender systems and how to leverage it. In: Proceedings of the 2008 ACM Conference on Recommender Systems, pp. 11–18. RecSys '08, New York, NY, USA (2008)
11. Ren, W., Wang, L., Liu, K., Guo, R., Lim, E., Fu, Y.: Mitigating popularity bias in recommendation with unbalanced interactions: A gradient perspective. In: IEEE International Conference on Data Mining, ICDM 2022, Orlando, FL, USA, November 28 - Dec. 1, 2022., pp. 438–447. IEEE (2022)
12. Rendle, S., Freudenthaler, C., Gantner, Z., Schmidt-Thieme, L.: BPR: Bayesian personalized ranking from implicit feedback. CoRR abs/1205.2618 (2012)
13. Singh, A., Joachims, T.: Fairness of exposure in rankings. In: Proceedings of the 24th ACM SIGKDD International Conference on Knowledge Discovery and Data Mining, pp. 2219–2228. KDD '18, New York, NY, USA (2018)
14. Steck, H.: Item popularity and recommendation accuracy. In: Proceedings of the Fifth ACM Conference on Recommender Systems, pp. 125–132. RecSys '11, New York, NY, USA (2011)
15. Wei, T., Feng, F., Chen, J., Wu, Z., Yi, J., He, X.: Model-Agnostic Counterfactual Reasoning for Eliminating Popularity Bias in Recommender System, pp. 1791–1800. New York, NY, USA (2021)
16. Xv, G., Lin, C., Li, H., Su, J., Ye, W., Chen, Y.: Neutralizing popularity bias in recommendation models. In: Proceedings of the 45th International ACM SIGIR Conference on Research and Development in Information Retrieval, pp. 2623–2628. SIGIR '22 (2022)
17. Zehlike, M., Bonchi, F., Castillo, C., Hajian, S., Megahed, M., Baeza-Yates, R.: Fa*ir: A fair top-k ranking algorithm. In: Proceedings of the 2017 ACM on Conference on Information and Knowledge Management, pp. 1569–1578. CIKM '17, New York, NY, USA (2017)
18. Zhang, Y., et al.: Causal intervention for leveraging popularity bias in recommendation. In: Proceedings of the 44th International ACM SIGIR Conference on Research and Development in Information Retrieval, pp. 11–20 (2021)
19. Zhao, Z., et al.: Popularity bias is not always evil: disentangling benign and harmful bias for recommendation. IEEE Trans. Knowl. Data Eng. 1–13 (2022)
20. Zheng, L., Noroozi, V., Yu, P.S.: Joint deep modeling of users and items using reviews for recommendation. In: Proceedings of the Tenth ACM International Conference on Web Search and Data Mining, pp. 425–434. WSDM '17, New York, NY, USA (2017)
21. Zheng, Y., Gao, C., Li, X., He, X., Li, Y., Jin, D.: Disentangling User Interest and Conformity for Recommendation with Causal Embedding, p. 2980–2991. New York, NY, USA (2021)
22. Zhu, Z., He, Y., Zhao, X., Zhang, Y., Wang, J., Caverlee, J.: Popularity-opportunity bias in collaborative filtering. In: Proceedings of the 14th ACM International Conference on Web Search and Data Mining, pp. 85–93. WSDM '21, New York, NY, USA (2021)
23. Zhu, Z., Hu, X., Caverlee, J.: Fairness-aware tensor-based recommendation. In: Proceedings of the 27th ACM International Conference on Information and Knowledge Management, pp. 1153–1162. CIKM '18, New York, NY, USA (2018)

Searching User Community and Attribute Location Cluster in Location-Based Social Networks

Yunzhe An, Chuanyu Zong(✉), Ruozhu Li, Tao Qiu, Anzhen Zhang, and Rui Zhu

School of Computer Science, Shenyang Aerospace University, 110136 Liaoning, China
{anyunzhe,zongcy,qiutao,azzhang,zhurui}@sau.edu.cn, liruozhu@stu.sau.edu.cn

Abstract. Community search is a fundamental problem in analyzing and managing graph data, which is searching for an optimal community based on query nodes. Attribute community search and geosocial community search have been extensively investigated, however, very few studies consider location-based social networks with *attributes* where users can get desired locations for some specified attributes. In this paper, we propose the Attribute Geosocial Community Search problem (AGCS) in a location-based social network with attributes, which aims to find a user community and a cluster of spatial locations with attributes that are densely connected simultaneously, while the community and the cluster should have high community metric based on attributes and check-in information. The AGCS can be used in many graph analysis applications, such as user and location recommendation, and geosocial data analysis. To solve this problem, we first defined a novel community metric based on attribute constraints and check-in information. Then, we explore three novel search algorithms: a basic algorithm based on k-core, a global algorithm for greedy deletion, and an improved global algorithm. Finally, we conduct comprehensive experimental studies, which demonstrate that our proposed solutions can efficiently searching user community and attribute location cluster in location-based social networks.

Keywords: Community search · User community · Location cluster with attributes

1 Introduction

Searching an internally densely connected, externally sparsely connected subgraph containing the query nodes is a fundamental problem in many graph analysis and management applications. In contrast to community detection, community search looks for communities in a graph that are connected by a set of query nodes and certain query restrictions [1]. Community search based only on user relationships can no longer fully satisfy the needs of community search

X. Yang et al. (Eds.): ADMA 2023, LNAI 14180, pp. 389–404, 2023.
https://doi.org/10.1007/978-3-031-46677-9_27

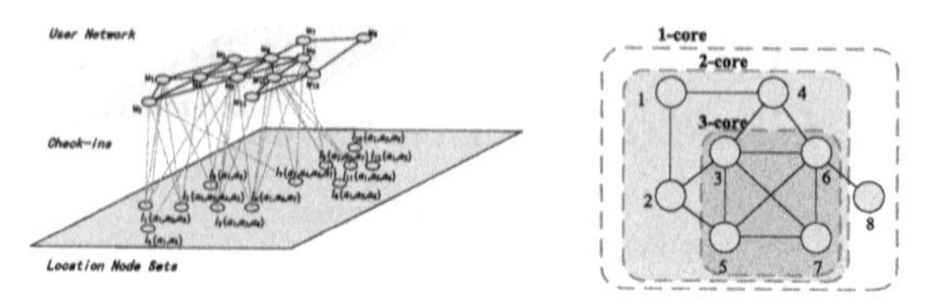

(a) Location-based social network with attributes (b) Example of k-core

Fig. 1. Location-based social network with attributes

as the diversity of data grows and users' requirements change, The emergence of attribute community search has enabled community search to add attribute constraints to user relationships, which searching for an attribute community that satisfies both structural and keyword cohesiveness. However, it may produce a community with a wide variety of spatial places. With the emergence of location-based social networks, the search criteria of community search can now take location relationships between users into account in addition to considering the social relationship between users. And thus, community search over location-based social networks emerges [2]. Searching community from a location-based social network yields a community of users and a cluster of social location that contains all the locations corresponding to the users' network, which are disorganised and do not all meet the actual requirements of the users.

Therefore, in order to find the users' intend communities that satisfy both the attribute and location constraints, we perform a community search with both location and attribute limitations on a location-based social network with attributes. We find that most previous location-based community searches have only yielded single user communities. And a few studies have explored the problem of searching user community and location cluster in location-based social networks, let alone the problem of searching location cluster based on user target attributes. In this paper, therefore, we explore the problem of Attribute Geosocial Community Search (AGCS), which focus on finding a densely connected user community and an attribute location cluster simultaneously with a high community metric from a location-based social networks with attributes.

A location-based social network with attributes is shown in Fig. 1(a), it includes user network, check-in information, and location node sets with attributes. A user is represented by each node in the user network, and a user's visited locations are represented by each node in the location node sets. The check-in edge (u_i, l_i) displays the user u_i has checked in at the location l_i, and a_i indicates the attribute or keyword associated with the location l_i. In Fig. 1(a), it is assumed that the user u_{11} wants to go to the movies and is looking for some suggestions on cinemas and which friends he/she can go to the movies with. In that situation, we need to find a user community and a cluster of spatial location with the specified attribute "cinema" simultaneously from the Location-based social networks with location attributes as detailed in Fig. 1(a).

To solve this type of community search problem for the user u_{11}, this paper proposes an attribute-constraint community search on a location-based social network with attributes: Attribute Geographical Community Search (AGCS). Given a location-based social network with attributes, a set of query nodes, and a set of query attributes, the purpose of AGCS is to search a subgraph containing the query nodes that satisfies structural, spatial, and attribute constraints. This subgraph contains a social community of users and an attribute location cluster, such that users in the same community often check in at the same location cluster with attribute constraints. Thus, they have closer social relationships. Additionally, the resulting attribute location clusters add to the richness of the community search results.

The main contributions of this paper are summarized as follows:

(1) A novel community metric for community search in location-based social networks based on attribute constraints and check-in information is defined.
(2) The geosocial community search problem in location-based social networks with attributes is proposed.
(3) Three algorithms are explored to solve the attribute geosocial community search problem effectively and efficiently.

2 Related Work

Community search with constraints other than structural constraints are discussed as below.

The first is community search with geographical location constraints. The Geosocial Community Search (GCS) proposed by Kim et al. [2] requires that the resulting community network and location clusters be not only closely connected in terms of user relationships but also geographically. Wang et al. [10] proposed a radius-constrained community search problem to find a set of spatially neighbouring groups on a location-based social network.

The second is community search with attribute constraints. Fang et al. [6] proposed the Attribute Community Query (ACQ) to return the attribute community (AC) for keyword cohesion. Zhang et al. [8] investigated keyword-centric community search (KCCS) on attribute graphs. The parameter-free contextual community model for attribute community search proposed by Chen et al. [9]. Liu et al. [12] proposed a vertex-centric attribute community (VAC) problem to solve the problem of difficult attribute setting and single query attribute types. Apon et al. [5] proposed the Top-k Flexible Socio-Spatial Keyword-aware Group Query (SSKGQ) problem based on social and spatial text.

The third is community search based on location and attribute. Guo et al. [7] proposed multi-attribute communities (MAC) Chen et al. [3] proposed the co-located community search (LCD) problem, in which the obtained communities satisfy the k-truss constraint and the user's spatial location constraint.

3 Problem Statement

Given an undirected graph G=(V, E), which contains a set of nodes V and a set of edges $E \subset V \times V$. Let H be a subset of nodes in V, $G[H]$=$(V[H], E[H])$ is denoted as the subgraph of G induced by H, i.e., $E[H]$=$\{(u, v) \in E|u, v \in H\}$. A bipartite graph G_C=(V_U, V_L, E_C) can represent a check in graph, where V_U is a set of user nodes, $V_L = \{(v_i, A(v_i))\}$ is a set of location nodes with attribute labels $A(v_i)$=$\{a_1, a_2, a_3, ..., a_j\}$, $(u, l) \in E_C$ is a check-in edge which stands for the user u has checked in the location l, and $W(u, l) \in R$ is the weight of the edge $(u, l) \in E_C$. The check-in weight W can be considered as the frequency of check-in. Finally, a Location-Based Social Network with Attributes (LBSNA for short) N=(G_U, V_L, G_C) is formed, which consists of a social network G_U, a set of location nodes V_L with attribute labels, and a check-in graph G_C.

There are many models to evaluate the structural cohesiveness of a community, such as k-core, k-truss, and k-clique [1], and the k-core is used as the measure of structural cohesiveness [11] in this paper. The number of nodes directly connected to the node v is denoted as $deg(v)$, i.e., the number of neighbours of node v. And $deg(v)$ is also known as the degree of a node. Additionally, given a subgraph $H = (V_H, E_H)$, $\delta(H)$ returns the minimum degree of any node $v \in V_H$.

Definition 1. ***k-Core***. *Given an integer $k(k \geqslant 0)$, a largest subgraph H_k of a graph G is said to be a k-core of the graph G when it satisfies $\forall v \in H_k$, $\delta(H_k) \geqslant k$, and H_k may be disconnected.*

Given two positive integers, i and j, $H_j \subseteq H_i$ if $i < j$. As shown in Fig. 1(b), the subgraph consisting of nodes $\{1, 2, 3, 4, 5, 6, 7\}$ is 2-core.

Definition 2. ***User Community***. *Let G_U be a social network and k be a degree threshold, a user community H_U is a connected subgraph $H_U = (V_{H_U}, E_{H_U})$ in G_U, which satisfies the degree constraint k, i.e., $\forall v \in V_{H_U}, deg(v) \geqslant k$.*

Definition 3. ***Distance Reachable***. *Let l_i and l_j be two location nodes in V_L, and r be a distance threshold, l_i and l_j are distance reachable if there exists a sequence of location nodes $l_1, l_2, l_3, ..., l_n$, where $l_i = l_1$, $l_j = l_n$ such that $dis(l_i, l_{i+1}) \leq r$ ($dis(l_i, l_j)$ returns the distance between l_i and l_j).*

Definition 4. ***Attribute Location Cluster***. *Given a location node set V_L, a distance threshold r, and a query attribute set $A_Q = \{a_1, a_2, .., a_n\}$, and a degree threshold k. A set of location nodes L_AC forms the location cluster if the following conditions are satisfied: (1) $\forall l_i$ and l_j in L_AC, l_i and l_j are distance reachable. (2) $\forall l_i \in L_AC$, $A_Q \subseteq A(l_i)$. (3) $\forall l_i \in L_AC$, There are at least k neighbours within r distances of l.*

The attribute location clusters need to be obtained from the attribute location network. By creating social links for nodes whose attributes satisfy the constraints and are distance reachable, we obtain the attribute location network.

In Fig. 2(a), an attribute location node set consisting of 12 location nodes $l_1 \sim l_{12}$ with different attribute labels. Figure 2(b) shows distances between

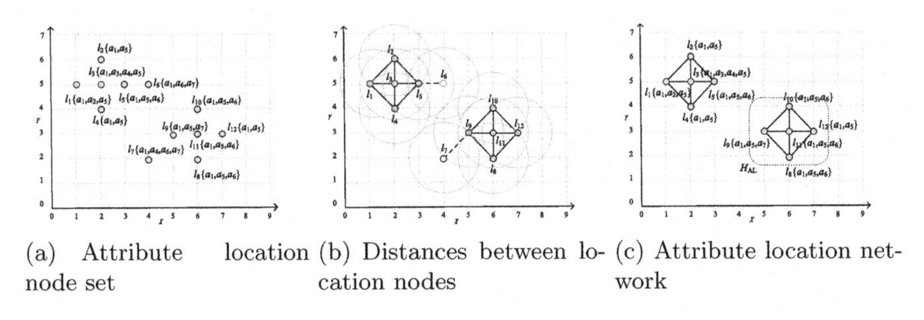

(a) Attribute location node set (b) Distances between location nodes (c) Attribute location network

Fig. 2. Attribute Location Clusters

location nodes in Fig. 2(a) when the distance threshold r is set as $r = \sqrt{2}$. It can be seen that the location network constructed based on location node sets and radius threshold r contains three connected components $\{l_1, l_2, l_3, l_4, l_5, l_6\}, \{l_7, l_8, l_9, l_{10}, l_{11}, l_{12}\}$, where any two nodes in any connected component are distance reachable. For the distance threshold $r = \sqrt{2}$ and the query attribute $A_Q = \{a_1, a_5\}$, since nodes l_6 and l_7 do not satisfy the attribute constraint, two attribute location networks$\{l_1, l_2, l_3, l_4, l_5\}$, and $\{l_8, l_9, l_{10}, l_{11}, l_{12}\}$ are obtained as seen in Fig. 2(c). For example, H_{AL} is a location attribute network as seen in Fig. 2(c).

Definition 5. ***Community Score.*** *According to the attribute constraints and check-in information, the community score of a community can be calculated by the following equation.*

$$S_G = x \cdot \frac{|V_t|}{|V_a|} + y \cdot \frac{\sum_{\mu \in H_U, \nu \in H_{AL}} W(\mu, \nu)}{\sum_{\mu \in H_U, \omega \in G_{AL}} W(\mu, \omega)} \tag{1}$$

In Eq. 1, $|V_t|$ stands for the number of location nodes in the searched attribute location cluster that satisfy the attribute constraints, $|V_a|$ represents the number of location nodes in the original location node set that satisfy the attribute constraints, $W(u, v)$ is the weight of the edge (μ, ν), $\sum_{\mu \in H_U, \nu \in H_{AL}} W(\mu, \nu)$ is the weight sum of the check-in edges from users in the user community to the attribute location cluster, $\sum_{\mu \in H_U, \omega \in G_{AL}} W(\mu, \omega)$ is the weight sum of the check-in edges from users in the user community to all location nodes in the original location node set, and the coefficients $x + y = 1$.

The magnitude of x indicates a focus on multiple choices of locations and the magnitude of y indicates a focus on how often users check-in at locations. For ease of representation, in this paper, $x = y = 1/2$ and the weights of the edges $W(u, v) = 1$. A geosocial community with location attributes is shown in Fig. 1(a) consists of $\{< u_{10}, u_{11}, u_{12}, u_{13} >, < l_{14}, l_{15}, l_{16}, l_{17}, l_{18} >\}$, and this geosocial community has a community score of $S_G = 1/2 \times 5/16 + 1/2 \times 8/18$.

Problem Definition: (Attribute Geosocial Community Search(AGCS)). Given a LBSNA $N = (G_U, V_L, E_C)$ with attributes, a query node set $Q \subseteq V_U \cup V_L$, a distance threshold r, a degree threshold k, and a query attribute set $A_Q =$

$\{a_1, a_2, ..., a_i\}$, the attribute geosocial community search problem aims to find a user community $H_U \subset G_U$ and an attribute location cluster L_{AC} that satisfy all query constraints, and the attribute geosocial community score consisting of the user community and the attribute location cluster is maximized.

The AGCS problem is an NP-hard problem, and the following heuristic algorithms are proposed to solve this problem;

Algorithm 1: Basic(N, Q, r, k)

Input: A graph $N = (G_U, V_L, G_C)$, a query node set $Q \subset (V_U \cup V_L)$, a query attribute set A_Q, a radius r, a degree threshold k

Output: A user community H_U, an attribtue location cluster L_{AC}, and the score *Score*

1 $C_U \leftarrow$ OCC(G_U, Q, k);
2 Delete all nodes that do not contain the query attribute from V_L ;
3 **for** $v \in V_L$ **do**
4     $N_L \leftarrow$Obtain possible neighbours of node v via R-tree;
5   Add node v to the attribute location network G_{AL};
6   **for** $n_j \in N_L$ **do**
7   Compute the Euclidean distance D_j between v and n_j;
8   **if** $D_j < r$ **then**
9   Add node n_j and edge (v, n_j) to G_{AL};
10 $C_L \leftarrow$ OCC(G_{AL}, Q, k);
11 $MaxScore \leftarrow 0$;
12 Remove candidate networks from C_U and C_L that do not contain Q ;
13 **for** $c_i \subset C_U$ **do**
14   **for** $c_j \subset C_L$ **do**
15   Calculate the score *Score* for the community consisting of c_i, c_j ;
16   **if** $Score > MaxScore$ **then**
17             $MaxScore \leftarrow Score$, $H_U \leftarrow c_i$, $H_{AL} \leftarrow c_j$;
18 **return** H_U, H_{AC}, *Score*;

---

4 Basic Algorithm

To solve this problem, firstly, a query attribute based search is performed on the set of attribute location nodes to find the location nodes that satisfy all the query attributes, and a radius r based attribute location network is constructed for all the location nodes that satisfy the attributes, and the edges are constructed between two nodes whose distance is less than r. To reduce the distance calculation, an R-tree is used to reduce the computational range. Then the two networks are searched for the connected components of k-core, and during the search process, the search k value is dynamically updated to avoid redundant k-core search. Until there are no connected components that can be searched, and

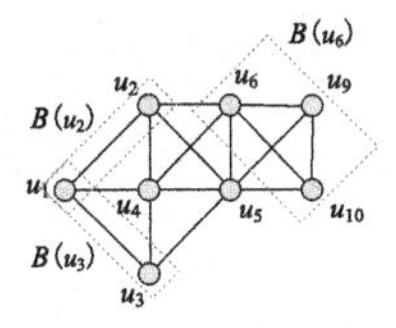

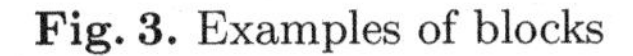
Fig. 3. Examples of blocks

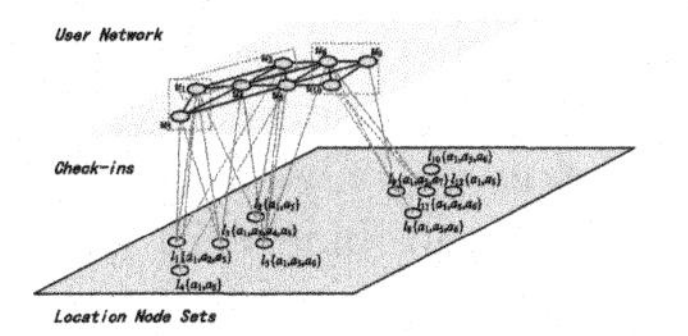

Fig. 4. Diagram of block check-in

then the community scores are calculated for the two types of connected components one by one, and the pair with the largest community score is obtained as the resultant community of the basic algorithm.

As shown in Algorithm 1, the set of candidate user connectivity components is first obtained (Line 1). The function *OCC(*G_u*, Q, k)* is used to obtain the set of nodes that satisfy the constraints. *OCC* removes all nodes of degree less than k, and also removes all nodes of degree less than k that are affected by the removed nodes, until the degree of all nodes is greater than or equal to k. Delete all nodes in the set of location nodes that do not satisfy the attribute constraints (Line 2). We use R-tree to extract the set of candidate location nodes within a certain range around the query nodes, and then further calculate the distance between the nodes to generate the location network (Lines 3–9). Obtain the set of candidate attribute location connectivity components (Line 10). The connected components that do not contain the query nodes are removed from the set of two connected components (Line 12), the community score is calculated for each pair of connected components (one user connectivity component and one location connectivity component), and the pair of connected components with the largest community score is the result obtained (Lines 13–18).

Time Complexity. The time complexity of the algorithm can be approximated as $O(|V_U| + |E_U| + |V_L| + |E_L| + n^2)$, where $|V_U|$, $|E_U|$, $|V_L|$, and $|E_L|$ are the number of nodes and edges in the user and location graphs, respectively, and n is the number of candidate connected components.

5 Global Algorithm

In order to maintain the community structure cohesiveness constraint, we use "block" as the deletion unit. The Global algorithm removes "block" one at a time in order to maintain the cohesiveness of the community as it removes nodes. By analysing the effect of "block" on the community score, we determine the "block score." The "block score" is used as a metric for deleting a block and as a basis for determining the greedy strategy of the Global algorithm.

Property 1. Given the post-pruning location network G_L, there is no $H_L \subseteq G_L$ that makes the community score higher.

Proof. From the community score equation in Definition 5, the deletion of a location block will result in a smaller $|V_t|$ and therefore a smaller community score S_G, so the current cluster of attribute locations is already optimal.

From Property 1, it is clear that the Global algorithm can directly obtain the attribute location network for the optimal community score. Therefore, the greedy strategy only needs to be executed for the user network, and we only need to generate blocks for the user nodes.

Definition 6. ***Block.*** *Given a network $G = (V, E)$ and a node $u \in V$, a block $B(u)$ is the concatenation of node u and the set of nodes affected by node u that leads to unsatisfied structural cohesion.*

Algorithm 2: GreedyDelete($G_U, G_{AL}, Q, k, Score$)

Input: A user network graph G_U and an attribtue location network G_{AL} , a query node set $Q \subset (V_U \cup V_L)$, a query attribute set A_Q, a degree threshold k and community score $Score$

Output: A user community H_U and a location cluster L_{AC}

1 **for** *node* $v \in G_U$ **do**

2   Generate block $B(v)$, determine if $B(v)$ is a deletable block ;

3   **if** $B(v)$ *is a deletable block* **then**

4   Add ¡block score of $B(v)$, $B(v)$¿ to B_U and ascending;

5 **while** B_U *is not empty* **do**

6   **for** $B(v) \in B_U$ **do**

7   Calculate the community score $S(v)$ after deleting block $B(v)$;

8   **if** $S(v) > Score$ **then**

9   Delete block $B(v)$ from the graph G_U;

10   **for** *node* $v \in G_U$ **do**

11   Regenerate block $B(v)$;

12   Remove B_v from B_U;

13 $H_U \leftarrow G_U$, $L_{AC} \leftarrow V_{AL} \in H_{AL}$;

14 **return** H_U, L_{AC};

If a block does not contain any query nodes and removing it from the graph does not produce multiple connected graph, we call this block a deletable block.

Definition 7. ***Block Score.*** *Given a graph $G_C = (V_U, V_L)$ and a deletable block $B(u) \subseteq V_U$, the block score $|W_U|/|v|$, where $|W_U|$ denotes the sum of the weights of the check-in from the nodes in the block to the network at the current location, and $|v|$ denotes the number of nodes in the block with check-ins.*

As shown in Algorithm 2, the Global algorithm first generates the corresponding blocks for the nodes in the current community, determines the blocks that can be deleted, and calculates the *block score* of the deletable blocks, deposit it in the block set B_U and sort them from highest to lowest *block score* (Lines 1–4). Pre-delete the first block, and if the community score increases after deletion, delete the current block from the result and repeat the process of generating blocks. If the score does not become larger after deletion, continue to pre-delete the next block and repeat the process until no blocks make the pre-deleted community score larger (Lines 5–12). The pair of connected components with the largest score is obtained as the result for the community (Lines 13–14).

Time Complexity. The time complexity of the algorithm can be approximated as $O(|G_U| + n(V + E))$, where $|G_U|$ is the number of nodes in the user network graph and n is the number of deletable blocks, $(V + E)$ is the cost of deleting the block and regenerating the corresponding block.

For the Global algorithm, the repetitive block generation and calculation affect the efficiency of the algorithm's execution. To speed up this process and improve the efficiency of the algorithm, we propose the Global$^+$ algorithm.

6 Global$^+$ Algorithm

The Global$^+$ algorithm optimises the repeated generation of blocks and the repeated calculation of community scores as part of the Global algorithm. To address the problem of duplicate calculations of scores, we first introduced the concept of impact factor, which is used to measure the impact of a block on the community score. We do not need to calculate the contribution of blocks with the same impact factor to the community score. In order to reduce the duplication of block generation, we determine the range of nodes that are affected by the deleted blocks. By reconstructing the block and the block generation strategy, we reduce the number of blocks that need to be generated repeatedly and the number of times the block score is repeatedly calculated.

For the above reasons, we refactor the block generation strategy after block deletion and proposed the concepts of impact factor and equivalent blocks.

Optimisation of Pre-deletion Judgement. In the Global algorithm, the computation process to determine whether a block should be deleted is performed once before each judgment, and if the block to be deleted cannot be found immediately, the computation process has to be repeated several times. In the case where different blocks contribute equally to the community score, we define an "equivalent block" based on the impact factor to determine whether the block scores contribute equally.

Definition 8. ***Impact Factor****. An impact factor $I(B(v))$ is defined to measure the impact of blocks on community scores.*

$$I(B(v)) = \frac{\sum_{\mu \in B(v), \nu \in H_{AL}} W(\mu, \nu)}{\sum_{\mu \in B(v), \omega \in G_{AL}} W(\mu, \omega)} \tag{2}$$

$\sum_{\mu \in B(v), \nu \in H_{AL}} W(\mu, \nu)$ is the sum of the check-in weights from the node in the block to the current location, and $\sum_{\mu \in B(v), \omega \in G_{AL}} W(\mu, \omega)$ is the sum of the check-in weights from the node in the block to all location nodes that satisfy the attribute constraints. In Fig. 4, $B(u_{16})$ has an impact factor of 5/6.

Definition 9. ***Equivalent Block.*** *$B(v_i), B(v_j)$ are blocks corresponding to nodes in the user's network, $B(v_i), B(v_j) \in B$, $v \in G_U$, if $I(B(v_i)) = I(B(v_j))$, then $B(v_i)$ and $B(v_j)$ are equivalent blocks.*

Theorem 1. *If the community score does not increase after a block is deleted, then, deleting its equivalent block will not make the community score greater.*

Proof. The impact factor for block B(a) is $I(B(a)) = \frac{\sum_{u \in B(a), v \in H_{AL}} W(u,v)}{\sum_{u \in B(a), v \in G_{AL}} W(u,v)}$, the community score after removing block B(a) from the current community is $S_{G/B(a)} = x \cdot \frac{|V_t|}{|V_a|} + y \cdot \frac{\sum_{u \in H_{U/B(a)}, v \in H_{AL}} W(u,v)}{\sum_{u \in H_U/B(a), v \in G_{AL}} W(u,v)}$. The impact factor for block B(b) is$I(B(b)) = \frac{\sum_{u \in B(b), v \in H_{AL}} W(u,v)}{\sum_{u \in B(b), v \in G_{AL}} W(u,v)}$,the community score after removing block B(b) from the current community is $S_{G/B(b)} = x \cdot \frac{|V_t|}{|V_a|} + y \cdot \frac{\sum_{u \in H_{U/B(b)}, v \in H_{AL}} W(u,v)}{\sum_{u \in H_U/B(b), v \in G_{AL}} W(u,v)}$. If $I(B(a)) = I(B(b))$ then $S_{G/B(a)} = S_{G/B(b)}$. Therefore, when the impact factors are equal, the blocks have the same impact on the community score, so if a block cannot make the community score larger, then its equivalent block cannot make the community score larger either. The impact factor for block B(a) is $I(B(a)) = \frac{\sum_{\mu \in B(a), \nu \in H_{AL}} W(\mu,\nu)}{\sum_{\mu \in B(a), \omega \in G_{AL}} W(\mu,\omega)}$, the community score after removing block B(a) from the current community is $S_{G/B(a)} = x \cdot \frac{|V_t|}{|V_a|} + y \cdot \frac{\sum_{\mu \in H_{U/B(a)}, \nu \in H_{AL}} W(\mu,\nu)}{\sum_{\mu \in H_U/B(a), \omega \in G_{AL}} W(\mu,\omega)}$. The impact factor for block B(b) is $I(B(b)) = \frac{\sum_{\mu \in B(b), \nu \in H_{AL}} W(\mu,\nu)}{\sum_{\mu \in B(b), \omega \in G_{AL}} W(\mu,\omega)}$, the community score after removing block B(b) from the current community is $S_{G/B(b)} = x \cdot \frac{|V_t|}{|V_a|} + y \cdot \frac{\sum_{\mu \in H_{U/B(b)}, \nu \in H_{AL}} W(\mu,\nu)}{\sum_{\mu \in H_U/B(b), \omega \in G_{AL}} W(\mu,\omega)}$. If $I(B(a)) = I(B(b))$ then $S_{G/B(a)} = S_{G/B(b)}$. Therefore, when the impact factors are equal, the blocks have the same impact on the community score, So if a block cannot make the community score larger, then its equivalent block cannot make the community score larger either.

Based on Theorem 1, when a block is known to be non-removable, we can obtain a number of blocks with the same impact factor as that block and they cannot be removed from the graph. We can remove these blocks from the candidate set and these blocks do not contribute to the community score.

Optimisation of Block Generation Strategy. During the Global algorithm, when a block is removed from the graph, it is necessary to regenerate the block for all the nodes that have not been removed. Through our research, we find that the set of regenerated blocks intersects with the set before regeneration, and by property 2 we know that a deleted block has limited "influence" on the blocks of the remaining nodes in the graph, so we only need to regenerate the blocks affected by the deleted block.

Property 2. After deleting block $B(u)$ in graph G_u, the nodes that may needs to be regenerate the block is the neighbouring nodes in the deleted block and the neighbouring nodes of those neighbouring nodes.

Algorithm 3: OptimisedGreedyDelete(G_U, G_{AL}, Q, k)

Input: A user network graph G_U and an attribtue location network G_{AL} , a query node set $Q \subset (V_U \cup V_L)$, a query attribute set A_Q, a degree threshold k and community score $Score$

Output: A user community H_U and a location cluster L_{AC}

```
1  deg_Max ← 0;
2  for v ∈ G_U do
3      Generate block B(v) and calculate the impact factor I(B(v));
4      if The max deg_{B(v)/v} > deg_Max then
5          deg_Max ← the max deg_{B(v)/v};
6      if B(v) is a deletable block then
7          Create block set B_U and ascending it;
8          for node m ∈ B(v) do
9              Deposit B(v) into the set of blocks containing the node m;
10 while B_U is not empty do
11     b_v ← B_U.pop();
12     if Community score after pre-deletion of b_v > Score then
13         N(b_v) ← neighbour nodes of b_v;
14         Remove the nodes in b_v, regeneration of partial blocks;
15         for node p ∈ the neighbour nodes set of b_v do
16             if deg(p) ⩽ deg_Max then
17                 Processing blocks according to Property 3;
18         Update B_U;
19     else
20         Delete all equivalent blocks;
21 H_U ← G_U;
22 return H_U, L_AC ← V_AL ∈ G_AL;
```

Property 3. According to the definition of a block, if a node is added to a block during block generation, it must be because the node does not satisfy the cohesiveness constraint. For $\forall v \in G_u$, we record the maximum value of the degree of all nodes that are added to the block as deg_{Max}. $\forall v$ which $deg(v) > deg_{Max}$ will not be joined to blocks of other nodes.

We determined by Property 3 that the blocks corresponding to neighbouring nodes whose deleted blocks have $deg(v) > deg_{Max}$ do not need to be regenerated.

The process of the optimised global algorithm is as follows. Blocks are generated for all nodes in the user network, the influence factor of the block is calculated and the maximum degree deg_{MAX} is recorded for all blocks (Lines 1–5). If the block is a deletable block, calculate the block score and put the blocks in ascending order. Record the set of all blocks containing this node corresponding to each node (Lines 6–9). All blocks in the block set are implicitly deleted in turn, and if the community score increases, the set of neighbouring nodes of the

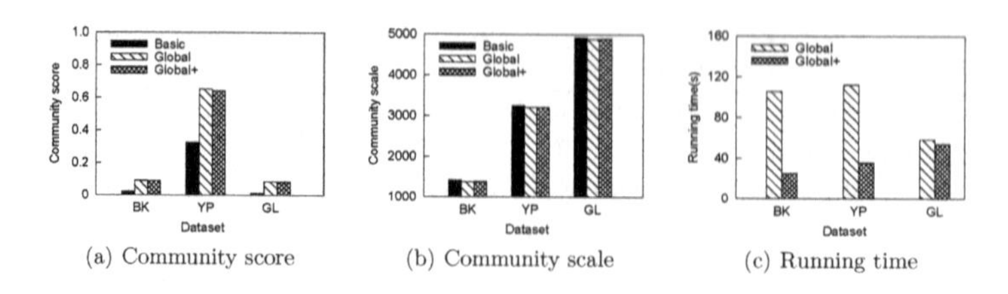

(a) Community score (b) Community scale (c) Running time

Fig. 5. Overall performance on different dataset

deleted block (called the current node set) is obtained. Remove this block from the graph and remove the node information in the deleted block from the rest of the block set. All nodes in the current node set regenerate the corresponding blocks (Lines 10–14). For neighbours $N(n)$ of node n of any degree less than deg_{Max} in $N(b_v)$, there is a block b_m composed of any node m in $N(n)$, if the number of nodes that b_m and $N(n)$ coincide is smaller than $deg(n)$ by k, then add n to b_m(Lines 15–17). After determining all nodes in the current node set, judge if the regenerated block is a deletable block and recalculate the score to update the block set (Line 18). If the community score does not increase after deleting, delete all blocks in the set of deletable blocks that are identical to this block and then pre-delete the next block (Lines 19–20). At last, we return the optimal user community H_U and location cluster L_{AC} (Lines 21–22).

7 Experiments

In this section, we discuss the experimental setup and the experimental results.

7.1 SetUp

Datasets. We conducted experiments on three datasets:

- **Yelp** [14]: The Yelp dataset comes from Yelp. We randomly extracted 100,000 user nodes, 177,969 merchants, and 748,188 check-ins.
- **Brightkite** [15]: Brightkite used to be a location-based social networking service provider. We randomly extracted 58,227 users and 100,000 location nodes with 183,384 check-ins.
- **Gowalla** [15]: Gowalla is a location-based social networking website where users share their locations by checking-in. We randomly extracted 100,000 users, 100,000 locations and 6,442,890 check-ins.

We set the default k-values to 40 for Yelp and Gawalla, 20 for Brightkite, one user node and one location node for two query nodes, and one query attribute by default, depending on the actual dataset, which will ensure that there is a k-core containing each query node and satisfying the query attribute.

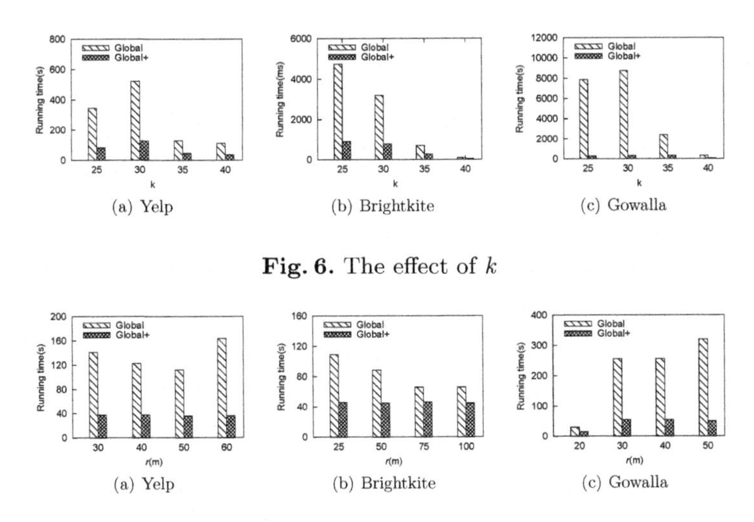

(a) Yelp (b) Brightkite (c) Gowalla

Fig. 6. The effect of k

(a) Yelp (b) Brightkite (c) Gowalla

Fig. 7. The effect of r

7.2 Experimental Results

Overall Performance on Different Datasets. As shown in Fig. 5, we first investigated the performance of algorithms on three datasets. The query constraints were set to their default values. As shown in Fig. 5(a), we find that the Global and Global+ algorithms significantly improve the community scores of the Basic algorithm on all three datasets. This is because the Global and Global+ algorithms removed some nodes that could have improved the community scores. Looking at Fig. 5(b), the significant improvement in execution efficiency of Global+ compared to Global demonstrates the effectiveness of several of our optimisation strategies. Combined with Fig. 5(c), we find that the reduction in community scale boosts the community score. This is due to the removal of low contribution nodes. By comparing community scores and community scale across the three datasets. We found that the amount of score increase and the scale of the score did not directly correlate with community scale. This is because the factors that influence community score are more related to the number of check-ins that users make to the location network.

The Effect of$k$. Next, we investigated the effect of the query condition k on the efficiency of the algorithm's execution. By setting k to 25–40, the other parameters are default values. As shown in Fig. 6, we find that the execution speed of both algorithms in general becomes faster as the value of k increases. This is because a stronger k constraint most of the time leads to a smaller number of nodes that the algorithm needs to process. Figure 6(a) and Fig. 6(c) show us two special cases. The algorithms become less efficient in execution when the k-bound changes from 25 to 30. Analyzing the experimental procedure, we find that this is because the user network processed at $k = 30$ is not a subgraph of the user network found at $k = 25$, i.e., the actual nodes processed grow instead as the

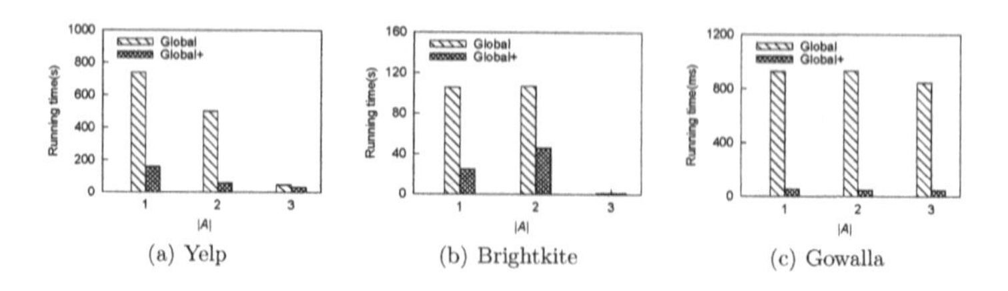

(a) Yelp (b) Brightkite (c) Gowalla

Fig. 8. The impact of query attributes

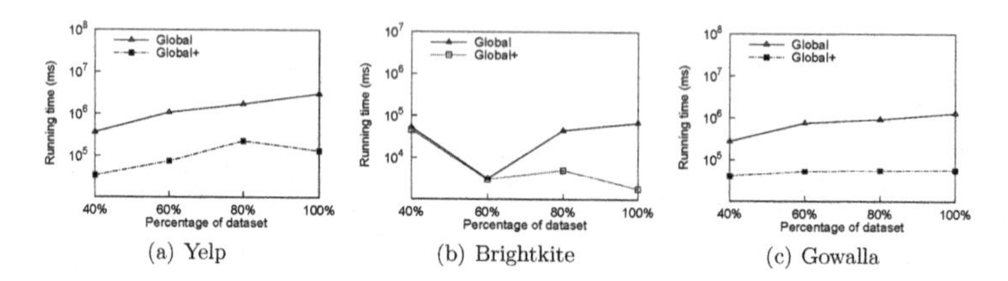

(a) Yelp (b) Brightkite (c) Gowalla

Fig. 9. The impact of node

user network we process changes due to the check-in count constraint. Overall, we can still see that the Global+ algorithm greatly improves the efficiency of the Global algorithm, especially when the Global algorithm is very slow.

The Effect of r. Due to the inconsistent structure of the different datasets, in Fig. 7 we set different ranges of r variation for each of the three datasets. One intuition is that as r increases, the graph that the algorithm needs to process becomes larger, and then the execution slows down. However, we actually observe that the execution time of the algorithm is inconsistently affected by r on different datasets. As shown in Fig. 7(a) and Fig. 7(b), the algorithm execution time is irregular on Yelp and Brightkite, and monotonically increases on the Gowalla dataset shown in Fig. 7(c). Observing the experimental process, we find that after the change in r leads to a change in the location network, the best user network found by the corresponding Basic may also change, which leads to irregularity in the size of each block and the number of removable blocks during the execution of the Global algorithm. Therefore the effect of changes in r on the execution time of the Global algorithm is irregular. Since the Global+ algorithm only partially updates the blocks after the first block generation, the Global+ algorithm execution time is not affected much overall.

The Impact of Query Attributes. We randomly select 1, 2, and 3 attributes as the set of query attributes. Figure 8 compares the running time of proposed algorithms for different numbers of query attributes. An increase in query attributes results in a decrease in the number of location nodes that satisfy the attribute constraints. This will reduce the cost of community score computation during algorithm execution and reduce execution time. In the previous section, we described that changes in the location network may result in irregular changes in the number of user network nodes to be processed. This affects the speed of algorithm execution. The increase in execution time of the Global+ algorithm when the number of attributes changes from 1 to 2 in Fig. 8(b) is due to the growth of user network nodes. However, overall, the algorithm execution time decreases with the increase in $|A|$. This is because the growth in the number of attributes leads to a massive reduction in the number of location network nodes, which results in the number of optimal user network nodes found by the corresponding Basic being reduced in most cases.

Scalability. Figure 9 compares the scalability at different data sizes and shows that the runtime of the Basic algorithm increases with the size of the data under the default query constraint. The Global algorithm is mainly influenced by the block generation process, and the Global+ is partially influenced by the block generation process, while the factor affecting the block generation time is the sparsity of the connections within the graph. If the graph is sparser, this will result in each block having a smaller area of influence. The smaller the number of blocks that need to be regenerated, the more the running time of the Global+ algorithm will be reduced. The Global+ algorithm is also affected by the number of check-in edges at the current data size. If the number of equivalent blocks at the current data size is higher, the running time of the Global+ algorithm will be reduced by more than the Global algorithm.

8 Conclusion

In this paper, we present the Attribute Geosocial Community Search Problem (AGCS). We define a new community metric based on attribute constraints and check-in information. We propose three algorithms to solve this problem. And we conduct extensively experiments on real and synthetic location-based social networks with attributes, which demonstrate the effectiveness and efficiency of the proposed algorithms. In the future, we will explore more effective community metrics and solutions for attribute geosocial community search.

Acknowledgements. This work is partially supported by the National Natural Science Foundation of China (Nos. 62002245, 62102271, 61802268), Natural Science Foundation of Liaoning Province (Nos. 2022-MS-303, 2022-BS-218).

References

1. Fang, Y., et al.: A survey of community search over big graphs. In: VLDB, pp. 353–392 (2020)
2. Kim, J., Guo, T., Feng, K., Cong, G., Khan, A., Choudhury, F.: Densely connected user community and location cluster search in location-based social networks. In: SIGMOD, pp. 2199–2209 (2020)
3. Chen, L., Liu, C., Zhou, R., Li, J., Yang, X., Wang, B.: Maximum co-located community search in large scale social networks. In: VLDB, pp. 1233–1246 (2018)
4. Liu, Y., Pham, T., Cong, G., Yuan, Q.: An experimental evaluation of point-of-interest recommendation in location-based social networks. In: VLDB, pp. 1010–1021 (2017)

5. Apon, H., et al.: Social-spatial group queries with keywords. In: ACM Transactions on Spatial Algorithms and Systems (TSAS), pp. 1–32 (2021)
6. Fang, Y., Cheng, R., Chen, Y., Luo, S., Hu, J.: Effective and efficient attributed community search. In: VLDB, pp. 803–828 (2017)
7. Guo, F., Yuan, Y., Wang, G., Zhao, X., Sun, H.: Multi-attributed community search in road-social networks. In: ICDE, pp. 109–120 (2021)
8. Zhang, Z., Huang, X., Xu, J., Choi, B., Shang, Z.: Keyword-centric community search. In: ICDE, pp. 422–433 (2019)
9. Chen, L., Liu, C., Liao, K., Li, J., Zhou, R.: Contextual community search over large social networks. In: ICDE, pp. 88–99 (2019)
10. Kai, W., et al.: Efficient radius-bounded community search in geo-social networks. In: IEEE, pp. 4186–4200 (2020)
11. Cui, W., Xiao, Y., Wang, H., Wang, W.: Local search of communities in large graphs. In: SIGMOD, pp. 991–1002 (2014)
12. Qing, L., et al.: Vertex-centric attributed community search. In: ICDE, pp. 937–948 (2020)
13. Sozio, M., Gionis, A.: The community-search problem and how to plan a successful cocktail party. In: KDD, pp. 939–948 (2010)
14. Liu, Y., Pham, T., Cong, G., Yuan, Y.: An experimental evaluation of point-of-interest recommendation in location-based social networks. In: VLDB, pp. 1010–1021 (2017)
15. Cho, E., Myers, S., Leskovec, J.: Friendship and mobility: user movement in location-based social networks. In: KDD, pp. 1082–1090 (2011)

Efficient Size-Constrained (k, d)-Truss Community Search

Chuanyu Zong[1(✉)], Pengcheng Gong[1], Xin Zhang[2] and Tao Qiu[1], Anzhen Zhang[1], and Meng-xiang Wang[3]

[1] School of Computer Science, Shenyang Aerospace University, Liaoning 110136, China
{zongcy,qiutao,azzhang}@sau.edu.cn, gongpengcheng@stu.sau.edu.cn
[2] Shenyang Aircraft Corporation, Liaoning 110850, China
[3] China National Institute of Standardization, Beijing 100191, China
wangmx@cnis.ac.cnand

Abstract. In recent years, finding a cohesive subgraph containing the user-given query vertices has been extensively explored as the community search problem. Most of the existing research ignores the size and diameter of the community and instead focuses only on how cohesive returned communities are. However, it has been a natural requirement for many applications that a community's number of vertices or members fall within a given range. In this paper, therefore, we investigate the problem of searching maximal (k, d)-truss community with a size constraint (denoted by SkdC) in a social network G: Given a size lower constraint l, a size upper constraint s, a query distance constraint d, an integer k, and a query vertex q, SkdC search problem aims to find a maximal (k, d)-truss H that contains the query vertex q, the query distance of each vertex in H from the query vertex q does not exceed d, the size of H (i.e., the total number of vertices in H) is no less than l and no more than s, and no subgraph H' has a size bigger than that of H. We prove that the SkdC search problem is NP-hard. To the best of our knowledge, this is the first work to find a size-constrained maximal (k, d)-truss community in a large graph. To address this issue, first, a practically efficient SkdC search solution is proposed. Then, we explore an improved searching algorithm by updating blocks locally. Finally, we conduct comprehensive experimental studies on several real social networks to evaluate both the efficiency and effectiveness of our proposed searching algorithms.

Keywords: $(k \cdot d)$-truss · Community search · Size constraint · Social network

1 Introduction

With the rapid development of information technology, graphs are widely used to model the relationships of entities, and graph data is ubiquitous in real-world

X. Yang et al. (Eds.): ADMA 2023, LNAI 14180, pp. 405–420, 2023.
https://doi.org/10.1007/978-3-031-46677-9_28

applications. Community search is one of the fundamental problems in graph analytics and managements since a community contains a set of internally tightly connected vertices. Currently, how to effectively find high-quality communities from large social networks is a hot research topic. The task of finding cohesive subgraphs with user-specified query vertex that may also satisfy other constraints [1] has practical applications in various fields such as team formation [2], social contagion modeling [3], and identification of protein functions [4].

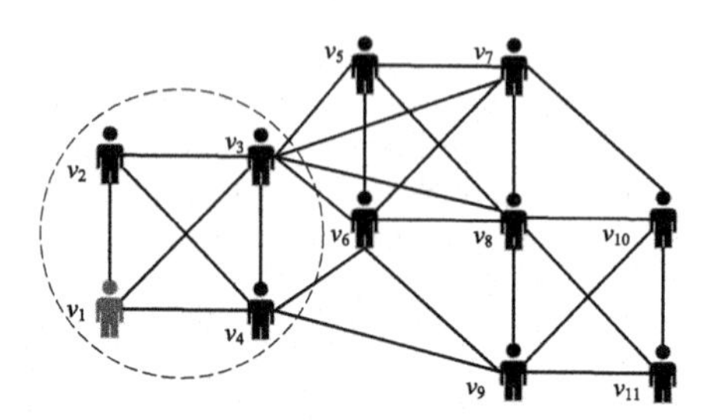

Fig. 1. An example of community search

In this paper, we follow the study of community search based on k-truss since it has a good cohesion [5]. The k-truss of a social network graph G is defined as the largest subgraph in which each edge is contained in at least k-2 triangles. Currently, the existing community models allow users to specify the query vertices to be included in the community. Based on this, in order to obtain a tighter and denser subgraph, a new community model (k, d)-truss [6] is proposed. This model requires that the query distance between each vertex of the community and the query vertex cannot exceed d. Due to functional or budget constraints in practical applications [7], a constraint is often imposed on the number of vertices in a community, and in addition to the size upper constraint s, a size lower constraint l on size is also required. This is because organizing an event often requires a minimum number of participants to ensure that the event is held successfully.

In social networks, users in a triangle usually have strong and stable relationships, because it means that these users have some friends in common. A user may want to organize a group activity, such as a party, and seek people with whom he has a close connection. To ensure the smooth conduct of the activity, there is a lower bound constraint on the number of participants. Apart from social connections, there is also an upper bound constraint on the size of the group due to various factors such as ticket availability or venue capacity [1]. Under the size constraints, organizers also aim to maximize the number of participants.

Therefore, we study the problem of searching maximal (k, d)-truss community with size constraint (SkdC) in this paper, Given a query vertex, the goal of the SkdC search is to find a maximal (k, d)-truss with size constraints. The (k, d)-truss algorithm primarily focuses on identifying the largest (k, d)-truss communities that satisfy the distance constraint, without considering the size limitations of the communities. On the other hand, SkdC incorporates a size constraint and aims to obtain communities within a specific size range. It also prioritizes communities that have close connections to the organizers and are larger in size. Consequently, SkdC is capable of discovering communities that exhibit higher cohesion and internal connectivity.

Figure 1 contains 11 users, where the edges between users indicate that two users are closely related and know each other. In Fig. 1, the whole graph is a k-truss with k=3. Assuming that v_1 is the query vertex, k=3, l=2, s=4, and d=2, then the SkdC search will return the community $\{v_1, v_2, v_3, v_4\}$.

Challenges. The size-constrained (k, d)-truss community search problem is proved to be NP-hardness. Existing studies, such as those presented in [8,9], have focused on computing all maximal k-trusses formed by triangular connections and establishing offline indexes for supporting online community searches. However, their indexing approaches are not directly applicable to the SkdC search problem due to the presence of query distance constraints. The goal of SkdC is to obtain the maximal number of (k, d)-truss communities with size constraints, which poses significant computational challenges, especially for large-scale social networks. Our proposed solution needs to efficiently search for the largest (k, d)-truss communities while ensuring that the identified communities satisfy the specified size constraints. Therefore, it is crucial to develop efficient algorithms and optimization techniques suitable for the SkdC search problem.

Our Solution. To overcome the above challenges, we propose the global idea of deleting vertices by "blocks". To further improve the search efficiency, we propose the concept of equivalent blocks, which can reduce some invalid searches and improve the efficiency of updating blocks to find SkdC faster. We observe that not all the remaining blocks in the graph need to be updated after we delete some blocks. Therefore, we propose a strategy to update "blocks" locally, so that after we delete a "block", only the blocks affected by the deletion are updated to search for SkdC faster.

The main contributions of this paper are as follows:

(1) We propose the maximal (k, d)-truss search problem with size constraints, and we prove that the SkdC search problem is NP-hard.

(2) We propose a global algorithm for block deletion that works as a solution for SkdC search.

(3) We explore an improved algorithm for local update of blocks to improve the efficiency of marking blocks after deleting them.

(4) We conduct some extensive experiments on several real graph datasets, and the results demonstrate the efficiency of our proposed algorithms.

2 Related Work

We review two research areas that are closely related to SkdC search below.

(k, d)-Truss. k-truss models possess various excellent properties for communities. For instance, k-truss is an (k-1)-core and an (k-1)-ECC that demonstrates high cohesiveness. However, when we concentrate on cohesion, it is essential to consider the communication cost among every vertex and the query vertex q in a graph. This approach is closer to reality and more consistent with sociological studies. Hence, a new definition of a compact substructure, (k, d)-truss, was introduced by Huang et al. [6]. It is based on the k-truss structure and necessitates that the query distance between each vertex in the subgraph and the query vertex should not exceed d. As k increases, the cohesiveness of (k, d)-truss also increases, while the proximity to the query vertex increases as d decreases. Like k-truss, (k, d)-truss inherits excellent structural properties such as k-edge connectivity and bounded diameter, allowing it to achieve a denser structure.

Size-Constrained Community Search. Size-constrained community search is a crucial type of community search due to budget or capacity restrictions in several practical applications. In [10], an online search algorithm was created to eliminate vertices from a large initial subgraph sequentially until a k-truss community that satisfies the constraints is obtained. In [11], the author suggests a branch degradation algorithm, SC-BRB, to solve the community search problem with the maximum and minimum degree of the size-constrained subgraph. In [7,12], researchers explored the search for the smallest community to further decrease the size of the returned community. [13] proposed an estimated solution for identifying the minimum k kernel. The previously mentioned algorithms do not possess definite thresholds to limit the size of each result and are not intended for community search with size thresholds. It is clearly impractical to apply their algorithms to our problem because they may overlook the (k, d)-truss community that we obtain the highest number of vertices under the size constraints.

3 Preliminaries

In this section, we define the problem and analyze the hardness of the problem.

3.1 Notations and Definitions

Given an undirected graph $G(V, E)$, where V is the set of vertices and E is the set of edges. A graph H is a subgraph of G if $V(H) \subseteq V(G)$ and $E(H) \subseteq E(G)$, denoted as $H \subseteq G$. For a vertex $v_i \in V$, the set of adjacent vertices (neighbors) of v_i in G is denoted by $N(v_i, G) = \{v_j | (v_i, v_j) \in E(G)\}$.

Definition 1. (Edge Support). *Given an edge (v_i, v_j) in G, the support of the edge (v_i, v_j) , denoted as $sup_G(v_i, v_j)$, returns the number of the triangles in G which contain (v_i, v_j), and it can be formalized as $sup_G(v_i, v_j) = | \{v_w \mid v_w \in V(G) \wedge \triangle(v_i, v_j, v_w) \in G\} |$.*

If the subgraph $H \subseteq G$ is a k-truss. Then the support of each edge in the subgraph H is greater than or equal to k-2. The support of an edge $e = (v_i, v_j) \in E$ in G, can be denoted by $sup_G(e)$. When the context is obvious, the subscript can be dropped and the support of e can be denoted as $sup(e)$. Since k-truss is allowed to be disconnected, which makes no sense for our search community, we define connected k-truss below.

Definition 2. (Connected k-truss) *[6]. Let G be a graph and k be a positive integer, a connected k-truss is a connected subgraph $(H \subseteq G)$, such that $\forall e \in E(H), sup_H(e) \leq k - 2$.*

Given two vertices $v_i, v_j \in G$, the length of the shortest path between v_i and v_j in G is denoted as $dist_G(v_i, v_j)$, if v_i and v_j are not connected, $dist_G(v_i, v_j) = +\infty$. The diameter of a graph G is defined as the maximal length of a shortest path in G, and it can be formalized as $\text{diam}(G) = \max_{v_i, v_j \in G} \{dist_G(v_i, v_j)\}$. The notion of graph query distance is defined below.

Definition 3. (Query Distance) *[6]. Let G be a graph, q be a query vertex, and $u(u \neq q)$ be a vertex in V, the query distance of vertex u is the maximal length of a shortest path from u to the query vertex q in G, and it can be defined as $dist_G(u, q)$. Let $H \subseteq G$ be a subgraph, then the graph query distance of H is denoted as $dist_H(H, q) = \max_{u \in H} dist_H(u, q)$.*

For the graph G in Fig. 1 and the query vertex q=v_1, the query distance of vertex v_6 is $dist_G(v_6, q) = dist_G(v_6, v_1) = 2$. And the query distance of the graph G is $dist_G(G, q) = dist_G(v_1, v_{11}) = 3$. The diameter of the graph is $diam(G) = 3$.

As mentioned in [8], a good community should have a low communication cost with a small graph query distance. Thus, based on the concept of connected k-truss and query distance, We present the concept of (k, d)-truss as follows.

Definition 4. ((k, d)-truss). *Let H be a graph, q be the query vertex, k and d be two numbers, then H is a (k, d)-truss iff the following conditions are satisfied: (1) H is a connected k-truss containing q;(2) $dist_H(H, q) \leq d$.*

Definition 5. (Size-constrained maximal (k, d)-truss community (SkdC). *Given a graph $G(V, E)$, a query vertex $q \in V$, a size constraint $[l, s]$, a positive integer k, and a query distance constraint d, H is a size-constrained maximal (k, d)-truss community (SkdC), such that (1) H is a (k, d)-truss containing q; (2) H satisfies the size constraint with $l \leq |V(H)| \leq s$; (3) There exists no subgraph $H' \subseteq G$ satisfying the above properties, such that $| V(H') | > | V(H) |$.*

Problem Definition. Given a graph G, a query vertex $q \in V(G)$, a size constraint $[l, s]$, a positive integer k, and a query distance constraint d, the SkdC search problem is to find a SkdC from G, i.e., a connected (k, d)-truss subgraph H of G such that H contains q and the number of vertices in H is not less than l and not greater than s.

3.2 Problem Hardness

In this subsection, we analysis the hardness of SkdC search problem.

Theorem 1. *The SkdC search problem is NP-hard.*

Proof. By reducing from the NP-complete k-clique problem, we can prove that the decision version of the SkdC problem is also NP-complete. The decision version of the SkdC problem involves a graph $G(V, E)$, a query vertex q, a size constraint $[l, s]$, and a positive integer k, and a query distance constraint d. The goal is to determine whether there exists a subgraph $R \subseteq G$ that satisfies the following conditions: (1) R is a (k, d)-truss containing q; (2) the number of vertices in R is between l and s (inclusive). It is evident that the decision problem belongs to NP.

To reduce the k-clique problem, which determines whether a graph $G = (V, E)$ contains a complete subgraph of k vertices, to the decision SkdC problem, we introduce a fictitious vertex v_i and connect it to every vertex in V, resulting in a new graph G' with $V(G') = V \cup \{v_i\}$ and $E(G') = E \cup \{(v_i, v_j)|v_j \in V\}$. The decision SkdC problem we create involves a query vertex v_i, a size constraint of $[k+1, k+1]$, and a minimum edge support threshold of k. It is evident that if $R \subseteq V$ is a k-clique, then $R \cup \{v_i\}$ is a solution to the decision SkdC problem. Therefore, the decision SkdC problem is NP-complete, and as a result, the optimization version of the SkdC problem is NP-hard.

4 Our Solution

Although the SkdC search problem is NP-hard, we explore the potential of finding an efficient exact solution as the initial research into the SkdC search. In the first part of this section, we present algorithms designed to solve the SkdC community search problem. We then propose a block local update strategy to enhance the efficiency of block deletion based on these algorithms.

4.1 The Framework

To solve the problem, we should obtain a result that fulfills the query distance d constraint in the (k, d)-truss structure as well as the k constraint in the k-truss structure, and that result is a connected subgraph. To begin, we use the query vertex q as our beginning point and process the graph in accordance with the query vertex q. That is, we must discover a (k, d)-truss containing q while satisfying the query distance constraint d. If a (k, d)-truss containing q cannot be identified under the query distance d restriction, we can certainly claim that there does not exist a qualified SkdC containing q.

The reason for using this framework is that SkdC requires us to discover the subgraph with the most vertices within the size constraints. We may significantly minimize the search space by first locating a qualified (k, d)-truss and then locating SkdC using this (k, d)-truss. This SkdC framework is formally introduced in the following sections.

Algorithm 1 gives the pseudo-code of the framework. Each edge's support in G is precomputed. To begin, we find all of the vertices in G whose query distance to the query vertex q is less than or equal to d (Line 1). Then, in the graph G, we obtain the edges of these vertices that are related to each other to get H' (Line 2). We remove all edges in H' with less than k-2 support (Line 3). Because removing edges may cause H' to no longer be a connected graph, the query vertice q exists independently. As a result, we get (k, d)-truss H from the query point in H' (Line 4). If H is an empty set or the number of vertices in H is less than l, the empty set is returned, indicating that a qualified SkdC containing q does not exist (Lines 5–6). If the number of vertices in H falls exactly inside the range $[l, s]$, H is a SkdC solution, and the result is returned (Lines 7–9). If the number of vertices in H is greater than s, we use the search algorithm (described in Sects. 4.2 and 4.3) to explore SkdC on H(Lines 10–12).

Algorithm 1: Framework

---

input : a graph G, the integer k, a query distance d, a size constraint $[l, s]$, a query vertex q
output: SkdC R

1 $V(H') \leftarrow$ obtain all vertices from G which $dist_{v \in V(G)}(v, q) \leq d$;
2 $H' \leftarrow$ obtain the interconnected edges in G through these vertices;
3 $H' \leftarrow$ remove edges e with $sup_{H'}(e) < k - 2$ from H';
4 $H \leftarrow$ obtain a (k, d)-truss from H';
5 **if** $H = \emptyset$ *or* $|V(H)| < l$ **then**
6   **return** $\emptyset$;
7 **if** $|V(H)| \in [l, s]$ **then**
8     $R \leftarrow H$;
9   **return** R ;
10 **else**
11     $R \leftarrow$ Search(H, k, q, d, l, s) ;
12   **return** R ;

4.2 Basic Algorithm

In this subsection, we present our basic algorithm aiming at obtaining a size constrained maximal (k, d)-truss. Our approach is based on the global idea of sequential deletion of vertices. After removing a vertex, the remaining graph must always satisfy both the cohesion constraint and the query distance constraint. To maintain these constraints in the remaining graph, we need to delete vertices that cannot satisfy the constraints at the same time. The concept of "block" is inspired by this idea.

Definition 6. **(k, d)-truss Violation Set.** *Let H be a (k, d)-truss, q be a query vertex, for $\forall v_i \in H$, after deleting the vertex from the H, in order to maintain the rest graph is still a (k, d)-truss, some extra vertices might need to be* *deleted. Iterating the above process, the set of extra deleted vertices is denoted by* $kdVS(v_i)$.

Definition 7. Block. *Let* H *be a* (k,d)*-truss and a vertex* $v_i \in V(H)$*, the union of vertex* v_i *and* $kdVS(v_i)$ *is denoted by block* $B(v_i)$.

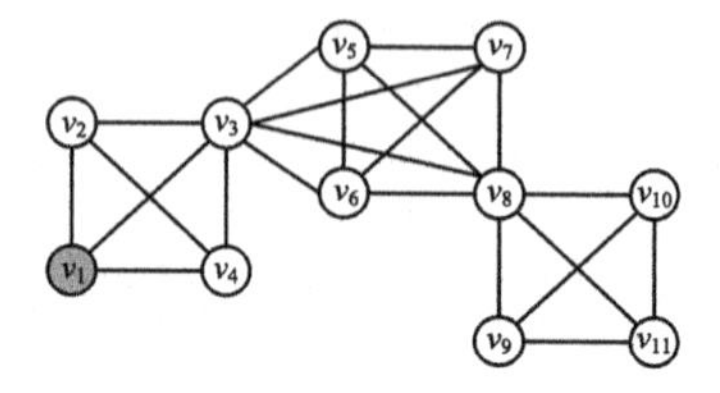

Fig. 2. A (k,d)-truss with q=v_1, k=4, d=3, l=2, and s=4

As shown in Fig. 2, the query vertex is set to v_1, the structural cohesion constraint k is set to 4, the query distance constraint d is set to 3, the size constraints l and s is set to 2 and 4, respectively. It can be seen that the graph represents a (k,d)-truss and each vertex corresponds to a block. If we delete vertex v_{11}, only vertex v_{11} is deleted under the satisfied constraint, i.e., $B(v_{11})$=$\{v_{11}\}$. If we delete vertex v_8, it causes vertices v_8, v_9, v_{10}, and v_{11} to be deleted, i.e., $B(v_8)$=$\{v_8, v_9, v_{10}, v_{11}\}$.

Our algorithm uses "block" as the unit of deletion. To achieve this, we compute and mark the block information for each vertex. We then delete blocks in the order of their size from the smallest to the largest to ensure that SkdC is not missed in the process of deletion. Whenever a "block" is deleted from the graph, the algorithm needs to recalculate the block information for the remaining vertices in the graph and mark them to see if they qualify for deletion.

Algorithm 2 presents the basic algorithm for the SkdC search. First, the block to be deleted is marked as empty (Line 1). When the number of vertices in H is greater than s, for each vertex v in H, its corresponding block information $B(v)$ is computed (Line 4). If the number of vertices in H minus the number of vertices in $B(v)$ is exactly equal to s, then it is determined that a SkdC has been obtained. The block in H is deleted, and R is returned (Lines 5–7). If the number of vertices in H minus the number of vertices in the block is greater than or equal to l, then deleting the block may obtain a SkdC, so it is marked with B_m. If B_m is empty, the block information is recorded (Lines 9–10). If the number of vertices in B_m is greater than the number of vertices in $B(v)$, the block information for B_m is updated because the blocks are deleted in order from the smallest to the largest (Lines 11–12). If B_m is still empty at this point, it means that none of the block information in H can be deleted, and the loop ends (Lines 13–14). Otherwise, the smallest block B_m marked in H is deleted (Line 15). If the number of vertices in H obtained is less than l or greater than s, it means there is no SkdC, and the algorithm returns an empty set (Lines 16–17). Otherwise, the result of H is output to obtain SkdC R (Lines 18–20).

Algorithm 3 displays the pseudo-code of Mark_Block, which is invoked at Line 4. It first pushes all the edges incident to u into a stack P (Lines 1–3). Then, for each edge (x, y) in P, Algorithm 3 removes it from H and reduces the edge support by one for those edges that form a triangle with (x, y) (Lines 4–8). If an edge's support is less than $k-2$, Algorithm 3 pushes it into P (Lines 9–10). After removing every edge whose support is less than k-2, the set of vertices in H whose distance from query vertex q is greater than d is recorded as the block corresponding to vertex v (line 12). Finally, Algorithm 3 returns the block of vertex v (Line 13). It should be noted that after each iteration of Algorithm 3, H remains unchanged. In H, we are just calculating the block information corresponding to vertex v. However, we are not removing the vertex v.

Algorithm 2: Basic Algorithm

```
input : a graph H, the integer k, the query distance d, the size constraint
        [l, s], the query vertex q
output: the miximal (k, d)-truss R with size constraint [l, s] if exists,
        otherwise ∅
1  B_m ← ∅ ;
2  while |V(H)| > s do
3      for each v ∈ V(H) do
4          B(v) ← Mark_Block(H, v);
5          if |V(H)| − |B(v)| == s then
6              R ← delete_block(H, B(v)) ;
7              return R ;
8          if |V(H)| − |B(v)| ≥ l then
9              if B_m == ∅ then
10                 B_m ← B(v);
11             if |B_m| > |B(v)| then
12                 B_m ← B(v);
13     if B_m == ∅ then
14         break;
15     H ← delete_block(H, B_m) ; // B_m ← ∅
16 if |V(H)| < l or |V(H)| > s then
17     return ∅ ;
18 else
19     R ← H ;
20     return R ;
```

4.3 Update Algorithm

During the search, we discover that if the number of vertices in the remaining graph after deleting a block is less than our size lower constraint l, we refer to the block as non-deletable, and the remaining blocks are deleteable blocks. Obviously, as a non-deletable block, we can disregard it because deleting a non-deletable block is not destined to get a SkdC. The block $B(v_3)$ corresponding to v_3 in Fig. 2 is a non-deletable block because if v_3 is deleted, a SkdC is not available, and the block $B(v_{11})$ corresponding to vertex v_{11} is a deleteable block.

When we compute blocks, we observe that there are numerous vertices in the (k, d)-truss H that correspond to a block. This will need us to execute numerous repeated computations, and calculating each block will take a significant amount of time. As a result, we propose the concept of equivalent blocks.

Definition 8. Equivalent Block. *There are two vertices u and $v \in V(H)$, their corresponding blocks in the (k, d)-truss H are $B(u)$ and $B(v)$. $B(u)$ and $B(v)$ are equivalent blocks if $B(u) = B(v)$.*

Algorithm 3: Update Algorithm

input : a graph H, the integer k, the query distance d, the size constraint $[l, s]$, the query vertex q
output: the maximal (k, d)-truss R with size constraint $[l, s]$ if exists, otherwise $\emptyset$

```
1  B_m ← ∅, β ← ∅, β_a ← ∅ ;
2  β ← record deleteable blocks not containing equivalent blocks in H ;
3  if β == ∅ then
4      return ∅ ;
5  while |V(H)| > s do
6      B_m ← iterate through all blocks in β and select the deleted block ;
7      β_a ← find the block in β affected by B_m according to property 2 ;
8      H ← delete_block(H, B_m) ;
9      β ← updated affected blocks β_a in β ;
10     if β == ∅ then
11         break ;
12     B_m ← ∅, β_a ← ∅ ;
13 if |V(H)| < l or |V(H)| > s then
14     return ∅ ;
15 else
16     R ← H ;
17     return R ;
```

Property 1. In all deletable blocks of (k, d)-turss H, if $B(u)$=$B(v)$, when we explore a block for deletion, we only need to record the block corresponding to one of the vertices.

Block Generation Strategy. When a block is deleted from the graph during the global algorithm, it is necessary to regenerate the block for all the vertices that were not deleted. We discover during our investigation that not all of the blocks that can be deleted need to be updated. Only a few blocks require updating. It would take a long time to update all blocks in H after each deletion. Based on this, we adopt the idea of updating blocks locally. After deleting a block, we only need to update the blocks affected by it.

As shown in Fig. 2, when we delete vertices v_9, v_{10}, and v_{11} respectively, the blocks corresponding to vertices v_9, v_{10}, and v_{11} are the same, so we call $B(v_9)$, $B(v_{10})$, and $B(v_{11})$ as equivalent blocks. When selecting a block to be deleted from the graph, we only need to determine whether the blocks corresponding to vertex v_9 meet the deletion conditions, and the blocks corresponding to vertices v_{10} and v_{11} are not need to be considered.

Property 2. After deleting the block $B(v)$ in (k, d)-truss H, the blocks that need to be updated are the deletable blocks that contain vertices in $B(v)$ and the deleteable blocks that contain neighbors of vertices in $B(v)$.

Proof. After removing a moveable block from the (k, d)-truss H, the blocks containing the vertices of the moveable block will definitely be affected and must be updated. Blocks containing neighbors of the movable block vertices may also be affected and should be taken into account when updating the blocks.

Example 1. *We show an example of an updated algorithm. The graph on the left in Fig. 3 is a (k, d) truss with k=4 and d=2. The query vertex is v_1, the SkdC lower constraint is 4, and the SkdC upper constraint is 5. Because the number of vertices in the graph exceeds the upper size constraint, the updated algorithm is invoked. First, we compute the candidate set of deletable blocks: $\beta = \{B(v_5), B(v_6), B(v_7), B(v_8)\}$, which excludes non-deletable and equivalent blocks. To obtain the middle graph in Fig. 3, we first need to eliminate the smallest block $B(v_5)$ from the graph. Only the blocks corresponding to vertex v_6, v_7, and v_8 need to be updated in β, according to Property 2. After taking away v_6, v_7, and v_8, we obtain $B(v_6) = B(v_7) = B(v_8)$ due to $k = 4$. Because of this, we alter β to be $\{B(v_6)\}$. After removing $B(v_6)$ from the graph, β has no updateable or deletable blocks. The number of vertices in the remaining graph is 4, satisfying the size constraints, returning the community consisting of $\{v_1, v_2, v_3, v_4\}$.*

Algorithm 4: Mark_Block

```
   input  : a graph H, the vertex v
   output: B(v)
1  P ← ∅ ;
2  for each u ∈ N(v, H) do
3  |  P.push(u, v) ;
4  while P ≠ ∅ do
5  |  (x, y) ← P.pop() ;
6  |  remove (x, y) from H ;
7  |  for each w ∈ N(x, H) ∩ N(y, H) do
8  |  |  sup_H(x, w) ← sup_H(x, w) − 1 ;
9  |  |  if sup_H(x, w) < k − 2 and (x, w) ∉ P then
10 |  |  |  P.push(x, w) ;
11 |  |  exchange x with y in lines 8-10 ;
12 B(v) ← the vertices whose query distance is greater than d in H ;
13 return B(v) ;
```

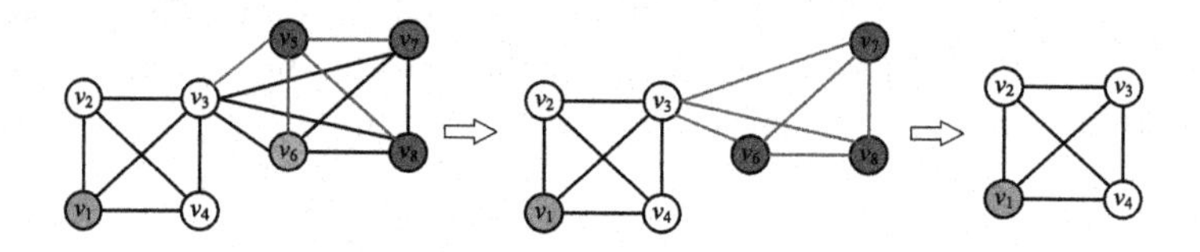

Fig. 3. A (k, d)-truss deletion block process with q=v_1, k=4, d=4, l=5, and s=4

Algorithm 4 shows the pseudo-code of the SkdC search update algorithm. We set the affected block β_a after block deletion, the block marked for deletion B_m, and the deleteable block β in H as empty (Line 1). Then record all the deleteable block information in H not containing equivalent blocks and put this block information into β. For a deleteable block $B(v)$ in H, if equivalent blocks exist, only one is need to be recorded in β (Line 2). All of the vertices in H are undeleteable if β is an empty set, and then the empty set is returned (Lines 3–4). When the number of vertices in H is greater than s, we enter the loop. If we delete a block, and the number of vertices in H is exactly equal to s, we directly output the result of H after deleting the block(Line 6). According to Property 2, we find the blocks in β that is affected by the block B_m(Line 7). Then, delete the block B_m in H (Line 8). In β, we update affected blocks β_a(Line 9). If there are no more blocks to be deleted after the update, the loop ends (Lines 10–11). Finally, both B_m and β_a are set to be empty. If the number of vertices in H is greater than s, the loop is continued; otherwise, end the loop (Line 12). If the number of vertices in H is less than l or greater than s, then there is no SkdC and the empty set is returned (Lines 13–14). Otherwise, the result of H is output to get SkdC R (Lines 15–17).

Table 1. Default parameter settings for each data set

Dataset			
Parameter	Youtube	DBLP	Amazon
k	18	21	4
d	3	4	5
l	200	50	20
s	300	150	50

5 Experiments

We evaluate the effectiveness and efficiency of our algorithm through conducting extensive performance studies in this section.

5.1 Experimental Setting

We conducted experiments on three datasets:
Youtube: Youtube is a video-sharing website that also incorporates a social network with 1134890 vertices and 2987624 edges.
DBLP: The DBLP creates a co-authorship network with 317080 vertices and 1049866 edges.
Amazon: Amazon is based on the Customers Who Bought This Item Also Bought feature of the Amazon website with 334863 vertices and 925872 edges.

Algorithms. As far as we know, there is no previous research that addresses the SkdC search problem. Our study focuses on evaluating two SkdC search algorithms, namely **Basic** and **Update**. The **Basic** consists of Algorithm 1 and Algorithm 2, while the **Update** comprises Algorithm 1 and Algorithm 4.

Parameters. We run experiments by adjusting the query distance constraint d, the constraint k, the size constraint l, the size constraint s, and the proportion of the datasets. The default parameter settings are shown in Table 1.

The algorithms covered in this paper are implemented in C++ and run on a machine with an Intel(R) Core(TM) i5-11400F CPU and 16 GB of memory.

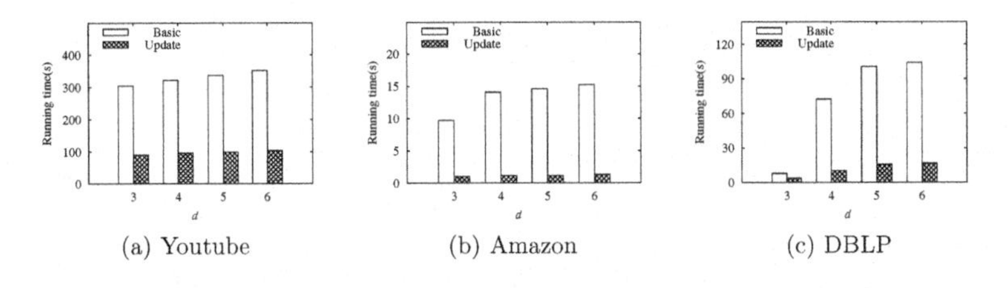

(a) Youtube (b) Amazon (c) DBLP

Fig. 4. The effect of d

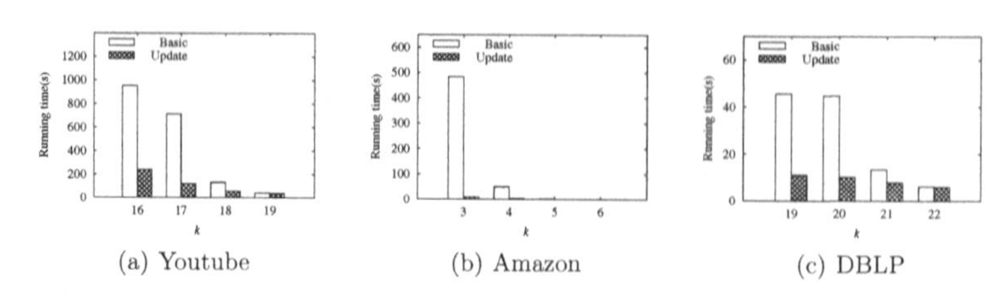

(a) Youtube (b) Amazon (c) DBLP

Fig. 5. The effect of k

5.2 Experimental Results

The Effect of d. We investigate the effect of the query condition d on algorithm efficiency. By setting d to 3–6 and leaving other parameters at their default values, as shown in Fig. 4, we see that as d grows, the execution performance of the two algorithms slows. This is due to the fact that the bigger the value of d, the more vertices there are in the (k, d)-truss H obtained in Algorithm 1, suggesting that we need to analyze more vertices. When the value of d reaches a certain threshold, increasing it further has little influence on the process because the outcome of Algorithm 1 no longer changes.

The Effect of k. Due to the inconsistent structure of different datasets, we change the value of k according to the actual situation, while leaving the other parameters at their default values. As seen in Fig. 5, both algorithms usually become faster as k grows. Because the number of vertices in the (k, d)-truss H obtained in Algorithm 1 reduces as k grows, we need to process fewer vertices. When k reaches a certain threshold, increasing it further may cause Algorithm 1 to return the empty set or not satisfy the deletion condition, which means we do not need to execute Algorithm 2 and Algorithm 4. Figure 5(c) shows a specific scenario in which the performance of the algorithm does not vary considerably when the value of k is changed from 19 to 20. We discover that after evaluating the experimental, the number of deleted blocks is lower when $k = 19$. This is because during the process deletion, there exists a block to obtain the result of SkdC directly. Overall, the Update algorithm significantly improves the efficiency of the SkdC search, particularly when Algorithm 1 gives a larger H.

The effect of l. Figure 6 compares the impact of different data sizes on the efficiency of the Basic and the Update under different values of the size constraint l. It is found that, under the default query constraints, changing the value of l does not significantly affect the efficiency of both algorithms. This is because, during the execution of the algorithm, l is only used to determine whether a block is a deleteable block. There are extremely few deleteable blocks in a big graph under other restriction circumstances, and it has essentially little effect on the algorithm's efficiency.

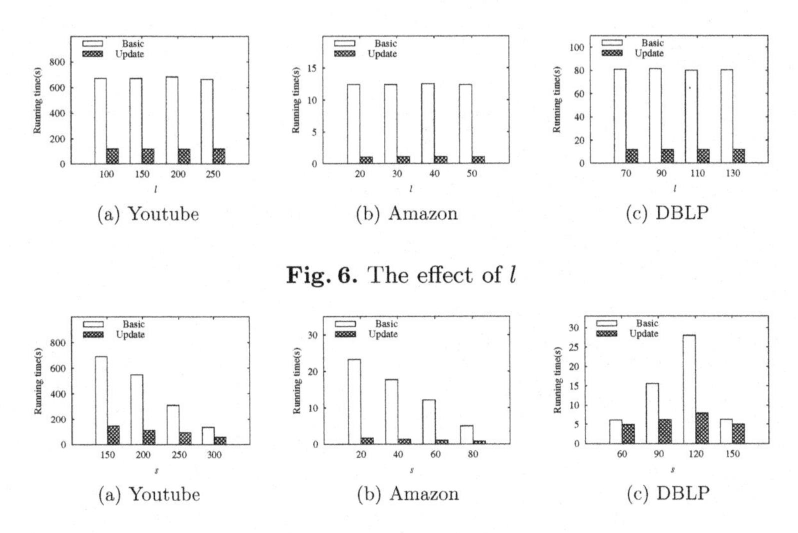

(a) Youtube (b) Amazon (c) DBLP

Fig. 6. The effect of l

(a) Youtube (b) Amazon (c) DBLP

Fig. 7. The effect of s

The effect of s. Figure 7 explores the influence of the query condition s on algorithm performance by modifying the size of s while leaving the other parameters at their default settings. As illustrated in Fig. 7, the execution speed of both algorithms tends to grow as s increases. This is due to the fact that the larger the value of s, the less times we need to delete blocks in Algorithm 2 and Algorithm 4. A specific situation is depicted in Fig. 7(c). We discover that there is a direct deletion of blocks in the process of removing blocks to end the algorithm after analyzing the testing process.

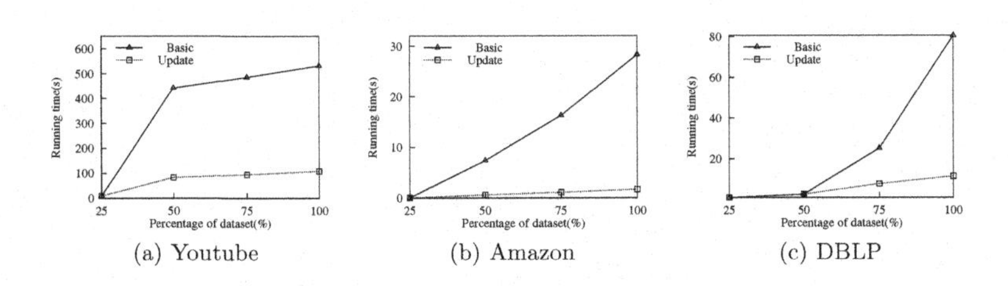

(a) Youtube (b) Amazon (c) DBLP

Fig. 8. Proportion of change in datasets

Scalability. Figure 8 compares the scalability of different datasets scale, whose result shows that the runtime of the Basic and Update increases as the datasets scale increases. The efficiency of the Basic is mainly influenced by the size of H returned by Algorithm 1 and the block generation process, while the Update is mainly influenced by the running time of Algorithm 1. When the number of blocks eliminated by the algorithm is modest, the Basic and Update algorithms' running times are mostly controlled by the running time of Algorithm 1.

6 Conclusion

We present the size-constrained (k, d)-truss community search problem (SkdC) in this paper, which seeks to find the largest (k, d)-truss satisfying size restrictions in a large graph. To tackle this problem, we propose two novel algorithms. To illustrate the usefulness and validity of the proposed algorithms, we conduct comprehensive experiments on real social networks. We will investigate more effective solutions for SkdC search problem in the future.

Acknowledgements. This work is partially supported by the National Natural Science Foundation of China (Nos. 62002245, 62102271, 61802268), Natural Science Foundation of Liaoning Province (Nos. 2022-MS-303, 2022-BS-218).

References

1. Fang, Y., et al.: A survey of community search over big graphs. In: VLDB, pp. 353–392 (2020)
2. Zhang, J., Yu, P., Lv, Y.: Enterprise employee training via project team formation. In: WSDM, pp. 3–12 (2017)
3. Ugander, J., Backstrom, L., Marlow, C., Kleinberg, J.: Structural diversity in social contagion. In: PNAS, pp. 5962–5966 (2012)
4. Dittrich, M., Klau, G., Rosenwald, A., Dandekar, T., Müller, T.: Identifying functional modules in protein-protein interaction networks: an integrated exact approach. In: ISMB, pp. i223–i231 (2008)
5. Huang, X., Cheng, H., Qin, L., Tian, W., Yu, J.: Querying k-truss community in large and dynamic graphs. In: SIGMOD, pp. 1311–1322 (2014)
6. Huang, X., Lakshmanan, L.: Attribute-driven community search. In: VLDB, pp. 949–960 (2017)
7. Barbieri, N., Bonchi, F., Galimberti, E., Gullo, F.: Efficient and effective community search. In: DMKD, pp. 1406–1433 (2015)
8. Huang, X., Lu, W., Lakshmanan, L.: Truss decomposition of probabilistic graphs: semantics and algorithms. In: SIGMOD, pp. 77–90 (2016)
9. Akbas, E., Zhao, P.: Truss-based community search: a truss-equivalence based indexing approach. In: VLDB, pp. 1298–1309 (2017)
10. Liu, B., Zhang, F., Zhang, W., Lin, X., Zhang, Y.: Efficient community search with size constraint. In: ICDE, pp. 97–108 (2021)
11. Yao, K., Chang, L.: Efficient size-bounded community search over large networks. In: VLDB, pp. 1441–1453 (2021)
12. Cui, W., Xiao, Y., Wang, H., Wang, W.: Local search of communities in large graphs. In: SIGMOD, pp. 991–1002 (2014)
13. Li, C., Zhang, F., Zhang, Y., Qin, L., Zhang, W., Lin, X.: Efficient progressive minimum k-core search. In: VLDB, pp. 362–375 (2020)

FSKD: Detecting Fake News with Few-Shot Knowledge Distillation

Jing Yuan[1], Chen Chen[1], Chunyan Hou[2](✉), and Xiaojie Yuan[1]

[1] College of Computer Science, TKLNDST, Nankai University, Tianjin, China
yuanjing@mail.nankai.edu.cn, {nkchenchen,yuanxj}@nankai.edu.cn
[2] School of Computer Science and Engineering, Tianjin University of Technology, Tianjin, China
chunyanhou@163.com

Abstract. The detection of fake news on social networks is highly desirable and socially beneficial. In real scenarios, there are few labeled news articles and a large number of unlabeled articles. One prominent way is to consider fake news detection as a few-shot learning task. However, existing few-shot learning methods suffer from two limitations when they are used to detect fake news. First, they may not directly learn linguistic knowledge from both labeled and unlabeled news articles. Second, few training articles usually lead to over-fitting of the model. In this paper, we propose a novel Few-Shot Knowledge Distillation (FSKD) model which is based on the student-teacher framework. BERT is fine-tuned on a few labeled news articles and used as the teacher model which provides soft labels of unlabeled articles. Then a few labeled news articles and a lot of unlabeled articles with soft labels are used to train a student model via knowledge distillation. An optimization algorithm is proposed to take advantage of the large-scale news articles and alleviate over-fitting simultaneously. Experimental results on real-world public datasets demonstrate that the proposed model can achieve superior performance.

Keywords: Fake news detection · Knowledge distillation · Few-shot learning

1 Introduction

The ever-increasing popularity of social networks has provided audiences with more convenient ways to access news than ever before. However, the social network is a double-edged sword for news consumption. It amplifies the exposure to a large number of fake news pieces that contain poorly checked or intentionally written false information. Fake news can be manipulated to mislead people for various commercial and political purposes, and even bring about detrimental effects on society. Therefore, fake news detection has attracted enormous attention in the past few years [12,13,20,26,27].

Existing studies on detecting fake news use a variety of machine learning algorithms that incorporate news characteristics in social networks. Approaches

X. Yang et al. (Eds.): ADMA 2023, LNAI 14180, pp. 421–436, 2023.
https://doi.org/10.1007/978-3-031-46677-9_29

based on traditional *machine learning* are applied in the early works for fake news detection. For example, Kwon *et al.* have proposed to use the Random Forest Classifier (RFC) to identify fake news with hand-crafted features related to the user, linguistic and structure [10]. Yang *et al.* have proposed to use the SVM classifier to identify rumors with hand-crafted features [26]. In contrast to these methods, researchers employed deep neural networks to learn efficient representation automatically [11,12,23]. For example, Ma *et al.* proposed gated recurrent unit network, which is an RNN-based model to learn the representation that captures the variation of contextual information of relevant posts [12]. The effectiveness of deep neural networks really depends on the amount and quality of training data. Building an effective deep neural network model requires a large amount of training data. Fake news annotation is a challenging task which is time-consuming and labor-intensive for human annotators to label news articles [20]. Fortunately, there are a large amount of unlabeled news articles in social networks and unsupervised learning has been proposed to detect fake news. Yang *et al.* proposed an unsupervised fake news detection algorithm by utilizing users' opinions on social network and estimating their credibilities [27]. However, it is still difficult for unsupervised models to outperform supervised models. Therefore, it is critical to design an effective fake news detection algorithm to learn with a few labeled articles.

Recently, Few-Shot Learning (FSL) has shown promising performances for learning with a limited number of labeled data such as translation [6], sentiment classification [28] and text classification [4,16]. Wang *et al.* conduct a thorough survey to understand few-shot learning [24]. In real scenarios, FSL is also a promising method to improve fake news detection. It meets the need to detect fake news as early as possible since a few news articles may be annotated in a short time to train a detective model. Existing few-shot learning methods, such as [4,22], suffer from two limitations when they are used to detect fake news. First, a few labeled articles affect model training because the model is easily over-fitted to such few articles. Second, they are designed to learn a model from a few labeled data rather than a lot of unlabeled data. Although it is difficult to learn a model from limited labeled news articles, a large amount of unlabeled articles is advantageous to learn rich linguistic knowledge such as a wide variety of topics, sentence structures and writing styles. This linguistic knowledge is helpful for fake news detection.

To address these issues, we propose a Few-Shot Knowledge Distillation (FSKD) model to detect fake news. The loss function of the FSKD model is designed to discriminate between the labeled and unlabeled news articles. The aim is to take advantage of a lot of unlabeled articles and alleviate the impact of soft labels during the model training. To distill the knowledge of BERT, BERT [2] is fine-tuned on labeled news articles and servers as the teacher model. The teacher model provides soft labels of a large number of unlabeled articles to guide the training of the detective model. We conduct experiments on two real-world and large-scale datasets. In addition, an optimization algorithm is also proposed to alleviate over-fitting and improve performance. Experimental results show that our proposed model can achieve substantial gains over state-of-the-art baseline models. The main contributes of this paper can be summarized as follows:

- We treat fake news detection as a few-shot learning task and propose a novel Few-Shot Knowledge Distillation (FSKD) model, which can take advantage of both labeled and unlabeled news articles to distill knowledge from fine-tuned BERT effectively and achieve the superior performance.
- We propose an optimization algorithm, which is able to make full use of the large-scale soft-labeled news articles and alleviate the over-fitting problem.
- We conduct extensive experiments on two real-world and large-scale datasets to demonstrate that our proposed model is more accurate in detecting fake news than previous works.

2 Related Works

Fake News Detection. Recently, deep neural networks have been widely used to detect fake news and achieve good performance in fake news detection [11,12,17,23]. Shu *et al.* present a comprehensive survey of detecting fake news on social media from a data mining perspective and point out that data-oriented fake news research is a promising direction [20]. Unsupervised learning has been proposed to detect fake news because there is a large amount of unlabeled data in social networks. Yang *et al.* proposed an unsupervised fake news detection algorithm by utilizing users' opinions on social networks and estimating their credibilities [27]. However, it is still difficult for unsupervised models to outperform supervised models. The major difference between prior works and ours is that we regard fake news detection as a few-shot learning task, and a few works are done to use few-shot learning for fake news detection. We use the few-shot learning based on labeled and unlabeled data to detect fake news.

Few-Shot Learning. Few-Shot Learning is proposed to tackle task which contains only a few examples with supervised information. Wang *et al.* conduct a thorough survey to understand few-shot learning [24]. They categorize existing few-shot learning into three perspectives: data, model and algorithm. From a data perspective, previous few-shot learning approaches, such as [4,22], usually focus on how to make full use of a few labeled data, and a few studies using both labeled and unlabeled data are applied in the vision domain [3,25]. For example, Wu *et al.* select some informative unlabeled samples with pseudo labels and update the model by the selected samples [25]. Our work differs from these studies because all unlabeled samples can be used by our model and an optimization algorithm is proposed to alleviate over-fitting. Recently, the use of few-shot learning has drawn attention in natural language processing, including translation [6], sentiment classification [28], text classification [4,16]. Relation network (RN) is a few-shot learning model that uses a neural network as the distance measurement and computes the class vectors [22]. The Induction network (IN) proposed by Geng *et al.* learns generalized class-wise representations by combining dynamic routing algorithm with a typical meta-learning framework, which is a recent state-of-the-art few-shot learning method [4]. Compared with these researches, we focus on fake news detection.

Knowledge Distillation. As a representative type of model compression and acceleration, knowledge distillation has received increasing attention from the research community. Generally, knowledge distillation has a teacher-student framework, that learns a small student model from a large teacher model. The large-scale deep models, such as BERT [2], have achieved great success in natural language processing. Thus, knowledge distillation is studied in the field of natural language processing to obtain lightweight, efficient and effective language models [8,15,18,21]. Gou *et. al.* provided a comprehensive survey of knowledge distillation from the perspectives of knowledge categories [5]. The most popular knowledge distillation for classification tasks is known as the soft labels [1,7]. Inspired by these studies, we fine-tune BERT on a few labeled data as a teacher model which provides soft labels for the unlabeled data. The major difference between prior works and ours is that an optimization algorithm is proposed to improve the generalization of knowledge distillation.

3 Knowledge Distillation Framework

3.1 Problem Formulation

The goal of this work is to classify fake/true news pieces accurately with limited labeled articles. Given a set of news articles $A = \{a_i\}_{i=1}^{M}$ and each article a has a textual content x, our goal is to learn a model $P(y|x;\theta)$ for the prediction of label $y(x) \in \{0,1\}$ based on the textual content x, where M denotes the number of news and θ is the model parameters. The label $y(x) = 1$ if the news article is fake and $y(x) = 0$ if it is true.

Let $D_L = \{(x_i, y_i)\}_{i=1}^{2K}$ denotes the labeled training set and $D_U = \{(x_i)\}_{i=1}^{N}$ is the unlabeled training set. There are K labeled news articles of each class for learning the model $P(y|x;\theta)$. Thus, fake news detection can be regarded as a K-shot learning problem. Note that K is small and N is far more than K here.

3.2 Few-Shot Knowledge Distillation

As shown in Fig. 1, our method has three stages. At stage 1, in order to distill knowledge of BERT, we use all of a few labeled news articles to fine-tune BERT as a teacher model. At stage 2, the teacher model generates soft labels of a large number of unlabeled articles. At stage 3, the labeled and soft-labeled articles are combined to distill knowledge from the teacher model to the student model (i.e. FSKD model). Softmax function with the parameter T is able to produce a softer probability distribution over classes. The loss function consists of distillation loss and student loss. Minimizing distillation loss can make FSKD similar to the teacher model and improve the generalization of the student model with soft-labeled news articles while the student loss can use the ground truth labels of labeled articles to keep knowledge distillation correctly. We introduce the teacher model and the student model respectively.

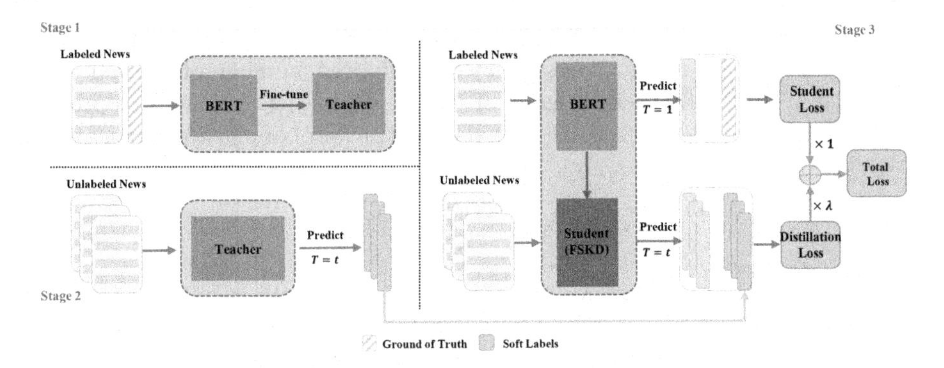

Fig. 1. An illustration of the FSKD framework.

Teacher Model. For fake news detection, the unlabeled data denotes the set of a lot of news articles without verified veracity while the labeled data is the set of a few news articles having the labels of veracity. Linguistic knowledge is required to detect the veracity of news articles. BERT consists of a multi-layer bidirectional transformer encoder, which can be pre-trained on a large unlabeled textual data and then be fine-tuned on a limited labeled textual data. It has shown excellent results in many natural language processing tasks. BERT benefits from learning deep representations which are full of rich linguistic knowledge. We use BERT to distill rich linguistic knowledge. A few labeled news articles are used to fine-tune pre-trained BERT as a teacher model. Then, the teacher model generates soft labels of a large amount of unlabeled articles.

Student Model. The main goal of knowledge distillation is to distill knowledge from the teacher model to the student model. A few labeled articles are not only used to fine-tune the teacher model, but also responsible for distilling the knowledge of BERT. Both the labeled and soft-labeled articles are used as the distillation set and the student model is exactly leveraged to detect fake news. Specifically, the labeled and soft-labeled articles are integrated into the training set to train FSKD model. As shown in Eq.(3), the loss function is designed for the training of the student model on the labeled and soft-labeled articles. Based on the soft-labeled articles, the distillation loss is to make the prediction of the student model similar to that of the teacher model. The student loss is responsible for using labeled articles to make the predictions of the student model close to the ground truth labels. The distillation loss and the student loss are combined as the total loss. An optimization algorithm is proposed to apply the loss function to train the student model. The softmax function with the temperature parameter T is used to compute the distillation loss because the higher value of T can generate a softer probability distribution over classes of news. The softmax function with $T = 1$ is utilized to provide the probability distribution for labeled articles.

In knowledge distillation, BERT serves as the teacher model. Pre-trained BERT is fine-tuned to be $P_{BERT}(y|x)$ on a limited number of labeled news articles D_L. To distill the knowledge of BERT, $P_{BERT}(y|x)$ is used to provide soft labels of unlabeled articles D_U. BERT's output probability for an unlabeled article x is formulated as:

$$P'_{BERT}(y|x) = \frac{\exp(v_x/T)}{\sum_{d \in D_U} \exp(v_d/T)} \tag{1}$$

where v_x is the logits before the softmax layer in the neural network of $P_{BERT}(y|x)$.

The predictive model for an unlabeled news article x is defined as

$$P(y|x;\theta) = \frac{\exp(z_x)}{\sum_{d \in D_U} \exp(z_d)}, P'(y|x;\theta) = \frac{\exp(z_x/T)}{\sum_{d \in D_U} \exp(z_d/T)} \tag{2}$$

where z_x is the logits before the softmax layer in the neural network of $P(y|x;\theta)$.

The loss function is as follows:

$$\begin{aligned} L(D_L, D_U, \theta) = -\tfrac{1}{|D_L|} \sum_{(x,y) \in D_L} (\log (P(y|x;\theta))) \\ + \tfrac{\lambda}{|D_U|} \sum_{x \in D_U} KL\left(P'_{BERT}(y|x), P'(y|x;\theta)\right) \end{aligned} \tag{3}$$

where T is a temperature parameter, and λ is a weighting parameter to balance the two parts of loss.

In Eq.(3), the former term is the student loss based on labeled news articles while the latter is the distillation loss on soft-labeled articles. Especially, the distillation loss use Kullback-Leibler (KL) divergence to measure the difference of probability distribution between fine-tuned BERT (i.e. teacher model) and FSKD (i.e. student model) model. The optimization objective is to obtain the optimal parameters θ^* of the model.

$$\theta^* = \underset{\theta}{\operatorname{argmin}}\, L(D_L, D_U, \theta) \tag{4}$$

3.3 An Optimization Algorithm

The training set consists of labeled and unlabeled news articles. The amount of labeled articles is much less than that of unlabeled ones. The challenge in the training procedure is to make full use of a large number of unlabeled articles, as well as alleviate over-fitting to a few labeled data simultaneously. To this end, an optimization algorithm is proposed. Algorithm 1 shows the pseudo code of the optimization algorithm in the training procedure. We use the adaptive gradient optimization algorithm Adam proposed by Kingma & Ba [9]. The Algorithm 1 begins with the random initialization of parameter vector θ_0 and the zero initialization of the first and second moment vectors m_0, v_0. Then, the batches of labeled and unlabeled news articles are extracted from the training set. The number of iterations is determined by T. In each iteration, ξ is calculated to control the size of the labeled articles which is used for the gradient

Algorithm 1. An Optimization Algorithm

Input: The set of labeled training data D_L; The set of unlabeled training data D_U; The number of iteration T; weight decay rate w; step size α; exponential decay rates β_1, β_2;
Output: The optimal parameters.

1: initialize θ_0, η_0
2: initialize $m_0 \leftarrow 0$
3: initialize $v_0 \leftarrow 0$
4: fill B_L with a batch of labeled data
5: fill B_U with a batch of unlabeled data
6: $t \leftarrow 0$
7: **while** $t <= T$ **do**
8: $\quad t \leftarrow t + 1$
9: $\quad \xi \leftarrow \frac{1}{2} + \frac{t}{T}\left(1 - \frac{1}{2}\right)$
10: $\quad B_Z = ((x, y)|(x, y) \in B_L \wedge P(y|x; \theta_{t-1}) < \xi)$
11: $\quad$ compute the gradient $\nabla_\theta L(B_Z, B_U, \theta)$
12: $\quad g_t \leftarrow \nabla_\theta L(B_Z, B_U, \theta) + w\theta_{t-1}$
13: $\quad m_t \leftarrow \beta_1 m_{t-1} + (1 - \beta_1) g_t$
14: $\quad v_t \leftarrow \beta_2 v_{t-1} + (1 - \beta_2) {g_t}^2$
15: $\quad \hat{m}_t \leftarrow m_t / (1 - \beta_1^t)$
16: $\quad \hat{v}_t \leftarrow v_t / (1 - \beta_2^t)$
17: $\quad \theta_t \leftarrow \theta_{t-1} - \eta_t \left(\alpha \hat{m}_t / \left(\sqrt{\hat{v}_t} + \varepsilon\right) + w\theta_{t-1}\right)$
18: $\quad$ fill B_L with the next batch of labeled data
19: $\quad$ fill B_U with the next batch of unlabeled data
20: **end while**
21: **return** optimized parameters θ_t

computation. The batch of labeled articles is filtered by the threshold ξ. The gradient is computed by the remaining articles. The weight decay term $w\theta_{t-1}$ is used to modify g_t. The vector m_t and v_t are maintained to be responsible for storing smoothed amplitudes of the gradients g_t and the squared gradients ${g_t}^2$. Specifically, Algorithm 1 updates exponential moving averages of the gradients g_t and the squared gradients ${g_t}^2$ while the hyper-parameters β_1, β_2 control the exponential decay rates of these moving averages. After the bias-corrected first and second moment estimation $\hat{m}_t, \hat{v}_t$ are computed, the parameter vector is updated as shown in line 17 of Algorithm 1. $\hat{m}_t$ is divided by $\sqrt{\hat{v}_t} + \varepsilon$ and then modified by the weight decay term $w\theta_{t-1}$. At the end of each iteration, the next batches of labeled and unlabeled data are extracted for the next iteration.

In brief, Algorithm 1 computes gradients and updates the model in each training iteration. The weight decay term plays a role in modifying the gradients and moments. In particular, as the model is trained on more unlabeled news articles, the number of labeled articles in each successive iteration is gradually increased to prevent the model from over-fitting. The labeled training data is selected as follows:

$$B_Z = ((x, y)|(x, y) \in B_L \wedge P(y|x; \theta_{t-1}) < \xi) \quad (5)$$

The value of ξ is given by Eq. (6)

$$\xi \leftarrow \frac{1}{C} + \frac{t}{T}\left(1 - \frac{1}{C}\right) \tag{6}$$

where C denotes the number of categories. In our experiments, $C = 2$ (i.e. fake or true), so, Eq. (6) can be simplified to Eq. (7). ξ increases linearly with t.

$$\xi \leftarrow \frac{1}{2} + \frac{t}{T}\left(1 - \frac{1}{2}\right) \tag{7}$$

As shown in Eq. (5), more labeled articles can be selected when ξ increases from 0.5 to 1. In this case, the model can alleviate over-fitting to a few labeled articles.

4 Experiments

4.1 Datasets

Table 1. Statistics of the datasets

Datasets	#fake	#true	#total	avg. length
FNDC	19,285	19,186	38,471	156
GossipCop	4,415	15,133	19,548	557

We conduct experiments on two large-scale datasets, FNDC and GossipCop, which are Chinese and English news articles respectively. The statistics of the two datasets are shown in Table 1. The FNDC (Fake News Detection Competition) dataset stems from the fake news detection competition in China[1]. FNDC has a total of 38,471 news articles, including 19,285 true news and 19,186 fake news. The GossipCop dataset comes from FakeNewsNet [19], which contains 19,548 news articles collected from a fact-checking website[2]. We observe that the distribution of labels in the GossipCop dataset is unbalanced, and the number of true news is about 3.5 times that of fake news. The average length of news articles in GossipCop is more than that of FNDC.

4.2 Baselines

We compare our model against the following baselines:

GRU [12]: The RNN-based model with the gated recurrent unit is used to learn the representation of high-level features. **GRU+SL** denotes training a GRU model on few labeled data, which is then used to predict labels for all unlabeled samples. Samples having the soft labels are then added to training data that is used to learn a new GRU model.

[1] https://www.biendata.xyz/competition/falsenews/.
[2] https://www.gossipcop.com.

RFC [10]: A Random Forest Classifier (RFC) that uses three parameters to fit the temporal properties and an extensive set of hand-crafted features related to the user, linguistic and structure characteristics. **RFC+SL** is in a similar way to **GRU+SL**.

RN [22]: Relation network (RN) is a few-shot learning model that uses a neural network as the distance measurement and computes the class vectors.

IN [4]: Induction network (IN) learns generalized class-wise representations by combining dynamic routing algorithm with a typical meta-learning framework, which is a recent state-of-the-art few-shot learning method.

BERT [2]: Bidirectional Encoder Representations from Transformers (BERT) is designed to pre-train deep bidirectional representations from unlabeled text. A few labeled news articles are used to fine-tune BERT. **BERT+SL** is in a similar way to **GRU+SL**.

VAT [14]: Virtual Adversarial Training (VAT), an algorithm that generates adversarial Gaussian perturbations on input. VAT uses both labeled and unlabeled data in the training for the purpose of consistent predictions.

4.3 Experimental Setup

We implement RFC with the scikit-learn machine learning library[3]. For BERT, VAT, GRU, RN and IN, we use Tensorflowfootnotehttps://www.tensorflow.org. for the implementation. We use accuracy and F1 measure as evaluation metrics. We denote our model as FSKD, which is also based on Tensorflow.

When the transformer is initialized with BERT, we use a batch size of 8 for labeled data and a batch size of 24 for soft-labeled data, so that the model can be trained on more soft-labeled data. We set the learning rate to 2e-5 and the number of training steps to 1K. In the optimization algorithm, we set β_1 to 0.9, β_2 to 0.999 and weight decay rate w to 0.01.

4.4 Results and Analysis

We treat fake news detection as the K-shot learning task. The training and testing sets are required to evaluate the performance of models in few-shot learning. We randomly select 20% from GossipCop and FNDC datasets as the testing sets. For the rest of data, we hold out $2K$ labeled data and the remaining data is used as unlabeled data.

[3] https://www.sklearn.org.

Table 2. Accuracy of K-shot fake news detection on FNDC.

K	5	10	15	20	25
RFC	0.514	0.628	0.662	0.718	0.742
GRU	0.531	0.677	0.720	0.759	0.772
BERT	0.634	0.692	0.725	0.766	0.795
RN	0.504	0.522	0.631	0.701	0.740
IN	0.517	0.551	0.617	0.642	0.710
RFC+SL	0.519	0.624	0.668	0.717	0.759
GRU+SL	0.523	0.670	0.741	0.755	0.762
BERT+SL	0.690	0.732	0.761	0.781	0.799
VAT	0.541	0.560	0.563	0.579	0.565
FSKD-ξ	0.712	0.761	0.780	0.796	0.805
FSKD	**0.757**	**0.791**	**0.800**	**0.811**	**0.815**

Table 3. Accuracy of K-shot fake news detection on GossipCop.

K	5	10	15	20	25
RFC	0.399	0.521	0.536	0.555	0.571
GRU	0.484	0.525	0.516	0.572	0.580
BERT	0.467	0.553	0.599	0.601	0.620
RN	0.507	0.513	0.510	0.526	0.486
IN	0.485	0.525	0.513	0.524	0.497
RFC+SL	0.372	0.501	0.531	0.549	0.563
GRU+SL	0.495	0.520	0.564	0.596	0.640
BERT+SL	0.505	0.508	0.510	0.560	0.639
VAT	0.549	0.569	0.587	0.571	0.591
FSKD-ξ	0.686	0.724	0.761	0.769	0.783
FSKD	**0.749**	**0.777**	**0.779**	**0.783**	**0.796**

Results with Different Sizes of Labeled Data. We evaluate the performance of the proposed model with different K. Taking $K = 10$ on FNDC as an example, 10 true news and 10 fake news are taken as supervised training examples, 20% samples were taken as test samples, and the remaining samples are used as unlabeled articles. Table 2 and Table 3 show the accuracy of the proposed model and that of baseline models on FNDC and GossipCop datasets respectively. In brief, our proposed method outperforms all baseline models on two datasets. Baselines can be divided into two groups. The one group, which includes RFC, GRU, BERT, RN and IN, uses only labeled news articles while both labeled and soft-labeled articles are leveraged by the other group, which consists of VAT,

RFC+SL, GRU+SL, BERT+SL and FSKD. FSKD-ξ uses Adam rather than the proposed optimization algorithm.

(1) As we can see, in Table 2, when K equals to 5, BERT achieves the best performance in the former group, which can obtain 0.634 accuracy, while the other baselines can only obtain about 0.5 accuracy. The reason might be that BERT is pre-trained on a large corpus and is stable while other baselines, such as RFC and GRU, are easy to over-fit to a few labeled news articles. Thus we use a few labeled news articles to fine-tune BERT Base version as the teacher model. When using FSKD-ξ to predict the veracity of news articles, the accuracy is increased from 0.634 to 0.712, and the improvement is 0.078. This shows that our proposed model successfully distills knowledge from the teacher model to the student model. In addition, after using the proposed optimization algorithm, the accuracy is up to 0.757, which demonstrates the effectiveness of our proposed algorithm.

(2) Among baselines in the former group, BERT can achieve the best performance because BERT is able to learn a variety of linguistic knowledge by the multi-head attention mechanism. With the increase of labeled news articles, the performance of FSKD has an increasing trend. The few-shot learning methods (i.e. RN and IN) are generally worse than GRU and RFC. One main reason is that RN and IN are unable to capture the language knowledge from limited labeled articles. RFC and GRU can learn knowledge better than RN and IN because RFC uses hand-crafted features related to linguistic characteristics and GRU is good at learning representation of articles.

(3) The increase of K can usually improve the performance of these baselines. When we compare baselines in the latter group, the soft-labeled articles are added into the training set and the accuracy may not be improved because of the noise in soft-labeled articles. VAT's performance is not consistent with the increase of K. The reason might be that although a large number of unlabeled articles can make VAT robust, a few labeled articles are not enough to guarantee the generalization of the VAT model. Although BERT+SL usually gains a competitive advantage over other baselines, our proposed FSKD model still outperforms it. The reason is that FSKD can give the higher weight to distillation loss with labeled articles and the lower weight to student loss on soft-labeled articles. It is helpful to distill linguistic knowledge in the few-shot scenario.

(4) Our optimization algorithm can effectively improve the performance of the model, especially when there are fewer labeled articles. Moreover, even when the label distribution is unbalanced, FSKD can still achieve better results than other baselines, which further demonstrates the robustness of FSKD.

In order to show more intuitively what knowledge is distilled from teacher model to student model, we show the distribution of soft labels when varying K from $\{5, 15, 25\}$ on two datasets respectively. According to Table 1, FNDC has a total of 38,471 news articles, including 19,285 true news and 19,186 fake news. The ratio of positive and negative samples is approximately $1:1$. In contrast, the distribution of labels in the GossipCop dataset is unbalanced, the number of true news is about 3.5 times that of fake news. As shown in Fig. 2, when $K = 5$,

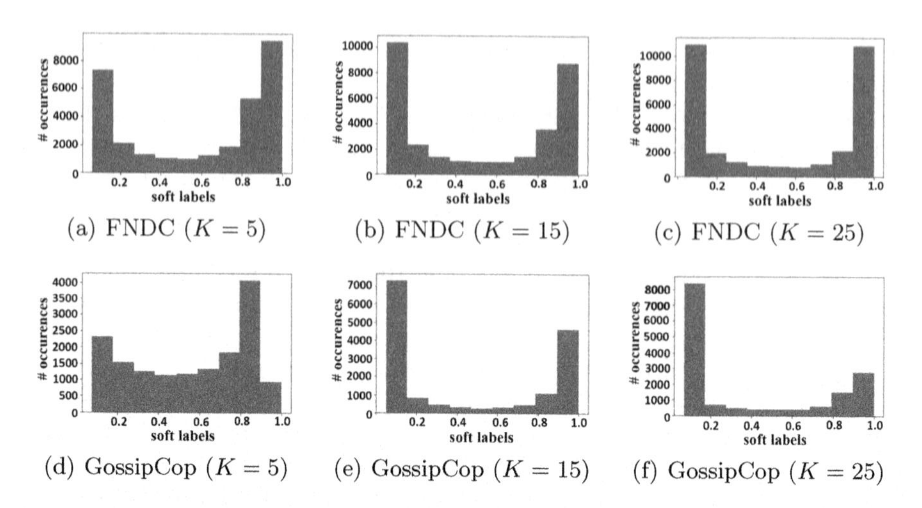

(a) FNDC ($K = 5$) (b) FNDC ($K = 15$) (c) FNDC ($K = 25$)

(d) GossipCop ($K = 5$) (e) GossipCop ($K = 15$) (f) GossipCop ($K = 25$)

Fig. 2. Soft labels distribution of FNDC and GossipCop with different numbers of K.

the predicted soft labels appear to be randomly distributed. With the increase of K, the distribution of soft labels is closer to that of the ground truth of labels, which demonstrates the higher the K, the more knowledge is distilled.

Results with Different Sizes of Soft-Labeled Data. To investigate the effect of the number of soft-labeled news articles, we conduct experiments with varied sizes of soft-labeled articles and the same size of labeled articles. K is set to 10 and we use the same testing set. BERT+SL and VAT are selected as baselines because they usually have good performance. We compare our method against BERT+SL and VAT with the increasing size of soft-labeled data on two datasets. The results are shown in Table 4 and Table 5. Two conclusions can be drawn from the results.

Firstly, BERT+SL outperforms VAT, and the increase of soft-labeled news articles does not consistently improve the accuracy of BERT+SL and VAT. This is perhaps because labeled articles are too limited to influence the effectiveness of the model and implicitly brings about more noise in the soft labels.

Secondly, Our proposed FSKD model is better than BERT+SL and VAT, and the improvement is consistent with the growth of soft-labeled data to a certain extent. When the size of soft-labeled data continues to increase, the performance gradually reaches the peak and keeps stable.

For understanding how wide of a gap the models can make up in using labeled data vs. soft-labeled data. We do an interesting experiment, which includes a "ceiling" model. As shown in Fig. 3, the first two columns indicate that the models BERT+SL and FSKD are trained using soft-labeled data. In contrast, we take the same number of labeled news articles to train BERT, which is regarded as the "ceiling" model. With the increase of soft-labeled data, the gap between FSKD and the "ceiling" model gradually decreases, which indicates the

Table 4. Accuracy on FNDC with different numbers of soft-labeled data.

#Soft Labeled	5000	10000	15000	20000
BERT+SL	0.682	0.708	0.726	0.730
VAT	0.538	0.556	0.550	0.561
FSKD	**0.760**	**0.780**	**0.782**	**0.785**

Table 5. Accuracy on GossipCop with different numbers of soft-labeled data.

#Soft Labeled	3500	7000	10500	14000
BERT+SL	0.514	0.605	0.577	0.627
VAT	0.546	0.574	0.555	0.563
FSKD	**0.587**	**0.685**	**0.768**	**0.768**

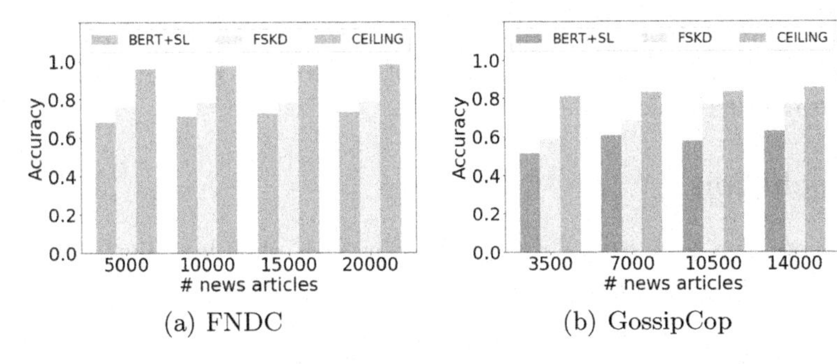

(a) FNDC (b) GossipCop

Fig. 3. An interesting experiment. The first two columns are trained using soft-labeled news items. The third column uses the same number of labeled news items to train BERT, which is regarded as the "ceiling" model.

effective distillation of knowledge from the teacher model to the student model. In addition, on GossipCop, the gap between FSKD and the "ceiling" model becomes smaller as the number of data increases, which indicates the robustness of the proposed model on unbalanced data.

4.5 Ablation Study

To illustrate the importance of each part of the loss function, we conduct an ablation study. We set $K = 10$ and use all soft-labeled articles on two datasets respectively. The accuracy and F1-score are used as the evaluation metrics.

The experimental results are shown in Table 6. **total loss** is to use both student loss and distillation loss. **w/o (student loss)** denotes to remove student loss from total loss while **w/o (distillation loss)** means to disregard distillation loss in total loss. The results show that both student loss and distillation loss significantly contribute to the effectiveness of FSKD. In contrast, distillation loss is more effective than student loss. Especially, distillation loss is able to

Table 6. Ablation study. K equals to 10 and all soft-labeled examples are used on FNDC and GossipCop. "w/o (student loss)" denotes to remove student loss from total loss while "w/o (distillation loss)" means not using distillation loss in total loss.

	FNDC		GossipCop	
	Acc	F1-Score	Acc	F1-Score
total loss	**0.791**	**0.805**	**0.777**	**0.873**
w/o (student loss)	0.727	0.715	0.695	0.736
w/o (distillation loss)	0.692	0.670	0.553	0.658

enhance the F1-score. The possible reason is that the number of soft-labeled articles is far more than that of labeled articles in the few-shot scenario and the detective model can benefit from linguistic knowledge distilled from BERT which is fine-tuned by a few labeled data. The model can achieve higher performance with both student loss and distillation loss. It is revealed that our proposed optimization algorithm enables two parts of loss to play an independent role in the improvement of fake news detection.

5 Conclusion

In this paper, we treat this detection task as a K-shot learning problem and propose a few-shot knowledge distillation model for fake news detection. We take BERT as the teacher model and use a few labeled articles to fine-tune the model to give soft labels for unlabeled articles. Then, we use a few labeled articles and a large number of unlabeled articles with soft labels to train a student model (i.e. FSKD) in the way of knowledge distillation. FSKD is exactly used to detect fake news. In addition, an optimization algorithm is proposed to make full use of large-scale unlabeled articles and alleviate over-fitting. The experimental results on real-world datasets show that the model has superior performance in detecting fake news in both Chinese and English, even if the data distribution is unbalanced.

In the future, when more information of news is available, we plan to combine news content and user profiles based on the existing model. In addition, how to improve the performance of unsupervised fake news detection could also be an interesting research direction.

Acknowledgement. This work is partially supported by NFSC-General Technology Joint Fund for Basic Research (No.U1936206) and the National Natural Science Foundation of China (No. 62172237, 62077031.). We thank the AC, SPC, PC and reviewers for their insightful comments on this paper.

References

1. Ba, J., Caruana, R.: Do deep nets really need to be deep? In: NeurIPS, pp. 2654–2662 (2014)

2. Devlin, J., Chang, M., Lee, K., Toutanova, K.: BERT: pre-training of deep bidirectional transformers for language understanding. In: NAACL-HLT, pp. 4171–4186 (2019)
3. Douze, M., Szlam, A., Hariharan, B., Jégou, H.: Low-shot learning with large-scale diffusion. In: CVPR, pp. 3349–3358 (2018)
4. Geng, R., Li, B., Li, Y., Zhu, X., Jian, P., Sun, J.: Induction networks for few-shot text classification. In: EMNLP-IJCNLP, pp. 3902–3911 (2019)
5. Gou, J., Yu, B., Maybank, S.J., Tao, D.: Knowledge distillation: a survey. Int. J. Comput. Vis. **129**(6), 1789–1819 (2021)
6. Gu, J., Wang, Y., Chen, Y., Li, V.O.K., Cho, K.: Meta-learning for low-resource neural machine translation. In: EMNLP, pp. 3622–3631 (2018)
7. Hinton, G.E., Vinyals, O., Dean, J.: Distilling the knowledge in a neural network. CoRR abs/1503.02531 (2015)
8. Jiao, X., et al.: Tinybert: Distilling BERT for natural language understanding. In: EMNLP, pp. 4163–4174 (2020)
9. Kingma, D.P., Ba, J.: Adam: A method for stochastic optimization. In: ICLR (2015)
10. Kwon, S., Cha, M., Jung, K., Chen, W., Wang, Y.: Prominent features of rumor propagation in online social media. In: ICDM, pp. 1103–1108 (2013)
11. Liu, Y., Wu, Y.B.: Early detection of fake news on social media through propagation path classification with recurrent and convolutional networks. In: AAAI pp. 354–361 (2018)
12. Ma, J., et al.: Detecting rumors from microblogs with recurrent neural networks. In: IJCAI, pp. 3818–3824 (2016)
13. Ma, J., Gao, W., Wong, K.: Rumor detection on twitter with tree-structured recursive neural networks. In: ACL, pp. 1980–1989 (2018)
14. Miyato, T., Maeda, S., Koyama, M., Ishii, S.: Virtual adversarial training: a regularization method for supervised and semi-supervised learning. IEEE Trans. Pattern Anal. Mach. Intell. **41**(8), 1979–1993 (2019)
15. Ren, X., Shi, R., Li, F.: Distill BERT to traditional models in Chinese machine reading comprehension (student abstract). In: AAAI, pp. 13901–13902 (2020)
16. Rios, A., Kavuluru, R.: Few-shot and zero-shot multi-label learning for structured label spaces. In: EMNLP, pp. 3132–3142 (2018)
17. Ruchansky, N., Seo, S., Liu, Y.: CSI: A hybrid deep model for fake news detection. In: CIKM, pp. 797–806 (2017)
18. Sanh, V., Debut, L., Chaumond, J., Wolf, T.: Distilbert, a distilled version of BERT: smaller, faster, cheaper and lighter. CoRR abs/1910.01108 (2019)
19. Shu, K., Mahudeswaran, D., Wang, S., Lee, D., Liu, H.: Fakenewsnet: a data repository with news content, social context and dynamic information for studying fake news on social media. CoRR abs/1809.01286 (2018)
20. Shu, K., Sliva, A., Wang, S., Tang, J., Liu, H.: Fake news detection on social media: a data mining perspective. SIGKDD Explor. **19**(1), 22–36 (2017)
21. Sun, S., Cheng, Y., Gan, Z., Liu, J.: Patient knowledge distillation for BERT model compression. In: EMNLP-IJCNLP, pp. 4322–4331 (2019)
22. Sung, F., Yang, Y., Zhang, L., Xiang, T., Torr, P.H.S., Hospedales, T.M.: Learning to compare: relation network for few-shot learning. In: CVPR, pp. 1199–1208 (2018)
23. Wang, W.Y.: "liar, liar pants on fire": a new benchmark dataset for fake news detection. In: ACL, pp. 422–426 (2017)
24. Wang, Y., Yao, Q., Kwok, J.T., Ni, L.M.: Generalizing from a few examples: a survey on few-shot learning. ACM Comput. Surv. **53**(3), 63:1–63:34 (2020)

25. Wu, Y., Lin, Y., Dong, X., Yan, Y., Ouyang, W., Yang, Y.: Exploit the unknown gradually: one-shot video-based person re-identification by stepwise learning. In: CVPR, pp. 5177–5186 (2018)
26. Yang, F., Liu, Y., Yu, X., Yang, M.: Automatic detection of rumor on sina weibo. In: SIGKDD Workshop (2012)
27. Yang, S., Shu, K., Wang, S., Gu, R., Wu, F., Liu, H.: Unsupervised fake news detection on social media: A generative approach. In: AAAI, pp. 5644–5651 (2019)
28. Yu, M., et al.: Diverse few-shot text classification with multiple metrics. In: NAACL-HLT, pp. 1206–1215 (2018)

STIP: A Seasonal Trend Integrated Predictor for Blood Glucose Level in Time Series

Weixiong Rao[1], Guangda Yang[1](✉), Qinpei Zhao[1], Yuzhi Liu[1], Hongming Zhu[1], Ming Li[2], Xuefeng Li[1], and Yinjia Zhang[3]

[1] Tongji University, Shanghai, China
{wxrao,2031550,qinpeizhao}@tongji.edu.cn
[2] Shanghai DianJi University, Shanhai, China
[3] Aalto University, Espoo, Finland

Abstract. Blood glucose prediction is important for managing diabetes, preventing hypoglycemia, optimizing insulin therapy, and improving the quality of life for people with diabetes. Because of the continuous glucose monitoring technique, the prediction models can be trained on the patient's historical blood glucose data in time series. In order to learn the seasonality and trend of the blood glucose data, we introduce a seasonal trend integrated predictor (*STIP*). Especially for the seasonality, the local and global patterns are captured by embedding and convolutions. The experimental results on different prediction methods indicate the performance of the introduced method.

Keywords: blood glucose prediction · seasonality and trend · time series · convolution · continuous glucose monitoring

1 Introduction

Diabetes is a common chronic diseases and its impact spans the globe. The number of adults with diabetes worldwide reached 463 million by 2019 [16], accounting for 9% of the global adult population. There are two main types of diabetes, which are Type 1 diabetes mellitus (T1DM) and Type 2 diabetes mellitus (T2DM). T1DM is a pathological condition in which blood glucose (BG) levels are elevated due to lack or insufficiency of insulin secretion. Patients with T1DM need exogenous insulin to compensate for the deficiency of endogenous insulin secretion. T2DM, on the other hand, results in hyperglycemia primarily due to insufficient circulating insulin to stimulate tissue uptake of glucose and is associated with obesity [7]. Oral hypoglycemic agents and lifestyle modifications through increased exercise and weight loss are priorities for most T2DM patients [11]. Diabetes patients must manage their BG levels by some necessary actions to avoid *hypoglycemia* and *hyperglycemia*. Both hyperglycemia and hypoglycemia were associated with severe complications of diabetes. Furthermore, Severe hypoglycemia could be associated with permanent brain damage and even be a lethal condition leading to mortality.

X. Yang et al. (Eds.): ADMA 2023, LNAI 14180, pp. 437–450, 2023.
https://doi.org/10.1007/978-3-031-46677-9_30

Much work has been done on BG predictions [18,21]. The review [21]compiled a concise guide on machine learning methods for T1DM blood glucose prediction. This review included 55 papers and presented their topics, input types, data sources, algorithms, prediction ranges and metrics, etc. A review [18] was conducted to examine models and algorithms for T1/T2DM hypoglycemia and blood glucose prediction using real data. In general, there are fewer studies on patients with T2DM, because the amount of data is relatively small. Another limiting factor is the low availability of free data [14].

Efficient and accurate prediction of future blood glucose levels could help diabetics reduce their risk of extremes, both hypoglycemia and hyperglycemia. Accurate prediction of BG levels is the basis for reliable BG control and helps physicians to better plan patients' insulin therapy [13]. In addition to traditional linear models, machine learning techniques have also been applied in BG prediction [21], such as Gaussian process (GP), support vector machine (SVM), deep neural network, recurrent neural networks, etc. Applications of these algorithms has made BG prediction a breakthrough.

Auto-regressive integrated moving average (ARIMA) is widely implemented for prediction which explains a given time series based on its own past value. An ARIMA model [2] with adaptive identification algorithm [12] by adjusting the model orders and simultaneously updating model parameters was introduced [22]. ARIMA models have weaknesses in the time series with nonlinear relationships. In addition, it is time-consuming to find the optimal hyper-parameters for ARIMA models.

Various machine learning algorithms have achieved success in BG prediction. Random Forest (RF) [5] is widely used for regression, classification and other tasks, which builds models by integrating a multitude of decision trees in a training process. Support Vector Regression (SVR) [15] is established by constructing a hyperplane which maximizes the margin to minimize the error. K-nearest neighbors (KNN) [3] algorithm uses the sequences most adjacent to the input sequence to calculate the future values. For the machine learning models, feature engineering is essential and important in time series prediction. However, feature engineering is usually domain knowledge needed and time-consuming.

Because neural networks can learn the complex patterns including temporal pattern and the nonlinear relationship of the sequence data, they are widely used in different fields. Recurrent neural networks (RNNs) design a unique structure to capture the sequential features of data effectively. An architecture of GluNet [6], a Sequence-to-Sequence neural network [17] and a multi-scale LSTM [23] were introduced for the BG forecasting based on the *OhioT1DM* dataset. [24] proposed a novel deep learning model for T1DM patients, which can predict BG levels in a multi-step forward manner.

Seasonality and trend are two important components of time series data. Trend is the general direction in which a time series changes over time. Seasonality, on the other hand, refers to the recurring patterns that occur in a time series over a fixed period of time. For the task of BG prediction by time series analysis, it is important to understand and account for the components of seasonality and trend when analyzing time series data. Meanwhile, the local feature and global correlations should also be considered. BG levels at a certain moment are not only affected by specific changes at different times of the day, but may also be related to general trends at different times of the week or month.

In this paper, we introduce a seasonality and trend integrated predictor for BG level forecasting, which is able to extract short-term local features and to capture long-term global correlations.

2 Methodology

Time series of blood glucose is usually expressed as $X = (x_1, x_2, ..., x_t) \in R^{n \times d}$, where n denotes the length and d denotes the dimension of feature. In blood glucose prediction, the whole time series is usually divided into several sub-series $X_t = (x_{t-\tau+1}, ..., x_t)$, where τ is the window size, and this sub-series is used for next prediction. It is defined as:

$$\widehat{x_{t+1}} = f(x_{t-\tau+1}, ..., x_t) \tag{1}$$

In time series analysis, seasonal and trend components are important factors that can affect the behavior of the data over time. A trend is a long-term pattern in a time series that shows the overall direction of the data. It describes the gradual increase or decrease in the data over time. A trend can be either positive (upward) or negative (downward) and can be linear or nonlinear. Seasonality refers to the repeating pattern of fluctuations in a time series that occur within a year or less. Seasonal patterns can be daily, weekly, monthly, quarterly, or yearly, and can be caused by factors such as weather, holidays, and other periodic events.

It's crucial to identify the trend and seasonal components of the data so that they can be separated from the random or irregular components. This separation allows for better modeling and forecasting of the data. Following the FEDformer [25], we decompose the data into its trend and seasonal components using the series decomposition modules, and then processing the trend and seasonal components using the convolutional and feedforward neural network layers. The local and global predictions of the seasonal components are then merged using the 2D convolutional layer (see Fig. 1).

$$X_t = Avg(AvgPool(Padding(X))_{kernel_i}) \tag{2}$$

$$X_s = X - X_t \tag{3}$$

where X_t denotes the trend component and X_s denotes the seasonality component. We employ several different average pooling ($AvgPool$) kernels to obtain different patterns.

For the seasonality part, it is embedded by three parts separately.

$$X_s' = \sum\{Time\_encoding, Pos\_encoding, Value\_encoding\} \tag{4}$$

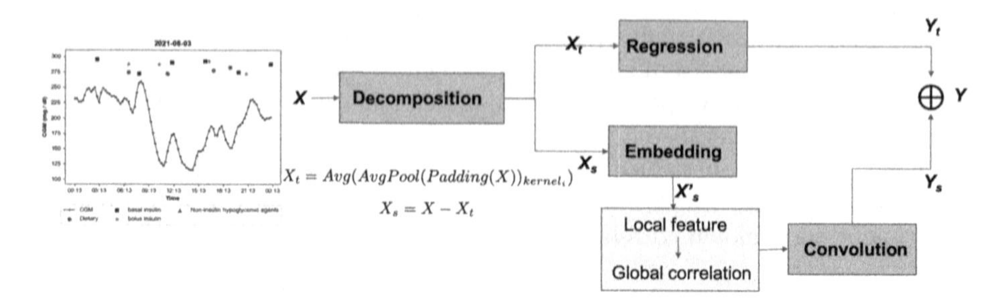

Fig. 1. The prediction on the seasonality part

where, the $Time\_encoding$ is to encode the time features such as minutes, days and months. For the value embedding, we complement zeros in order to adapt to the prediction length, i.e., $X_s = Concat(X_s, X_0)$.

With the embedding X_s', the local and global feature extraction is performed, which consists of multiple convolutional and deconvolutional layers, as well as a series decomposition block and a feedforward neural network.

Require: Input time series data X_s'

1: Apply isometric convolution [20] layers (a set of 1D convolutions) with different kernel sizes to X_s'
2: Apply downsampling convolution layers (a set of 1D convolutions) with different kernel sizes to the output of the isometric convolution layers
3: Apply upsampling convolution layers (a set of 1D transposed convolutions) with the same kernel sizes as the downsampling convolutions to the output of the downsampling convolution layers
4: Decompose the input time series data into trend and residual components using moving average filters with different kernel sizes
5: Process the output of the series decomposition blocks with a feedforward neural network layer to extract local and global features
6: Merge the output of the downsampling convolution layers into a single feature map
7: Apply layer normalization and a *tanh* activation function to the output of the feedforward neural network layer

Ensure: Output Y_s

With the trend prediction Y_t and the seasonality prediction Y_s, the final prediction result is $Y = Y_t + Y_s$.

3 Experiments

3.1 Blood Glucose Data

We conducted a comprehensive study of both real and simulated patient data for T1DM. The simulator can conveniently provide data on the diet and treatment strategies of virtual diabetic patients.

Table 1. Some statistics of the datasets studied for blood glucose level prediction models.

Datasets	Study period (days)	Monitoring interval (minutes)	Number of patients	Total CGM measurements
ShanghaiT1DM	4–14	15	12	15,695
ShanghaiT2DM	3–14	15	100	112,475
SimulatorT1DM	112	3	30	1,612,800
OhioT1DM	56	5	12	191,605

SimulatorT1DM was collected from UVA/Podova1 type diabetes patient simulator [8]. This simulator was first proposed in 2009, and with the deepening of research, more perfect subsequent versions have been proposed successively [19]. The simulator constructs 30 virtual patients, including 10 children, 10 adolescents and 10 adults. The user can control the blood sugar of the virtual patient by choosing the corresponding amount of insulin injection and carbohydrate intake. Blood glucose of the patient. The virtual patient received input every five minutes for the actions, including insulin injection and carbohydrate intake, and then was given information about the physical status of the virtual patient after performing the actions, including key information such as blood sugar. The virtual patient received input every three minutes for the actions, including insulin injection and carbohydrate intake, and then was given information about the physical status of the virtual patient after performing the actions, including key information such as blood sugar. In this paper, blood glucose data are collected through an 16-week simulation experiment with 30 virtual patients provided by the simulator.

OhioT1DM is a proposed dataset from Ohio University [9,10] that includes physiological sensor data, life event data, insulin injection and blood glucose monitoring data of 12 T1DM patients for up to 8 weeks. All participating diabetic patients received insulin pumps and wore continuous glucose monitoring (CGM) devices. They wore a Medtronic 530G or 630G insulin pump and used a Medtronic Enlite CGM sensor to monitor blood glucose during the 8 weeks of data collection. They recorded life events on a custom-made smartphone app and wore wristbands to further record relevant data.

The *ShanghaiT1DM* contains CGM readings from 10 type 1 Chinese patients. On the other hand, the datasets for T2DM are quite rare. The dataset *ShanghaiT2DM* was introduced in [14] with 100 type 2 diabetic patients in Shanghai, China. The dataset contains CGM readings over a 3 to 14 day period along with daily dietary information. All patients used the CGM device (Freestyle Libre HAbbott Diabetes Care) for 7–14 days. The CGM device collects glucose data at 15-minute intervals and automatically stores it on a sensor. We collected 681–1344 glucose values from T1DM patients and 1344 glucose values from T2DM patients.

Detailed information about the datasets was shown in Table 1. As shown in Fig. 2, the trend and seasonality parts from the four datasets have similar pattern with the original data. Therefore, the decomposition of two parts are reasonable.

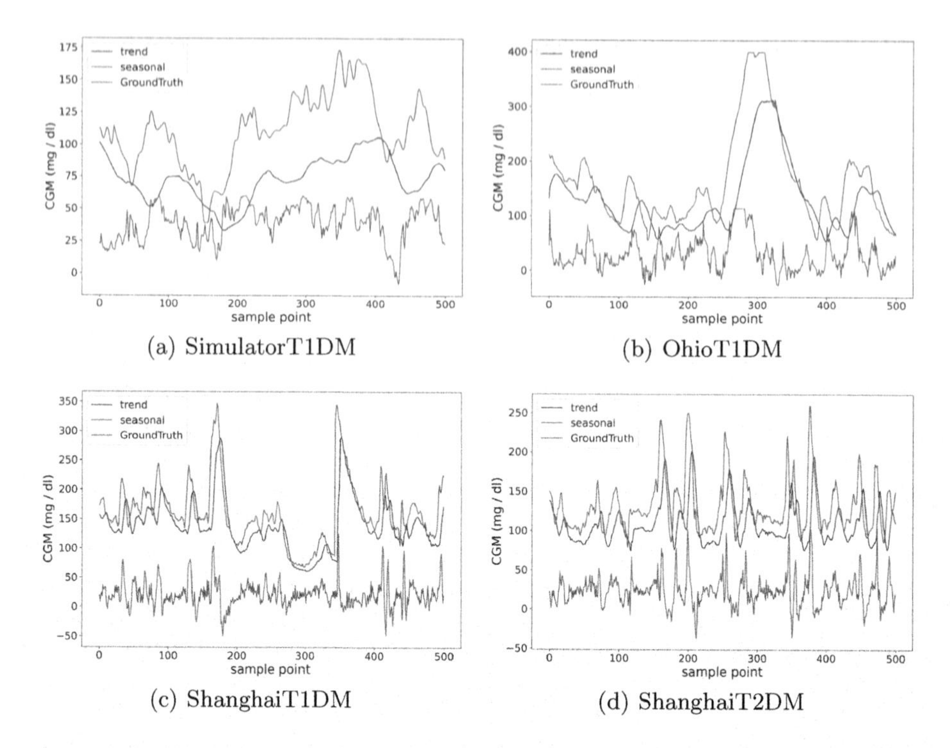

(a) SimulatorT1DM (b) OhioT1DM

(c) ShanghaiT1DM (d) ShanghaiT2DM

Fig. 2. The seasonality, trend parts of four BG data, together with the original time series.

3.2 Experimental Settings

We divide the data of each patient in the four datasets to facilitate model training and effect verification. Due to the obvious sequentiality of time series data, the random split of the dataset will involve future information, thus affecting the model. Therefore, we choose to split the dataset sequentially, splitting each patient's data in a 6:2:2 ratio into training, validation, and testing sets. The model is constantly updated based on the error on the training set. The validation set is used to test the performance of the model on unknown data during training, so as to select a model with stronger generalization ability. The final performance of the model is tested on the testing set.

Root mean square error (RMSE) is used as the evaluation metric of the blood glucose prediction model. It is defined as follows:

$$RMSE = \sqrt{\frac{1}{n}\sum_{i=1}^{n}(y_i - \hat{y})^2} \tag{5}$$

where y denotes the actual blood glucose value, $\hat{y}$ denotes the predicted blood glucose value, n denotes the length of time series. A model with a low RMSE represents a good performance of blood glucose prediction.

However, in the scenario of blood glucose prediction, the model not only needs to consider the accuracy of the prediction, but also has certain requirements for security. The security refers to whether the prediction model can accurately predict the occurrence of hypoglycemia or hyperglycemia events when the actual blood glucose value is hypoglycemia or hyperglycemia. For example, when the actual blood glucose is hypoglycemia, the prediction value given by the prediction model should be less than or equal to the actual blood glucose value, which means high security. We can find that RMSE can only express the accuracy of the model, but cannot evaluate the security point.

Therefore, we introduce a security metric. The Surveillance Error Grid [4] was developed by integrating the opinions of 206 clinicians and 28 non-clinicians. Each of them was asked to divide the Grid into five areas according to the measures they needed to take, and gave their own Error Grid. Then the Error grids of all the people were combined together, and different areas were represented with different colors. Finally, the Surveillance Error Grid was formed (Table 2).

Table 2. A summary of the time series prediction models involved in this paper.

Model	Category	Description
ARIMA [22]	Statistical Models	It models the future value of a variable as a linear combination of past observations and errors. It applies finite difference of the data points to eliminate the non-stationary of time series.
Random Forest [5] SVR [15] KNN [3]	Machine Learning Models	They are the models that can effectively solve nonlinear problems. Because of the good interpretability and low data dependency, they are applied in many time series prediction problems.
GRU [1]	Deep Learning Models	GRU is a variant of the RNN, which can effectively model sequence and solve the problem of gradient vanishing and exploding
Seq2Seq [17]		Seq2Seq is a special structure that trains models to convert one sequence into another. This model can captures dependencies between output values

In this paper, we want to compare the effects of the STIP and other blood glucose prediction algorithms on four datasets as comprehensively as possible, so we selected some commonly used models as our experimental baseline models. Because the models used in this paper involve many different types of blood glucose prediction models, we used various libraries in Python to implement these models. ARIMA model is implemented by PMDARIMA library. The models Random Forest, SVR and KNN are implemented by Scikit-Learn library. The GRU and Seq2Seq models are implemented by Pytorch. We use Python for data preprocessing, model building, training, and evaluation. For the ARIMA model, we use the auto_arima function from the PMDARIMA library to automatically select the best hyper-parameters for the ARIMA model. For the Random Forest model, the maximum depth is 10 and the iteration is 100. For the SVR model, we set the kernel function as RBF kernel. For KNN, we iterate over numbers between 10 and 15 to find the optimal K.

For the deep learning model, the number of nodes for each layer is 512 in the GRU model. The batch_size is set to 512 in the experiment. The Adam is used to optimize the objective function. For each patient in the dataset, we train the model using 1000 epochs. In order to have better convergence and prevent overfitting, we use an early stop strategy, where the training would stop early when the performance of the model does not increase for 30 consecutive epochs on the validation set.

For the STIP, the neural network layer is set as 512 nodes with batch size as 64. The learning rate is 0.005. The convolution kernel for downsampling and upsampling is 12 and 16 respectively. The input sequence length and initial feature length are both 24 for the OhioT1DM and SimulatorT1DM. Those are 4 for the ShanghaiT1DM and ShanghaiT2DM.

3.3 Experimental Results

We plot 500 samples of the results of the learned trend and seasonality parts in Fig. 2. The ground-truth time series is also plotted. It shows that the regression of the trend part model has already model the BG level well, especially for the ShanghaiT1DM and ShanghaiT2DM.

The performance in RMSE values of different models and strategies on the four datasets are shown in Table 3. The 30 min and 60 min in the table indicate the predicted blood glucose after 30 min and 60 min, respectively. When we train the model using data from all patients in the same dataset, we observe that $GRU_{allData}$ generally outperforms the GRU on single patient. This shows that the size of data impacts the performance of deep learning models. When the amount of data is too small, deep learning cannot fully play its performance and effectively capture the information in the data set, so the performance of model is not good. However, with the increase of the amount of data, the performance of deep learning model is effectively improved.

Comparing the effect of deep learning model under recursive strategy and multi-output strategy, we can find from SimulatorT1DM, OhioT1DM and ShanghaiT1DM that GRU_m performs better than GRU_r in 60 min. In addition, in SimulatorT1DM and OhioT1DM,GRU_m was also better than GRU_r at 30 min. This reflects the difference between the recursive strategy and the multi-output strategy. The recursive strategy only considers the error of the prediction at the next moment when training the model, so the error will accumulate with the increase of the prediction step, thus affecting the prediction result. While the multi-output strategy considers the error of multiple output values when training the model.In the process of training, the multi-output strategy keeps iterating to

Table 3. The prediction model results of recursive policy and multi-output policy model in four data sets, measured in RMSE.

Strategy	Models	SimulatorT1DM		OhioT1DM		ShanghaiT1DM		ShanghaiT2DM	
		30min	60min	30min	60min	30min	60min	30min	60min
Recursive strategy	ARIMA	33.44	77.99	54.20	68.01	17.31	33.96	16.58	29.88
	RF	38.10	64.22	28.62	40.02	27.04	34.12	20.15	28.02
	SVR	22.83	42.89	25.83	37.28	35.92	41.57	27.88	31.62
	KNN	22.37	41.14	27.01	37.32	25.99	33.96	26.62	31.45
	GRU_r	21.57	44.33	20.50	34.71	22.82	37.92	16.79	25.11
	$\mathrm{GRU}_{allData}$	21.99	45.28	19.82	33.88	**13.18**	24.95	12.02	21.45
Multiple output strategy	GRU_m	21.14	40.74	20.31	32.73	30.67	35.23	25.67	27.64
	Seq2Seq	21.01	40.45	20.89	33.07	28.33	34.65	23.76	26.86
	GRU_{meal}	13.02	23.12	20.49	32.90	24.32	33.22	21.39	25.35
	STIP	7.45	11.51	**13.70**	**21.79**	14.56	**24.12**	11.12	18.38
	STIP_{mv}	**7.30**	**11.42**	14.24	22.34	14.62	24.67	**11.08**	**18.18**

reduce the error in the multiple output values. Therefore, the multi-output strategy takes the long-term prediction error into account in the training to avoid the accumulation of errors in the recursive strategy. Since the sampling interval of SimulatorT1DM and OhioT1DM is 3 and 5 min, the prediction steps at 30 min is 10 and 6, which are longer, so the multi-output strategy works better. The sampling interval of ShanghaiT1DM and ShanghaiT2DM is 15 min which means the prediction step at 30 min is 2, so the recursive strategy works better.

To study the influence of diet on BG prediction, we introduce GRU_{meal} model based on multi-output strategy in the comparison, and include dietary information into the input variables of the model. As can be seen from the table, GRU_{meal} model performs better than GRU_m on most datasets. Although the food information in ShanghaiT1DM and ShanghaiT2DM only includes information about whether to eat or not, it still improves the performance of the model. The $STIP_{mv}$ is a multi-variate version which includes the meal, insulin as other input variables. The result on the $STIP_{mv}$ indicates that the other factors may affect the prediction results in some cases.

It is quite obvious that the learning from trend and seasonality parts separately in the STIP works in most cases. The $GRU_{allData}$ beats the STIP only on the ShanghaiT1DM's 30min prediction. For the other cases, the STIP outperforms the others, which illustrates the superiority of the STIP under both multi-output and recursive strategy.

We test the RMSE errors of different blood glucose prediction models on each dataset. However, the security of blood glucose prediction models cannot be measured simply by RMSE. In this section, we further evaluate the model used in the basic experiments and use the monitoring error grid to judge the security of the model.

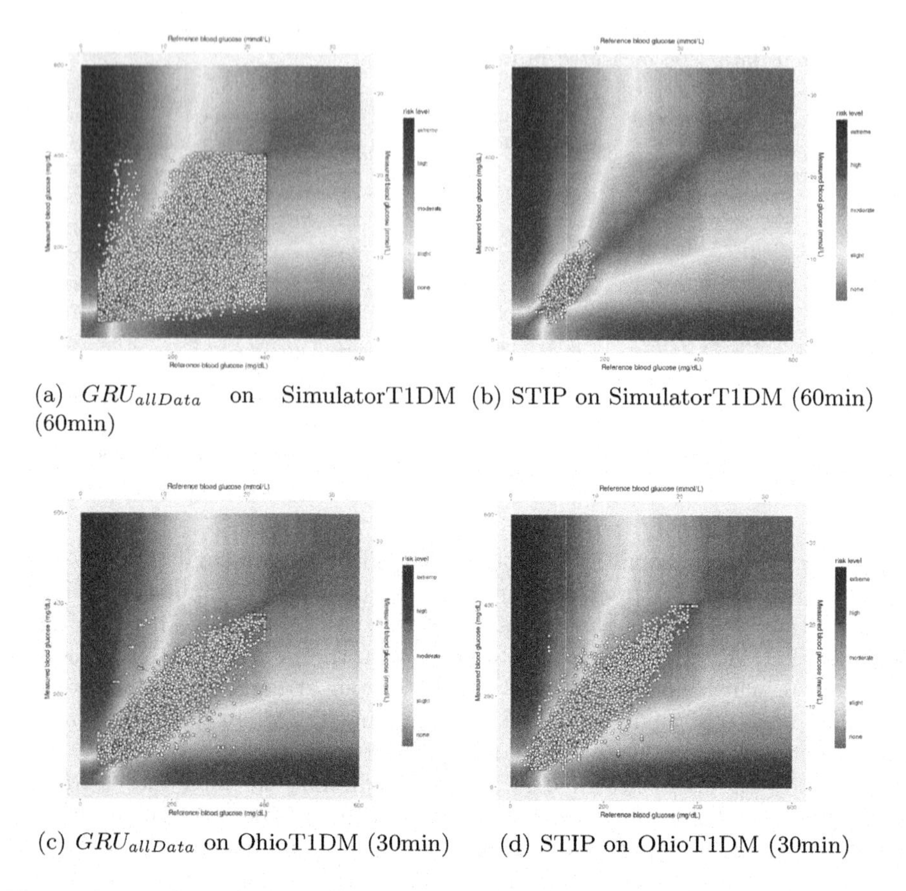

(a) $GRU_{allData}$ on SimulatorT1DM (60min)
(b) STIP on SimulatorT1DM (60min)

(c) $GRU_{allData}$ on OhioT1DM (30min)
(d) STIP on OhioT1DM (30min)

Fig. 3. Comparison of the security of $GRU_{allData}$ and STIP on the SimulatorT1DM and ShanghaiT1DM for either 30-minute or 60-minute prediction.

We choose $GRU_{allData}$ as the comparison baseline on the security experiment. The prediction time span will also affect the effect of the model prediction, so we conduct security experiments on the 30-minute and 60-minute spans both.

The error grid plots of the $GRU_{allData}$ and STIP models either with 30-minute or 60-minute prediction span on the four datasets are shown in Fig. 3 and Fig. 4. The abscissa of each graph represents the actual BG value, and the ordinate is the predicted BG value. The background color in the figure represents the risk value of the point, which is the risk value of the predicted BG pair (actual BG value and predicted BG value). Green represents low risk, while red represents high risk. According to the definition of coordinates, it is easy to find that the point close to the upper left corner represents the serious overestimation of BG value and the point close to the lower right corner represents the serious underestimation of BG value, which will both pose a great risk. According to different risk scores, the regions in the monitoring error grid can be divided into five levels: A, 0-0.5; B, 0.5-1.0; C, 1.0-2.0; D, 2.0-3.0; E, higher than 3.0. We can more intuitively compare the safety of the models by comparing the ratio of predicted BG value to each of the five grades. The proportions of different risk grades in the 30-minute and 60-minute prediction results of the $GRU_{allData}$ and STIP models on the four data sets are listed in Table 4. To better compare the difference between RMSE and security, the corresponding RMSE values are also presented in the table.

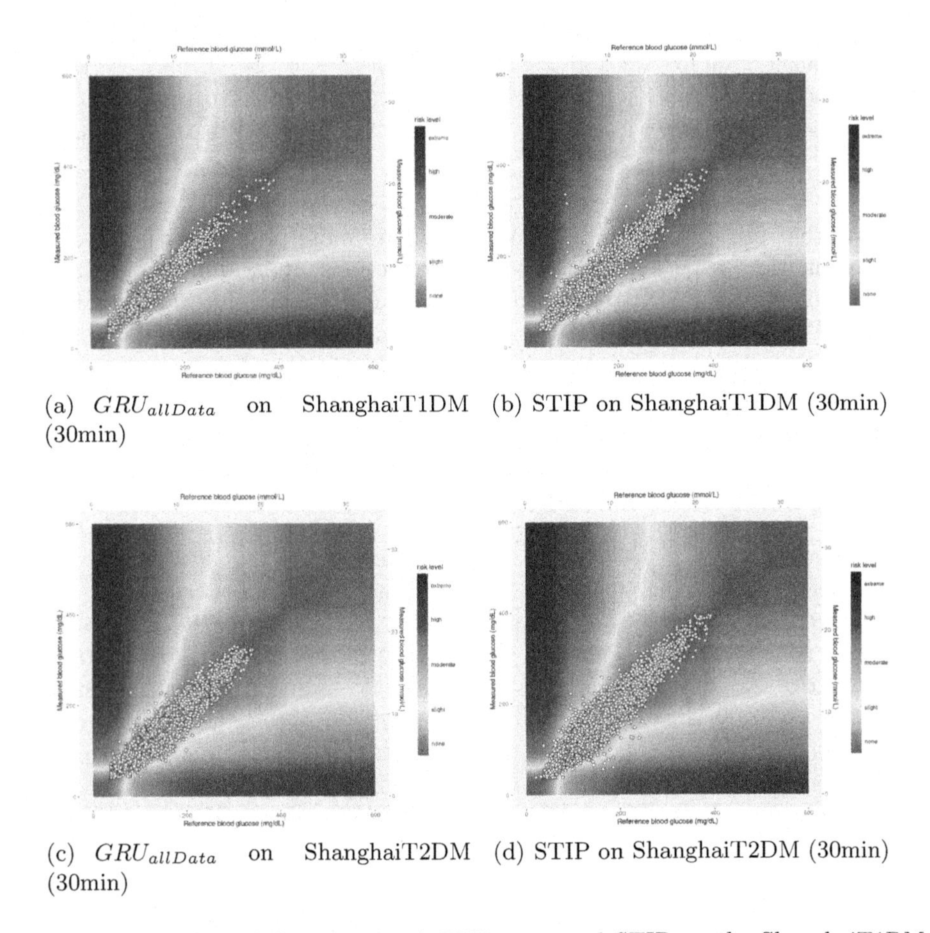

(a) $GRU_{allData}$ on ShanghaiT1DM (30min)
(b) STIP on ShanghaiT1DM (30min)

(c) $GRU_{allData}$ on ShanghaiT2DM (30min)
(d) STIP on ShanghaiT2DM (30min)

Fig. 4. Comparison of the security of $GRU_{allData}$ and STIP on the ShanghaiT1DM and ShanghaiT2DM for 30-minute prediction.

Figure 3, Fig. 4, Table 4 and Table 5 compare the security of the $GRU_{allData}$ and the STIP on the four datasets. At 30-minutes prediction, the $GRU_{allData}$ overestimates hypoglycemia as much as it underestimates hyperglycemia, and at 60-minutes prediction, the $GRU_{allData}$ underestimates hyperglycemia more. In the STIP model, the underestimation of hyperglycemia is also observed. This

Table 4. The proportion of $GRU_{allData}$ and STIP risk levels in the 30-minute prediction and 60-minute prediction on SimulatorT1DM and OhioT1DM dataset. The comparison of the RMSE and model security.

	SimulatorT1DM				OhioT1DM			
	30 min		60 min		30 min		60 min	
	$GRU_{allData}$	STIP	$GRU_{allData}$	STIP	$GRU_{allData}$	STIP	$GRU_{allData}$	STIP
RMSE	21.99	7.45	44.33	11.51	19.82	13.7	33.88	21.79
A	83.90%	95.30%	68.10%	90.10%	88.70%	90.20%	75.40%	76.60%
B	15.60%	4.70%	28.20%	9.80%	11.00%	9.40%	22.80%	22.10%
C	0.50%	0.10%	3.50%	0.10%	0.30%	0.40%	1.70%	1.20%
D	0.00%	0.00%	0.10%	0.00%	0.00%	0.00%	0.00%	0.00%
E	0.00%	0.00%	0.00%	0.00%	0.00%	0.00%	0.00%	0.00%

suggests that projections on SimulatorT1DM are more prone to hyperglycemia underestimation, especially in long-term projections. It is found from Table 5 that although the RMSE of the STIP on the 30-minute forecast of ShanghaiT2DM is lower than that of the $GRU_{allData}$, the latter is more secure than the former.

Table 5. The proportion of $GRU_{allData}$ and STIP risk levels in the 30-minute prediction and 60-minute prediction on ShanghaiT1DM and ShanghaiT2DM dataset. The comparison of the RMSE and model security.

	ShanghaiT1DM				ShanghaiT2DM			
	30 min		60 min		30 min		60 min	
	$GRU_{allData}$	STIP	$GRU_{allData}$	STIP	$GRU_{allData}$	STIP	$GRU_{allData}$	STIP
RMSE	13.18	14.56	24.95	24.12	12.02	11.12	21.45	18.38
A	88.60%	90.60%	72.60%	75.70%	94.10%	93.20%	83.70%	82.50%
B	11.20%	8.90%	24.30%	21.90%	5.80%	6.60%	15.80%	16.80%
C	0.20%	0.50%	3.00%	2.30%	0.10%	0.10%	0.40%	0.70%
D	0.00%	0.00%	0.00%	0.10%	0.00%	0.00%	0.00%	0.00%
E	0.00%	0.00%	0.00%	0.00%	0.00%	0.00%	0.00%	0.00%

It can be found that the $GRU_{allData}$ and STIP models do not show a significant tendency to underestimate or overestimate blood glucose in the 30-minute and 60-minute prediction on three real datasets. By comparing the ratios of different risk grades of the three models, we can find that, in general, on the same data set, the RMSE of the model and the security of the model show a similar trend. The low RMSE represents the high security of the model. However, when comparing the relationship between RMSE and security of models on different datasets, lower RMSE of models does not mean higher security. For example, the security of the STIP model on the ShanghaiT1DM is higher than that of the $GRU_{allData}$. However, the RMSE of the $GRU_{allData}$ is lower than that of the STIP. Similarly on the ShanghaiT2DM, the two models behave contrary. This indicates the importance of safety verification of blood glucose prediction model, and RMSE is not perfectly suitable for evaluating the effect of blood glucose prediction model.

4 Conclusion

The accurate prediction of blood glucose levels is critical for effective diabetes management, and the use of continuous glucose monitoring techniques has enabled the development of prediction models based on time series data. In this paper, we propose a novel predictor, the Seasonal Trend Integrated Predictor (STIP), which incorporates embedding and convolutions to capture both local and global patterns in the data, including seasonality and trend. The experimental results show the good performance of the STIP model compared to other prediction methods. It suggests that the STIP model can be a valuable tool for more accurately predict the blood glucose level. Further research is needed to explore the potential for incorporating additional variables and real-time feedback to improve the accuracy and usefulness of blood glucose prediction models.

Acknowledgement. This work is partially supported by National Key R&D Program of China (No. 2022YFE0208000, 2021YFE204500, 2021YFC3340601), National Natural Science Foundation of China (No. 61972286), the Shanghai Science and Technology Development Funds (No. 22410713200, 20ZR1460500), the Shanghai Municipal Science and Technology Major Project (2021SHZDZX0100), and Shanghai Key Lab of Vehicle Aerodynamics and Vehicle Thermal Management Systems, and the Fundamental Research Funds for the Central Universities.

References

1. Cho, K., et al.: Learning phrase representations using RNN encoder-decoder for statistical machine translation. arXiv preprint arXiv:1406.1078 (2014)
2. Doherty, S.T., Greaves, S.P.: Time-series analysis of continuously monitored blood glucose: the impacts of geographic and daily lifestyle factors. J. Diabetes Res. **2015**, 1–6 (2015)
3. Hidalgo, J.I., Colmenar, J.M., Kronberger, G., Winkler, S.M., Garnica, O., Lanchares, J.: Data based prediction of blood glucose concentrations using evolutionary methods. J. Med. Syst. **41**(9), 1–20 (2017)
4. Klonoff, D.C., et al.: The surveillance error grid. J. Diabetes Sci. Technol. **8**(4), 658–672 (2014)
5. Li, J., Fernando, C.: Smartphone-based personalized blood glucose prediction. ICT Express **2**(4), 150–154 (2016)
6. Li, K., Liu, C., Zhu, T., Herrero, P., Georgiou, P.: Glunet: a deep learning framework for accurate glucose forecasting. IEEE J. Biomed. Health Inform. **24**(2), 414–423 (2019)
7. M, P., S, P., A, B., A., D.G.: A comparison among three maximal mathematical models of the glucose-insulin system. PloS one **16**(9), e0257789 (2021)

8. Man, C.D., Micheletto, F., Lv, D., Breton, M., Kovatchev, B., Cobelli, C.: The uva/padova type 1 diabetes simulator: new features. J. Diabetes Sci. Technol. **8**(1), 26–34 (2014)
9. Marling, C., Bunescu, R.: The OhioT1DM dataset for blood glucose level prediction: Update 2020. In: CEUR workshop proceedings. vol. 2675, p. 71. NIH Public Access (2020)
10. Marling, C., Bunescu, R.C.: The ohiot1dm dataset for blood glucose level prediction. In: KHD@ IJCAI (2018)
11. Marín-Peñalver, J., Martín-Timón, I., Sevillano-Collantes, C., Del Cañizo-Gómez, F.: Update on the treatment of type 2 diabetes mellitus. World J. Diabetes **7**(17), 354–95 (2016)
12. Novara, C., Pour, N.M., Vincent, T., Grassi, G.: A nonlinear blind identification approach to modeling of diabetic patients. IEEE Trans. Control Syst. Technol. **24**(3), 1092–1100 (2015)
13. Oviedo, S., Vehi, J., Calm, R., Armengol, J.: A review of personalized blood glucose prediction strategies for t1dm patients. Int. J. Num. Methods Biomed. Eng. **33**(6), e2833 (2017)
14. Q. Zhao, J. Zhu, X.S.e.a.: Chinese diabetes datasets for data-driven machine learning. Sci Data **10**(35) (2023)
15. Reymann, M.P., Dorschky, E., Groh, B.H., Martindale, C., Blank, P., Eskofier, B.M.: Blood glucose level prediction based on support vector regression using mobile platforms. In: 2016 38th Annual International Conference of the IEEE Engineering in Medicine and Biology Society (EMBC), pp. 2990–2993. IEEE (2016)
16. Sun, H., et al.: IDF diabetes atlas: global, regional and country-level diabetes prevalence estimates for 2021 and projections for 2045. Diabetes Res. Clin. Pract. **183**, 109119 (2022)
17. Sutskever, I., Vinyals, O., Le, Q.V.: Sequence to sequence learning with neural networks. In: Advances in Neural Information Processing Systems 27 (2014)
18. Vfa, B., Nmga, B., Npa, B., Im, C.: Data-based algorithms and models using diabetics real data for blood glucose and hypoglycaemia prediction: a systematic literature review. Artif. Intell. Med. **118**, 102120 (2021)
19. Visentin, R., Campos-Náñez, E., Schiavon, M., Lv, D., Vettoretti, M., Breton, M., Kovatchev, B.P., Dalla Man, C., Cobelli, C.: The UVA/padova type 1 diabetes simulator goes from single meal to single day. J. Diabetes Sci. Technol. **12**(2), 273–281 (2018)
20. Wang, H., Peng, J., Huang, F., Wang, J., Chen, J., Xiao, Y.: MICN: multi-scale local and global context modeling for long-term series forecasting (2023)
21. Woldaregay, A.Z., et al.: Data-driven modeling and prediction of blood glucose dynamics: machine learning applications in type 1 diabetes. Artif. Intell. Med. **98**, 109–134 (2019)
22. Yang, J., Li, L., Shi, Y., Xie, X.: An arima model with adaptive orders for predicting blood glucose concentrations and hypoglycemia. IEEE J. Biomed. Health Inform. **23**(3), 1251–1260 (2018)
23. Yang, T., et al.: Multi-scale long short-term memory network with multi-lag structure for blood glucose prediction. In: KDH@ ECAI, pp. 136–140 (2020)
24. Zaidi, S.M.A., Chandola, V., Ibrahim, M., Romanski, B., Mastrandrea, L.D., Singh, T.: Multi-step ahead predictive model for blood glucose concentrations of type-1 diabetic patients. Sci. Rep. **11**(1), 24332 (2021)
25. Zhou, T., Ma, Z., Wen, Q., Xue Wang, L.S., Jin, R.: Fedformer: frequency enhanced decomposed transformer for long-term series forecastings. In: International Conference on Machine Learning (2022)

Window-Controlled Sepsis Prediction Using a Model Selection Approach

Shiyan Su[1], Su Lan[1], Zhicheng Zhang[1], and Anjie Zhu[2](✉)

[1] The University of Queensland, 4072 Brisbane, QLD, Australia
[2] University of Electronic Science and Technology of China, Sichuan, China
anjie.zhu@uq.edu.au

Abstract. Sepsis is a major risk to patient in the Intensive Care Unit (ICU) and is associated with substantial treatment expenditure. As most cases of sepsis are acquired during the ICU stay, timely identification and intervention play a crucial role in enhancing the survival rate of septic patients and reducing the financial burden of treatment. Prior research has established that machine learning approaches surpass conventional scoring systems in predicting sepsis. However, these existing machine learning methodologies exhibit limitations when predicting sepsis with flexible window settings. Their performance is heavily reliant on the selection of prediction and feature windows, which restricts their practical applicability in clinical settings. This paper aims to overcome this challenge by introducing a model selection approach for sepsis prediction, utilizing a window-controlled strategy. Experimental results demonstrate that our proposed model outperforms existing models and exhibits enhanced stability across various prediction and feature windows.

Keywords: Sepsis prediction · Window-Controlled strategy · Model selection approach · Neural Architecture Search · Machine learning

1 Introduction

Sepsis, a life-threatening organ dysfunction caused by a dysregulated host response to infection, is among the leading causes of mortality in the Intensive Care Unit (ICU) and is significantly costly for treatment [1,21,28]. Previous research suggests that treatment delays are associated with increased mortality and the optimal time window for initiating antimicrobial therapy is within one hour of sepsis onset [19,25]. Consequently, the ability to predict sepsis onset before any clinical manifestations can offer valuable warning signals for clinicians, guiding them to make informed decisions regarding the timely commencement of appropriate treatment.

Existing approaches for sepsis prediction can be primarily classified into two categories: conventional scoring systems and machine learning models. The conventional scoring systems generally employ a set of clinical and biochemical criteria, with representative examples including Systemic Inflammatory Response

X. Yang et al. (Eds.): ADMA 2023, LNAI 14180, pp. 451–465, 2023.
https://doi.org/10.1007/978-3-031-46677-9_31

Syndrome (SIRS), Sequential Organ Failure Assessment (SOFA), quick Sequential Organ Failure Assessment (qSOFA) and Modified Early Warning System (MEWS) [29]. It is worth noting that these scoring systems have laid the groundwork for identifying septic patients and their criteria have been further utilized for feature selection in machine learning prediction models. However, previous research has demonstrated that existing machine learning models consistently outperform conventional scoring systems in the early recognition of septic and non-septic patients [13].

Current machine learning approaches are typically designed to address two types of sepsis prediction tasks: the visit level early diagnosis (Left-aligned) and the event level early prediction (Right-aligned) (see Fig. 1) [11,15]. Left-aligned models utilize a fixed period as the feature window, aiming to recognize whether patients will develop sepsis at any point within this predetermined window. In contrast, right-aligned models partition the duration of a patient's ICU stay before sepsis onset into two consecutive time windows, namely the feature window and prediction window. By utilizing data within the feature window, these models aim to predict whether or not the given patients will develop sepsis following the prediction window. In the clinical context, right-align models are more valuable as they can provide early warning signals to clinicians before the actual onset of sepsis, thereby prompting timely treatment interventions [11]. As a result, our proposed approach seeks to enhance the performance of sepsis prediction under the concept of right-aligned models.

As previously discussed, while current machine learning approaches consistently surpass traditional scoring systems in predicting sepsis, right-aligned models still have some limitations. The majority of existing right-aligned models are based on neural network models (NNMs) including recurrent neural networks (RNNs), long short-term memory (LSTM) and gated recurrent unit (GRU) [22]. Since all these approaches use an NNM with fixed neural architectures to handle various settings of prediction windows and feature windows, their prediction performance is subject to variance when faced with different window settings. As shown in Fig. 2, it has been observed that as the length of the feature window decreases and the length of the prediction window increases, the prediction performance of these models tends to worsen. Nevertheless, in clinical settings, it is crucial for right-aligned models to maintain precise prediction performance consistently across various window settings, given that a significant number of sepsis cases develop during the early stages of ICU stay [26]. In addition, from clinicians' perspective, the length of the feature window should be determined for each patient individually based on the clinical manifestations and symptoms to allow more personalised diagnosis [10]. However, existing models are unable to provide that kind of flexibility in practical applications as they cannot guarantee the consistency of the model performance under different window settings.

To address the above challenges, we proposed a model selection approach by utilizing the neural architecture search method to achieve window-controlled sepsis prediction. This method is classified as the right-aligned model, which takes time-series data from each patient within the feature window as input

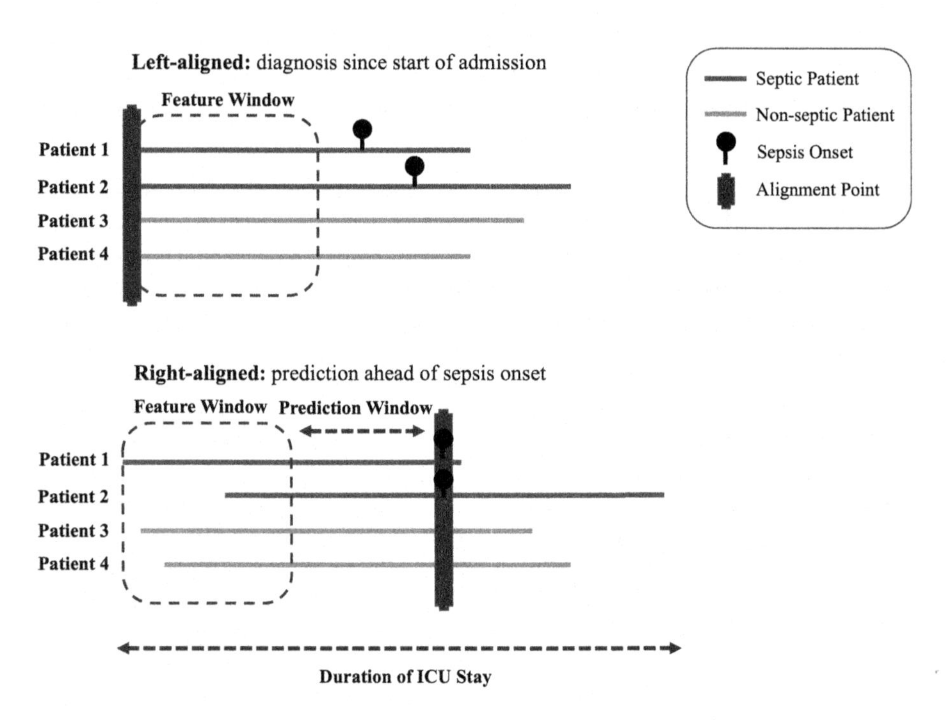

Fig. 1. Left-aligned vs. Right-aligned. Patients' data during the feature window is utilized for performing prediction.

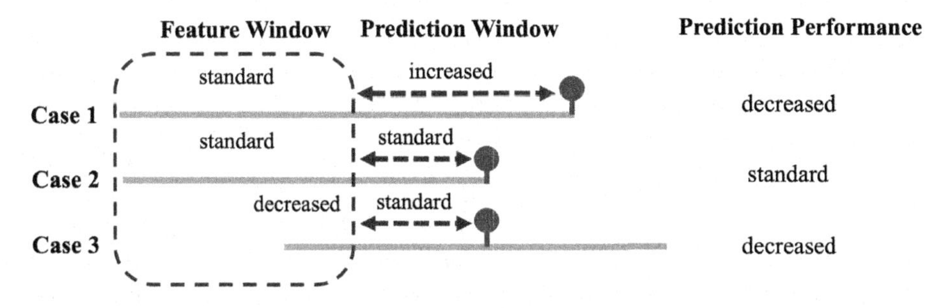

Fig. 2. Prediction performance comparison of existing right-aligned models on various window settings.

and generates classification results by the selected model which has optimal neural architectures to predict whether the patient will develop sepsis after the prediction window. We utilize the Medical Information Mart for Intensive Care III (MIMIC-III) database for the development and evaluation of our model, comparing its performance to other baseline models. The main contributions of our work can be summarized as follows:

1. We propose a model selection approach for achieving window-controlled sepsis prediction. This model selection approach is achieved by utilizing the neural architecture search, a method that has seen sparse application in the realm of sepsis prediction. This approach not only enhances the stability of prediction performance across a diverse range of prediction window and feature window settings compared to other right-aligned models but also surpasses other right-aligned models under the same window settings.
2. We introduce a novel concept for right-aligned models, known as window-controlled sepsis prediction. By incorporating this concept, the right-aligned model becomes more applicable in clinical settings as it can be applied to all patients in ICU, irrespective of their admission duration. Additionally, this approach allows clinicians to leverage their domain knowledge when utilizing the prediction model by flexibly adjusting window settings. This added flexibility ensures that the model is better suited to individual patient needs and can further improve sepsis prediction outcomes in diverse clinical scenarios.

2 Related Work

2.1 Conventional Scoring Systems

Over the past few decades, the definition of sepsis has evolved several times. In accordance with these evolving definitions, multiple conventional scoring systems have been introduced as evaluation criteria. Sepsis (Sepsis-1) has been initially defined and categorized into various severity levels including severe sepsis and septic shock in 1991 [3]. Subsequently, the Systemic Inflammatory Response Syndrome (SIRS) criteria have been established to classify Sepsis-1. In 2001, Sepsis-2 has been introduced where novel signs and symptoms of sepsis have been expanded based on the previous definition. Despite these advancements, the SIRS criteria have still been used for diagnosing Sepsis-2 [16]. However, subsequent research on sepsis diagnosis reveals that SIRS lacked specificity and exhibited excessive sensitivity when identifying sepsis cases [4]. In response to growing criticism of the low specificity of the SIRS criteria, an update of the sepsis definition and criteria has been conducted in 2016. The latest definition of sepsis (Sepsis-3) describes it as a life-threatening organ dysfunction caused by a dysregulated host response to infection [28]. Concurrently, the Sequential Organ Failure Assessment (SOFA) has been introduced, which includes a set of clinical and biochemical criteria for diagnosing Sepsis-3 [20]. In addition, a modified version of the SOFA called quick SOFA (qSOFA) has been proposed in 2016 to facilitate easier identification of potential sepsis risks in patients [20].

In contrast to SIRS criteria, SOFA and qSOFA have demonstrated improved specificity but poor sensitivity which would result in many sepsis cases being unnoticed [7].

The Modified Early Warning Score (MEWS) is a bedside early warning scoring system based on physiological parameters and it is not disease-specific [6]. Previous studies have also demonstrated the capability of MEWS for identifying sepsis [29].

Detailed information regarding the criteria of these scoring systems is provided in Table 1. Although the prediction performance for these conventional scoring systems has been proven that is worse than existing machine learning-based approaches, they establish a benchmark for sepsis prediction and provide valuable insights for feature selection.

2.2 Right-Aligned Models

For right-aligned models, modelling time-series patients' data is the most significant task. As previous research has demonstrated that NNMs generally outperform other approaches in predicting clinical events [9,18], the majority of existing right-aligned models for sepsis prediction are implemented by using NNMs [22]. A recurrent neural network (RNN) based approach which uses the gated recurrent unit (GRU) as the hidden layer architecture has been proposed [27]. This approach exhibits superior performance compared to other related works using NNMs. However, the performance of this approach is highly reliant on an elongated look back (feature window) and it shows a noticeable decrease with increasing prediction window and decreasing feature window. LSTM is a type of RNN which is more effective at accessing long-term dependencies than basic RNNs [12]. There are several existing approaches which are implemented based on LSTM [2,14,17]. Compared to the basic LSTM approach, the Convolutional-LSTM based approach and bidirectional LSTM based approach show improved prediction performance [2,17]. However, the fixed structure of these proposed models leads to unstable prediction performance for various feature windows and prediction windows.

2.3 Neural Architecture Search

Neural Architecture Search (NAS) is the process of automating architecture engineering which aims to output a neural architecture that can obtain the best performance with the minimum human intervention [24]. Previous research has demonstrated that NAS methods have achieved better performance than manually designed neural architectures, including image classification, object detection and semantic segmentation [8]. However, the application of NAS is currently limited in the field of disease prediction, particularly in the context of sepsis prediction.

3 Proposed NAS Method

In this section, we introduce a model selection approach by using NAS. The framework of this proposed NAS model is illustrated in Fig. 3. A NAS method can be categorized into three dimensions that are search space, search strategy and performance estimation strategy (see Fig. 4) [8]. By testing and evaluating architectures across the search space using a specific search strategy, our method is able to output a model with optimized architectures automatically based on the performance estimation strategy in response to various window settings. Therefore, the limitations of existing right-aligned models we mentioned in Sect. 1 can be addressed. The detailed implementation of three dimensions will be demonstrated in the following subsections.

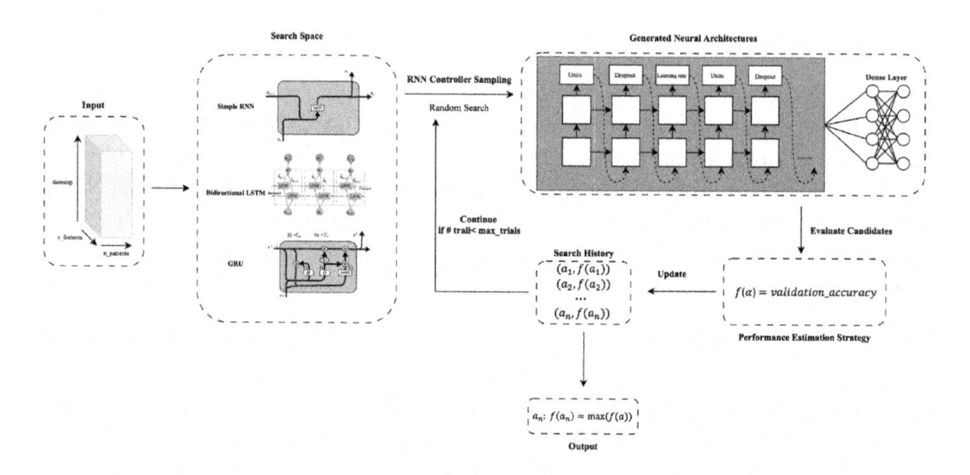

Fig. 3. The framework of our proposed NAS method.

3.1 Search Space

The search space provides all the necessary components of the optimal neural architecture that our method aims to generate. Based on prior knowledge, we realized that certain RNNs have demonstrated exceptional performance in sepsis prediction including simple RNN, bidirectional-LSTM and GRU. Therefore, we introduced an RNN controller consisting of a simple RNN layer, a bidirectional-LSTM layer and a GRU layer in the search space. To reduce the size of the search space, the value range of units, dropout rate and learning rate are defined based on experiment results on each layer individually. This search space is constructed as a custom hypermodel and the high-fidelity structure of this hypermodel is shown in Fig. 3.

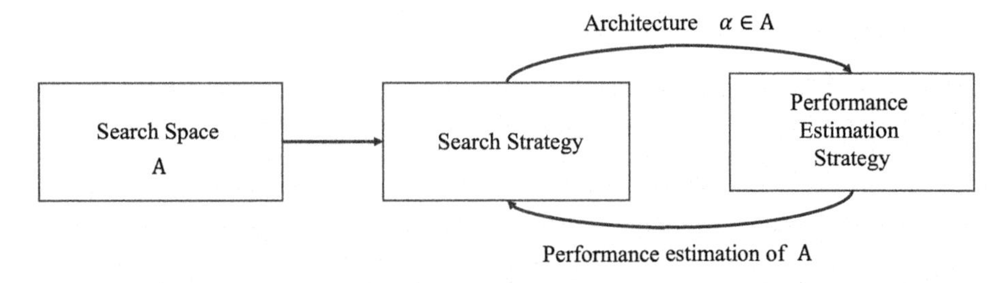

Fig. 4. Abstract illustration for 3 dimensions of Neural Architecture Search methods.

3.2 Search Strategy

The search strategy for the NAS model aims to find an architecture which obtains optimal performance by exploring the search space. Given that the size of the search space for our model is relatively small and computational resources are generally limited in clinical settings, we have employed random search as the search strategy for our model. Throughout the search process, the model will randomly sample different combinations of hyperparameters for each architecture and train the architecture for each combination. This approach allows for efficient exploration of the search space while remaining computationally feasible in the context of clinical settings.

3.3 Performance Estimation Strategy

In our study, we have a relatively large dataset with balanced classes, making the holdout validation approach a suitable performance estimation strategy for the NAS model. In addition, this approach requires less computation compared to other strategies such as k-fold cross-validation and leave-one-out validation, making it more applicable in clinical practice. During the search process, the model trains each neural architecture selected from the predefined search space and estimates its performance by evaluating it on a separate validation set, which is created by splitting a portion of the training data. We utilized the validation accuracy as the objective function for this performance estimation strategy to determine how well each architecture generalizes to new data and to select the optimal architecture accordingly.

4 Experiments and Results

In this section, we present comprehensive information regarding the experiment settings and evaluation results of our proposed model compared with other baseline methods. This includes details on dataset preparation, model training, and performance metrics used for evaluation.

4.1 Settings

Datasets. We utilized the Medical Information Mart for Intensive Care III (MIMIC-III) database for training and evaluating our model. The version of MIMIC-III database employed in our study is 1.4, which contains 58,976 admissions for 46,520 patients in critical care units of the Beth Israel Deaconess Medical Center between 2001 and 2012. The MIMIC-III employs the 9th edition of the International Classification of Disease (ICD-9) for the disease diagnosis where sepsis is defined by codes 995.91, 995.92 and 788.52. Considering the specificity of sepsis in the juvenile population, we extracted data for all adult patients who were diagnosed with sepsis during their first admissions. To maintain a class-balanced dataset, we then randomly selected an equal number of adult patients who were not diagnosed with sepsis during their first admissions. Figure 5 illustrates the process of preparing the final data collection for experiments.

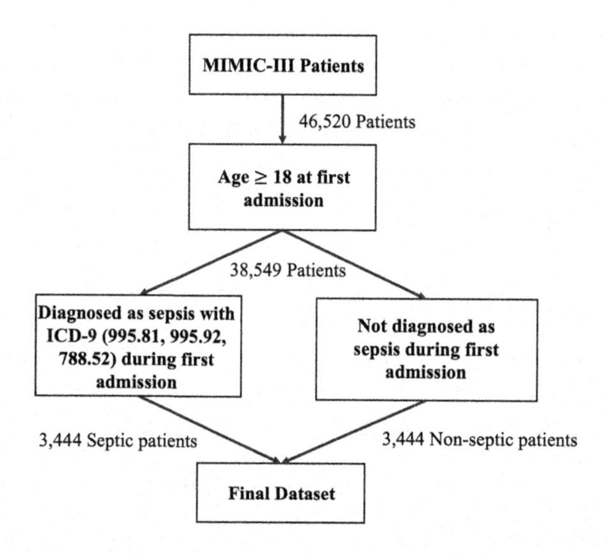

Fig. 5. Process of patient extraction.

Definition of Sepsis Onset. As the time of diagnosis for a disease is not explicitly defined in MIMIC-III, identifying the onset time of sepsis for extracted septic patients is a critical task for right-aligned models. All state-of-the-art methods, as well as our proposed method, utilize the gold standard proposed by [5] to define the onset time of sepsis. The gold standard consists of two criteria

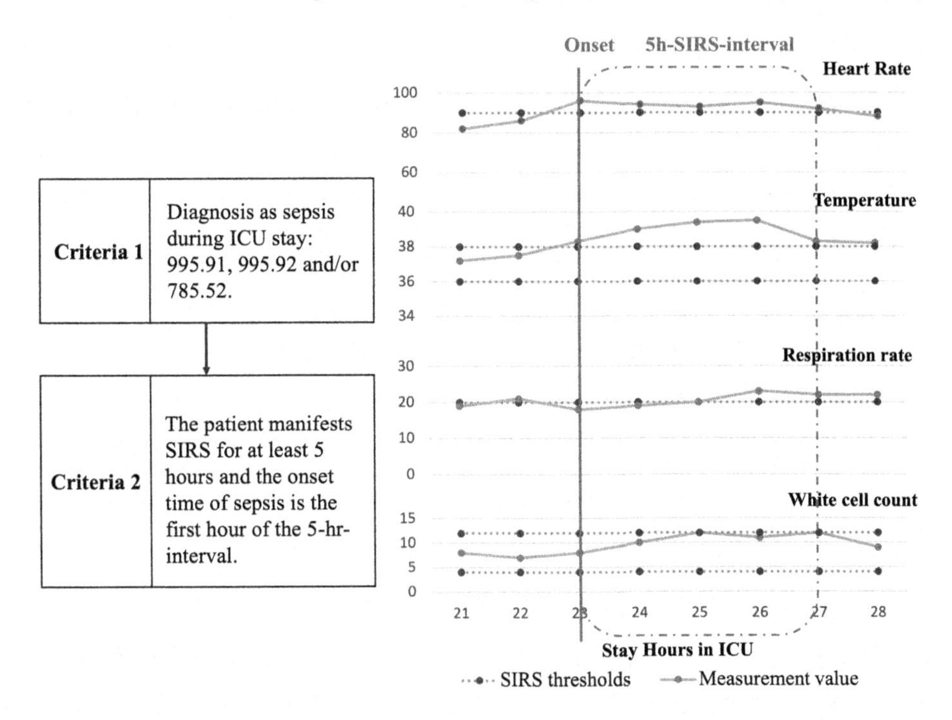

Fig. 6. Criteria of the gold standard (left) and the process of defining sepsis onset for a sepsis patient extracted from MIMIC-III database by using the gold standard (right).

as shown in Fig. 6. It designates the onset time of sepsis as the beginning of 5-h-SIRS-Interval. Figure 6 also illustrates the process of defining the sepsis onset time by employing the gold standard for a specific patient extracted from the MIMIC-III dataset, who was diagnosed with sepsis with ICD-9 995.91.

Evaluation Metrics. To evaluate the prediction performance of our proposed method in comparison to other baseline models, we utilized the Area Under the Receiver Operating Characteristic (AUROC) curve as it is widely used for right-aligned models. The AUROC is a plot of the true positive rate (sensitivity) against the false positive rate (1 - specificity) at various decision thresholds. It ranges from 0 to 1 with higher values indicating better classification performance. Since the optimal decision threshold for sepsis prediction is unknown, the AUROC is more suitable to use than F1-score as it is a threshold-independent metric which measures the overall ability of the classifier to distinguish between positive and negative instances. To evaluate the stability of prediction performance across various window settings, the variance of the AUROC is measured. The model with lower variance is considered more stable as it exhibits consistent performance across various time windows.

4.2 Baselines

To evaluate the performance of our proposed method, we compare it with several state-of-the-art models.

1. **RNN-GRU Approach** [27]: This approach implements an RNN comprising two hidden layers with 40 neurons each, utilizing the GRU as the hidden layer architecture. It demonstrates improved performance in sepsis prediction compared to existing approaches that use simple RNNs.
2. **Bidirectional LSTM Approach** [2]: This method introduces a modified Bidirectional LSTM model to overcome the limitation of LSTM that is only capable of using inputs it has seen from the past.

The aforementioned approaches serve as suitable representatives for right-aligned models by using NNMs, which were evaluated across various window settings and exhibited strong results for multiple time windows. Both of these models were trained and evaluated by using the MIMIC-III database.

4.3 Implementation Details

Feature Extraction. The features we extracted for preparing the final dataset are mainly based on conventional scoring systems including SIRS, SOFA, qSOFA

Table 1. Extracted Features compared with features of SIRS, SOFA, qSOFA and MEWS where 'o' represents the feature is selected.

	SIRS	SOFA	qSOFA	MEWS	Extracted Features
Age					o
Temperature	o			o	o
Heart rate	o			o	o
Respiratory rate	o	o	o		o
$PaCO_2$	o				o
White Cell Count	o				o
PaO_2		o			o
FiO_2		o			
Platelets		o			
Glasgow Coma Scale		o	o		o
Bilirubin		o			
Creatinine		o			
Mean Arterial Pressure		o			
Vasopressors		o			
Systolic Blood Pressure			o	o	o
AVPU Score				o	

and MEWS. Features with more than 80% missing values were eliminated at this stage. Additionally, we extracted several demographic features based on other baseline methods to evaluate the results under consistent experimental settings. Table 1 presents all the extracted features.

Data Pre-processing. For data pre-processing, we mainly conducted four steps. The initial step involved partitioning records of all measurements for each patient into consecutive 1-hour intervals. Within a given time interval, we utilized the maximum value for each type of measurement. The second step focused on imputing missing values. We hypothesized that if a vital sign had not been measured during a specific time period in clinical settings, it would likely remain unchanged since the last measurement. As a result, we first imputed missing values by carrying forward to locate the nearest non-missing values. Then, for any remaining missing values, we applied backward filling. After imputing all missing values, we determined the onset time for each sepsis patient by using the gold standard, as outlined in Sect. 4.1. Finally, we organized the data for each patient according to the feature window and prediction window settings to accommodate the input requirements of our model.

Model Implementation. Our source code is implemented based on TensorFlow. The framework of our model consists of three dimensions which are the search space, search strategy and performance estimation strategy. This model takes input which is sequences of feature vectors for patients and then outputs a trained neural network with the best architecture found during the search process. The shape of input is (num_patients, time_steps, num_features) where the time_steps represents the length of the feature window. The search space of our model comprises three recurrent layers, including Bidirectional LSTM, SimpleRNN, and GRU. Additionally, we incorporated a dense layer into the search space, enabling the searched neural network to perform classification tasks. Softmax is used as the activation function for this dense layer. For the search strategy, the random search is implemented. Lastly, as the performance estimation strategy, we utilized validation accuracy based on the holdout validation approach.

Parameter Settings. The parameters we set for three dimensions of our NAS model are as follows:

1. **Search Space:** For each layer type, the number of hidden units is a tunable parameter specified by the units hyperparameter, with a minimum value of 32 and a maximum value of 128. The dropout rate for the layers is specified using the dropout_rate hyperparameter, which spans from 0.0 to 0.5 with a step of 0.1. The learning rate for the Adam optimizer is specified by the learning_rate hyperparameter. It varies between 1e-4 and 1e-2, and is sampled on a logarithmic scale.
2. **Search Strategy:** The random search tuner is configured to execute a maximum of 50 trials, with the primary objective of maximizing validation accuracy. To guarantee reproducibility, the random seed is set to 42.

3. **Performance Estimation Strategy:** The hyperparameter validation_split is set to 0.1 which indicates that 10% of the training data will be held out for validation purposes. In addition, the number of training epochs is set to 50.

4.4 Comparisons with State-of-The-Art

In this section, we conduct an assessment of the reliability and stability of the prediction performance for our proposed method in comparison to other baseline methods. We initiate our comparison with the Bidirectional LSTM model as shown by Table 2, aiming to examine the model's performance under the identical feature window settings (6 h) while modulating the prediction windows (from 1 to 6 h). This exercise aims to discern the impact of extending prediction windows on the accuracy and reliability of the prediction performance.

We further evaluate our model with the RNN-GRU approach under an extensive array of window configurations (see Table 3). For this comparative analysis, three distinct settings are established for both the prediction window and the feature window, yielding nine unique window configurations.

To ensure that the experimental variables are consistent, we utilize the same patients' measurement features with each baseline model for conducting the experiment. The AUROC score is calculated based on 5-fold cross validation to provide a more accurate measure of models' performance.

Reliability Evaluation. In this section, we evaluate the performance of our model compared other two baseline methods under each identical window configuration, with the objective of demonstrating the superior reliability of our model. As depicted in Table 2, our model outperforms the Bidirectional LSTM model across all window settings, with particularly noteworthy dominance in the last two test scenarios where the prediction window is significantly elongated. Furthermore, when evaluated against the RNN-GRU methodology (see Table 3), our model presents impressive dominance across all given window configurations, further solidifying the reliability of our approach.

Stability Evaluation. To demonstrate the stability of our proposed model, we compare the variance of the AUROC values with other baseline methods across identical window settings. The results, presented in Table 2 and Table 3, indicate that our model obtains enhanced stability for predicting under various window configurations.

Subsequently, we delve into the model's performance across certain specific window configurations. As evidenced in the relevant table, a situation where the feature window remains static while the prediction window is expanded elicits a substantial downturn in the performance of the Bidirectional LSTM model. This marked degradation in performance is the primary factor contributing to the large variance in the AUROC values observed across the test cases for this particular model. Furthermore, for the RNN-GRU method, when the prediction window is amplified in conjunction with a diminution in the feature window, its performance also manifests a decline leading to large variance.

Table 2. Comparisons of AUROC results for various window settings and the variance between bidirectional LSTM method with the proposed method. Values in boldface indicate better performances.

Methods	Window Settings (prediction, feature window)						Variance
	1hr, 6hr	2hr, 6hr	3hr, 6hr	4hr, 6hr	5hr, 6hr	6hr, 6hr	
Bidirectional LSTM	91.7	90.5	89.2	89.8	87.1	85.7	4.944
Proposed NAS	**92.7**	**91.8**	**91.7**	**91.9**	**92.0**	**92.2**	**0.131**

Table 3. Comparisons of AUROC results for various window settings (in hour) and the variance between RNN-GRU method with the proposed method. Values in boldface indicate better performances.

Methods	Window Settings (prediction, feature window)									Variance
	3, 5	3, 10	3, 15	6, 5	6, 10	6, 15	12, 5	12, 10	12, 15	
RNN-GRU	71.2	74.3	78.7	71.9	74.1	78.1	71.8	73.1	76.8	7.870
Proposed NAS	**91.8**	**90.9**	**91.2**	**90.5**	**92.2**	**91.9**	**90.4**	**90.1**	**89.7**	**0.755**

5 Conclusion

In this work, we introduced a NAS method for performing window-controlled sepsis prediction, a more practically applicable right-aligned prediction task that is overlooked by existing approaches. This model selection approach addresses the challenges encountered by previous right-aligned models, which were unable to consistently maintain accurate prediction performance across various window settings. Compared to existing right-aligned models, our proposed model exhibits enhanced stability in predicting sepsis across diverse window settings and surpasses their performance under the same window configurations. Therefore, our method offers increased reliability and encourages clinicians to apply it to patients with window settings determined based on each patient's unique situation, ultimately enhancing the potential for timely and effective sepsis detection and intervention.

It is reasonable to point out that our proposed approach substantially contributes to disease prediction via the application of machine learning models, particularly in the context of disease prediction with variable window sizes. Nevertheless, despite the considerable merits of this approach, it confronts some limitations such as the extensive computational costs associated with our NAS model. This may impede the efficiency and practicality of this proposed NAS model in real-world applications. For future work, alternative neural architectures with the potential to reduce computational cost and improve performance can be considered such as the transfer learning with transformer, where time series can be treated as language tokens [23].

References

1. Angus, D.C., Linde-Zwirble, W.T., Lidicker, J., Clermont, G., Carcillo, J., Pinsky, M.R.: Epidemiology of severe sepsis in the united states: analysis of incidence, outcome, and associated costs of care. Crit. Care Med. **29**(7), 1303–1310 (2001)
2. Baral, S., Alsadoon, A., Prasad, P., Al Aloussi, S., Alsadoon, O.H.: A novel solution of using deep learning for early prediction cardiac arrest in sepsis patient: enhanced bidirectional long short-term memory (lstm). Multimedia Tools Appl. **80**, 32639–32664 (2021)
3. Bone, R.C., et al.: Definitions for sepsis and organ failure and guidelines for the use of innovative therapies in sepsis. Chest **101**(6), 1644–1655 (1992)
4. Brink, A., et al.: Predicting mortality in patients with suspected sepsis at the emergency department; a retrospective cohort study comparing qsofa, sirs and national early warning score. PLoS ONE **14**(1), e0211133 (2019)
5. Calvert, J.S., et al.: A computational approach to early sepsis detection. Comput. Biol. Med. **74**, 69–73 (2016)
6. Çıldır, E., Bulut, M., Akalın, H., Kocabaş, E., Ocakoğlu, G., Aydın, ŞA.: Evaluation of the modified meds, mews score and charlson comorbidity index in patients with community acquired sepsis in the emergency department. Intern. Emerg. Med. **8**, 255–260 (2013)
7. Dykes, L.A., Heintz, S.J., Heintz, B.H., Livorsi, D.J., Egge, J.A., Lund, B.C.: Contrasting qSOFA and sirs criteria for early sepsis identification in a veteran population. Fed. Pract. **36**(Suppl 2), S21 (2019)
8. Elsken, T., Metzen, J.H., Hutter, F.: Neural architecture search: a survey. J. Mach. Learn. Res. **20**(1), 1997–2017 (2019)
9. Esteban, C., Staeck, O., Baier, S., Yang, Y., Tresp, V.: Predicting clinical events by combining static and dynamic information using recurrent neural networks. In: 2016 IEEE International Conference on Healthcare Informatics (ICHI), pp. 93–101. IEEE (2016)
10. Evans, T.: Diagnosis and management of sepsis. Clin. Med. **18**(2), 146 (2018)
11. Fleuren, L.M., et al.: Machine learning for the prediction of sepsis: a systematic review and meta-analysis of diagnostic test accuracy. Intensive Care Med. **46**, 383–400 (2020)
12. Hochreiter, S., Schmidhuber, J.: Long short-term memory. Neural Comput. **9**(8), 1735–1780 (1997)
13. Islam, M.M., Nasrin, T., Walther, B.A., Wu, C.C., Yang, H.C., Li, Y.C.: Prediction of sepsis patients using machine learning approach: a meta-analysis. Comput. Methods Programs Biomed. **170**, 1–9 (2019)
14. Kaji, D.A., Zech, J.R., Kim, J.S., Cho, S.K., Dangayach, N.S., Costa, A.B., Oermann, E.K.: An attention based deep learning model of clinical events in the intensive care unit. PLoS ONE **14**(2), e0211057 (2019)
15. Khoshnevisan, F., Ivy, J., Capan, M., Arnold, R., Huddleston, J., Chi, M.: Recent temporal pattern mining for septic shock early prediction. In: 2018 IEEE International Conference on Healthcare Informatics (ICHI), pp. 229–240. IEEE (2018)
16. Levy, M.M., et al.: 2001 SCCM/ESICM/ACCP/ATS/sis international sepsis definitions conference. Intensive Care Med. **29**, 530–538 (2003)
17. Lin, C., et al.: Early diagnosis and prediction of sepsis shock by combining static and dynamic information using convolutional-LSTM. In: 2018 IEEE International Conference on Healthcare Informatics (ICHI), pp. 219–228. IEEE (2018)

18. Lipton, Z.C., Kale, D.C., Elkan, C., Wetzel, R.: Learning to diagnose with LSTM recurrent neural networks. arXiv preprint arXiv:1511.03677 (2015)
19. Liu, V.X., et al.: The timing of early antibiotics and hospital mortality in sepsis. Am. J. Respir. Crit. Care Med. **196**(7), 856–863 (2017)
20. Marik, P.E., Taeb, A.M.: Sirs, qSOFA and new sepsis definition. J. Thorac. Dis. **9**(4), 943 (2017)
21. Martin, G.S., Mannino, D.M., Eaton, S., Moss, M.: The epidemiology of sepsis in the united states from 1979 through 2000. N. Engl. J. Med. **348**(16), 1546–1554 (2003)
22. Moor, M., Rieck, B., Horn, M., Jutzeler, C.R., Borgwardt, K.: Early prediction of sepsis in the ICU using machine learning: a systematic review. Front. Med. **8**, 607952 (2021)
23. Raffel, C., et al.: Exploring the limits of transfer learning with a unified text-to-text transformer. J. Mach. Learn. Res. **21**(1), 5485–5551 (2020)
24. Ren, P., et al.: A comprehensive survey of neural architecture search: challenges and solutions. ACM Comput. Surv. (CSUR) **54**(4), 1–34 (2021)
25. Rhodes, A., et al.: Surviving sepsis campaign: international guidelines for management of sepsis and septic shock: 2016. Intensive Care Med. **43**, 304–377 (2017)
26. Sakr, Y., et al.: Sepsis in intensive care unit patients: worldwide data from the intensive care over nations audit. In: Open Forum Infectious Diseases, vol. 5, p. ofy313. Oxford University Press US (2018)
27. Scherpf, M., Gräßer, F., Malberg, H., Zaunseder, S.: Predicting sepsis with a recurrent neural network using the mimic iii database. Comput. Biol. Med. **113**, 103395 (2019)
28. Singer, M., et al.: The third international consensus definitions for sepsis and septic shock (sepsis-3). JAMA **315**(8), 801–810 (2016)
29. Van der Woude, S., Van Doormaal, F., Hutten, B., Nellen, F., Holleman, F.: Classifying sepsis patients in the emergency department using sirs, qSOFA or mews. Neth. J. Med. **76**(4), 158–166 (2018)

An Effective Pre-trained Visual Encoder for Medical Visual Question Answering

Yefan Huang, Xiaoli Wang(✉), and Jinsong Su

School of Informatics, Xiamen University, Xiamen, China
yefanhuang@stu.xmu.edu.cn, {xlwang,jssu}@xmu.edu.cn

Abstract. Medical Visual Question Answering (Med-VQA) is a domain-specific task that answers a given clinical question regarding a radiology image. It requires sufficient prior medical knowledge, resulting in additional challenges compared to general VQA tasks. However, the lack of well-annotated large-scale datasets makes it hard to learn sufficient medical knowledge for Med-VQA. To address the challenge, this paper employs a large-scale medical multi-modal dataset to pre-train and fine-tune an effective model, denoted by ROCOGLoRIA. The model can locate semantic-rich regions implied in medical texts and extract local semantic-focusing visual features from the image. We propose to combine the global visual features with the weighted local visual features, for capturing fine-grained semantics in the image. We further incorporate ROCOGLoRIA as the visual encoder into baselines, to investigate whether it benefits Med-VQA. We conduct extensive experiments on three benchmark datasets and the results show that the method using ROCOGLoRIA as a pre-trained visual encoder outperforms strong baselines in the overall accuracy.

Keywords: Medical visual question answering · pre-trained visual encoder · fine-tuning · transfer learning

1 Introduction

Medical visual question answering (Med-VQA) aims to answer a given question based on a medical image. It has become a hot research topic in computer vision and natural language processing, which processes multi-modal information of visual images and textual language. Med-VQA task has great potential to benefit medical practice. It may aid doctors in interpreting medical images to obtain more accurate diagnoses with responses to closed-ended questions or help patients with urgent needs get timely feedback on open-ended questions raised. Different from general VQA, Med-VQA requires substantial prior domain-specific knowledge to thoroughly understand the contents and semantics of medical visual questions.

Inspired by general VQA, Med-VQA systems have witnessed great successes in recent years while still hindered by challenges [9]. These systems generally need

X. Yang et al. (Eds.): ADMA 2023, LNAI 14180, pp. 466–481, 2023.
https://doi.org/10.1007/978-3-031-46677-9_32

to be trained on well-annotated large-scale datasets to learn enough domain-specific knowledge for understanding medical visual questions. Although benchmark datasets [10,13,19,23] have been published, such as VQA-RAD, SLAKE, and PathVQA, we still suffer from the problem of data limitation on Med-VQA [7]. To tackle the problem, a native solution is data augmentation. VQAMix [9] focused on generating new Med-VQA training samples. However, it may incur noisy samples that affect the performance of models [9]. Recent studies adopted deep encoders to interpret the image and question for predicting an answer. They typically contain four main components: visual feature extraction, textual feature extraction, multi-modal fusion, and answer prediction [22]. Several studies adopted meta-learning to pre-train a visual encoder [6,24]. However, they did not consider the impact of linguistic representation on Med-VQA [9], or pre-trained the visual encoder for specific body regions, limiting the generalization ability to other settings [7]. Current studies adopt transfer learning to pre-train a visual encoder on external medical image-text pairs to capture suitable visual representations for subsequent cross-modal reasoning [4,7,9,20]. These approaches have been particularly successful by performing pre-training using large-scale medical image-text pairs without additional manual annotations. Following such studies, we investigate to what extent using external medical multi-modal datasets without any manual annotation can contribute to Med-VQA.

In this paper, we propose an effective and generic model, denoted by ROCOGLoRIA, to extract useful visual features for Med-VQA. We employ a large-scale medical multi-modal dataset, named by ROCO [26], to train ROCOGLoRIA. ROCO includes a diverse range of body organs and image modalities. Especially, we perform multi-task learning to pre-train an effective visual encoder by fine-tuning the GLoRIA model [12], which has been pre-trained on a medical dataset. We first fine-tune it using ROCO and output semantic-oriented visual features, which are then used as the input of the R2Gen [2] model to generate a medical report about the medical image. We consider both losses of GLoRIA and R2Gen to pre-train the final visual encoder and output both global and local weighted visual features in the medical image. These features are combined and fused with question features to predict a final answer. We further investigate the performance gains when incorporating our ROCOGLoRIA as a visual encoder into state-of-the-art Med-VQA methods. Overall, our contributions are summarized as follows:

- We propose an effective model to perform multi-task learning for pre-training an effective visual encoder for Med-VQA. The model is trained using a large-scale medical multi-modal dataset without additional manual annotations, which can overcome the data limitation problem for Med-VQA.
- Different from previous visual encoders used in Med-VQA, ROCOGLoRIA is pre-trained using medical images from a diverse range of body regions, which can achieve better generalization ability to a wide range of Med-VQA datasets.
- ROCOGLoRIA can locate semantic-rich regions implied in medical texts, and extract local semantic-focusing visual features from the medical image. This

can help understand the contents and semantics of the medical image for predicting a correct answer to the clinical question.
- We conduct extensive experiments on three public datasets. The results show that incorporating ROCOGLoRIA as the visual encoder into baselines can benefit Med-VQA. Our model also shows the new state-of-the-art performance over all three benchmark datasets, which may indicate a better generalization ability.

2 Related Work

2.1 Medical Visual Question Answering

Current Med-VQA systems mainly consist of four components [21]: visual feature extraction, textual feature extraction, multi-modal fusion, and answer prediction.

To address the challenge of lacking well-annotated large-scale datasets for Med-VQA, a native solution is to use data augmentation (e.g. VQAMix [9]). However, such methods may generate noisy training samples which hinder the Med-VQA models from learning effective visual representations. Therefore, recent studies adopted deep encoders to interpret the image and question for predicting an answer. As the textual feature extraction module was reported to have less impact on the Med-VQA task, most existing studies focused on improvements of the visual feature extraction [7].

Medical image features are crucial for Med-VQA. Early studies attempted to use pre-trained models based on large-scale natural scene image datasets (e.g. ImageNet [30]) as the visual extractor. However, they cannot achieve satisfactory performance on Med-VQA due to the data limitation problem [7]. Several studies improved the visual feature extraction module by adopting meta-learning, such as MEVF [24] and MMQ [6]. However, they did not consider the impact of linguistic representation on Med-VQA [9]. Besides, some studies pre-trained the visual encoder for specific body regions. For example, CPRD [22] utilized a large number of unlabeled radiology images to train three teacher models and a student model for the body regions of the brain, chest, and abdomen through contrastive learning, limiting the generalization ability to other settings [7]. Recent studies employed transfer learning by using medical image-text pair datasets to fine-tune the cross-modal pre-trained models [4,7,9,20]. For example, Eslami et al. [7] fine-tuned the CLIP [27] model using the ROCO dataset [26], demonstrating its effectiveness on Med-VQA. These approaches had been particularly successful by performing pre-training using large-scale medical image-text pairs without additional manual annotations. Along this line, we further investigate what extent using external medical multi-modal datasets without any manual annotation can contribute to Med-VQA.

With the effective visual encoder, existing systems employed multi-modal fusion models to generate representations for final answer prediction. SAN [32] and BAN [17] are two representative models, which are effective for Med-VQA.

However, they only support two modalities. MTPT-CMSA [8] introduces a cross-modal self-attention model to capture the long-range contextual relevance from more modalities. While the multi-modal fusion module in M^{3}AE [3] consists of two transformer models.

Several studies are focusing on the improvements of model framework and model reasoning. For example, MMBERT [16] used the transformer framework instead of the typical deep learning framework and employed the image features of Med-VQA for mask language modeling. This work also used ROCO [26] for pre-training. CR [33] proposed a novel conditional reasoning framework, which aimed to automatically learn effective reasoning skills for various Med-VQA tasks. Its research ideas had been adopted in many studies to improve the accuracy of Med-VQA models [7,8,22].

2.2 Transfer Learning

As Med-VQA suffers from the general lack of well-annotated large-scale datasets for training, most studies adopted transfer learning. For multi-modal tasks in the general domain, many studies have proposed multi-modal self-supervised models, such as CLIP [27], simCLR [1] and DALLE [28], to overcome the challenge of requiring additional annotated data. These models can be transferred to most tasks and are competitive with fully supervised methods. Tasks such as VQA, image captioning, and image-text retrieval, can benefit from these models. Inspired by this, current studies have emerged to pre-train effective visual representations. conVIRT [34] contrasts the image representations with the paired descriptive texts via a bidirectional objective between two modalities. GLoRIA [12] is a framework for jointly learning multi-modal global and local representations of medical images by contrasting attention-weighted image regions with words in the paired reports. In this paper, we use ROCO [26] to fine-tune GLoRIA [12] pre-trained on CheXpert [14]. Specifically, we perform the multi-task pre-training based on GLoRIA. The original GLoRIA can learn more image types after fine-tuning, and hence it can be applied to various datasets while ensuring the effectiveness of the extracted visual features.

3 Methodology

This paper proposes an effective model, denoted by ROCOGLoRIA, to extract semantic-oriented visual features for Med-VQA. Figure 1 provides an overview of our model architecture. We first perform multi-task pre-training to extract visual features by fine-tuning the GLoRIA model [12] and then incorporate our ROCOGLoRIA as a visual encoder into state-of-the-art Med-VQA methods for improving the answer prediction accuracy.

3.1 Multi-task Pre-training

The GLoRIA used for fine-tuning is trained based on CheXpert [14] which contains a total of 224,316 chest radiographs from 65,240 patients. We fine-tune it

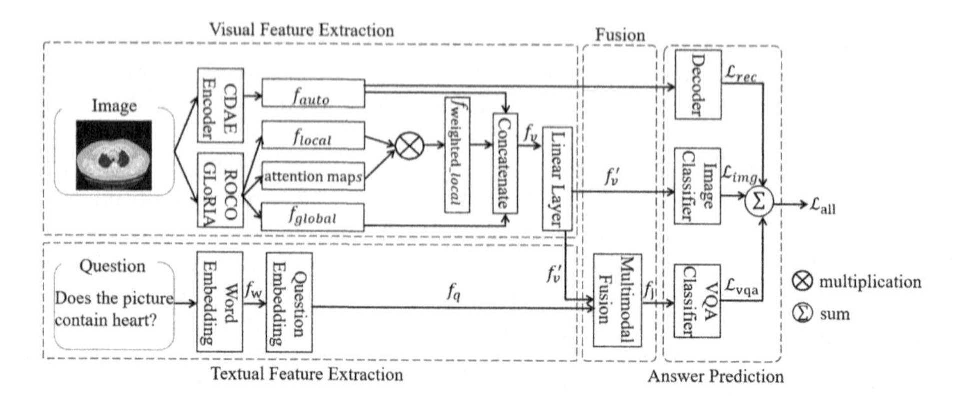

Fig. 1. Model architecture

using the ROCO dataset [26]. As shown in Fig. 2, during multi-task pre-training, we have two pre-training objectives: GLoRIA-based text-image pair similarity calculation [12] and R2Gen-based image report generation [2].

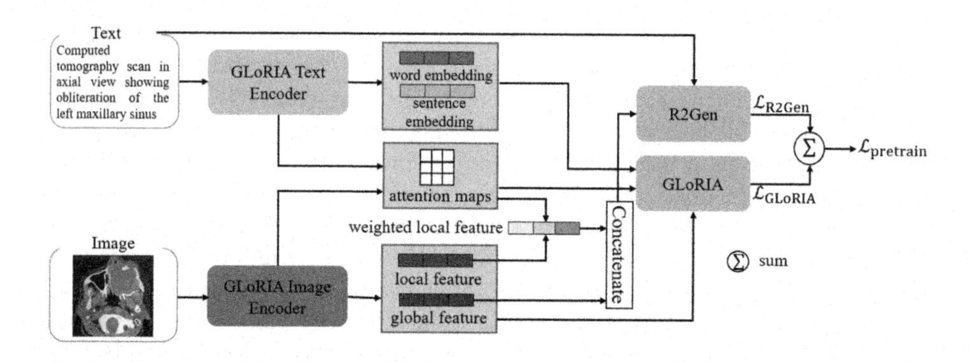

Fig. 2. Multi-task pre-training

GLoRIA-Based Similarity Calculation. Texts and images from ROCO are respectively encoded by the text encoder and the image encoder in GLoRIA. For texts, GLoRIA extracts the word embeddings denoted by f_{word} and the sentence embeddings denoted by f_s. For images, GLoRIA extracts the global features denoted by f_{global} from the final adaptive average pooling layer of ResNet50, and the local features denoted by f_{local} from an intermediate convolution layer. The local features are vectorized to the C-dimensional feature map for each of M image sub-regions. The generated f_{word}, f_s, f_{global}, and f_{local} are used in the GLoRIA model for similarity calculation between text-image pairs. At the same time, in the process of fine-tuning GLoRIA, we output word-based attention maps, which are subsequently multiplied with the image local features of f_{local} to obtain the attention-weighted local features denoted by $f_{weighted\_local}$.

R2Gen-Based Image Report Generation. This task is designed to further ensure the effectiveness of the visual features outputted by fine-tuned GLoRIA. The visual features are obtained by concatenating the global features of f_{global} and the weighted local features of $f_{weighted\_local}$ produced by GLoRIA. We feed them into R2Gen to generate a report about the medical image and calculate the loss denoted by $\mathcal{L}_{R2Gen}$. Specifically, we replace the visual extractor in R2Gen with visual features generated by GLoRIA. Formally, the multi-task loss function is defined as:

$$\mathcal{L}_{pretrain} = \mathcal{L}_{GLoRIA} + \mathcal{L}_{R2Gen}. \tag{1}$$

It is worthwhile to continuously fine-tune the GLoRIA model by using the reports generated from R2Gen. We use the fine-tuned GLoRIA model as a visual feature extractor in our proposed ROCOGLoRIA. The results in Sect. 4.6 also verify the effectiveness of our fine-tuned version, compared with the original GLoRIA model.

3.2 Our Med-VQA Model

In this paper, we evaluate the effectiveness of our proposed ROCOGLoRIA by incorporating it as a visual encoder into three advanced Med-VQA methods: CR [33], VQAMix [9] and M^3AE [3]. In Fig. 1, Our Med-VQA model consists of four main components: visual feature extraction for capturing visual features of the medical image, textual feature extraction for generating the embedding of the given question, fusion module for visual-textual feature fusion, and answer prediction.

We follow existing studies (e.g., [6,8,24]) to define the Med-VQA task as a classification task. Given an image denoted by v and an associated question denoted by q, the goal is to select a correct answer from a set of candidate answers denoted by A. F denotes our model, and α denotes the attention maps. Thus, the predicted answer denoted by $\hat{a}$ is formulated: $\hat{a} = \arg\max_{a \in A} F(a|v, q, \alpha)$.

Visual Feature Extraction. Most existing studies obtained the global visual features by averaging the visual feature vectors of different local regions in the medical image [6,24,29]. Existing studies found that the semantic-oriented medical content for general VQA may only lie in a particular area of the image [4]. Thus, the need arises to focus on local semantic-rich features in the medical image. However, it is much more challenging for Med-VQA, as the range of semantic-oriented regions associated with various types of questions can be very different. Single-use of global features or semantic-rich local features cannot achieve a satisfactory result. To tackle this problem, we propose ROCOGLoRIA, which can simultaneously extract global features and weighted local features of the medical image. The extracted features are then combined to capture useful visual information, which can help understand the contents and semantics of the medical image for predicting a correct answer to the clinical question.

Given an image-text pair, the original GLoRIA [12] model considers each word in the text to generate attention maps denoted by $\alpha \in \mathbb{R}^{L\times H\times W}$, L denotes the length of the text. Considering that the lengths of questions in a Med-VQA dataset are different, we need to ensure the dimensions of the weighted local features generated later are consistent. Thus, we convert raw attention maps to average attention maps denoted by $\alpha^{'} \in \mathbb{R}^{1\times H\times W}$, where $H \times W$ is the shape of the input medical image. The weighted local features can be calculated as $f_{weighted\_local} = \alpha^{'} f_{local}^{\mathrm{T}}$, where $f_{weighted\_local} \in \mathbb{R}^{1\times C}$, $f_{local} \in \mathbb{R}^{C\times H\times W}$, and C is the image feature dimension.

In this way, we exploit the attention maps in the question to localize the semantic-oriented regions in the medial image. We concatenate the global features of f_{global}, and the weighted local features of $f_{weighted\_local}$ to generate f_{gloria}, which can be concatenated with the features extracted by CDAE [24] denoted by f_{auto} to generate the final visual features: $f_v = [f_{gloria}; f_{auto}]$, where $f_{auto} \in \mathbb{R}^{1\times C}$ and $f_{global} \in \mathbb{R}^{1\times C}$. Finally, a linear layer is used to convert f_v to $f_v^{'}$ for multi-modal fusion.

Textual Feature Extraction and Multi-modal Fusion. We follow the MEVF [24] to perform textual feature extraction and multi-modal fusion. Specifically, each word of the question is represented as a 600-D vector, and then the word embedding denoted by f_w is fed into a 1024-D LSTM [31] to produce the question embedding denoted by $f_q \in \mathbb{R}^{12\times 1024}$. We do not directly use the text encoder in ROCOGLoRIA to generate question embeddings as the text encoder is trained on lengthy medical reports which may incur noisy information for embedding questions in Med-VQA. Subsequently, question embedding of f_q and visual features of $f_v^{'}$ are fed into the BAN [17] to generate the joint feature denoted by f_j. Noted that for M^{3}AE, we keep other modules in it and only replace its visual encoder with ROCOGLoRIA.

Multi-task Learning. The joint features of f_j are fed into a classifier for Med-VQA answer prediction. Existing studies showed that introducing multi-task learning into Med-VQA can effectively improve the accuracy of the model [4,8]. In this paper, to ensure the effectiveness of the visual features generated by the visual extraction component, we use the generated visual features of $f_v^{'}$ to perform the task of image classification for predicting the image type in parallel. Meanwhile, we also incorporate the CDAE into Med-VQA to adjust the loss function [24], and finally, our multi-task loss function is defined:

$$\mathcal{L}_{all} = \mathcal{L}_{vqa} + \beta\mathcal{L}_{img} + \mathcal{L}_{rec}, \tag{2}$$

where $\mathcal{L}_{vqa}$ is a cross-entropy loss for VQA classification, $\mathcal{L}_{rec}$ stands for the reconstruction loss of CDAE, and $\mathcal{L}_{img}$ represents the loss for image type classification. β is a hyperparameter for balancing the three loss terms. Noted that the $\mathcal{L}_{img}$ in the loss function is not used in our subsequent experiments in the PathVQA dataset [10] as no image type label is provided in this dataset.

4 Experiments

4.1 Datasets

For pre-training, we use a large-scale publicly available medical dataset, named ROCO [26]. It contains image-text pairs that are collected from PubMed articles, which cover a variety of imaging modalities such as X-Ray, MRI, angiography, etc. It also widely covers diverse body regions, such as the head, neck, teeth, etc. We select 87,952 non-compound radiological images with the associated captions, which provide rich semantic information about the content of images. This paper follows the training and validation data splits from the original paper [4].

Three benchmark Med-VQA datasets of VQA-RAD [19], SLAKE [23], and PathVQA [10] are used to train and evaluate our ROCOGLoRIA model. The VQA-RAD dataset contains 315 images and 3,515 corresponding questions posted and answered by clinicians. In this paper, we utilize the English subset of SLAKE dataset, denoted by SLAKE-EN. The SLAKE-EN dataset comprises 642 images and more than 7,000 question-answer pairs. The PathVQA dataset consists of 32,799 question-answer pairs generated from 1,670 pathology images collected from two pathology textbooks and 3,328 pathology images collected from the PEIR digital library[1]. For a fair comparison, we use the same data splits by following previous studies to partition the VQA-RAD dataset [6,24], the SLAKE-EN dataset [7], and the PathVQA dataset [6].

Medical visual questions are usually divided into two types: closed-ended and open-ended questions. Closed-ended questions are typically answered with "yes/no" or other limited choices; while open-ended questions do not have a restrictive structure and can have multiple correct answers.

4.2 Metrics

To quantitatively measure the performance of Med-VQA models, we use accuracy as the evaluation metric by following previous studies [6,8,24]. Let P_i and L_i denote the prediction and the label of sample i in the test set, and T represents the set of samples in the test set. The accuracy is calculated as: $accuracy = \frac{1}{|T|}\sum_{i \in T} l(P_i = L_i)$.

4.3 Competitors

We incorporate ROCOGLoRIA into three strong baselines for Med-VQA: CR [33], VQAMix [9] and M^{3}AE [3]. Our adapted methods are named **ROCOGLoRIA+BAN+CR**, **ROCOGLoRIA+BAN+VQAMix** and **ROCOGLoRIA+M^{3}AE**. We replace the visual encoder in these three methods using our

[1] https://peir.path.uab.edu/library/.

model. Our competitors are as follows:
MEVF [24] leverages the meta-learning MAML and deploys the auto-encoder CDAE for image feature extraction to overcome the data limitation problem.
CR [33] proposes a question-conditioned reasoning module and a type-conditioned reasoning module to automatically learn effective reasoning skills for various Med-VQA tasks.
CPRD [22] leverages large amounts of unannotated radiology images to pre-train and distill a lightweight visual feature extractor via contrastive learning and representation distillation.
MMBERT [16] is pre-trained using the ROCO dataset with masked language modeling using image features for Med-VQA.
MMQ [6] increases metadata by auto-annotation, dealing with noisy labels, and output meta-models which provide robust features for Med-VQA.
PubMedCLIP [7] is a fine-tuned version of CLIP in the medical domain based on PubMed articles.
MTPT-CMSA [8] reformulates image feature pre-training as a multi-task learning paradigm.
VQAMix [9] combines two training samples with a random coefficient to improve the diversity of the training data instead of relying on external data.
MKBN [11] is a medical knowledge-based VQA network that answers questions according to the images and a medical knowledge graph.
M^{3}AE [3] learns cross-modal domain knowledge by reconstructing missing pixels and tokens from randomly masked images and texts.

4.4 Implementation Details

All the models in this paper are implemented based on the PyTorch library [25] and run on a Ubuntu server with NVIDIA GeForce RTX 3090 Ti GPUs. The detailed training process is described below.

Pre-training. The pre-training model in this paper is implemented based on the ViLMedic library [5]. GLoRIA [12] was pre-trained on CheXpert [14] with ResNet50 as the visual feature extractor and R2Gen [2] was pre-trained on MIMIC-CXR [15]. The model is trained by the Adam optimizer [18] with an initial learning rate of 5e-5, and the learning rate is decreased by loss plateau decay.

Med-VQA Model. When adopting CR, the model is trained by the Adam optimizer [18] with an initial learning rate of 0.0009 on both VQA-RAD and SLAKE-EN, and an initial learning rate of 0.002 on PathVQA. When adapting VQAMix, we follow the original experimental settings for VQA-RAD and PathVQA. For SLAKE-EN, we set the epoch to 80 and the learning rate to 0.02. When adapting and reproducing M^{3}AE on PathVQA, the batch size is set to 16.

4.5 Performance

Table 1 summarizes the results of our model compared against the competitors on three benchmark datasets. We provide overall accuracy along with accuracy in answering open-ended and closed-ended questions. Our proposed model can achieve the best overall accuracy on three datasets, and yield the highest accuracy for open-ended and closed-ended questions. Other observations are summarized as follows:

(1) The performance of CR, VQAMix, and M^3AE is improved when adopting our ROCOGLoRIA as the pre-trained visual encoder. This verifies the effectiveness of ROCOGLoRIA for Med-VQA. And as shown in Table 1, the improvements in our model are effective for both open-ended and closed-ended questions.
(2) Our ROCOGLoRIA+BAN+CR achieves better overall accuracy than our ROCOGLoRIA+BAN+VQAMix on SLAKE-EN and PathVQA. On VQA-RAD, ROCOGLoRIA+BAN+VQAMix achieves better overall accuracy than ROCOGLoRIA+BAN+CR. This suggests that there may be underlying differences in the question type distribution in these datasets. It also verifies that VQAMix can give full play to its advantages on the small labeled dataset of VQA-RAD, by using data augmentation. While on large datasets like SLAKE-EN and PathVQA, this strategy may introduce noise to degrade the performance. Anyway, our ROCOGLoRIA can further enhance VQAMix by capturing both global and weighted local information, which helps to distinguish meaningless pairs and avoid introducing noise during training.
(3) Existing Med-VQA systems have shown too limited performance on PathVQA, as the pathological images contained in PathVQA are quite different from the clinical images contained in VQA-RAD and SLAKE-EN. The superior advantage of our model on PathVQA also indicates that our model can achieve better generalization ability to a wide range of Med-VQA datasets.
(4) Our model achieves the most significant improvements on open-ended questions in PathVQA. We achieve 289%, 38.8% and 9.5% absolute open-ended questions accuracy gain on CR, VQAMix, and M^3AEon PathVQA respectively.
(5) ROCOGLoRIA not only far outperforms classical MEVF, but also compares very well to the multi-stage pre-trained model M^3AE [3] and CPRD [22]. CPRD is pre-trained on head, abdomen, and chest images. In contrast, the pre-training images we use target a larger range of body parts. In other words, our model can achieve good results on more datasets. The pre-training of CPRD is more tailor-made for datasets of the three parts of the head, abdomen, and chest. Taking the SLAKE dataset as an example, the head, abdomen, and chest account for 88% of the total.
(6) Compared to advanced models that improve Med-VQA using ROCO, such as PubMedCLIP [7]), ROCOGLoRIA achieves 3.5% and 3% absolute overall accuracy gain on VQA-RAD and SLAKE-EN respectively. This verifies the

Table 1. Comparisons with competitors on three benchmark datasets. The highest and the second-highest accuracy are respectively marked in bold and by an underline. * indicates the reproduced results. The proposed methods yield the highest overall accuracy on every dataset and achieve the highest accuracy for each kind of questions in the datasets. Differences between the highest and second-highest accuracy are shown in brackets.

Dataset	Models	VQA accuracy (%)		
		Open	Closed	Overall
VQA-RAD	MEVF+SAN	49.2	73.9	64.1
	MEVF+BAN	49.2	77.2	66.1
	MMQ+SAN	46.3	75.7	64.0
	MMQ+BAN	53.7	75.8	67.0
	CPRD+BAN	52.5	77.9	67.8
	PubMedCLIP	48.9	76.7	65.5
	MTPT-CMSA	61.5	80.9	73.2
	MMBERT	63.1	77.9	72.0
	PubMedCLIP+CR	58.4	79.5	71.1
	MEVF+BAN+CR	60.0	79.3	71.6
	CPRD+BAN+CR	61.1	80.4	72.7
	MEVF+SAN+VQAMix	60.0	77.2	70.4
	MEVF+BAN+VQAMix	62.4	81.2	73.8
	M^3AE	67.2	83.5	77.0
	ROCOGLoRIA+BAN+CR	60.6	82.3	73.6
	ROCOGLoRIA+BAN+VQAMix	62.6	**83.8**(+0.1)	75.4
	ROCOGLoRIA+M^3AE	**67.6**(+0.4)	83.7(-0.1)	**77.4**(+0.4)
SLAKE-EN	MEVF+SAN	75.3	78.4	76.5
	MEVF+BAN	77.8	79.8	78.6
	CPRD+BAN	79.5	83.4	81.1
	PubMedCLIP	76.5	80.4	78.0
	MKBN_MGE	77.7	85.1	80.6
	PubMedCLIP+CR	78.4	82.5	80.1
	MEVF+BAN+CR	78.8	82.0	80.0
	CPRD+BAN+CR	81.2	83.4	82.1
	MEVF+SAN+VQAMix*	76.3	79.6	77.6
	MEVF+BAN+VQAMix*	77.4	79.1	78.0
	M^3AE	80.3	87.8	83.3
	ROCOGLoRIA+BAN+CR	**81.7**(+0.5)	83.7	82.5
	ROCOGLoRIA+BAN+VQAMix	80	84.9	81.9
	ROCOGLoRIA+M^3AE	81.1	**88.5**(+0.7)	**84.0**(+0.7)
PathVQA	MEVF+SAN	6.0	81	43.6
	MEVF+BAN	8.1	81.4	44.8
	MMQ+SAN	9.6	83.7	46.8
	MMQ+BAN	11.8	82.1	47.1
	MEVF+BAN+CR*	7.3	84.0	45.8
	MEVF+SAN+VQAMix	12.1	84.4	48.4
	MEVF+BAN+VQAMix	13.4	83.5	48.6
	M^3AE*	26.3	90.3	58.4
	ROCOGLoRIA+BAN+CR	28.4(-0.4)	85.5	57.1
	ROCOGLoRIA+BAN+VQAMix	18.6	83.3	51.1
	ROCOGLoRIA+M^3AE	**28.8**(+0.4)	**90.8**(+0.5)	**59.9**(+1.5)

effectiveness of our proposed pre-training and fine-tuning model that may capture more useful semantic information from Med-VQA.

Table 2. Ablation study on ROCOGLoRIA+BAN+CR. "GLoRIA" means using the original Gloria model as the visual encoder. "global" means using only global features. "local" means only using weighted local features. "w/o img classify" means not performing the image type classification.

	VQA-RAD	SLAKE-EN	PathVQA
ROCOGLoRIA+BAN+CR	**73.6**	**82.5**	**57.1**
GLoRIA	72.3	79.7	54.0
global	69.8	81.1	50.5
local	72.3	78.8	54.5
w/o img classify	69.8	81.5	–

4.6 Ablation Study

To verify the effectiveness of each component in our model, we conducted an ablation study based on our adapted method of ROCOGLoRIA+BAN+CR. The results are shown in Table 2. We observe that:

(1) Adopting ROCOGLoRIA as the pre-trained visual encoder in existing Med-VQA systems can further boost the overall accuracy of 1.3%, 2.8%, and 3.1% on VQA-RAD, SLAKE-EN, and PathVQA, respectively, compared to those by adopting the original GLoRIA.
(2) We evaluate the impact of global features and weighted local features on Med-VQA. Using only global features in our model, the performance drops by 3.8%, 1.4%, and 6.6% on VQA-RAD, SLAKE-EN, and PathVQA, respectively; while using only weighted local features, the performance of our model drops by 1.3%, 3.7%, and 2.6%, respectively. The most frequently asked questions in VQA-RAD are about the presence of an abnormality in the images. This requires the visual encoder to detect local features and abnormalities in the image. In this case, our model with better visual localization outperforms that using only the global features. Similarly, the frequently asked questions in PathVQA are about the abnormality in the images. Hence our model using only the local features also performs better in this dataset than that with global ones. However, on SLAKE-EN, the most frequently asked questions ask about the presence of organ type in the image. In this case, the visual encoder needs the overall understanding of the image content, and thus our model using only the global visual features achieves better accuracy than that using only the local ones.

(3) Employing image type classification in our multi-task learning scheme helps to boost the performance of Med-VQA. The results are consistent with previous studies [4,8]. After removing the image type classifier, the performance of our model drops by 3.8% and 1% on VQA-RAD and SLAKE-EN, respectively. No result is shown in PathVQA as no image type label is provided in this dataset.

4.7 Qualitative Analysis

We provide a qualitative comparison of our model with two competitors. Our goal is to illustrate the performance of the original MEVF and VQAMix in comparison with VQAMix when adopting ROCOGLoRIA as the visual encoder for Med-VQA. Examples from three datasets in Fig. 3 show that the MEVF model fails to answer Med-VQA questions. It cannot correctly understand the content of the medical image. For example, in the second left image in Fig. 3, we observe that the given image is from the lung, but the predicted answer from MEVF relates to the heart. In the same image, the original VQAMix model provides answers that located the correct organ to the given image. However, it fails to

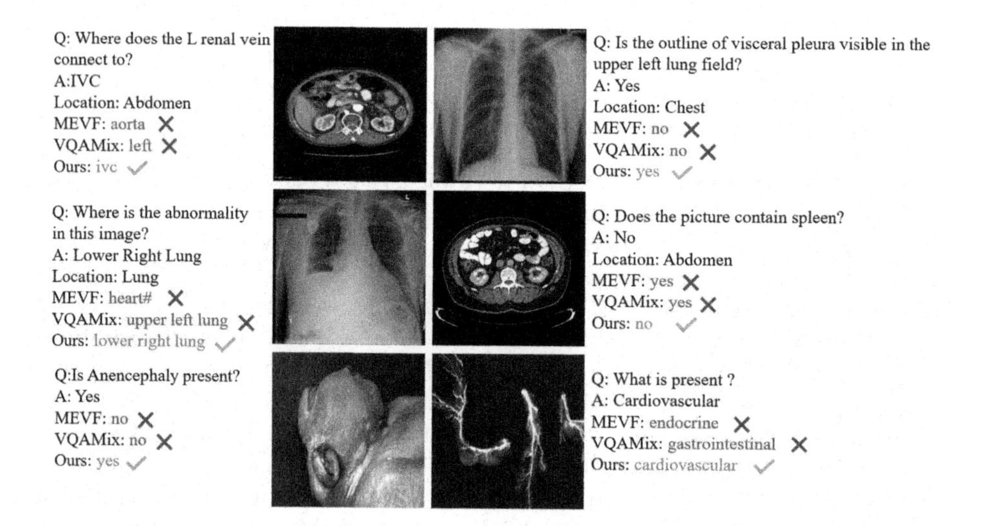

Fig. 3. Examples from VQA-RAD, SLAKE-EN, and PathVQA

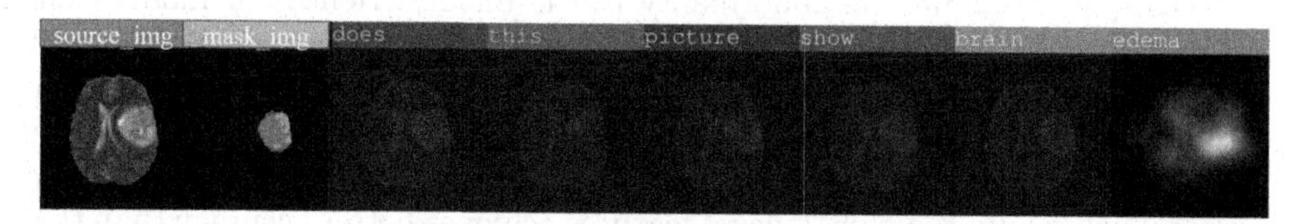

Fig. 4. Visualization of attention maps

predict a correct answer by pointing to a wrong region. In contrast, adopting ROCOGLoRIA as the visual encoder in VQAMix shows an improvement and results in providing answers that are correct throughout all examples. This also indicates that our model not only correctly captures image content holistically but also understands regions of interest as specifically associated with the questions to provide correct answers.

Moreover, we select an image from SLAKE-EN to visualize the attention maps of α generated by ROCOGLoRIA. In Fig. 4, we ask one question "does this picture show brain edema" for the left first medical image namely source_img. ROCOGLoRIA can correctly identify the semantic focusing regions of the image corresponding to each word in the question, and the region identified by "edema" is basically consistent with the annotated region of the image namely mask_img provided by SLAKE-EN. Correct attention maps guarantee the correctness of weighted local features of $f_{weighted\_local}$, which helps to predict a correct answer.

5 Conclusion

To address the problem of data limitation problem on Med-VQA, this paper employs a large-scale medical multi-modal dataset to fine-tune an effective model, denoted by ROCOGLoRIA, which can be used as a pre-trained visual encoder for enhancing existing Med-VQA systems. The model can locate semantic-rich regions implied in medical texts. Experimental results show that our model outperforms the state-of-the-art models. Considering global features and weighted local features at the same time can ensure our model has a better generalization ability.

Acknowledgments. The project was supported by the Natural Science Foundation of Fujian Province of China (No. 2021J01003).

References

1. Chen, T., Kornblith, S., Norouzi, M., Hinton, G.: A simple framework for contrastive learning of visual representations. In: ICML, pp. 1597–1607. PMLR (2020)
2. Chen, Z., Song, Y., Chang, T.-H., Wan, X.: Generating radiology reports via memory-driven transformer. ArXiv, abs/2010.16056 (2020)
3. Chen, Z., et al.: Multi-modal masked autoencoders for medical vision-and-language pre-training. In: Wang, L., Dou, Q., Fletcher, P.T., Speidel, S., Li, S. (eds.) Medical Image Computing and Computer Assisted Intervention - MICCAI 2022. MICCAI 2022. LNCS, vol. 13435, pp. 679–689. Springer, Cham (2022). https://doi.org/10.1007/978-3-031-16443-9_65
4. Cong, F., Xu, S., Guo, L., Tian, Y.: Caption-aware medical VQA via semantic focusing and progressive cross-modality comprehension. In: MM, pp. 3569–3577 (2022)
5. Delbrouck, J.-B., et al.: ViLMedic: a framework for research at the intersection of vision and language in medical AI. In: ACL, pp. 23–34 (2022)

6. Do, T., Nguyen, B.X., Tjiputra, E., Tran, M., Tran, Q.D., Nguyen, A.: Multiple meta-model quantifying for medical visual question answering. In: de Bruijne, M., et al. (eds.) MICCAI 2021. LNCS, vol. 12905, pp. 64–74. Springer, Cham (2021). https://doi.org/10.1007/978-3-030-87240-3_7
7. Eslami, S., de Melo, G., Meinel, C.: Does clip benefit visual question answering in the medical domain as much as it does in the general domain? ArXiv, abs/2112.13906 (2021)
8. Gong, H., Chen, G., Liu, S., Yu, Y., Li, G.: Cross-modal self-attention with multi-task pre-training for medical visual question answering. In: ICMR, pp. 456–460 (2021)
9. Gong, H., Chen, G., Mao, M., Li, Z., Li, G.: VQAMix: Conditional triplet mixup for medical visual question answering. TMI **41**, 3332–3343 (2022)
10. He, X., Zhang, Y., Mou, L., Xing, E.P., Xie, P.: Pathvqa: 30000+ questions for medical visual question answering. ArXiv, abs/2003.10286 (2020)
11. Huang, J., et al.: Medical knowledge-based network for patient-oriented visual question answering. IP&M **60**(2), 103241 (2023)
12. Huang, S.-C., Shen, L., Lungren, M.P., Yeung, S.: Gloria: a multimodal global-local representation learning framework for label-efficient medical image recognition. In: ICCV, pp. 3922–3931 (2021)
13. Huang, Y., Wang, X., Liu, F., Huang, G.: OVQA: a clinically generated visual question answering dataset. In: SIGIR, pp. 2924–2938 (2022)
14. Irvin, J., et al.: ChExpert: a large chest radiograph dataset with uncertainty labels and expert comparison. In: AAAI, vol. 33, pp. 590–597 (2019)
15. Johnson, A.E.W., et al.: Mimic-CXR: a large publicly available database of labeled chest radiographs. ArXiv, abs/1901.07042 (2019)
16. Khare, Y., Bagal, V., Mathew, M., Devi, A., Priyakumar, U.D., Jawahar, C.: MMBERT: multimodal bert pretraining for improved medical VQA. In: ISBI, pp. 1033–1036 (2021)
17. Kim, J.-H., Jun, J., Zhang, B.-T.: Bilinear attention networks. NeurIPS **31**, 1571–1581 (2018)
18. Kingma, D.P., Ba, J.: Adam: a method for stochastic optimization. abs/1412.6980 (2015)
19. Lau, J.J., Gayen, S., Ben Abacha, A., Demner-Fushman, D.: A dataset of clinically generated visual questions and answers about radiology images. Sci. Data **5**(1), 1–10 (2018)
20. Li, P., Liu, G., Tan, L., Liao, J., Zhong, S.: Self-supervised vision-language pre-training for medical visual question answering. ArXiv, abs/2211.13594 (2022)
21. Lin, Z., et al.: Medical visual question answering: a survey. ArXiv, abs/2111.10056 (2021)
22. Liu, B., Zhan, L.-M., Wu, X.-M.: Contrastive pre-training and representation distillation for medical visual question answering based on radiology images. In: de Bruijne, M., et al. (eds.) MICCAI 2021. LNCS, vol. 12902, pp. 210–220. Springer, Cham (2021). https://doi.org/10.1007/978-3-030-87196-3_20
23. Liu, B., et al.: Slake: A semantically-labeled knowledge-enhanced dataset for medical visual question answering. In: ISBI, pp. 1650–1654 (2021)
24. Nguyen, B.D., Do, T.-T., Nguyen, B.X., Do, T.K.L., Tjiputra, E., Tran, Q.D.: Overcoming data limitation in medical visual question answering. In: MICCAI, pp. 522–530 (2019)
25. Paszke, A., et al.: PyTorch: an imperative style, high-performance deep learning library. vol. 32, pp. 8024–8035 (2019)

26. Pelka, O., Koitka, S., Rückert, J., Nensa, F., Friedrich, C.M.: Radiology objects in COntext (ROCO): a multimodal image dataset. In: Stoyanov, D., et al. (eds.) LABELS/CVII/STENT -2018. LNCS, vol. 11043, pp. 180–189. Springer, Cham (2018). https://doi.org/10.1007/978-3-030-01364-6_20
27. Radford, A., et al.: Learning transferable visual models from natural language supervision. In: ICML, pp. 8748–8763. PMLR (2021)
28. Ramesh, A., et al.: Zero-shot text-to-image generation. In: ICML, pp. 8821–8831. PMLR (2021)
29. Ren, F., Zhou, Y.: CGMVQA: a new classification and generative model for medical visual question answering. IEEE Access **8**, 50626–50636 (2020)
30. Russakovsky, O., et al.: ImageNet large scale visual recognition challenge. IJCV **115**, 211–252 (2014)
31. Sundermeyer, M., Schlüter, R., Ney, H.: LSTM neural networks for language modeling. In: Interspeech (2012)
32. Yang, Z., He, X., Gao, J., Deng, L., Smola, A.: Stacked attention networks for image question answering. In: CVPR, pp. 21–29 (2015)
33. Zhan, L.-M., Liu, B., Fan, L., Chen, J., Wu, X.-M.: Medical visual question answering via conditional reasoning. In: MM, pp. 2345–2354 (2020)
34. Zhang, Y., Jiang, H., Miura, Y., Manning, C.D., Langlotz, C.: Contrastive learning of medical visual representations from paired images and text. ArXiv, abs/2010.00747 (2020)

BBAC: Blockchain-Based Access Control Scheme for EHRs with Data Sharing Support

Peng Qin, Tong Zhang, Canming Fang, and Lina Wang(✉)

Key Laboratory of Aerospace Information Security and Trusted Computing, Ministry of Education, School of Cyber Science and Engineering, Wuhan University, Wuhan 430072, China
lnwang@whu.edu.cn

Abstract. With the advancement of modern medicine, human whole genome sequencing technology has become more and more efficient and accurate. Genomic data are characterized by large data volume, privacy, and ease of being tampered with. Genomic data are usually stored in different data centers, and it is not easy to share the data. We propose a blockchain-based dual-verifiable cloud storage solution BBAC with the features of traceability and non-comparability of blockchain. First, we use homomorphic encryption technology to encrypt data and upload it to the cloud to ensure transmission security and data privacy protection. At the same time, the aggregated ciphertext and uploader information is stored on the blockchain to avoid the risk of data tampering by illegal users effectively and enable the traceability of malicious users, realizing double verification of data integrity in the cloud. Security analysis proves BBAC is more secure and reliable than similar schemes.

Keywords: blockchain · verifiable calculation · privacy protection · cloud storage

1 Introduction

With the development of sequencing technology, the cost of sequencing genomes has decreased dramatically, generating large amounts of genomic data in high dimensions [14,15]. Genomic data have been widely used in scientific research, healthcare, legal and forensic, and direct-to-consumer services [6]. However, genomic data uniquely identifies individuals and is associated with sensitive information such as diseases and bloodlines. There is a need to protect the privacy information of genomic data.

Genomic data is notable for its vast data volume, extensive private information, and susceptibility to manipulation. Genomic data are usually stored in different data centers, and it is challenging to share the data. In addition, the data-sharing process may also lead to user privacy leakage. We propose a

X. Yang et al. (Eds.): ADMA 2023, LNAI 14180, pp. 482–494, 2023.
https://doi.org/10.1007/978-3-031-46677-9_33

blockchain-based solution for the secure storage and sharing of genomic data with the help of blockchain features such as traceability and tamper-evident. The data is encrypted and uploaded to the cloud using homomorphic encryption technology to ensure transmission security and data privacy protection [18]. The aggregated ciphertext and uploader information is stored on the blockchain, eliminating the risk of data tampering by illegal users and verifying data integrity in the cloud. Our proposed scheme is more secure and reliable than several other competing approaches.

In the past decades, many academic and industrial researchers have explored extensively and intensively the problems related to verifiable computing. Verifiable computing schemes mainly involve application security, computer theory, and cryptography [21]. Meanwhile, verifiable computing is widely used in various scenarios such as billing systems, outsourced computing, and electronic voting systems. In the application security domain, most schemes are based on auditing and various security coprocessor tools from the application perspective. Most schemes in computer theory rely on probabilistically checkable proofs [1]. However, due to the extremely high computational cost of PCPs, they are challenging to apply in real life. The main cryptographic tools are homomorphic encryption [2], obfuscation circuits, attribute-based encryption, and homomorphic signatures [3].

Blockchain and verifiable computing work together for genomic data privacy protection, facilitating distributed storage of genomic data while testing the integrity of genomic data in the cloud. The above scheme can effectively control the security problems caused by genomic data vandalism. It also facilitates effective regulation by regulators to track malicious input in scenarios such as outsourced computing.

Recently, blockchain-based cloud storage solutions for verifiable genomic data have been proposed. However, the following three problems still need to be solved. The availability of genomic data should be considered while ensuring the security and confidentiality of genomic data during transmission. Existing schemes need to enable the validation of comprehensive genomic data and individual genomic data in a cloud storage service environment. Tracking users who maliciously modify and corrupt genomic data is not possible.

This paper proposes a blockchain-based dual verifiable cloud storage solution (BBAC) to address the above challenges.

- We Proposed a blockchain-based system model of a dual verifiable cloud storage scheme with an analysis of the adversary's malicious behavior. Consistent genomic data storage and tampering prevention are achieved through blockchain technology. At the same time, it increases the difficulty and cost of genomic data forgery on the chain. It thus can eliminate most users' concerns about genomic data credit.
- BBAC applying hash functions with homomorphic nature, the cloud server can aggregate the signatures of multiple users and perform overall verification of the aggregated genomic data, solving the problems of high communication burden and low computational efficiency. When the malicious users

who upload false genomic data need to be traced, separate verification is performed, which solves the problem of tracing malicious users in distributed computing and cloud storage and effectively avoids the vicious circle.
- Security analysis shows that our scheme is efficient and verifiable when encrypting and aggregating genomic data. It is safer and more secure than similar schemes as it can trace malicious users who uploaded incorrect genomic data while ensuring genomic data privacy.

2 Related Work

Based on blockchain technology and combined with the security protocol of verifiable computing, the researchers implement various functions such as privacy protection of genomic data, data deposition, and verification. According to the number of participants in verifiable computation, the schemes can be divided into one-to-one and multi-party verifiable computation schemes.

In a one-to-one verifiable computing scheme, Wang et al. [20] proposed a blockchain-based personal health record-sharing scheme that verifies the integrity of genomic data. This scheme achieves the security of genomic data in sharing personal health records. It improves the privacy of sensitive genomic data of patients. However, the scheme has a high computational burden, a single application scenario, and needs to be more scalable. Guo et al. [9] proposed a new dynamic single sign-on scheme based on blockchain technology to achieve a reliable search of encrypted genomic data and design an encrypted checklist through intelligent contracts. A cryptographic checklist was designed through a smart contract, which allows secure result verification within the blockchain. Chen et al. [5] proposed a lightweight CP-ABE protocol scheme with verifiable outsourced decryption. This scheme adds verification algorithms to the decryption process. It uses the ciphertext's verifiability to verify the user device's correctness. However, this scheme does not satisfy public verifiability. It thus cannot solve the problems arising from the correctness of genomic data between the user and the cloud service platform.

Dimitriou et al. [7] proposed a blockchain-based voting system in a multi-party verifiable computing scheme. This system ensures coercion resistance and relentlessness through random number generator tokens and is suitable for large-scale elections. The blockchain structure ensures the solution is verifiable and scalable while increasing the election's credibility. However, it is not resistant to conspiracy attacks. Wang et al. [20] combine blockchain technology for fair payments, solving the problem of servers returning incorrect results, using smart contracts to store certain indexes, and performing searches. Dor et al. [7] designed a proof-based verifiable computational fairness protocol scheme that provides blockchain systems such as Bitcoin with public verifiability for generating transactions in their network. The scheme makes extensive use of bilinear mappings leading to reduced efficiency and increased cost of script execution. Zhang et al. [22] temporarily freeze deposits in the blockchain to ensure that users receive search results and service fees as long as they perform honestly. This scheme achieves a decentralized and fair payment scheme for fees. However, it requires a large amount of signature verification computation on the user side to verify the correctness of the results, which has a high user overhead.

The management and use of large-scale genomic data can be achieved by cloud computing due to its scalability [17] and the availability of low-cost computing resources. Cloud computing provides high-quality on-demand services for genomic data storage and analysis. However, it also poses an increasingly severe problem of privacy leakage during network access and genomic data upload.

3 Proposed Approach

3.1 System Model

The system model of the blockchain-based dual verifiable cloud storage scheme proposed in this paper is shown in Fig. 1. The genomic data owner transmits the ciphertext and signature to the cloud. Genomic data users can apply to the miner node to verify the integrity of the overall genomic data and individual genomic data and enable malicious users' traceability. There are five types of participating entities: CS (cloud server), DO (data owner), DU (data user), MN (miner node), and blockchain. The roles of each entity and the functions they belong to are described below.

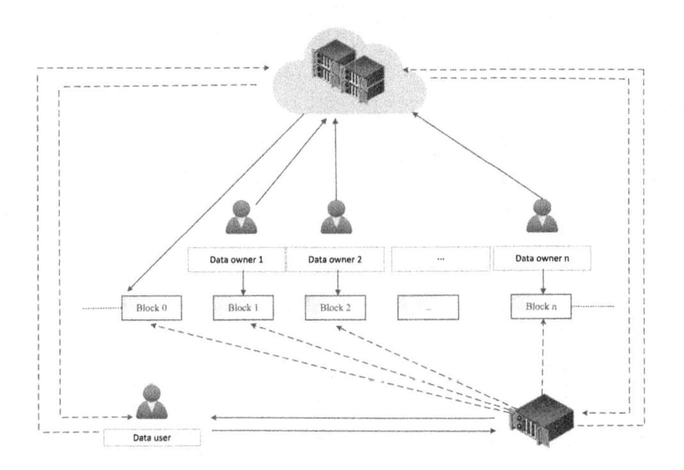

Fig. 1. System overview.

- Cloud Servers. The cloud server is a third-party untrusted storage server with mighty computing power and huge storage space to process and maintain the genomic data owner's genomic data. In the model of this paper, it is the specific executor of function computation, which is mainly responsible for storing, aggregating, and computing the encrypted files and signatures

uploaded by the genomic data owners and then sending the computation and verification results to the genomic data users.
- Genomic data owner. The genomic data owner encrypts its original genomic data using the stream key encryption algorithm in the genomic data encryption phase. In the signature generation phase, the genomic data owner generates a signature on the file with his private key and random numbers, uploads the cipher text and signature to the cloud server for storage, and records the corresponding uploader information into the blockchain.
- Genomic data users. Genomic data users, as legitimate users, can request integrity verification of the overall genomic data and individual genomic data from the miner nodes in the blockchain. Du obtains the correct aggregation result and pays the service fee to the cloud server to complete the transaction.
- Miner nodes. The miner node is used to validate the overall and individual genomic data. Also, the validation results are returned for DO and CS to ensure the integrity and correctness of the genomic privacy data. When the miner node performing the validation task fails, a new node recurs from the standby node.
- Blockchain. The blockchain is mainly responsible for storing on the chain the signatures uploaded by the users of the genomic data and the aggregated results uploaded by the cloud server. Only the address of the encrypted file is stored on the blockchain to trace the encrypted file and the owner of the genomic data who uploaded this encrypted file. Since the blockchain is public, all users can browse and access it, effectively avoiding illegal modification of genomic data and ensuring genomic data security.

3.2 Threat Model

The cloud server in this scenario is honest and curious. It will comply with the protocol and return computation and verification results for the user. However, it may also use the intermediate or final results to infer the user's private genomic data through malicious means, posing a threat to their privacy. Genomic data owners are only partially trustworthy and may forge or provide the wrong ciphertext for some purposes and dishonestly upload genomic data to the cloud. Also, the data owner may be subject to malicious tampering and spoofing attacks by adversaries during uploading. Blockchain is entirely trustworthy and can be introduced to help users of genomic data to verify the correctness of the results returned by the cloud server and the ciphertext uploaded by the data owner. The blockchain only records the uploader's corresponding information and aggregated results, and no original genomic data can be uploaded to the blockchain. The miner node is honest and trustworthy and will strictly perform the verification according to the algorithm in this paper and send the verification results back to the genomic data users.

3.3 System Overview

The verifiable cloud storage scheme timing is shown in Fig. 2: The data owner encrypts and signs the genomic data using homomorphic encryption. Multiple users upload the ciphertext CC and signatures to the cloud server. The cloud server records the uploader information corresponding to each ciphertext uploaded into block 1 to block n. It uploads the ciphertext aggregation result to block 0. The genomic data user requests the genomic data result from the cloud server, and the cloud server returns the computation result. The genomic data user issues a task to the miner node to verify the total genomic data.

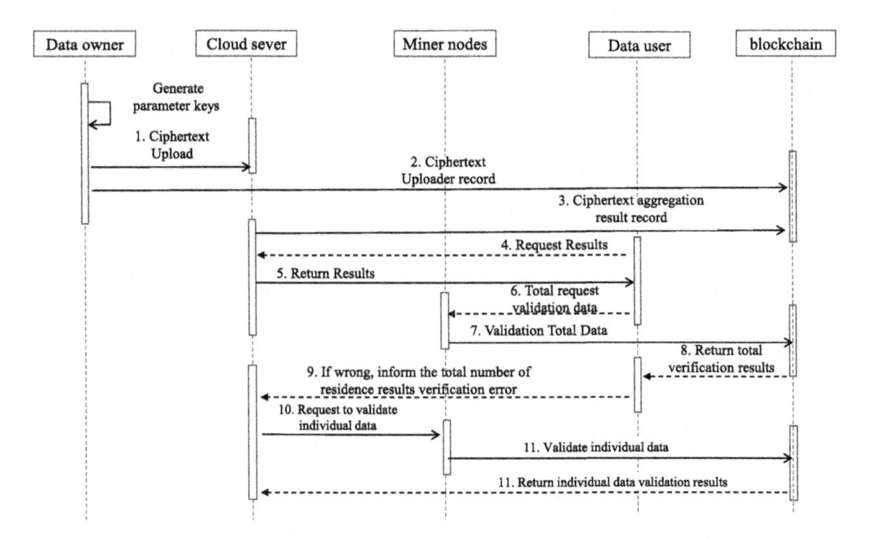

Fig. 2. Blockchain-based access control scheme.

The miner node receives the verification task, performs the overall result verification at the blockchain head, and returns the total genomic data verification result. The cloud server gets the service fee and completes the transaction if the genomic data is correct. If the validated genomic data is incorrect, the result is returned to the cloud server, and the transaction is terminated.

After the cloud server receives the error result feedback, it sends the task of validating individual genomic data information to the miner node again. After receiving the verification task, the miner node visits block 1 to block n, respectively. It then returns the single genomic data verification result to the cloud server. The traced malicious user will be penalized, and the user can refuse to pay for the service.

3.4 Dual Verifiable Cloud Storage Algorithm

This section will describe the system and threat model of the blockchain-based dual verifiable cloud storage algorithm.

Ciphertext Aggregation Processing Phase. This scheme uses the stream key encryption algorithm to homomorphically encrypt the raw genomic data because the algorithm is fast and easy to implement. Moreover, in this scheme, each genomic data user does not share the same key with the cloud server.

$$\mathrm{CAGG} = c_1 + c_2 + \cdots + c_V = \mathrm{Enc}\left(m_1\right) + \mathrm{Enc}\left(m_2\right) + \cdots + \mathrm{Enc}\left(m_v\right) \tag{1}$$

1) Key distribution. The system broadcasts a key k to each data user in the Key distribution phase. Key k is generated based on a variable-length cluster of stream encryption algorithms.

2) Counter t. The cloud server needs to verify whether the genomic data uploaded by the genomic data owner is the latest genomic data. Each genomic data owner generates a counter t and initializes it to ensure real-time genomic data and resist replay attacks.

3) Genomic data encryption. The genomic data owner L encrypts the original genomic data mL and then uploads it to the cloud server to ensure the privacy of the genomic data. Its encryption algorithm is.

Cloud Server Aggregation Ciphertext. The cloud server aggregates the ciphertexts uploaded by V genomic data owners by the encryption algorithm. Since we use additive homomorphic encryption, we do not need to decrypt the ciphertexts. We can directly perform the related operations on the uploaded genomic data. The above operations reduce the computational overhead and increase the scheme's security, reducing the risk of privacy leakage.

$$\mathrm{Dec(CAGG)} = \mathrm{CAGG} - \mathrm{rt} \tag{2}$$

Ciphertext Aggregation Processing Stage. The cloud server aggregates the ciphertexts uploaded by V genomic data owners by the encryption algorithm. Since we use additive homomorphic encryption, we do not need to decrypt the ciphertexts. We can directly perform the related operations on the uploaded genomic data. The above operations reduce the computational overhead and increase the scheme's security, reducing the risk of privacy leakage. Genomic data decryption phase. After receiving the cipher text uploaded by the genomic data owner, the cloud server decrypts it, and its decryption function is N.

$$F = \{m_1, m_2, \cdots, m_V\} \tag{3}$$

Overall Verification. Verifying each user's ciphertext leads to excessive traffic in a verifiable computing scheme. So, the cloud server needs to aggregate the ciphertext and signature of each user. The specific algorithm is described as follows.

4 Experiments

4.1 Security Analysis

Content Privacy. The ciphertext of homomorphic encryption is operable when the genomic data information is stored in this scheme. Each genomic data owner uses different public keys to encrypt the genomic data information through cryptography homomorphically. After the above processing, it is stored in the cloud server, enhancing the security of transmission and storage of genomic data. The aggregated ciphertext and uploader information is also stored in the block. The above operations further ensure the privacy of the shared content.

Verifiability and Traceability. Existing verifiable cloud storage solutions suffer from the security risk of participants' conspiracy to tamper with genomic data due to the lack of trusted third parties. The scheme in this paper utilizes the decentralization of blockchain to provide a trusted environment for verifiable computing, which can effectively avoid the risk of private tampering with genomic data by third-party organizations or malicious users. We perform homomorphic hashing of genomic data and store the signatures distributed on the blockchain. Genomic data users can send requests to the blockchain for double-verifying genomic data in the cloud through the miner node with guaranteed security. Finally, the review result is obtained, and the signature information is used to trace and punish malicious users.

Anti-conspiracy Attack. The blockchain-based dual verifiable cloud storage scheme proposed in this paper can avoid conspiracy attacks. Blockchain provides a trustworthy environment for verifiable computing through cryptography. The cloud server stores the aggregated cryptographic results of genomic data owners on the chain, effectively preventing conspiracy among participants and enabling each participant to execute the protocol honestly. At the same time, conspirators who are traced are subject to a high-security deposit for violating the agreement. The participant's compliance with the contract provides higher benefits than conspiracy and eliminates the incentive for conspiracy.

Scalability. The information consensus generation blocks are time-stamped, traceable, and cannot be modified. However, the limited block storage makes it impossible to store all the information in complete blocks. We store the complete cipher text in the cloud server to ensure fast verification of the correctness of the stored genomic data. The on-chain blocks record uploader information, store aggregation results for easy verification and traceability, and solve the limited block storage capacity problem.

4.2 Performance Evaluation

Computational Cost. The generation validation and the genomic data validation phase mainly generate the computational cost in this scheme. The computational cost is shown in Table 1, where V is the number of validated genomic data owners, M is the multiplication over G1, E is the exponent, P is the bilinear mapping, and H is the monomial hash function.

Table 1. Computation overhead

Verification	Generate verification phase	Data validation phase
Overall validation	$2H + (2Vc - 1)M + VcE$	$VcH + (Vc + V)M + (Vc + V)E$
Single verification	$1H + (2c - 1)M + cE$	$(c + 1)H + (c + 1)M + (c + 1)E$

To check the integrity of L, the user sends a validation request to the miner node. After receiving the verification request, the miner node generates a random subset A and B containing c elements. For each S, the miner node randomly selects P, generates a challenge message I, and sends it to the cloud server.

Communication Volume. The communication volume The scheme in this paper needs to transmit A during the verification process. In contrast, the scheme in the literature [13] needs to transmit R. The scheme in this paper reduces the communication volume between the cloud server and the mining nodes. As the number of verified genomic data blocks increases, the algorithm in this paper also dramatically reduces the average communication per block for verification (Fig. 3).

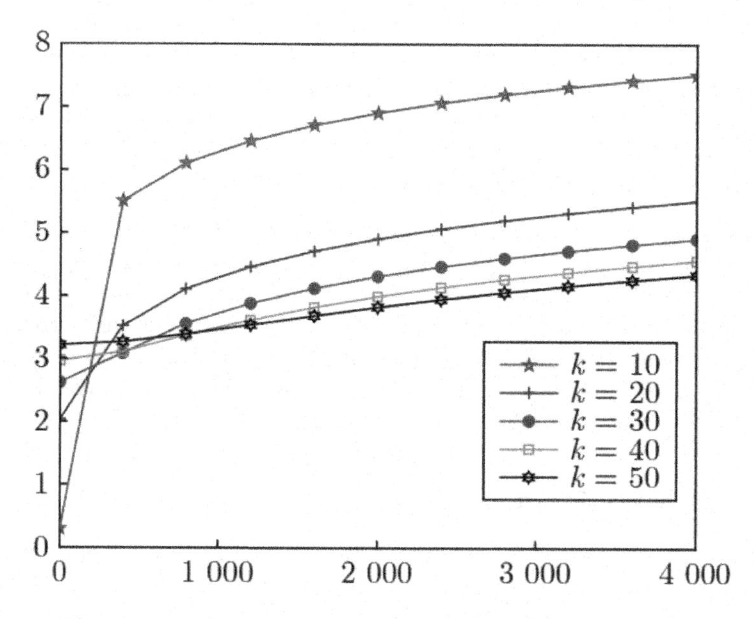

Fig. 3. communication volume.

Validation Algorithm. The validation algorithm in this paper is based on a probabilistic model. Let b blocks be tampered with, and c blocks of n blocks of genomic data are validated simultaneously. The random variable X denotes the number of blocks of genomic data detected to be tampered with, then P. The probability of detecting genomic data tampered with in one validation is at least Q. Let the random variable Z denote the number of blocks in which genomic data are detected as tampered with in validation times. R, a total of w = ac blocks are challenged, i.e., at least W probability of detecting genomic data as tampered.

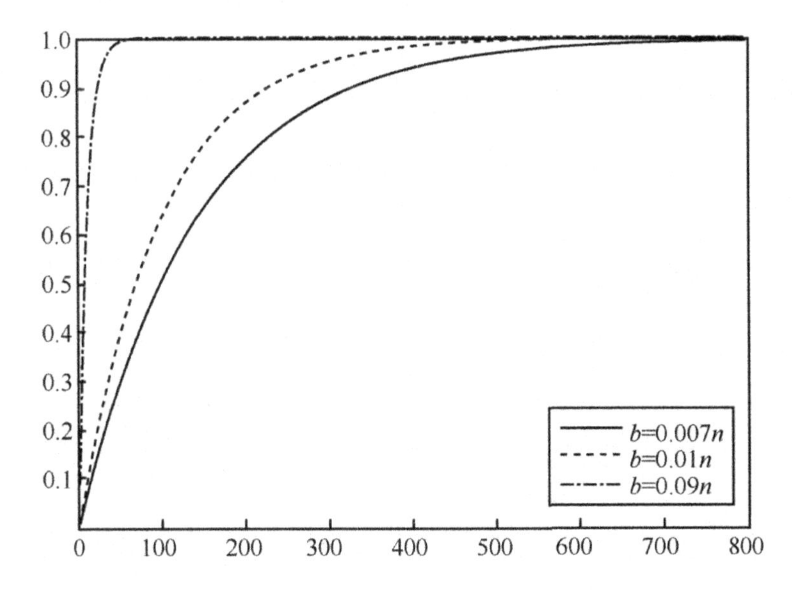

Fig. 4. Verify the validity of the algorithm.

Assume that 0.7% of the n blocks of genomic data are tampered with W. Similarly, we get $bn = 0.01$ and $bn = 0.09$. The validity of the verification algorithm is shown in Fig. 4, where $PZ \geq 1$ is independent of n. The algorithm in this paper also proves 100% genomic data completeness when each piece of data is verified.

We obtained the following findings from a comparative analysis of existing verifiable computing schemes. The literature [16,19] schemes are all blockchain-based verifiable computation studies that do not rely on trusted third parties, but none of them are traceable. The literature [9] scheme achieves the security of genomic data while sharing personal health records. However, the scheme has a high computational burden and needs to be more scalable.

The literature [10,16] schemes all achieve data verifiability. Buccafurri et al. [4] proposed a deterministic approach to verify query results in a cloud environment, which implements user verification of the integrity of the results. The genomic data owner verifies the first token's time value, calculates each element's MAC attribute, and compares it with the value returned by the cloud server to verify each link starting from the first token. This scheme reduces the complexity of insertion space and time to some extent. Wu et al. [20] proposed a verifiable outsourced computation scheme for matrix products. However, none of the schemes in the literature [8,12] is scalable and traceable.

The literature [11] proposed a privacy-preserving framework to store patient health records in a cloud-assisted environment. Its performance is based on performance metrics such as adaptability, privacy, and utility, but it cannot verify the correctness of genomic data and trace malicious users.

Table 2. Scheme Comparison

solution	Cloud Storage	Verifiability	Privacy	Scalability	Collusion-Resistant	Blockchain
[11]	×	✓	✓	×	–	✓
[8]	×	✓	✓	✓	×	✓
[12]	✓	✓	✓	–	×	×
[4]	✓	✓	✓	×	–	×
[10]	✓	×	✓	–	–	×
BBAC	✓	✓	✓	✓	✓	✓

The scheme in this paper ensures genomic data confidentiality and transmission security and achieves the privacy protection of genomic data. We utilize the tamper-evident and traceability of blockchain to realize the double verification of genomic data integrity in the cloud. At the same time, we can effectively avoid the risk of genomic data being tampered with by illegal users and traceable to malicious users. The scheme comparison is shown in Table 2.

5 Conclusion

Blockchain can provide extremely high transparency, distributed verifiability, and immutability. Using blockchain technology and verifiable computing technology, the integrity of genomic data uploaded to the cloud can be checked, effectively avoiding the risk of genomic data being tampered with by illegal users. Based on the above functions of blockchain, the scheme in this paper adopts homomorphic encryption technology to encrypt genomic data information to ensure the confidentiality of genomic data and the security of transmission and realize the privacy protection of genomic data. Moreover, users of genomic data can double-verify genomic data in the cloud under the premise of ensuring security and can trace the malicious users. The combination of blockchain's characteristics of tamper-evident, decentralization, and anonymity with verifiable computing can be well used in specific applications such as billing systems and medical sensitive information sharing. Combining blockchain with verifiable computing and applying it to specific real-world scenarios to solve existing critical problems is the focus of the following research work.

Acknowledgement. This work was supported in part by the National Key Research and Development Program of China (No. 2020YFB1805400); in part by the National

Natural Science Foundation of China (No. 42071431); in part by the Provincial Key Research and Development Program of Hubei, China (No. 2020BAB101).

References

1. Akavia, A., Leibovich, M., Resheff, Y.S., Ron, R., Shahar, M., Vald, M.: Privacy-preserving decision trees training and prediction. ACM Trans. Priv. Secur. **25**, 1–30 (2022)
2. Bai, J., Song, X., Cui, S., Chang, E.C., Russello, G.: Scalable private decision tree evaluation with sublinear communication. In: Proceedings of the 2022 ACM on Asia Conference on Computer and Communications Security (2022)
3. Barhoun, R.: A trust and activity based access control model for preserving privacy and sensitive data in a distributed and collaborative system: application to a healthcare system. Int. Rev. Comput. Softw. (IRECOS) (2021)
4. Buccafurri, F., Lax, G., Nicolazzo, S., Nocera, A.: Range query integrity in cloud data streams with efficient insertion. In: Cryptology and Network Security (2016)
5. Chen, Y.C., Chang, C.C., Hung, C.C., Lin, J.F., Hsu, S.Y.: SecDT: privacy-preserving outsourced decision tree classification without polynomial forms in edge-cloud computing. IEEE Trans. Signal Inf. Process. Netw. **8**, 1037–1048 (2022)
6. Davis, S., et al.: Standardized health data and research exchange (share): promoting a learning health system. JAMIA Open **5** (2022)
7. Dimitriou, T.: Efficient, coercion-free and universally verifiable blockchain-based voting. IACR Cryptol. ePrint Arch. **2019**, 1406 (2020)
8. Ding, Y., Sato, H.: Derepo: a distributed privacy-preserving data repository with decentralized access control for smart health. In: 2020 7th IEEE International Conference on Cyber Security and Cloud Computing (CSCloud)/2020 6th IEEE International Conference on Edge Computing and Scalable Cloud (EdgeCom), pp. 29–35 (2020)
9. Gan, Q., Wang, X., Li, J., Yan, J., Li, S.: Enabling online/offline remote data auditing for secure cloud storage. Clust. Comput. **24**, 3027–3041 (2021)
10. Iezzi, M.: Practical privacy-preserving data science with homomorphic encryption: an overview. In: 2020 IEEE International Conference on Big Data (Big Data), pp. 3979–3988 (2020)
11. Kim, D., Kim, K.S.: Privacy-preserving public auditing for shared cloud data with secure group management. IEEE Access **PP**, 1 (2022)
12. Li, X., Zhou, R., Zhou, T., Liu, L., Yu, K.: Connectivity probability analysis for green cooperative cognitive vehicular networks. IEEE Trans. Green Commun. Netw. **6**, 1553–1563 (2022)
13. Jie Lu, W., Huang, Z., Hong, C., Ma, Y., Qu, H.: PEGASUS: bridging polynomial and non-polynomial evaluations in homomorphic encryption. In: 2021 IEEE Symposium on Security and Privacy (SP), pp. 1057–1073 (2021)
14. Müller, S.: Is there a civic duty to support medical AI development by sharing electronic health records? BMC Med. Ethics **23** (2022)
15. Pankhurst, T., et al.: Introducing SNOMED-CT* coding into an electronic health record: Impact on clinicians, data sharing and research potential *systemised nomenclature of medicine clinical terminology (preprint) (2021)
16. Song, B.K., Yoo, J.S., Hong, M., Yoon, J.W.: A bitwise design and implementation for privacy-preserving data mining: from atomic operations to advanced algorithms. Secur. Commun. Netw. **2019**, 3648671:1–3648671:14 (2019)

17. Subbayamma, M.S.S.: The secured client-side encrypted data with public auditing in cloud storage (2021)
18. Sun, Y., Zhang, R., Wang, X., Gao, K., Liu, L.: A decentralizing attribute-based signature for healthcare blockchain. In: 2018 27th International Conference on Computer Communication and Networks (ICCCN), pp. 1–9 (2018)
19. Vinoth, R., Deborah, L.J.: An efficient key agreement and authentication protocol for secure communication in industrial IoT applications. J. Ambient. Intell. Humaniz. Comput. **14**, 1431–1443 (2021)
20. Wang, J., Shi, D., Chen, J., Liu, C.C.: Privacy-preserving hierarchical state estimation in untrustworthy cloud environments. IEEE Trans. Smart Grid **12**, 1541–1551 (2021)
21. Yu, Y., Liu, S., Yeoh, P.L., Vucetic, B., Li, Y.: Layerchain: a hierarchical edge-cloud blockchain for large-scale low-delay industrial internet of things applications. IEEE Trans. Industr. Inf. **17**(7), 5077–5086 (2021). https://doi.org/10.1109/TII.2020.3016025
22. Zhang, J., Lu, R., Wang, B., Wang, X.A.: Comments on "privacy-preserving public auditing protocol for regenerating-code-based cloud storage." IEEE Trans. Inf. Forensics Secur. **16**, 1288–1289 (2021)

Cross-Genre Retrieval for Information Integrity: A COVID-19 Case Study

Chaoyuan Zuo[1(✉)], Chenlu Wang[2], and Ritwik Banerjee[2,3]

[1] School of Journalism and Communication, Nankai University, Tianjin, China
zuocy@nankai.edu.cn

[2] Department of Computer Science, Stony Brook University, Stony Brook, USA
{chenlwang,rbanerjee}@cs.stonybrook.edu

[3] Institute for AI-Driven Discovery and Innovation, Stony Brook University, Stony Brook, USA

Abstract. Ubiquitous communication on social media has led to a rapid increase in the proliferation of unreliable information. Its ill-effects have perhaps been seen most obviously during the COVID-19 pandemic, and have rightfully raised concerns about the integrity of shared information. This work focuses on *derivative* Twitter posts (tweets), *i.e.*, posts that re-transmit preexisting content. We acknowledge that a considerable number of such tweets do not provide a source of information, which undoubtedly undermines the integrity of the information and poses difficulties in fact-checking. To address this concern, we propose an *ad hoc* information retrieval (IR) task to identify the support for claims made in tweets from reputable news outlets. We demonstrate the feasibility of such cross-genre IR by presenting experiments on a COVID-19 dataset of 11K pairs of tweets and news articles. We describe a two-step methodology: (i) selecting the most relevant candidates from 57K pandemic-related news articles, and (ii) a final re-ranking of this selection. Our method achieves significant improvements over the classical token-based approach using BM25 as well as a state-of-the-art transformer-based language model pretrained on COVID-19 tweets. Our findings demonstrate the viability of cross-genre IR across news and social media in safeguarding the integrity of information disseminated through social media.

Keywords: Dataset · COVID-19 · Twitter · Information retrieval

1 Introduction

The COVID-19 outbreak was a global public health emergency with widespread effects. Major disease outbreaks such as Ebola, Zika, and Yellow Fever, have underscored the importance of combating false information [16,34,40,41]. Misinformation related to the COVID-19 pandemic on social media, particularly on Twitter, has been a pressing social issue. It has, for instance, motivated racial violence [10], caused destruction of infrastructure due to belief in conspiracy

X. Yang et al. (Eds.): ADMA 2023, LNAI 14180, pp. 495–509, 2023.
https://doi.org/10.1007/978-3-031-46677-9_34

Table 1. Derivative tweets with (✓) and without (✗) a reference to the claim source.

(✓) JPMorgan Chase is investigating potential misuse of pandemic relief programs by its workers and customers, saying some conduct "may even be illegal" *Cited News: JPMorgan investigates employees over potential misuse of PPP loans* [www.cnn.com/2020/09/08/business/jpmorgan-covid-relief-misuse]
(✗) I read a WSJ article that said anxiety is also a symptom of Coronavirus....we're all doomed aren't we?

theories[1], and discouraged people from seeking medical assistance and receiving vaccines [37]. Numerous studies have examined the occurrence and spread of such misinformation [6,17]. One study found that around 70% tweets contained medical or public health information, and nearly 25% and 17.5% of the tweets were found to contain misinformation and unverifiable information, respectively [26]. In addition to the direct harm caused by misinformation, these findings are particularly concerning because users are more vulnerable to misinformation after prior exposure, leading to a ripple effect where false information spreads faster than accurate and verifiable news [3,53].

Notable studies on the proliferation of false information have distinguished posts containing original content from those that re-transmit preexisting content (*i.e.*, "derivative" posts) [3,5,42]. This is indeed important, since derivative posts are often shared together with reference to external sources such as renowned news outlets, to establish their own credibility [15]. In Table 1, we see two derivative posts propagating claims related to COVID-19, attributing the respective claims to articles from CNN while providing a reference (✓) and the Wall Street Journal without any reference (✗). The verifiability of information by ordinary users and professional fact-checkers alike is greatly impeded by social media posts of the latter kind, where no source is cited. It is unfortunate that such posts, where information is presented without a clear reference to its source, are quite common. Clearly, this demonstrates the need for an intelligent and automated system for the retrieval of news articles (if any such article exists) that support the information in a tweet.

The unique characteristics of social media posts, including their brevity, informal language, and non-standard grammar and spelling, present significant challenges for natural language processing (NLP) tasks. As a result, Twitter and other short-form social media content have been treated as a linguistic genre distinct from traditional newswire [19,45]. To effectively retrieve information from news articles for social media posts, a retrieval system must be robust and adaptable to the specific nuances of both genres. This necessitates a thorough comprehension of the characteristics that set social media posts apart from conventional news articles and the capacity to modify retrieval techniques accord-

[1] UK phone masts attacked amid 5G-coronavirus conspiracy theory. https://www.theguardian.com/uk-news/2020/apr/04/uk-phone-masts-attacked-amid-5g-coronavirus-conspiracy-theory.

ingly. Developing an approach that can bridge this gap is thus crucial to ensure the integrity of information presented in contemporary social media for consumption by the global populace. To that end, we commence by reviewing relevant literature (Sect. 2) before describing our contributions:

(a) a dataset of 11, 444 tweets and $57K$ news articles related to the COVID-19 pandemic (Sect. 3);
(b) an *ad hoc* cross-genre IR task to retrieve news articles from reputable sources that corroborate the information presented in a tweet related to COVID-19. To this end, we present a two-stage pipeline comprising (i) candidate selection from the large set of news articles, and (ii) re-ranking using a transformer-based cross-encoder (Sect. 4); and
(c) an in-depth analysis of our findings (Sect. 5), demonstrating the viability of cross-genre information retrieval for information integrity.

2 Related Work

Fact-checking and identifying misinformation has a long and rich history in journalism [18]. With the rise of user-generated content and social media platforms, however, the last two decades have seen a massive surge in automated computational (or at least computer-aided and partially automated) means of fact-checking and the detection of various forms of misinformation (see, for example, the multilingual architecture proposed by Martin et al. [33] or the recent survey by Guo et al. [20]). Two threads of research are particularly relevant to our work: (i) identifying misinformation on social networks, and (ii) information retrieval across different genres. The contributions we present in subsequent sections grow both these directions of research, discussed hereunder.

Misinformation Detection on Social Media Platforms. News consumption habits have undergone a drastic change since the advent of social media, especially Twitter. There are manifold reasons for it, leading to a large fraction of the populace consuming news and information on social media. For example, nearly half of all Americans use social media as a news source at least sometimes[2]. A manual verification of information being impossible due to the deluge of data and rate at which information spreads, computational means have been increasingly explored. Some techniques involve the retrieval of posts associated with misinformation identified *a priori* by the researchers [24,49]. These are, by design, helpful in the study of the propagation of misinformation in a network, but not in their initial identification. To identify falsehoods, methods can broadly be categorized into three stages: claim detection, evidence discovery, and claim verification. The first of these often relies on a claim being worth checking and it being checkable [23,25]. The discovery of evidence is a more recent

[2] News Consumption Across Social Media in 2021. Pew Research Center, www.pewresearch.org/journalism/2021/09/20/news-consumption-across-social-media-in-2021/.

research thrust, in contrast to earlier work on misinformation, which did not use external evidence beyond the immediate text [14,52]. Recently, Dougrez-Lewis et al. [13] introduced a dataset of tweets along with external evidence retrieved from the Web, and showed how the inclusion of external evidence can benefit rumor identification models. Their work, however, was restricted to five pre-identified rumors. On the other hand, Haouari [21] proposed a method to verify tweets immediately upon posting by evidence retrieval from multiple sources, but these sources were limited to other tweets. For a more in-depth survey of misinformation detection in social media, we point the reader to Shu et al. [47].

With regard to COVID-19, misinformation had a particularly deleterious effect in many ways. The extent of misinformation has been studied by many. Kouzy et al. [26], for example, found in a small sample of 673 tweets, that 25% contained misinformation, and 17.4% propagated unverifiable information. We thus use COVID-19 as a timely case study that distinguishes itself from this literature by not only looking for external evidence in a specialized domain, but also doing so across two different genres.

Information Retrieval Methods and Datasets. The majority of modern *ad hoc* IR systems are based on bag-of-words representations with term-weighting approaches such as the BM25 algorithm or its variants [31,44]. BM25 is a ranking function commonly used for text document retrieval tasks, which determines the relevance of a document to a query based on the frequency of query terms in the document and other factors. These methods have traditionally been employed for query-based news retrievals that focused on specific parts of an article [8]. With the success of BERT [12] and its successors [27,29,46] in NLP, however, more recent IR research on evidence retrieval has incorporated these language models [48]. Of particular relevance to this work is that some applications of transformer-based encoders show the value of contextual information in re-ranking candidate documents retrieved by models based on BM25 [39]. With this body of work as our motivation, we design a two-stage pipeline for information retrieval across news and social media wherein classical term-weighting as well as transformer-based cross-encoders are used.

Most IR datasets of sufficient size are able to provide a reasonably accurate representation of a specific domain or topic, making them effective benchmarks for the evaluation and comparison of various IR models. Notable examples include the various TREC collections[3], the NTCIR task datasets[4], and the CLEF initiative task datasets[5]. These are largely confined to single genres and document types. Despite several cross-lingual IR (CLIR) and a few other cross-genre IR datasets (*e.g.*, [50,56]), there is a comparative dearth of this latter category, even more so for retrieval tasks related to COVID-19. In this work, we present a dataset that comprises two distinct linguistic genres (tweets and news articles) as a much needed contribution not just to help mitigate the effects of misin-

[3] trec.nist.gov/data.html.
[4] research.nii.ac.jp/ntcir/data/data-en.html.
[5] www.clef-initiative.eu.

Table 2. News-related keywords: the 22 keywords used to filter Tweets.

bloomberg, cbs, cnbc, cnn, forbes, nypost, reuters, sfgate, theatlantic, wsj, abc news, chicago tribune, fox news, new york post, new york times, ny times, the atlantic, the daily beast, the guardian, us news, usa today, washington post

formation during health crises, but also as a potential benchmark for future IR tasks that query across such distinct linguistic genres.

3 Dataset

Our work starts by analyzing a subset of a large open dataset of COVID-19 related tweets developed and made publicly available by Banda *et al.* [7]. Specifically, our work focuses on 46.86 million tweets collected from March to May. Although this dataset pertains to COVID-19, it required significant pre-processing steps to ensure its suitability for our tasks. This includes the application of rigorous filtering and data cleaning procedures, such as the removal of retweets and non-English posts. Furthermore, as our work is focused on derivative posts that transmit information from reputable news agencies, we implement a more stringent criterion to establish the relevance of each post to the corresponding news article. Specifically, we employ a set of 22 carefully chosen keywords (shown in Table 2) and retain only those tweets containing at least one.

After applying the aforementioned filtering steps, we are left with a total of 876,325 Tweets. Among these derivative posts, we observe that a significant proportion of 23% (197,464) tweets have no hyperlink to any information source. This indicates a potential loss of information integrity and underscores the need to establish an information retrieval system to identify the sources of information for these Tweets. To build such an IR system, we then create a IR dataset for training purposes.

3.1 IR Dataset Creation

Our IR Dataset is based on a large corpus of check-worthy tweets about COVID-19, developed by Zuo *et al.* [57]. It offers about 30K tweets collected from March through May 2020, where each tweet contains a factual claim and is deemed worth checking for factual accuracy[6]. Moreover, each tweet refers to the external source of its information by providing a hyperlink to a news article from a reputable news publisher. This relation between a tweet and a news article provides the ground truth for strong relevance labels for our task of retrieving the article that supports the claim made in a tweet.

[6] The work by Zuo *et al.* [57] follows the notion of *check-worthiness* set forth by a large body of work in NLP research (Arslan *et al.* [4] and Hassan *et al.* [22], among others).

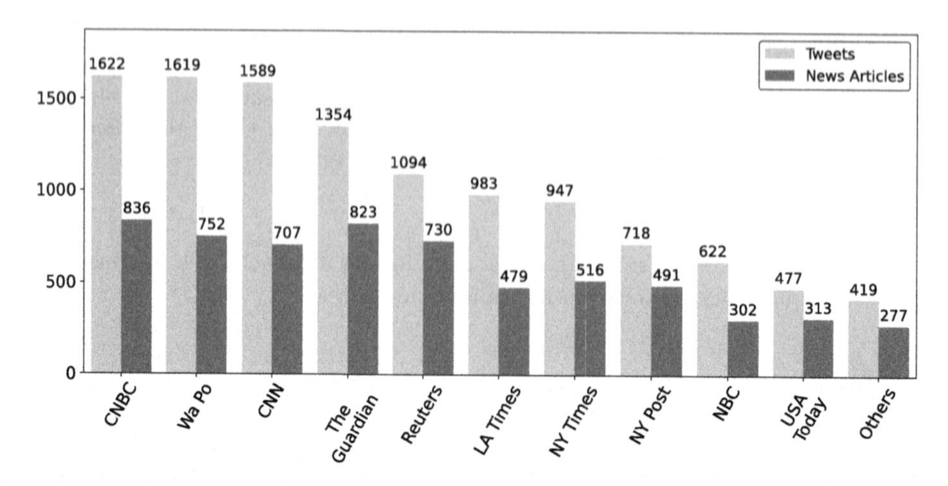

Fig. 1. Distribution of news articles and tweets citing them, across the ten most frequent news agencies (plus "Others", which comprises the combination of all news enterprises that have fewer than 400 tweets).

We notice that many tweets are mere retweets of a news article, and the linguistic content of the tweet already contains the article headline. The inclusion of such instances will make the IR task of this work overly and impractically simple, since any standard search will immediately find the exact match. Thus, we discard such tweets, retaining a corpus of 11,444 tweets and 6,226 news articles. The number of unique news articles is understandably smaller, since multiple tweets often cite the same article from a well-known news publication. The distribution of tweets and the linked news articles, over the news agencies, is shown in Fig. 1. Our final dataset[7] consists of tuples of the form $(t, \{h, b\})$, where t is the tweet with its hyperlinks removed, and $\{h, b\}$ is the headline-body pair from the news article cited by that tweet. Moreover, we consider a realistic scenario for a lay human reader who wants to verify the information in a tweet. This reader will usually need to search for a news report from a vast collection. To imitate this scenario, we add a collection of 51,003 news articles from the RSS feeds of several popular news websites over a period of three years (2020–2022). We collect these news articles based on a set of six keywords related to COVID-19: *corona, coronavirus, covid, covid-19, pandemic, quarantine*. In our entire dataset, the full text and the headline of all the articles are retained, while images and videos are discarded.

4 Experiments

Given a tweet propagating a claim pertaining to COVID-19, we present the cross-genre IR task of retrieving news articles that support it. Along the lines of other

[7] https://github.com/chzuo/adma2023_tweet_IR.

Table 3. Results of candidate selection for different models and K values when only the title of news was searched and when both title and body were searched. The best results for each K value are in bold. MiniLM* indicates that the model is pretrained on several corpora, while models with † are fine-tuned on the MS MARCO dataset.

Model	Title				Title + Body			
	R@1	R@10	R@100	R@500	R@1	R@10	R@100	R@500
BM25	63.5	75.0	81.5	86.3	58.3	**79.5**	**89.3**	**93.2**
DistilBERT†	37.5	60.8	76.0	84.1	58.0	69.6	77.7	83.0
MiniLM†	29.0	52.7	71.0	81.1	56.4	68.3	76.4	81.9
MiniLM*	40.4	67.2	82.6	89.3	**59.7**	73.0	81.9	87.3
CT-BERT	**64.1**	**79.9**	**88.5**	**92.2**	1.1	2.6	5.7	12.1

ad hoc pipelines [11,32], this comprises two stages: a retrieval system to obtain a large candidate list, and the re-ranking of this list by a transformer-based cross-encoder. Furthermore, we notice that tweets may often simply modify the news headline. Thus, using only the news headline for retrieval may boost the results as well as improve efficiency. For this reason, we conduct two sets of experiments: by considering (i) only the headline of each news article, and (ii) the headline as well as the body of the article.

4.1 Candidate Selection

In this first stage, we aim to reduce the search space for the final re-ranking task. For this, we consider the classical lexical IR approach of token-based bag-of-words models, *i.e.*, BM25 algorithm, which has remained an extremely viable model for information retrieval [30] for nearly two decades. Additionally, we also use transformed-based Bi-Encoders as semantic search. It encodes the query (*i.e.*, the tweet post) into vector space and retrieves the news embeddings that are nearby in the same space.

For the BM25 approach, proper preprocessing steps are added before retrieval, which comprises converting the words into lowercase, removing function words, and stemming. We also include text cleaning procedures designed for tweet posts, including removing Twitter user handles, emojis, and hash symbols for hashtags (the term is retained, *e.g.*, "#quarantine" to "quarantine").

As part of the Bi-Encoder approach, we use two pretrained models from Sentence-BERT [43]: MiniLM [54] and DistilBERT [46]. Those models encode the tweet t, the news headline h, and the full news (headline and body) $h{+}b$ respectively. We then obtain the ranked list of news pertinent to the tweet based on cosine similarity. Given that a claim may bear some semantic similarity to the evidence supporting it [2,35], the pretrained models are fine-tuned on several corpus, including the Natural Language Inference (NLI), the Semantic Textual Similarity (STS) benchmark datasets [9] and *etc.*. Considering the newswire article is significantly longer than the given query, we also use the pretrained models tuned on the large-scale information retrieval corpus, *i.e.*, MS MARCO Passage Ranking Dataset [38]. Moreover, since our task involves COVID-19 tweets, we use the COVID-Twitter-BERT(CT-BERT) [36], which is pretrained on 97M tweets.

For evaluation, since each tweet in our dataset cites only one newswire article, precision is not an important measure for this task. Instead, we measure recall@k ($k = 1, 10, 100, 500$) to report whether the hyperlinked news is in the top-K selection. As it has been argued in other two-stage IR pipelines [48], a high recall is crucial in this case because the correct news report will otherwise be excluded from the final re-ranking. Recall@K would also be useful in a scenario where a reader needs to verify tweets in real-time. Our retrieval system could return a shortlist of up to ten news items that are pertinent to the tweets, and the reader could rapidly scan them for verification.

Table 3 shows the results of experiments across the BM25 and Bi-Encoders, which compare the fraction of times the correct document was found in the first K ranked documents. When K is equal to 1, BM25 could locate the correct document more than 63% of the time using only the title. By increasing the value of K to 500, this reaches 86.3%. The search results get worse for smaller K values when the body of the article is also included. However, including the body gives us better results when we have a more considerable K value. One explanation could be some tweets refer to a specific part of the news that is not included in the title, but the algorithm cannot rank them on top of the list; when the K value increases, those documents are retrieved. When the title is used, CT-BERT outperforms all other algorithms in terms of results for all K values, which shows that domain-specific knowledge is crucial in this task. When adding the news body, however, the model's performance drops significantly, as the news body may be too extensive for it to embed the core information. All models, with the exception of CT-BERT, achieve similar performance when the news body is included for lower values of K, while the MiniLM model pretrained on a large corpus performs marginally better. However, the BM25 model outperforms others for $K > 1$, demonstrating this token-based approach is still a hard-to-beat baseline for asymmetric semantic search with long documents and short queries.

4.2 Re-Ranking

We keep 6,000 tweet-news pairs for training, 2,444 for development, and 3,000 for testing. We first use the BM25 approach (on the full news) with highest score in Recall@500 to generate a list of 500 newswire articles for each tweet. It is possible that the correct news was not retrieved during candidate selection. In that case, we add it back to the list. To train the Cross-Encoder, we pair the tweet t with the newswire article with headline h and body b. These concatenated strings serve as training data for our task. The ground-truth label is 1 for an input $t + h$ (or $t + h + b$, for experiments on the full news), where the article is indeed cited by the tweet. For other inputs, which are created by random sampling from the BM25 candidate list, the label is 0. As discussed by Thakur *et al.* [51] and Zuo *et al.* [56], using the right sampling strategy to create negative samples is crucial to achieving performance improvement. We randomly sample 4 news

Table 4. Re-ranking results. The best results for each K value are in bold. BM25 and CT-BERT without training serve as the baseline. MiniLM* indicates that the model is pretrained on several corpus, while models with † are pretrained as Cross-Encoder on the MS MARCO dataset.

Model	Title				Title + Body			
	R@1	R@10	R@100	R@MRR	R@1	R@10	R@100	R@MRR
Baseline								
BM25	63.5	75.7	81.3	17.8	57.8	79.7	89.4	17.2
CT-BERT	64.4	79.7	88.8	18.4	1.3	2.5	5.6	0.5
MiniLM*	68.3	85.3	97.5	74.4	67.7	88.9	99.6	75.4
MiniLM†	70.1	88.3	98.4	76.5	76.9	92.1	99.4	82.5
TinyBERT†	67.9	83.7	97.6	73.5	73.6	88.7	98.1	79.0
CT-BERT	**73.2**	**93.5**	**99.5**	**80.5**	**77.6**	**95.0**	**99.8**	**83.9**

articles from the top 50–100 candidate list retrieved by the BM25 approach. Although random selection typically results in dissimilar pairs, these articles from the candidate list may share similarities with the tweet someway, making them strong negative samples to prevent the classification process from being overly simple. We use this labeled data to tune pretrained transformer-based models. During prediction, we use the softmax probabilities of the classification scores to re-rank the news article for each tweet and calculate recall@k for $k = 1, 3, 5, 20$, as well as the mean reciprocal rank (MRR).

As part of our experiments, we train different models – MiniLM from sentence-BERT, MiniLM (tuned as Cross-Encoder on the MS MARCO dataset), TinyBERT (tuned as Cross-Encoder on MS MARCO dataset), and CT-BERT. All models are trained for 2 and 3 epoch, batch sizes of 16 and 24, and maximum sequence lengths of 256 and 512 tokens. The final hyperparameters are manually chosen based on MRR achieved on the development set.

The results of re-ranking are shown in Table 4. As a benchmark for candidate selection, BM25 is difficult to surpass, but token-based lexical search makes mistakes when words from the newswire article do not match those in the tweet, which frequently happens when the same or similar meanings are expressed across two different genres. A Cross-Encoder based re-ranker can perform attention across the query and the document, improving the final results with higher performance. Following training, CT-BERT performs better than competitors, and it may include the right news story in the Top 10 list more frequently than 93% of the time when only the title is used. Even if a slight improvement could result from including the news body, relying just on the news title would be a more efficient way for this IR task. Also, for MiniLM, fine-tuning on the MS MARCO dataset significantly improves, indicating the importance of task-specific training (Table 5).

Table 5. The tweet disseminates information pertaining to three newswire articles. In the retrieval results generated by CT-BERT, the cited news ranks third.

Tweet: JC Penney files for bankruptcy during coronavirus pandemic #retailers #retail #retailbankruptcy

Cited News: Long-struggling JC Penney files for bankruptcy as coronavirus crushes hopes for a quick turnaround (Source: CNBC)

Retrieval results

1. *JCPenney files for bankruptcy as the coronavirus hammers retail (Source: NBC)*
2. *JC Penney could join a growing list of bankruptcies during the coronavirus pandemic (Source: CNBC)*

5 Discussion

5.1 News Event Clustering

Our dataset contains a set of 57k covid-19-related news reports, and a subset of 6k newswire articles ($\{n_i\}$) are linked by the tweets. It is also worth pointing out that our evaluation is based on relevance labels from these hyperlinks. However, as noted in the by Liu *et al.* [28] regarding the redundancy of news ("*a news event or story is likely to be reported and discussed by multiple publishers*"), we may encounter the case where several newswire articles regarding the same event are retrieved when we perform information retrieval for the given tweet. It is possible that some documents that are given higher rankings actually support the tweet, but are judged as irrelevant because they are not cited (as shown in Table 4). This will impact our evaluation results of the models. In fact, many IR benchmark datasets – *e.g.*, MS MARCO [38] – do not offer strong non-relevance labels. Instead, one could consider the results as a lower bound in this general evaluation setup (*i.e.*, With exhaustive ground-truth labels of non-relevance, the true performance are better, not worse.)

Additionally, we run same-event news searches to take a second look at the findings. We characterize a group of documents as referring to the same event when they have a simultaneous publication date, have semantic similarity, and share comparable named entities. These criteria are similar to the definition of an event in other previous work [1,55]. We perform a three-step pipeline to obtain such sets. First, for each news n_i linked by tweets, we identify a set of candidate news articles ($\{N_i\}$) from the overall news collection that was published five days before and after it. In the second step, we employ sentence-BERT to measure the semantic similarity between the news n_i and the candidates N_i among the headlines, and only keep the articles N_i above a given threshold. Then, we use the named entity recognition (NER) algorithm to obtain the named entities from the full news (title and body) and compare the intersection between the retained news candidates N_i and news n_i. Only news articles that exceed the predetermined threshold are kept. For articles N_i retained after this step, it is related to the same event with the given news n_i. Two independent readers who each received a random sample of 50 such sets noticed this. They both agreed that the final candidate lists for each set present the same event with the provided news. Finally, for 1,461 news articles linked by tweet (23.5%), we obtained a set of reports on the same news story separately. To adjust the evaluation results, for each tweet, if the linked news n_i has a same-event news set $\{N_i\}$, we additionally assign the relevant label to the news N_i. We find that the MRR for CT-BERT trained on title and body increased slightly, from 83.9 to 85.0, while Precision@1(equal to Recall@1 without same-event news searching) improves from 77.6 to 78.8.

5.2 Error Analysis

We conduct a thorough analysis of instances where the CT-BERT model fails to accurately identify the news in the Top 5 list. Our findings reveal that 10% of the errors were due to inaccurate or dirty data, as these tweets do not contain any factual information. Additionally, over 40% of the mistakes are Type I errors, where the tweets are only relevant to the latter half of a lengthy news document, making it challenging for the Cross-Encoder model to fully encode the information due to its token length restriction. The remaining 50% of errors are Type II errors: the tweet contains an additional statement that is not in the news, even though the content is relevant.

This has prompted us to contemplate the utilization of the model. Firstly, for posts on social media platforms that lack reference sources, we can leverage the model to identify relevant authoritative news articles to aid in verifying the authenticity of the original post. Secondly, with regard to tweets that have listed information sources, we can also conduct IR research to uncover additional news reports that pertain to the news event, thus mitigating the potential for any bias inherent in a singular news report. Additionally, by evaluating the similarity score generated by the model between the tweet and the news, we can make an assessment of the extent to which the original tweet is supported by the cited news. If the similarity score is comparatively low, this indicates that the tweet includes information not present in the news, and a more rigorous fact-checking process for the tweet would be necessary.

6 Conclusion

In this work, we provide a novel dataset by linking tweets related to COVID-19 to reliable news articles. We use this linked dataset to then present a pipeline for the retrieval of news related to the tweeted claim(s). The findings of our investigation show that cross-genre information retrieval is practical for confirming the veracity of information about the epidemic, based on the support such information finds in journalistic organizations of repute. Furthermore, our findings emphasize the importance of incorporating domain-specific knowledge in the information retrieval process. This highlights the need for responsible and informed social media usage, particularly in times of crisis, where access to accurate information is of paramount importance. By utilizing reliable sources of information, we can help ensure the integrity of such information across genres, as we demonstrated through our experiments across traditional and social media.

Acknowledgment. This work was supported by the Fundamental Research Funds for the Central Universities (No. 63232115).

References

1. Allan, J., Carbonell, J.G., Doddington, G., Yamron, J., Yang, Y.: Topic detection and tracking pilot study final report. Technical report, Carnegie Mellon University (1998). https://doi.org/10.1184/R1/6626252.v1
2. Alonso-Reina, A., Sepúlveda-Torres, R., Saquete, E., Palomar, M.: Team GPLSI. Approach for automated fact checking. In: Proceedings of the Second Workshop on Fact Extraction and VERification (FEVER), pp. 110–114 (2019). https://doi.org/10.18653/v1/D19-6617
3. Arif, A., Shanahan, K., Chou, F., Dosouto, Y., Starbird, K., Spiro, E.S.: How information snowballs: Exploring the role of exposure in online rumor propagation. In: Proceedings of the 19th Conference on Computer-Supported Cooperative Work & Social Computing, 2016, pp. 465–476 (2016). https://doi.org/10.1145/2818048.2819964
4. Arslan, F., Hassan, N., Li, C., Tremayne, M.: A benchmark dataset of check-worthy factual claims. In: Proceedings of the International AAAI Conference on Web and Social Media, vol. 14, no. 1, pp. 821–829 (2020). https://doi.org/10.1609/icwsm.v14i1.7346
5. Badawy, A., Ferrara, E., Lerman, K.: Analyzing the digital traces of political manipulation: the 2016 Russian interference twitter campaign. In: ASONAM, pp. 258–265 (2018). https://doi.org/10.1109/asonam.2018.8508646
6. Balakrishnan, V., Ng, W.Z., Soo, M.C., Han, G.J., Lee, C.J.: Infodemic and fake news - a comprehensive overview of its global magnitude during the COVID-19 pandemic in 2021: a scoping review. Int. J. Disaster Risk Reduct. **78**, 103144 (2022)
7. Banda, J.M., et al.: A large-scale COVID-19 Twitter chatter dataset for open scientific research-an international collaboration. Epidemiologia **2**(3), 315–324 (2021). https://doi.org/10.3390/epidemiologia2030024
8. Catena, M., Frieder, O., Muntean, C.I., Nardini, F.M., Perego, R., Tonellotto, N.: Enhanced news retrieval: passages lead the way! In: SIGIR, SIGIR 2019, pp. 1269–1272 (2019). https://doi.org/10.1145/3331184.3331373
9. Cer, D.M., Diab, M.T., Agirre, E., Lopez-Gazpio, I., Specia, L.: SemEval-2017 task 1: semantic textual similarity multilingual and crosslingual focused evaluation. In: SemEval, pp. 1–14 (2017)
10. Chong, M., Froehlich, T.J., Shu, K.: Racial attacks during the COVID-19 pandemic: politicizing an epidemic crisis on longstanding racism and misinformation, disinformation, and misconception. Proc. Assoc. Inf. Sci. Technol. **58**(1), 573–576 (2021). https://doi.org/10.1002/pra2.501
11. Dai, Z., Callan, J.: Deeper text understanding for IR with contextual neural language modeling. In: SIGIR, pp. 985–988 (2019). https://doi.org/10.1145/3331184.3331303

12. Devlin, J., Chang, M., Lee, K., Toutanova, K.: BERT: pre-training of deep bidirectional transformers for language understanding. In: NAACL-HLT, pp. 4171–4186 (2019). https://doi.org/10.18653/v1/n19-1423
13. Dougrez-Lewis, J., Kochkina, E., Arana-Catania, M., Liakata, M., He, Y.: PHEMEPlus: enriching social media rumour verification with external evidence. In: Proceedings of the Fifth Fact Extraction and VERification Workshop (FEVER) (2022)
14. Dungs, S., Aker, A., Fuhr, N., Bontcheva, K.: Can rumour stance alone predict veracity? In: COLING, pp. 3360–3370 (2018)
15. Fogg, B.J., Cuellar, G., Danielson, D.: Motivating, influencing, and persuading users: an introduction to captology. In: The Human Computer Interaction Handbook: Fundamentals, Evolving Technologies and Emerging Applications, pp. 159–172. CRC Press (2007). https://doi.org/10.1201/9781410615862
16. Fung, I.C.H., et al.: Social media's initial reaction to information and misinformation on Ebola, August 2014: facts and rumors. Public Health Rep. **131**(3), 461–473 (2016). https://doi.org/10.1177/003335491613100312
17. Gabarron, E., Oyeyemi, S.O., Wynn, R.: Covid-19-related misinformation on social media: a systematic review. Bull. World Health Organ. **99**, 455-463A (2021)
18. Graves, L.: Deciding What's True: The Rise of Political Fact-Checking in American Journalism. Columbia University Press (2016). https://doi.org/10.7312/grav17506
19. Gui, T., et al.: Transferring from formal newswire domain with hypernet for twitter POS tagging. In: EMNLP, pp. 2540–2549 (2018). https://doi.org/10.18653/v1/d18-1275
20. Guo, Z., Schlichtkrull, M., Vlachos, A.: A survey on automated fact-checking. Trans. Assoc. Comput. Linguist. **10**, 178–206 (2022). https://doi.org/10.1162/tacl_a_00454
21. Haouari, F.: Evidence-based early rumor verification in social media. In: ECIR 2022, pp. 496–504 (2022). https://doi.org/10.1007/978-3-030-99739-7_61
22. Hassan, N., Arslan, F., Li, C., Tremayne, M.: Toward automated fact-checking: Detecting check-worthy factual claims by claimbuster. In: SIGKDD, pp. 1803–1812 (2017). https://doi.org/10.1145/3097983.3098131
23. Hassan, N., Li, C., Tremayne, M.: Detecting check-worthy factual claims in presidential debates. In: CIKM, pp. 1835–1838 (2015). https://doi.org/10.1145/2806416.2806652
24. Jin, Z., Cao, J., Guo, H., Zhang, Y., Wang, Yu., Luo, J.: Detection and analysis of 2016 US presidential election related rumors on Twitter. In: Lee, D., Lin, Y.-R., Osgood, N., Thomson, R. (eds.) SBP-BRiMS 2017. LNCS, vol. 10354, pp. 14–24. Springer, Cham (2017). https://doi.org/10.1007/978-3-319-60240-0_2
25. Konstantinovskiy, L., Price, O., Babakar, M., Zubiaga, A.: Toward automated factchecking: developing an annotation schema and benchmark for consistent automated claim detection. Digit. Threats **2**(2) (2021). https://doi.org/10.1145/3412869
26. Kouzy, R., et al.: Coronavirus goes viral: quantifying the COVID-19 misinformation epidemic on Twitter. Cureus **12**(3), e7255 (2020)
27. Lan, Z., Chen, M., Goodman, S., Gimpel, K., Sharma, P., Soricut, R.: ALBERT: a lite BERT for self-supervised learning of language representations. In: ICLR (2020)
28. Liu, J., Liu, T., Yu, C.: NewsEmbed: modeling news through pre-trained document representations. In: SIGKDD, pp. 1076–1086 (2021). https://doi.org/10.1145/3447548.3467392
29. Liu, Y., et al.: RoBERTa: a robustly optimized BERT pretraining approach. CoRR abs/1907.11692 (2019)

30. Lv, Y., Zhai, C.: Adaptive term frequency normalization for BM25. In: CIKM, pp. 1985–1988 (2011). https://doi.org/10.1145/2063576.2063871
31. Lv, Y., Zhai, C.: When documents are very long, BM25 fails! In: SIGIR, pp. 1103–1104 (2011). https://doi.org/10.1145/2009916.2010070
32. MacAvaney, S., Yates, A., Cohan, A., Goharian, N.: CEDR: contextualized embeddings for document ranking. In: SIGIR, pp. 1101–1104 (2019). https://doi.org/10.1145/3331184.3331317
33. Martín, A., Huertas-Tato, J., Álvaro Huertas-García, Villar-Rodríguez, G., Camacho, D.: FacTeR-Check: semi-automated fact-checking through semantic similarity and natural language inference. Knowl.-Based Syst. **251**, 109265 (2022). https://doi.org/10.1016/j.knosys.2022.109265
34. Miller, M., Banerjee, T., Muppalla, R., Romine, W., Sheth, A.: What are people tweeting about Zika? An exploratory study concerning its symptoms, treatment, transmission, and prevention. JMIR Public Health Surveill. **3**(2), e38 (2017)
35. Mohtarami, M., Baly, R., Glass, J.R., Nakov, P., Màrquez, L., Moschitti, A.: Automatic stance detection using end-to-end memory networks. In: NAACL-HLT, pp. 767–776 (2018). https://doi.org/10.18653/v1/n18-1070
36. Müller, M., Salathé, M., Kummervold, P.E.: COVID-Twitter-BERT: a natural language processing model to analyse COVID-19 content on twitter. CoRR abs/2005.07503 (2020)
37. Muric, G., Wu, Y., Ferrara, E.: COVID-19 vaccine hesitancy on social media: building a public twitter data set of antivaccine content, vaccine misinformation, and conspiracies. JMIR Public Health Surveill. **7**(11), e30642 (2021)
38. Nguyen, T., et al.: MS MARCO: a human generated machine reading comprehension dataset. In: NeurIPS. CEUR Workshop Proceedings, vol. 1773 (2016)
39. Nogueira, R., Cho, K.: Passage Re-ranking with BERT (2020). https://doi.org/10.48550/arXiv.1901.04085
40. Ortiz-Martínez, Y., Jiménez-Arcia, L.F.: Yellow fever outbreaks and Twitter: rumors and misinformation. Am. J. Infect. Control **45**(7), 816–817 (2017)
41. Oyeyemi, S.O., Gabarron, E., Wynn, R.: Ebola, Twitter, and misinformation: a dangerous combination? BMJ **349**, g6178 (2014)
42. Piergiorgio, C., Giulia, A., Riccardo, G., Eugenia, P., Manlio, D.D.: The voice of few, the opinions of many: evidence of social biases in Twitter COVID-19 fake news sharing. R. Soc. Open Sci. **9**(220716) (2022). https://doi.org/10.1098/rsos.220716
43. Reimers, N., Gurevych, I.: Sentence-BERT: sentence embeddings using Siamese BERT-networks. In: EMNLP-IJCNLP, pp. 3980–3990 (2019). https://doi.org/10.18653/v1/D19-1410
44. Robertson, S.E., Zaragoza, H.: The probabilistic relevance framework: BM25 and beyond. Found. Trends Inf. Retr. **3**(4), 333–389 (2009). https://doi.org/10.1561/1500000019
45. Rosenthal, S., Ritter, A., Nakov, P., Stoyanov, V.: SemEval-2014 task 9: sentiment analysis in twitter. In: SemEval@COLING, pp. 73–80 (2014). https://doi.org/10.3115/v1/s14-2009
46. Sanh, V., Debut, L., Chaumond, J., Wolf, T.: DistilBERT, a distilled version of BERT: smaller, faster, cheaper and lighter. CoRR abs/1910.01108 (2019)
47. Shu, K., Sliva, A., Wang, S., Tang, J., Liu, H.: Fake news detection on social media: a data mining perspective. SIGKDD Explor. Newsl. **19**(1), 22–36 (2017). https://doi.org/10.1145/3137597.3137600

48. Soleimani, A., Monz, C., Worring, M.: BERT for evidence retrieval and claim verification. In: Jose, J.M., et al. (eds.) ECIR 2020. LNCS, vol. 12036, pp. 359–366. Springer, Cham (2020). https://doi.org/10.1007/978-3-030-45442-5_45
49. Starbird, K., Maddock, J., Orand, M., Achterman, P., Mason, R.M.: Rumors, false flags, and digital vigilantes: misinformation on Twitter after the 2013 Boston marathon bombing. In: iConference 2014 Proceedings (2014). https://doi.org/10.9776/14308
50. Sun, S., Duh, K.: CLIRMatrix: a massively large collection of bilingual and multilingual datasets for Cross-Lingual Information Retrieval. In: EMNLP, pp. 4160–4170 (2020). https://doi.org/10.18653/v1/2020.emnlp-main.340
51. Thakur, N., Reimers, N., Daxenberger, J., Gurevych, I.: Augmented SBERT: data augmentation method for improving bi-encoders for pairwise sentence scoring tasks. In: NAACL-HLT, pp. 296–310 (2021). https://doi.org/10.18653/v1/2021.naacl-main.28
52. Volkova, S., Shaffer, K., Jang, J.Y., Hodas, N.: Separating facts from fiction: linguistic models to classify suspicious and trusted news posts on Twitter. In: ACL, pp. 647–653 (2017). https://doi.org/10.18653/v1/P17-2102
53. Vosoughi, S., Roy, D., Aral, S.: The spread of true and false news online. Science **359**(6380), 1146–1151 (2018)
54. Wang, W., Wei, F., Dong, L., Bao, H., Yang, N., Zhou, M.: MINILM: deep self-attention distillation for task-agnostic compression of pre-trained transformers. In: NeurIPS (2020)
55. Yang, Y., Carbonell, J.G., Brown, R.D., Pierce, T., Archibald, B., Liu, X.: Learning approaches for detecting and tracking news events. IEEE Intell. Syst. **14**(4), 32–43 (1999). https://doi.org/10.1109/5254.784083
56. Zuo, C., Acharya, N., Banerjee, R.: Querying across genres for medical claims in news. In: EMNLP, pp. 1783–1789 (2020). https://doi.org/10.18653/v1/2020.emnlp-main.139
57. Zuo, C., Banerjee, R., Chaleshtori, F.H., Shirazi, H., Ray, I.: Seeing should probably not be believing: the role of deceptive support in COVID-19 misinformation on twitter. J. Data Inf. Quality **15**(1) (2022). https://doi.org/10.1145/3546914

PBCI-DS: A Benchmark Peripheral Blood Cell Image Dataset for Object Detection

Shuyao You[1], Mingshi Li[1], Wanli Liu[1], Hongzan Sun[2], Yuexi Wang[2], Marcin Grzegorzek[3], and Chen Li[1](✉)

[1] Microscopic Image and Medical Image Analysis Group, College of Medicine and Biological Information Engineering, Northeastern University, Shenyang, China
lichen@bmie.neu.edu.cn

[2] Shengjing Hospital, China Medical University, Shenyang, China
sunhz@sj-hospital.org

[3] Institute of Medical Informatics, University of Luebeck, Luebeck, Germany
marcin.grzegorzek@uni-luebeck.de

Abstract. Blood testing has always been one of the important methods for disease diagnosis, but currently, blood testing instruments face the problems of long time consumption, complex processes, and limited detection types. Therefore, a blood cell dataset for artificial intelligence is necessary. *Peripheral Blood Cell Image Dataset* (PBCI-DS) provides a total of 17092 images of 8 categories of peripheral blood cells and corresponding cell labeling files. The purpose of PBCI-DS is to serve as a training model for object detection. The experiment uses four YOLO series models (YOLO-v5s, YOLO-v5l, YOLO-v6, YOLO-v7) and SSD models of deep learning methods to train the database, and compares and evaluates the results to prove the effectiveness and usability of PBCI-DS.

Keywords: Peripheral Blood Cell Dataset · Object Detection · Deep Convolution Neural Network

1 Introduction

1.1 Peripheral Blood Cells

In the diagnosis of many diseases, blood test is an important basis for judgment. At present, blood cell analyzer and flow cytometry are usually used in various hospitals for blood testing. However, the accuracy and efficiency are not ideal [10]. In this paper, a benchmark *Peripheral Blood Cell Image Dataset* (PBCI-DS) for Object Detection is introduced, which can be used to develop a peripheral blood cell images object detection system based on convolutional neural networks [1]. PBCI-DS contains a total of 17,092 images of cells, divided into the following eight categories: neutrophils, eosinophils, basophils, lymphocytes, monocytes, immature granulocytes (metamyelocytes, myelocytes and promyelocytes), erythroblasts, and platelets (thrombocytes). The image information of the dataset is shown in Table 1.

X. Yang et al. (Eds.): ADMA 2023, LNAI 14180, pp. 510–519, 2023.
https://doi.org/10.1007/978-3-031-46677-9_35

Table 1. Images information of the dataset.

Class/Dataset	Train	Test	Total
Neutrophils	2658	671	3329
Eosinophils	2456	661	3117
Basophils	987	231	1218
Lymphocytes	953	261	1214
Monocytes	1114	306	1420
Immature Granulocytes (Metamyelocytes, Myelocytes, Promyelocytes)	2295	600	2895
Erythroblasts	1225	326	1551
Platelets (Thrombocytes)	1849	499	2348

These images are captured using CellaVision DM96 [20]. The images are in Joint Photographic Experts Group (JPG) format and 360 × 360 pixels in size and are marked by clinical pathologists at the hospital clinic. The peripheral blood cell dataset images are of high quality and cover a wide range of normal peripheral blood cell types. This dataset can be used to perform training and testing of machine and deep learning models to achieve automatic classification of peripheral blood cells [4]. The dataset can also be used for model benchmarking and comparison, acting as a common canonical image set. In addition, this dataset can be used as a model weight initializer, which can be further trained to classify other types of abnormal cells [2].

1.2 Purpose of Blood Test

Clinical blood tests can provide measurements for red blood cell count (RBC), white blood cell count (WBC), white blood cell classification count (DC), and other parameters, offering valuable feedback on various aspects of blood composition. The complete blood count (CBC) is an important and powerful diagnostic tool [6]. It can be used to monitor response to treatment, measure the severity of disease or differential diagnosis as a starting point for developing a checklist. Doctors can see what categories of cells and what state they are. Many human health conditions can be predicted through blood tests. For example, liver blood tests that measure bilirubin, platelet count and prothrombin time (PT) in blood can effectively help detect abnormalities [18]. A serum microRNA labeling test can be a useful tool for early lung cancer screening in high-risk groups [17]. Blood tests can also be combined with PET-CT to screen for cancer and guide the feasibility of intervention [12].

1.3 Image Processing and Analysis

With the rapid development of computer science, machine learning has significantly improved the ability of medical industry to assist diagnosis. Digital image processing, machine learning and deep learning are among the hottest technologies, and machine learning can be used for everything from face recognition to autonomous driving. The prevalence of computer aids diagnostic techniques. These technologies can also be applied to the recognition and classification of blood cells in medicine, and computer technology can also be attempted to solve the problem of blood cell detection. Alam et al. [3] proposed a detection method using YOLO algorithm to identify and count blood cells. Finally, the accuracy of red blood cells (RBCS) and white blood cells (WBCS) [5] is calculated using the iterative circle method. The average accuracy of the final WBC is 98.4% and the average accuracy of RBC is 95.3% in 100 training images.

1.4 Contribution

At present, there is a limited availability of open-source datasets suitable for scientific research. In comparison to other blood cell datasets, PBCI-DS stands out with its extensive variety of cell types and images. Furthermore, unlike other datasets, the microscopic images in PBCI-DS are magnified by the object, which greatly enhances the accuracy of feature extraction for target cells. Additionally, PBCI-DS provides high-quality object label files specifically designed for deep learning algorithms. These label files can be directly utilized for multiple object detection and analysis, and have yielded excellent training results in this regard. The significant contributions of PBCI-DS can be summarized as follows:
(1) PBCI-DS, as an open source dataset available for researchers to use, contributes to the development of peripheral blood cell detection.
(2) PBCI-DS contains high-quality corresponding object label files that can be used for algorithm and model evaluation. This label file can also be directly used for the detection and analysis of multiple objects.
(3) The performance analysis of the multi object detection model on PBCI-DS is provided, which helps further mechanical learning and deepens the auxiliary application of object detection in the medical field.

2 Dataset Information of PBCI-DS

PBCI-DS contains a total of 17092 cell microscopic images, including eight types: Neutrophils [16], Eosinophils, Basophils [15], Lymphocytes [11], Monocytes [8], Immature Granulocites [7], Erythroblasts [22], Platelets [9]. Among them, there are three types of Immature Granulocytes: Metamyelocytes, Myeloccytes, and Promyelocytes, while Platelets also include Thrombocytes. Two medical scientists manually prepare the object file in ".XML" in a format corresponding to 17092 original images (from 06-2022 to 02-2023). In order to more intuitively present the labeling situation of PBCI, detailed information for each type of cell is listed in Table 2. Figure 1 shows an example of cell labeling for specific categories. The labeling file image of PBCI-DS is manually labeled based on the following two rules:

Rule A: If there are two or more target cells exposed, more than 80% of the cell volume can recognize PBCI. In all images, over 98% of the images are labeled with category labels corresponding to 8 categories.
Rule B: Except for the central cells, other types of cells and impurities in the image background are not labeled.

PBCI-DS is released free of charge for non commercial purposes on: https://www.data-in-brief.com/article/S2352-3409(20)30368-1/fulltext

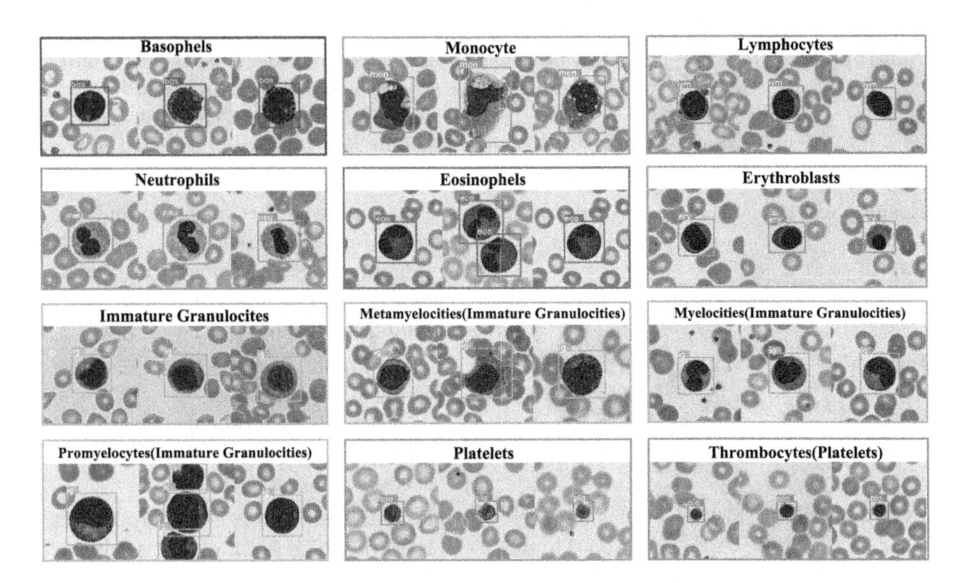

Fig. 1. An example of PBCI-DS annotation

3 Object Detection Methods for PBCI-DS Evaluation in This Paper

3.1 YOLO-v5

YOLO-v5 is consisting of three main components: the backbone network, neck, and head. Based on the depth and width of the network architecture, YOLO-v5 is divided into four versions: YOLO-v5s, YOLO-v5m, YOLO-v5l, and YOLO-v5. The architecture of YOLO-v5 is illustrated in the diagram. The CBS module is the fundamental building block of YOLO-v5, incorporating convolutional (Conv) layers, batch normalization (BN) layers, and SiLU activation layers. The original backbone of YOLO-v5 is CSPDarknet53, which includes two types of CSP modules. Additionally, the Spatial Pyramid Pooling (SPP) module provides different receptive fields to enhance feature representation. The activation functions used in YOLO-v5 are Leaky ReLU and Sigmoid. YOLO-v5 selects Stochastic Gradient

Descent(SGD) [13] as the optimization function. The classification loss and confidence loss in the Loss function are binary Cross entropy losses, while the location loss is based on CIOU [21].

Table 2. Introduction to Database Cell Types in Medicine.

Name	Dataset quantity	Source	Function Description
Neutrophils	3329	Multipotent Progenitor Cells	Invasive microorganisms react, and changes in their quantity and function can reflect the immune status and inflammation level of the body
Eosinophils	3117	Myeloblasts	Participate in the initiation and reproduction of various inflammatory reactions in the body, as well as regulators of innate and adaptive immunity
Basophils	1218	Promyelocytes	The number and functional changes of alkaline particles in the cytoplasm of basophils can indicate parasitic infections, allergic diseases, and certain malignant tumors
Lymphocytes	1214	Bone Marrow and Thymus	It is divided into B cells and T cells, both of which are antigen specific lymphocytes. Play important immune functions
Monocytes	1420	Monoblasts	Monocyte can recognize and ingest pathogens, such as bacteria, viruses, fungi, parasites and other pathogens to eliminate, and limit the spread of infection
Immature Granulocytes	2895	Hematopoietic Stem Cells	Although relatively immature in function, it may still participate in the immune response to increase its ability to phagocytose and kill pathogens
Erythroblasts	1551	Hematopoietic Stem Cells	Synthesis and differentiation into mature red blood cells
Platelets (Thrombocytes)	2348	Megakaryocytes	Participate in blood coagulation and hemostasis processes, activate coagulation cascade reactions. It can also play a regulatory role in the activation and function of immune cells

YOLO-v5s strikes a balance between speed and model size, achieving a certain level of accuracy and real-time requirements for object detection. YOLO-v5l has a larger network structure and parameter size, providing higher detection accuracy and better object detection performance.

3.2 YOLO-v6

The Backbone module of YOLO-v6 utilizes the more efficient EfficientNet as its backbone network. The Neck part of the network is built based on Rep and PAN, forming Rep-PAN. The Head module has been decoupled, separating the processes of bounding box regression and class classification. This decoupling not only speeds up the convergence but also reduces the complexity of the Head module. YOLO-v6 introduces a more efficient decoupled Head structure while still using the anchor-free approach. In addition to the model structure, a new bounding box regression loss function called SIOU is introduced [24]. The calculation formula of SIOU Loss function is: $SIOU = IoU \times ShapeSimilarity$. Among them, Shape Similarity is used to adjust the value of IoU to consider the shape factor of the bounding box.

3.3 YOLO-v7

The YOLO-v7 algorithm network consists of three convolutional modules: LSTM module, Dropout module, and Softmax output module. The LSTM module is mainly detected through connection and full activation, while the Dropout module is mainly added after the first two networks to prevent overfitting. The output part of Softmax adopts convolutional processing [19]. Dropout is a commonly used regularization technique used to reduce overfitting problems in neural networks.

The backbone network design of YOLO-v7 is mainly based on Shufflenet or MobileNet. In addition to architecture optimization, the method proposed by YOLO-v7 also focuses on optimizing the training process to improve the accuracy of the detector without increasing the duration of inference consumption [23]. The convolution layer consisting of convolution, BN layer and ReLU Activation function in the main part is called CBS for short.

3.4 SSD

SSD is a deep learning network structure for object detection, proposed by Wei Liu et al. in 2016 [14]. The design of the SSD model structure aims to detect multiple targets in the image during a single forward propagation process. For each position, SSD uses multiple anchor boxes of different scales to detect targets of different sizes. The basic network architecture of SSD is the VGG16 network. SSD using Conv, FC7 and gradually smaller pyramid layers are used for prediction, and each of these layers can generate a set of detection predictions using a set of 3×3 convolutional filters.

4 Experiment

PBCI-DS is divided into the training set and the test set according to the ratio of 4:1 by a 5-fold crossover experiment. After training four YOLO network models and SSD models, the experiment records the model and the data obtained during the training process, including the training time and the memory size occupied by the model. Next, the training results are used to predict the PBCI-DS test set. In this study, Precision (P), Recall (R), and mean average precision (mAP) are used as evaluation indicators to evaluate the accuracy and stability of the model. Finally, the prediction time of a single image is recorded as indicator for determining the detection speed of the model. All indicator data obtained are recorded in Table 3.

Table 3. Comparison of test results of PBCI-DS on different models.

Models	P (%)	R (%)	mAP (%)	mAP variance	Memory Cost (MB)	Training Time (h)	Single Test Time (ms)
YOLO-v5s	98.6	97.8	97.8	5.21	598	40.6	9.8
YOLO-v5l	97.7	98.0	98.0	4.41	667	42.0	10.4
YOLO-v6	98.3	94.3	98.2	5.14	602	37.5	11.2
YOLO-v7	97.0	97.2	98.8	2.43	726	21.7	11.5
SSD	98.5	97.0	98.9	2.13	783	25.5	11.2

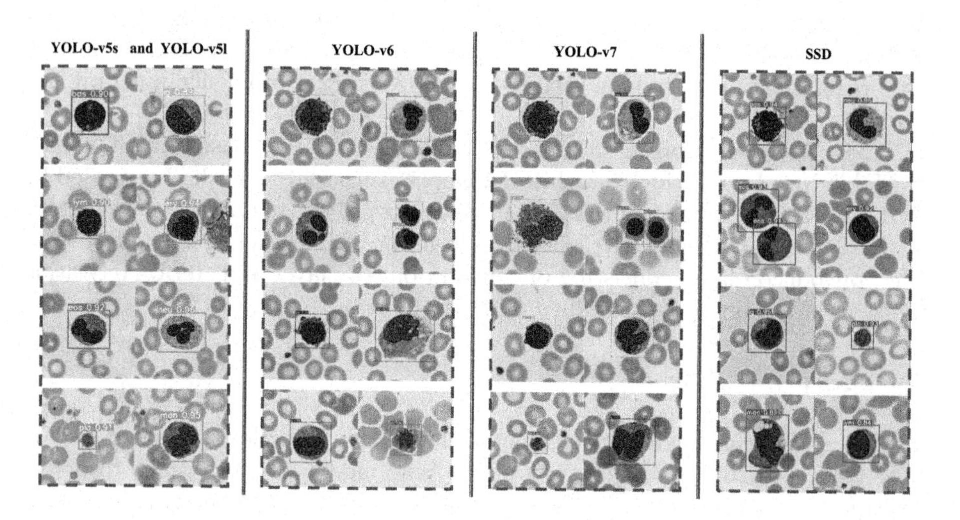

Fig. 2. PBCI-DS prediction results.

Based on the comparison presented in Table 3, it is evident that the YOLO-v5s and YOLO-v5l models exhibit high similarity. However, due to differences in depth, the YOLO-v5l model demonstrates higher mAP indicators compared to the YOLO-v5s, albeit with longer training and detection times. On the other hand, the YOLO-v6 model, despite employing a more efficient EfficientRep module than YOLO-v5, does not exhibit significant performance improvement, resulting in a mere 0.2% increase in accuracy. In terms of performance, both YOLO-v7 and SSD models outperform the aforementioned models, achieving final accuracy rates of 98.8% and 98.9%, respectively, while also requiring approximately 10 h less training time. Notably, the SSD model combines the VGG16 large network, which boasts the highest detection accuracy but occupies the most memory space. In contrast, the YOLO-v5s offers the fastest detection speed, at 9.8 ms per image, and occupies the smallest space, measuring only 598 MB. Figure 2 shows the predicted results of five models.

5 Conclusion and Future Work

PBCI-DS is a dataset for object detection divided into eight categories of cell images, with a total of 17092 images and labeled files. This article verifies its effectiveness by using multiple deep learning network models. The cell images of different types in PBCI-DS, including damaged cells, multiple object cells, and images mixed with unrelated cells, can effectively complete model training and prediction. PBCI-DS, with its large sample size and high feature extraction ability requirements, demonstrates improved training effectiveness using YOLO series networks as observed in the experimental results. Meanwhile, SSD networks with deeper depths achieve higher accuracy, albeit at the cost of slower detection speed and increased memory usage. PBCI-DS proves valuable in testing and comparing the performance of different models, yielding significant comparative effects across various types of model training. It possesses a discerning capability to distinguish between models and provides reliable data for comparing model performance.

In the future, we will aim to broaden the scope of PBCI-DS and expand the category of blood cells involved in disease detection. This expansion will enable PBCI-DS to directly diagnose specific diseases through object detection. Our goal is to enhance the functionality of PBCI-DS, allowing it to achieve even more capabilities in the future.

Acknowledgements. We thank B.A. Qiuqi from Foreign Studies College of Northeastern University, China, for her professional English proofreading in this paper. We thank Miss Zixian Li and Mr. Guoxian Li for their important discussion. This work is support by "National Natural Science Foundation of China" (No. 82220108007). PBCI-DS is open available at: https://figshare.com/articles/figure/PBCI-DS_A_Benchmark_Peripheral_Blood_Cell_Image_Dataset_for_Object_Detection/24417049.

References

1. Acevedo, A., Alférez, S., Merino, A., Puigvl, L., Rodellar, J.: Recognition of peripheral blood cell images using convolutional neural networks. Comput. Methods Programs Biomed. **180**, 105020 (2019)

2. Acevedo, A., Merino, A., Alférez, S., Molina, Á., Boldú, L., Rodellar, J.: A dataset of microscopic peripheral blood cell images for development of automatic recognition systems. Data Brief **30** (2020)
3. Alam, M.M., Islam, M.T.: Machine learning approach of automatic identification and counting of blood cells. Healthc. Technol. Lett. **6**(4), 103–108 (2019)
4. Alférez, S., Merino, A., Bigorra, L., Mujica, L., Ruiz, M., Rodellar, J.: Automatic recognition of atypical lymphoid cells from peripheral blood by digital image analysis. Am. J. Clin. Pathol. **143**(2), 168–176 (2015)
5. Alomari, Y.M., Sheikh Abdullah, S.N.H., Zaharatul Azma, R., Omar, K.: Automatic detection and quantification of WBCS and RBCS using iterative structured circle detection algorithm. Comput. Math. Methods Med. **2014** (2014)
6. Barger, A.M.: The complete blood cell count: a powerful diagnostic tool. Vet. Clin. Small Anim. Pract. **33**, 1207–1222 (2003)
7. Fujimoto, H., Sakata, T., Hamaguchi, Y., Shiga, S., Tohyama, K.: Flow cytometric method for enumeration and classification of reactive immature granulocyte populations. Cytometry **42**(6), 371–378 (2000)
8. Geissmann, F., Manz, M.G., Jung, S., Sieweke, M.H., Merad, M., Ley, K.: Development of monocytes, macrophages, and dendritic cells. Science **327**(5966), 656–661 (2010)
9. Kaplan, Z.S., Jackson, S.P.: The role of platelets in atherothrombosis. Hematology **2011**(1), 51–61 (2011)
10. Kratz, A., et al.: Digital morphology analyzers in hematology: ICSH review and recommendations. Int. J. Lab. Hematol. **41**(4), 437–447 (2019)
11. LeBien, T.W., Tedder, T.F.: B lymphocytes: how they develop and function. Blood **112**(5), 1570–1580 (2008)
12. Lennon, A.M., Buchanan, A.H., Kinde, I., Warren, A., Honushefsky, A.: Feasibility of blood testing combined with PET-CT to screen for cancer and guide intervention. Science **369**(6499), eabb9601 (2020)
13. Li, R., Xiao, X., Ni, S., Zheng, H., Xia, S.: Byte segment neural network for network traffic classification. In: 2018 IEEE/ACM 26th International Symposium on Quality of Service (IWQoS), pp. 1–10. IEEE (2018)
14. Lu, X., Kang, X., Nishide, S., Ren, F.: Object detection based on SSD-ResNet. In: 2019 IEEE 6th International Conference on Cloud Computing and Intelligence Systems (CCIS), pp. 89–92. IEEE (2019)
15. Marone, G., Lichtenstein, L.M., Galli, S.J.: Mast cells and basophils (2000)
16. Mayadas, T.N., Cullere, X., Lowell, C.A.: The multifaceted functions of neutrophils. Annu. Rev. Pathol. **9**, 181–218 (2014)
17. Montani, F., Marzi, M.J., Dezi, F., Dama, E., Carletti, R.M.: miR-Test: a blood test for lung cancer early detection. JNCI J. Natl. Cancer Inst. **107**(6), djv063 (2015)
18. Newsome, P.N., Cramb, R., Davison, S.M., Dillon, J.F., Foulerton, M.: Guidelines on the management of abnormal liver blood tests. Gut **67**(1), 6–19 (2018)
19. Olorunshola, O.E., Irhebhude, M.E., Evwiekpaefe, A.E.: A comparative study of YOLOv5 and YOLOv7 object detection algorithms. J. Comput. Soc. Inform. **2**(1), 1–12 (2023)
20. Piaton, E., Fabre, M., Goubin-Versini, I., Bretz-Grenier, M.F., Courtade-Saïdi, M.: Recommandations techniques et règles de bonne pratique pour la coloration de may-grünwald-giemsa: revue de la littérature et apport de l'assurance qualité. In: Annales de Pathologie, vol. 35, pp. 294–305. Elsevier (2015)

21. Shuo, Y., Hui, L., Fangjun, G., Yanqi, Y., Chenyang, L.: Real-time detection algorithm of mask wearing based on YOLOv5 in complex scenes. Comput. Meas. Control **29**(12), 188–194 (2021)
22. Tordjman, R., Delaire, S., Plouët, J., Ting, S., Gaulard, P., Fichelson, S.: Erythroblasts are a source of angiogenic factors. Blood **97**(7), 1968–1974 (2001)
23. Wang, C.Y., Bochkovskiy, A., Liao, H.Y.M.: YOLOv7: trainable bag-of-freebies sets new state-of-the-art for real-time object detectors. In: Proceedings of the IEEE/CVF Conference on Computer Vision and Pattern Recognition, pp. 7464–7475 (2023)
24. Wei, J., Qu, Y.: Lightweight improvement of YOLOv6 algorithm for small target detection (2023)

Skilled Task Assignment with Extra Budget in Spatial Crowdsourcing

Yunjun Zhou[1], Shuhan Wan[1], Detian Zhang[1(✉)], and Shi-ting Wen[2]

[1] Institute of Artificial Intelligence, School of Computer Science and Technology, Soochow University, Suzhou, China
{yjzhou1015,20195227042}@stu.suda.edu.cn, detian@suda.edu.cn
[2] Ningbo Institute of Technology, Zhejiang University, Ningbo, China
wensht@nit.zju.edu.cn

Abstract. With the prevalence of mobile devices and ubiquitous wireless networks, spatial crowdsourcing has attracted much attention from both academic and industry communities. On spatial crowdsourcing platforms, task requesters can publish spatial tasks and workers need to move to destinations to perform them. This paper formally defines a new problem, i.e., the Skilled Task Assignment with Extra Budget (STAEB), which aims to maximize total platform revenue. In the STAEB problem, the complex task needs more than one worker to satisfy its skill requirement and has the extra budget to subsidize extra travel costs of workers to attract more workers. We prove that the STAEB problem is NP-complete. Therefore, two approximation algorithms are proposed to solve it, including greedy and game-theoretic approaches. Extensive experiments on both real and synthetic datasets demonstrate the efficiency and effectiveness of our proposed approaches.

Keywords: Spatial Crowdsourcing · Task Assignment · Extra budget

1 Introduction

With the development of smart devices and high-speed wireless networks, Spatial Crowdsourcing (SC), which assigns moving workers to location-based tasks, has recently gained significant attention. In SC, workers must physically move to specified locations to accomplish tasks published by task requesters. SC has found applications in various fields, including online taxi-calling services (e.g., DiDi and Uber), food delivery services (e.g., Grubhub, Eleme, and Meituan), traffic monitoring (e.g., Waze), and geographical data generation (e.g., OpenStreetMap). However, some complex tasks require workers with specific skills and also need to ensure that workers are within a certain distance to ensure reasonable efficiency.

Previous works [4–6,26] on spatial crowdsourcing have mainly focused on assigning complex tasks to multiple workers who possess the necessary skills while ignoring the additional travel costs that may be incurred by workers. However, in reality, workers need to physically move to the location of the task, which

X. Yang et al. (Eds.): ADMA 2023, LNAI 14180, pp. 520–535, 2023.
https://doi.org/10.1007/978-3-031-46677-9_36

incurs travel costs. Therefore, workers may only accept tasks that offer travel reimbursements within a certain budget to ensure their actual earnings are reasonable. Take Fig. 1 for example, there are three workers with different skills (denoted by different colors), and two task requesters that require appropriate skills to complete (the required skills are also denoted by the different colors). The fixed range constraints of the task are shown with solid lines, and the extra range constraints of the task are shown with dashed lines. We can see, worker w_2 can well meet the skill requirement of task requester t_2 and does not exceed the distance limit. However, if task requester t_1 wants to find a worker to complete his task, no worker can answer it, because the worker (i.e., w_1) who meets the skill requirement exceeds his distance limit, while the skill of the worker (i.e., w_3) who within the distance limit do not meet the requirements. It results that task requester t_1 may never be assigned. In practice, task requester t_1 may have an extra budget [23,25] to subsidize the extra travel cost for worker w_1 so that the task can be completed, which can also facilitate a reasonable online recommendation.

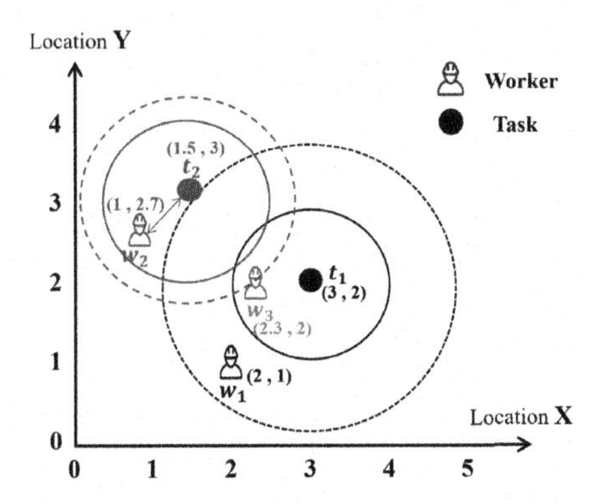

Fig. 1. Example of distance constraint in STAEB problem.

Therefore in this paper, we investigate the task assignment of SC under such a problem setting, namely Skilled Task Assignment with Extra Budget (STAEB). To be more specific, given a set of workers and a set of tasks, it aims to assign multi-skilled workers to complex tasks with extra budget.

To summarize, we make the following contributions to this paper:

- We formally define the Skilled Task Assignment with Extra Budget (STAEB) problem and prove it is NP-complete.
- We propose two batch-based approximation algorithms to solve the STAEB problem, i.e., greedy and game-theoretic approaches.
- We conduct extensive experiments on real and synthetic datasets to prove the effectiveness and efficiency of our algorithms.

The rest of this paper is organized as follows. Section 2 reviews some related work. The STAEB problem is formally defined in Sect. 3. Section 4 gives the greedy algorithm. The game-theoretic algorithm is proposed in Sect. 5. Extensive experiments on real and synthetic datasets are presented in Sect. 6. Finally, Sect. 7 concludes this work.

2 Related Work

Spatial crowdsourcing is an essential topic in match-based services that has garnered significant attention from scholars in the mobile Internet and sharing economy domains. In SC, we can classify task assignment into single-worker planning [7,9,17] and multi-worker planning [5,6,21] based on the complexity of the task.

Most of the previous works mainly focus on assigning tasks to a single worker. Den et al. [8] firstly study maximizing the number of performed tasks under travel budget and deadline constraints using an exact algorithm and several approximation algorithms based on the greedy heuristic. To better achieve the trade-off between efficiency and effectiveness, the beam search heuristic (BSH) is proposed [9], which extends the base of the candidate set to a given threshold in the nearest-neighbor heuristic (NNH) algorithm. Tong et al. [22] first consider the online scenario of task assignment and propose threshold-based algorithms with theoretical guarantees to maximize the total utility of the assignment. Cheng et al. [7] propose a cross-online matching that enables the platform to borrow some unoccupied workers from other platforms. Zhao et al. [17] first consider reducing the average waiting time of users, and many complete tasks so as to improve the user experience.

In practice, some complex tasks require the collaboration of groups of workers as a single worker may not possess all the necessary skills. Consequently, researchers have focused on multi-worker planning, where multiple workers collaborate on a single complex task. Similarly, the planning modes are to maximize general utility, e.g., satisfaction scores [11,19], payoffs [2,12] and distance [3,13]. In detail, Shin et al. [12] propose a local ratio-based algorithm, to maximize the reward of the performed tasks. Cheng et al. [6] propose the multi-skill spatial crowdsourcing (MS-SC) problem, which finds an optimal worker-and-task assignment strategy. Cheng et al. [5] consider the cooperative relationship of multiple workers, to maximize the total cooperative quality income of tasks. Considering the practical application of multi-worker planning, Song et al. [20] first proposed an online multi-worker assignment algorithm. It uses the Online-Greedy algorithm for fast computing task assignments. Recently, gaming-theoretic algorithms became popular. For example, Zhao et al. [26] designed gaming-theoretic algorithms to form worker coalitions in order to maximize the overall rewards.

However, these methods do not address the urgent need for tasks with extra budget (as depicted in Fig. 1), which may lead to long wait times or even task abandonment. Some studies have considered the issue of extra budget in SC, such as the work of Wan et al. [23], who propose two improved greedy algorithms to minimize extra budget in maximizing worker-task matching. In another paper by Jin et al. [14], the authors consider how extra budget affects worker-task assignment satisfaction and proposed a threshold-based algorithm to maximize satisfaction. However, the problems studied in these papers do not consider the skilled task assignment with extra budget. Therefore, in this paper, we propose two task assignment algorithms, i.e., a greedy algorithm and a game-theoretic algorithm to solve the skilled task assignment with extra budget (STAEB) problem.

3 Problem Definitions

In this section, we formally define the Skilled Task Assignment with Extra Budget (STAEB) problem. Assume that $S = \{s_1, s_2 ... s_k\}$ is a set of k skills. Each worker has one or multiple skills in S and each task needs one or multiple skills in S. Different skills require task requesters to pay different fees p_s.

Definition 1. *(Skilled Task) A skilled task, denoted by* $t =< l_t, a_t, r_t, b_t, S_t >$, *is released on the platform with location* l_t *in the 2D space at time* a_t. *In addition,* r_t *is the radius of* t *which is the fixed range constraint of* t, b_t *is its provided extra budget and* t *needs skills* $S_t \subseteq S$.

Definition 2. *(Worker) A worker, denoted by* $w =< l_w, a_w, S_w >$, *appears on the platform at time* a_w *and at location* l_w *in the 2D space. Moreover, each* w *has a series of skills* $S_w \subseteq S$ *to meet the needs of the task.*

Definition 3. *(Travel cost) The travel cost, denoted by* $cost(t, w)$, *is determined by the travel distance from* l_t *to* l_w.

Travel distance can be measured by any type of distance such as Euclidean distance or road network distance. In this paper, we use Euclidean distance as the travel distance and take it as the travel cost directly for simplicity.

Definition 4. *(Extra travel cost) Extra travel cost, denoted by* e_t, *is the actual travel cost exceeding the fixed range constraint, i.e.,* $e_t = cost(t, w) - r_t$.

Definition 5. *(Valid worker set) The set of multiple workers who can complete the task is called the valid worker set, denoted by* $W_v(t)$, *and the skill set of workers in a valid worker set can cover the skills that the task requires.*

Furthermore, each worker in a valid worker set must have at least one skill that the task needs but other workers do not have, so each valid worker set has two restrictive conditions:

(1) Each skill of the task can be covered by the skills in the set, i.e., $s \in \bigcup_{w \in W_v(t)} S_w, \forall s \in S_t$.
(2) The total extra travel cost of workers in the set should be less than the extra budget of the task, i.e., $\sum_{w \in W_v(t)} e(t, w) \leq b_t$.

Definition 6. *(Platform revenue) Given a task t and a worker w, the platform revenue is denoted as:*

$$p_{(t,w)} = \begin{cases} \alpha \sum\limits_{s \in S_w \cap S_t} p_s & , \quad cost(t,w) < r_t, \\ \alpha \sum\limits_{s \in S_w \cap S_t} p_s - \beta e(t,w) & , \quad cost(t,w) \geq r_t. \end{cases} \quad (1)$$

where $\sum_{s \in S_w \cap S_t} p_s$ represents the total revenue of providing skills the task needs by the worker. $\alpha (0 < \alpha < 1)$ is the percentage parameter that the platform shares from the skill fee. In practice, workers' earnings need to be increased when they exceed a fixed distance. Therefore, we introduce parameter $\beta (0 < \beta < 1)$ to control the subsidy of the platform for the workers' extra travel costs.

Consequently, the platform revenue is equal to skill revenue minus extra travel costs, which is equal to the revenue of the workers assigned to the task, denoted as:

$$P_{<t,W_v(t)>} = \sum_{w \in W_v(t)} p_{(t,w)} \quad (2)$$

Definition 7. *(Skilled Task Assignment with Extra Budget (STAEB) problem) Given a set of tasks T and set workers W. The STAEB problem is to find a feasible matching result M that can achieve the goal of maximizing total platform revenue P_M:*

$$P_M = \sum_{<t,W_v(t)> \in M} P_{<t,W_v(t)>} \quad (3)$$

where M is a feasible matching result, consisting of a set of < task, valid worker set >, and any $\cap_{i=1}^{|n|} W_v(t_i) = \emptyset$.

Theorem 1. *The Skilled Task Assignment with Extra Budget problem is an NP-complete problem.*

Proof. We prove that the Skilled Task Assignment with Extra Budget problem is NP-complete by reducing it to a maximum weighted independent set problem. In our problem, we can first initialize each available $W_v(t)$ of all $t \in T$ to be a vertex in the graph. Then let all $W_v(t)$ vertices belonging to the same t be connected one by one. Finally, we connect the $W_v(t)$ vertices with intersections (i.e., the same workers). At this point, any set of $W_v(t)$ vertices that can be assigned to T must be an independent set (otherwise there would be at least one t that has more than one $W_v(t)$, or two $W_v(t)$ with the same worker). Since there is revenue for each $W_v(t)$, the revenue of the platform is also the sum of the vertex weights on the independent subset, the STAEB problem is reduced successfully to the maximum-weighted independent set problem. Because the maximum weight-independent subset problem is an NP-complete problem, the STAEB problem is also an NP-complete problem.

Algorithm 1. Greedy algorithm

Input: A set of tasks T, a set of workers W
Output: the matched pair set M

1: $M \leftarrow \emptyset$, $W^{'} \leftarrow W$;
2: Sort the tasks in descending order according to the size of $\frac{\sum_{s \in S_t} p_s}{|number\ of\ task\ skills|}$;
3: **for** each task $t \in T$ **do**
4: $\quad W_v(t) \leftarrow \emptyset$;
5: $\quad$ **while** $S_t \neq \emptyset$ **do**
6: $\quad\quad R_t = r_t + b_t$;
7: $\quad\quad W^* = \{w | S_w \cap S_t \neq \emptyset, cost(t, w) \leq R_t\}$;
8: $\quad\quad$ **if** $W^* == \emptyset$ **then**
9: $\quad\quad\quad W \leftarrow W + W_v(t)$;
10: $\quad\quad\quad W_v(t) \leftarrow \emptyset$;
11: $\quad\quad\quad$ Break;
12: $\quad\quad$ **end if**
13: $\quad\quad w = argmax_{w \in W^*} |S_w \cap S_t|$;
14: $\quad\quad W_v(t) = W_v(t) \cup \{w\}$;
15: $\quad\quad b_t = b_t - e(t, w)$;
16: $\quad\quad S_t = S_t - \{S_w \cap S_t\}$;
17: $\quad\quad W = W - \{w\}$;
18: $\quad$ **end while**
19: $\quad$ **if** $W_v(t) \neq \emptyset$ **then**
20: $\quad\quad M \leftarrow M \cup \{< t, W_v(t) >\}$;
21: $\quad$ **end if**
22: **end for**
23: **return** M;

4 Greedy Algorithm

This section presents a proposed greedy algorithm [24] for task assignment in the STAEB problem. The main concept of the algorithm is to sort tasks in descending order of the average fee for the required skills and then identify the smallest set of workers to cover the required skills for each task. In practice, the fees for different skills are not the same, and workers prefer tasks that offer higher fees for skills to obtain higher revenue. Additionally, workers prefer tasks that satisfy the most skills to save on travel costs.

Algorithm 1 outlines the steps of the Greedy algorithm in detail. In line 1, the matched pair set and the unmatched worker set are initialized. In line 2, the tasks are sorted in descending order according to the average fee of skills. Then, appropriate workers are assigned to each task t in turn (lines 4–21). If the unmatched skills of a task are not empty, the algorithm assigns workers to the task as follows: in lines 6–7, the total range constraint of the task is determined as the sum of the fixed range constraint and the extra range constraint. And the set of workers who can complete the task is denoted as W^*. If the set of workers who are capable of completing the task is empty, it indicates that no worker possesses the remaining skills required to complete the task. Therefore, the task cannot be completed, and the workers in the valid worker set $W_v(t)$ are added to the worker set $W^{'}$. Then in lines 13–17, we select and update the worker w in W^* who has the most skills required by the task and add w to the valid worker set of the task. In lines 19–20, the task and its valid worker set are added to the matched pair set if the valid worker set is not empty, as it means the valid worker set can cover the skills required by the task. The task is then removed from the task set. Finally, in line 23, the algorithm returns the matched pair set.

Complexity Analysis. The time complexity of sorting tasks is $O(|S_t| \cdot |T| + |T| \cdot \ln |T|)$ and finding valid worker sets for each task is $O(|T| \cdot |S_t|^2 \cdot |W|)$, so the total time complexity is $O(|S_t| \cdot |T| + |T| \cdot \ln |T| + |T| \cdot |S_t|^2 \cdot |W|)$.

5 Game-Theoretic Algorithms

The greedy algorithm can effectively find results. However, it only considers that workers will complete the tasks based on the platform's assignments and ignores competition between different workers. The STAEB problem is fundamentally such that each task requires multiple workers who satisfy the skill requirements, meaning that the choice of workers is affected by decisions made by other workers. Each worker desires to choose a task with high revenue, and this interdependent decision can be modeled through game theory, where workers can be regarded as independent players in the game. Thus, the STAEB problem can be formalized as a multiplayer game. Building on existing game-theoretic models [10,16,18], we propose a game theory method to find valid worker sets for tasks until the Nash equilibrium is achieved in this section.

5.1 Game Formulation

Our STAEB problem can be formulated as an $n-$player strategic game $\mathcal{G} =< W, \mathbb{Y}, \mathbb{U} >$, which consists of players, the overall strategies, and utility functions. It is formulated as follows:

(1) $W = \{w_1, w_2, ..., w_n\} (n \geq 2)$ is a limited set of workers as game players. In the rest of the paper, we will use player and worker interchangeably.
(2) $\mathbb{Y} = \cup_{i=1}^{n} Y_i$ is the overall strategies for the players. Y_i is the finite set of strategies that worker w_i can choose.
(3) $\mathbb{U} = \cup_{i=1}^{n} U_i$ denotes the utility functions of all the players. The value of U_i depends on the player w_i and other players' strategies. $U_i : \mathbb{Y} \rightarrow \mathbb{P}$ is the utility function of player w_i. For every joint strategy $\boldsymbol{y} \in \mathbb{Y}$, $U_i(\boldsymbol{y}) \in \mathbb{P}$ represents the utility of player w_i, which can be calculated as follows:

$$U_i(\overrightarrow{y}) = p_{(t,w_i)} - p_{(t_0,w_i)} \tag{4}$$

where $p_{(t,w_i)}$ is the platform revenue of assigning player w_i to task t. $p_{(t_0,w_i)}$ is the as platform revenue of assigning player w_i to task t_0.

Let y_i be the strategy of player w_i in the joint strategy Y_i and y_{-i} be all other players joint strategies except for player w_i. A strategic game has a pure Nash equilibrium $Y^* \in \mathbb{Y}$ if and only if for every player $w_i \in W$ satisfies the following conditions:

$$U_i({y_i}^*, {y_{-i}}^*) \geq U_i(y_i, {y_{-i}}^*), \forall y_i \in Y_i, y_i^* \in Y^* \tag{5}$$

In our problem, since each worker needs to have a deterministic strategy, i.e., selecting a task or doing nothing, which means that the probability of a multiplayer worker w_i can choose from Y_i is 1, while the probabilities of the remaining strategies in Y_i are 0. Then, we prove that the STAEB game is an Exact Potential Game (EPG) [15] which has at least one pure Nash Equilibrium. In short, no player can improve his utility by unilaterally changing his strategy in the current situation means achieving a Nash equilibrium. We first introduce the theory of the Exact Potential Game.

Definition 8 (Exact Potential Game). *A strategic game* $\mathcal{G} =< W, \mathbb{Y}, \mathbb{U} >$ *is called an exact potential game if and only if there exists a potential function* $\phi : \mathbb{Y} \to \mathbb{P}$*, such that for all* $\boldsymbol{y} \in \mathbb{Y}$ *and* $w_i \in W$*:*

$$U_i(y_i, y_{-i}) - U_i(y_i', y_{-i}) = \phi(y_i, y_{-i}) - \phi(y_i', y_{-i}), \forall y_i, y_i' \in Y_i \tag{6}$$

where y_i and y_i' are the strategies that worker w_i can choose, y_{-i} is the joint strategy of other workers except for worker w_i.

Theorem 2. *Our STAEB problem is an Exact Potential Game(EPG).*

Proof. The total platform revenue of all tasks in T is represented by the potential function $\phi(y) = \sum_{t\in T} P_{<t,W_v(t)>}$. Worker w_i can choose between y_i and y_i' strategies, while y_{-i} is the strategy of all other workers except worker w_i. The task chosen in strategies y_i' is denoted by t_j and t_k. Then we have:

$$\begin{aligned}
&\phi(y_i, y_{-i}) - \phi(y_i', y_{-i}) \\
&= p_{<t_j,W_v(t_j)>} + p_{<t_k,W_v(t_k)-w_i>} + \sum_{t\in T-t,_j-t_k} p_{<t,W_v(t)>} - (p_{<t_k,W_v(t_k)>} \\
&+ p_{<t_j,W_v(t_j)-w_i>} + \sum_{t\in T-t_j-t_k} p_{<t,W_v(t)>}) \\
&= p_{<t_j,W_v(t_j)>} + p_{<t_k,W_v(t_k)-w_i>} - (p_{<t_k,W_v(t_k)>} + p_{<t_j,W_v(t_j)-w_i>}) \\
&= p_{<t_j,W_v(t_j)>} - p_{<t_j,W_v(t_j)-w_i>} - (p_{<t_k,W_v(t_k)>} - p_{<t_k,W_v(t_k)-w_i>}) \\
&= p_{(t_j,w_i)} - p_{(t_k,w_i)} \\
&= (p_{(t_j,w_i)} - p_{(t_0,w_i)}) - (p_{(t_k,w_i)} - p_{(t_0,w_i)}) \\
&= U_i(y_i, y_{-i}) - U_i(y_i', y_{-i})
\end{aligned} \tag{7}$$

Thus, the strategic game of the STAEB problem is an exact potential game, according to the definition.

□

5.2 The Game Theoretic Approach

To improve the effectiveness of finding a solution to the STAEB problem, we propose the Extra Budget-aware Game-Theoretic method (EBGT) in this section.

Algorithm 2. Extra Budget-aware Game-Theoretic (EBGT) Approach

Input: A set of tasks T, a set of workers W
Output: the matched pair set M
1: // **Step 1: Initialize the strategy.**
2: $M \leftarrow \emptyset$, $T' \leftarrow T$, $W' \leftarrow W$;
3: **for** each task $t \in T$ **do**
4: $W_v(t) \leftarrow \emptyset$;
5: **for** each task skill $s \in S_t$ **do**
6: $R_t = b_t$;
7: $W^* = \{w | s \in S_w \text{ and } cost(t, w) \leq R_t\}$;
8: **if** $W^* = \emptyset$ **then**
9: $W' \leftarrow W' + W_v(t)$;
10: $W_v(t) \leftarrow \emptyset$;
11: Break;
12: **end if**
13: $w = argmax_{w \in W^*} P_{(t,w)}$;
14: $W_v(t) = W_v(t) \cup \{w\}$, $b_t = b_t - e(t, w)$ $W' = W' - \{w\}$;
15: **end for**
16: **if** $W_v(t) \neq \emptyset$ **then**
17: $M \leftarrow M \cup \{< t, W_v(t) >\}$, $T' = T' - \{t\}$;
18: **end if**
19: **end for**
20: // **Step 2: Find the Nash equilibrium.**
21: **repeat**
22: $M' \leftarrow M$;
23: **for** each worker $w \in W$ **do**
24: find the best-response task t^* for w_i;
25: **if** t^* not exists **then**
26: Continue;
27: **else**
28: **if** $t^* \in T'$ **then**
29: obtain $W_v(t^*)$ containing w_i;
30: $M \leftarrow M \cup \{< t^*, W_v(t^*) >\}$, $T' = T' - \{t^*\}$;
31: **else**
32: $W' \leftarrow W' + W_v(t^*)$;
33: $M \leftarrow M - \{< t^*, W_v(t^*) >\}$;
34: $W_v(t^*) \leftarrow \emptyset$;
35: obtain $W_v(t^*)$ containing w_i;
36: $M \leftarrow M \cup \{< t^*, W_v(t^*) >\}$, $T' = T' - \{t^*\}$;
37: **end if**
38: **end if**
39: compare M and M';
40: **end for**
41: **until** Nash equilibrium
42: Update M;
43: **return** M;

This method is based on the best response framework to find the Nash equilibrium joint strategy of the strategic game $\mathcal{G}$. Since the STAEB game has a pure Nash equilibrium, we assign each worker to their "best" task in the algorithm to obtain a higher total platform revenue. The details of the game theory method are described in two steps in Algorithm 2.

Step 1: Initialize the strategy. In line 2, the matching set is initialized to empty, and the initial unmatched task and worker set are set to the inputs of task and worker, respectively. Then, we use the same steps of the Greedy algorithm to initialize each task t (lines 3–19), which is described in Sect. 4.

Step 2: Find the Nash equilibrium. The algorithm iteratively adjusts each worker's strategy to obtain the best response strategy based on the current joint strategy of other workers until no one changes their strategy. This maximizes the platform revenue. In each iteration, only one worker is allowed to choose the best game. The algorithm records the matching results of the previous round in line 22. Then, the best response task t^* for each worker $w_i \in W$ in the current situation is found (line 24), which can be calculated as follows:

$$t^* = \mathbf{argmax}_{t \in \{t | S_{w_i} \cap S_t \neq \emptyset \ and \ cost(t, w_i) \leq R_t\}} (p_{<t, W_v(t)>} - p_{<t, W_v(t) - \{w_i\}>} - (p_{<t_O, W_v(t_O)>} - p_{<t_O, W_v(t_O) - \{w_i\}>})) \quad (8)$$

The worker w_i will not change his strategy when there is no best response task (line 25). Or else if w_i has the best response task and the valid worker set of the best response task is empty, we obtain the valid worker set which includes w_i (lines 28–30). Otherwise, the current valid worker set of the task is cleared, and the valid worker set which includes w_i is obtained (lines 31–38). Then t^* and a valid worker set to the matched set are added. Finally, we update M according to the Nash equilibrium.

5.3 Analysis of the Game Theoretic Approach

As the STAEB problem is an exact potential game with a limited strategy set S, a pure Nash equilibrium can be reached after workers change strategies in a limited number of rounds. To simplify the analysis, we consider a scaled version of the problem to prove the upper bound of the total rounds required to achieve a pure Nash equilibrium.

To verify the upper bound of the total rounds required to achieve a pure Nash equilibrium, we explore a scaled version of the issue where the objective function is scaled by a positive integer factor. We suppose that a comparable game with potential function exists: $\phi_{\mathbb{Z}}(S) = d \cdot \phi(S)$, in which d is a positive multiplicative factor such that $\phi_{\mathbb{Z}}(S) \in \mathbb{Z}, \forall S \in \mathbb{S}$. Then our EBGT performs at most $\phi_{\mathbb{Z}}(S^*)$ rounds using this scaled potential function, where S^* is the optimal strategy that workers can select in this potential STAEB game. $\phi_{\mathbb{Z}}(S^*)$ is the product of the positive multiplier factor d and the optimal value of the objective function P_M.

Lemma 1. *The upper bound on the number of rounds to converge to a pure Nash equilibrium is $d \cdot \phi(S^*)$ in each batch with the EBGT algorithm.*

Proof. The EBGT approach converges when no worker deviates from his current strategy, which means that there is at least one worker who deviates from his current strategy in each round. Because $\phi_{\mathbb{Z}}(S) \in \mathbb{Z}$, each worker w_i is changed from its current strategy $s_i{}'$ to a better strategy s_i, which will increase the scaled potential function by at least 1, i.e., $U_i(s_i, s_{-i}) - U_i(s_i{}', s_{-i}) \geq 1$. Thus the upper bound of the number of rounds to converge to pure Nash equilibrium is the maximum value $\phi_{\mathbb{Z}}(S^*) = d \cdot \phi(S^*)$.

Similarly, the upper bound of the platform revenue of the task t_j in our best joint strategy is:

$$\overline{p}_{t_j} = \sum_{w_i \in W_v(t_j)} p_{(t_j, w_i)} \quad (9)$$

where $W_v(t_j)$ are the workers assigned to task t_j, p_i is the maximum platform revenue got from the worker completing the task. So the upper bound of the platform revenue is $d \cdot \sum_{t_j \in T} p_{<t_j, W_v(t_j)^*>} = d \cdot \overline{p}_{t_j}$.

Complexity Analysis. The time complexity is $O(|T| \cdot |S_t| \cdot |W|^2 + |W|^2 \cdot |S_t| \cdot d \cdot \overline{p}_{t_j})$, where $|T|$ is the number of tasks, $|W|$ is the number of workers, $|S_t|$ is the maximum number of skills, $d \cdot \overline{p}_{t_j}$ is the number of iterations which adjusts the best response strategy of each worker until the Nash equilibrium is reached.

6 Experimental Study

6.1 Experiment Setup

We use two datasets in our experiment. For the real dataset, we use the taxi data from Didi Chuxing [1], which contains order data in Chengdu from November 1 to November 30, 2016. The order data has information on pick-ups and drop-offs, the start and the end of billing time. Specifically, the pick-up location and the starting time of billing are used as the location information and arrival time of the task respectively. Since the worker becomes available again after the passenger gets off the car, the drop-off location is also used as the location of the worker and the end billing time is used as the arrival time of the worker. For the synthetic dataset, we randomly generate task requests and workers in a rectangular area of Chengdu. The arrival time distribution of workers and task requests follows the uniform distribution in a day. Table 1 depicts our experimental settings, where the default values of parameters are in bold font.

Table 1. Experiments settings

Parameter	Setting
Fixed range constraint r	600, 800, **1000**, 1200, 1400
Extra range constraint b	(400,600),600,800),**(800,1000)**, (1000,1200), (1200,1400)
Number of skills S	10, 11, **12**, 13, 14
Number of tasks T	800, 900, **1000**, 1100, 1200
Number of workers W	2400, 2700, **3000**, 3300, 3600

Compared Algorithms. We evaluate the performance of the representative algorithms, i.e., Random algorithm (RAN), Greedy algorithm (GRY), and Extra budget-aware Game-Theoretic algorithm (EBGT). All the algorithms are implemented in Java and run on a machine with Intel(R) Core (TM) i7-7700 CPU @ 3.60 GHz and 16 GB RAM.

6.2 Results on the Real Dataset

Effect of the Fixed Range Constraint. As shown in Fig. 2(a), the running time of algorithms increases with the larger fixed range constraint. This is because more workers will be located in the fixed range constraint of each task which needs longer running time to be processed. Similar to existing works [24–27], EBGT runs slower than RAN and GRY because it needs multiple rounds of iteration to find the optimal solution of each round continuously. As the platform revenue results are shown in Fig. 2(b), it increases as the fixed range constraint grows. EBGT gains the highest platform revenue for it finds the best-response task for each worker.

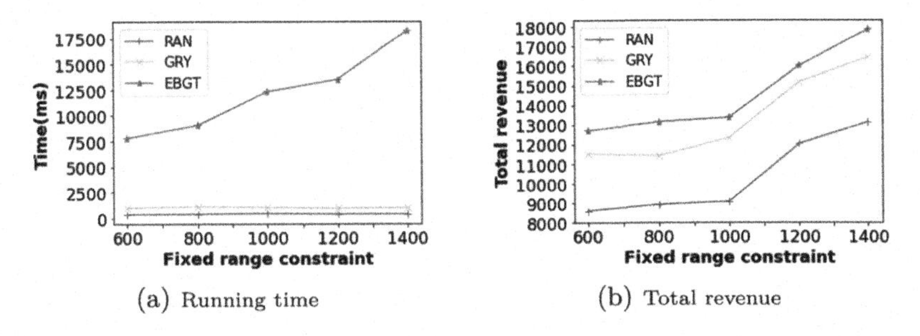

(a) Running time (b) Total revenue

Fig. 2. Effect of the fixed range constraint on real dataset

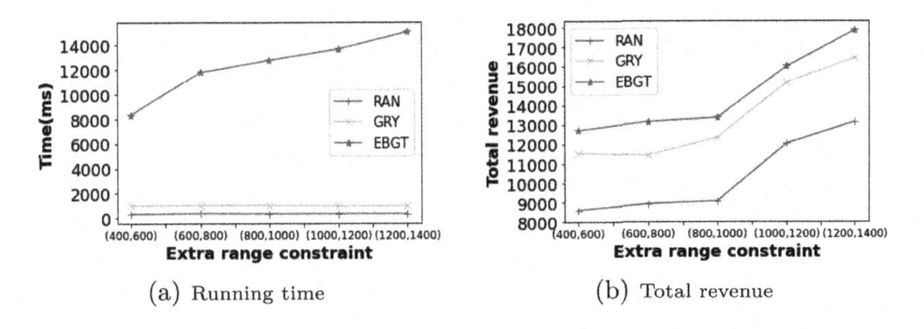

(a) Running time (b) Total revenue

Fig. 3. Effect of the extra range constraint on real dataset

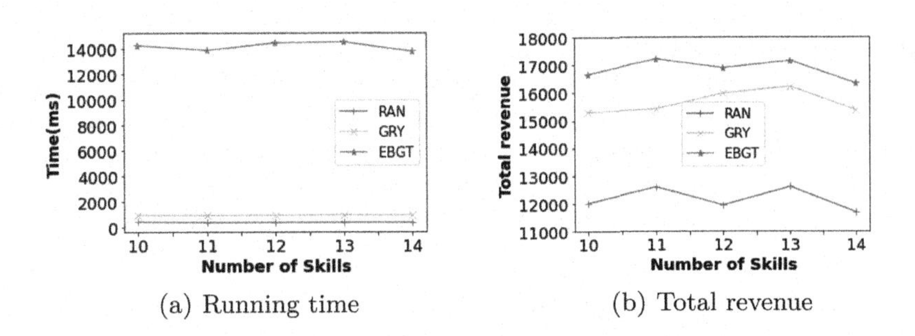

(a) Running time (b) Total revenue

Fig. 4. Effect of the number of skills on real dataset

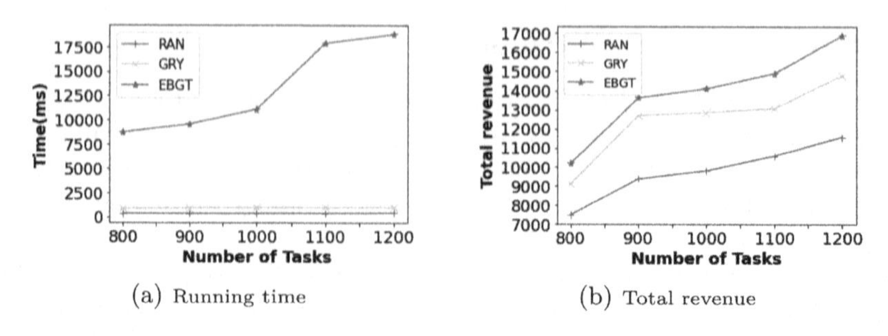

(a) Running time

(b) Total revenue

Fig. 5. Effect of the number of tasks on synthetic dataset

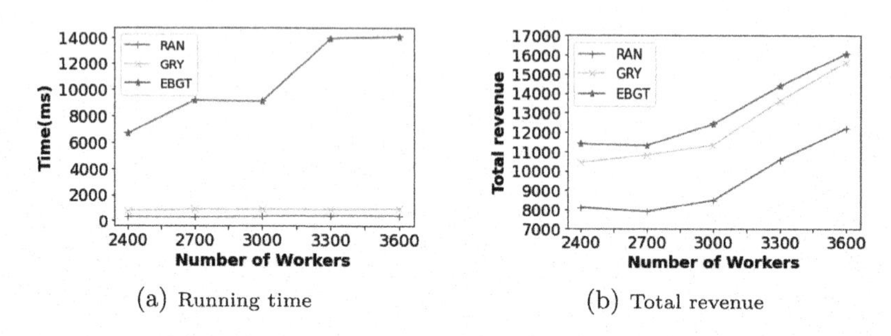

(a) Running time

(b) Total revenue

Fig. 6. Effect of the number of workers on synthetic dataset

Effect of the Extra Range Constraint. As depicted in Fig. 3(a), when the extra range constraint becomes larger, the running time of algorithms increases. This is because more workers will be located in each task's extra range constraint, which needs longer running time. From Fig. 3(b) we can see that the platform revenue of all the algorithms increases with the larger extra range constraint. This is because more pairs will be matched. EBGT gains keep owning the highest platform revenue as it finds the best-response task for each worker.

Effect of the Number of Skills. As shown in Fig. 4(a), the running time of algorithms with the number of skills changes little. Because with the number of skills required for a task increases, the number of tasks that can be satisfied decreases, then reduces the number of iterations, which in turn makes the running time stable. EBGT still runs longer than the other two algorithms because it needs multiple rounds of iteration to find the optimal solution of each round continuously. From Fig. 6(b), we can see that the total revenue of all three algorithms changes a little. EBGT gains the highest platform revenue and RAN gains the least platform revenue.

6.3 Results on the Synthetic Dataset

Effect of the Number of Tasks. As we can see from Fig. 5(a), the running time of algorithms increases with the increment of the number of tasks. EBGT still runs longer than RAN and GRY. In Fig. 5(b), the platform revenue increases with the increment of the number of tasks. The reason is that more tasks can be completed, generating more platform revenue. EBGT still gains the most platform revenue. In summary, the experimental results are similar to those on the real data set above.

Effect of the Number of Workers. In Fig. 6(a), the running time of algorithms increases with the increment of the number of workers. EBGT runs longer than RAN and GRY because it needs multiple rounds of iteration to find the optimal solution. From Fig. 6(b), the platform revenue increases with the increment of the number of workers. The reason is that more workers can complete tasks, generating more platform revenue. EBGT gains the most platform revenue. In conclusion, the results of the experiment are also similar to the above on the real data set.

7 Conclusion

In this paper, we study the problem of the Skilled Task Assignment with Extra Budget (STAEB) in spatial crowdsourcing. Then two approximation algorithms are proposed, including greedy and game-theoretic approaches. Specifically, the greedy approach sorts tasks in order of average fee of skills and greedily assigns fewer workers to cover the skills required by tasks. In addition, the game-theoretic approach uses game theory to further increases the total platform revenue. Extensive experiments on real and synthetic datasets show that our proposals achieve good efficiency and scalability.

Acknowledgment. Detian Zhang is supported by the Collaborative Innovation Center of Novel Software Technology and Industrialization, the Priority Academic Program Development of Jiangsu Higher Education Institutions. Shi-ting Wen is supported by the Ningbo Science and Technology Special Innovation Projects with Grant Nos. 2022Z095 and 2022Z235, and Opening Foundation of Jiangsu Intelligent Big Data Engineering Lab with Grant Nos. SDG2156.

References

1. Didi chuxing data. https://gaia.didichuxing.com
2. Asghari, M., Deng, D., Shahabi, C., Demiryurek, U., Li, Y.: Price-aware real-time ride-sharing at scale: an auction-based approach. In: Proceedings of the 24th ACM SIGSPATIAL International Conference on Advances in Geographic Information Systems, pp. 1–10 (2016)
3. Asghari, M., Shahabi, C.: On on-line task assignment in spatial crowdsourcing. In: 2017 IEEE International Conference on Big Data (Big Data), pp. 395–404. IEEE (2017)
4. Chen, Z., Cheng, P., Zeng, Y., Chen, L.: Minimizing maximum delay of task assignment in spatial crowdsourcing. In: ICDE, pp. 1454–1465. IEEE (2019)

5. Cheng, P., Chen, L., Ye, J.: Cooperation-aware task assignment in spatial crowdsourcing. In: 2019 IEEE 35th International Conference on Data Engineering (ICDE), pp. 1442–1453. IEEE (2019)
6. Cheng, P., Lian, X., Chen, L., Han, J., Zhao, J.: Task assignment on multi-skill oriented spatial crowdsourcing. TKDE **28**(8), 2201–2215 (2016)
7. Cheng, Y., Li, B., Zhou, X., Yuan, Y., Wang, G., Chen, L.: Real-time cross online matching in spatial crowdsourcing. In: ICDE, pp. 1–12. IEEE (2020)
8. Deng, D., Shahabi, C., Demiryurek, U.: Maximizing the number of worker's self-selected tasks in spatial crowdsourcing. In: Proceedings of the 21st ACM Sigspatial International Conference on Advances in Geographic Information Systems, pp. 324–333 (2013)
9. Deng, D., Shahabi, C., Demiryurek, U., Zhu, L.: Task selection in spatial crowdsourcing from worker's perspective. GeoInformatica **20**(3), 529–568 (2016)
10. Fudenberg, D., Tirole, J.: Game theory. Economica **60**(238), 841–846 (1992)
11. Gao, D.T.: Team-oriented task planning in spatial crowdsourcing. In: Web and Big Data (2017)
12. He, S., Shin, D.H., Zhang, J., Chen, J.: Toward optimal allocation of location dependent tasks in crowdsensing. In: IEEE INFOCOM 2014-IEEE Conference on Computer Communications, pp. 745–753. IEEE (2014)
13. Huang, Y., Jin, R., Bastani, F., Wang, X.S.: Large scale real-time ridesharing with service guarantee on road networks. arXiv preprint arXiv:1302.6666 (2013)
14. Jin, L., Wan, S., Zhang, D., Tang, Y.: Extra budget-aware online task assignment in spatial crowdsourcing. In: Web Information Systems Engineering-WISE 2022: 23rd International Conference, Biarritz, France, 1–3 November 2022, Proceedings, pp. 534–549. Springer, Heidelberg (2022). DOI: https://doi.org/10.1007/978-3-031-20891-1_38
15. Monderer, D., Shapley, L.S.: Potential games. Games Econ. Behav. **14**(1), 124–143 (1996)
16. Myerson, R.B.: Game Theory: Analysis of Conflict. Harvard University Press, Cambridge (1997)
17. Peng, W., Liu, A., Li, Z., Liu, G., Li, Q.: User experience-driven secure task assignment in spatial crowdsourcing. WWW **23**(3), 2131–2151 (2020)
18. Rasmusen, E.: Games and information: an introduction to game theory (1990)
19. She, J., Tong, Y., Chen, L.: Utility-aware social event-participant planning. In: Proceedings of the 2015 ACM SIGMOD International Conference on Management of Data, pp. 1629–1643 (2015)
20. Song, T., Xu, K., Li, J., Li, Y., Tong, Y.: Multi-skill aware task assignment in real-time spatial crowdsourcing. GeoInformatica **24**, 153–173 (2020)
21. To, H., Shahabi, C., Kazemi, L.: A server-assigned spatial crowdsourcing framework. ACM Trans. Spatial Algor. Syst. (TSAS) **1**(1), 1–28 (2015)
22. Tong, Y., She, J., Ding, B., Wang, L., Chen, L.: Online mobile micro-task allocation in spatial crowdsourcing. In: ICDE, pp. 49–60. IEEE (2016)
23. Wan, S., Zhang, D., Liu, A., Fang, J.: Extra-budget aware task assignment in spatial crowdsourcing. In: WISE (2021)
24. Wangze, N., Peng, C., Lei, C., Xuemin, L.: Task allocation in dependency-aware spatial crowdsourcing. In: 2020 IEEE 36th International Conference on Data Engineering (ICDE) (2020)
25. Xu, Y., Xiao, M., Wu, J., Zhang, S., Gao, G.: Incentive mechanism for spatial crowdsourcing with unknown social-aware workers: a three-stage stackelberg game approach. IEEE Trans. Mobile Comput. **22**, 4698–4713 (2022)

26. Zhao, Y., Guo, J., Chen, X., Hao, J., Zhou, X., Zheng, K.: Coalition-based task assignment in spatial crowdsourcing. In: 2021 IEEE 37th International Conference on Data Engineering (ICDE), pp. 241–252. IEEE (2021)
27. Zhao, Y., Zheng, K., Cui, Y., Su, H., Zhu, F., Zhou, X.: Predictive task assignment in spatial crowdsourcing: a data-driven approach. In: ICDE, pp. 13–24. IEEE (2020)

Synchronous Prediction of Asset Prices' Multivariate Time Series Based on Multi-task Learning and Data Augmentation

Jiahao Li, Qinghua Zhao, Simon Fong, and Jerome Yen(✉)

Faculty of Science and Technology, University of Macau, Macau SAR, China
{mc05504,mc05505,ccfong,jeromeyen}@um.edu.mo

Abstract. Multi-task Learning (MTL) makes a positive difference in many fields by improving the prediction effects of correlated tasks among multiple related data sets. Some financial Multivariate Time Series (MTS) also have a high correlation, but applications like price synchronous prediction based on MTL still lack enough attention from researchers. The future values of a certain price are not only related to its own historical values, but also related to other correlated price sequences, and this is a suitable condition for applying the MTL model. This paper constructs an MTL model for synchronous learning and predicting price time series based on the Sequence-to-Sequence (Seq2Seq) model. To obtain enough data for modeling, the Weighted Soft-dtw Barycentric Averaging (wDBA) is used as the Data Augmentation (DA) method to generate more time series data for each Forex pair based on its original OHLC bid quotes. On the testing of 8 Forex pairs' minute-level quote data from 2020 to 2022, our model, Seq2Seq with DA, outperforms baseline models including the single Long Short-term Memory (LSTM) and Seq2Seq without DA. During the experiment, to provide a comprehensive evaluation on such a long time sequence, more than 1 million minutes, we design the Chronological Randomly-sampling Walk-forward (CRSWF) Validation for a quick evaluation. As a result, when the DA degree is 125%, on the MAE, NMAE, RMSE, and NRMSE, our model respectively reduces by 95.23%, 97.24%, 94.07%, and 95.99% than LSTM, and also reduces by 7.87%, 4.84%, 6.05%, and 1.71% than the Seq2Seq without DA.

Keywords: Forex Forecast · Synchronous Prediction · Multivariate Time Series Analysis · Multi-task Learning · Data Augmentation

1 Introduction

Price forecasting has long been one of the most challenging tasks in time series research. Accurately predicting prices can reduce investors' investment risks and effectively improve investment returns. There are deep learning models having

J. Li and Q. Zhao contributed equally to this work.

X. Yang et al. (Eds.): ADMA 2023, LNAI 14180, pp. 536–551, 2023.
https://doi.org/10.1007/978-3-031-46677-9_37

some achievements in price forecasting but most of them only focus on the prediction for a single asset but not on multiple assets' prices. More application values will be found if prices of a group of assets with strong correlations are forecasted synchronously because it is conducive to obtaining a more comprehensive market analysis and drawing a more timely arbitrage or hedging strategy. For example, in the Forex market, scholars propose the concept, Currency Baskets, based on the correlation and interaction between various currencies' exchange rates to reduce the risk of single-piece fluctuations [17].

There are research demands for high-dimensional time series analysis in the financial field. MTL is a feasible deep learning mode to realize a synchronous predictor of asset prices' MTS, which aims at improving the generalization performance of each inner task by using other related tasks [12]. Not only the pattern of the time series itself could be learned, but also the interaction between the time series should be captured, so as to maximize the use of all the sequences' information. Many price time series have mutual influence, for instance, the bid quotes of a Forex pair depend not only on its own trend but are also closely related to the one of other Forex pairs [10]. Among the models for MTL of MTS, Seq2Seq is a promising one, especially emerges as an effective paradigm for dealing with variable-length input and output [8]. It is proven to be good at long-term time series forecasting while having the ability to learn from the high-dimension data [18].

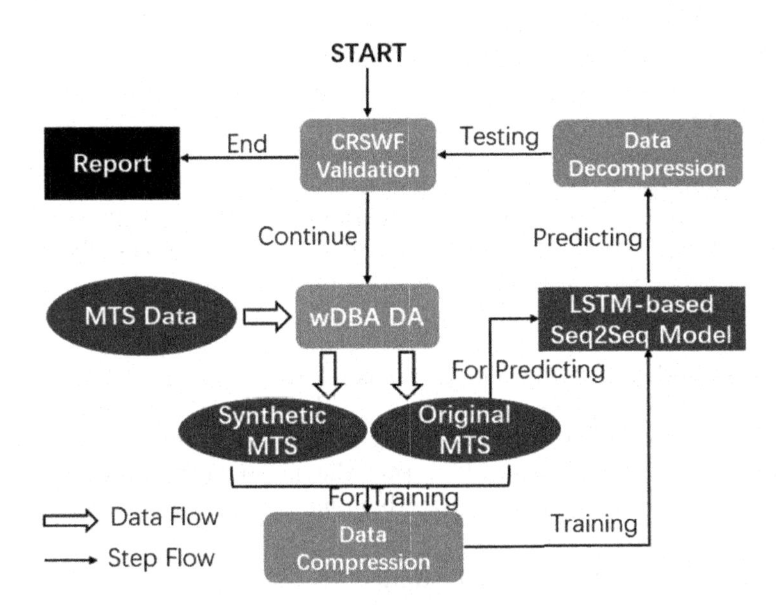

Fig. 1. Overall flowchart.

Furthermore, deep learning models rely heavily on big data, but the time series of strong-correlated financial prices are seriously inadequate. DA is an effective way to extend data samples, primarily to prevent overfitting, when datasets are small [14]. DA has shown its effectiveness in many applications, such as computer vision, speech processing, and time series research. DA is beneficial for MTL when the newly generated time series have enough correlation with the existing time series. In this research, the wDBA Algorithm is conducted to extend the training dataset.

The rest of the paper is organized as follows: related works about MTS forecasting based on the MTL and DA are presented in Sect. 2. In Sect. 3, as preliminaries, the Seq2Seq model and wDBA algorithm are introduced. Then the modeling methodology of these two concepts and the CRSWF validation are explained in Sect. 4, and the overall flowchart of the model constructed in this paper is shown in Fig. 1. Finally, the experimental setup and the results of the regression are described in Sect. 5, while the conclusion and future work are presented at the end of this paper.

2 Related Work

2.1 Time Series Prediction Enhanced by Multi-task Learning

Multi-task Learning [3] is a sub-field of machine learning, in which multiple learning tasks are solved at the same time, while overfitting is reduced by sharing representations, and fast learning is performed by using auxiliary information. Compared with the single-task model, it can improve the learning efficiency and prediction accuracy of the task-specific model [5].

The Sequence-to-Sequence model [4,16] maps an input sequence to a variable-length output sequence via recurrent neural networks. The model was originally designed to improve machine translation techniques and can also be used for multi-step forecasting of time series and MTS analysis tasks. As early as 2015, Minh-Thang Luong et al. explored the performance of the Seq2Seq framework in multitasking learning [12]. Then in 2020, Du et al. successfully performed multivariate time series predictions through an attention-based encoder-decoder framework [8]. Furthermore, ZihAo Gao et al. used the Seq2Seq model based on the attention model for stock forecasting and found it superior to the LSTM and AMAR models [9]. So far, Seq2Seq has not been applied to the synchronous prediction field of multiple assets' prices. In the financial portfolio adjustment, it is very necessary to predict the timing of multiple assets at the same time, which is a task about MTL of MTS, thus this paper builds up a simultaneous predictor for 8 well-known Forex pairs' bid quotes based on the Seq2Seq model.

2.2 Time Series Prediction Enhanced by Data Augmentation

Time series data augmentation algorithms can be mainly divided into basic methods and advanced methods [19]. Basic methods can be divided into the time, frequency, and time-freq domain, of which common approaches are Window Cropping or Slicing, Window Warping, Flipping, DTW Barycentric Averaging, etc. Advanced methods derive from decomposition-based, model-based, and learning-based models, of which common approaches are STL, Gaussian Tree, Generative Adversarial Network, etc. The above two DA algorithms are usually respectively suited for classification and regression. For classification, DBA is shown to be useful in some UCR datasets [13]. For regression, Lee S W et al. propose a novel regularization method, random column-wise shuffling, as the data augmentation technique of their convolutional neural networks to prevent the overfitting problem, which is proved to be beneficial for their model to forecast stock market indexes [13]. Moreover, Demir S et al. devise a novel time series augmentation method using generative models, resulting in a statistically significant improvement in the accuracy of electricity price forecasting [7]. Kasun Bandara et al. proved DA can improve the accuracy of regression prediction of the Global Forecasting Model called the MTL model in this paper through experiments, in which the comprehensive improvement effect of DBA was the most obvious [1], and that's why this paper uses the DBA-based algorithm as the DA method.

3 Preliminaries

3.1 Sequence to Sequence (Seq2Seq) Model

Seq2Seq is an important implementation of encoder-decoder architecture, of which the input and output are both sequences data. Moreover, different from Recurrent Neural Network (RNN), it can solve the problem of indefinite-length sequences pairing, which is very suitable for the financial time series prediction because more future ticks could be predicted each time accurately, there will be more time for portfolio managing in advance. The function of the encoder is to represent the pattern information of the whole visible sequence by the context vector C, and the decoder interprets the context vector and decoder's hidden state vectors based on the weight matrix obtained by training, to complete the pairing between the input and output sequences. Since RNN is characterized by accumulatively considering the information of each previous timestep, C can theoretically include all the information of the input sequence, and its mathematical representation is as follows:

$$C = h_t \bullet X_t \tag{1}$$

where h_t and X_t respectively are the hidden state vector and the input value, which are corresponding to the last timestamp t of the input sequence. Based on this, the conditional probability of the model output Y_t is expressed as:

$$P\left(Y_t \left| Y_{t-1}, Y_{t-2}, \ldots, Y_1, C\right.\right) = g(h_t, Y_{t-1}, C) \tag{2}$$

where g is always the softmax function. According to the different ways of inputting C into the decoder, the Seq2Seq architecture can be divided into two types, namely the parallel structure proposed by Cho et al. [4], in which C is inputted into each RNN cell of the decoder, and the serial structure proposed by

Sutskever et al. [16], in which C is only used as the input of the first RNN cell, and in sequence, the hidden state vector output by the previous cell is used as the input of the next cell. There are differences between the models' mathematical expression, and the parallel structure is:

$$P(Y_t) = f(h_t, Y_{t-1}, C) \tag{3}$$

and the serial structure is:

$$P(Y_t) = f(h_t, Y_{t-1}) \tag{4}$$

where f represents the mapping function of RNN like LSTM.

Depending on the different definitions of the input Y of the decoder, there are two decoding strategies as shown in Fig. 2.

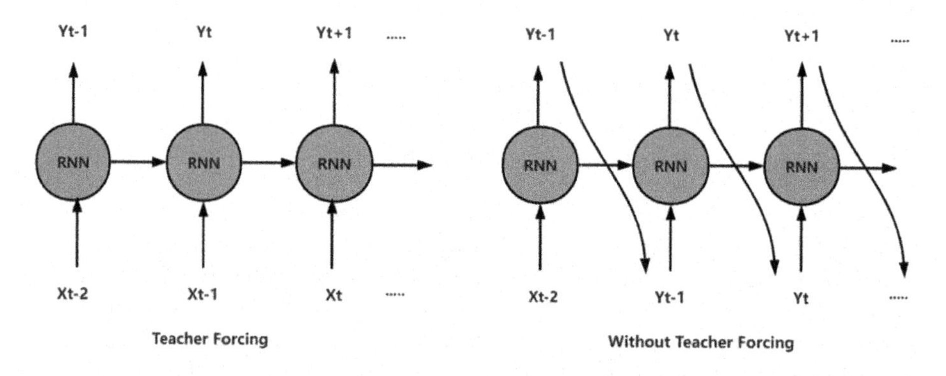

Fig. 2. Decoding strategy.

Teacher Forcing strategy means taking the past one target value, the ground truth, as the input for the next RNN cell, and the Without Teacher Forcing strategy means using the past one predicted value as the input for each RNN cell of the decoder. The former is suitable for the training phase when the target value is fully visible, while the latter is suitable for the testing phase when the target value is not visible. LSTM is adopted as the RNN cell in this paper because the traditional RNNs cannot handle the problem of exponential explosions of weights or disappearance of gradients as they recursion, and the improved LSTM can solve this problem by introducing dropout layers.

3.2 Weight Soft-DTW Barycentric Averaging (wDBA) Algorithm

The modified DBA proposed by Seto S et al. [15] is adopted as the foundation of the data augmentation method in this paper, which is a weighted version of the DTW barycentric averaging (DBA) [14] algorithm. Different from this modified DBA, the Soft-dtw [6] rather than the classical DTW is adopted as the distance measurement in wDBA. For two time series S and K, the Soft-dtw distance is defined as follows:

$$DTW\left(S,K\right)=\sqrt{\sum_{(i,j)\in P}\left(S_i-K_j\right)^2} \tag{5}$$

where i and j are time indices for points on the DTW path, and S_i and K_j are the matched-computing values, of which i is not equal to j necessarily, which means that it does not have to compute the sequences' difference by comparing S and K at each same timestamp. Once the set of points (i, j) on this path has been established, the Soft-dtw distance can be calculated as the square root of the summary of all the point pairs' squared differences. There is always a path P which could minimize the Soft-dtw distance, which is subjected to two constraints: Firstly, the first (i, j) point pair must be $(0, 0)$, the start of both time series, and the last (i, j) point pair must be (n, n), where n is the shared length of the two sequences; Secondly, the i and j are allowed to change by one step or stay the same when going from one timestamp point on the path to the next.

It turns out that the constrained optimization problem of finding the path P that minimizes $DTW(S, K)$ can be efficiently solved using a technique called dynamic programming [3]. The primary idea of wDBA is that the synthetic time series of any quantity could be generated by weighted (barycentre) averaging based on two or more time series. For a set of time series, the related synthetic sequence is their weighted average (barycentre) which is the S subjecting to the following objective function:

$$\min_{S\in R}\sum_{i=1}^{N}\lambda_i DTW\left(S,K_i\right) \tag{6}$$

DA to generate a synthetic sequence is the minimum optimization of the above formula according to the known sequence set K and the DBA algorithm. This function represents the minimum value of the sum of the weighted Soft-dtw distances between S and each time series K_i in K, where the value space of the optimization problem is R, while N is the number of sequences of K, and λi are the weight factors. The λi are designed to be subjected to $\sum \lambda_i = 1$, and there is an infinite quantity of combinations satisfying it, which gives the diversity of generating synthetic sequences by paying different attention to various K_i.

4 Methodology

After the data collection and exploratory data analysis, to alleviate the lack of training data for MTL of MTS analysis, the wDBA is applied for DA during the modeling of Seq2Seq consisting of LSTM as RNN cells. Simultaneously, for effectively evaluating the prediction ability of the improved model on long-term time series, we propose the Chronological Randomly-sampling Walk-Forward (CRSWF) Validation.

Table 1. Data description.

Forex	XAU, EUR, JPY, GBP, AUD, CHF, CAD, NZD
Price	Open, High, Low and Close minutely bid quotes
Time	2020-01-01 18:01:00 to 2022-12-30 16:57:00
Minute	1,041,238 min (exclude missing values)

4.1 Data Acquisition and Processing

The OHLC bid quotes of 8 major Forex pairs are collected as research objects from the HistData website shown in Table 1, namely XAU/USD (XAU), EUR/USD (EUR), JPY/USD (JPY), GBP/USD (GBP), AUD/USD (AUD), CHF/USD (CHF), CAD/USD (CAD), NZD/USD (NZD).

For each bid quote of each Forex pair, the crawled data ranges from 2020-01-01 18:01:00 to 2022-12-30 16:57:00, which is at the minute level, with 1,041,238 timestamps. There are 4,164,952 relevant bid quote data for each Forex pair (four quotes), and a total of 33,319,616 data for the dataset (eight Forex pairs) is involved in this study.

To understand the synchronicity between the bid quotes of various Forex pairs, this paper standardizes the close prices and calculates their Pearson correlations. The standardization does not cause data leakage because they are only used to observe if quotes are relevant, but they are not used for modeling and experiments.

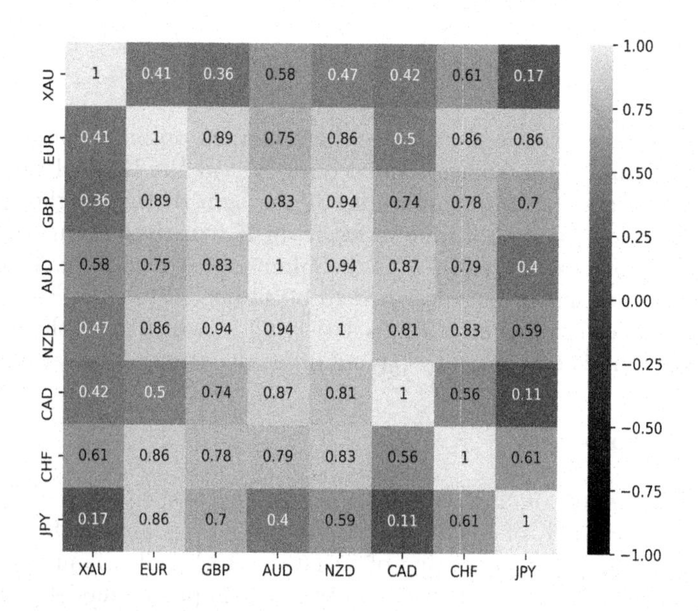

Fig. 3. Correlation matrix upon Forex pairs' standardized close bid quotes.

As shown in Fig. 3, most of the absolute values of Pearson correlations are above 0.5, and some are even higher than 0.8, which indicates that there are some common patterns of these quote sequences, giving feasibility to apply MTL.

In data preprocessing, the use of the $log1p(x)$ function can avoid the problem of outliers, and the function only takes the current value as the parameter. The function of $log1p(x)$ is:

$$log1p(x) = \log(x+1) \tag{7}$$

In econometrics, the price time series has been proved many times that it is not stationary, and consequently, the data after the above processing still needs to be differentially processed by subtracting the mean, as follows:

$$x' = log1p\,(x) - \bar{x} \tag{8}$$

where x' and $\bar{x}$ are respectively the inputting data and the historical mean. The whole process above does not need to introduce any future information, so it will not lead to data leakage. For reference purposes, this process is called data compression.

The output data obtained from prediction after modeling based on the compressed data has no longer realistic readability, so it is necessary to recover them through the inverse operation by the formula:

$$expm1\,(x',\ \bar{x}) = \ exp\,(x' + \bar{x}) - 1 \tag{9}$$

In the following, this process is called data decompression. By making the input data more obedient to the Gaussian distribution, these two steps of processing lead the MTL model to a better-predicting result.

4.2 Seq2Seq Modeling with wDBA Algorithm

The whole modeling process of Seq2Seq is divided into the training phase and the testing phase. One of the obvious differences is that for the former, all the data is accessible at the beginning, while the latter requires that the input data obey the chronological order. And this makes the Seq2Seq architectures and decode strategies of the two phases quite different.

In the training phase, the decoding strategy is Teaching Forcing, and the Seq2Seq architecture is a parallel structure shown in Fig. 4 (upper), which avoids overfitting by transferring the fixed context vector rather than the past one decoder cell's hidden state vector. Unlike the classical parallel architecture introduced in the previous section, the model adopted in this paper only uses the context vector as the state information, and the objective function is also adjusted to

$$P(Y_t) = f(Y_{t-1}, C). \tag{10}$$

In this architecture, X_i, Y_i, and T_i are respectively the historical values, prediction values, and target values. Different from the general Seq2Seq model using

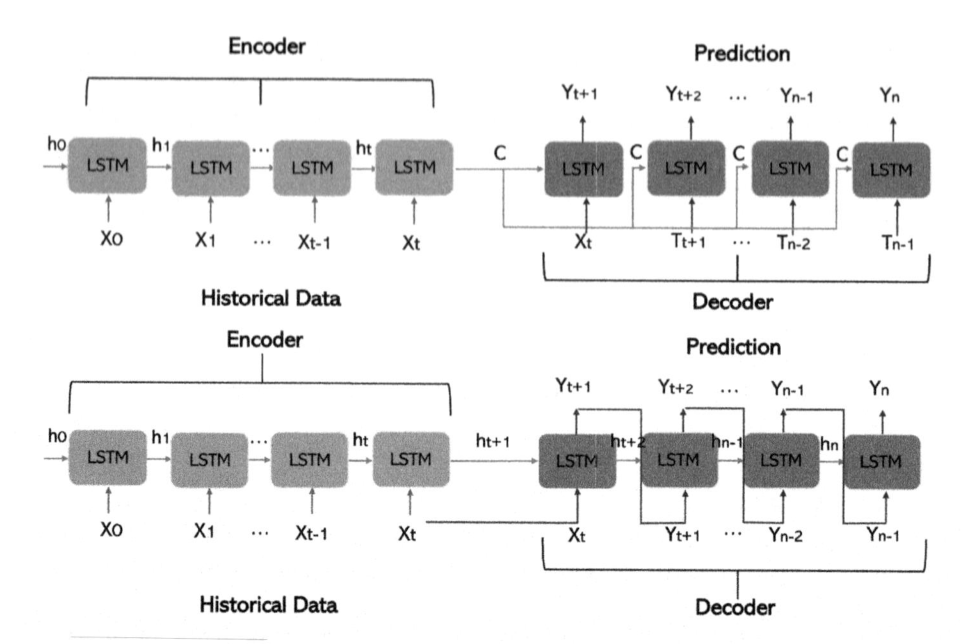

Fig. 4. The parallel structure of Seq2Seq for training (upper); the serial structure of Seq2Seq for predicting (bottom).

a specific [STRAT] signal as the input of the first cell of the decoder, our model adopts the last value of the inputted sequence, which is the same between both architectures. And for the rest cells of the decoder in the parallel structure, every Y_i can be calculated out only based on the context vector C and every ground truth T_{i-1}.

In the testing phase, the decoding strategy is Without Teaching Forcing, while the Seq2Seq architecture is the serial structure shown in Fig. 4 (bottom), which chronologically transports the accumulated information from the past to the next LSTM cell by feeding the past output $Y_{(i-1)}$ and the hidden state vector h_i into the decoder cell for providing the probability distribution of Y_i.

For the training process, the input of the model is the parallel combination based on the same timestamp, from 32 raw quote time series of 8 Forex pairs, and the synthetic time series based on the wDBA algorithm. The number of synthetic time series depends on the data augmentation degree (DA degree) which means the percent of the number of synthetic time series to the number of the original time series. And the wDBA algorithm is as follows:

1) Select randomly a time series instance i for each type of Forex bid quote from the training data;
2) Select randomly an integer from 2 to 4 as the hyperparameter k of the K-NearestNeighbor (KNN) model that is used to determine how much the nearest time series to the i needs to be found out;
3) Figure out Soft-dtw distances $DTW(i, j)$ of the k nearest neighbors (j);

4) Define the normalized weight vectors base on $\lambda_i = e^{\log(0.5)DTW(i,j)/DTW_{nn}}$, in which the DTW_{nn} is the Soft-dtw distance between i and its nearest neighbor in (j), and $\log(0.5)$ makes sure the weight increases as the distance decreases;
5) Generate a synthetic instance by taking the barycenter average with Soft-dtw and the defined weight vectors;
6) Repeat the above steps until the number of synthetic instances meets the DA degree.

It should note that the process of generating the synthetic time series uses only the data from the training set, thus eliminating data leakage and the modeling bias caused by peeping the distribution of the test set.

4.3 Chronological Randomly-Sampling Walk-Forward (CRSWF) Validation

It is essential to appropriately partition the time series into encoding and decoding intervals and the Walk-forward validation is widely adopted in time series forecasting to make full use of the data. Considering that the training set and the testing set of time series cannot be scrambled, and the traditional Hold-out or K-fold cross-validation cannot be avoided from resulting in the future data being used to predict the data of the past, which is obviously the data leakage.

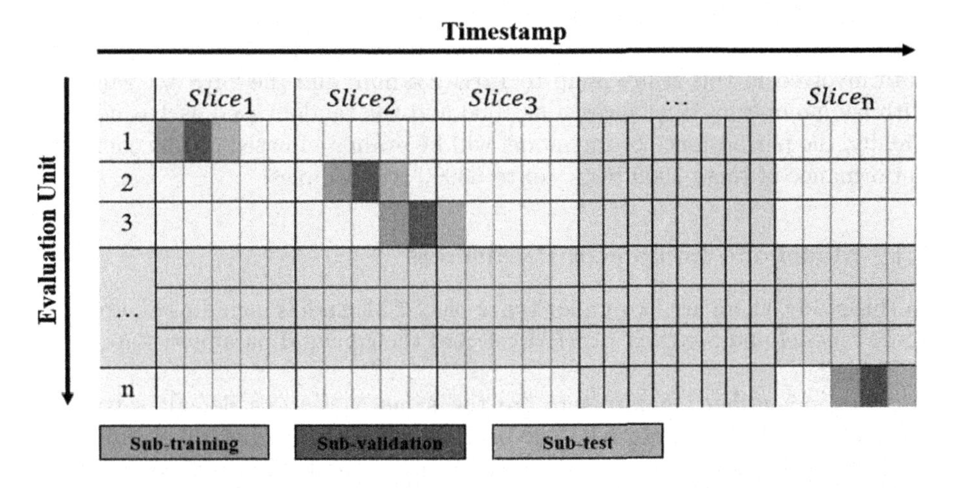

Fig. 5. The Chronological Randomly-sampling Walk-forward (CRSWF) Validation.

The Walk-forward validation divides the whole time series into evaluation units of equal length including the sub-training period, sub-validation period, and sub-test period. The final evaluation of the model can be obtained by concatenating all sub-test data for testing. In this approach, the difference between the start times of any two temporally adjacent sub-test periods equals the length of the sub-test period, and all the timestamps are considered in the testing except the first sub-training and sub-validation set. However, since the time span of the Forex bid quote data is too long to test every timestamp, in order to improve the experimental efficiency, a new adjusted version of the Walk-forward validation, the Chronological Randomly-sampling Walk-forward (CRSWF) Validation is proposed in this paper as shown in Fig. 5. For the sake of presentation, all three subsets in the graph contain only one timestamp, which is hardly accepted in practical evaluations, but this assumption helps to explain how the method randomly selects the start timestamp of each Sub-test set. We first divide the whole data into a number n of length-equal time slices $Slice_i$, and chronologically randomly select a timestamp, the left border of each green square, in every slice as the start time of every sub-test period. It should be noted that random selection is conditional: the end time of the Sub-test set of each evaluation unit, the right border of the green square, will not exceed the last minute of the corresponding slice. Then we concatenate every sub-test data for loss calculation to make an unbiased estimate of the model performance, in which start timestamps are evenly distributed on the timeline of the entire dataset.

5 Evaluation

This paper conducts comparative experiments on the LSTM, Seq2Seq without DA, and Seq2Seq with DA, which are based on the OHLC data of 8 Forex pairs, a total of 32 price time series, and in each evaluation unit, we use the historical data of the past 6 h to predict the bid quote in the next 1 h. The time span of the data involved in this study is up to 1,041,238 min, and the CRSWF validation with the number of time slices n as 1000 and the random seed as 1 is adopted. Finally, the performance of the model will be evaluated based on the compound performance of these 1000 tests, up to 60000 test samples.

5.1 Model and Equipment Parameters

In this study, there are two major types of LSTM models including the baseline LSTM model and Seq2Seq's LSTM cell and their pivotal parameters are shown in Table 2.

The unmentioned parameters are the same as Keras's default settings of TensorFlow-GPU version 2.5.0. The baseline LSTM is designed in a rolling prediction way for the comparability to the Seq2Seq model, in which the sequence to be predicted is divided into 60 steps while Seq2Seq's each prediction step is 60 min, and each one of these 60 targets for the former is concatenated behind the 6 h of data preceding it to form the data used for one round of training-test (totally 60 rounds), so there are consequently the same as 60 samples (minutes) for every evaluation unit of baseline LSTM.

In addition, our experimental machine is equipped with a CPU: AMD Ryzen 9 4900HS with Radeon Graphics, and a GPU: NVIDIA GeForce RTX 2060 with Max-Q Design GDDR6.

Table 2. Summary of LSTM models' parameters

	LSTM (baseline)	LSTM Cell (Seq2Seq)
Layer number	1	1
Latent unit number	50	50
Dropout rate	0.25	0.25
Batch size	4	4
Epoch	30	30
Optimizer	Adam	Adam
Total (trainable) parameters	82,251	20,851

5.2 Indicators

This paper focuses on the regression task, so the Mean Absolute Error (MAE), Root Mean Square Error (RMSE), Normalized Mean Absolute Error (NMAE), and Normalized Root Mean Square Error (NRMSE) are chosen as evaluation metrics. Suppose m is the number of samples used for the test, and the target values are $y = (y_1, y_2 \dots y_m)$, while the predicted values are $\hat{y} = (\hat{y}_1, \hat{y}_2 \dots \hat{y}_m)$. The MAE and RMSE ARE used to measure the difference between the target value and the predicted value with formulas:

$$MAE = \frac{1}{m} \sum_{n=1}^{m} |y_n - \hat{y}_n| \tag{11}$$

$$RMSE = \sqrt{\frac{1}{m} \sum_{n=1}^{m} (y_n - \hat{y}_n)^2} \tag{12}$$

For the comparability between various Forex pairs, the NMAE and NRMSE are introduced as percentages, in which smaller values indicate less residual variance, and $\bar{y}$ is the mean of target values.:

$$NMAE = \frac{MAE}{\bar{y}} \tag{13}$$

$$NRMSE = \frac{RMSE}{\bar{y}} \tag{14}$$

5.3 Experiments

There are two comparative experiments in this paper. One is to compare the regression performance of the single LSTM model and the LSTM-based Seq2Seq model, for figuring out whether MTL is beneficial for MTS prediction. The other one is about comparing different DA degrees' impacts on the predicting ability of Seq2Seq models to find out the potential pattern.

All shown in Table 3 are mean values of the regression indicators of 8 Forex pairs. The 'LSTM' stands for the baseline model, the 'Seq2Seq' represents the LSTM-based Seq2Seq model, and the '%wDBA' means that the DA Degree of wDBA is X, and this degree indicates the percentage of the number of newly generated time series accounting to the original series.

Table 3. Summary of models' prediction performance (records are rounded approximations)

Models	MAE	NMAE	RMSE	NRMSE
LSTM (baseline)	7.2866	42.79%	8.4878	42.93%
Seq2Seq - 0%wDBA	0.3774	1.24%	0.5358	1.75%
Seq2Seq - 25%wDBA	0.3432	1.36%	0.4874	1.92%
Seq2Seq - 50%wDBA	0.3228	1.26%	0.4801	1.89%
Seq2Seq - 75%wDBA	0.3239	1.33%	0.4711	1.85%
Seq2Seq - 100%wDBA	0.3298	1.27%	0.4762	1.83%
Seq2Seq - 125%wDBA	0.3477	1.18%	0.5034	1.72%
Seq2Seq - 150%wDBA	0.3513	1.31%	0.5137	1.86%

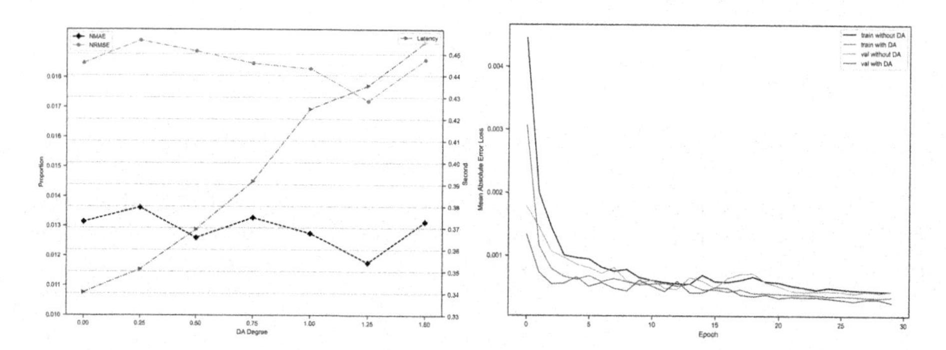

Fig. 6. Visualization of models' prediction performance (left); training loss functions of Seq2Seq model without DA and the one with DA degree as 125% (right).

As shown in Table 3 and Fig. 6 (left), our MTL model based on Seq2Seq respectively achieve average improvements of 95.3%, 97.01%, 94.16% and 95.73% in MAE, NMAE, RMSE, and NRMSE, compared with the LSTM-based baseline model. In addition, by comparing different DA Degrees, the error first decreases and then rebounds. The MAE and RMSE reach the lowest level under the condition of DA degree as 75%, while NMAE and NRMSE reach the lowest level when the DA degree is 100%. The comparison between the Seq2Seq without DA with the Seq2Seq with DA shows that an appropriate degree of data augmentation can improve the prediction ability of the model. But when the proportion of synthetic data exceeds $[75\%, 100\%]$ of the raw data, the prediction error of the model becomes higher for introducing so much generated data that the input of the model becomes fuzzier.

In addition to analyzing the predicted performance, we should also analyze the differences between the Seq2Seq model without DA and those with DA, from the model training process and latency. The Loss functions of the Seq2Seq models with or without the DA are shown in Fig. 6 (right). It is clearly observed that DA allows the model to converge faster by giving it more samples, but DA also brings longer latency. As the DA degree increases, the average latency used to predict each price time series also gradually increases, but they are still less than the baseline LSTM's latency of 3.4 s per time series.

To further understand the gap between the models, we take the close bid quotes of EUR/USD and XAU/USD as examples and intuitively recognized the improvement of the Seq2Seq model relative to the LSTM model by visualizing the prediction results in Fig. 7. It is also easy to observe that the predicted value of the LSTM model is basically the lag value of the true value, and the model cannot learn the information in the historical quotes. It is efficiently solved in the Seq2Seq model after using data augmentation.

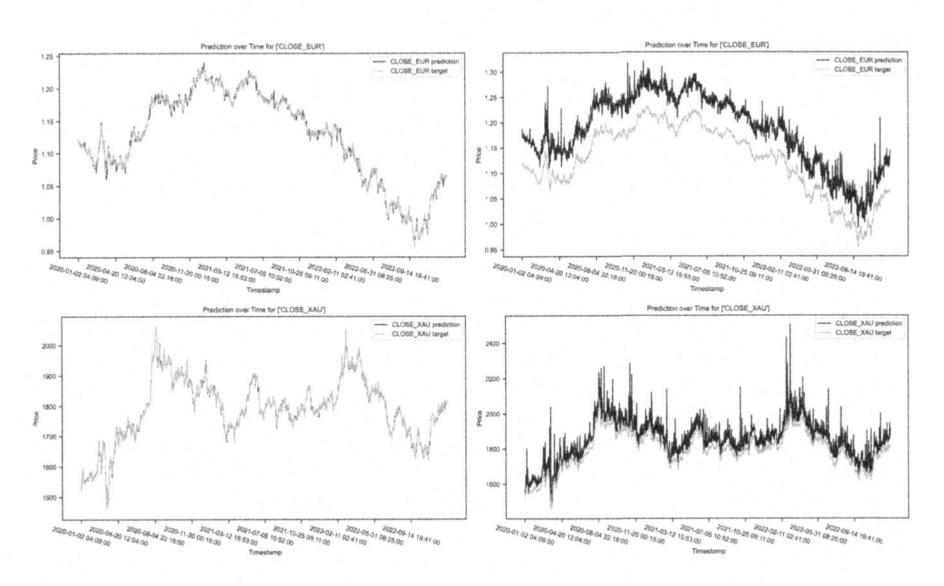

Fig. 7. EUR's prediction result of Seq2Seq - 125%wDBA model (upper left) and baseline LSTM model (upper right); XAU's prediction result of Seq2Seq - 125%wDBA model (bottom left) and baseline LSTM model (bottom right).

6 Conclusions and Future Work

In this paper, an LSTM-based Seq2Seq MTL model for MTS with a high correlation in assets' price is designed, which has been proven to have better regression prediction ability than the LSTM-based single-task learning model. In the process, to alleviate the lack of time series samples, we utilize the wDBA algorithm for DA to improve the predictive ability of the model. The experimental results show that the Seq2Seq - 125%wDBA model performs best, reaching 0.3477, 1.18%, 0.5034, and 1.72% on MAE, NMAE, RMSE, and NRMSE.

For future work, we are considering designing a specific attention mechanism for price time series, optimizing the decoder strategy of the current best model, using data other than Forex quotes to further verify the generalization of the model, and exploring the difference of various DA algorithms in the synchronous prediction of asset prices' multivariate time series.

ACKNOWLEDGMENT. We deeply appreciate the funding from Grant no. 2021GH10, Grant no: 2020GH10, and Grant no: EF003/FST-FSJ/2019/GSTIC by Guangzhou Development Zone Science and Technology; Grant no. 0032/2022/A and 0091/2020/A2, by Macau FDCT; Grant no. MYRG-GRG2022 and Collaborative Research Grant (MYRG-CRG) - CRG2021-00002-ICI, by University of Macau.

References

1. Bandara, K., Hewamalage, H., Liu, Y.H., Kang, Y., Bergmeir, C.: Improving the accuracy of global forecasting models using time series data augmentation. Pattern Recogn. **120**, 108148 (2021)
2. Bellman, R.: The theory of dynamic programming. Bull. Am. Math. Soc. **60**(6), 503–515 (1954)
3. Caruana, R.: Multitask learning. Mach. Learn. **28**, 41–75 (1997)
4. Cho, K., et al.: Learning phrase representations using RNN encoder-decoder for statistical machine translation. arXiv preprint arXiv:1406.1078 (2014)
5. Crawshaw, M.: Multi-task learning with deep neural networks: a survey. arXiv preprint arXiv:2009.09796 (2020)
6. Cuturi, M., Blondel, M.: Soft-DTW: a differentiable loss function for time-series. In: International Conference on Machine Learning, pp. 894–903. PMLR (2017)
7. Demir, S., Mincev, K., Kok, K., Paterakis, N.G.: Data augmentation for time series regression: applying transformations, autoencoders and adversarial networks to electricity price forecasting. Appl. Energy **304**, 117695 (2021)
8. Du, S., Li, T., Yang, Y., Horng, S.J.: Multivariate time series forecasting via attention-based encoder-decoder framework. Neurocomputing **388**, 269–279 (2020)
9. Gao, Z.: Stock price prediction with arima and deep learning models. In: 2021 IEEE 6th International Conference on Big Data Analytics (ICBDA), pp. 61–68. IEEE (2021)
10. Katsiampa, P., Corbet, S., Lucey, B.: Volatility spillover effects in leading cryptocurrencies: a BEKK-MGARCH analysis. Financ. Res. Lett. **29**, 68–74 (2019)
11. Lee, S.W., Kim, H.Y.: Stock market forecasting with super-high dimensional time-series data using ConvLSTM, trend sampling, and specialized data augmentation. Expert Syst. Appl. **161**, 113704 (2020)

12. Luong, M.T., Le, Q.V., Sutskever, I., Vinyals, O., Kaiser, L.: Multi-task sequence to sequence learning. arXiv preprint arXiv:1511.06114 (2015)
13. Petitjean, F., Forestier, G., Webb, G.I., Nicholson, A.E., Chen, Y., Keogh, E.: Dynamic time warping averaging of time series allows faster and more accurate classification. In: 2014 IEEE International Conference on Data Mining, pp. 470–479. IEEE (2014)
14. Petitjean, F., Ketterlin, A., Gançarski, P.: A global averaging method for dynamic time warping, with applications to clustering. Pattern Recogn. **44**(3), 678–693 (2011)
15. Seto, S., Zhang, W., Zhou, Y.: Multivariate time series classification using dynamic time warping template selection for human activity recognition. In: 2015 IEEE Symposium Series on Computational Intelligence, pp. 1399–1406. IEEE (2015)
16. Sutskever, I., Vinyals, O., Le, Q.V.: Sequence to sequence learning with neural networks. In: Advances in Neural Information Processing Systems, vol. 27 (2014)
17. Wang, P., Wang, P.: Assessment on estimations of currency basket weights-with coefficient correction for common factor dominance. Int. J. Financ. Econ. **27**(1), 1401–1418 (2022)
18. Wang, X., Cai, Z., Luo, Y., Wen, Z., Ying, S.: Long time series deep forecasting with multiscale feature extraction and seq2seq attention mechanism. Neural Process. Lett. **54**(4), 3443–3466 (2022)
19. Wen, Q., et al.: Time series data augmentation for deep learning: a survey. arXiv preprint arXiv:2002.12478 (2020)

MGCP: A Multi-View Diffusing Graphs Based Traffic Congestion Prediction for Roads Around Factory

Wei Zhao[1], Yingzi Shen[1], Jiali Mao[1,2](✉), and Lei Cheng[1]

[1] East China Normal University, Shanghai 200062, China
{52195100008,51255903061,71205901085}@stu.ecnu.edu.cn
[2] Shanghai Engineering Research Center of Big Data Management, Shanghai, China
jlmao@dase.ecnu.edu.cn

Abstract. Traffic congestion easily occurs on the roads around factories due to limited road space, which will be worsen especially when many trucks stay at the roadside of the same road. Actually the truck's movement or staying at the road depends on the phase of its implementing cargo-loading task, e.g., the truck may strand in the road outside the factory due to waiting for loading cargoes, or move toward the road near the gate of factory when receiving loading notification. Thus, to predict traffic jams of the roads around the factory, it is necessary to consider the influence of the truck's cargo-loading task phase on road traffic situation. However, the influences of different task phases that the trucks are on traffic situation of the roads are not the same, and such influences may change with the transition of the trucks' task phases, which brings severe challenges for precise prediction. In this paper, we put forward a multi-view task diffusing graphs based traffic congestion prediction method for Roads around Factory, called *MGCP*. To capture discrepant impacts of task phases on traffic situations of the roads, we present a task phase based multi-view diffusing graphs generating method. In addition, we leverage a *Markov* process to denote the transition of each task phase, and build a transition matrix of task phases to extract the transited task phases in the future. Experimental results on real steel logistics data sets demonstrate that our proposed method outperforms the existing prediction approaches in terms of prediction accuracy.

Keywords: Bulk logistics · Task phase · Traffic congestion · Multi-view diffusing graphs

1 Introduction

With the increasing demand for bulk freight transportation, the roads around large-scale enterprises or logistic parks are easily congested. This also intensifies the traffic pressures of their adjacent roads such as the city roads and highways. It is reported that, on Aug. 2021, there are often heavy traffic jams in the logistics

X. Yang et al. (Eds.): ADMA 2023, LNAI 14180, pp. 552–568, 2023.
https://doi.org/10.1007/978-3-031-46677-9_38

park in *Ningbo* city, *Zhejiang* province, China[1]. Likewise, a similar situation prevails in the roads near Benxi Iron and Steel Group on Apr. 2020[2]. The analysis shows that traffic congestion occurred on the roads around the factory or logistics park (or called *FactRoad* for short) is caused by thousands of trucks going to the factory or logistics park for loading (or unloading) the cargoes. Due to no sufficient amount of parking lots, most of trucks have to stay at the side of roads until receiving the transport task instruction for entering the factory or the warehouses. Generally, each road within *FactRoad* is about 16-m wide, while each truck is about 13.5-m long and 4-m wide. The trucks' long-staying behaviors seriously decrease the traffic capacity of the roads where they are on, and further worsen the road congestion problem. To maintain smooth traffic in *FactRoad*, it necessitates an appropriate traffic congestion prediction method incorporating with the effect of trucks' long-time staying behaviors.

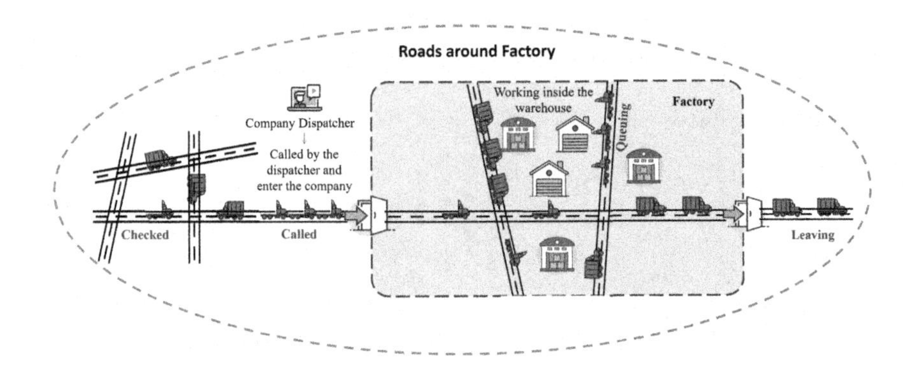

Fig. 1. Illustration of cargo loading procedure of a truck in a steel mill

Recently, although the traffic prediction issue has attracted wide attentions from academia and industry, their proposed solutions mainly focus on the traffic of the city roads not having long-time stranding vehicles [1–4], thus they cannot be directly leveraged to tackle the issue of traffic congestion prediction for *FactRoad*. What factors lead to long stays of trucks on *FactRoad*? Take a steel mill in *Shandong* province, China as an example, each truck's transport workflow on *FactRoad* (highlighted in a red ellipse) can be divided into four task phases: check-in at the steel mill (or *Checked* for short), called to enter the mill (or *Called* for short), queue around warehouses to wait for loading or unloading of cargoes (*Queuing* for short), and leaving the mill (or *Leaving* for short) after finishing their work. The above task phases keep an linearly unidirectional transition process as shown in Fig. 1, i.e. *Checked* → *Called* → *Queuing* → *Leaving*. Each type of task phase has an unique probability of transitioning to subsequent phases at the next timestamp, which varies depending on the duration of the

[1] http://www.bl.gov.cn/art/2021/8/3/art_1229044479_59034428.html.
[2] https://gaj.benxi.gov.cn/gadt/gzdt/content_462692.

phase. Different task phases may lead to a truck's moving or staying on the road. Figure 2 shows the speed distribution of trucks with the four task phases. We can observe that trucks with *Checked* or *Queuing* may strand on the roads outside or inside the factory for a long time, but trucks with *Called* and *Leaving* are in motion state. To predict traffic congestion in *FactRoad*, it needs to analyze the diffusion of traffic situations among roads resulting from different moving behaviors of trucks that are in various task phases.

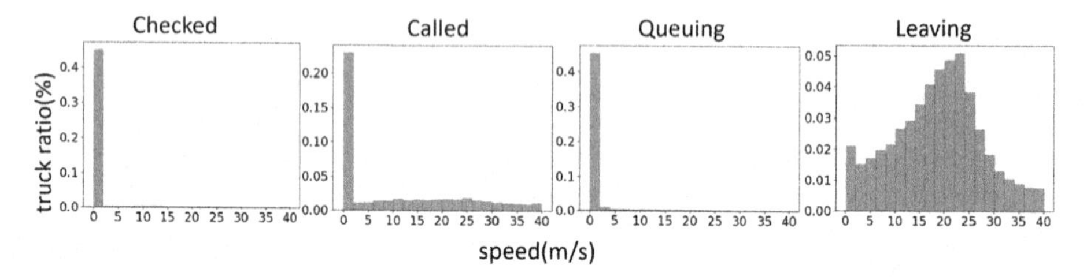

Fig. 2. Speed distribution of trucks in different task phases

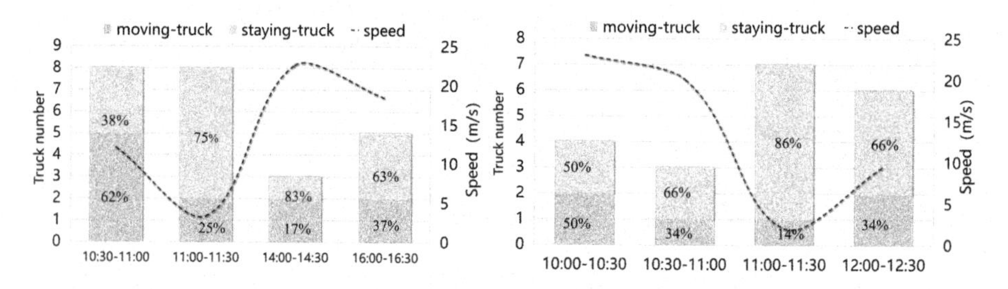

Fig. 3. The correlations between average speed of moving trucks and the amount of moving-truck or staying-truck on two specified roads during different time periods of one day.

However, there are two challenges needed to be tackled for predicting traffic congestion in *FactRoad*.

- *The influences of different task phases on the traffic situation of roads vary significantly.* As discussed above, each task phase corresponds to a different moving behaviors of trucks, which can affect the traffic situation of the roads differently. This is evident in Fig. 3(a) and Fig. 3(b), where even when the total number of trucks on two roads is similar, the higher number of staying-trucks results in a lower average speed of moving trucks. Moreover, these unique moving behaviors of trucks result in varying tendencies of truck diffusion among roads, making it challenging to predict the future traffic situation by simply aggregating the task phases into a single representation for each road.
- *The transition of task phases leads to changes in the traffic situation of roads.* When the moving behaviors of numerous trucks change, the fluctuations may occur in the traffic situation of the roads that the trucks are moving on. It can be verified by an obvious gap in the average speed of moving trucks during the time intervals of 10:00–10:30 and 11:00–11:30 in Fig. 3(a). Inspired

by this, we can perceive the variation in traffic situation of the road by capturing the transitive process of trucks' task phases. As different task phases have varying durations, it is challenging to construct an uniform transition evaluation module fit for all types of task phases.

To generate an accurate traffic representation of roads based on task phases, we first construct a graph-structured diffusion network for trucks with each type of task phase in the road network, and then use an attention network to capture highly related task phase-based diffusing graphs for traffic situation diffusing mining. In addition, we recognize that task phases transition unidirectionally and linearly, so we model it as a Markov process and introduce a transition probability matrix for each task phase to capture these transitions. Finally, we construct a formula for evaluating traffic congestion on *FactRoad* by combining predicted traffic situation representations of roads from a sequence forecasting model. The contributions of our work are summarized as follows:

- To predict traffic congestion in *FactRoad*, we propose a Multi-View Diffusing Graphs based Traffic Congestion Prediction architecture, called *MGCP*, which incorporates the traffic situation of roads influenced by the moving behaviors of trucks with different task phases.
- To capture different impacts of trucks' task phases on traffic situations, we design a task phase-based multi-view graph-structured diffusion network and leverage an attention network to mine highly related task phase-based diffusion graphs for traffic situation analysis.
- In view of the impact of the transitive process of trucks' task phases on traffic situations of roads, we build a transition probability matrix and employ a *Markov process*-based transiting strategy to update the task phase representations of roads.
- We conduct extensive experiments on real data sets to demonstrate that *MGCP* outperforms the existing approaches in accurately predicting traffic congestion of *FactRoad*.

2 Related Works

2.1 Attention Mechanism Based Approach

In recent years, attention mechanism has been employed in traffic prediction solutions to extract important factors affecting traffic situation from multiple temporal dimensions such as recent, daily and weekly periods [3]. For instance, Yao et al. introduced a periodically shifted attention mechanism to handle the issue of long-term periodic temporal shifting [4]. Some solutions yet designed spatial-temporal attention mechanisms to extract traffic flow variations, and based on that, transformed historical traffic flow into predictive traffic flow [2]. To improve traffic prediction precision, a few solutions fused external factors extracted from different domains (e.g., sensor readings, meteorological data, and spatial data) into traffic prediction models by employing multi-level attention-based recurrent neural networks [5]. Inspired by them, we employ attention mechanism to obtain trucks' task phases diffusion from spatial-temporal dimension in order to capture the traffic situation diffusion of the roads.

2.2 Graph Neural Network Based Approach

Owing to that graph is an useful topology to reflect spatial dependency of traffic situations in road networks, it has been widely used in traffic prediction solutions. In the literature [6], spectral convoluted neural network that considered graph signals with multiple channels was leveraged to extract different features reflecting traffic situations. Some approaches also designed graph learning models (e.g., dynamic graph convolutional network) using traffic data to capture spatial or temporal feature of traffic situations [7,8]. Lately, a few prediction approaches incorporated graph convolutional network with gated recurrent unit (GRU) to simultaneously capture spatial and temporal dependencies among features influencing traffic situation [9,10]. Further, some researches designed multi-graphs based solutions according to spatio-temporal features and external ones like weather to proliferate traffic prediction accuracy [11,12]. Nevertheless, they cannot be directly applied to predict traffic congestion among *FactRoad*, because they do not consider disparate impacts of the trucks' task phases on traffic situations of the roads which the trucks are in.

3 Preliminaries

Definition 1 Roads around Factory. Roads around the factory, denoted as *FactRoads*, can be modeled as a weighted directed graph $G = (V, E, W)$, where V is a set of vertices representing the roads, and we have the total number of roads as $N = |V|$. E is a set of weighted directed edges that connect the vertices, representing the traversing correlations among the roads. The weight of each edge $W_{i,j}$ denotes the traversing possibility from road i to road j.

Traversing correlations can be classified into two types: *in-traversing* and *out-traversing*. *In-traversing* refers to the traffic situation or trucks with different task phases diffusing from other roads into the current road, while *out-traversing* refers to the traffic situation or trucks with different task phases diffusing from the current road to other roads.

Definition 2 Transitive Probability Matrix of Task Phases. Given a set C of task phases $S^k, k \in C$, and a predefined maximum number of timestamps T^d, transition probability matrix of task phase $\Gamma \in R^{C \times T^d}$ is used to record transitive probabilities between each pair of task phases extracted from historical data.

Definition 3 Traffic Congestion among *FactRoad*. Traffic congestion among road *FactRoad* refers to the traffic congestion degree $J \in R^{Q\times N}$. For each road j at each timestamp i, $J_{j,i}$ is calculated based on Eq. (1):

$$J_{j,i} = (1 - \frac{AS_{j,i}}{\overline{AS}_j}) * ((1-\theta_j) * \frac{AMT_{j,i}}{\overline{AMT}_j} + \theta_j * \frac{AST_{j,i}}{\overline{AST}_j}) \quad (1)$$

The calculation of $J_{j,i}$ takes into account the following factors, including the average speed of moving trucks $AS_{j,i}$, the number of moving trucks $AMT_{j,i}$, and the number of staying trucks $AST_{j,i}$. Specifically, the maximum values of these variables in historic records are denoted as $\overline{AS}_j, \overline{AMT}_j$, and $\overline{AST}_j$, separately. Besides, the predefined parameter $\theta_j \in [0,1]$ reflects the effect of the number of staying trucks on the average speed of moving trucks of each road j.

Problem Studied. Given traffic situation input signal $X \in R^{P\times N}$ (e.g., the average speed of moving trucks, the number of moving trucks, or the number of staying trucks) of N roads during previous P timestamps and the task phases input signal $S \in R^{P\times N\times C\times T^d}$, our goal is first to construct a traffic situation prediction model $f(\cdot)$ that forecasts the traffic situation $X' \in R^{Q\times N}$ during next Q timestamps, and then evaluate the traffic congestion $J \in R^{Q\times N}$ using the evaluation function g as defined in Eq. 1.

$$J = g(f^{AS}(X^{AS}, S), f^{AST}(X^{AST}, S), f^{AMT}(X^{AMT}, S)) \quad (2)$$

4 Methodology

To begin, we model task phase transitions using a Markov process and encode the task phases with their duration. To generate a suitable representation for each road based on traffic situation, we introduce a time-difference attention based graph module to capture the task phase diffusion process among roads from one timestamp to the next. Using these task phase diffusion graphs, we identify the traffic situation variation patterns of each road in each timestamp. Finally, we use two self-attention based *GRU* networks to separately capture the traffic situation and task phase variation patterns of each road in the temporal dimension. To improve the precision of our traffic predictions, we incorporate auxiliary prediction tasks with double weighting of the coefficient of variation approach(DCVW) that aim to estimate the amounts of different task phases during each duration contained in the roads. After obtaining the predictions for future traffic situations on the roads, the traffic congestion levels can be evaluated using a predefined formula 1.

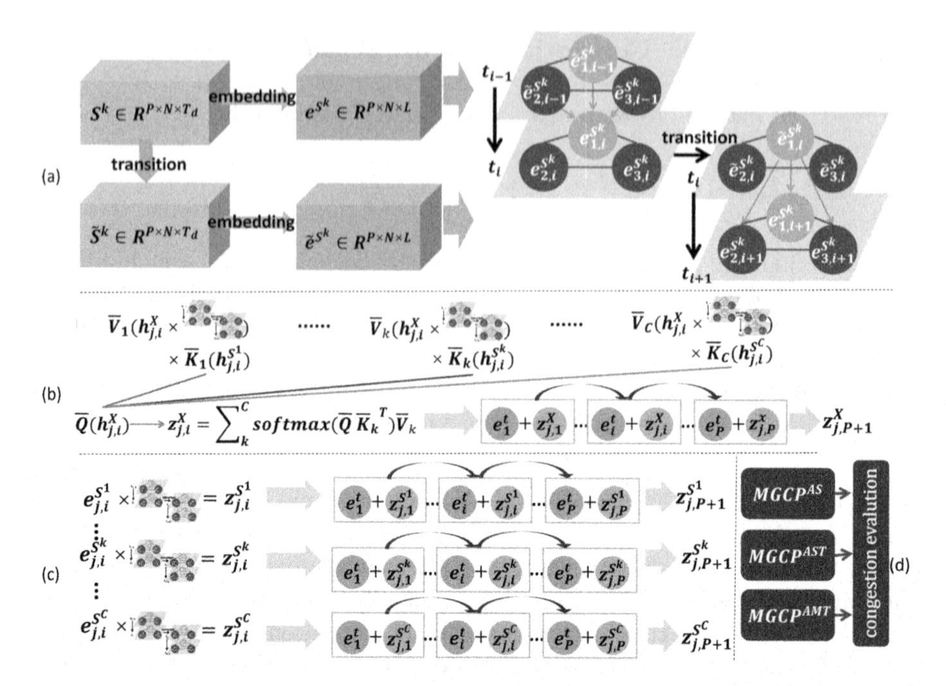

Fig. 4. *MGCP* architecture. (a) A task phase-based graph-structured diffusion network module for each type of task phase. (b) A multi-view task phase diffusion-based traffic situation prediction module. (c) A graph-structured diffusion-based task phase prediction module. (d) The traffic congestion evaluation based on the three predicted traffic situations of roads.

4.1 Task Phase Based Graph-Structured Diffusion Network Construction

Markov Process Based Task Phase Transition. Since the transition of task phases is influenced by their duration, we use the task phase matrix S together with the task phase transition matrix Γ to compute the transitive expectation of the task phase with different duration into next phase.

For each type of task phase in road j during timestamp i, $S^k_{j,i} \in R^{T_d}$ records the amount of trucks with the task phase S^k in each duration. To obtain the transited task phase, we first compute the expected transited amount from phase S^k to S^{k+1}, $\Delta S^{k+1}_{j,i,1} = \sum_t^{T_d} S^k_{j,i,t} \times \Gamma^k_t$, where Γ^k_t is the transition probability of phase S^k at duration t. It should be noted that these transitions are updated into the first duration of the next task phase, hence the duration index of $\Delta S^{k+1}_{j,i,1}$ is set to 1.

After updating the task phase at timestamp $i+1$, the expected amount of transition should be subtracted from each duration of $S^k_{j,i}$ to obtain the updated task phase S^k at timestamp i. Additionally, each duration count in $S^k_{j,i}$ should be shifted one timestamp forward to account for the passage of time. The transition process can be summarized by the following equation:

$$\tilde{S}_{j,i} = \|_{k=1}^{C} \begin{cases} \|_{t=1}^{T_d} S_{j,i,t+1}^{k} \leftarrow S_{j,i,t}^{k} \times (1 - \Gamma_t^k) \\ S_{j,i,1}^{k+1} \leftarrow S_{j,i,1}^{k+1} + \Delta S^{k+1} j, i, 1 \end{cases} \tag{3}$$

where $\tilde{S}_{j,i}$ is the transited task phase matrix, $\|$ is a concatenation operation for each vector. In order to avoid the influence of transition processes between different type of task phases, the transition process of each type of task phase is parallel execution, and the transited task phases are fused into the new task phase matrix $\tilde{S}_{j,i}$.

Duration Based Task Phase Encoding. Consider that the transition of task phases is influenced by their duration, to capture more detailed moving behaviors of trucks, we generate a unified representation e^{S^k} for each task phase S^k using a duration-based embedding layer $embed(\cdot)$ defined as $e^{S^k} = \sum_{t=1}^{T_d} embed(S_t^k)$. Here, $S_t^k \in R^{P \times N}$, $e^{S^k} \in R^{P \times N \times L}$, and L represents the output dimensionality of the duration-based embedding layer. We also use the same embedding layer to encode the transited task phase matrix $\tilde{S}^k$ and generate the unified representation $\tilde{e}^{S^k}$. These embeddings are then used to analyze the task phase diffusion among roads.

Time-Difference Attention Based Task Phase Diffusion Graph Construction. Trucks exhibit individual moving behaviors with each type of task phase, which makes it useful to extract traffic situation variation patterns by analyzing the diffusion of task phases among roads. However, due to the existing of task phase transitions, task phase diffusion across different roads may change in different timestamps, making it unreasonable to mine spatial dependencies in just one timestamp. To address this issue, we propose a task phase diffusion graph construction method based on the time-difference attention. This method captures the task phase diffusion during two consecutive timestamps by considering the transition of task phases using attentive operations. As shown in the right of Fig. 4(a), we construct the *in-traversing* matrix $D_{j \leftarrow j'}$ by considering the task phase diffusion from timestamp $i-1$ to i (as shown in Fig. 4), and the *out-traversing* matrix $D_{j \rightarrow j'}$ by considering the task phase diffusion from timestamp i to $i+1$ (as shown in Eq. 5).

$$u_{j \leftarrow j',i}^{k} = \frac{exp(\overline{Q}^k(e_{j,i}^{S^k}) \times (\overline{K}^k(\tilde{e}_{j',i-1}^{S^k}))^{\mathrm{T}})}{\sum_{n}^{V_I^k(j)} exp(\overline{Q}^k(e_{j,i}^{S^k}) \times (\overline{K}^k(\tilde{e}_{n,i-1}^{S^k}))^{\mathrm{T}})} \tag{4}$$

$$u_{j \rightarrow j',i}^{k} = \frac{exp(\overline{Q}^k(\tilde{e}_{j,i}^{S^k}) \times (\overline{K}^k(e_{j',i+1}^{S^k}))^{\mathrm{T}})}{\sum_{n}^{V_O^k(j)} exp(\overline{Q}^k(\tilde{e}_{j,i}^{S^k}) \times (\overline{K}^k(e_{n,i+1}^{S^k}))^{\mathrm{T}})} \tag{5}$$

where $V_I^k(j)$ and $V_O^k(j)$ denotes the high dependent *in-traversing* and *out-traversing* roads set of road j for trucks with task phase S^k. The transforms

$\overline{Q}^k$, $\overline{K}^k$, and $\overline{V}^k$ represent the *Query*, *Key*, and *Value* operations for task phase S^k in the attentive computing process.

The traffic situation in a road is often affecting or affected by the traffic situations in its neighboring roads, which are highly correlated. Therefore, it is useful to capture the diffusion of traffic situation among roads by analyzing their connectivity. However, due to the demands of trucks' task phases in the *FactRoad*, some geographically adjacent roads may not be relevant in reality. Thus, we focus more on the highly dependent roads for each task phase S^k when generating $V_I^k(j)$ and $V_O^k(j)$, which are identified by calculating the average frequency of trucks' movements from one road to another based on the traffic volume of trucks with each task phase S^k. This approach helps to prioritize the roads with high traffic volume and task phase relevance, resulting in a more accurate construction of the task phase diffusion graph.

4.2 Multi-view Task Phase Diffusion Based Traffic Situation Prediction

To extract the traffic situation diffusion among roads, we utilize the generated multi-view task phase based *in-traversing* and *out-traversing* graphs. These graphs reflect the traversing correlations of trucks with different task phases among each road, which can be used to model the variation patterns of traffic situations. However, as the ratio of trucks with different task phases varies among roads, we use an attention mechanism to focus on the specific task phase-based graphs that have a greater impact on the traffic situation of each road:

$$z_{j,i}^X = \sum_k^C \alpha_{j,i}^k \Big(\sum_{j'}^{V_I^k(j)} u_{j \leftarrow j',i}^k \cdot \overline{V}_I(h_{j',i-1}^X \| \tilde{e}_{j',i-1}^{S^k}) - \sum_{j'}^{V_O^k(j)} u_{j \rightarrow j',i}^k \cdot \overline{V}_O(h_{j',i+1}^X \| \tilde{e}_{j',i+1}^{S^k}) \Big) \tag{6}$$

where h^X is the encoded representation of traffic situation X, obtained applying a fully-connected layer. $\overline{V}_I$ and $\overline{V}_O$ are used as the *Value* operation in the attentive computing process to capture the *in-traversing* and *out-traversing* of each road's traffic situation caused by the trucks with task phase S^k, respectively. α^k denotes the underlying attentive relevance between the traffic situation and the task phase, which is formally estimated as follows:

$$\alpha_{j,i}^k = \frac{exp(\overline{Q}^k(h_{j,i}^X) \times (\overline{K}^k(e_{j,i}^{S^k}))^{\mathrm{T}})}{\sum_c^C exp(\overline{Q}^k(h_{j,i}^X) \times (\overline{K}^k(e_{j,i}^{S^c}))^{\mathrm{T}})} \tag{7}$$

Through the attentive operations, we can identify the traffic situation on roads which are more related with that of other roads, and learn the spatial dependency of traffic situation diffusion among roads.

The traffic situation diffusion among roads are correlated with their previous observations. Encoding traffic situation representation z^X with its previous observations can enhance its representation of task phase and reduce the impact of abnormal data. To accomplish this, we construct a self-attentive network to enhance the traffic situation representations z^X from historical temporal dimension. The network is built upon the scaled dot-product multi-head attention architecture, as shown in the Eq. 8 and 9:

$$\beta_{i,i'}^{S^k,(w)} = \sigma(T_i)(\overline{Q}^w(z_{j,i}^X \| e_i^t) \cdot (\overline{K}^w(z_{j,i'}^X \| e_{i'}^t))^{\mathrm{T}}) \tag{8}$$

$$\tilde{z}_{j,i}^X = \|_{w=1}^W \{ \sum_{i' \in P} \beta_{i,i'}^{S^k,(w)} \cdot \overline{V}^w(z_{j,i'}^X) \} \tag{9}$$

where $\sigma(T_i)$ is a normalized function by considering a set of timestamps T_i before timestamp i and $i' \in T_i$. The transforms $\overline{Q}^w$, $\overline{K}^w$, and $\overline{V}^w$ represent the *Query*, *Key*, and *Value* operations in the wth head attention, respectively, producing $l = L/W$ dimensional outputs. In the generating process of attention score $\beta_{i,i'}^{S^k,(w)} \in [0, 1]$, e_i^t is the temporal representation of each timestamp i [2].

After generating the more comprehensive traffic situation representation of each road with the weighted *in-traversing* and *out-traversing* graphs, the paper utilizes a *GRU* model for sequence prediction to model the relationship between the historical traffic situation representations of each road and forecast its future traffic situation representation. Finally, a fully connected layer is utilized to decode the learned representations into predicted traffic situations for the roads.

4.3 Iteration Based Multi-step Traffic Situation Prediction

To avoid errors accumulating in the multi-timestamp prediction process of traffic situation representations due to the lack of task phase transition prediction in the future, we use an iterative prediction process with a fixed-size historical time window P that covers the historical timestamps.

This process is illustrated in Fig. 5. The green part represents the historical timestamps, while the yellow and gray parts represent the current predictive timestamp and unpredictable timestamp, respectively. At each iteration, the predicted traffic situation representations are used as inputs to the next iteration, along with the traffic situation representations of the last $P - 1$ timestamps in the time window. After the traffic situation prediction in one timestamp is finished, the historical time window moves forward with one timestamp to cover the current predicted timestamp.

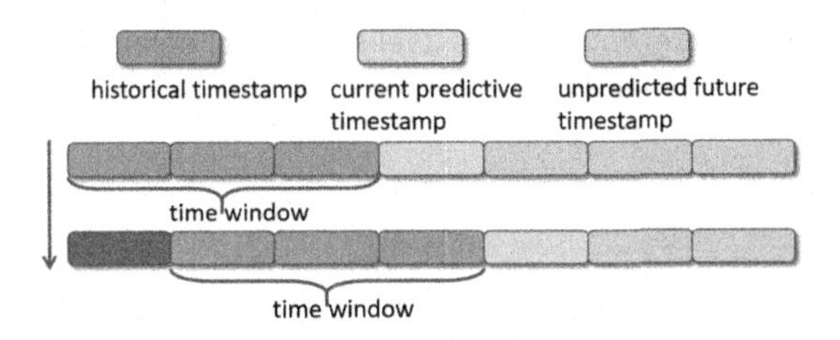

Fig. 5. The iterative predictive process

Before advancing the time window, the task phases in the current predicted timestamp need to be predicted. As the diffusion of trucks with each task phase has been captured from the task phase-based diffusion graphs, the spatial traversing correlations-based task phase representations can be accomplished by updating the task phase-based representation of each road using the graph, as illustrated in Fig. 4(c) and Eq. 10:

$$z_{j,i}^{S^k} = \sum_{j'}^{V_i^k(j)} u_{j \leftarrow j',i}^k \times \overline{V}(\tilde{e}_{j',i-1}^{S^k}) - \sum_{j'}^{V_o^k(j)} u_{j \rightarrow j',i}^k \times \overline{V}(\tilde{e}_{j',i+1}^{S^k}) \tag{10}$$

Using the enhanced task phase representations generated above, we construct a task phase sequence prediction model based on self-attention and *GRU*. This model is designed to forecast the future task phase representations of each road, similar to the process used to predict future traffic situation representations. Subsequently, a fully connected layer is utilized to decode these representations into the task phases that are contained in each road.

4.4 *DCVW* Based Multi-loss Module

There are two types of prediction missions in the MGCP model, including traffic situation prediction of roads and prediction of task phases contained in roads. For the traffic situation prediction loss and the task phase prediction loss, we use the mean squared error (MSE) to compute the traffic situation prediction loss $L_{ts} = (J - \tilde{J})^2$, and the task phase prediction loss $L_{tp} = \sum_k^C (S^k - \tilde{S}^k)^2$. However, due to the varying gradient ranges of the different predictive tasks, the overall learning process may deviate towards the predictive task with a larger gradient range. To address this issue, the DCVW approach firstly scales the losses L with hyper-parameters λ_i, where i denotes the i-th prediction task in the model, such that $L_i = \lambda_i * L_i$.

The second step is to design reasonable weights to control the learning emphasis during training, based on the training progress [13]. Tasks with complicated and changeable losses may need more attention in the learning process, and should therefore be allocated larger weights. Conversely, tasks with stable gradients may reach the end of the learning process and should be allocated smaller weights. The fluctuation of gradients in the paper is evaluated by computing the variation between the mean and standard deviation of the loss, using $a_{i,t} = \frac{1}{b_t} \frac{\sigma_{i,t}}{\mu_{i,t}}$ and $b_t = \sum_i \frac{\sigma_{i,t}}{\mu_{i,t}}$, where $\sigma_{i,t}$ and $\mu_{i,t}$ represent the standard deviation and mean of the loss L_i at training step t, respectively. $a_{i,t}$ is the computed weight for the i-th training task at training step t.

5 Experiments

5.1 Dataset Description

We evaluate the performance of *MGCP* on a real world data set from a steel factory mill in Shandong province factory, which contains 4 months of trucks' trajectories and tasks data on the observed road network from January 1st, 2021 to April 31st. The trucks' trajectory dataset includes more than 30,000,000 trajectory points, and each trajectory point contains the attributes of timestamp, instant speed, angle, latitude as well as longitude. The angle refers to the gap between truck's moving direction and road segment's main direction. Task data contains nearby 200,000 task records about trucks' task phases in an entire working process. We use 70% of the dataset generated for training, 10% for validation, and 20% for testing.

5.2 Experimental Settings

Hyperparameters. We use historical traffic situation of roads and trucks' task phases during past P (P is set as 12) timestamps (about 1 h) to predict the future traffic congestion levels of the next Q (Q is set as 12) timestamps (about 1 h) in the experiment, and the timestamp is set as 5 min. We train our model using Adam optimizer with an initial learning rate of 0.001. Here the learning rate will be automatically decayed by 0.1 times every 15 training epochs.

The value of θ_j for each road j is determined by analyzing the impact of the number of moving trucks and the number of staying trucks on the average speed of moving trucks from the historical records, whose computing process is as follows:

$$\theta_j = \frac{\sum_i AST_{j,i}/AS_{j,i}}{\sum_i (AMT_{j,i}/AS_{j,i} + AST_{j,i}/AS_{j,i})} \tag{11}$$

Baselines. We compared our model with several baselines, including: (1) *LSTM* [14], which uses memory units for short-term traffic situation prediction; (2) *SSGRU* [15], a hybrid stacked GRU-based traffic volume prediction approach for road networks; (3) *STSGCN* [16], which captures localized spatial-temporal correlations through an elaborately designed modeling mechanism; (4) *STGCN* [17], which formulates the traffic prediction problem on graphs and is built with complete convolutional structures; (5) *ASTGCN* [3], which adopts two parts to capture dynamic spatial-temporal correlations and common standard convolutions separately for traffic prediction; (6) *TGCN* [9], which leverages graph convolutional networks to capture spatial dependence and gated recurrent units to capture temporal dependence when predicting traffic situation; (7) *GMAN* [2], which designs an encoder-decoder architecture using multiple spatial-temporal attention blocks to model the impact of spatial-temporal factors on traffic conditions.(8) *DuETA* [18], which designs an ETA model that considers traffic congestion propagation patterns among roads as the basic factor affecting travel time. In the paper, we only compared the traffic congestion mining module of *DuETA*, including the congestion-sensitive graph and the graph transformer to generate the representation of roads in each timestamp.

Table 1. The comparison of traffic congestion prediction results

Models	MAE	MSE	MAPE	Reference
LSTM	1.168	1.642	0.994	IET2017
SSGRU	0.311	0.341	0.219	CC2020
ASTGCN	0.259	0.269	0.199	AAAI2019
STGCN	0.289	0.253	0.238	arXiv2017
STSGCN	0.321	0.282	0.229	AAAI2020
TGCN	0.301	0.334	0.211	ITS2019
GMAN	0.264	0.255	0.198	AAAI2020
DuETA	0.259	0.250	**0.193**	CIKM2022
MGCP	**0.258**	**0.224**	0.194	-

5.3 Comparison with Baseline Methods

Table 1 presents the results of our traffic congestion prediction model, *MGCP*, and the other baseline methods on the *RIZHAO* dataset. As shown in the table, *MGCP* outperforms all the other methods that do not take into account trucks' task phases. In particular, *LSTM* and *SSGRU* perform poorly, likely due to error accumulation in predicting traffic congestion across multiple timestamps without considering the transition of trucks' phases. Moreover, we observe that *STGCN* and *STSGCN* have an increment due the application of *GCN*, but still have a subpar performance as they do not fully explore for the influence of trucks' task phases on the road network feature. On the other hand, *DuETA*, *ASTGCN* and *GMAN* perform relatively better, thanks to the incorporation of Attention mechanism for capturing important related information in traffic condition prediction, thereby mitigating the problem of error propagation. Moreover, *DuETA* models the traffic congestion propagation among different order neighbors, which captures the propagation features caused by the demands of different task phases in some degree.

Table 2. Performance comparison of three traffic situation

Models	*AS*		*AMT*		*AST*		Reference
	MAE	MSE	MAE	MSE	MAE	MSE	
LSTM	2.932	77.165	0.384	0.530	0.184	0.131	IET2017
SSGRU	3.080	74.884	0.353	0.505	0.172	0.133	CC2020
ASTGCN	3.554	66.550	0.371	0.497	0.196	0.130	AAAI2019
STSGCN	**2.787**	98.096	0.445	0.548	0.266	0.209	AAAI2020
STGCN	3.698	79.038	**0.338**	0.506	0.191	0.150	arXiv2017
TGCN	3.535	66.249	0.349	0.461	0.171	0.120	ITS2019
GMAN	3.411	65.959	0.358	0.466	0.172	0.121	AAAI2020
DuETA	3.351	65.876	0.355	0.463	0.174	0.119	CIKM2022
MGCP	3.035	**64.526**	0.341	**0.447**	**0.171**	**0.109**	–

Table 2 presents the performance comparison of our *MGCP* method and other baseline approaches on three different traffic situations. Our method outperforms all other baseline methods in terms of *MSE*. However, some baseline methods have lower *MAE* than our method. It is worth noting that *MSE* is more sensitive to outliers such as congestion data, which is the main focus of our study. On the other hand, *MAE* reflects the overall volatility of the model's performance. Therefore, we can conclude that our method has a better ability to handle congestion data, but may exhibit slightly more instability than some of the baseline methods.

5.4 Predicted Results Analysis on Different Timestamps

In the paper, we focus on the traffic congestion prediction results of 12 timestamps (one hour) in the future, thus it is necessary to analyze the robustness of models on the 12 timestamps. We choose the seven baselines with the best performance in the above comparisons, and the compared results on *MSE* are shown in the Fig. 6. In this paper, we focus on predicting traffic congestion 12 timestamps (one hour) into the future, so it's important to assess the robustness of models for this time horizon. We selected the seven baselines with the best performance from the previous comparisons and compared their performance on the *MSE* metric, as shown in Fig. 6.

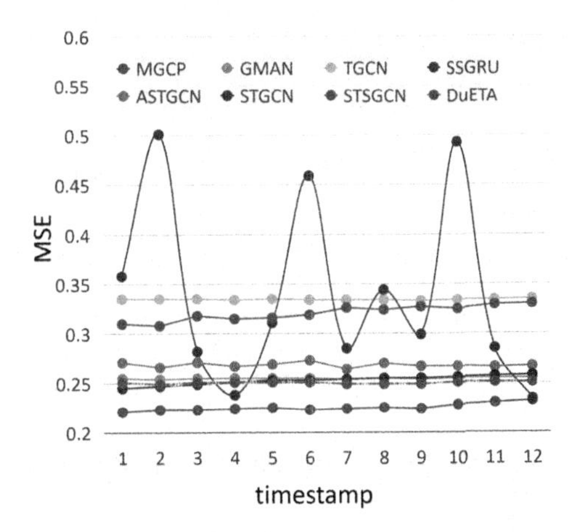

Fig. 6. predicted results of traffic congestion analysis on different timestamps

The results indicate that our proposed method, *MGCP*, outperforms the other baseline methods in all 12 timestamps, demonstrating its robustness in long-term traffic congestion prediction around the factory. Furthermore, the *MGCP* model performs best in the first predicted timestamp, but its performance declines as the timestamps progress. This is due to the fact that the task phase of trucks is known in the first timestamp prediction, whereas the proportion of task phase predicted by *MGCP* increases as the timestamps progress, and the biased prediction results of task phases have a certain impact on the performance of traffic congestion prediction.

Table 3. Ablation of *MGCP* on several important modules

Ablation of module	MAE	MSE	MAPE
$g_o^{Checked}$	0.274	0.248	0.199
g_o^{Called}	0.314	0.306	0.295
$g_o^{Queuing}$	0.266	0.258	0.213
$g_o^{Leaving}$	0.261	0.235	0.206
$TPTP_o$	0.287	0.269	0.203
IP_o	0.296	0.243	0.271
$DCVW_o$	0.273	0.267	0.219
MGCP	**0.258**	**0.224**	**0.194**

5.5 Ablation of *MGCP* on Several Important Modules

To verify the effectiveness of each module proposed in the paper, a series of ablation experiments are designed, and the evaluation results are shown in the Table 3.

The variants of *MGCP* are introduced as follows:

- $g_o^{Checked}$: The variant removes task phase *Checked* from the multi-view task phase diffusion graph
- g_o^{Called}: The variant removes task phase *Called* from the multi-view task phase diffusion graph
- $g_o^{Queuing}$: The variant removes task phase *Queuing* from the multi-view task phase diffusion graph
- $g_o^{Leaving}$: The variant removes task phase *Leaving* from the multi-view task phase diffusion graph
- $TPTP_o$: The variant replaces the transition probability matrix of task phase with a fully-connected layer
- IP_o: The variant replaces the iterative predictive process with a parallel predictive process
- $DCVW_o$: The variant replaces the *DCVW* module with a simple summation based loss

The experimental results in Table 3 demonstrate that *MGCP* outperforms all the ablation variants. Among the first four results, the performance of g_o^{Called} shows a more significant decrement due to the regularity of the *Called* trucks that move towards their target warehouses through relatively fixed roads. Conversely, it is not easy to capture the moving behaviors of the *Leaving* trucks since they have high movability and their target directions are not stable. Therefore, $g_o^{Leaving}$ shows a less significant decrement. The $TPTP$ module is a key component in $MGCP$, and the experimental results of $TPTP_o$ confirm its usefulness in mining the transition of task phases. The importance of task phases in the overall prediction process is demonstrated by the IP_o variant, which proves that a parallel predictive process may result in a lack of exact task phase information in the future. Finally, the results of $DCVW_o$ highlight the significance of suitable design for the multi-predictive mission loss function, with $DCVW$ showing a great improvement in the prediction results.

6 Conclusion

We proposed a prediction method incorporated with the trucks' task phase for forecasting traffic congestion among the roads around factory, called *MGCP*. It consists of a task phase embedding and updating module, *Markov process* based transition prediction module, multi-view traffic situation prediction module, and multi-predictive mission loss function module. Experiments on real data sets show that *MGCP* outperforms state-of-the-art baselines in prediction accuracy. Further, ablation experiments also confirm the effectiveness of each component of *MGCP*. In the future, we will apply *MGCP* to improve the efficiency in more emerging fields such as delivering food and packages.

Acknowledgment. This research was supported by NSFC (Nos.62072180, U1911203 and U1811264).

References

1. Zhang, X., et al.: Traffic flow forecasting with spatial-temporal graph diffusion network. In: AAAI, vol. 35, pp. 15008–15015 (2021)
2. Zheng, C., Fan, X., Wang, C., Qi, J.: Gman: a graph multi-attention network for traffic prediction. In: AAAI, vol. 34, pp. 1234–1241 (2020)
3. Guo, S., Lin, Y., Feng, N., Song, C., Wan, H.: Attention based spatial-temporal graph convolutional networks for traffic flow forecasting. In: AAAI, vol. 33, pp. 922–929 (2019)
4. Yao, H., Tang, X., Wei, H., Zheng, G., Li, Z.: Revisiting spatial-temporal similarity: a deep learning framework for traffic prediction. In: AAAI, vol. 33, pp. 5668–5675 (2019)
5. Liang, Y., Ke, S., Zhang, J., Yi, X., Zheng, Yu.: Geoman: multi-level attention networks for geo-sensory time series prediction. In: IJCAI, vol. 2018, pp. 3428–3434 (2018)
6. Estrach, J.B., Zaremba, W., Szlam, A., LeCun, Y.: Spectral networks and locally connected networks on graphs. In: ICLR, vol. 2014 (2014)
7. Dai, R., Xu, S., Gu,Q., Ji, C., Liu, K.: Hybrid spatio-temporal graph convolutional network: improving traffic prediction with navigation data. In: SIGKDD, pp. 3074–3082 (2020)

8. Guo, S., Lin, Y., Wan, H., Li, X., Cong, G.: Learning dynamics and heterogeneity of spatial-temporal graph data for traffic forecasting. IEEE Trans. Knowl. Data Eng. **34**(11), 5415–5428 (2021)
9. Zhao, L., et al.: T-GCN: a temporal graph convolutional network for traffic prediction. IEEE Trans. Intell. Transp. Syst. **21**(9), 3848–3858 (2019)
10. Zhu, J., et al.: KST-GCN: a knowledge-driven spatial-temporal graph convolutional network for traffic forecasting. IEEE Trans. Intell. Transp. Syst. **23**(9), 15055–15065 (2022)
11. Ni, Q., Zhang, M.: STGMN: a gated multi-graph convolutional network framework for traffic flow prediction. Appl. Intell. **52**(13), 15026–15039 (2022)
12. Li, F., Yan, H., Jin, G., Liu, Y., Li, Y., Jin, D.: Automated spatio-temporal synchronous modeling with multiple graphs for traffic prediction. In: CIKM, pp. 1084–1093 (2022)
13. Groenendijk, R., Karaoglu, S., Gevers, T., Mensink, T.: Multi-loss weighting with coefficient of variations. In: WACV, pp. 1469–1478 (2021)
14. Zhao, Z., Chen, W., Wu, X., Chen, P.C.Y., Liu, J.: LSTM network: a deep learning approach for short-term traffic forecast. IET Intell. Transp. Syst. **11**(2), 68–75 (2017)
15. Sun, P., Boukerche, A., Tao, Y.: SSGRU: a novel hybrid stacked GRU-based traffic volume prediction approach in a road network. Comput. Commun. **160**, 502–511 (2020)
16. Song, C., Lin, Y., Guo, S., Wan, H.: Spatial-temporal synchronous graph convolutional networks: a new framework for spatial-temporal network data forecasting. In: AAAI, vol. 34, pp. 914–921 (2020)
17. Yu, B., Yin, H., Zhu, Z.: Spatio-temporal graph convolutional networks: a deep learning framework for traffic forecasting. In: IJCAI, pp. 3634–3640 (2018)
18. Huang, J., et al.: Dueta: traffic congestion propagation pattern modeling via efficient graph learning for eta prediction at baidu maps. In: CIKM, pp. 3172–3181 (2022)

M2MTR: Reposition Idle Taxis in the Many-to-Many Manner with Multi-agent Reinforcement Learning

Hao Yu[1,2], Xi Guo[1,2](✉), Jie Chen[5], and Xiao Luo[3,4]

[1] School of Computer and Communication Engineering, University of Science and Technology Beijing, No. 30 Xueyuan Road, Beijing 100083, China
[2] Beijing Key Laboratory of Knowledge Engineering for Materials Science, No. 30 Xueyuan Road, Beijing 100083, China
xiguo@ustb.edu.cn
[3] College of Transportation Engineering, Tongji University, No.4800 Caoan Road, Shanghai 201804, Shanghai, China
luo.xiao@tongji.edu.cn
[4] Urban Mobility Institute, Tongji University, No.1239 Siping Road, 200092 Shanghai, China
[5] Network Product Line, New H3C Technologies Co., Ltd., No.466 Changhe Road, 310052 Hangzhou, China
13452131581@sina.cn

Abstract. Ride-hailing apps, such as Didi and Uber, allow people to easily request a ride by inputting their desired origin and destination locations. Due to transportation system complexity and vast city areas, uneven distribution of vehicle supply versus rider demand frequently occurs. This can lead to overcrowded areas with insufficient taxis or sparsely populated zones with abundant empty taxis. To balance supply and demand, many studies have proposed taxi-repositioning methods. However, recent studies limit to repositioning taxis in the one-to-one manner. In this paper, we propose the M2MTR method that can reposition idle taxis in the many-to-many manner. We define the reposition task as a partially observable Markov decision process and define the optimization objectives. To find good reposition strategies, we propose the M2MTR method that is a variation of a multi-agent cooperative A2C method. To make models converge quickly, we design the rewards delicately. To update the policy networks efficiently, we design a local reward combiner. We build an environment simulator to train and evaluate M2MTR. Extensive experiments on real datasets show that M2MTR outperforms other three baseline algorithms. The reposition strategies obtained from M2MTR can make supply and demand more balance, can increase the response rates, and can reduce response time of taxis.

Keywords: Taxi repositioning · Multi-agent reinforcement learning · A2C model · Partially observable Markov decision process · Balance supply and demand

X. Yang et al. (Eds.): ADMA 2023, LNAI 14180, pp. 569–583, 2023.
https://doi.org/10.1007/978-3-031-46677-9_39

1 Introduction

Currently, ride-hailing platforms are very popular, such as Didi, Uber, etc. As long as the passenger sends the starting and ending locations of his (or her) trip to the platform, the platform will automatically match an appropriate taxi to pick up the passenger. The platform monitors the locations of taxis in real time and answers the ride-hailing requests as fast as possible. Since the urban transportation system is complicated and the geographical area of a city is quite large, the imbalance between supply and demand is a common problem that platforms often encounter. There are many idle vehicles in some areas, but there are few demands for taxis. Or there are few idle vehicles in some areas, but there are many demands for taxis. In the oversupplied area, the drivers have to compete for several ride-hailing orders, while the passengers have to wait for a long time to get on a taxi in the area where demand is exceeding supply. For example, in Fig. 1, the region Q_1 has many idle taxis but no passengers. But the regions Q_2, Q_3, and Q_4 have some passengers but no taxis. The supply and demand is imbalance. To attain balance, we should reposition the taxis in Q_1 in an effective way by considering the different amounts of demands in Q_2, Q_3, and Q_4.

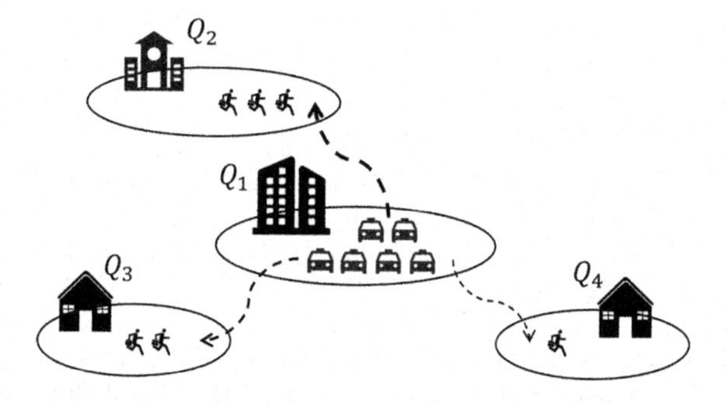

Fig. 1. The problem of repositioning idle taxis.

To balance supply and demand, many studies ([1–3], etc.) consider the taxi-repositioning problem as a network flow problem and solve it by using operations research algorithms. Since such algorithms are time-consuming, some recent studies ([4,5], etc.) consider the reposition task as a Markov decision process and use multi-agent reinforcement learning methods to train policy models. Assuming the entire city is divided by a grid of hexagonal cells, the policy models trained can guide the idle taxis from one cell to another cell. However, in multi-agent reinforcement learning, it is difficult to make a model converge. So the recent studies have to reposition taxis in the one-to-one manner, i.e., all of the idle taxis staying at the current cell or flocking into another cell. Specifically, HTRS [4] guides taxis to flock into a neighbor cell and META [5] guides taxis to flock into a nearby cell that may not be a neighbor cell. To make the idle taxis move independently rather than in groups, we propose the M2MTR method, which can reposition the taxis in the many-to-many manner.

In our taxi-repositioning problem, the agents are the idle taxis. Each agent has seven actions including staying at the current cell and moving to the six neighbor cells. The environment is the taxi transportation system of the entire city. Our goal is to balance the supply and demand in the environment. We propose M2MTR to solve the problem, and M2MTR is an variation of a multi-agent cooperative A2C method [6]. It consists of an actor part and a critic part. The actor part has k policy networks and each cell corresponds to a policy network. A taxi decides its action using the policy network with respect to (w.r.t.) the cell where it is. The critic part has a value network and a target network. The two networks incorporate each other in order to evaluate the state of the environment after agents executing actions. To train M2MTR, we also design and implement a simulator of the environment. We conduct experiments by using a real dataset that is collected from taxis in Shanghai, 2015. The experimental results show that M2MTR outperforms three baseline methods. This paper has the following contributions.

- We define the taxi-repositioning task as a partially observable Markov decision process where the global rewards and local rewards are designed delicately in order to make models converge quickly.
- We propose the M2MTR method, which can reposition the taxis in a many-to-many manner rather than in an one-to-one manner like recent studies. In the training procedure, we propose a local reward combiner in order to update the policy networks efficiently.

The rest of the paper is organized as follows. We introduce our problem settings in Sect. 2. We present the M2MTR method in Sect. 3. Specially, we introduce the training details in Sect. 3.1 and introduce the local reward combiner in Sect. 3.2. We report the experimental settings and results in Sect. 4. Specially, we introduce the environment simulator in Sect. 4.1. We give a literature review in Sect. 5 and conclude the paper in Sect. 6.

2 Problem Statement

To balance the supply and demand of taxis for different regions in a large city, the taxi dispatching system should have an effective strategy to reposition idle taxis. It is challenging to design the strategy because demand and supply are changing over time. So in this paper we use a multi-agent reinforcement learning method to form effective strategies incrementally. We define the repositioning problem as a multi-agent version of the partially observable Markov decision process (POMDP) [7]. The environment is the geographical area of a city, which is divided by a hexagonal grid. $\mathcal{C} = \{c^1, c^2, \ldots, c^k\}$ indicates the set of cells in the grid and c^j indicates a single cell.

Agents. Each taxi is an agent. $\mathcal{G} = \{g^1, g^2, \ldots, g^m\}$ indicates the set of agents and g^i indicates a single taxi (or agent).

Actions. At a certain time t, a taxi falls (g^i) into a cell (c^j). The taxi g^i can execute an action ($a_t^i \in \mathcal{A}^j$). The action set $\mathcal{A}^j$ contains seven different actions, i.e., $\{0,1,2,3,4,5,6\}$. In $\mathcal{A}^j$, 0 denotes the taxi stays where it is (i.e., c^j), and 1 (or 2, 3, 4, 5, 6) denotes the taxi moves to a neighbor cell c^{j_1} (or c^{j_2}, c^{j_3}, c^{j_4}, c^{j_5}, c^{j_6}). We use $\{c^{j_1}, c^{j_2}, \ldots, c^{j_6}\}$ to indicate the six neighbor cells of c^j. In our problem settings, the taxis falling into the same cell (c^j) share the same action set ($\mathcal{A}^j$). There are k action sets in total and the global action space is a combination of all actions sets, i.e., $\mathcal{A} = \mathcal{A}^1 \times \mathcal{A}^2 \times \ldots \times \mathcal{A}^k$. At a certain time t, the global action is $\boldsymbol{a}_t = \left[a_t^1, a_t^2, \ldots, a_t^m\right]$ that is an m-dimensional vector.

States. In the grid, to describe the state of an individual cell (c^j), we use a three-dimensional vector $\boldsymbol{s}^j = \left[s_{dem}^j, s_{sup}^j, s_{ser}^i\right]$, where s_{dem}^j denotes the number of demands in c^j, s_{sup}^j denotes the number of supplies (i.e., idle taxis) in c^j, and s_{ser}^i denote the number of taxis in-service in c^j. At a certain time t, the **global state** $\boldsymbol{s}$ consists of all the states of cells, i.e.,

$$\boldsymbol{s}_t = \boldsymbol{s}_t^1 \circ \boldsymbol{s}_t^2 \circ \cdots \circ \boldsymbol{s}_t^k, \tag{1}$$

where $\circ$ means concatenating vectors. Thus, considering the actions, the state-transition function is

$$p(\boldsymbol{s}_{t+1}|\boldsymbol{s}_t; \boldsymbol{a}_t) = \mathbb{P}\left[S_{t+1} = \boldsymbol{s}_{t+1}|S_t = \boldsymbol{s}_t, A_t = \boldsymbol{a}_t\right]. \tag{2}$$

In our problem settings, a taxi g^i in cell c^j does not have a global vision. It only has a **local observation** $\boldsymbol{o}^j$ that consists of the states of c^j and the states of c^j's six neighbors, i.e.,

$$\boldsymbol{o}_t^j = \boldsymbol{s}_t^j \circ \boldsymbol{s}_t^{j_1} \circ \boldsymbol{s}_t^{j_2} \circ \cdots \boldsymbol{s}_t^{j_6}. \tag{3}$$

Since the taxis in the same cell c^j share the same partial observations, without ambiguity, we use $\boldsymbol{o}_t^j$ to denote the partial observations of all the taxis in c^j. In the next section, the global state $\boldsymbol{s}$ will be used to train the critic part of the multi-agent reinforcement learning model, and the local observation $\boldsymbol{o}^i$ will be used to train the policy network w.r.t. the cell c^j in the actor part.

Policy Functions. We aim to design effective repositioning policies for idle taxis (i.e., agents). At a certain time t, a taxi g^i in cell c^j executes an action a_t^i by observing $\boldsymbol{o}_t^j$ according to its policy function π^j, i.e.,

$$a_t^i \sim \pi(\cdot \mid \boldsymbol{o}_t^j; \boldsymbol{\theta}^j). \tag{4}$$

In this equation we use π instead of π^j for simple. In our problem settings, we consider the taxis in the same cell c^j use the same policy function π^j. So there are k policy functions in total. We use a neural network with parameters $\boldsymbol{\theta}^j$ to approximate the complicated π^j. Without ambiguity, we use $\boldsymbol{\theta}^j$ to denote the policy network for simple.

Rewards. After executing an action, the taxi (g^i) obtains rewards from the environment. Each cell c^j can provide two types of rewards, namely, $\dot{r}^j$ and $\ddot{r}^j$. The $\dot{r}^j$ is

$$\dot{r}^j = 1 - \frac{|s^j_{sup} - s^j_{dem}|}{\max(s^i_{sup}, s^i_{dem})}, \quad (5)$$

and it is a decimal between 0 to 1. If $\dot{r}^j = 1$, the supply and demand of c^j attains balance. If $\dot{r}^j$ is close to 0, the supply and demand is imbalance. The $\ddot{r}^j$ is defined by Eq. 6, i.e.,

$$\ddot{r}^j = 1 - \frac{s^j_{sup}}{s^j_{dem}}. \quad (6)$$

If $\ddot{r}^j = 0$, the supply and demand attains balance. If $\ddot{r}^j$ is negative, there are more supplies. If $\ddot{r}^j$ is positive, there are more demands. Using the first type of rewards ($\dot{r}^j$'s), we define the **global reward** that is obtained after all taxis executing their actions at a certain time t. The global reward $r^\star_t$ is

$$r^\star_t = \dot{r}^1_t + \dot{r}^2_t + \cdots + \dot{r}^k_t, \quad (7)$$

where $\dot{r}^j_t$ $j \in [1..k]$ is the first type of reward w.r.t. cell c^j at time t. The global reward is used to train the critic model. Using the second type of reward ($\ddot{r}^j$'s), we define the **local reward** that is obtained after a taxi executing its action a^i_t at a certain time t, i.e.,

$$r^i_t = \begin{cases} \omega^j \cdot \ddot{r}^j_t & a^i_t = 0 \\ \omega^j \cdot \ddot{r}^j_t - \omega^{j_{nbr}} \cdot \ddot{r}^{j_{nbr}}_t & a^i_t = nbr \; nbr \in [1..6] \end{cases} \quad (8)$$

The reward $\ddot{r}^j_t$ is provided by the cell c^j where the taxi g^i falls into. The reward $\ddot{r}^{j_{nbr}}_t$ is provided by the cell $c^{j_{nbr}}$ which is neighboring to c^j. In Eq. 8, we assign the reward to a taxi in two different ways. If the action of the taxi is staying where it is ($a^i_t = 0$), we use the first formula to calculate the reward. If the action of the taxi is moving to a neighbor cell ($a^i_t = nbr$), we use the second formula to calculate the reward. The taxi executing a moving action should collect the reward $\ddot{r}^j_t$ as well as the reward $\ddot{r}^{j_{nbr}}_t$, because the taxi can influence the supply and demand situations of the starting cell (c^j) and the ending cell ($c^{j_{nbr}}$). In Eq. 8, we use ω^j and $\omega^{j_{nbr}}$ to weighing the rewards from cell c^j and cell $c^{j_{nbr}}$.

Optimization Problem. To evaluate the long-term reward in the future, we use γ as the discount factor. The global discounted return is

$$U_t = R_t + \gamma \cdot R_{t+1} + \gamma^2 \cdot R_{t+2} + \gamma^3 \cdot R_{t+3} + \cdots, \quad (9)$$

where R_t is a random variable indicating the global reward at t. The global action-value function is

$$Q(\boldsymbol{s}_t, \boldsymbol{a}_t) = \mathbb{E}\left[U_t | S_t = \boldsymbol{s}_t, A_t = \boldsymbol{a}_t\right], \quad (10)$$

and the global state-value function is

$$V(\boldsymbol{s}_t) = \mathbb{E}_A \left[Q(\boldsymbol{s}_t, A) \right], \tag{11}$$

where A is a random variable indicating the global action at t. The optimization problem is

$$\max_{\boldsymbol{\theta}^1, \boldsymbol{\theta}^2, \ldots, \boldsymbol{\theta}^k} \mathbb{E}_S \left[V(S) \right], \tag{12}$$

where S is a random variable indicating the global state at t, and $\boldsymbol{\theta}^1, \boldsymbol{\theta}^2, \ldots, \boldsymbol{\theta}^k$ indicate the parameters of k policy networks. In other words, the problem is to minimize the expectation in Eq. 12 by adjusting the parameters of k policy networks.

Table 1. Mathematical Symbols

Mathematical Symbols	Explanation
g^i $(i \in \{1, 2, \ldots, m\})$	an agent (i.e., an idle taxi)
c^j $(j \in \{1, 2, \ldots, k\})$	a cell in the city grid
a_t^i	the action of agent g^i
$\boldsymbol{s}_t^\star$	the global state from the environment
$\boldsymbol{o}_t^j$	the local observation of the agents in cell c^j
$r_t^\star$	the global reward received from the entire environment
r_t^i	the local reward received from the cells influenced by g^i
$\boldsymbol{\theta}^j$	the parameters of the policy network w.r.t. c^j
$\boldsymbol{\omega}$	the parameters of the value network
$\boldsymbol{\omega}^-$	the parameters of the target network

The important mathematical symbols and their meanings are summarized in Table 1. The last two symbols $\boldsymbol{\omega}$ and $\boldsymbol{\omega}^-$ will be introduced in Sect. 3.1.

3 Method

We propose M2MTR model to solve the optimization problem. The M2MTR is a multi-agent cooperative A2C method [6] and is consistent with "the centralized training with decentralized execution" framework. As Fig. 2 shows, the M2MTR model consists of the critic part and the actor part. The critic part has a value network and a target network. The two networks cooperate with each other in order to predict the value of the global state at a certain time as accurate as possible. The actor part has k policy networks. Each cell maintains a policy network. A taxi makes decisions by using the policy network w.r.t. the cell where it is. When training the critic part, we use the global rewards and global states from the environment. When training the actor part, we use the predicted state values from the critic part, the local rewards and the local states from the environment. We will show the details of training and decision-making in Sect. 3.1. To update the policy networks faster, we design a local reward combiner to aggregate the local rewards (Sect. 3.2). We also design and implement an environment simulator and we will introduce it in Sect. 4.1.

3.1 Training and Decision Making

We train the M2MTR model according to the following steps. Figure 3 shows the information flow in the M2MTR model while training.

(1) **Execute actions.** Each idle taxi g^i in cell c^j executes the action a_t^j which is sampled randomly according to the probability distribution given the partial observation o_t^j, i.e.,

$$a_t^i \sim \pi(\cdot \mid o_t^j; \boldsymbol{\theta}_{now}^j), \tag{13}$$

where θ_{now}^j is the parameter of the policy network sharing by the taxis in the cell c^j. The green arrows in Fig. 3 illustrate the information flow in this step.

(2) **Obtain rewards.** After all the idle taxis executing actions, we obtain the global reward $r_t^\star$ (Eq. 7), the local rewards $\{r_t^1, r_t^2, \ldots, r_t^k\}$ (Eq. 8), and the new global state $\boldsymbol{s}_{t+1}$ (Eq. 1) from the environment.

(3) **Update the critic part.** We predict the state value $(\hat{v}_t)$ at time t using the value network, i.e., $\hat{v}_t = v(\boldsymbol{s}_t; \boldsymbol{w}_{now})$, and predict the state value $(\hat{v}_{t+1}^-)$ at time $t+1$ using the target network, i.e., $\hat{v}_{t+1}^- = v(\boldsymbol{s}_{t+1}; \boldsymbol{w}_{now}^-)$. The value network is updated as follows.

$$\boldsymbol{w}_{new} \leftarrow \boldsymbol{w}_{now} - \alpha \cdot \delta_t^\star \cdot \nabla_{\boldsymbol{w}} v(\boldsymbol{s}_t; \boldsymbol{w}_{now}), \tag{14}$$

where α is the learning rate, $\nabla_{\boldsymbol{w}}$ denotes the gradients of v, and $\delta_t^\star$ is the global TD-error. The $\delta_t^\star$ is calculated as follows.

$$\delta_t^\star = \hat{v}_t - (r_t^\star + \gamma \cdot \hat{v}_{t+1}^-), \tag{15}$$

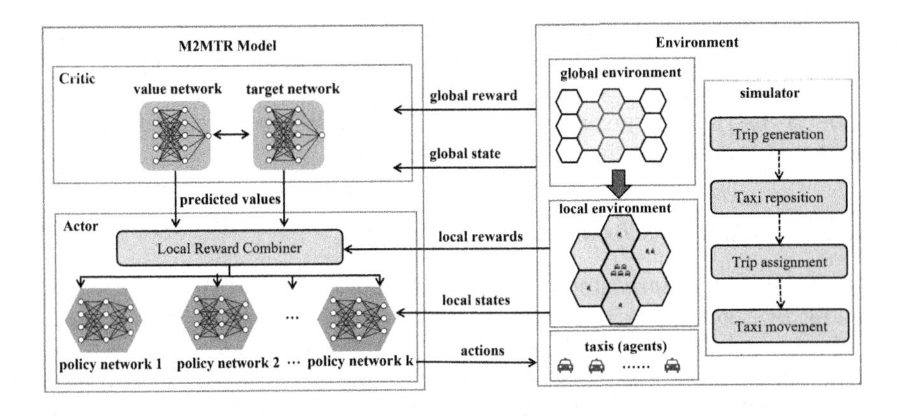

Fig. 2. M2MTR model and the environment

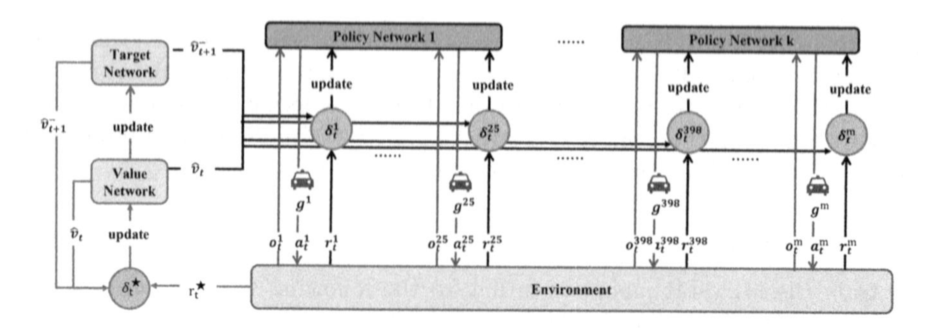

Fig. 3. Information flow when training M2MTR

where $r_t^\star$ is calculated by Eq. 7. The target network is updated by using the new parameters in the critic network, i.e.,

$$\boldsymbol{w}_{new}^{-} \leftarrow \tau \cdot \boldsymbol{w}_{new} + (1 - \tau) \cdot \boldsymbol{w}_{now}^{-}, \tag{16}$$

where τ is a weight. The orange arrows in Fig. 3 illustrate the information flow in this step.

(4) **Update the policy part.** We update the policy network π^j of each cell c^j by using the local TD-error δ_t^i, i.e.,

$$\boldsymbol{\theta}_{new}^{j} \leftarrow \boldsymbol{\theta}_{now}^{j} - \beta \cdot \delta_t^i \cdot \nabla_{\boldsymbol{\theta}^j} \ln \pi(a_t^i | \boldsymbol{o}_t^j; \boldsymbol{\theta}_{now}^j), \tag{17}$$

where $i \in \{1, \ldots, m\}$, $j \in \{1, \ldots, k\}$, and δ_t^i is calculated as follows.

$$\delta_t^i = \hat{v}_t - (r_t^i + \gamma \cdot \hat{v}_{t+1}^{-}). \tag{18}$$

Here r_t^i is calculated by Eq. 8. There are k policy networks $\{\boldsymbol{\theta}^1, \boldsymbol{\theta}^2, \ldots, \boldsymbol{\theta}^k\}$ in total. We update them according to Eq. 17, respectively. The blue arrows in Fig. 3 illustrate the information flow in this step.

After completing the training procedure, the agents (i.e., taxis) use the policy networks for decision making. At a decision making point t, assuming an idle taxi g^i is at cell c^j, it should execute the action a_t^i that is sampled according to the probability distribution given by $\pi(\cdot \mid o_t^j; \boldsymbol{\theta}^j)$. Here $\boldsymbol{\theta}^j$ is the learned parameters of the network and o_t^j is the partial observation.

3.2 Local Reward Combiner

In the training procedure, step (4) updates each policy network $(\boldsymbol{\theta}^j)$ m^j times, if there are m^j idle taxis at time t. To make this step more efficiently, we propose a local reward combiner which aims to update $\boldsymbol{\theta}^j$ one time. The local reward combiner divides m^j idle taxis into seven groups according to their difference actions, i.e., $\{\mathcal{G}^0, \mathcal{G}^1, \mathcal{G}^2, \ldots, \mathcal{G}^6\}$, where $\mathcal{G}^0$ denotes the taxis staying at c^j, and $\mathcal{G}^1$

to $\mathcal{G}^6$ denotes the taxis moving to the neighbor cells $\{c^{j_1}, c^{j_2}, \ldots, c^{j_6}\}$. According to Eq. 8 and Eq. 18, the taxis in the same group share the same local TD-error,

$$\delta_t^{j.\mathcal{G}^0}, \delta_t^{j.\mathcal{G}^1}, \ldots, \delta_t^{j.\mathcal{G}^6}, \tag{19}$$

where superscripts $j.\mathcal{G}^0$, $j.\mathcal{G}^1$, ..., $j.\mathcal{G}^6$ denote the seven taxi groups in c^j. Thus, we aggregate the TD-errors and update $\boldsymbol{\theta}^j$ as follows.

$$\begin{aligned}\boldsymbol{\theta}^j_{new} \leftarrow \boldsymbol{\theta}^j_{now} & -(m^j \cdot p_t^0) \cdot \beta \cdot \delta_t^{j.\mathcal{G}^0} \cdot \nabla_{\boldsymbol{\theta}^j} \ln \pi(0|\boldsymbol{o}_t^j; \boldsymbol{\theta}^j_{now}) \\ & -(m^j \cdot p_t^1) \cdot \beta \cdot \delta_t^{j.\mathcal{G}^1} \cdot \nabla_{\boldsymbol{\theta}^j} \ln \pi(1|\boldsymbol{o}_t^j; \boldsymbol{\theta}^j_{now}) \\ & - \cdots \\ & -(m^j \cdot p_t^6) \cdot \beta \cdot \delta_t^{j.\mathcal{G}^6} \cdot \nabla_{\boldsymbol{\theta}^j} \ln \pi(6|\boldsymbol{o}_t^j; \boldsymbol{\theta}^j_{now})\end{aligned} \tag{20}$$

where $p_t^0, p_t^1, \ldots, p_t^6$ are the probabilities of the seven actions predicted by $\boldsymbol{\theta}^j_{now}$, and $(m^j \cdot p_t^0)$, $(m^j \cdot p_t^1)$, ..., $(m^j \cdot p_t^6)$ are the numbers of taxis executing the seven different actions, respectively. In the training procedure, we use Eq. 20 to update the policy networks instead of using Eq. 17.

4 Experiments

4.1 Environment Simulator

Different from supervised learning methods, reinforcement learning methods interact with environments in the process of training networks. When using reinforcement learning methods to solve vehicle dispatching problems, a common way is to design and implement a simulator that can simulate the traffic environment [8]. In this paper, we design and implement an environment simulator that can simulate the four key procedures of urban transportation, namely, trip generation, taxi reposition, trip assignment, and taxi movement, as Fig. 2 shows. We employ the simulator both in training and in evaluating the M2MTR model.

In our simulator, we set a timestamp every 10 min. It means that the 24 h is divided into 144 intervals. At a timestamp t, the simulator conducts the four key procedures sequentially.

(1) **Trip generation.** New trips are generated chronologically according to the real dataset. We will introduce the real dataset in Sect. 4.2.
(2) **Taxi reposition.** Each idle taxi executes an action according to the decision-making result of the policy network. After repositioning, the idle taxis appear at new locations. Subsequently, the simulator assigns trips considering the new locations. In this step, the simulator also collects the number of supplies and the number of demands. The number of supplies of a cell c^j is the number of idle taxis falling into c^j at this time stamp. The number of demands of c^j is the number of trips that are still waiting for pick-ups at this time stamp.

(3) **Trip assignment.** When a trip request is issued, if there are taxis nearby, the simulator assigns one taxi to the trip, otherwise, the simulator makes the trip waiting for a taxi. In our simulator, if the distance between a trip and a taxi is less than 5 km, we consider the taxi can pick up the trip. In addition, the simulator decides a trip is expired, if the trip has been waiting for a long time (10 min).
(4) **Taxi movement.** When a trip is assigned to a taxi, the taxi should move the the starting location of the trip and drive to the destination.

4.2 Experimental Settings

We conduct experiments by using the taxi trajectory data collected in Shanghai, 2015. The data set for each day contains about 130 million trajectory points. The GPS collects the information of taxi every 5 s, which includes its position, speeds, status (operating, available, suspended), and so on. We extract trips from this raw trajectory data. We sort the trips chronologically in order to enable the simulator to use the real data.

We divide the geographical area of Shanghai with a hexagonal grid. The grid contains 41 hexagonal cells with 5 km width. The grid excludes the cells that have no trajectory data.

All the algorithms are implemented with Python 3.11.1, Pytorch 2.0.0, and trained with a single NVIDIA 3090 GPU. We implement the policy networks, the value network, and the target network by using the multi-layer perceptrons (MLPs).

We use three indices for evaluating reposition performances, i.e., .

- Reward is the global reward that is defined in Eq. 7. A larger Reward means the method can make the supply and demand more balance.
- TRR (%) is the response rate of trips in a whole day throughout the city, i.e.,

$$\mathsf{TRR} = \frac{|n_{trip} - n_{expired}|}{n_{trip}}, \tag{21}$$

  where n_{trip} is the total number of trips and $n_{expired}$ is the number of trips expired. A larger TRR means the method can make more trips picked up.
- AWT (seconds) is the average waiting time of a trip before it is picked up. A smaller AWT means the method can make passengers waiting for a shorter time range.

We compare the proposed M2MTR to three baseline methods according to the three indices, i.e.,

- Random. This method randomly repositions idle taxis to neighbor cells.
- Diffusion. Considering the supply and demand, this method diffuses idle taxis to the neighbor cells proportionally in order to attain balance [1]. We implement a simplified version of this method.
- Shared-A2C. This method trains a tradition A2C model and all agents use the same policy network. It is a simplified version of M2MTR. The main difference is M2MTR has k policy networks for agents in different cells.

4.3 Experimental Results

To evaluate the reposition performances of M2MTR, we compare M2MTR with the three baseline methods. Table 2 summarizes experimental values of the three indices, i.e., Reward, TRR, AWT. To evaluate the scalability of the methods, we conduct experiments under different fleet sizes, i.e., 5000 taxis, 6500 taxis, and 8000 taxis. Diffusion performs better than Random, because Diffusion encourages taxis to leave the cells where the density of taxis is high and in this way it can alleviate the imbalance more effectively. Shared-A2C and M2MTR are both reinforcement learning (RL) algorithms in A2C framework. Our M2MTR performs better than Shared-A2C because it has k policy networks that can make decisions more effectively by observing local environments. Generally speaking, the RL algorithms (Shared-A2C and M2MTR) performs better than the Non-RL algorithms (Random and Diffusion). The RL algorithms perform well when the number of taxis is sufficient.

Table 2. Comparing M2MTR with three baseline methods

Method	5000 Taxis			6500 Taxis			8000 Taxis		
	Reward	TRR	AWT	Reward	TRR	AWT	Reward	TRR	AWT
Random	4717.5	68.1%	268.6	5263.7	77.5%	201.2	5533.9	83.3%	159.5
Diffusion	5034.9	76.2%	215.7	5665.9	87.7%	126.6	5982.5	90.7%	102.7
Shared-A2C	5246.4	76.4%	214.3	6132.3	91.0%	96.9	6351.4	96.9%	37.7
M2MTR	**5374.1**	**79.1%**	**196.8**	**6605.8**	**93.3%**	**79.6**	**6810.9**	**98.6%**	**16.7**

To further evaluate the RL-algorithms, namely, M2MTR and Shared-A2C, we draw their training curves in Fig. 4. In the experiments, we train the models using a fleet of 8000 taxis. Figure 4 shows how Reward, TRR, and ATW change when the number of epochs increases. On the one hand, the experimental results tell us that the Reward curve is roughly consistent with the TRR curve, and it is also roughly consistent with the ATW curve if we flip it down. It means that the global reward (Eq. 7) we designed can meet the goals of taxi repositioning. On the other hand, comparing the curves of M2MTR and Shared-A2C, we can see that when the number of epochs increases, the Reward and the TRR obtained by M2MTR become larger than Shared-A2C, and the ATW obtained by M2MTR becomes smaller than Shared-A2C. It means that M2MTR performs better than Shared-A2C in the training procedures.

Figure 5 illustrates two taxi repositioning scenarios at 8:00 a.m. and at 23:30 p.m. in the same area. This area is divided by hexagon cells. We use different color depths to illustrate different demand amounts. A dark cell indicates there is a large demand in this cell. We also show the exact numbers (in black) of demands in some cells. The blue arrows illustrate the taxi repositioning strategies that is provided by our M2MTR model. The numbers in blue denote the numbers of idle taxis that should move to the neighbor cells. In the figure, there are many idle

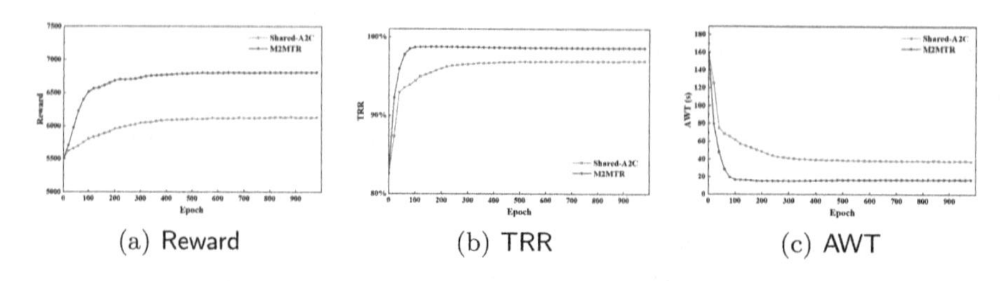

(a) Reward (b) TRR (c) AWT

Fig. 4. The training curves of M2MTR and Shared-A2C.

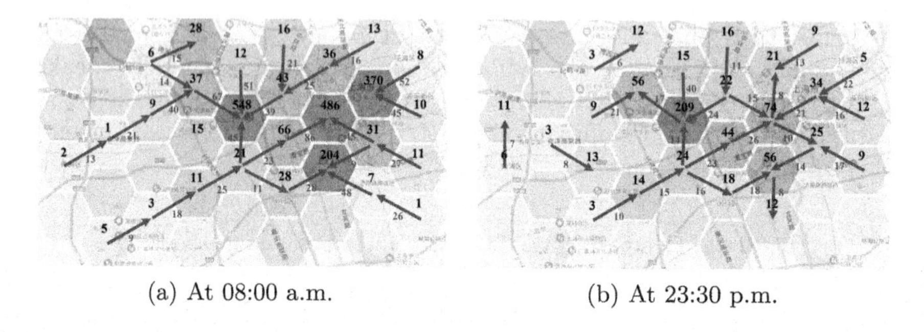

(a) At 08:00 a.m. (b) At 23:30 p.m.

Fig. 5. The scenario of taxi repositioning using M2MTR

taxis moving from low-demand cells to high-demand cells. To make the figure clearer, we only show a portion of the movements.

5 Related Work

The problem of repositioning taxis is an important issue in fleet management, and it has positive implications for balancing the supply and demand of taxis. A mainstream research direction is to relate the problem of repositioning taxis to a Markov decision process model and train the model using reinforcement learning methods. It has been proved that reinforcement learning methods based on Markov decision process models are more effective than traditional model-based methods [1–3,9]. Lin et al. proposed a contextual multi-agent reinforcement learning framework to balance supply and demand of in different regions [4]. There are two algorithms, contextual deep Q-learning and contextual multi-agent actor-critic. Jin et al. proposed a joint order dispatching and fleet management framework named CoRide, which models ride-hailing as a massively parallel ranking problem and investigates the joint decision task [10]. Li et al. proposed a mean field multi-agent reinforcement learning algorithm to simplify local interactions by taking an average action among neighborhoods [11]. Liu et al. proposed a context-aware taxi dispatching approach named COX, which integrates rich contexts into DRL modeling for more efficient taxi dispatching [12]. Liu et al. proposed a city-wide taxi repositioning framework named META, which builds two components to address the gap collaboratively. Each region is treated as an agent, and taxis within that region can make two different actions [5]. Xu et al. modeled order matching and vehicle repositioning as a uniform Markov decision process, and addressed the competition between the huge state space and the driving factors [13].

Another related research direction is multi-agent reinforcement learning [14], including homogeneous and heterogeneous multi-agent reinforcement learning. Homogeneous agents take the same action in the same state, while heterogeneous agents take different actions separately. Tan et al. compared independent Q-learning and cooperative Q-learning in different environments and found that cooperative Q-learning can improve the learning speed [15]. Tampuu et al. extended independent Q-learning to DRL and let agents cooperate and compete with each other through rewards [16]. Lowe et al. proposed a general algorithm named MADDPG for learning continuous action deterministic policies and open-sourced their simulation environment, which includes various scenarios such as cooperation, competition and communication [17]. Foerster et al. proposed a multi-agent actor-critic algorithm called COMA, which uses a counterfactual baseline to marginalize the actions of individual agents while keeping the actions of other agents fixed [18]. Nguyen et al. studied large-scale multi-agent planning for fleet management and explicitly modeled the expected counts of agents [19,20].

6 Conclusions

In this paper, we propose the many-to-many taxi repositioning method called M2MTR, to balance supply and demand. The M2MTR is a variation of the multi-agent cooperative A2C model. To train the model, we design and implement an environment simulator and design a local reward combiner. We conduct experiments to evaluate M2MTR by using real-world datasets. The experimental results show M2MTR can repositioning taxis efficiently. In the future, we will study how to recommend good reposition routes for taxi drivers.

Acknowledgment. This work was partially supported by National Natural Science Foundation of China (61602031), Shanghai Natural Science Foundation (21ZR1466600), Fundamental Research Funds for the Central Universities (FRF-IDRY-19-023), and Fundamental Research Funds for the Central Universities (2022-5-YB-03). This work was also supported by International Cooperation and Exchanges from the National Natural Science Foundation of China (No. 72061137071) and the General Project of the National Natural Science Foundation of China (No. 52372340).

References

1. Zhe, X., et al.: When recommender systems meet fleet management: practical study in online driver repositioning system. Proc. Web Conf. **2020**, 2220–2229 (2020)
2. Li, W.: A fleet manager that brings agents closer to resources: GIS cup. In: Proceedings of the 28th International Conference on Advances in Geographic Information Systems, pp. 655–658 (2020)

3. Ming, L., Hu, Q., Dong, M., Zheng, B.: An effective fleet management strategy for collaborative spatio-temporal searching: GIS cup. In: Proceedings of the 28th International Conference on Advances in Geographic Information Systems, pp. 651–654 (2020)
4. Lin, K., Zhao, R., Xu, Z., Zhou, J.: Efficient large-scale fleet management via multi-agent deep reinforcement learning. In: Proceedings of the 24th ACM SIGKDD International Conference on Knowledge Discovery and Data Mining, pp. 1774–1783 (2018)
5. Liu, C., Chen, C.-X., Chen, C.: Meta: a city-wide taxi repositioning framework based on multi-agent reinforcement learning. IEEE Trans. Intell. Transp. Syst. **23**(8), 13890–13895 (2021)
6. Zhang, Z., Wang, S., Li, Y.: Deep reinforcement learning. Posts and Telecom Press (2022)
7. Littman, M.L.: Markov games as a framework for multi-agent reinforcement learning. In: Machine learning proceedings 1994, pp. 157–163. Elsevier (1994)
8. Maciejewski, M., Nagel, K.: The influence of multi-agent cooperation on the efficiency of taxi dispatching. In: Wyrzykowski, R., Dongarra, J., Karczewski, K., Waśniewski, J. (eds.) PPAM 2013. LNCS, vol. 8385, pp. 751–760. Springer, Heidelberg (2014). https://doi.org/10.1007/978-3-642-55195-6_71
9. Qu, M., Zhu, H., Liu, J., Liu, G., Xiong, H.: A cost-effective recommender system for taxi drivers. In: Proceedings of the 20th ACM SIGKDD International Conference on Knowledge Discovery and Data Mining, pp. 45–54 (2014)
10. Jin, J., et al.: Coride: joint order dispatching and fleet management for multi-scale ride-hailing platforms. In: Proceedings of the 28th ACM International Conference on Information and Knowledge Management, pp. 1983–1992 (2019)
11. Li, M., et al.: Efficient ridesharing order dispatching with mean field multi-agent reinforcement learning. In: The World Wide Web Conference, pp. 983–994 (2019)
12. Liu, Z., Li, J., Kaishun, W.: Context-aware taxi dispatching at city-scale using deep reinforcement learning. IEEE Trans. Intell. Transp. Syst. **23**(3), 1996–2009 (2020)
13. Xu, M., et al.: Multi-agent reinforcement learning to unify order-matching and vehicle-repositioning in ride-hailing services. Int. J. Geograph. Inform. Sci. 1–23 (2022)
14. Gronauer, S., Diepold, K.: Multi-agent deep reinforcement learning: a survey. Artif. Intell. Rev. 1–49 (2021). https://doi.org/10.1007/s10462-021-09996-w
15. Tan, M.: Multi-agent reinforcement learning: independent vs. cooperative agents. In: Proceedings of the Tenth International Conference on Machine Learning, pp. 330–337 (1993)
16. Tampuu, A., et al.: Multiagent cooperation and competition with deep reinforcement learning. PLoS ONE **12**(4), e0172395 (2017)
17. Lowe, R., Wu, Y.I., Tamar, A., Harb, J., Pieter Abbeel, O., Mordatch, I.: Multi-agent actor-critic for mixed cooperative-competitive environments. Adv. Neural Inform. Process. Syst. **30**, 6379–6390 (2017)
18. Foerster, J., Farquhar, G., Afouras, T., Nardelli, N., Whiteson, S.: Counterfactual multi-agent policy gradients. In: Proceedings of the AAAI Conference on Artificial Intelligence, vol. 32, (2018)

19. Nguyen, D.T., Kumar, A., Lau, H.C.: Collective multiagent sequential decision making under uncertainty. In: Proceedings of the AAAI Conference on Artificial Intelligence, vol. 3 (2017)
20. Nguyen, D.T., Kumar, A., Lau, H.C.: Policy gradient with value function approximation for collective multiagent planning. In: Advances in Neural Information Processing Systems, 30 (2017)

Difficulty-Controlled Question Generation in Adaptive Education for Few-Shot Learning

YuChen Wang and Li Li(✉)

School of Computer and Information, Southwest University, Chongqing, China
ainloy9@email.swu.edu.cn, lily@swu.edu.cn

Abstract. Adaptive education aims to achieve common educational goals by implementing targeted education based on differences in student status. However, existing intelligent teaching methods cannot be applied in data-limited environments. This paper aims to address a difficulty-controlled question generation task in adaptive education in data-limited settings, where a knowledge tracing model is used to judge the difficulty of new questions and generate problems adaptively based on the student's mastery of knowledge. Based on this, a knowledge tracing and question generation model based on few-shot learning is proposed to predict the difficulty of new questions and generate adaptive questions as needed. The proposed method improves the performance of few-shot knowledge tracing by adjusting the design of input and output and adding a self-control loss function during fine-tuning. In terms of difficulty-controllable question generation, the controlled generation is achieved by splitting the Bayesian formulation into a probabilistic model for generation model and control sampling. The knowledge tracking model is used as a probabilistic model for control sampling thus solving the problem of the inability to generate questions with difficulty differentiation in the few-shot case. Experimental results show that the proposed method can generate questions with difficulty distinctions in a few-shot setting based on the given difficulty. In this way, the proposed method provides a broader application range for modern intelligent personalized education.

Keywords: Adaptive education · Question generation · Knowledge tracing · Few-shot learning

1 Introduction

Adaptive education is an instructional approach that emphasizes the unique needs of individual students and aims to achieve common educational objectives through the implementation of targeted teaching strategies that align with their learning status and needs. A crucial element in achieving this personalized approach is the generation of controlled-difficulty questions, which are specifically tailored to match the knowledge levels of the students. These questions enable

X. Yang et al. (Eds.): ADMA 2023, LNAI 14180, pp. 584–598, 2023.
https://doi.org/10.1007/978-3-031-46677-9_40

educators to design exams with varying levels of difficulty that are suitable for educational purposes [15].

In recent years, the development of natural language processing has led to an increasing interest in difficulty-controllable question generation techniques. Researchers have explored various approaches to generate questions of different difficulty levels, Gao et al. [7] achieved text generation of questions with varying levels of difficulty by inputting context and answers of reading comprehension and generating questions of specified difficulty. Kumar et al. [14] developed automatic generation of multi-hop complex questions based on knowledge graphs. Cheng et al. [4] redefined question difficulty as reasoning steps required to answer the question and rewrote the questions into more flexible and controllable levels of difficulty.

Traditional difficulty-controlled question generation methods typically define the difficulty level of the question in advance and generate questions accordingly, which limits their ability to adapt to individual learners' needs. To address this issue, the LM-KT model [23] proposes a novel approach that combines difficulty-controlled question generation with the pre-trained autoregressive language model GPT-2 [18]. The proposed method not only enables the difficult judgment of a specified student facing a new question but also generates questions of specified difficulty to match the learner's current level.

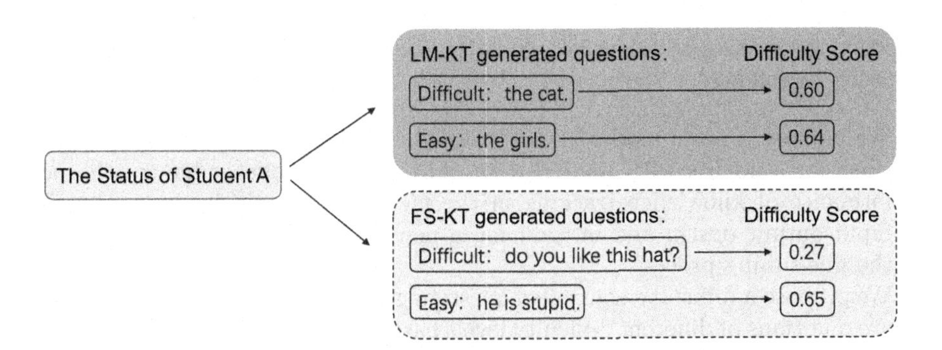

Fig. 1. With limited data, the performance of LS-KT and FS-KT models on generating questions of different difficulty for the same student A in the translation task. Given labels of difficult and easy, the generated questions evaluate the difficulty score reference [23] (the score interval is (0,1), and the higher the score the easier the question is)

However, these methods demand a substantial amount of data, and when we try to use educational questions in other languages or specialized fields with only a few dozen student data, the performance of the model drops sharply, and the difficulty of generating questions cannot be controlled. To address these challenges, this paper proposes a few-shot learning method for difficulty-controlled question generation in adaptive education. In the few-shot setting, the proposed method can predict the difficulty of new questions for different students' acquired knowledge status and generate a question that will achieve a target difficulty.

Specifically, we introduce the knowledge tracking model FS-KT to predict the difficulty of a new problem for students (e.g., the target phrase to be translated). FS-KT predicts the difficulty of a new question by composing an autoregressive text sequence of student states and the new question into the model and improves the few-shot performance by adding a self-control loss function to the fine-tuning process. Furthermore, to generate questions that match the target difficulty, we incorporate the student state and difficulty into a pre-trained model. During the generation of the question, control generation is achieved by splitting the Bayesian formulation ($P(x|a)$) into generation model ($P(x)$) and control sampling ($P(a|x)$), the sampling method is improved by adding the FS-KT model as a discriminator for difficulty-controlled sampling to the traditional Top-p sampling method [11]. With this approach, we can effectively improve the difficulty prediction performance of the problem for data-limited and solve the problem of not being able to generate questions based on difficulty level for data-limited. The controlled generation is achieved by splitting the Bayesian formulation into model generation and control sampling. The proposed approach adopts the same input-output design as the FS-KT model and improves the sampling method by incorporating the FS-KT model for difficulty control sampling in the traditional Top-p sampling method.

To the best of our knowledge, this is the first work on difficulty-controlled question generation in adaptive education for few-shot learning, our contributions are as follows:

1. We propose a novel model FS-KT to predict the difficulty level of new questions for students with limited data. The FS-KT model enhances the performance of knowledge tracking in the few-shot scenario by improving the input-output design and introducing a new self-control loss function during the fine-tuning process.
2. We propose a difficulty-controlled question generation method that can generate questions of different difficulty levels based on students' current knowledge states when data are scarce. This approach addresses the limitations of the scope of use of previous question generation methods in adaptive education.
3. We perform comprehensive experiments that demonstrate the effectiveness of our proposed approach for difficulty-controlled question generation in adaptive education for few-shot learning.

2 Related Work

In recent years, significant research efforts have been devoted to adaptive education, which utilizes Item Response Theory (IRT) to model students' abilities based on their responses to various questions, thereby creating a mathematical model to find out the degree of knowledge mastery based on the objective function of the answers [9,16], while Computerized Computerized Adaptive

Testing (CAT), which is based on IRT, uses computer technology as a tool to build a question bank using item response theory and the computer automatically selects questions based on the test taker's ability level to estimate the test taker's ability [21,24,25]. These testing methods have disadvantages such as slow updating of the question bank and inability to track student status, rendering them unsuitable for modern educational contexts. We address the shortcomings by tracking the evolving knowledge state of each student and generating questions of controlled difficulty, addressing the limitations of traditional adaptive education methods [23]. Furthermore, by combining the popular few-shot learning technique with the above methods, we apply adaptive education to more realistic scenarios.

2.1 Knowledge Tracing

Knowledge tracing (KT) is a critical component in achieving adaptive education, where learners' knowledge mastery state is modeled based on their learning behaviors and performance [6]. Over the years, KT has undergone significant development, with Bayesian knowledge tracing (BKT) [27] emerging as a popular method. BKT models the knowledge state as a set of binary variables and leverages students' real-time interactions to predict changes in their knowledge mastery through Hidden Markov Models(HMM).

Despite its effectiveness, BKT has some limitations, particularly in modeling complex learning patterns. To address this, Deep knowledge tracing (DKT) [17] was introduced, which combines Recurrent Neural Networks (RNN) with knowledge tracing for improved results over previous probabilistic models. DKT has shown success in capturing complex learning dynamics, including knowledge decay and growth, and has been applied to a wide range of educational contexts.

Recently, the SAINT+ model [22] was introduced, which is a transformer-based knowledge tracing model that leverages temporal features to improve performance and show state-of-the-art performance on the popular EdNet dataset [5], demonstrating its effectiveness in modeling student learning.

2.2 Controllable Text Generation

Controlled text generation enables the generation of text with desired properties. Ghazvininejad et al. [8] proposed a method to control the style of generated poems by adjusting the sampling weights. CTRL model [12], based on the Transformer architecture, can generate text with desired attributes in terms of a specific domain, style, topic, date, entity, inter-entity relationship, slot, and task-related code. PPLM model [26] introduced an attribute discriminator, which can be fine-tuned with a small number of parameters to achieve effectively controlled text generation. GeDi model [13] solves the inefficiency of decoding caused by multiple backpropagations of PPLM by adding an attribute discriminator directly to the decoding stage. CoCon model [3] improves the training efficiency of controlled text generation by fine-tuning the CoCon layer inserted into the pre-training model of GPT-2 [18]. CBART model [10] controls text generation by using the output of the encoder to judge the retention of different tokens, which in turn constrains the output of the decoder.

2.3 Few-Shot Learning

With the swift advancement of pre-trained models in the field of natural language processing, there has been a growing interest in utilizing the acquired knowledge of such models to address the few-shot problem. While the GPT3 model [1] has shown exceptional performance in few-shot and zero-shot cases owing to its vast number of parameters and pre-training data, its high cost restricts its access to a limited group of researchers. In contrast, the PET model [19] has adopted prompt-learning to transform text classification into an ideal fill-in-the-blank task, thereby enabling it to achieve favorable performance with few-shot. By aligning the downstream task with the pre-training task, it can better utilize the knowledge gained in the pre-training phase. Similarly, the FewshotQA framework [2] improves the performance of QA scenarios by aligning only the inputs, targets, and outputs of the pre-training and fine-tuning phases without any model modification, this approach has yielded a significant boost in few-shot performance.

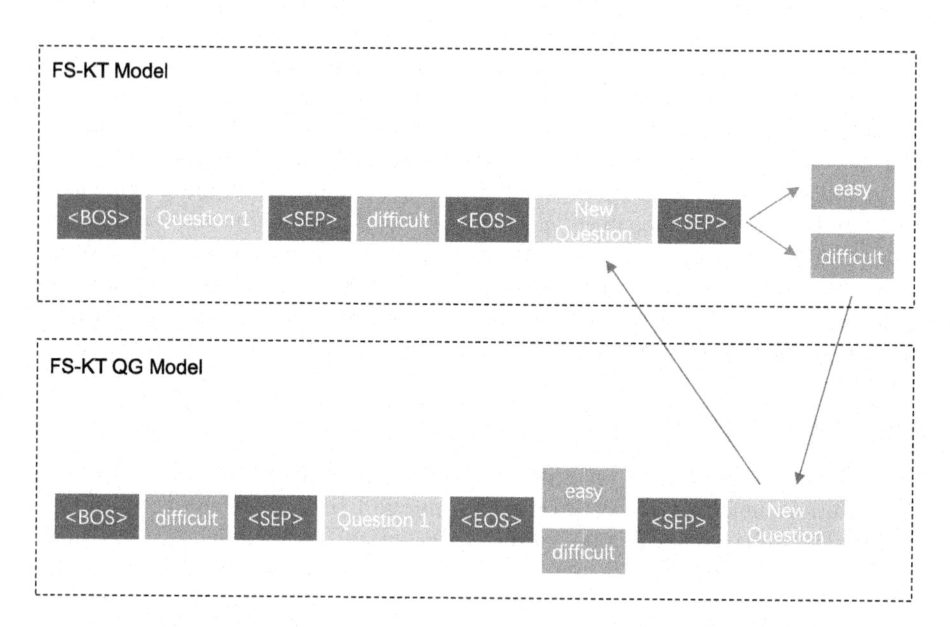

Fig. 2. Illustrating the input-output design for the FS-KT model and FS-KT QG model.

3 Method

Our method utilizes any autoregressive language model, such as GPT-2 [18]. To simulate a realistic educational scenario in situations where data are scarce, we limit the number of training examples to 50 or 25, corresponding to the learning progress of 50 or 25 students over time.

We first fine-tune the FS-KT model ($P\theta_{KT}$) to complete the knowledge tracking task by inputting a student's learning status to predict whether the student can answer the next question correctly, allowing us to gauge their mastery of the material. The student's learning status comprises the text of the questions that the student has answered, as well as whether the questions were answered right or wrong in chronological order. Secondly, we put the right and wrong of the questions done in the student state in front of the questions, and then input the difficulty of the questions that we want to generate in the model($P\theta_{QG}$) to generate the corresponding questions needed.

3.1 FS-KT Model for Knowledge Tracking

Input-Output Design: Our knowledge-tracking task is to determine the difficulty of a new question (the text of a question) for this student (the student's learning status), rather than the question ID or hand-extracted feature labels used in standard knowledge-tracking tasks, and the single training example input to the task consists of the student's current status S^i for the ith student and the question Q^i to be judged, where the student's current status S^i consists of the questions the student has done over time and whether the student answered correctly or not:

$$Input_i = S^i Q^i \tag{1}$$

$$S^i = q_1^i d_1^i q_2^i d_2^i \dots q_n^i d_n^i \tag{2}$$

q_j^i denotes the jth question answered by the ith student and d_j^i denotes whether the ith student answered the jth question correctly. Unlike previous common practices, we represent the correct and incorrect answers with the labels "*easy*" and "*difficult*" respectively. Notably, we did not use special tokens such as $< Y >$ and $< N >$ for the labels. This decision was motivated by the scarcity of training data, which makes it challenging for pre-trained models to learn where to output special tokens. Additionally, we did not use simple *Yes* and *No* lexical symbols to map the labels, as our experimental tests showed that this approach led to the model outputting words with similar lexical meanings and composition, such as *Yescid* and *Not*. This, in turn, decreased the model's performance, possibly because *Yes* and *No* are widely applicable in English and can indicate various lexical meanings beyond affirmation and negation.

To address this issue, we set the labels to "*easy*" and "*difficult*" lexical symbols, which is consistent with the labels used in question generation. To further enhance the model's performance, we utilize the GPT-2 model's own special lexical symbols in the input to separate the different questions and difficult level labels. This approach is particularly effective in improving the model's performance with limited training examples. Specifically, we use the separator $< SEP >$ to split the questions and difficult level labels and the terminator $< EOS >$ to indicate the end of each question. The overall input of the FS-KT model for the knowledge tracking task is illustrated in Fig. 2.

Self Control Loss: The autoregressive language model is a one-way generation model that generates words from left to right, with each word tag being based on the preceding text. Our knowledge tracking is to input the text of the question and the difficult label in chronological order to the model at the same time. Consequently, the model learns the question and the difficult label sequentially in a left-to-right order during fine-tuning. When a large amount of data is available for fine-tuning, the model can learn to generate output effectively. However, in the case of sparse data, the text of the questions dominates the data input to the model, while the difficult label labels are only a few. Consequently, when fine-tuning with the same loss function, the model focuses on tuning parameters for the question text, and the difficult labels are neglected. As shown in Fig. 3.

To overcome this issue, we propose a segmented, self-controlled loss function for the few-shot case. During the fine-tuning process, we exclude the question and the difficult labels from the input student state S^i used to compute the loss function:

$$\begin{aligned} L_{control} = & -\alpha \sum_{i=t^q}^{l} \log p_{\theta,\varphi}(x_i^q | \{x_1, \cdots, x_i\}) \\ & - \beta \sum_{i=t^d}^{l} \log p_{\theta,\varphi}(x_i^d | \{x_1, \cdots, x_i\}). \end{aligned} \tag{3}$$

This enables us to control the amount of parameter tuning applied to the text of the questions and ensures that the difficult labels receive sufficient attention during fine-tuning. By adopting this approach, we aim to improve the FS-KT model's performance in knowledge tracking tasks where training data is limited.

3.2 Question Generation

Input-Output Design: Given the limited availability of training examples, we approach the question generation task with controlled difficulty as a categorical text generation task. In this regard, the input to the model comprises the student state and the corresponding difficulty label of the question that requires generation. The output, on the other hand, is the generated question with the corresponding difficulty. It is worth noting that the student state in this task is distinct from that in the knowledge tracking task, we put the right and wrong of the questions done in the student state in front of the questions, then after combining the student status S^j with the desired difficulty, in the input is shown below:

$$Input_i = S^j D^j \tag{4}$$

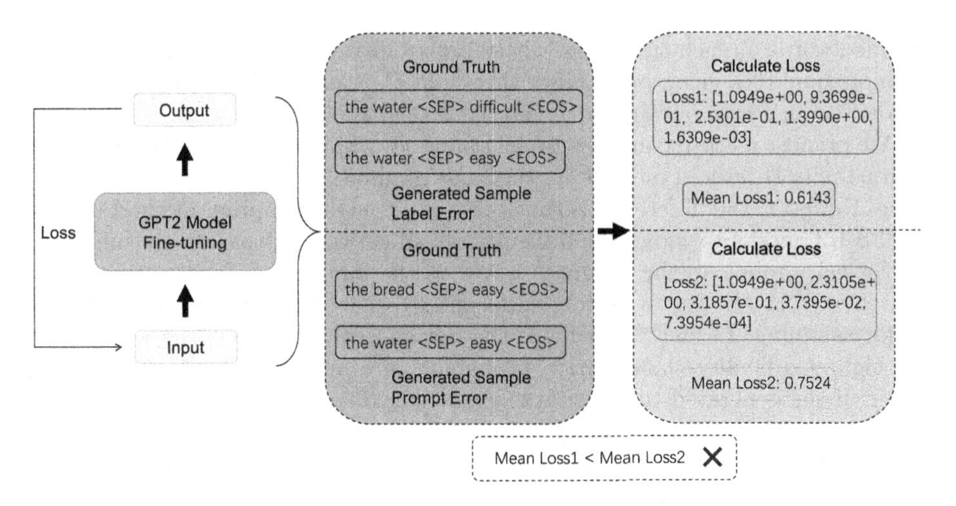

Fig. 3. In knowledge-tracking tasks where the input text sequence is ordered chronologically from left to right and includes question templates and labels, the loss of errors in hint templates can be greater than that of label errors. This can impede our ability to obtain accurate labels, especially in scenarios with limited data.

$$S^j = d_1^j q_1^j d_2^j q_2^j \dots d_n^j q_n^j \tag{5}$$

Once the user inputs the difficulty label D for the question they wish to generate, the model will proceed to generate questions with corresponding difficulty labels up to the $< EOS >$ terminator. The input configuration is illustrated in Fig. 2.

Top-P Control Sampling: Previous pre-trained models have yielded favorable outcomes in text generation utilizing extensive unsupervised data training. However, controlling the generated text has proven to be challenging. To address this issue, controllable text generation is the probabilistic model $P(x|a)$, which means that the text x is generated based on some attribute a. We decompose it as follows:

$$P(x|a) \propto P(a|x)P(x) \tag{6}$$

where $P(x)$ is directly output by pre-trained language model, and $P(a|x)$ is obtained through the FS-KT model. Due to the potential negative impact of adjusting the model structure when dealing with a limited number of examples, we have chosen to improve the sampling method rather than modify the model itself.

Specifically, we enhance the sampling process, as the currently favored technique. The current popular sampling method is mostly the Top-p sampling method [11], which considers that within each time step, the probability distribution of decoded words may have a long-tail distribution, which means that the occurrence probability of a few words at the head has occupied most of the probability space, Top-p sampling involves selecting a minimum set V of decoded words at each time step, whose probability sum of occurrence is greater than or equal to a given probability threshold P. A probability sampling is then conducted on V_P to obtain the generated lexical symbols.

We propose a top-p control sampling method, which involves first establishing a probability threshold p to obtain the same minimum set V_P as with top-p sampling. However, the difference is that the top-p control sampling method utilizes the FS-KT model to judge the difficulty classification probability by combining the lexical symbols in the minimum set with the input in the generative model, and then from the set of words conforming to the difficulty is removed from V_d for sampling. In fact, To achieve the optimal control effect, the words with the highest difficulty classification probability are selected for sampling in this paper. If the generated text wants to get better diversity, the generative model can also set up an interval when sampling, and probabilistically sample in the interval. The method is shown in Fig. 4.

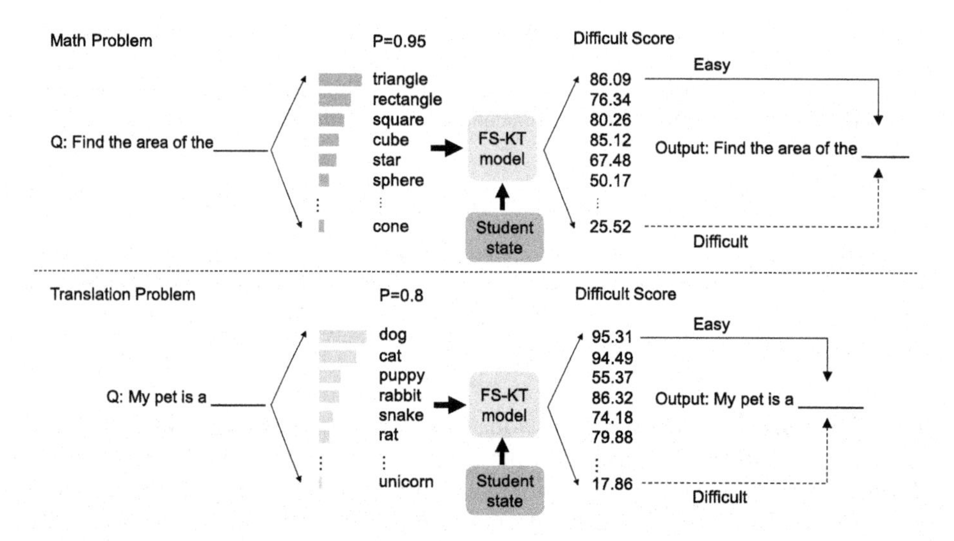

Fig. 4. An example pertains to a mathematical question(top), while the other involves a translation question(bottom). Both were generated using controlled sampling during the generation process, by which questions of specified difficulty can be generated.

4 Experiments

4.1 Datasets

Our approach can be applied to any educational task that can be represented using text sequences. Due to the paucity of publicly available datasets in this domain, we focus on translation educational tasks, which have specialized publicly available datasets. Specifically, we concentrate on tasks that involve translating phrases from the native language into the second language being learned.

To this end, we employ the 2018 Duolingo Second Language Acquisition Modeling Shared Task dataset [20], a language learning tool software that comprises questions and answers from users of the Duolingo software during the first 30 d of learning a second language. We use the translation learning task question dataset from this corpus. The original purpose of this dataset was to identify word-level errors. To achieve our research objectives, we leverage the already improved dataset in LM-KT [23], which is designed for two types of users learning Spanish and French. The dataset decomposes the original word-level error identification task into an overall correct or wrong label for each question and creates a split between the training, validation, and test sets.

To ensure the robustness of our experiments, we perform the experiments on the dataset using three different random seeds. Each experiment employs a distinct dataset containing 50 and 25 students, ensuring that the extracted data are unique for each experiment. We report the mean of the experimental results to reduce the impact of random variations and to provide a reliable measure of performance.

4.2 Experimental Details

To closely approximate real-world educational scenarios, we restricted the number of training examples to 50 and 25, which corresponds to the capacity of an average classroom, and represents the learning data of 50 students or 25 students. We evaluated students' knowledge mastery using the knowledge tracking model and subsequently generated difficult questions using the question generation model, which allowed us to complete an adaptive educational task.

The model used in the knowledge tracking experiment is called the student model, while the model used in the question generation experiment is called the question model. Both models we fine-tuned using the GPT-2 autoregressive pre-trained language model. We use the same randomly extracted training set with few examples, the test set for the student model is the entire test set in the dataset, while the test set for the question model consists of 200 randomly extracted student's given 100 "*difficult*" lexical symbol labels and 100 "*easy*" lexical symbol labels to generate the questions. To better reflect the difficulty of the generated questions, we put the generated questions and the previous student status into the LM-KT model [23] trained on the full training set, which is currently the optimal performance model for this task using the full data set, and we only use this model to judgment the difficulty level of the generated problem. We use the average of the scores of the different level labels output by the model as a reflection of the difficulty of the generated questions (the score interval is (0,1), and the higher the score the easier the question is).

Our experiments were conducted using an RTX 3090 GPU. In the FS-KT model for knowledge tracking, we used 24 training cycles, with a learning rate of 5e-5. Additionally, we set the hyperparameters α in the self-control loss function to 1, and β to 1.1(50 examples) and 1.05(25 examples). For the other comparative model parameters, we obtained the optimal results through experimentation. In the question generation model, we used 24 rounds of training cycles with a learning rate of 5e-5 and p to 0.9.

4.3 Few-Shot Results for Knowledge Tracking

We tested the FS-KT model using the knowledge-tracking traditional ROC curve area, or AUC for short. The model needs to predict whether a student can correctly answer a new question in a test with a set of real student responses. Table 1 shows the results of compared the performance of the FS-KT model against other models, where unseen indicates that the IDs of questions in the test set that are not included in the Duolingo training set (not seen ever for any student). Among them, the LM-KT model introduces the specific text of the question, so it has the best performance after fine-tuning using the full training set compared to the standard DKT model (using only the question ID) for scenarios with a large amount of data. However, the performance of the previous model degrades when entering the limited data scenario, and the FS-KT model achieves a better performance.

To gain further insights into the FS-KT model's performance, we conducted ablation experiments. Our findings indicate that the model's performance improved after using an input-output design suitable for a few-shot scenario. Moreover, we observed that the model's performance was further enhanced by introducing a self-control loss function. These results suggest that the autoregressive language model's generalization can be improved in the data-limited case by using a segmented loss function that helps the model to better learn the difference between the prompt and the location that needs to be generated.

4.4 Few-Shot Results for Question Generation

The present study aimed to evaluate the ability of a question generation model to produce questions of corresponding difficulty. In assessing the fluency of the generated text, we opted not to use Perplexity (PPL) as a measure, because this method is used to measure the ability of the language model and our sampling method cannot be demonstrated by this evaluation metric, So we relied on the judgments of human experts. Five individuals with some knowledge of natural language processing and English reading ability evaluated the fluency of the generated text using a five-point scale, with scores ranging from 1 (not fluent or grammatically incorrect) to 5 (very fluent or grammatically error-free).

Since the LM-KT model controls the difficulty of the generated questions through difficulty scores, the 200 students we gave the test set were given difficulty scores from 0.1 in increments of 0.1 to 1 each time (the higher the difficulty score, the easier the question), with 20 students for each difficulty score, which allowed us to test the model's ability to generate questions of varying difficulty with few-shot.

Table 2 presents the results of the difficulty-controlled question generation experiments on the dataset. Our findings indicate that, after modifying the input-output design of the question generation model, the model's ability to

Table 1. Experimental results of knowledge tracking in datasets.

Full data				
Model	*Spanish*		*French*	
	AUC(seen)	AUC(unseen)	AUC(seen)	AUC(unseen)
LM-KT	**0.75 ± .0001**	**0.76 ± .001**	**0.73 ± .0002**	**0.71 ± .002**
Standard DKT	0.72±.0001	0.70±.001	0.70±.0001	0.65±.002
Question only	0.67±.0001	0.58±.002	0.58±.0002	0.62±.001
50 students				
FS-KT	**0.690 ± .0002**	0.655±.003	**0.680 ± .0002**	**0.681 ± .003**
∟ w/o $L_{control}$	0.684±.0002	**0.661 ± .003**	0.672±.0002	0.678±.003
LM-KT	0.672±.0002	0.647±.003	0.664±.0003	0.673±.003
25 students				
FS-KT	**0.644 ± .0004**	**0.667 ± .003**	**0.675 ± .0002**	**0.641 ± .002**
∟ w/o $L_{control}$	0.638±.0003	0.661±.003	0.665±.0002	0.640±.002
LM-KT	0.627±.0003	0.659±.003	0.655±.0002	0.639±.003

generate questions of varying difficulty remained limited. Specifically, the pre-trained model performed poorly in controlled generation based on PROMPT when working with few examples. In contrast, the LM-KT model successfully controlled difficulty by embedding the difficulty scores through the layers in full data, but could not achieve a good fit with the limited data. Our model, with the addition of the Top-p control sampling method, substantially improves the ability of difficulty-controlled question generation with differentiation of the generated questions of different difficulties, although it loses some fluency.

Table 2. Experimental results of question generation in datasets.

50 students						
Model	Spanish			*French*		
	Easy	Difficult	Fluency	Easy	Difficult	Fluency
FS-KT QG	**0.76 ± .002**	**0.49 ± .002**	2.32	**0.81 ± .001**	**0.28 ± .003**	2.46
∟ w/o $Top - p_{control}$	0.73 ± .002	0.74 ± .002	**2.47**	0.74 ± .002	0.60 ± .003	**2.55**
LM-KT QG	0.70±.002	0.68 ± .002	2.34	0.65 ± .002	0.66±.002	2.48
25 students						
FS-KT QG	**0.76 ± .002**	**0.62 ± .001**	2.29	**0.80 ± .001**	**0.59 ± .002**	2.37
∟ w/o $Top - p_{control}$	0.75 ± .002	0.80 ± .001	**2.39**	0.75 ± .002	0.64 ± .002	**2.43**
LM-KT QG	0.66 ± .002	0.67 ± .002	2.32	0.65 ± .002	0.67 ± .002	2.38

Table 3. Comparison of our model with the LS-KT model for example questions(translation questions) generated by a French learner.

FS-KT	
Difficult	Easy
do you want a red apple?	it is ugly.
twenty years is a very quick transition.	he loves you.
i grab my hat.	the cat.
LM-KT	
Difficult(0.3)	**Easy(0.7)**
yes or no?	it breaks
the ducks.	where?
who eats?	the dog is dirty

5 Case Study

Table 3 provides a comparison of some examples of different models for generating questions for translation tasks. LS-KT model we use a difficulty factor of 0.3 and 0.7 to generate difficult and easy questions, respectively. Our model performs similarly to the comparison model in generating easy questions, but generates more difficult questions that are longer and harder to translate.

For the examples, when generating more difficult questions, we observe that the comparison model tends to generate phrases or words with a smaller number of tokens, while our model generates a complete sentence, and for the translation task, often sentences with a larger number of words are more difficult to translate. Specifically, by top-p control sampling, the model does not select tokens that usually have a greater probability, but rather selects tokens that may generate more difficult questions, and then uses the tokens as input to generate questions autoregressively, so that when generating more difficult questions, a single long sentence is often generated. In generating easy questions, top-p control sampling then selects tokens that are likely to generate easy questions, thus generating simple phrases or words. However, one drawback of our approach is that it may generate more repetitive questions, this issue is also present in the comparison model, and it is due to the limited amount of available data.

6 Conclusions

Our work aims to apply artificial intelligence and adaptive education to a broader range of educational scenarios by combining currently popular few-shot learning with previous question generation in adaptive education [23].

Our approach is limited by the number of parameters of the pre-trained model and the amount of data learned from pre-training. Currently, GPT-3 and ChatGPT have been developed accordingly, and we will investigate how to use and improve more advanced models to improve the performance for few-shot learning in the future. Moving forward, our future direction is to learn to generate logical mathematics problems with less data, and how to control the specific direction and knowledge points of problem generation.

Acknowledgments. This work was supported by NSFC (grant No.61877051). Li is the corresponding author of the paper.

References

1. Brown, T., et al.: Language models are few-shot learners. Adv. Neural. Inf. Process. Syst. **33**, 1877–1901 (2020)
2. Chada, R., Natarajan, P.: Fewshotqa: a simple framework for few-shot learning of question answering tasks using pre-trained text-to-text models. In: Proceedings of the 2021 Conference on Empirical Methods in Natural Language Processing, pp. 6081–6090 (2021)
3. Chan, A., Ong, Y.S., Pung, B., Zhang, A., Fu, J.: Cocon: a self-supervised approach for controlled text generation. In: International Conference on Learning Representations
4. Cheng, Y., et al.: Guiding the growth: difficulty-controllable question generation through step-by-step rewriting. In: Proceedings of the 59th Annual Meeting of the Association for Computational Linguistics and the 11th International Joint Conference on Natural Language Processing (Volume 1: Long Papers), pp. 5968–5978 (2021)
5. Choi, Y., et al.: EdNet: a large-scale hierarchical dataset in education. In: Bittencourt, I.I., Cukurova, M., Muldner, K., Luckin, R., Millán, E. (eds.) Artificial Intelligence in Education: 21st International Conference, AIED 2020, Ifrane, Morocco, July 6–10, 2020, Proceedings, Part II, pp. 69–73. Springer International Publishing, Cham (2020). https://doi.org/10.1007/978-3-030-52240-7_13
6. Corbett, A.T., Anderson, J.R.: Knowledge tracing: modeling the acquisition of procedural knowledge. User Model. User-Adap. Inter. **4**, 253–278 (1994)
7. Gao, Y., Bing, L., Li, P., King, I., Lyu, M.R.: Generating distractors for reading comprehension questions from real examinations. In: Proceedings of the AAAI Conference on Artificial Intelligence. vol. 33, pp. 6423–6430 (2019)
8. Ghazvininejad, M., Shi, X., Priyadarshi, J., Knight, K.: Hafez: an interactive poetry generation system. In: Proceedings of ACL 2017, System Demonstrations. pp. 43–48 (2017)
9. Hambleton, R., Jodoin, M.: Item response theory: models and features. Encyclopedia of psychological assessment, pp. 510–515 (2003)
10. He, X.: Parallel refinements for lexically constrained text generation with BART. In: Proceedings of the 2021 Conference on Empirical Methods in Natural Language Processing, pp. 8653–8666 (2021)
11. Holtzman, A., Buys, J., Du, L., Forbes, M., Choi, Y.: The curious case of neural text degeneration. In: International Conference on Learning Representations
12. Keskar, N.S., McCann, B., Varshney, L.R., Xiong, C., Socher, R.: CTRL: A conditional transformer language model for controllable generation. arXiv preprint arXiv:1909.05858 (2019)
13. Krause, B., et al.: GEDI: generative discriminator guided sequence generation. In: Findings of the Association for Computational Linguistics: EMNLP 2021, pp. 4929–4952 (2021)

14. Kumar, V., Hua, Y., Ramakrishnan, G., Qi, G., Gao, L., Li, Y.F.: Difficulty-controllable multi-hop question generation from knowledge graphs. In: The Semantic Web-ISWC 2019: 18th International Semantic Web Conference, Auckland, New Zealand, October 26–30, 2019, Proceedings, Part I, pp. 382–398 (2019)
15. Kurdi, G., Leo, J., Parsia, B., Sattler, U., Al-Emari, S.: A systematic review of automatic question generation for educational purposes. Int. J. Artif. Intell. Educ. **30**, 121–204 (2020)
16. Lord, F.M.: Applications of item response theory to practical testing problems. Routledge (1980)
17. Piech, C., et al.: Deep knowledge tracing. In: Advances in Neural Information Processing Systems 28 (2015)
18. Radford, A., Wu, J., Child, R., Luan, D., Amodei, D., Sutskever, I., et al.: Language models are unsupervised multitask learners
19. Schick, T., Schütze, H.: Exploiting cloze-questions for few-shot text classification and natural language inference. In: Proceedings of the 16th Conference of the European Chapter of the Association for Computational Linguistics: Main Volume, pp. 255–269 (2021)
20. Settles, B., Brust, C., Gustafson, E., Hagiwara, M., Madnani, N.: Second language acquisition modeling. In: Proceedings of the Thirteenth Workshop on Innovative use of NLP for Building Educational Applications, pp. 56–65 (2018)
21. Machine learning-driven language assessment: Settles, B., T. LaFlair, G., Hagiwara, M. Trans. Assoc. Comput. Linguist. **8**, 247–263 (2020)
22. Shin, D., Shim, Y., Yu, H., Lee, S., Kim, B., Choi, Y.: Saint+: integrating temporal features for EdNet correctness prediction. In: LAK21: 11th International Learning Analytics and Knowledge Conference, pp. 490–496 (2021)
23. Srivastava, M., Goodman, N.: Question generation for adaptive education. In: Proceedings of the 59th Annual Meeting of the Association for Computational Linguistics and the 11th International Joint Conference on Natural Language Processing (Volume 2: Short Papers), pp. 692–701 (2021)
24. Thissen, D., Mislevy, R.: Testing algorithms//h. wainer (ed.) computerized adaptive testing: A primer (2000)
25. Weiss, D.J., Kingsbury, G.G.: Application of computerized adaptive testing to educational problems. J. Educ. Meas. **21**(4), 361–375 (1984)
26. Xu, C., Zhao, J., Li, R., Hu, C., Xiao, C.: Change or not: a simple approach for plug and play language models on sentiment control. In: Proceedings of the AAAI Conference on Artificial Intelligence. vol. 35, pp. 15935–15936 (2021)
27. Yudelson, M.V., Koedinger, K.R., Gordon, G.J.: Individualized Bayesian knowledge tracing models. In: Lane, H.C., Yacef, K., Mostow, J., Pavlik, P. (eds.) AIED 2013. LNCS (LNAI), vol. 7926, pp. 171–180. Springer, Heidelberg (2013). https://doi.org/10.1007/978-3-642-39112-5_18

A Multi-truth Discovery Approach Based on Confidence Interval Estimation of Truths

Xiu Fang[1], Chenling Shen[1], Quan Z. Sheng[2], Guohao Sun[1(✉)], Yating Tang[1], and Haiyan Zhuo[1]

[1] Donghua University, Shanghai, China
{xiu.fang,ghsun}@dhu.edu.cn, {2212626,2212646,2222698}@mail.dhu.edu.cn
[2] Macquarie University, Sydney, Australia
michael.sheng@mq.edu.au

Abstract. The rapid development of the Internet makes it easier to spread and obtain data. However, conflicting descriptions of an object from different sources make identifying trustworthy information challenging. This is known as the truth discovery task. In truth discovery, an object may have multiple values, such as a book written by multiple authors. Existing multi-truth discovery methods primarily focus on the probability of each candidate value being correct and provide a point estimate. However, practical applications face the problem of unbalanced object distribution, where a single point estimate may overlook critical confidence information. Additionally, ambiguous terms like "etc." and "et. al" can lead to estimation deviations. To address these issues, we propose MTD_VCI, an optimization model for confidence perception of multiple truths to detect truth from unbalanced data distribution. MTD_VCI estimates the credibility score of each candidate value and considers the confidence interval to reflect the unevenness distribution, improving decision-making. Additionally, the number of values claimed by ambiguous sources is re-estimated using other sources as a reference. Experiment results on real-world and simulated datasets demonstrate that MTD_VCI produces better results and effective confidence intervals for each value.

Keywords: Multi-Truth discovery · Unbalanced distribution · Confidence interval

1 Introduction

The internet's rapid development has made it easier for people to acquire and share data through various channels like social networking, mobile devices, and crowdsourcing platforms [4,9,15]. However, in the massive data of unknown reliability, different sources may provide conflicting descriptions of the same object due to errors, missing records, and typos. The spread of erroneous information

X. Yang et al. (Eds.): ADMA 2023, LNAI 14180, pp. 599–615, 2023.
https://doi.org/10.1007/978-3-031-46677-9_41

can lead to misunderstandings, harm, or worse. Hence, truth discovery is essential for obtaining accurate information. Majority voting is the simplest approach to handling conflicting values from different sources, where the value with the most occurrences is taken as the true value. However, majority voting assumes that the reliability of all sources is consistent, which is contrary to the difference in the reliability of different sources in the real life, failing to identify the correct value when low-quality sources are in the majority. To solve this problem, various truth discovery methods [3,8,10,18–20] have been proposed. These methods consider various influencing factors such as source reliability, source relationship, object difficulty, and the number of truths. Though these methods differ in many ways, they all follow the principle that a source is reliable if it regularly provides trustworthy information and a piece of information is trustworthy if it is supposed by reliable sources.

Most existing methods [6,7,9,16–18] assume that every object has only one true value. However, in reality, many objects have multiple truths, such as a book co-authored by multiple authors. In this case, different sources have different performances. Some sources may be cautious and only claim values that they believe to be correct, leading to high precision but potentially incomplete information. Other sources may claim all possible correct values to improve recall, sacrificing accuracy. While some multi-truth discovery methods [5,8,10,11,19] have been proposed, they all provide a single point estimate for each value, and select one or more values according to the ordering of the credibility scores. However, such a mere estimation of confidence score ignores important confidence information. For example, if book A may receive 1000 claims, while book B only receives one, despite both receiving the same credibility score of their values, the confidence levels of those scores could be very different due to the unbalanced distribution of collected data. Obviously, with more supportive source, the estimated confidence score of A should be more reliable. Therefore, an estimated score confidence interval is preferable to a point estimate.

In addition, in multi-truth discovery, the estimation of the number of truth is an important clue for the selection of true values. The existing methods [5,10,19] mostly capture two types of errors (false positive and false negative) to estimate two aspects of source reliability. Wang et al. [11] proposed a model to adapt the existing single-truth discovery methods to multi-truth scenario by additionally estimating truth number for each object. However, they all ignore the ambiguous claims provided by rigorous sources. In practice, some sources tend to be more precise and well-rounded, by adding words such as "et al", "etc." to their claims, to show that they know the object has more values, but they can only name part of them. Simply ignoring these ambiguous words would underestimate the number of true values for each object and lead to low recall.

In this paper we propose a confidence-aware optimization algorithm model MTD_VCI, which not only considers the unbalanced distribution of data objects in multi-value scenarios, but also the negative impact of ambiguous claims from some rigorous sources. For ambiguous claims from some rigorous sources, we calculate the expected values based on the number of values provided by other sources to update their claims, while converting the values of the claims provided by sources into two forms: the number of values and the vector of values, and performing truth discovery on the number of values and value vectors, respectively, to iteratively estimate the source weights and the truths (i.e., the true number of values and the specific true values). In addition, confidence intervals for the confidence scores of the number of values and the specific values are constructed by randomly selecting the sources and their claims through a bootstrap sampling method. The contributions of this paper are as follows:

(i) We consider the unbalanced data distribution in multi-truth scenario, and propose an optimization model framework of confidence perception, which incorporates the estimation of confidence interval into multi-truth discovery.
(ii) We recognize the negative impact of ignoring ambiguous words in sources' claims. We propose a method that takes sample expectations to update ambiguous source claims on the number of values for each object.
(iii) Experimental results show that our method has advantages on both real-world and simulated datasets, and achieves the desired effect.

2 Related Work

Most truth discovery algorithms, such as TruthFinder [18] and CATD [3], assume that each object has only one truth. However, practical applications often involve multiple truths. In 2012, Zhao et al. [19] proposed LTM, the first algorithm to address multi-truth discovery using a probabilistic graphical model to model the false positives and false negatives factors. Pochampally et al. [8] used Bayesian analysis to estimate the probability that a triplet is true according to the accuracy and recall of the source, assuming independence between sources. Wang et al. [12] proposed a probabilistic model to balance positive and negative claims and incorporate the meaning of value co-occurrence in the same claim. Wang et al. [10] proposed MBM, a comprehensive Bayesian method that considers the different meanings of mutual exclusion between values and the relationship between sources. They reconstructed the problem model based on the mapping between sources and value sets to improve efficiency. Subsequently, they also proposed three models to improve truth discovery results by using the number of truth as an important clue [11]. Lin et al. [5] integrated domain expertise into multi-truth discovery to achieve more accurate source reliability estimation. Fang et al. [2] proposed SmartVote, a graph-based method that models and quantifies two types of source relationships to accurately estimate source reliability and detect malicious agreements between sources. Azzalini et al. [1] proposed STORM, a domain-aware data fusion algorithm based on Bayesian inference. STORM detects replication relationships between sources and determines the trustworthiness of a source based on its authority. Existing multi-truth discovery methods focus on point estimation of the credibility score of a value, but miss important confidence information at this point.

Being aware of the unbalanced data distribution, some recent works are proposed. Li et al. [3] proposed the CATD method to address truth discovery with long-tail phenomena by introducing source reliability estimation based on confidence intervals. In 2016, Xiao et al. [13] proposed ETCIBoot, a new truth discovery method based on bootstrapping that constructs confidence interval estimation to identify the truth. This framework has been applied to large-scale truth discovery tasks under distributed conditions [14]. For the feature of unbalanced data distribution, the existing methods assume that there is only one truth for an object. However, in the multi-value scenario, the above methods cannot be directly used to estimate the confidence interval.

3 Problem Setting

3.1 Notation Definition

In the truth discovery problem, there are three inputs: (i) objects, the things of interest and their truths are to be discovered. (ii) sources, which provide the candidate values for objects. (iii) claims, the values provided by sources.

Suppose a set of sources, denoted as S, providing claims on a set of objects, denoted as O. Each object $o \in O$ may have multiple true values, the number of the values is unknown a priori. For each object $o \in O$, the set of sources which provide claims for object o is denoted by S_o. S_o is divided into two categories, S_o^a(sources provide ascertain claims) and S_o^u(sources provide uncertain claims), according to whether a source provides uncertain claim or not. The set of claims on object o provided by sources is denoted by C_o, $c_o^s \in C_o$, where c_o^s is the claim provided by source s for object o.

For each object $o \in O$, we treat the values in each claim individually, and collect all potential values from all claims into a set, denoted as U_o. Then, we preprocess each claim into two representations, i.e., the number of values, denoted as N_o, and the vector of values, denoted as V_o. Let $n_o^s \in N_o$ be the number of claimed values provided by source s for object o. N_o is divided into two categories, N_o^a(the number provided by source in S_o^a) and N_o^u(the number provided by source in S_o^u). The vector of values is the representation of specific values in a source's claim, which is coded as a 0–1 vector, with the length equals to $|U_o|$. If a source provides one of the candidate values, the corresponding value is 1, if not, the value is 0. Let $v_o^s \in V_o$ be the vector provided by source s for object o.

For example, in Table 1, the set of candidate values for object 0201609215 U_o={*Foley, James D.; Dam, Andries Van; James D. Foley; Foley James; Vandam Andries; Van Dam, Andries; Feiner, Steven K.; Hughes, John F.*}, $S_o^u = \{s_1, s_2\}$, $S_o^a = \{s_3, s_4, s_5, s_6, s_7, s_8\}$, $N_o^u = \{>2, >1\}$, $N_o^a = \{2, 1, 3, 5, 1, 3\}$, $c_{0201609215}^{s_1}$ = {*Foley, James D.; Dam, Andries Van*}, $n_{0201609215}^{s_1} = 2$, $v_{0201609215}^{s_1}$ = [1, 1, 0, 0, 0, 0, 0, 0].

3.2 Multi-truth Discovery Task

In the multi-truth scenario, each object may have multiple values. Therefore, the truths of an object contain two implications, i.e., the true number of values

Table 1. The example of book authors

source	object	source's claim	#truth
s_1	0201609215	Foley, James D.; Dam, Andries Van; etc	>2
s_2	0201609215	James D. Foley; et al.	>1
s_3	0201609215	Foley James; Vandam Andries	2
s_4	0201609215	Foley, James D	1
s_5	0201609215	Foley, James D.; Van Dam, Andries; Feiner, Steven K	3
s_6	0201609215	Foley, James D.; Van Dam, Andries; Feiner, Steven K.; Hughes, John F.; Phillips, Richard L	5
s_7	0201609215	FOLEY, JAMES D	1
s_8	0201609215	Foley, James D.; Van Dam, Andries; Feine	3

and the specific true values. Accordingly, we split the multi-truth discovery task into two sub-tasks: truth discovery on the number of values and truth discovery on the vector of values. Given an object o, in true number discovery, we need to estimate the number of values, denoted as N_o^e, which should be as close to the true number of truths N_o^* as possible. Besides, for any $\alpha \in (0, 1)$, we also provide a level-α bilateral confidence interval for the true number of each object.

Similarly, in the truth discovery on the vector of values, we need to identify the estimated vector V_o^e, which should be as close to the true vector V_o^*, and construct a level-α bilateral confidence interval for the credibility score of each value.

4 Multi-truth Discovery Based on CATD

4.1 Source Reliability Estimation

In our model, the weight of the source consists of two parts, i.e., the reliability of its claim on the number of values of each object, and the reliability of its claim on specific values of each object, which are represented by w_s^n and w_s^v, respectively. Source weights play an important role in truth discovery, and the basic principle is: If a source often provides correct claims, it should have a larger weight.

In the truth discovery on the number of values, we can reflect the source reliability by the error between the value provided by the source and the truth. The reliability of the source is inversely proportional to the error. If a source is unreliable, the error variance of the source will be larger and the error distribution will be wider. Therefore, an optimization model can be established to make the error variance as small as possible:

$$\begin{aligned} &\min_{w_s^n} \textstyle\sum_{s \in S} {w_s^n}^2 {\sigma_s^n}^2 \\ &\text{s.t.} \textstyle\sum_{s \in S} w_s^n = 1, w_s^n \geq 0, \forall s \in S, \end{aligned} \tag{1}$$

where ${\sigma_s^n}^2$ is the error variance of the source in calculating the number of values, which is an unknown variable. The real variance is estimated by the sample variance, i.e. $(\hat{\sigma}_s^n)^2 = \frac{1}{|O_s|}\sum_{o \in O}(n_o^s - n_o^e)^2$. $|O_s|$ is the number of objects provided by source s. However, when the source provides few claims, the value of ${\sigma_s^n}^2$ is not accurate due to the lack of sufficient data support. Therefore, a point estimate for each source may have some deviation. A range of values can be considered as an estimate of the real variance. We assume that the error of the source follows a normal distribution. Since the sum of normal distribution has chi-squared distribution, the following formula is satisfied:

$$\frac{\sum_{o \in O_s}(n_o^s - n_o^e)^2}{{\sigma_s^n}^2} = \frac{|O_s|(\hat{\sigma}_s^n)^2}{{\sigma_s^n}^2} \sim \chi^2(|O_s|). \tag{2}$$

We use $(1 - \alpha)$ confidence interval for ${\sigma_s^n}^2$, where α is the significance level and is a small number, such as 0.05. According to the significance level, the confidence interval can be calculated, i.e. $(\frac{\sum_{o \in O_s}(n_o^s - n_o^e)^2}{\chi^2_{(1-\alpha/2,|O_s|)}}, \frac{\sum_{o \in O_s}(n_o^s - n_o^e)^2}{\chi^2_{(\alpha/2,|O_s|)}})$. We use the upper bound of the confidence interval as the estimator of the real variance of the source. Since the source reliability is inversely proportional to the error, the calculation formula of the source reliability is as follows:

$$w_s^n \propto \frac{\chi^2_{(\alpha/2,|O_s|)}}{\sum_{o \in O_s}(n_o^s - n_o^e)^2}. \tag{3}$$

Similarly, in the truth discovery of the value vector, the weight of the source is calculated as follows:

$$w_s^v \propto \frac{\chi^2_{(\alpha/2,|O_s|)}}{\sum_{o \in O_s}(v_o^s - v_o^e)^2}. \tag{4}$$

where w_s^v is the source reliability in calculating the vector of values.

4.2 Truth Inference

The basic principle of estimating value credibility score is that if a value is provided by a high-reliability data source, the probability of this value becoming a truth is higher. In multi-truth discovery, we need to calculate the truth value of the number of values and the truth value of the vector of values respectively.

The truth of the number of values can be calculated by:

$$n_o^e = \frac{\sum w_s^n n_o^s}{\sum w_s^n}. \tag{5}$$

The truth of the vector of values can be calculated by:

$$v_o^e = \frac{\sum w_s^v v_o^s}{\sum w_s^v}. \tag{6}$$

5 Methodology

5.1 Ambiguous Claims Elimination

When estimating the number of truth, some sources provide ambiguous claims that contain words like "et al", "etc.", such as s_1 and s_2 in Table 1. If we ignore these words, we may underestimate the number of values, and result in low recall. Taking the data in Table 1 as an example, if ambiguous words as "et al" are ignored, the number of values of object 0201609215 obtained through majority voting is 1, which is less than the ground truth 3. By adopting our method to deal with the uncertain claims, we update n_1^1 to 4 and n_1^2 to 3. Based on this, we could get the true number of values, i.e., 3, by majority voting. Therefore, it can be seen that such ambiguous words do have an impact on the results. We need to estimate the number of truth by taking those knowledge into consideration.

We assume that S_o^a is a sample, and use the expectation of the sample to estimate and update the uncertain claims provided by S_o^u. Specifically, we create a dictionary, in which the key is the unique value in N_o^a, and the corresponding value is the number of its occurences. Continue with the example in Table 1, $D_o = \{1 : 2, 2 : 1, 3 : 2, 5 : 1\}$. Given a conditional constraint in N_o^u, say $> n_o^s$, we find out all the keys in D_o that meet with this constraint. Then, based on the keys and values, we calculate the expectation, and update $> n_o^s$ as the expectation. The number of claims for a source in S_o^a is calculated as follows:

$$n_o^{s*} = \sum_{n > n_o^s} n \times \frac{D_o[n]}{\sum_{n > n_o^s} D_o[n]}, \tag{7}$$

where n_o^{s*} is an estimation of the number of values provided by source s in S_o^u. Still take Table 1 as an example, the estimated number of values provided by s_1 is $n_1^{1*} = 3 \times (2/(2+1)) + 5 \times (1/(2+1)) = 11/3 \approx 4$, the estimated number of values provided by s_2 is $n_1^{2*} = 2 \times (1/(1+2+1)) + 3 \times (2/(1+2+1)) + 5 \times (1/(1+2+1)) = 13/4 \approx 3$.

5.2 Confidence Intervals Construction

The confidence interval can better reflect the unbalanced distribution of dataset. If the width of the confidence interval is smaller, it means that more sources provide support for the object, and the estimated value is more reliable. In this part, we construct the corresponding confidence intervals for the number of values and the vector of values, respectively.

Bootstrap Sampling Strategy. Bootstrap has stronger robustness and can better estimate the truth value when it is integrated into the truth discovery process, so we propose a bootstrap sampling strategy, which samples the number of values and the value vector respectively, and finds their truth values.

The bootstrap sampling strategy for the number of values is as follows: In the i-th sampling, for object o, we randomly select with place back $|S_o|$ sources from the set S_o to form the set S_o^i, and select the number of values claimed by the corresponding source $N_o^i = \{n_o^s\}_{s \in S_o^i}$. Based on the sampled data, we calculate the estimated number of values $\hat{\theta}(N_o^i)$ by Eq. (5).

The sampling strategy for the vector of values is similar to the number of values. For an object o, we sample on each candidate value $u_o \in U_o$. We randomly select the source set $S_{u_o}^i$ from the set S_{u_o}, where S_{u_o} is the set of sources that claim the specific value u_o. Then, for $o \in O$, we recombine the specific values of the source claimed into a new value vector V_o^i. Based on the sampled data, we calculate the estimated the vector of values $\hat{\theta}(V_o^i)$ by Eq. (6).

Confidence Interval Construction. Next, we introduce the process of constructing the bilateral confidence intervals of the number of values and the credibility score of each candidate value in the value vector.

Confidence interval construction method of the number of values: we will take object o as an example to describe, and calculate the confidence interval of other objects by analogy. For convenience, we omit the subscript o in this part, and use $\hat{\theta}(N)$ to represent the number of values estimated based on the sampling data, and $\hat{Var}(N)$ to represent the variance of the number of values estimated based on the sampling data.

After M times of sampling, we can obtain M estimated values. In order to obtain the confidence interval of the number of values, we need to construct a statistic T_n related to n_o^e and estimate the cumulative density function of $T_n \sim F(t)$, where

$$T_n = \frac{\hat{\theta}(N) - n_o^e}{[\hat{Var}(N)]^{0.5}/\sqrt{|S_o|}}. \tag{8}$$

This formula measures the error between the number of values obtained by each sampling and the estimated number of values.

We only need to determine the distribution of T_n, and use $T_n^{(\alpha)}$ to represent the $(100 \cdot \alpha)$th percentile of T_n. According to $P(T_n^{(\alpha/2)} \leq \frac{\hat{\theta}(N)-n_o^e}{[\hat{Var}(N)]^{0.5}/\sqrt{|S_o|}} \leq T_n^{(1-\alpha/2)}) = \alpha$, we can calculate the confidence interval of n_o^e.

Since $T_n^{(\alpha/2)}$ and $T^{(1-\alpha/2)}$ are priori unknown, they need to be estimated according to the results obtained by M sampling.

According to the M sampling data, we can calculate the number of values $\hat{\theta}(N^i)$ and the variance $\hat{Var}(N^i)$ of M results. Based on this, we can estimate the statistic $\hat{T}^i$ after each sampling, and it is calculated by:

$$\hat{T}^i = \frac{\hat{\theta}(N^i) - n_o^e}{[\hat{Var}(N^i)]^{0.5}/\sqrt{|S_o|}}. \tag{9}$$

Then, the estimate of $T_n^{(\alpha)}$ is defined as follows:

$$T_n^{(\alpha)} = \sup\{t \in \{\hat{T}^1, \cdots \hat{T}^M\} : \frac{\#(\hat{T}^i) \leq t}{M} \leq \alpha\}. \tag{10}$$

Finally, the confidence interval of the number of values is obtained:

$$(n_o^e - \frac{T_n^{(1-\alpha/2)}[\hat{Var}(N)^{0.5}]}{\sqrt{|S_o|}}, n_o^e - \frac{T_n^{(\alpha/2)}[\hat{Var}(N)^{0.5}]}{\sqrt{|S_o|}}). \tag{11}$$

Similar to the method of constructing confidence interval for the number of values, the confidence interval we construct for the vector of values is

$$(v_o^e - \frac{T_v^{(1-\alpha/2)}[\hat{Var}(V)^{0.5}]}{\sqrt{|S_v|}}, v_o^e - \frac{T_v^{(\alpha/2)}[\hat{Var}(V)^{0.5}]}{\sqrt{|S_v|}}). \tag{12}$$

5.3 MTD_VCI Algorithm

After introducing the methods of source weight updating, credibility score estimation, and confidence interval construction, we combine them to propose a new method of multi-truth discovery, which can calculate the truths according to the confidence intervals. The pseudo code of our proposed MTD_VCI algorithm is shown in algorithm 1. In each iteration, lines 3–5 estimate the two aspects of source weights w_s^n and w_s^v. Lines 7 estimates the number of values and the credibility score of each value in the value vector. Line 8–13 use the bootstrap strategy to sample to estiamte the stastic. Lines 14–15 construct confidence intervals for the estimated number of values and the credibility score of the values. Line 16 judges the truth according to the number of values, the vector of values, and their corresponding confidence intervals. If there are values with similar credibility scores, the one with a smaller confidence interval is selected. Then denote the vector of values to 0–1 vector.

6 Experiment

6.1 Experimental Setup

Datasets. In this part, we introduce the two real-world datasets, i.e., Book Dataset and Movie Dataset used in [19], and generate simulated datasets in the experiment.

- Book Dataset: The book dataset is crawled from abebooks.com, including 33,158 records. Each record represents the statement of the bookstore to the author of a book. We deleted duplicates to make the problem more challenging. The final dataset includes 875 sources (i.e. bookstores) and 1,263 objects (i.e. books).

Algorithm 1 MTD_VCI

Input: set of the number of values N_o, set of the vector of values V_o, confidence level α and the sampling times M
Output: N_o^e, V_o^e, confidence intervals $CI_o^n\{\alpha\}$ and $CI_o^v\{\alpha\}$

```
1: Initialize the number of values N_o^(0) and the vector of values V_o^(0) as average;
2: while the convergence condition is not satisfied do
3:     for each source s ∈ S do
4:         update source weights w_s^n and w_s^v according to Eq. (3) and (4);
5:     end for
6:     for each object o ∈ O do
7:         estimate the truth of n_o^e and v_o^e according to Eq.(5) and (6);
8:         while i ≤ M do
9:             sample N_o^i and V_o^i from N_o and V_o;
10:            calculate θ̂(N_o^i) and θ̂(V_o^i) according to Eq.(5) and (6);
           //to be consistent with Sect. 5, the subscript o of T̂_o^i is omitted;
11:            calculate the stastic T̂^i according to Eq. (9);
12:            i = i + 1;
13:        end while
14:        calculate T^(α/2), T^(1−α/2) based on T̂^i according to Eq. (10);
15:        construct confidence intervals CI_o^n{α} and CI_o^v{α} by Eq. (11) and (12);
16:        select top-n values as truth according to n_o^e, v_o^e and confidence intervals,
           then denote v_o as 0-1 vector.
17:    end for
18: end while
```

- Movie Dataset: The movie dataset contains 1,134,329 records. Each record contains statements from different websites about the director and the year of the movie. Since the director of a movie may have multiple values, we choose the director of each movie as the experimental object. The final dataset includes 15 sources (i.e. websites) and 403,526 objects (i.e. movies).
- Simulated Dataset: We simulated the different error variance distribution of sources. Each dataset contains 31 objects and 100 sources. The generation process will be described later.

Baseline Methods. We choose several classical truth discovery methods and Majority Voting for comparison, including single-truth methods and multi-truth methods.

- TruthFinder [18] iteratively update the source weight and truth, and calculate the probability that the claim is correct based on the reliability of source.
- AverageLog [7] estimates the trustworthiness of all sources that whether claim a value. If the trustworthiness of claiming a value is larger than another, the value is considered to be true.
- LTM [19] uses a probability graph model to calculate the source weight and reliability of truth, which is the first method for multi-truth discovery.
- MBM [10] groups sources and values, and considers both false positive and false negative errors in source reliability.

- MTD_VCI_et is our improved version of MTD_VCI that deals with uncertain claims before estimating the number of truths.

Besides, there are other multi-truth discovery methods, such as Precrec [8], SmartVote [2], DART [5] and STORM [1]. Precrec condiders a wide correlation between sources, while we assume that the sources are independent. SmartVote uses a graph model to detect malicious protocols between sources, which we do not assume in our method. The DART and STORM incorporate the domain into the truth discovery, which is not significant in our model. Thus, we do not compare our method with Precrec, SmartVote, DART and STORM.

Metrics. For multi-value problems, we use three metrics to measure the accuracy of estimates.

• Precision and Recall: In multi-truth discovery, precision is used to represent the proportion of accurate prediction among all estimated truth. Recall is used to represent the proportion of accurate prediction among ground truth.

• F1-score: We use F1-score to measure the overall accuracy of multi-truth discovery, which takes into account both accuracy and recall.

6.2 The Results on Real-World Dataset

We present the results of MTD_VCI and baseline methods in Table 2. Because our proposed MTD_VCI method considers more information in the confidence interval, while other baseline methods only make a single point estimation, the results show that MTD_VCI can achieve good results on the two datasets.

Table 2. Comparison of different algorithms on real-world datasets: the best performance values are in bold.

Method	Book Dataset			Movie Dataset		
	Precision	*Recall*	*F1-score*	*Precision*	*Recall*	*F1-score*
Majority Voting	**0.949**	0.791	0.863	0.832	0.663	0.738
AverageLog	0.900	0.701	0.788	0.847	0.744	0.792
TruthFinder	0.933	0.756	0.835	0.868	0.716	0.785
LTM	0.864	0.879	0.872	0.879	0.783	0.826
MBM	0.824	**0.931**	0.874	0.849	0.780	0.827
MTD_VCI	0.900	0.865	0.882	0.878	**0.801**	**0.838**
MTD_VCI_et	0.902	0.867	**0.884**	**0.880**	0.800	**0.838**

Experimental results indicate that the book dataset outperforms the movie dataset, possibly due to the presence of more unreliable sources in the movie dataset. Moreover, the performance of MTD_VCI_et on the movie dataset is similar to that of MTD_VCI, and the improvement is not as obvious as that of the book dataset. This may be because there are fewer statements containing uncertain words in the movie dataset, which has little impact on the results.

Due to the large number of objects and values in the datasets, it is hard to display the confidence intervals corresponding to the values of all objects in the paper. So we randomly selected some objects and one truth for each selected object to display the corresponding confidence interval. The results are shown in Fig. 1(a) and (b) for the book and movie datasets, respectively. The horizontal axis represents the object index, the vertical axis represents the endpoints of the confidence interval, the shaded area represents the width of the confidence interval, and the point represents the credibility score of the estimated truth.

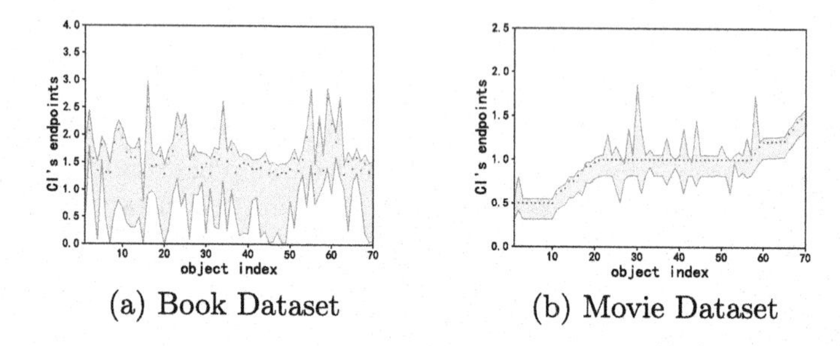

(a) Book Dataset (b) Movie Dataset

Fig. 1. The confidence interval of values on real-world datasets.

From the two figures, we can find that in most cases, the confidence interval provided by MTD_VCI contains the credibility score of values, but some are not included in the confidence interval. This may be because for these values, the amount of sources which claims the value is a small number. The credibility score of the value provided by these sources differs greatly from the real situation. In this particular case, other truth discovery methods also fail to correctly identify the truth. Besides, the width of confidence interval in movie dataset is smaller than book dataset, the possible reason is that data in book dataset is unbalanced, which leads to the results uncertain.

6.3 The Results on Simulated Dataset

Datasets Generation. The method of generating simulated dataset proposed in [14] is only applicable to single-truth scenarios, so we propose a new generation method for multi-truth scenarios on this basis.

First, we give a collection of the number of objects claimed by the source N. Next, for each $n_i \in N$, there are o_i objects receive n_i claims, where $o_i = e^7 n_i^{-1.5}$. This function ensures that the generated data has a long-tail phenomenon. Therefore, there are $\sum_i o_i$ objects and $\max\{n_i\}$ sources in simulated datasets. Considering the special case of multi-truth, for each source, we randomly generate two groups of errors of the source claim on the number of values and the vector of values respectively. We assume that these two groups of errors obey the same distribution function, i.e. $\sigma_s^{n2}, \sigma_s^{v2} \sim F$, where F is a predefined distribution function. Therefore, we use σ_s^2 to simplify the description of the two errors. Then, according to the two sets of errors, the source statements are generated from $Normal(0, \sigma_s^2)$, where the larger the σ_s^2, the lower the reliability of the source.

Experiment. We set the number of claims between 70 and 100. So we generate 31 objects and 100 sources. In order to reduce the randomness of the experiment, we repeat the experiment 100 times and take the average results. We simulated 4 situations of distribution of sources' reliability.

- $\sigma_s^2 \sim Uniform(0, 1)$. In this situation, the reliability of the sources are evenly distributed between 0 and 1.
- $\sigma_s^2 \sim Gamma(1, 3)$. In this situation, most of the sources are reliable, and only a small number of sources have low reliability.
- $\sigma_s^2 \sim Beta(1, 0.5)$. In this situation, the reliability of the source is higher than other situations. The value is distributed in (0,1).
- $\sigma_s^2 \sim FoldedNormal(1, 2)$. The distribution is a long-tailed distribution, in which situation there will be a few unreliable sources.

Table 3. Comparison of different algorithms on simulated datasets: the best performance values are in bold.

Method	Uniform(0,1)			Gamma(1,3)			Beta(1,0.5)			FoldedNormal(1,2)		
	Precision	*Recall*	*F1-score*	*Precision*	*Recall*	*F1-score*	*Precision*	*Recall*	*F1-score*	*Precision*	*Recall*	*F1-score*
Majority Voting	0.839	0.757	0.795	0.856	0.783	0.818	0.890	0.871	0.880	0.901	0.829	0.863
AverageLog	0.852	0.814	0.833	0.903	0.802	0.850	0.892	0.867	0.879	0.883	0.841	0.861
TruthFinder	0.843	0.783	0.812	0.875	0.798	0.835	0.901	0.852	0.876	0.899	0.847	0.872
LTM	0.793	**0.874**	0.832	0.873	0.849	0.861	0.896	0.874	0.885	0.900	0.894	0.897
MBM	**0.881**	0.825	0.852	**0.912**	0.851	0.880	0.906	0.885	0.895	0.898	**0.902**	**0.900**
MTD_VCI	0.872	0.853	**0.862**	0.897	**0.893**	**0.895**	**0.916**	**0.891**	**0.903**	**0.907**	0.893	**0.900**

Results Analysis. MTD_VCI and baseline methods were evaluated on four simulated datasets, and the results are shown in Table 3. The results demonstrate that MTD_VCI outperforms the other baseline methods. In addition, the experimental results obtained in the third and fourth situations are better than those in other situations, possibly due to higher source reliability, leading to better results. In other words, the more reliable sources, the better results.

We also randomly selected a value for each object and compared the confidence intervals to a theoretical normal distribution. Results for four different situations are presented in Fig. 2, 3, 4 and 5. The figures indicate that the width of the confidence intervals estimated by our method is smaller for all simulated datasets. A smaller confidence interval indicates a more reliable estimated truth. Therefore, our proposed method is effective in determining the confidence interval of the credibility score and provides a basis for judging the truth.

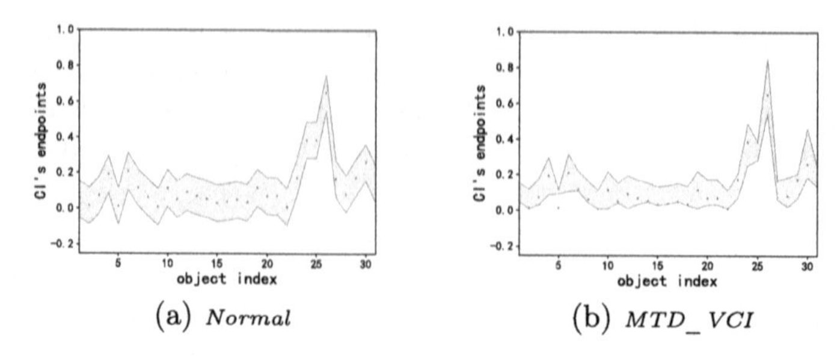

(a) *Normal* (b) *MTD_ VCI*

Fig. 2. The confidence interval of values in the first situation:$Uniform(0, 1)$.

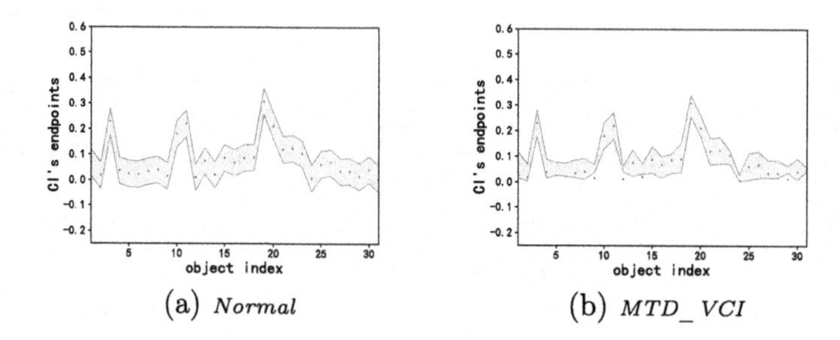

(a) *Normal* (b) *MTD_ VCI*

Fig. 3. The confidence interval of values in the second situation:$Gamma(1, 3)$.

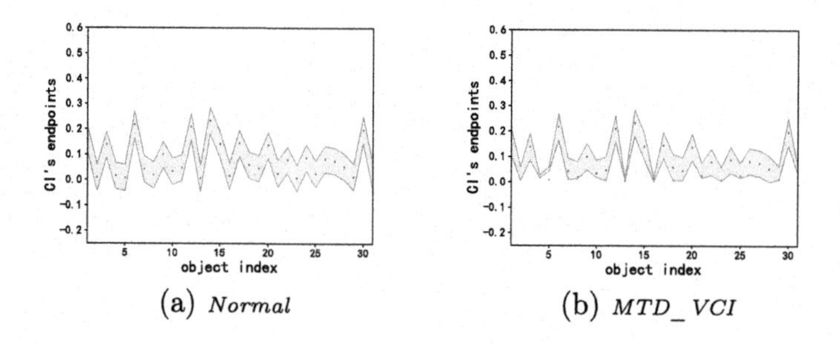

(a) *Normal* (b) *MTD_ VCI*

Fig. 4. The confidence interval of values in the third situation:$Beta(1, 0.5)$.

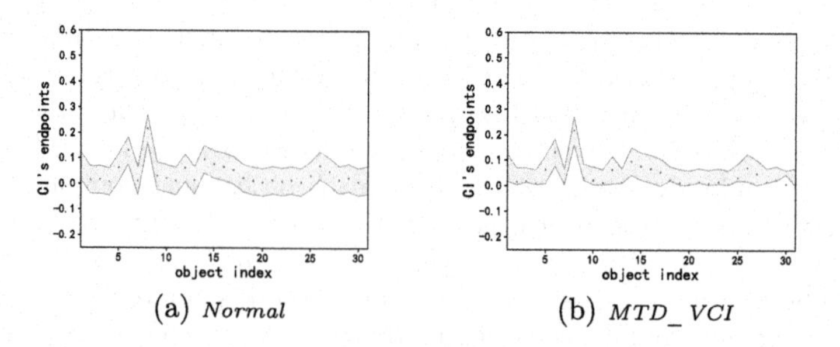

(a) *Normal* (b) *MTD_ VCI*

Fig. 5. The confidence interval of values in the fourth situation:$FoldedNorm(1, 2)$.

Additionally, we observed that in most cases, the width of the confidence interval in situations two to four was smaller than in the uniform distribution situation. This was mainly because that the sources in the last three situations were more reliable, leading to less uncertainty in the estimated truth values.

6.4 Influence of Ambiguous Words

We selected the book dataset, which contains a considerable number of ambiguous words, and varied the proportion of ambiguous words in the dataset to investigate their impact. Figure 6 illustrates the effect of different proportions of ambiguous words on the results, where the horizontal axis indicates the proportion of ambiguous words and the vertical axis represents the F1-score.

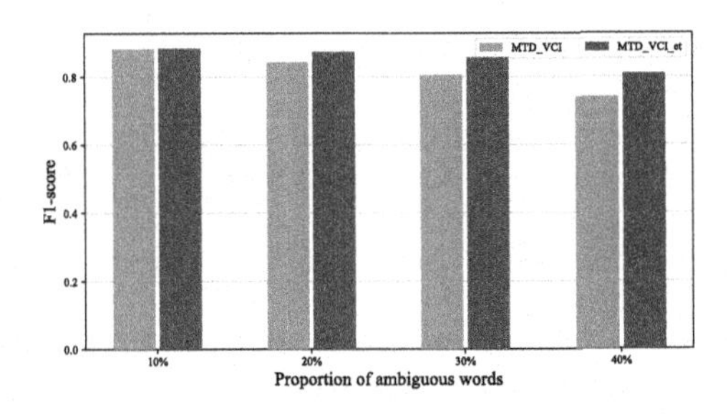

Fig. 6. The influence of different proportion of ambiguous words.

Figure 6 shows that when there are fewer ambiguous words, the results of two methods are relatively close. When the number of ambiguous words increases, the result of MTD_VCI_et become more accurate compared to MTD_VCI. This is because MTD_VCI_et takes into account the influence of ambiguous words by re-estimating the number of values provided by sources that use uncertain words. But when the number of ambiguous words continues to increase, both MTD_VCI and MTD_VCI_et produce less accurate results due to the reduced number of determined values, leading to fewer references available for estimation.

7 Conclusion

In this paper, we consider the problem of unbalanced data distribution in multi-truth discovery and also the negative impact of sources with ambiguous words. We propose a confidence-aware optimization model MTD_VCI. For sources with ambiguous words, we calculate the expected values based on the number of values provided by other sources to re-estimate them. Then we estimate the truth of the number of values and the vector of values of each object in multi-truth scenario based on the improved CATD method. Next, we construct confidence intervals for the number of values and the vector of values by bootstrap sampling methods to reflect the unbalanced distribution. Experimental results demonstrate that MTD_VCI can effectively improve the results of truth discovery, and provide an effective confidence interval.

Acknowledgements. This work was supported by Fundamental Research Funds for the Central Universities (No. 23D111204, 22D111210), Shanghai Science and Technology Commission (No. 22YF1401100), and National Science Fund for Young Scholars (No. 62202095).

References

1. Azzalini, F., Piantella, D., Rabosio, E., Tanca, L.: Enhancing domain-aware multi-truth data fusion using copy-based source authority and value similarity. VLDB **32**, 1–26 (2022)
2. Fang, X.S., Sheng, Q.Z., Wang, X., Chu, D., Ngu, A.H.: SmartVote: a full-fledged graph-based model for multi-valued truth discovery. WWW **22**(4), 1855–1885 (2019)
3. Li, Q., et al.: A confidence-aware approach for truth discovery on long-tail data. VLDB **8**(4), 425–436 (2014)
4. Li, Y., et al.: A survey on truth discovery. SIGKDD **17**(2), 1–16 (2016)
5. Lin, X., Chen, L.: Domain-aware multi-truth discovery from conflicting sources. VLDB **11**(5), 635–647 (2018)
6. Lyu, S., Ouyang, W., Shen, H., Cheng, X.: Truth discovery by claim and source embedding. In: CIKM, pp. 2183–2186 (2017)
7. Pasternack, J., Roth, D.: Knowing what to believe (when you already know something). In: COLING, pp. 877–885 (2010)
8. Pochampally, R., Das Sarma, A., Dong, X.L., Meliou, A., Srivastava, D.: Fusing data with correlations. In: SIGMOD, pp. 433–444 (2014)
9. Shao, H., et al.: Truth discovery with multi-modal data in social sensing. TC **70**(9), 1325–1337 (2020)
10. Wang, X., Sheng, Q.Z., Fang, X.S., Yao, L., Xu, X., Li, X.: An integrated Bayesian approach for effective multi-truth discovery. In: CIKM, pp. 493–502 (2015)
11. Wang, X., et al.: Empowering truth discovery with multi-truth prediction. In: CIKM, pp. 881–890 (2016)
12. Wang, X., et al.: Truth discovery via exploiting implications from multi-source data. In: CIKM, pp. 861–870 (2016)
13. Xiao, H., et al.: Towards confidence in the truth: a bootstrapping based truth discovery approach. In: SIGKDD, pp. 1935–1944 (2016)
14. Xiao, H., et al.: Towards confidence interval estimation in truth discovery. TKDE **31**(3), 575–588 (2018)
15. Yan, L., Yang, K., Yang, S.: Reputation-based truth discovery with long-term quality of source in internet of things. IOT **9**(7), 5410–5421 (2021)
16. Yang, J., Tay, W.P.: An unsupervised Bayesian neural network for truth discovery in social networks. TKDE **34**(11), 5182–5195 (2022)
17. Ye, C., et al.: Constrained truth discovery. TKDE **34**(1), 205–218 (2020)
18. Yin, X., Han, J., Yu, P.S.: Truth discovery with multiple conflicting information providers on the web. In: SIGKDD, pp. 1048–1052 (2007)

19. Zhao, B., Rubinstein, B.I., Gemmell, J., Han, J.: A Bayesian approach to discovering truth from conflicting sources for data integration. VLDB **5**(6), 550–561 (2012)
20. Zhi, S., Yang, F., Zhu, Z., Li, Q., Wang, Z., Han, J.: Dynamic truth discovery on numerical data. In: ICDM, pp. 817–826 (2018)

Development and Application of Flight Parameter Data Analysis Based on Multi-text Timescale Alignment

Pengbo Li(✉), Liang Zhang, Chuhan Cai, and Kuang Li

AVIC Shenyang Aircraft Design & Research Institute, Shenyang, China
lpb1120@163.com

Abstract. Flight parameter data plays a crucial role in comprehensively capturing the operational characteristics of various aircraft components and recording essential flight status information. This data holds significant value and finds extensive application in the domains of aircraft repair, maintenance, support, troubleshooting, and system optimization. However, due to the vast number of parameters present in flight record bus data and the generation of massive datasets at the gigabyte level, direct extraction becomes challenging. Therefore, it is necessary to filter the flight parameter data to enable efficient utilization for targeted problem analysis. This paper proposes an auxiliary method for analyzing flight parameter data using multi-text clustering and multi-timescale alignment. Additionally, an automated tool is designed and developed to address the complexities of synchronizing and concatenating parameters across multiple data buses, facilitating the effective extraction of flight parameter data and unlocking its true value.

Keywords: Flight Parameter Data · Multi-text Clustering · Multi-timescale Alignment · Data Cropping

1 Introduction

With the rapid advancements in modern science and technology, information technology, including artificial intelligence, network communication, and data mining, has made significant strides. The aviation industry has been deeply influenced by the processes of informatization, digitalization, and automation, permeating every aspect of its operations. In the realm of flight training, the objective evaluation method centered around flight parameter monitoring and analysis has gained wide-spread recognition and research attention within the aviation community [1]. Flight parameter data encompasses a comprehensive range of flight status information, including aircraft position parameters, aircraft attitude parameters, engine status parameters, aircraft control parameters, and the deflection angle of aircraft control surfaces, among others. It also captures critical details concerning the working status of the control system, engine, and other subsystems and equipment onboard the aircraft. This wealth of flight parameter data proves invaluable in providing reliable and objective support to technical and maintenance personnel engaged in tasks such as system repairs, troubleshooting, and optimization [2].

X. Yang et al. (Eds.): ADMA 2023, LNAI 14180, pp. 616–622, 2023.
https://doi.org/10.1007/978-3-031-46677-9_42

The flight record bus contains a vast amount of detailed data, comprising numerous parameters and recording cycles that vary across different systems [3]. In some cases, the event-driven recording frequencies can reach as high as tens of Hz, resulting in a massive volume of data at the GB level. Analyzing such raw data directly would be time-consuming and inefficient. Moreover, technical personnel do not need to analyze the entire dataset as their focus is often on specific problem areas. Therefore, it becomes essential to extract and filter the relevant flight parameter data required for addressing specific issues. This approach enables accurate data integration, reduces the burden of manual analysis, and enhances the efficiency and quality of problem analysis. To address these challenges, this paper proposes a method for analyzing flight parameter data using multi-text clustering and multi-timescale alignment. Simultaneously, a specialized tool is designed and developed to support this methodology. The paper is structured as follows: Sect. 2 provides a detailed description of the multi-text timescale alignment method. Section 3 focuses on the development and application of the proposed approach. Finally, Sect. 4 presents the concluding remarks of this study.

2 Method of Multi-text Timescale Alignment

2.1 Multi-text Clustering

Multi-text clustering is an effective approach for organizing, summarizing, and navigating textual information. Traditional text clustering algorithms are classified into four categories: partition-based, hierarchy-based, density-based, and model-based. However, these algorithms often have limitations, such as the requirement of predefined cluster trees and lack of self-organization. To overcome these shortcomings, this paper employs an ant colony text clustering algorithm, which enables self-organization [4].

The algorithm begins by constructing a two-dimensional spatial grid with dimensions $Z \times Z$, where Z represents the grid width. The value of Z is determined by the number of texts, denoted as $\sqrt{N}$. Next, the algorithm initializes parameters such as the number of texts N and the number of ants K. The texts are then randomly distributed within the grid, with each grid cell containing at most one text. Subsequently, a certain number of ants are randomly positioned within the grid. Each ant autonomously moves and selects a text, evaluating the probability of 'picking up', 'putting down' or further processing the text based on similarity calculations with all texts in the local area. Through multiple iterations of this process, the texts in the grid are clustered based on their similarity [5].

Similarity $f(d_i)$ between text and local regions is also known as the group similarity function. It can be represented by:

$$f(d_i) = \frac{1}{s^2} \sum\nolimits_{d_j \in Neighs \times s(r)} \left| 1 - \frac{1 - \mathrm{sim}\left(d_i, d_j\right)}{\alpha} \right| \tag{1}$$

The value of α is in the range of 1 to 10. When α approaches 1, the resulting value of $f(d_i)$ will be smaller, leading the ant to subdivide similar texts and breaking down the original aggregated large class into several smaller classes. Consequently, this increases the number of generated clusters and slows down the convergence speed of the algorithm. Conversely, when α tends towards 10, the value of $f(d_i)$ will be larger, causing the ant to aggregate dissimilar texts into a single cluster. This reduces the number of clusters generated and accelerates the convergence speed of the algorithm. In the equation, $s \times s$ denotes the area of unit r, which directly influences the movement range of the ants. The distance $\mathrm{sim}(d_i, d_j)$ between text d_i and text d_j is typically calculated using cosine distance.

The probability conversion function is determined by $f(d_i)$. The conversion rule of the probability $P_p(d_i)$ of the text d_i picked up by ant is as follows: when the current ant is in an empty state, it decides whether to pick up the text based on $f(d_i)$. The conversion rule of the probability $P_p(d_i)$ of the text d_i picked up by ant is as follows: when the current ant is in a loaded state, the ant decides whether to pick up the text d_i based on $f(d_i)$. $P_p(d_i)$ and $P_d(d_i)$ can be represented by:

$$P_p(d_i) = \left(\frac{k_1}{k_1 + f(d_i)}\right)^2 \tag{2}$$

$$P_d(d_i) = \begin{cases} 2f(d_i), \; if\,(f(d_i) < k_2) \\ 1, \quad otherwise \end{cases} \tag{3}$$

where k_1 and k_2 represent the threshold constants, which are typically determined based on empirical values. In the case of $P_p(d_i)$, as $f(d_i)$ approaches k_1, the value tends towards 1 indicating the highest probability of picking up the text. Conversely, when (d_i) is significantly far greater than k_1, $P_p(d_i)$ tends to zero, resulting in the lowest probability of picking up the text.

2.2 Multi-timescale Alignment

The recording system for flight parameters has found extensive use in military applications, scientific and technological research, aviation and space technology, air passenger transportation, communication satellites, and more [6]. As aircraft technology continues to advance, the complexity of flight parameter recording systems has increased, necessitating the recording of more parameters and achieving greater time scale accuracy. These systems collect various parameter indicators from different aircraft systems, encompassing multiple types of interfaces and signals such as analog signals, discrete signals, and bus signals [7]. Since the data from these interfaces lack a unified time scale, it is crucial to establish a consistent relative time reference for conducting meaningful aircraft state analysis [8]. Ensuring the accuracy of the relative time scale between different flight parameters is of utmost importance. By adopting optimized time scales, the aircraft's relevant flight instructions can be significantly improved, thereby assisting technicians in accomplishing their flight-related tasks and facilitating scientific research projects. This paper also incorporates multi-timescale alignment processing for flight parameter data, with the specific algorithm outlined in Fig. 1.

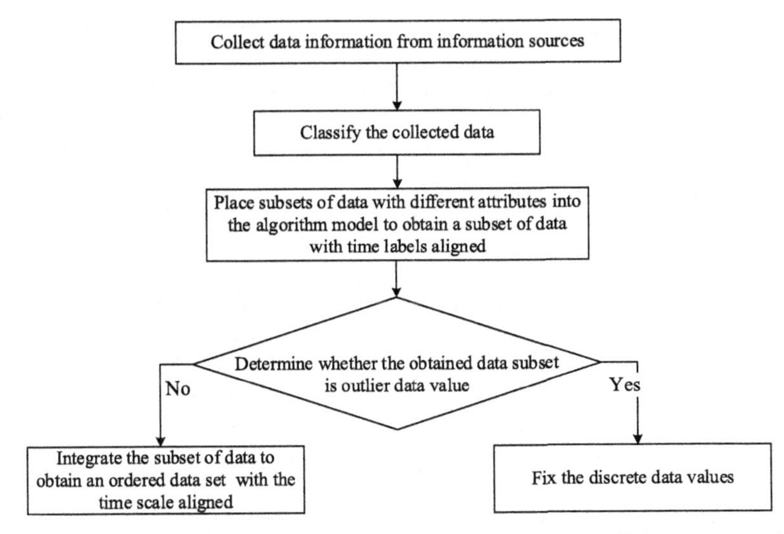

Fig. 1. Flow chart of the multi-timescale alignment algorithm for flight parameter data.

3 Development and Application

3.1 Tool Implementation

Utilizing the ant colony multi-text clustering algorithm and multi-timescale alignment algorithm as the foundation, an automated processing tool is developed for extracting flight parameter data. This tool enables the loading of multiple parsing texts in '.csv' format, which contain bus data into the memory. Through the implementation of the multi-text clustering and multi-timescale alignment techniques, a parameter list is extracted, allowing users to select their desired parameters. Selected parameters are retained, while irrelevant parameter sets are discarded. Additionally, the tool offers data item cropping functionality, allowing for the streamlined processing of data in seconds or minutes, depending on the specific work objectives. Once the data is organized and stored in memory according to the user's requirements, it can be further structured and stored in a new text file. This new file contains intuitive, clear, and lightweight key parameter data, facilitating quick completion of data preparation tasks. The efficiency of technical personnel in repair, maintenance, support, troubleshooting, and system optimization design is significantly enhanced as a result. Furthermore, the tool resolves the challenges associated with manually synchronizing and concatenating parameters from multiple data buses. The working principle of the tool is shown in Fig. 2.

3.2 Tool Application

This tool has been developed through a combination of practical design analysis and field support, resulting in successful applications in flight parameter data analysis, coolant over-temperature investigation, and other related tasks. Specifically, in the context of investigating coolant over-temperature problems, the tool plays a crucial role. The initial

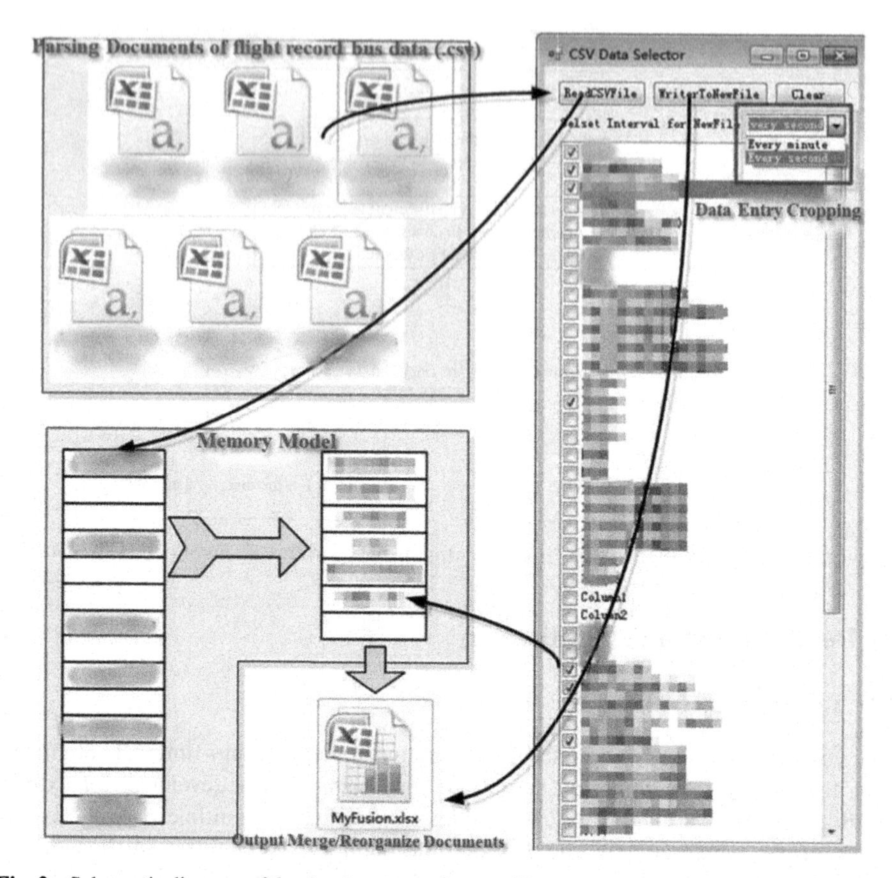

Fig. 2. Schematic diagram of the development of an auxiliary analysis tool for multi-text timescale alignment of flight parameter data.

step involves utilizing the tool to read the parsing text set, formatted as '.csv' files, which contain flight record bus data, as illustrated in Fig. 3. Each text within the set has different parameter types and associated time information. When the texts are read, the tool processes the text set using the multi-text clustering and multi-timescale alignment technique, thereby generating a comprehensive dataset where all parameter types are sorted chronologically. Simultaneously, the tool presents a display of all the parameter types within the dataset, allowing users to filter and select the relevant data of interest. Finally, the tool outputs a new text file containing the parameter types sorted by time, precisely as required by the users, as exemplified in Fig. 4. This clearly exhibits a two-dimensional distribution, showcasing the time dimension and the parameter type dimension that users are specifically concerned with. Consequently, the processed data can be directly employed for the analysis of coolant over-temperature issues and their corresponding problems.

Following extensive practical implementation, the tool developed in this research paper demonstrates reliable performance on the Windows platform when coupled with

Fig. 3. Diagram of the application scenario for troubleshooting coolant over-temperature issues (pre-processing flight parameter data).

	A	C	D	E	F	G	H	I	J	K	R	S	T	U	V	W
1	Time															
2	16:32:22	0	6500	3925	61	70	45	39	0	0	68.44	70.19	1.75	0	67.25	66.56
3	16:32:23	0	6500	3925	61	70	45	38	0	0	68.38	70.06	1.688	0	67.313	66.56
4	16:32:24	0	6500	3925	61	70	45	39	0	0	68.44	70.06	1.625	0	67.25	66.43
5	16:32:25	0	6500	3925	61	70	45	39	0	0	68.38	70.13	1.75	0	67.25	66.43
6	16:32:26	0	6500	3925	61	70	45	38	0	0	68.31	69.94	1.625	0	67.25	66.43
7	16:32:27	0	6500	3925	61	70	45	38	0	0	68.44	70.13	1.688	0	67.125	66.43
8	16:32:28	0	6500	3925	61	70	45	38	0	0	68.38	70.06	1.688	0	67.125	66.43
9	16:32:29	0	6500	3925	61	69	45	38	0	0	68.31	69.94	1.625	0	67.25	66.43
10	16:32:30	0	6500	3925	61	70	45	38	0	0	68.44	70.06	1.625	0	67.125	66.43
11	16:32:31	0	6500	3925	61	70	45	38	0	0	68.31	69.94	1.625	0	67.25	66.43
12	16:32:32	0	6500	3925	61	69	45	38	0	0	68.31	69.94	1.625	0	67.25	66.43
13	16:32:33	0	6500	3925	61	69	45	39	0	0	68.31	70.06	1.75	0	67.125	66.37
14	16:32:34	0	6500	3925	61	69	45	39	0	0	68.19	69.88	1.688	0	67.125	66.37
15	16:32:35	0	6500	3925	61	69	45	39	0	0	68.31	69.94	1.625	0	67.125	66.37
16	16:32:36	0	6500	3925	61	69	45	39	0	0	68.19	69.94	1.75	0	67.125	66.37
17	16:32:37	0	6500	3925	61	69	45	39	0	0	68.19	69.94	1.75	0	67.125	66.37

Fig. 4. Diagram of the application scenario for troubleshooting coolant over-temperature issues (processed flight parameter data).

the Excel software. Notably, the tool exhibits remarkable versatility and broad applicability, efficiently and accurately organizing flight data based on user requirements. Its efficacy remains consistent in both single-system parameter analysis scenarios and multi-system parameter coupling analysis scenarios. With the aid of the tool, more than 100 sorties' flight data conversion was completed per person per day, when it needs about 50 people per day previously. The utilization of this tool brings about notable improvements in operational efficiency. The tool's capability to crop data records with varying time steps not only expedites the data preparation process but also eliminates redundant data, resulting in a lightweight data output. As a result, users can dedicate their focus to interpretation and analysis, significantly enhancing efficiency when addressing specific issues. One of the notable advantages of this tool is its ability to save users between

90% and 99% of the time typically required for data organization. Moreover, it is compatible with various data file formats featuring parameter column names and recorded in time slices. With its wide applicability, high operational efficiency, practicality, and substantial value in various applications, this tool holds significant promise.

4 Conclusion

In this paper, an auxiliary flight parameter data analysis tool is designed and developed, encompassing the essential functionalities of multi-text clustering, multi-timescale alignment, and data cropping. It can efficiently filter crucial parameters dispersed across different buses and data packets, consolidate multiple data texts from the same flight record, synchronize time scales, and crop records at intervals of either one second or one minute. These features facilitate both high-level and detailed analyses of flight parameters. By incorporating this tool, analysts gain access to enhanced flexibility, efficiency, and cost-effectiveness. It expands the scope of parameter correlation and facilitates in-depth exploration of independent parameter sensitivity. Importantly, it overcomes the limitations of existing bus data analysis tools by providing comprehensive subsequent data processing capabilities. As a result, it serves as a valuable auxiliary tool for analyzing flight bus recorded data while streamlining data organization and preparation processes. Ultimately, the tool enables the efficient extraction of flight parameter data, facilitating the optimal utilization of flight recorded bus data and unlocking its true value.

Acknowledgement. This paper is funded by the 1912 project of the pre research team of AVIC Shenyang Aircraft Design and Research Institute.

References

1. Li, Y., Zhang, D., Zhu, L.: Application and development prospects of flight parameter data. Metrol. Test. Technol. **36**(1), 10–11 (2009)
2. Chaojiang, H., et al.: Aircraft Flight Reference System and its Application, pp. 21–98. National Defense Industry Press, Beijing (2012)
3. Wang, Z., Zhao, Z., Liu, K., et al.: Application of flight parameter data in aircraft usage and maintenance. China Sci. Technol. Inf. **17**, 30–31 (2021)
4. Zhang, H.: Research on Text Clustering Based on Text Dimension Reduction and Ant Colony Algorithm, pp. 21–23. Anhui University, Anhui (2015)
5. Fang, Z.: Research and Application of Ant Colony Text Clustering Algorithm. Xi'an University of Electronic Science and Technology, Xi'an (2013)
6. Jin, M., Hehai, S.: Application of flight parameter data in fault analysis. Sci. Technol. Innov. **23**, 56–57 (2019)
7. Li, H., Zhang, X.: Flight parameter calculation method of multi-projectiles using temporal and spatial information constraint. Defence Technol. **19**, 63–75 (2023)
8. Zhang, L., Tang, Z., Lei, T.: Time scale design and application of flight parameter recording system. Mil. Autom. **32**(03), 60–62 (2013)

LFD-CD: Peripheral Blood Cells Detection Using a Lightweight Cell Detection Model with Full-Connection and Dropconnect

Mingshi Li[1,5], Shuyao You[1], Wanli Liu[1], Hongzan Sun[2], Yuexi Wang[3], Marcin Grzegorzek[4], and Chen Li[1(✉)]

[1] College of Medicine and Biological Information Engineering, Microscopic Image and Medical Image Analysis Group,, Northeastern University, Shenyang, China
lichen@bmie.neu.edu.cn
[2] Shengjing Hospital, China Medical University, Shenyang, China
[3] Fourth Hospital, China Medical University, Shenyang, China
[4] Institute of Medical Informatics, University of Luebeck, Luebeck, Germany
[5] Department of Knowledge Engineering, University of Economics, Katowice, Poland

Abstract. Blood testing is an important basis for the diagnosis of many diseases. However, the accuracy of detection instruments in hospitals is often not high, and manual detection of blood samples is sometimes necessary. The whole detection process is not only time-consuming, but also limited in terms of the types of detection. The addition of Deep Convolution Neural Network (DCNN) can quickly and accurately detect blood cells. This study designs a *Lightweight Cell Detection Model with Full-connection and Dropconnect* (LFD-CD) to achieve the detection task of the peripheral blood cell dataset. In the detection of 8 categories, the model is optimized by adding the full connection layer and Dropconnect model, and the accuracy of the LFD-CD results achieves 99.3%. In the comparative experiment, LFD-CD demonstrates higher detection accuracy and faster detection speed than other independent DCNN models. Moreover, the space required for LFD-CD is only 395 MB, which is half the size of the YOLO-v7 model used in the comparison experiment.

Keywords: Peripheral Blood cell · Object detection · Lightweight model · Deep Convolution Neural Network

1 Introduction

Blood testing is an important judgment basis for the diagnosis of many diseases. Clinical blood tests can provide a count of blood cells, feedback on certain parameters, and enable observation of the species and status of the cells. The assay of blood can help to speculate on many human health conditions, such as liver blood test [18], early lung cancer detection [17], and pulmonary tuberculosis in young children [3].

X. Yang et al. (Eds.): ADMA 2023, LNAI 14180, pp. 623–633, 2023.
https://doi.org/10.1007/978-3-031-46677-9_43

At present, tests in various large hospitals mainly apply blood cell analyzer and flow cytometer to realize automated blood cell detection. However, the accuracy of the hematology analyzer is only around 80% [11]. In case of abnormalities, blood samples are manually examined under a microscope. The detection of these two kinds of cell detection instruments takes longer, the price of the instrument is also quite expensive, and the number of detecting cell species is less [12]. Manual microscopic examination, although highly accurate, is very time-consuming and not suitable for extensive work.

Nowadays, with the rapid development of computer science, machine learning technology has significantly improved the auxiliary diagnosis ability of the medical industry. Many applications such as face recognition and cell detection can be achieved using machine learning. Additionally, incorporating lightweight techniques can greatly reduce the resource usage of deep learning models, especially the parameters in neural networks [9]. It can help effectively implement model compression as well as reduce inference time [4].

In this study, *Lightweight Cell Detection Model with Full-connection and Dropconnect* (LFD-CD) are builded. The results obtained from training this model on the peripheral blood cell dataset are applied to the cell detection framework. The final test accuracy is 97.7%.

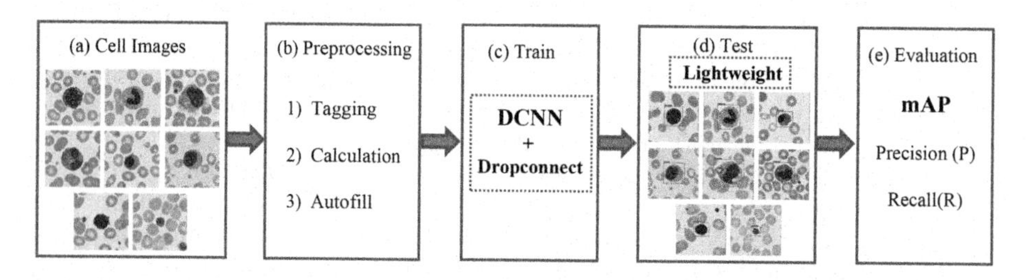

Fig. 1. Workflow of the proposed LFD-CD method.

In Fig. 1, the cells demonstrated in (a) are from the training set and are selected from a peripheral blood cell dataset comprising a total of eight classes of cells [1]. (b) For preprocessing of data. First, the images need to be tagged by adding labeling classification. The size of the images needs to be calculated before training, followed by auto-fill processing. (c) This stage is the training part, where the preprocessed data input is trained and generalized by the training model. Shown in (d) is a plot of assay results for the eight classes of cells and sets this module to light quantification. In (e), the mean Average Precision(mAP), precision, and recall rate are calculated as the performance of the object detection model.

The main contributions of this study are as follows: 1) the design of a lightweight Deep Convolutional Neural Networks (DCNN) model for detecting eight classes of peripheral blood cells. 2) The designed training network is generalized by adding Dropconnect. The detection network is lightweight, which is the innovation of this study. The processed network has the advantages of a small model and a short reasoning time. The average accuracy of the results also reaches 97.7%.

2 Related Work

2.1 Progress in Blood Cell Detection

During a time when early science and technology were underdeveloped, computers were unable to distinguish objects from complex images. For blood cell images that are prone to overlap, and where morphological feature distinctions are not obvious, one could only rely on professionally trained physicians for human identification classification. But this method is susceptible to subjective awareness and is not efficient due to a large number of cells, which can lead to low accuracy rates and unsatisfactory work efficiency. In [20], this study proposes a method for localizing leukocytes in microscopic blood samples and partitioning them into nuclear and cytoplasmic regions. The study concludes that the average correction rate of the linear classifier is higher than that of the Bayesian classifier. Both offer high precision and are powerful and efficient. Another study proposes a method for differentiating white blood cells (WBC) from red blood cells (RBC) using H and S components. This work is applied to a Wright Giemsa stain image dataset in [26]. In [16], a method of counting blood cells using the nearest neighbor and support vector machines (SVM) is proposed, which involves manually cropping each cell. The final mean accuracy of RBC is lower than that of WBC in the Acute Lymphoblastic Leukaemia image database (ALL-IDB). Cruz et al. In [6], it proposes a basin analysis method for RBC counting, which involves spot analysis and color phase saturation (HSV). In [22], this study first proposes a deep learning semantic segmentation method to segment RBC and WBC using gram Schmidt orthogonalization to enhance the color of the nucleus. The study then utilizes morphological enhancement to locate the nucleus.

In recent years, there has been a continuous development of computer technology, and more models and technologies are being applied in medical detection. An assay is proposed to identify and count blood cells using the YOLO algorithm, which is trained using a modified Blood Cell Count Dataset (BCCD) of blood smear images, to automatically identify and count RBC, WBC, and platelets [2]. In [21], this study proposes a region-based faster convolutional neural network (fast R-CNN) focusing not only on RBC but also on variants of leukocytes. The aim of this study is to have a fast and reliable system that can help the field of medicine classify red and white blood cells. In [19], the method proposes a simple circular representation for medical object detection and introduces the anchorless detection framework circlet. The experiment uses whole slide images captured from kidney biopsies and annotated as experimental data.

2.2 Related Methods of Object Detection

For the object detection in this experiment, the feature selection method is the selection of variables or attributes in the data, which is the process of choosing a subset of unique functions (variables, prediction indicators) to be used in building machine learning and data science models. In the training process, the rectified linear unit (ReLU) function is used to calculate the activation of all convolution-extracted features [8]. ReLU is typically assigned to the output of each hidden unit in the volume accumulation layer and the fully connected layer. In the space pooling process, the DCNN structure used in the experiment uses maximum pooling. The maximum pooling operator extracts patches from the input characteristic graph, outputs the maximum value of each patch, and discards all other values. The fully connected layer connects each node with all nodes of the previous layer and integrates the extracted features. In this study, there is also a regularization method. Dropconnect is applied to the full connection layer. The regularization method discards the connection between neurons but still retains the calculated output unchanged [23]. Dropconnect can reduce over-fitting, increase the generalization ability of the model and improve the network detection performance. Finally, this study employs quantization processing, which converts the training file into Open Neural Network Exchange (ONNX) format [4], and stores the weight of the neural network structure through the Protobuf data structure.

3 Methodology

3.1 Structure of LFD-CD

To accomplish the object detection task for blood cells, this study designs a training model and detection framework, the structure of which is shown in Fig. 2.

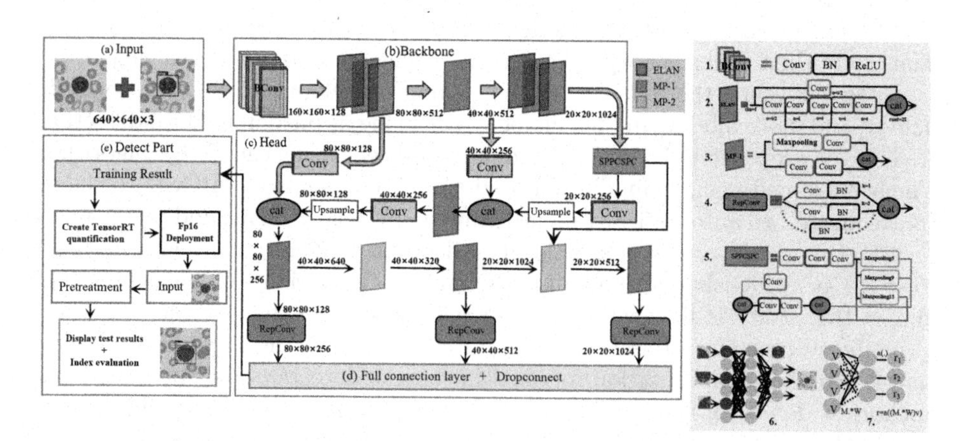

Fig. 2. The structure of the object detection model. (1: Conv, BN layer and ReLU (BConv); 2: Emulated Local Area Network (ELAN); 3: Maxpool and Conv (MP); 4: RepConv; 5: Spatial Pyramid Pooling (SPP); 6: Full connection layer; 7: Dropconnect)

As shown in Fig. 2, the object detection units in the picture are first annotated and classified. The preprocessed data is alternately halved in length and width and doubled in the channel through multiple convolution layers of the backbone layer. After feature extraction, the training results are obtained by entering the head layer. Finally, for the detection part, this study converts the trained results into a format for lightweight processing, output, and save the detection results.

3.2 Backbone and Head Parts for Object Detection

The size of the input image is 640×640 pixels. The training model is pre-trained using the cell images from the training set. The model can be grouped into three parts: Input Part for images input, Backbone Part for extracting features and Head Part for prediction. In the backbone part of Fig. 2(b), Convolution layer composed of Conv, BN layer and ReLU activation function is called BConv for short. Through the convolution layer of backbone, the length and width are reduced by half, the channel is doubled, and the feature is extracted. The feature map of the images is changed from $640 \times 640 \times 3$ to $80 \times 80 \times 128$, $40 \times 40 \times 256$ and $20 \times 20 \times 1024$. In the head layer of Fig. 2(c), the three feature maps are output through three repvgg_block (RepConv) and volume layer output three unprocessed prediction results of different sizes.

3.3 Dropconnect and Lightweight Parts for Model Optimization

A full connection layer is added at the end of the model, and a Dropconnect is added to the full connection layer. The main function of the fully connected layer is to map the feature space calculated from the previous layer to the sample label space. Its advantage is to reduce the impact of feature location on classification results and improve the robustness of the whole network. Each connection of Dropconnect is set to a probability of $1 - p = 0.5$ to be deleted, not every output unit. It weakens the joint adaptability of neuron nodes and enhances the generalization ability of training results.

The detection part in Fig. 2 shows the object detection framework flowchart. Convert from the trunk and prediction part of the training model to the Open Neural Network Exchange (ONNX) format. Then transfer the ONNX file to the Team Sports Scheduling System Report Template (TRT) file of TensorRT as the engine file, and adjust the precision from Full-recision floating-point 32 (Fp32) to Half-precision floating-point 16 (Fp16) during the serialization construction. The TRT model established is applied to the detection framework as calibration data, and the space occupied is reduced from 714 MB to 71.2 MB. 3555 cell images from the test dataset are input into the detection frame for detection. For the lightweight operation of the network, this step effectively reduce the space occupied by the storage of the object detection network and reduce the detection inference time. The performance of the object detection DCNN model is evaluated by calculating mAP, precision, recall and inference time. These are commonly used evaluation indexes, with mAP being the most important one.

4 Experiments and Analysis

4.1 Data Setting

The peripheral blood cell dataset used in this study consists of a total of 17,092 single normal cell images, which are acquired using analyzers at the core laboratory of the hospital clinic of Barcelona [1]. The dataset is composed of eight categories of cells: neutrophils (neu), eosinophils (eos), basophils (bas), lymphocytes (ly), monocytes (mon), immature granulocytes (ig), erythroblasts (ery) and platelets (pla). 17092 images of the cells in the database are used in this study and Fig. 3 are some examples. Table 1 is the situation of data allocation.

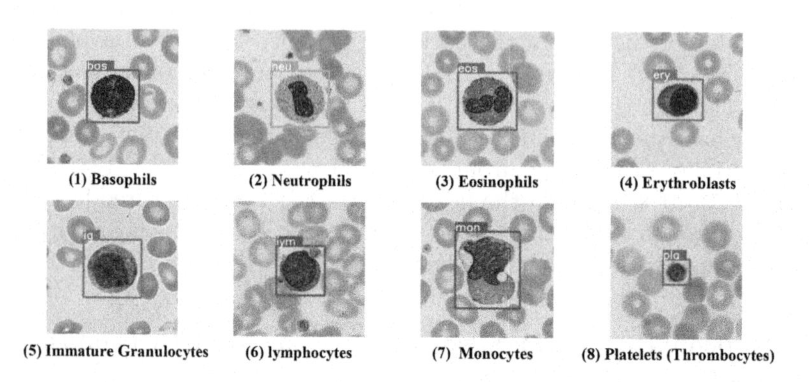

Fig. 3. Examples of peripheral blood cell datasets.

For the purpose of conducting a 5-fold cross-validation experiment, the eight categories in the dataset are divided into five parts, four of which are used as training sets and the rest as test set. Table 1 shows the data distribution.

Table 1. Table captions should be placed above the tables.

Dataset/Class	bas	neu	eos	ery	ig	ly	mon	pla	Total
Train	987	2658	2456	1225	2295	953	1114	1849	13537
Test	231	671	661	326	600	261	306	499	3555
Total	1218	3329	3117	1551	2895	1214	1420	2348	17092

4.2 Experimental Environment and Evaluation Indicators

The experimental environment uses a windows 11 system, and the local memory of the computer is 32 GB. GPU is NVIDIA GeForce RTX 3060 ti. CPU is 12th Gen Intel (R) core (TM) i5-12600kf. This study uses Pycharm and Pytorch for programming the experimental code. The trained epoch is 200 and the patch size is 5. The Dropconnect rate value added to the training full connected layer is 0.5.

The Mean Average Precision (mAP) is the average value of average precision (AP), which is the main evaluation index of object detection algorithm. The precision (P) is the true positive sample among the positive predicted samples. The recall rate (R) is the sample that is predicted to be positive among the real positive samples. True Positive (TP) represents the number of positive samples correctly detected, TN represents the number of negative samples correctly detected, False Positive (FP) represents the number of negative cases wrongly classified as positive, and False Negative (FN) represents the number of positive cases wrongly classified as negative. The value of precision is $\frac{TP}{(TP+FP)}$. The recall rate is $\frac{TP}{(TP+FN)}$ [7]. The calculation of the value of mAP is to draw the P-R curve by constructing the coordinate system of precision and recall rate, and calculate the mean value of the area values under various P-R curves [10].

4.3 Experimental Results and Analysis

To evaluate the effectiveness of LFD-CD, this experiment compares it with four DCNN models: YOLO-v5s [28], YOLO-v5l [25], YOLO-v6 [13], and YOLO-v7 [24]. The remaining experimental settings are kept the same as in the main experiment. Through the test results of the two cells listed in Fig. 4, the advantages and disadvantages of the five models can be obviously compared.

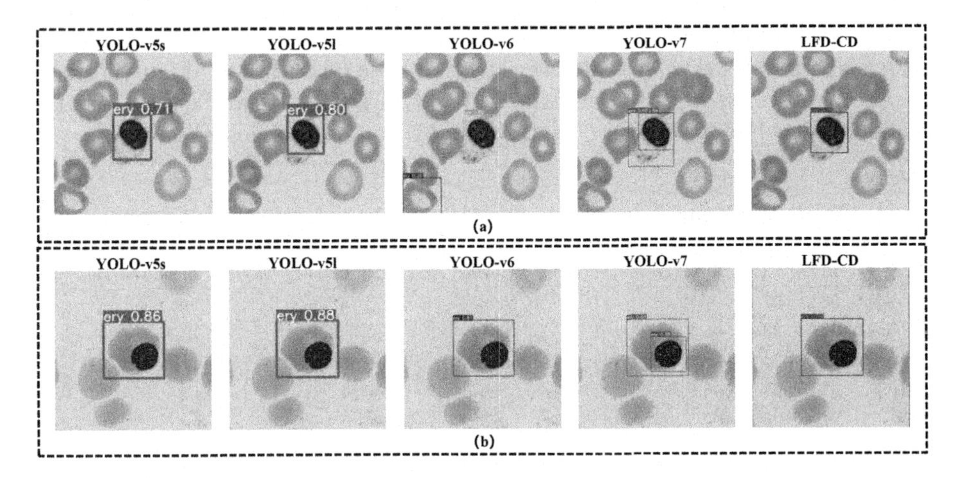

Fig. 4. LFD-CD is compared with YOLO-v5s [28], YOLO-v5l [25], YOLO-v6 [13] and YOLO-v7 [24]. The two cell images in (a) (b) are from the class of erythroblasts. In these images, the cells circled in the box represent the detected object cells. The label on the box indicates the category and confidence level of the object.

The cell image in Fig. 4(a) is fuzzy. The lowest confidence level of the YOLO-v6 model is only 66%, and there are false detection results of unrelated cells.

Compared with YOLO-v5s and YOLO-v5l models, the YOLO-v5l network with higher network width performs better at 80% confidence, 9% higher than the YOLO-v5s model. The confidence level of YOLO-v7 is 68%, but due to the small nuclear and cytoplasmic ratio of the cell [27], the network repeats the detection of the nucleus and the whole cell. LFD-CD obtains a 76% confidence level and avoids repeated detection. In Fig. 4(b), the effects of the three models of YOLO-v5s, YOLO-v5l, and YOLO-v6 have little difference. The confidence levels are 86%, 88%, and 87% respectively. However, YOLO-v7 has a repeated detection confidence of 88%. The detection results obtained by LFD-CD still have no duplicate detection and the highest confidence is 89%.

4.4 Extended Experiment

After recording and analyzing the test results, LFD-CD is compared to the other four models based on indicators. The results are shown in Table 2. It can be seen that the maximum mAP of the lightweight object detection model is 99.3%, 1.5% higher than YOLO-v5s, 1.3% higher than YOLO-v5l, and 1.1% higher than YOLO-v6. The mAP variance calculated using the value of each training in the 5-fold experiment is shown in Table 2. The mAP performance of LFD-CD in the 5-fold experiment has the smallest difference and the lowest variance, only 1.45. Therefore, it has a more stable training mode. Compared with the other four models, the memory cost of LFD-CD is only 395 MB, about half of YOLO-v7. The model optimization of the lightweight part effectively compresses LFD-CD and improves the detection speed. In addition, LFD-CD has the highest detection speed of a single cell image, with an average detection time of 9.6 ms. This is about 2 ms shorter than the detection time of YOLO-v7. Figure 5 shows the eight categories of cell detection results of LFD-CD. It can be seen that the confidence level of the model detection is very good, achieving about 95%.

Table 2. Comparison results of models on the test set of the dataset.

Models	P(%)	R(%)	mAP (%)	mAP variance	Memory Cost (MB)	Training Time (h)	Single Test Time (ms)
Yolov5s	98.6	97.8	97.8	5.21	598	40.6	9.8
Yolov5l	97.7	98.0	98.0	4.41	667	42.0	10.4
Yolov6	98.3	94.3	98.2	5.14	602	37.5	11.2
Yolov7	97.0	97.2	98.8	2.43	726	21.7	11.5
LFD-CD	97.4	98.3	99.3	1.45	395	19.7	9.6

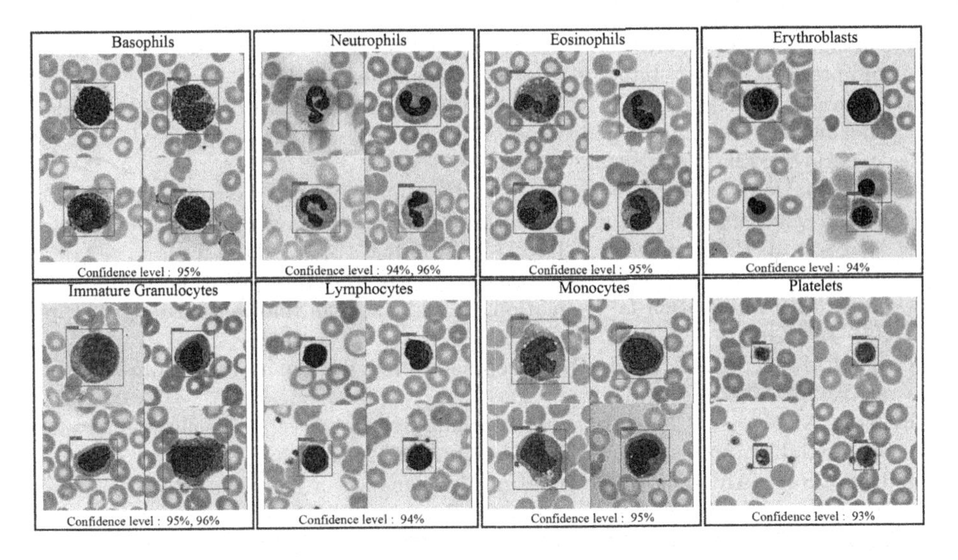

Fig. 5. Detection results of LFD-CD.

5 Conclusion and Future Work

This study designs a *Lightweight Cell Detection Model with Full-connection and Dropconnect* (LFD-CD) to perform the task of blood cell object detection. The model tests 8 categories of peripheral blood cells, and the accuracy of LFD-CD reaches 99.3%. LFD-CD has been lightened, and Dropconnect has been added to the training to enhance the generalization ability. In the comparison experiment, LFD-CD outperforms the single DCNN model in terms of accuracy, memory cost and detection speed. In future research, more models and functions may be combined to improve the training, reasoning and detection speed and reduce the model's space occupation. The experiment can also explore other methods to preprocess or normalize the data. In addition, the LFD-CD method could be used in other cell analysis work, such as cervical cells [14,15] and sperm cells [5].

Acknowledgements. This work is supported by the "National Natural Science Foundation of China" (No. 82220108007). We thank Miss Zixian Li and Mr. Guoxian Li for their important discussion. We thank B.A. Yingying Hou for her proof reading.

References

1. Acevedo, A., Merino, A., Alférez, S., et al.: A dataset of microscopic peripheral blood cell images for development of automatic recognition systems, vol. 30 (2020)
2. Alam, M.M., Islam, M.T.: Machine learning approach of automatic identification and counting of blood cells. Healthcare Technol. Lett. **6**(4), 103–108 (2019)
3. Bergamini, B.M., Losi, M., Vaienti, F., et al.: Performance of commercial blood tests for the diagnosis of latent tuberculosis infection in children and adolescents. Pediatrics **123**(3), 419–424 (2009)

4. Chang, D., Lee, J., Heo, J.: Lightweight of onnx using quantization-based model compression. J. Inst. Internet Broadcast. Commun. **21**(1), 93–98 (2021)
5. Chen, A., Li, C., Zou, S., et al.: SVIA dataset: a new dataset of microscopic videos and images for computer-aided sperm analysis. Biocybernetics Biomed. Eng. **42**, 204–214 (2022)
6. Cruz, D., Jennifer, C., Valiente, et al.: Determination of blood components (WBCS, RBCS, and platelets) count in microscopic images using image processing and analysis. In: Proceedings of ICHNITCCEM 2017, pp. 1–7 (2017)
7. Goutte, C., Gaussier, E.: A probabilistic interpretation of precision, recall and f-score, with implication for evaluation. In: Advances in Information Retrieval, pp. 345–359 (2005)
8. He, J., Li, L., Xu, J., et al.: Relu deep neural networks and linear finite elements. arXiv: 1807.03973 (2018)
9. Jiang, X., Wang, N., Xin, J., et al.: Learning lightweight super-resolution networks with weight pruning. Neural Netw. **144**, 21–32 (2021)
10. Khan, S.A., Ali Rana, Z.: Evaluating performance of software defect prediction models using area under precision-recall curve (AUC-PR). In: Proceedings of ICACS 2019. pp. 1–6 (2019)
11. Kratz, A., Lee, S.h., Zini, G., et al.: Digital morphology analyzers in hematology: ICSH review and recommendations. Int. J. Labor. Hematol. **41**(4), 437–447 (2019)
12. Lapsley, M.I., Wang, L., Huang, T.J.: On-chip flow cytometry: where is it now and where is it going? Biomark. Med. **7**(1), 75–78 (2013)
13. Li, C., Li, L., Jiang, H., et al.: Yolov6: A single-stage object detection framework for industrial applications. ArXiv abs/2209.02976 (2022)
14. Liu, W., Li, C., Rahaman, M.M., et al.: Is the aspect ratio of cells important in deep learning? A robust comparison of deep learning methods for multi-scale cytopathology cell image classification: from convolutional neural networks to visual transformers. Comput. Biol. Med. **141**, 105026 (2022)
15. Liu, W., Li, C., Xu, N., et al.: CVM-Cervix: a hybrid cervical pap-smear image Classi fication framework using CNN, visual transformer and multilayer perceptron. Pattern Recogn. **130**, 108829 (2022)
16. Loddo, A., Putzu, L., Di Ruberto, C., et al.: A computer-aided system for differential count from peripheral blood cell images, pp. 112–118 (2016)
17. Montani, F., Marzi, M.J., Dezi, F., et al.: miR-test: a blood test for lung cancer early detection. JNCI: J. Natl. Can. Inst. **107**(6), djv063 (2015)
18. Newsome, P.N., Cramb, R., Davison, S.M., et al.: Guidelines on the management of abnormal liver blood tests. Gut **67**(1), 6–19 (2018)
19. Nguyen, E.H., Yang, H., Deng, R., et al.: Circle representation for medical object detection. IEEE Trans. Med. Imaging **41**(3), 746–754 (2022)
20. Prinyakupt, J., Pluempitiwiriyawej, C.: Segmentation of white blood cells and comparison of cell morphology by linear and naïve bayes classifiers. BioMed. Eng. OnLine **14**, 63 (2015)
21. Tobias, R.R., Carlo De Jesus, L., Mital, M.E., et al.: Faster R-CNN model with momentum optimizer for RBC and WBC variants classification. In: Proceedings of LifeTech 2020, pp. 235–239 (2020)
22. Tran, T., Kwon, O.H., Kwon, K.R., et al.: Blood cell images segmentation using deep learning semantic segmentation. In: Proceedings of ICECE 2018, pp. 13–16 (2018)
23. Wan, L., Zeiler, M., Zhang, S., et al.: Regularization of neural networks using dropconnect. In: Proceedings of ICML 2013, vol. 28, pp. 1058–1066 (2013)

24. Wang, C.Y., Bochkovskiy, A., Liao, H.Y.M.: Yolov7: trainable bag-of-freebies sets new state-of-the-art for real-time object detectors. ArXiv abs/2207.02696 (July 2022)
25. Wang, S., Ke, Y., Liu, Y., et al.: Establishment and clinical validation of an artificial intelligence yolov51 model for the detection of precancerous lesions and superficial esophageal cancer in endoscopic procedure. Zhonghua Zhong liu za zhi [Chinese Journal of Oncology] **44**(5), 395–401 (2022)
26. Wei, X., Cao, Y., Fu, G., et al.: A counting method for complex overlapping erythrocytes-based microscopic imaging. J. Innov. Opt. Health Sci. **44**(06), 1550033 (2015)
27. Zhang, M.L., Guo, A.X., VandenBussche, C.J.: Morphologists overestimate the nuclear-to-cytoplasmic ratio. Cancer Cytopathol. **124**(9), 669–677 (2016)
28. Zhu, X., Lyu, S., Wang, X., et al.: Tph-yolov5: improved yolov5 based on transformer prediction head for object detection on drone-captured scenarios. In: Proceedings of ICCV 2021, pp. 2778–2788 (2021)

A Spatio-Temporal Attention-Based GCN for Anti-money Laundering Transaction Detection

Hengdi Huang[1], Pengwei Wang[1](✉), Zhaohui Zhang[1], and Qin Zhao[2]

[1] School of Computer Science and Technology, Donghua University, Shanghai, China
{wangpengwei,zhzhang}@dhu.edu.cn

[2] Department of Computer, Shanghai Normal University, Shanghai, China
zhao@shnu.edu.cn

Abstract. Money laundering is a serious financial crime that has significant implications for a country's economy, finance, and political stability. Machine learning and deep learning methods have been used to identify instances of money laundering, with some notable successes. However, most studies focus on static structure of transaction graph, ignoring the dynamic of transactions over time. In this study, we propose a novel approach called TemporalGAT that leverages temporal and spatial attention mechanisms to improve the accuracy and efficiency of money laundering detection. Specifically, we employ multi-head attention mechanisms to perform node embedding on spatial structure of graph, extract features from transaction data, and introduce a dynamic update mechanism that enables the LSTM to adaptively update graph convolutional network parameters over time. This approach allows the model to capture dynamic changes and spatial correlations in transaction data. We evaluate the proposed method on the publicly available Elliptic dataset for node (transaction entity) classification tasks, and the experimental results demonstrate that TemporalGAT outperforms existing methods in money laundering transaction detection.

Keywords: Graph convolutional network · Time series · Attention mechanism · Anti-money laundering

1 Introduction

Money laundering is a pervasive financial crime with long-standing and profound implications for the global economy. Annually, a substantial amount of criminals

Pengwei Wang is the corresponding author. This work was partially supported by the National Natural Science Foundation of China (NSFC) under Grant 61602109, DHU Distinguished Young Professor Program under Grant LZB2019003, Shanghai Science and Technology Innovation Action Plan under Grant 22511100700, Fundamental Research Funds for the Central Universities, the Key Innovation Group of Digital Humanities Resource and Research of Shanghai Municipal Education Commission.

X. Yang et al. (Eds.): ADMA 2023, LNAI 14180, pp. 634–648, 2023.
https://doi.org/10.1007/978-3-031-46677-9_44

exploit financial services to disguise illegal funds as legitimate assets, once these illegal funds enter the financial system, criminal activities involving these funds are difficult to detect because their sources are difficult to trace [1]. According to statistics from the United Nations, the global amount of money laundered each year ranges from 800 billion to 2 trillion [2]. In today's era of economic globalization and financial integration, money laundering activities pose a severe threat to a country's economic and social stability. Therefore, effectively detecting abnormal financial activities has become a significant challenge for both governments and financial institutions [3].

To address this challenge, researchers have explored various methods for detecting suspicious financial activities. Some have used rule-based expert systems to detect suspicious transactions, but these rules can be easily circumvented by criminals. To overcome the limitations of expert systems, other researchers have employed supervised learning or clustering methods that consider transaction features to identify illegal entities. While these methods can achieve good classification results, the complex feature engineering process is time-consuming, and they often overlook the impact of the network of transaction entities on detection performance. To consider the influence of transaction entities on each other, graph neural network models can be used to explore this problem [4]. However, few studies have simultaneously considered the impact of both graph spatial structures and temporal information on classification model performance, and even fewer studies have investigated the impact of dynamic changes in transaction subgraphs over time on detection model performance. Therefore, it is necessary to not only consider the information of current transaction subgraph but also analyze the impact of information on historical time step on current subgraph.

In this study, we propose a temporal and spatial attention-based GCN model (named TemporalGAT) that considers the impact of spatial structure of graph and temporal information of transaction data on detection performance. Validating the performance of the proposed model by predicting illegal transaction entities in Elliptic dataset.

The main contributions of this paper are:

- We proposed a novel spatial-temporal attention-based GCN model named TemporalGAT, which integrates spatial and temporal information of transaction data to detect suspicious trading entities.
- We devised a new scheme to aggregate the nodes features of the graph spatial structure by introducing a multi-head attention mechanism.
- We developed a new dynamic update mechanism to enable the model to perceive changes in graph structure over time.
- The benchmark analysis results showed that TemporalGAT outperforms other models in terms of precision, recall, and F1-score on Elliptic dataset.

The paper is structured as follows: Sect. 2 describes the related work. Section 3 provides a detailed description of the proposed methods, including problem definition, model architecture, task, and model training. Section 4 describes the experimental results and comparative analysis. Finally, Sect. 5 summarizes this paper.

2 Related Work

In response to the harm caused by money laundering, financial institutions have implemented a range of anti-money laundering measures. For instance, FATF regulations allow financial institutions to report suspicious transactions to national financial regulatory agencies for further investigation. In the past, detecting money laundering involved manually screening massive transaction records, with the complexity of interrelated and seemingly harmless transactions making it difficult to identify money laundering behavior. However, today, many researchers have developed anti-money laundering solutions using advanced technologies such as autonomous and controllable big data and artificial intelligence to assist in monitoring money laundering transactions. These solutions have significantly reduced the cost of manual anti-money laundering reviews, improved the timeliness and accuracy of anti-money laundering, and made it easier to identify money laundering behavior.

Researchers have extensively investigated various machine learning approaches to detect money laundering transactions in the financial system. Some studies [5–7] have shown that tree-based classifiers can efficiently detect money laundering activity. For example, Senator et al. [7] noted the potential of tree-based models in detecting traditional money laundering transactions. Ketenci et al. [8] used temporal frequency analysis to obtain a new feature set and employed a random forest classifier to achieve good classification performance. Le-Khac et al. [9] proposed a clustering and neural network-based data mining framework for analyzing transaction datasets to detect suspicious transactions. Tai and Kan [10] proposed a two-stage intelligent identification method based on machine learning and data analysis techniques to identify suspicious money laundering accounts from transaction data. Jamshidi et al. [11] proposed an adaptive neuro-fuzzy inference system based on adaptive neuro-fuzzy inference system (ANFIS) based novel intelligent multi-objective identification method for identifying money laundering in banking and foreign exchange transactions. Chen et al. [12] proposed a set of unsupervised deep learning techniques based on autoencoder (AE), variational autoencoder (VAE) and generative adversarial networks for implementing AML and fraud detection. Desrousseaux et al. [13] proposed a SOM+fzART prediction framework that reduces the number of false alerts from expert systems. Wang and Dong [14] proposed a novel money laundering detection algorithm based on improved minimum spanning tree clustering and a new variance metric to detect suspicious money laundering transactions in financial applications.

Traditional methods for money-laundering detection are typically inadequate in identifying money laundering activities from a relational network perspective. In recent years, there has been significant interest in artificial intelligence solutions that use graph computing. Graph-based techniques offer distinctive solutions for investigating financial crimes [15]. By utilizing graph neural networks to more accurately describe the relationships between transaction entities, suspicious transaction can be effectively detected. Li et al. [16] proposed a new multi-view semi-supervised learning method, Co-GCN, which introduces graph convolutional network into multi-view learning and adaptively utilizes graph information from multiple views using a combined Laplacian operator. Experimental results on real datasets show that Co-GCN outperforms the latest multi-view semi-supervised methods. Weber et al. [4] discussed the excellent performance of the random forest model compared to the graph convolutional network model in classifying legal and illegal transactions derived from the Bitcoin blockchain. Pareja et al. [17] introduced EvolveGCN, which is composed of graph convolutional network and recursive neural networks such as GRU and LSTM. This study revealed that EvolveGCN outperforms the graph convolutional network model used by Weber et al. [4].

3 Problem Definition and Proposed Model

In this section, we will provide a detailed description of the problem definition for classifying money laundering transaction entities, perform data analysis on the dataset, outline the architecture of the classification model, and discuss the tasks of the proposed model.

3.1 Problem Definition

We can represent a complex transaction network as a graph structure, where each node V in graph G represents a transaction entity and each edge E represents a transaction between nodes. Thus, the entire transaction network can be represented by $G = (V, E)$, where each transaction has a corresponding time step indexed by t. The entire transaction network can be partitioned into subgraphs for each time step, with graph structure of transaction subgraph at time step t represented by $G_t = (V_t, E_t)$, where $t \in \{1, 2, 3, , T\}$ and $|V| = n$ represents the number of nodes in graph. The entire transaction network can be represented by $G = \{G_1, G_2, , G_T\}$. The adjacency matrix for graph G_t can be represented as $A_t \in R^{n*n}$, where $x_t \in R^d$ is the node feature vector with d dimensions, and y_t represents the node label indicating whether it is a legal or illegal transaction entity. For a transaction subgraph G_t with n nodes and time step t, the node feature matrix is $X_t \in R^{n*d}$ and the node label matrix is $Y_t \in R^n$, represented as $G_t = (X_t, Y_t)$. A sliding window is used for dynamic updates, represented as $W = (G_1, G_2, , G_i) : i \in \{1, 2, 3, , w\}$, where w represents the size of the sliding window, and the sliding step size is represented by S.

3.2 Data Analysis

After data analysis, we observed significant differences in the distribution of transaction feature values between legal (0) and illegal (1) transaction entities at the same timestamp. As illustrated in Fig. 1, trans_feat_4 and trans_feat_6 of illegal are significantly smaller than those of legal, which can effectively distinguish between legal and illegal transactions. This suggests that temporal information of transaction is an important factor to identify legal and illegal transaction entities.

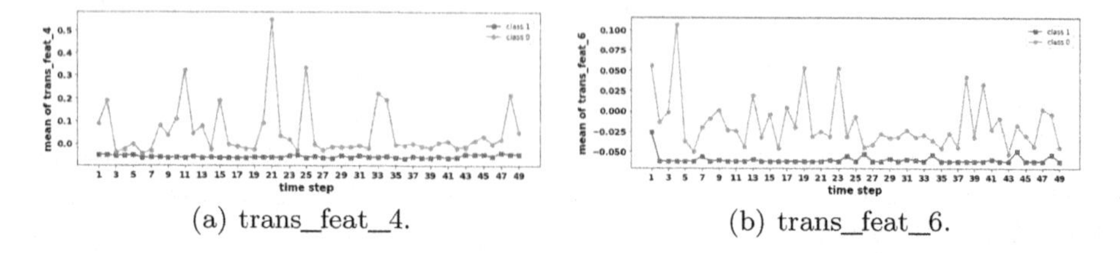

(a) trans_feat_4. (b) trans_feat_6.

Fig. 1. Distribution of trans_feat_4 and trans_feat_6 of legal and illegal entities at different time steps.

3.3 Model Architecture

The proposed model both considers spatial structures and temporal information from transaction subgraphs, by using spatial convolutions with multi-head attention performs feature extraction, and a dynamic update mechanism is introduced to account for the impact of changing spatial structures on model performance over time. The dataset used in this paper spans 49 time steps, with a transaction subgraphs formed at each time step. A sliding window of length 6 is created on the time series of these subgraphs, starting from time t_1, and the stride size is 1. Each subgraph within the window, a two layers spatial convolution with multi-head attention is required for them. The weight for spatial convolution at the sixth time step are updated based on the weights from the previous five time steps, enabling the model to consider features in historical transaction records and adapt to dynamic changes in graph structure. As illustrated in Fig. 2, the model performs feature aggregation and extraction on transaction subgraphs at each time step using the multi-head attention spatial convolution layer and the dynamic update mechanism, allowing it to adapt to changes in graph structure.

Multi-head attention mechanism assigns different weights to the first-order neighbors of the central node, thereby aggregating the features of nodes that have a greater impact on the central node. To capture the temporal dynamics of the transaction subgraphs, dynamic updating mechanism utilizes LSTM to adaptively update the weight of spatial convolution layer in the network, enabling the model to adapt to dynamic graph structures and fuse neighbor information at each node's current and historical time points. Finally, the extracted transaction subgraph features vector is fed into an MLP with two linear hidden layers, which performs binary classification of nodes in the transaction subgraph.

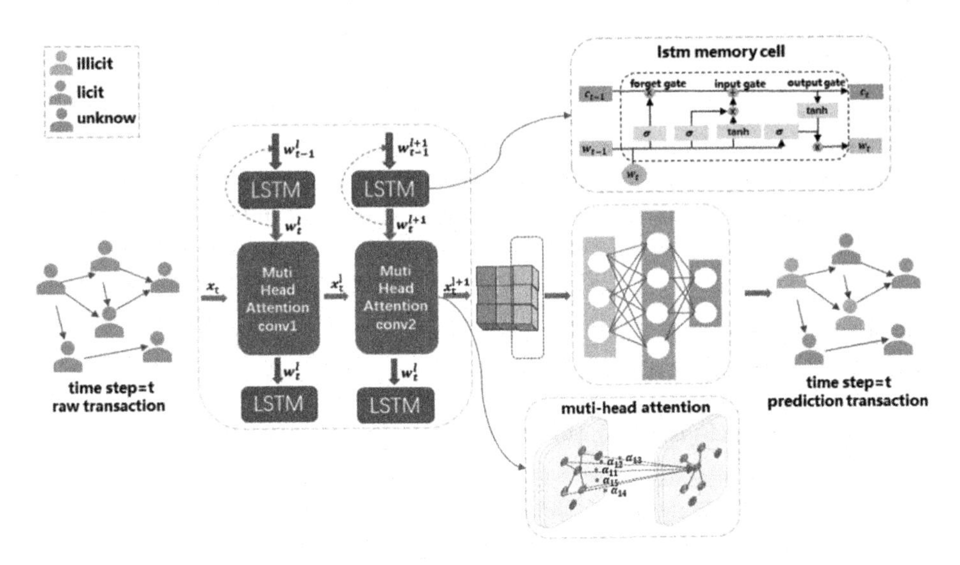

Fig. 2. TemporalGAT model architecture.

(1) Spatial Convolution of Multi-headed Attention Mechanisms

By taking advantage of the graph attention network (GAT), the spatial structure is extracted from the graph data of the transaction network by correlating and learning information about the transaction features of graphs on adjacent time steps. In GAT, we aggregate neighbor node information by assigning different attention coefficients to same-order neighbor nodes. For a graph $G_t = (V_t, E_t)$ at time step t, the input of the first layer of GAT is the node feature matrix X_t, where each row represents the d-dimensional features of nodes in the subgraph. We use l as the convolutional layer index of GAT, where the input of layer l is the output vector X_t^{l-1} of layer $l-1$, and the output is the updated node feature vector. The operation of layer l can be represented as:

$$X_t^l = F(x_t^{l-1}, W_t^{l-1}) = \sigma(\sum_{j \in N_i} \alpha_{ij} X_t^{l-1} W_t^{l-1}) \tag{1}$$

$$\alpha_{ij} = softmax(e_{ij}) = \frac{exp(e_{ij})}{\sum_{j \in N_i} exp(e_{ij})} \tag{2}$$

$$e_{ij} = LeakyRelu(a^T [W X_i || W X_j]) \tag{3}$$

The formula l represents graph convolutional layer with attention mechanism in the l-th layer, and t represents the t-th time step. W_t^l is the weight matrix at time step t, α_{ij} is the weight coefficient calculated by the attention mechanism, and σ is the ReLU non-linear activation function. The attention coefficient α_{ij} is obtained from the influence coefficient e_{ij} of neighbor node j on center node i. The relevance between the two is represented by the weight parameters a of a feedforward network layer. When considering graph structure information, center node i only calculates the relevance e_{ij} of its same-order neighboring nodes $j \in N_i$ to the target node. In order to better allocate weights between different nodes, we use the softmax activation function to normalize the relevance calculated between the target node and all neighbors.

In this paper, we use the multi-head attention mechanism to improve the learning efficiency of self-attention. We call K groups of independent attention mechanisms on Formula (1) and then concatenate the output results, as shown in Formula (4), where $||$ denotes the concatenation operation and α_{ij}^k is the weight coefficient calculated by the k-th group attention mechanism. To reduce the dimensionality of the feature vector, the concatenation operation can also be replaced with the average operation, as shown in Formula (5).

$$X_t^l = F(K, X_t^{l-1}, W_t^{l-1}) = ||_{k=1}^K \sigma(\sum_{j \in N_i} \alpha_{ij}^k X_t^{l-1} W_t^{l-1}) \tag{4}$$

$$X_t^l = F(K, X_t^{l-1}, W_t^{l-1}) = \sigma(\frac{1}{K} \sum_{k=1}^{K} \sum_{j \in N_i} \alpha_{ij}^k X_t^{l-1} W_t^{l-1}) \tag{5}$$

After L layers of graph convolutional layers, the feature matrix outputted at each time step has aggregated information from same-order neighboring nodes.

(2) Dynamic Update Mechanism

To consider the dynamic changes of transaction subgraphs at different time steps, a dynamic update mechanism was added to static graph attention network to form dynamic graph convolutional layer. As time passes, the transaction subgraphs at different time steps also change. To adapt to this change, the weight parameters of the convolutional operation should also be dynamically updated. The dynamic graph convolutional layer uses LSTM [19] to update the weight of graph attention network. For each time step $t \in \{1, 2, 3, , T\}$ and each graph convolutional layer $l \in \{1, 2, 3, , L\}$, LSTM takes the weight parameters W_{t-1}^l of the l-th graph convolutional layer at time step t-1 as the hidden state and the randomly initialized weight parameters W_t^l at time step t as the input to update the weight parameters W_t^l of the l-th graph convolutional layer at time step t. Modeling the dynamic update mechanism with LSTM which has a long-term memory function for time series.

The weight updating process of the lth graph convolutional layer at time step t is as follows:

$$W_t^l = LSTM(W_t^l) = H_t = f(Wt^l) \tag{6}$$

where H_t is the output of the LSTM, as follows:

$$H_t = f(X_t) = O_t \odot tanh(C_t) \tag{7}$$

$$O_t = \sigma(W_O \cdot [X_t, H_{t-1}] + B_O \tag{8}$$

$$C_t = F_t \odot C_{t-1} + I_t \odot \tilde{C}_T \tag{9}$$

$$F_t = \sigma(W_F \cdot [X_t, H_{t-1}] + B_F \tag{10}$$

$$I_t = \sigma(W_I \cdot [X_t, H_{t-1}] + B_I \tag{11}$$

$$\tilde{C}_T = tanh(W_C \cdot [X_t, H_{t-1}] + B_C \tag{12}$$

In the above formula, O_t, C_t, F_t,I_t, and $\tilde{C}_T$ represent the output gate, memory cell, forget gate, input gate, and predicted memory cell, respectively. The weights W_O, W_F, W_l, and W_C correspond to the gates and predicted memory cell mentioned above. H_{t-1} is the hidden state from the previous time step (i.e., the weight parameters from the previous time step). The randomly initialized weight matrix at time t is used as input to the LSTM, and the LSTM unit aggregates historical information and current time step information to obtain the weight matrix W_t^l at time step t.

In order to aggregate the features of neighbor nodes from the spatial construction of the transaction subgraph, we use multi-head attention graph convolution. The model is then given a dynamic update mechanism by updating the weight on every transaction time step using LSTM, which enables the initial static spatial convolution to dynamically aggregate the feature data of neighbors with the same order. The following formula can be used to express this dynamic spatial convolution:

$$X_t^{l+1} = F(K, X_t^l, W_t^l) = F(K, X_t^l, LSTM(W_t^l)) \tag{13}$$

3.4 Model Tasks and Training

To evaluate the classification performance of our model, we conducted experiments on the Elliptic Bitcoin Transaction dataset. The model utilizes the multi-head attention mechanism for spatial convolution and the dynamic update mechanism to obtain the transaction features at the current time step. Then, the extracted features are fed into MLP to obtain the final classification results. For a node v_t in transaction graph at time t, its predicted label is y_t^v, and its feature vector is denoted as x_t. $P(y_t^v|x_t)$ represents the probability distribution of node v_t on model prediction, as follows:

$$y_t^v = softmax(P(y_t^v|X_t)) \tag{14}$$

In this model, if the predicted label y_t^v of node v_t is 0, it means that the node is a legitimate transaction entity; if y_t^v is 1, it means that the node is a suspected money laundering transaction entity. We use the cross-entropy loss function as the training loss of the model, as follows:

$$Loss = -\frac{1}{N}\alpha \sum_{i=1}^{N} p(y_t^i) logq(y|X_t^i) \tag{15}$$

In this loss function, $p(y_t^i)$ represents the true probability distribution of node i, while $q(y|X_t^i)$ represents the probability distribution predicted by the model based on the input features X_t^i of node i. Due to the problem of imbalanced label distribution during training, we introduce a hyperparameter α to increase the weight of samples in the imbalanced class.

4 Experimental Evaluation

4.1 Dataset

The dataset used in this article is the Elliptic dataset [4], which was published by the cryptocurrency compliance company Elliptic. The dataset includes 203,769 transaction entities and 234,355 transaction records. 2% (4,545) of the transaction entities are labeled as illegal and 21% (42,019) are labeled as legal, with the remaining unlabeled. Each transaction entity has 166 transaction features, with the first 94 features reflecting the transaction itself. The remaining 72 features are aggregated features based on transaction information obtained from neighboring entities of the central transaction entity. Each transaction entity is associated with a time step that measures the time it is broadcast to the Bitcoin network. The time step ranges from 1 to 49.

During model training, only the first 94 local transaction features of each transaction entity were used, and the dataset was divided according to time steps. Data from the first 31 time steps were used to train, 32–36 were used to validate, and 37–49 were used to test (Table 1).

Table 1. Division of dataset

Nodes	Edge	Illicit nodes / licit nodes	Time Setp (Train / Val / Test)
203769	234355	2% / 21%	31 / 5 /13

4.2 Setup

To train the TemporalGAT model, we selected Ubuntu18.04.1 as the experimental environment and utilized tools such as Intel(R) Xeon(R) Silver 4214 CPU @ 2.20GHz, python3.7, pytorch1.7.1, dgl0.6.1, and cuda11.4. The number of hidden layers in the multi-head attention layer was set to 2, with 3 heads and hidden layer unit numbers of 128 and 384. The Adam optimizer was used to update the model parameters, with epoch set to 50 and a fixed learning rate of 0.001. Additionally, to address the impact of sample imbalance on model training, we set the loss_weight parameter to 0.35 and 0.65, and use precision, recall and F1-score to evaluate the proposed model performance.

4.3 Compared Methods

GCN [4]: is a static graph convolutional model that embeds a central node by aggregating information from its neighbors. Since there is a transaction subgraph for each time step, we train the same GCN model on transaction subgraphs at different time steps, without considering the dynamic nature of graph over time.

GAT [18]: is a static graph convolutional model that aggregates information from neighboring nodes to the central node by utilizing attention mechanisms.

Like GCN, GAT does not consider the temporal dynamics of graph structure, and therefore, we train the same GAT model on different transaction subgraphs at different time steps.

EvolveGCN-O [17]: The model trains graph convolution operation in a bottom-up manner along the convolutional layers, and uses LSTM for dynamic modeling in the time dimension, the model obtains the new node representations.

EvolveGCN-H [17]: In contrast to EvolveGCN-O, EvolveGCN-H utilizes GRU for dynamic modeling of graph convolutional network.

All of the above graph models have been used in anti-money laundering research.

4.4 Results

By conducting experiments on the same dataset, comparing the TemporalGAT model with existing research methods, and analyzing classification evaluation metrics, ablation experiments, and Hyperparameters experiments, we have validated the effectiveness of the TemporalGAT model. Specifically, we have conducted comprehensive experimental comparisons and validated the significant advantages of the TemporalGAT model on multiple metrics, providing strong support for the application and promotion of this model. The TemporalGAT model proposed in this paper achieves superior performance in various classification evaluation metrics compared to existing methods when using only 94 local transaction features on the same training, validation, and test sets. Compared to existing research that employs 94 local transaction features and 71 manually engineered aggregate features, this paper only considers the first 94 local transaction features, significantly improving detection efficiency. As feature engineering is a time-consuming and laborious task, if a model that can obtain good classification results only through simple data processing is more practical. The TemporalGAT model aggregates transaction features by considering both graph spatial information of the transaction subgraph and the dynamic nature of graph over time, and performs binary classification using an MLP. Additionally, even under imbalanced sample conditions, TemporalGAT also shows significant advantages over other models in terms of precision, recall and F1-scores by increasing the weight of unbalanced samples in the loss function. Specifically, the TemporalGAT classification model achieves precision, recall, and F1-score of

Table 2. Node classification results using only local features of elliptic dataset

Methods	Precision (%)	Recall (%)	F1-score (%)
GCN	71.15	42.05	52.86
GAT	74.28	44.59	55.72
EvolveGCN-H	26.38	51.61	34.92
EvolveGCN-O	63.83	51.04	56.72
TemporalGAT	**79.65**	**51.84**	**62.81**

79.65%, 51.84%, and 62.81%, respectively, in detecting illegal transaction entities. As shown in Table 2, compared to the static GCN and GAT models that do not consider temporal information, as well as two versions of the EvolveGCN model that consider graph dynamics over time, the TemporalGAT model performs the best. Compared to other methods, the TemporalGAT model shows the most significant improvement in precision, followed by F1-score.

(1) Ablation Experiments
Compared with GCN and GAT which only take static graph structure into consideration, GAT, which uses attentional spatial convolution to assign different weights to first-order neighbors for feature aggregation, provides more accurate results. GAT outperforms GCN on three commonly used evaluation metrics. Compared to the static GAT, TemporalGAT introduces a dynamic updating mechanism to consider the dynamic changes of transaction subgraphs at different time steps. Specifically, by using LSTM to dynamically model the attention-based graph convolution weights, the classification model at the current time step can take into account the features from past time steps, thus further improving the classification performance. We observed through F1-score on the test set that prior to the 43rd time step, TemporalGAT consistently achieved the best performance, while all methods were affected after the 43rd time step. It is known that the 43rd time step is when the black market closed, which led to insufficient training data on previous time steps to meet the needs of identifying current transaction entities, resulting in performance degradation. Models that do not consider the temporal dynamics of graph, such as GCN, were more affected. Although the proposed model, which considers temporal dynamics of graph, can partially alleviate the impact of sudden events, it still cannot achieve stable classification performance because historical data is insufficient for training the model to recognize features of current transaction entities after such events (Figs. 3 and 4).

(2) Hyperparametric Experiments
The effect of sliding window size on classification results. To investigate the impact of sliding window size on experimental results, we conducted experiments on the TemporalGAT model by varying the size of the sliding window to observe its effect on the model's dynamic updating mechanism. We used three evaluation metrics to describe the model's classification performance. The results showed that different sliding window sizes can result in different classification performances of the model. When the window size is smaller than 5, the dynamic updating mechanism of the model fails to fully exert its effect, leading to poor classification performance because the model learns less historical information from a too short time span. Conversely, when the window size is too large (exceeding 6), the model needs to consider too much historical information, making it less suitable for frequently changing transaction subgraphs. With other parameters being constant, when the sliding window size is set to 5 or 6, the dynamic updating mechanism has the most significant effect on the model,

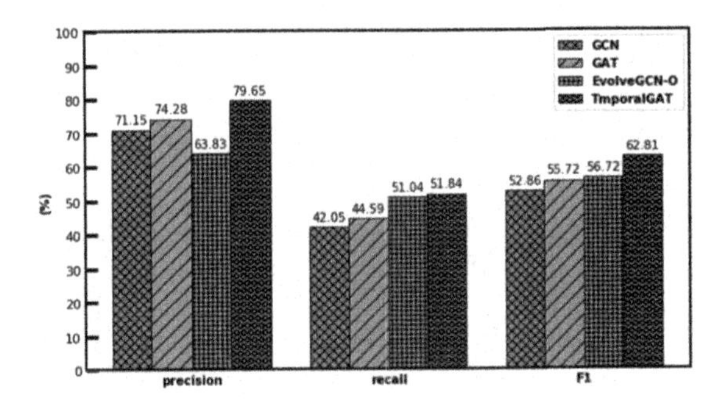

Fig. 3. Ablation studies with different models.

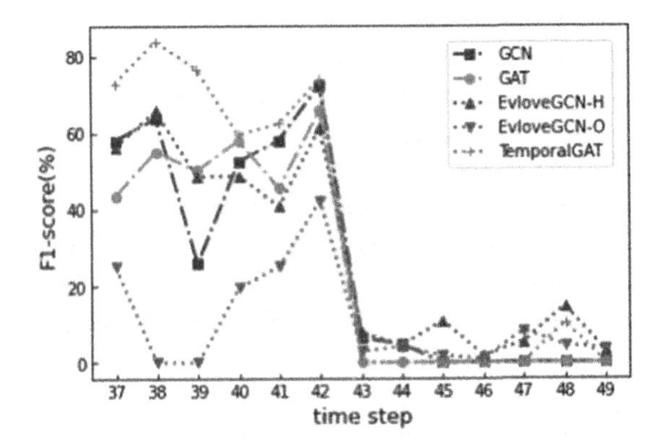

Fig. 4. The F1-score of node classification over time.

and the window size of 6 yields the best classification performance, with the precision, recall, and F1-score reaching 79.65%, 51.84%, and 62.8% respectively.

The effect of the number of heads of multi-headed attention on classification results. When aggregating neighbor node features through spatial convolution, we use a multi-head attention mechanism to assign different weights to first-order neighbor nodes, so as to achieve targeted feature aggregation of central nodes. To stabilize the attention process, we adopt multi-head attention, which can use multiple independent attention to perform feature transformation and connection. By changing the value of the attention head number k, we study its impact on the classification performance of the model. When $k = 1$, self-attention is used, and when $k \geq 1$, multi-head attention is used. Experimental results show that when $k = 3$, the classification performance of the model reaches its optimal level (Figs. 5 and 6).

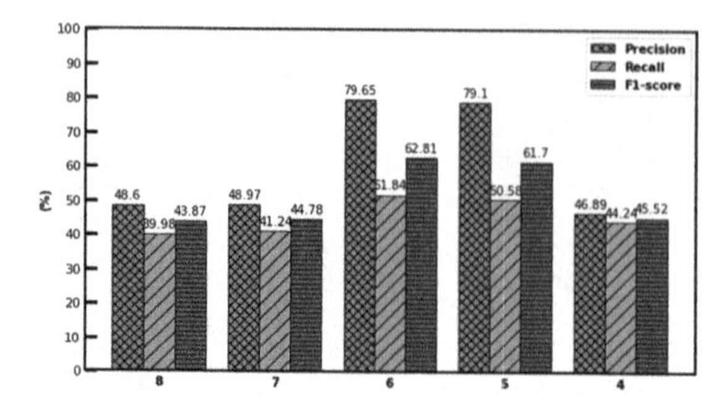

Fig. 5. Comparison of classification performance with changing of window sizes.

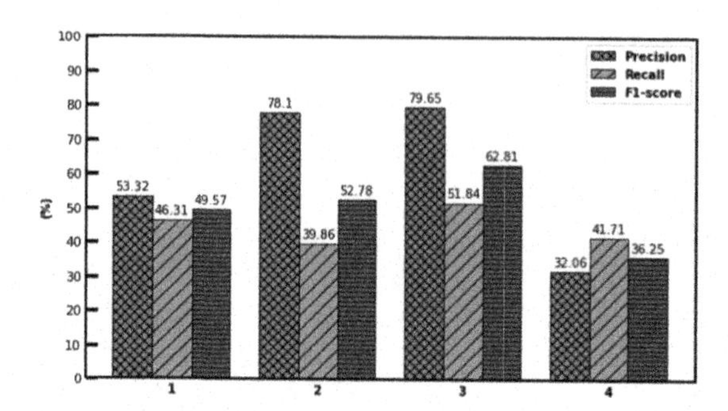

Fig. 6. Comparison of classification performance with changing of heads.

The effect of loss weight on classification results. By analyzing the dataset, it is evident that there is a data imbalance problem where the number of negative samples (representing legal transactions) greatly outnumber the number of positive samples (representing illegal transaction entities). In the learning process, this may cause the model to focus more on the negative samples, leading to overfitting them and underfitting the positive samples, thereby reducing the model's generalization ability. Therefore, when dealing with imbalanced data, one can assign higher weights to the minority class to alleviate this issue, so that the model is penalized more heavily when predicting the positive samples and thus improving its generalization ability. This can be achieved by setting sample weights in the cross-entropy loss function.

Through observing and comparing the model's classification performance under different weights assigned to positive samples, we can clearly see that in this experiment, setting the weight of positive samples to 0.65 achieves the best classification performance. The experimental results also show that changing the weight of positive samples does not significantly affect the recall evaluation metric, but has a significant improvement on the precision evaluation metric. Additionally, we can also observe that setting the loss weight of positive samples to 0.75 can achieve better recall, but at the cost of sacrificing more precision (Fig. 7).

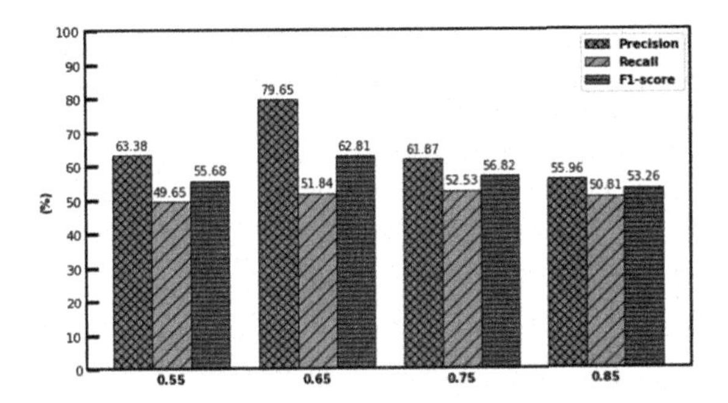

Fig. 7. Comparison of classification performance with changing of loss weight.

5 Conclusion

This paper proposes a semi-supervised graph attention neural network model called TemporalGAT for detecting illicit entities in financial transaction networks for anti-money laundering. The model integrates spatial and temporal information of the transaction data graph by aggregating neighbor features using a graph convolutional network with multiple attention mechanisms and dynamically updating the weights of graph convolutional layers with an LSTM to adapt to the changing graph structure at different time steps. Compared to existing research, the proposed model shows significant improvements in various classification metrics under the same experimental settings, with an accuracy of up to 79.65%. The ablation experiments demonstrate that considering both spatial structure and temporal information of the transaction graph is beneficial to achieve better classification performance. Hyperparameter experiments further determine the optimal sliding window size, number of attention heads, and loss weight for the model's best performance. In future research, we plan to incorporate an active learning framework into the model to further address the impact of imbalanced data on model performance. Additionally, we will attempt to extend the TemporalGAT model to other graph-structured datasets related to anti-money laundering, in order to improve the generalizability of the model.

References

1. Denny, C., Shihab, M.R.: Bank account classification for gambling transactions. In: 2021 3rd East Indonesia Conference on Computer and Information Technology, pp. 302–308 (2021)
2. Chen, Z., Van Khoa, L.D., Teoh, E.N., et al.: Machine learning techniques for anti-money laundering (AML) solutions in suspicious transaction detection: a review. Knowl. Inf. Syst. **57**(2), 245–285 (2018)
3. Li, X., Cao, X., Qiu, X., et al.: Intelligent anti-money laundering solution based upon novel community detection in massive transaction networks on spark. In: 2017 Fifth International Conference on Advanced Cloud and Big Data, pp. 176–181 (2017)

4. Weber, M., Domeniconi, G., Chen, J., et al.: Anti-money laundering in bitcoin: experimenting with graph convolutional networks for financial forensics. arXiv preprint arXiv:1908.02591 (2019)
5. Harlev, M.A., Yin, S.H., Langenheldt, K.C., et al.: Breaking bad: de-anonymising entity types on the bitcoin blockchain using supervised machine learning. In: Proceedings of The 51st Hawaii International Conference on System Sciences, pp. 1–10 (2018)
6. Savage, D., Wang, Q., Zhang, X., et al.: Detection of money laundering groups: supervised learning on small networks. In: Workshops at the Thirty-First AAAI Conference on Artificial Intelligence, pp. 24–34 (2017)
7. Senator, T.E., Goldberg, H.G., Wooton, J., et al.: The FinCEN artificial intelligence system: identifying potential money laundering from reports of large cash transactions. In: Proceedings of the 7th Conference on Innovative Applications of AI, pp. 156–170 (1995)
8. Ketenci, U.G., Kurt, T., Onal, S., et al.: A time-frequency based suspicious activity detection for anti-money laundering. IEEE Access **9**, 59957–59967 (2021)
9. Le-Khac, N.-A., Markos, S., et al.: A data mining-based solution for detecting suspicious money laundering cases in an investment bank. In: 2010 Second International Conference on Advances in Databases, Knowledge, and Data Applications, Menuires, pp. 235–240 (2010)
10. Tai, C.H., Kan, T.J.: Identifying money laundering accounts. In: International Conference on System Science and Engineering, pp. 379–382 (2019)
11. Jamshidi, M.B., Gorjiankhanzad, M., Lalbakhsh, A, et al.: A novel multiobjective approach for detecting money laundering with a neuro-fuzzy technique. In: 2019 IEEE 16th International Conference on Networking, Sensing and Control, pp. 454–458 (2019)
12. Chen, Z., Soliman, W.M., Nazir, A., et al.: Variational autoencoders and Wasserstein generative adversarial networks for improving the anti-money laundering process. IEEE Access **9**, 83762–83785 (2021)
13. Desrousseaux, R., Bernard, G., Mariage, J.J.: Predicting financial suspicious activity reports with online learning methods. In: 2021 IEEE International Conference on Big Data (Big Data), pp. 1595–1603 (2021)
14. Wang, X., Dong, G.: Research on money laundering detection based on improved minimum spanning tree clustering and its application. In: 2009 Second International Symposium on Knowledge Acquisition and Modeling, pp. 62–64 (2009)
15. Jin, Y., Qu, Z.: Research on anti-money laundering hierarchical model. In: 2018 IEEE 9th International Conference on Software Engineering and Service Science, pp. 406–411 (2018)
16. Li, S., Li, W.T., Wang, W.: Co-GCN for multi-view semi-supervised learning. Proc. AAAI Conf. Artif. Intell. **34**(4), 4691–4698 (2020)
17. Pareja, A., Domeniconi, G., Chen, J., et al.: EvolveGCN: evolving graph convolutional networks for dynamic graphs. Proc. AAAI Conf. Artif. Intell. **34**, 5363–5370 (2020)
18. Velikovi, P., Cucurull, G., Casanova, A.: Graph attention networks. arXiv preprint arXiv:1710.10903 (2017)
19. Hochreiter, S., Schmidhuber, J.: Long short-term memory. Neural Comput. **9**(8), 1735–1780 (1997)

Author Index

X. Yang et al. (Eds.): ADMA 2023, LNAI 14180, pp. 649–651, 2023.
https://doi.org/10.1007/978-3-031-46677-9

Printed in the United States
by Baker & Taylor Publisher Services